Surveyor Reference Manual

Sixth Edition

George M. Cole, PhD, PE, PLS

Professional Publications, Inc. • Belmont, California

Benefit by Registering This Book with PPI

- Get book updates and corrections.
- Hear the latest exam news.
- Obtain exclusive exam tips and strategies.
- Receive special discounts.

Register your book at **ppi2pass.com/register**.

Report Errors and View Corrections for This Book

PPI is grateful to every reader who notifies us of a possible error. Your feedback allows us to improve the quality and accuracy of our products. You can report errata and view corrections at **ppi2pass.com/errata**.

SURVEYOR REFERENCE MANUAL
Sixth Edition

Current printing of this edition: 1

Printing History

edition number	printing number	update
5	3	Minor corrections.
5	4	Minor corrections. Copyright update.
6	1	New edition. Copyright update.

Printed in the United States of America.

PPI
1250 Fifth Avenue, Belmont, CA 94002
(650) 593-9119
ppi2pass.com

ISBN 978-1-59126-485-9

Library of Congress Control Number: 2014953831

Topics

Table of Contents

Preface and Acknowledgments

As we hurtle full-tilt through the digital age, our knowledge and technical capability seem to increase at a faster pace every year. Advances in surveying technology, together with changes in associated areas such as law and computer science, seem to follow one upon the other at an almost exponential rate. Such changes are certainly good in that they have made surveying processes easier and surveying results more accurate and usable than ever before. On the other hand, such an amazing rate of change requires significant time and effort on the part of practicing surveyors and exam candidates to keep up with the rapidly changing scope of the profession.

For this sixth edition of the *Surveyor Reference Manual*, I have expanded the chapter on aerial mapping, including a discussion of the growing use of LiDAR in hydrographic mapping, and added a new chapter on laser scanning. Other information has been updated and corrected.

This edition is dedicated to Andrew L. Harbin, the original author of this manual. Mr. Harbin obtained a bachelor of science degree from Texas A&M University in civil engineering. He taught mathematics to high school students, surveying and related courses at Texas State Technical Institute, and developed a correspondence course for surveyors. Well known to surveyors, he was the sole author of the first three editions of the *Surveyor Reference Manual,* and his well written and comprehensive coverage of the broad field of surveying serves as the basis for this edition. Sincere appreciation is also expressed to my colleague Allen K. Nobles, PLS, for his helpful comments regarding the new material.

In addition, let me express my appreciation to Heather Turbeville, associate project manager; Scott Marley, lead editor; Tom Bergstrom, production associate and technical illustrator; Kate Hayes, production associate; Cathy Schrott, production services manager; Sarah Hubbard, director of product development and implementation; and Jenny King, associate editor-in-chief.

Finally, when you have finished going through this book, I encourage you to assist in its continuing improvement by notifying PPI of any comments or errors through the online errata submission form at **ppi2pass.com/errata**. Revisions to this edition were greatly influenced by readers' comments on the previous edition, and revisions to the next edition will be determined just as surely by your comments on this one. Please take the time to help those who will follow you in the surveying profession by letting us know what you liked and disliked about this book, and what we can do to make the next edition even better. Thank you.

George M. Cole, PhD, PE, PLS

Introduction

HOW TO USE THIS BOOK

This book serves multiple purposes.

First, as the title suggests, it is intended to serve as a comprehensive reference for practicing surveyors. If you are a practicing surveyor, you will find it to be an invaluable and frequently used addition to your library.

Second, because of its comprehensive coverage, it is an ideal resource for people preparing for the Fundamentals of Surveying (FS) exam. (It is also an ideal textbook for instructors teaching review courses for the FS exam.) This book may also be used as a study aid for the Principles and Practice of Surveying (PS) exam, but since it is intended as a comprehensive manual of fundamental principles rather than an in-depth analysis of any particular area of surveying, it may not be suitable as a stand-alone study guide for all portions of the PS.

THE FS AND PS EXAMS

The FS and PS exams are standardized tests prepared by the National Council of Examiners for Engineering and Surveying (NCEES) to ensure that only qualified people are legally allowed to practice as surveyors. To ensure the reliability and validity of the tests, the FS and PS exams are based on input from committees of professional surveyors and educators throughout the United States.

The FS exam tests general entry-level surveying principles you are expected to have gained through academic study. In most states, passing the FS exam is a requirement for registration as a surveyor-in-training or surveyor intern. The PS exam tests your ability to apply those principles to the kinds of problems typically encountered in professional practice. Passing the PS exam is required in most states for full licensure as a professional surveyor. This may be necessary if your company requires licensure for employment or advancement, if your state requires registration before you may use the title "Surveyor," if you wish to be an independent consultant, or for other reasons. The PS exam is usually administered in conjunction with an exam on surveying practices and regulations specific to the state administering the exam.

Although both the FS and PS exams are prepared by NCEES, they are administered under the direction of the licensing boards of the various states, which usually also require you to have attained a certain level of education and (for the PS exam) experience before being allowed to take the exams. For information regarding these requirements in the state in which you plan to become licensed, and to apply for licensure, contact that state's board. Current addresses and phone numbers for each state board may be obtained at PPI's website, **ppi2pass.com/stateboards**.

THE FS EXAM

Structure

The FS exam is a computer-based test that concentrates on the fundamentals and basics of surveying. You may take it at any Pearson VUE test center. The exam contains 110 multiple-choice problems given in two sessions with a break between. Each session contains approximately 55 problems. During either session, you cannot view or respond to problems in the other session.

Each problem has four possible answer options, labeled (A), (B), (C), and (D). Only one problem and its answer options are given onscreen at a time. The exam is not adaptive (i.e., your response to one problem has no bearing on the next problem you are given). Even if you answer the first five mathematics problems correctly, you'll still have to answer the sixth problem.

Your exam will include a limited (and unknown) number of problems (known as "pretest items") that will not be scored and will not have an impact on your results. NCEES does this to evaluate potential problems for future exams. You won't know which problems are pretest items. They are not identifiable and are randomly distributed throughout the exam.

The FS exam is 6 hours long and includes an 8-minute tutorial, a 25-minute break, and a brief survey at the conclusion of the exam. The total time you'll actually have to answer the exam problems is 5 hours and 20 minutes. This works out to slightly less than 3 minutes per problem. However, the exam does not pace you. You may spend as much time as you like on each question. Within a session, you may work through the problems in any sequence. If you want to go back and check your answers before you submit a session for grading, you may. However, once you submit a session you are not able to go back and review it.

You can divide your time between the two sessions any way you'd like. That is, if you want to spend 4 hours on the first section, and 1 hour and 20 minutes on the second section, you can do so. Or, if you want to spend 2 hours and 10 minutes on the first section, and 3 hours and 10 minutes on the second section, you can do that instead. Between sessions, you can take a 25-minute break. (You can take less, if you like.) You cannot work through the break, and the break time cannot be added to the time permitted for either session. Once each session begins, you can leave your seat for personal reasons, but the "clock" does not stop for your absence. Unanswered problems are scored the same as problems answered incorrectly, so you should use the last few minutes of each session to guess at all unanswered problems.

The NCEES Nondisclosure Agreement

At the beginning of the FS exam, a nondisclosure agreement will appear on the screen. In order to begin the exam, you must accept the agreement within two minutes. If you do not accept within two minutes, your exam appointment will end, and you will forfeit your appointment and exam fees. The nondisclosure agreement is discussed in the section titled "Subversion After the Exam." The nondisclosure agreement, as stated in the NCEES Examinee Guide, is as follows.

> This exam is confidential and secure, owned and copyrighted by NCEES and protected by the laws of the United States and elsewhere. It is made available to you, the examinee, solely for valid assessment and licensing purposes. In order to take this exam, you must agree not to disclose, publish, reproduce, or transmit this exam, in whole or in part, in any form or by any means, oral or written, electronic or mechanical, for any purpose, without the prior express written permission of NCEES. This includes agreeing not to post or disclose any test questions or answers from this exam, in whole or in part, on any websites, online forums, or chat rooms, or in any other electronic transmissions, at any time.

Your Exam Is Unique

The exam that you take will not be exactly the same exam taken by the person sitting next to you. NCEES says that, for each examinee, its computer-based testing (CBT) system randomly selects different but equivalent probems from its database using a linear-on-the-fly (LOFT) algorithm. Each examinee will have a unique exam, and all exams will be of equivalent difficulty.

The Exam Interface

The onscreen exam interface contains only minimal navigational tools. Onscreen navigation is limited to selecting an answer, advancing to the next problem, going back to the previous problem, and flagging the current problem for later review. The interface also includes a timer, the current problem number (e.g., 45 of 110), a pop-up scientific calculator, and access to an onscreen version of the *FS Reference Handbook*.

During the exam, you can advance through the problems in sequence, but you cannot jump to any specific problem, whether or not it has been flagged. After you have completed the last problem in a session, however, the navigation capabilities change, and you are permitted to review problems in any sequence and navigate to flagged problems.

Knowledge Areas and Problem Distribution

The FS exam includes problems in 13 knowledge areas. Each area is listed as follows with the number of problems relating to this area that you can expect to see on your exam and a list of topics included in this area.

I. Mathematics (13–20 problems): algebra, trigonometry, basic geometry, spherical trigonometry, linear algebra, matrix theory, analytic geometry, and calculus

II. Basic sciences (5–8 problems): geology, dendrology, cartography, and environmental sciences

III. Spatial data acquisition and reduction (6–9 problems): vertical measurement, distance measurement, angle measurement, unit conversions, redundancy, knowledge and utilization of instruments and methods, and understanding of historical methods and instruments.

IV. Survey computations and computer applications (19–29 problems): coordinate geometry, traverse closure and adjustment, area and volume, horizontal and vertical curves, spirals, and spreadsheets

V. Statistics and adjustments (6–9 problems): mean, median, mode, variance, standard deviation, error analysis, least squares adjustment, measurement and positional tolerance, and relative, network, and positional accuracy

VI. Geodesy (5–8 problems): basic theory, satellite positioning, gravity, coordinate systems, datums, and map projections

VII. Boundary and cadastral survey law (13–20 problems): controlling elements, gathering and identifying evidence, records research, legal descriptions, case law, riparian rights, public land survey system, metes and bounds, simultaneously created parcels, easements, and encumbrances

VIII. Photogrammetry and remote sensing (4–6 problems): interpretation and analysis, project and flight planning, quality control, ground control, and LiDAR

IX. Survey processes and methods (11–17 problems): land development (principles, standards, and regulations), boundary location, mapping, cartography, topography, construction, riparian surveys, route surveying, and control surveys

X. Geographic information systems (GIS) (5–8 problems): feature collection and integration, database concepts and design, accuracy and use, and metadata

XI. Graphical communication and mapping (6–9 problems): plans and specifications, contours and slopes, scales, planimetric features and symbols, land forms, digital terrain modeling, digital elevation modeling, and survey maps, plats, drawings, and reports

XII. Professional communication (4–6 problems): oral and written communication, alternative forms of communication, documentation, and recordkeeping

XIII. Business concepts (3–5 problems): contracts, liability, risk management, financial practices, leadership and management principles, personnel manage ment principles, project planning and design, ethics, and safety

Exam content is subject to change. Consult PPI's website (**ppi2pass.com/survfaq**) for current specifications.

THE PS EXAM

Structure

The PS exam consists of a four-hour morning section containing 67 problems and, following a one-hour break, a two-hour afternoon section containing 33 problems. Machine-scored sheets are provided for recording answers.

Knowledge Areas and Problem Distribution

The PS exam includes problems in five knowledge areas. These areas are listed here with the percentage of problems from each area that you can expect to see on your exam.

I. Standards and specifications (12% of problems): Federal statutes, laws, rules and regulations; state and local statutes, laws, rules and regulations; monumentation laws and ordinances; U.S. Public Land Survey System; American Land Title Association/American Congress on Surveying and Mapping (ALTA/ACSM) surveys; geodetic control network accuracy standards; Federal Geographic Data Committee (FGDC) standards (digital mapping); U.S. National Map Accuracy Standards (analog mapping); Federal Emergency Management Agency (FEMA)

II. Legal principles (26% of problems): common law and case law boundary principles, sequential and simultaneous conveyances, U.S. Public Land Survey System, controlling elements in legal descriptions, riparian and littoral rights, property title issues (e.g., encumbrances, interpretation, deficiencies), sovereign land rights (e.g., navigable waters, eminent domain), prescriptive rights/adverse possession, easement rights, and parol evidence

III. Professional survey practices (26% of problems): public and private record sources; project planning (e.g., photogrammetric, geodetic, boundary); control datums; encumbrances (e.g., easements, rights of way, mineral rights, subsurface rights); control network accuracy standards; supervision of and responsibility for field procedures, including instrument operations and usage, monumentation (e.g., identification, classification, perpetuation), vegetation identification (e.g., wetlands, bearing/corner trees, first line of vegetation, aquatic and upland species), survey control (e.g., boundary, topographic, photogrammetric), GPS operations, and construction surveying; supervision of and responsibility for the application of surveying principles and computations, including mapping methods and/or projections, graphical terrain representations, orthometric heights (geoid or ellipsoid), state plane and other coordinate systems, GPS data reduction and analysis, control networks (calculations, analysis, and adjustments), bearings and azimuths, area and volume calculations, horizontal and vertical alignment calculations, construction surveying calculations (e.g., plan interpretation), and data preparation for importation into geographical information systems (GIS); grading and site preparation; survey maps and plats; survey reports; and descriptions

IV. Business/professional practices (20% of problems): project planning (e.g., parameters, costs, budgeting); contracts, risk management (e.g., liability, safety procedures, insurance); ethics; communications (oral, written, graphical); quality assurance procedures; and activities, background, and skills of related professions (e.g., engineers, lawyers, architects, planners)

V. Types of surveys (16% of problems): American Land Title Association/American Congress on Surveying and Mapping (ALTA/ACSM) surveys, control and geodetic surveys, construction surveys (e.g., construction calculations and staking), hydrographic surveys (e.g., elevations of submerged surfaces), boundary surveys, route and right-of-way surveys, topographic surveys (e.g., scanning, photogrammetry, LiDAR, field), condominium surveys, subdivision surveys, and record drawing (as-built) surveys

Exam content is subject to change. Consult PPI's website (**ppi2pass.com/survfaq**) for current specifications.

TYPICAL PROBLEM FORMAT FOR THE FS AND PS EXAMS

The multiple-choice problems on the FS exam are typically short, straightforward, and designed to test your knowledge of the fundamentals of surveying and mapping.

The PS exam is designed to test your ability to apply surveying fundamentals to typical problems encountered in surveying practice. For example, a series of land descriptions from deeds might be provided, followed by multiple-choice problems requiring you to analyze the descriptions, establish certain boundaries or corner positions, and treat encroachments.

Both the FS and PS exams test in customary U.S. units. Therefore, the majority of this book also utilizes U.S.

units. However, in some cases where SI units are commonly used in practice, dual units are given. For a complete list of unit conversions, see Apps. D, E, and F.

EXAM SCORING

Neither the FS exam nor the PS exam is graded on a curve since a certain minimum competency must be demonstrated to safeguard the public welfare. Nevertheless, it is recognized that the tests may vary slightly in difficulty, depending upon the problems selected for a particular exam. Therefore, problems are reviewed by committees of practicing surveyors before the exams. These committees evaluate the difficulty of each problem in order to develop a "standard of minimum competency," or recommended passing score for each exam. However, the individual state boards have the authority to determine the passing score in their respective states. Credit is given for each correct answer and no points are deducted for incorrect answers. The sum of the correct answers is scaled so that the grade of 70 reflects the standard minimum competency.

USE OF CALCULATORS AND COMPUTERS IN THE EXAMS

The exams require use of a scientific calculator. However, it may not be obvious that you should also bring a spare calculator with you. It would be unfortunate not to be able to finish because your calculator was dropped or stolen or stopped working for some unknown reason.

NCEES has banned communicating and text-editing calculators from the exam site. Only select types of calculators are permitted. Check the current list of permissible devices at PPI's website (**ppi2pass.com/calculators**). All the listed calculators have enough functionality for the exam.

The exams have not been optimized for any particular brand or type of calculator. In fact, for most calculations, a $15 scientific calculator will produce results as satisfactory as those from a $200 calculator. There are definite benefits to having built-in statistical functions, graphing, unit-conversion, and equation-solving capabilities. However, these benefits are not so great as to give anyone an unfair advantage.

You may not share calculators with other examinees. Be sure to take your calculator with you whenever you leave the exam room for any length of time.

Laptop computers are not permitted in the exam. You may not use a walkie-talkie, cell phone, or other communications device during the exam.

THE NCEES REFERENCE HANDBOOKS

Both the FS exam and the PS exam are "closed book." No references may be used except for the *FS Reference Handbook* or *PS Reference Handbook*, which will be supplied to you at the exam site.

If you are taking the FS exam, the computer you use will have a 24-inch monitor that is large enough to display at the same time both the exam problems and a searchable PDF version of the *FS Reference Handbook*. The search function is capable of finding anything in the *FS Reference Handbook*, even individual variables, but it can find only precise search terms (e.g., searching for "non-annual compounding" will not locate "nonannual compounding").

If you are taking the PS exam, you will be provided with a bound copy of the *PS Reference Handbook*. You may not write in it or remove pages from it, and you must return it at the end of the exam.

Whichever exam you are preparing for, you should download a copy of the appropriate *Reference Handbook* from the NCEES website and use it in your studies. Become familiar with its contents and organization, so that during the exam you can find equations and tables quickly when you need them. You may download and print out either *Reference Handbook* for your personal use, but you may not bring your personal copy to the exam site. For this reason, though you can make notes in your printed copy if you want to, it's important to know well how to find the information you need in an unannotated copy.

The *PS Reference Handbook* contains the following statement.

> The *Handbook* does not contain all the information required to answer every question on the exam. Some of the basic theories, conversions, formulas, and definitions examinees are expected to know have not been included in the supplied references. When appropriate, NCEES will provide information in the question statement itself to assist you in solving the problem.

Some basic formulas and conversion factors not in the *PS Reference Handbook*, then, may be needed to finish the exam. To be well prepared, you should know many of the basic formulas. Although the *FS Reference Handbook* doesn't contain a similar statement, the same may be assumed to be true of the FS exam.

CHEATING AND EXAM SUBVERSION

The proctors are well trained to insure that cheating does not occur. Obviously, you should not talk to other examinees during the exam, nor should you pass notes back and forth. To prevent discussion, the number of people permitted to use the restrooms at the same time will typically be limited.

The NCEES regularly reuses good problems from previous exams. Therefore, exam security is a serious issue with NCEES, which goes to great lengths to prevent copying of problems. You may not keep your exam booklet, enter text of problems into your calculator, or copy problems into your own material.

The proctors are especially concerned about exam subversion, which generally means any activity that might invalidate the exam or the exam process. The most common form of exam subversion involves trying to copy exam problems for future use.

PREPARING FOR YOUR EXAM

Plan Your Approach

You should consider preparation for the FS or PS exam to be a long-term project and plan carefully. The exams are both comprehensive and fast paced; rapid recall, discipline, stamina, and mastery of the subject areas covered are all essential to success. Development of these qualities may require months of preparation in addition to the years of academic study and work practice you needed to qualify. Therefore, it is important to plan your preparation for the exam as you would plan for a large surveying and mapping project.

These steps can help prepare you.

(1) Review the list of subject areas earlier in this Introduction to gain insight into the nature and content of the exams.

(2) Answer the practice problems at the end of each chapter of this book. For future reference, prepare a concise outline as you work through each area. Your review should be on a rigorous schedule to help you develop the discipline and stamina necessary to do well on the exams. Table 1 at the end of this Introduction is a fill-in-the-dates schedule you can use in planning your study.

(3) For any areas in which you are not comfortable, read additional reference material as you work through each chapter. Tab pages where frequently used or hard-to-find information is located. Consider taking continuing education courses in problem areas.

(4) Take a sample exam, such as *Fundamentals of Surveying Sample Exam* or *Principles and Practice of Surveying Sample Exam* (both available from PPI), to evaluate your readiness for the exams.

(5) Work on any weak areas revealed by the sample exam.

(6) Conduct a final review of your notes.

Learning to use your time wisely is one of the most important things that you can do during your review. You will undoubtedly encounter review problems that take much longer than you expect. You may cause some delays yourself by spending too much time looking through the *Reference Handbook* for the information you need. Other problems will just entail too much work. Learning to recognize such situations more quickly will help you make intelligent decisions during the exams.

Additional Reference Material

You will find that this book is an excellent starting point for preparing for your exam. However, additional references may be helpful, especially in areas in which you are uncomfortable. There are countless texts available that cover the various topics in depth. Listed here are several personal favorites which offer coverage of the areas to be tested on the exams.

Legal Principles

Brown, Robillard, and Wilson. *Boundary Control and Legal Principles*, John Wiley & Sons.

Brown, Robillard, and Wilson. *Evidence and Procedures for Boundary Location*, John Wiley & Sons.

Bureau of Land Management. *Manual of Instruction for Surveys of Public Lands*, Government Printing Office.

Cole, George M. *Water Boundaries*, John Wiley & Sons.

Wattles, W. C. *Land Survey Descriptions*, Gordon Wattles Publications.

Measurement and Computation Theory and Practice

Bureau of Land Management. *Manual of Instruction for Surveys of Public Lands*, Government Printing Office.

Cole, George M. *Water Boundaries*, John Wiley & Sons.

Davis, Foote, Anderson, and Mikhail. *Surveying Theory and Practice*, McGraw-Hill.

Hickerson, Thomas F. *Route Location and Design*, McGraw-Hill.

Wolf and Ghilani. *Adjustment Computations*, John Wiley & Sons.

Geodesy (including GPS) and Survey Astronomy

Buckner, R. B. *A Manual on Astronomic and Grid North*, Landmark Enterprises.

Smith, James R. *Introduction to Geodesy*, John Wiley & Sons.

Van Sickle, Jan. *GPS for Land Surveyors*, Ann Arbor Press.

Geographic Information Systems and Photogrammetry

Clarke, Keith C. *Getting Started with Geographic Information Systems*, Prentice Hall.

Wolf, Paul R., and Bon Dewitt. *Elements of Photogrammetry (with Applications in GIS)*, McGraw-Hill.

Land Development

Colley, Barbara C. *Practical Manual of Land Development*, McGraw-Hill.

Business Law, Management, Economics, and Finance

Denny, Milton E. *Surveyors and Engineers Small Business Handbook*, CED Technical Services.

Lindeburg, Michael R. *Engineering Economic Analysis: An Introduction*, Professional Publications, Inc.

Practice Problems

Cole, George M. *Fundamentals of Surveying Sample Examination*, Professional Publications, Inc.

Cole, George M. *Principles and Practice of Surveying Sample Examination,* Professional Publications, Inc.

Van Sickle, Jan. *1001 Solved Surveying Fundamentals Problems*, Professional Publications, Inc.

Last-Minute Preparation

A week or so before your exam, conduct an intensive review of the outlines you prepared during your study. However, do not attempt to cram the night before the exam.

During the last week or so before the exam, arrange for child care and transportation. Since the exam does not always start or end at the designated time, make sure such arrangements are flexible. If convenient, visit the exam site ahead of time to locate the building, parking areas, exam rooms, and restrooms.

Take a backup calculator to your exam. If your spare calculator is not the same type as your primary one, spend some time familiarizing yourself with it. Make sure that you have correct replacement batteries for both calculators. In addition, you should prepare a kit of items to take to the exam.

Take the day before the exam off from work to relax. If you live far from the exam site, consider getting a hotel room in which to spend the night. Calculate your wake-up time, and set two alarms. Select and lay out your clothing and breakfast items, and make sure that you have gas in your car and money in your wallet.

TAKING YOUR EXAM

What to Take to Your Exam

Your exam kit should contain items you need, such as your photo ID and your calculator, as well as items pertaining to your personal comfort. However, you may bring only certain items into the testing room, and the list of permitted items is different for the FS and PS exams.

You will need your photo ID in order to be admitted to the test site for either exam. Your ID must be government issued and must include

- your name
- your date of birth
- a recognizable photo of you
- your signature (except for U.S. military IDs)
- an expiration date (not past)

The FS Exam

For the FS exam, your exam kit might contain the following items. Some of these, however, you may not bring into the testing room and must leave in a small locker that will be provided to you. You may access the locker during the break between sessions, but not during either session.

- an acceptable form of photo ID (essential)
- a printed copy of your appointment confirmation letter (strongly recommended)
- your primary calculator, with fresh batteries installed
- your backup calculator (left in locker)
- spare batteries for your calculators (left in locker)
- eyeglasses (case left in locker)
- eyeglasses repair kit, including a small screwdriver for fixing glasses or removing batteries from your calculator (left in locker)
- contact lens wetting solution (left in locker)
- a light sweater or jacket (pockets empty)
- cough drops (unwrapped and not in a bottle or container; a clear plastic bag is acceptable)
- aspirin or other pills (unwrapped and not in a bottle or container; a clear plastic bag is acceptable)
- eyedrops
- a pillow or cushion (if you need one in order to sit comfortably through the exam)
- several dollars in loose change (left in locker)
- an extra set of car keys (left in locker)
- something to eat during the break (left in locker)

You will be provided with earplugs, noise-cancelling headphones, tissues, and a reusable booklet and marker for scratch work, so you don't have to bring your own; if you do, you will not be permitted to bring them into the testing room. You must leave your wallet, purse, wristwatch, car keys, cell phone, and other personal items in your locker.

You may also bring essential medicines, medical devices, and mobility devices. These items will be inspected visually before you may bring them into the testing room. These include

- bandages
- braces (neck, back, wrist, leg, or ankle)
- cane
- crutches
- casts, slings, and other injury-related items that cannot be removed
- eye patches

- handheld magnifying glass (not electronic; case left in locker)
- hearing aids or cochlear implants
- inhaler
- insulin pump (or other medical device attached to your body)
- medical alert bracelet
- medical/surgical face mask
- motorized scooter or chair
- oxygen tank
- walker
- wheelchair

For medical-related items not on this list, you must get approval in advance of your exam day.

The PS Exam

For the PS exam, your exam kit might contain any or all of the items in the FS exam kit, as well as the following items, which may be brought into the testing room.

- snacks, such as hard candies, that you can eat without disturbing other examinees (unwrapped and in a clear plastic bag)
- bottled water or other nonalcoholic drinks
- a wristwatch or small clock
- a handkerchief, or tissues in a clear plastic bag
- two straightedges, such as a ruler, scale, triangle, or protractor

You will be provided with a pencil with an eraser for use during the exam, and the exam booklet contains blank areas for scratch work. Leave your wallet, purse, briefcase, backpack, cell phone, cigarettes, eyeglasses case, magnifying glass case, writing implements, writing tablets, and other personal items in your locked car.

What to Do at the Exam

Arrive at least 30 minutes before your exam is scheduled to begin. This will allow you to find a convenient parking place, get to the exam room, and calm down. Be prepared, though, to find that the exam room is not open or ready at the designated time.

All the procedures typically associated with timed, proctored, computer-graded assessment tests will be in effect when you take the PS exam. The proctors will distribute the exam booklets and answer sheets. However, you should not open the booklets until instructed to do so.

Listen carefully to everything the proctors say. Do not ask the proctors any technical questions. Even if knowledgeable in engineering, they are not permitted to answer your questions. They will guide you through the process of putting your name and other biographical information on the material. Time for these instructions and for initializing the answer sheets is not part of the timed exam period.

The common suggestions to "completely fill the bubbles and erase completely" apply here. Mechanical pencils with erasers are provided by NCEES.

On both the FS and PS exams, every problem is worth the same number of points, so it is a good idea to answer all of the problems that you can within a reasonable time before attempting to solve problems that will take a disproportionate amount of time. If time allows, you can go back to those difficult problems.

Many points are lost due to carelessness. Therefore, it is a good idea to read each problem twice before solving. Check to make sure that you used all of the given data and made the appropriate conversion of units. While the exam problems are not tricky, you may find the results of commonly made mistakes are represented among the available answer choices. Thus, just because there is an answer matching your results does not mean that you have obtained the correct results.

Both the FS and PS exams are multiple choice. Credit is given for correct answers, but no credit is deducted for wrong answers. Therefore, it is in your best interest to answer every question. It is a good idea to use the last five minutes of each session to guess at any remaining unsolved multiple-choice problems. You will be successful with about 25% of your guesses, and those points will more than make up for the few points you might earn by working during the last five minutes.

After Your Exam

People react quite differently to the exam experience. Some people are energized and need to unwind by talking with other examinees, describing every detail of their experience and dissecting every exam problem. However, most people are completely exhausted and need a lot of quiet space and a hot tub in which to soak and sulk. Since everyone who took the exam has seen it, you will not be violating your "oath of silence" if you talk about the details with other examinees. It is difficult not to ask how someone else approached a problem that had you completely stumped. However, it is also very disquieting to think you did well on a problem, only to have someone else tell you where you went wrong.

Waiting for your exam results is its own form of mental torture. There is no predictable pattern to the release of the results. Exam results are not released by NCEES to all states simultaneously. They are not released alphabetically by state or examinee name. The people who failed are not notified first or last. Your co-worker might receive his or her notification today, and you might have to wait another three weeks. It all depends on when the entire process is complete. Some states have to have the results approved at a board meeting. Some prepare certificates before sending out notifications. Some states are more highly automated that others. The number of examinees also varies by state, as do numerous other factors. Therefore you just have to wait patiently.

You will typically receive your results within 10 weeks for the PS exam and within 7 to 10 days for the FS exam. Your licensing board will contact you with your results. If you passed the exam, you will receive a letter that states you passed. If you failed, you will receive notice of this and get a diagnostic report that shows your strengths and weaknesses.

Now that you know all there is to know about the exams and about how to prepare for them, the rest is up to you. Plan your approach, and get to work. The very best of luck to you!

Table 1 *Schedule for Self-Study*

chap. no.	subject	date to start	date to finish
1	Algebra		
2	Basic Geometry		
3	Dimensional Equations		
4	Systems of Units		
5	Perimeter and Circumference		
6	Area		
7	Volume		
8	Trigonometry		
9	Rectangular Coordinate System		
10	Analytical Geometry		
11	Taping		
12	Electronic Distance Measurement		
13	Leveling		
14	Compass Survey		
15	Traverse		
16	Area of a Traverse		
17	Partitioning of Land		
18	Horizontal Curves		
19	Vertical Alignment		
20	Astronomic Observations		
21	Global Positioning System		
22	Map Projections and Plane Coordinate Systems		
23	History and Origins of Title		
24	Transfer of Ownership		

chap. no.	subject	date to start	date to finish
25	Water Boundaries		
26	Riparian and Littoral Rights		
27	Public Land Survey System		
28	Restoration of Lost and Obliterated Public Land Survey Corners		
29	Land Descriptions		
30	Colonial History and the U.S. Legal System		
31	Subdivisions		
32	Residential Planning		
33	Topographic Surveying and Mapping		
34	Geographic Information Systems		
35	Aerial Mapping		
36	Laser Scanning		
37	Construction Staking		
38	Earthwork		
39	Hydrographic Surveying		
40	Computer Hardware		
41	Data Structure and Programming		
42	Job Costing		
43	Economic Analysis		
44	Ethics for Surveyors		

Topic I: Mathematics Basics

Chapter

1. Algebra
2. Basic Geometry
3. Dimensional Equations
4. Systems of Units
5. Perimeter and Circumference
6. Area
7. Volume
8. Trigonometry
9. Rectangular Coordinate System
10. Analytical Geometry

1 Algebra

1. VARIABLES

Letters of the alphabet used to represent numbers are called variables (or *literal numbers*). The letters a, b, c, x, y, and z are commonly used to represent a number, but any letter of the alphabet can be used. Variables are called *general numbers* because they do not represent a specific number. For example, the area of a certain room that is 12 ft long and 10 ft wide is equal to the product 12×10. To express the area of any rectangle, it can be written as $A = LW$, where A is area in square measure, L is length in linear measure, and W is width in linear measure.

2. USING VARIABLES

Addition, subtraction, multiplication, and division using literal numbers are performed in the same manner as in arithmetic. If a and b represent any two numbers, their sum is $a+b$; their difference is $a-b$; their product is $a \times b$, $a{\cdot}b$, $(a)(b)$, or ab; and their quotient is $a \div b$ or a/b. Expressing the product as $a{\cdot}b$, or ab prevents confusing the letter x with the multiplication sign $\times$.

3. DEFINITIONS

A *term* is an algebraic expression not separated within itself by a plus or minus sign, such as $4xy$, $5x^2y$, or $2ab^2c$.

A *product* implies multiplication. The term $5xy$ means $5 \times x \times y$.

A *monomial* consists of one term, such as $4a^2b$; a *binomial* consists of two terms, such as $3x^2-5xy$; a *trinomial* consists of three terms; and a *polynomial* consists of any number of terms more than one.

The quantities multiplied together to form a product are called the *factors* of the product. The factors of $3xy$ are 3, x, and y.

The numerical factor in a monomial is known as the *numerical coefficient* or simply the *coefficient.* The coefficient of $3x^2y$ is 3, and the coefficient of $-5ab$ is -5.

Like or *similar terms* are terms that have the same literal factors. Terms that do not have the same literal factors are called *unlike* or *dissimilar* terms. $3x^2y$ and $5x^2y$ are like terms; $5x^2y$ and $5x^2y^2$ are unlike terms.

4. HORIZONTAL ADDITION AND SUBTRACTION OF MONOMIALS

Horizontal addition and subtraction of monomials is carried out according to the rules of addition and subtraction of signed numbers. Like terms can be combined; unlike terms cannot be combined.

Example 1.1

(a) $11ab + 4ab - 10ab - 8ab =$

(b) $5x^3 + 2x^2y - 8x^2y + 2x^3 =$

(c) $2ab^2 - 3ab + 2 - 4ab^2 + 3ab =$

Solution

(a) $-3ab$

(b) $7x^3 - 6x^2y$

(c) $-2ab^2 + 2$

5. EXPONENTS

As shown previously, 3 squared equals 9, meaning that 3 multiplied by itself equals 9. This can be written as $3{\cdot}3 = 9$ or $3^2 = 9$. It is also true that 2 cubed equals 8, meaning that $2{\cdot}2{\cdot}2 = 8$. This can be written as $2^3 = 8$. The small number 3 is known as the *exponent*; the number 2 is known as the *base*. 4^6 means that six 4's are to be multiplied. The exponent is the power to which a number is to be raised. If a number does not have an exponent, it is of the first power and the exponent is considered to be 1. To raise a literal number to a power, use an exponent. Thus, x^5 means $x{\cdot}x{\cdot}x{\cdot}x{\cdot}x$ and is read "x raised to the fifth power" or "x to the fifth."

6. EXPONENTS USED IN MULTIPLICATION

Multiplication in algebra follows the same rules, or laws, as multiplication in arithmetic. Exponents simplify the process. To multiply $x^3{\cdot}x^4$,

$$x^3{\cdot}x^4 = (x{\cdot}x{\cdot}x)(x{\cdot}x{\cdot}x{\cdot}x) = x^7$$

However, it is much easier to say that

$$x^3{\cdot}x^4 = x^{3+4} = x^7$$

This can be expressed in the form of a rule.

rule: The exponent of the product of powers with the same base is the sum of the exponents of the factors.

The rule holds for either positive or negative exponents, and the base may be an arithmetic number. Thus,

$$(2)^5(2)^{-2} = (2)^3 = 8$$

The rule for multiplication holds only for the product of powers of the same base; however, multiplication involving powers of more than one base may be simplified. For example,

$$(x^3y^4z^2)(x^2y^{-2}z) = x^5y^2z^3$$
$$(2)^4(3)^3(4)^2 = 16{\cdot}27{\cdot}16 = 6912$$

7. EXPONENTS USED IN DIVISION

Division is the inverse of multiplication, so it is logical to conclude that in division the exponent of the divisor is subtracted from the exponent of the dividend.

rule: The exponent of the quotient of two powers with the same base is the difference of the exponent of the dividend and the exponent of the divisor.

Example 1.2

Divide the following numbers.

(a) $\frac{x^5}{x^2}$

(b) $\frac{(3)^5}{(3)^2}$

(c) $a^2 \div a^{-3}$

(d) $\frac{x^3}{x^5}$

Solution

(a) $\frac{x{\cdot}x{\cdot}x{\cdot}x{\cdot}x}{x{\cdot}x} = x{\cdot}x{\cdot}x = x^3$

(b) $3^{5-2} = (3)^3 = 27$

(c) $a^{2+3} = a^5$

(d) $x^{3-5} = x^{-2}$

8. EXPONENT OF THE POWER OF A POWER

The exponent of a power is the product of the exponents of the powers.

Example 1.3

(a) $(x^3)^2$

(b) $(-5x^2y^3)^2$

(c) $\frac{a^3(b^2)^4}{(ab^3)^2}$

Solution

(a) $(x{\cdot}x{\cdot}x){\cdot}(x{\cdot}x{\cdot}x) = x^{3{\cdot}2} = x^6$

(b) $(-5)^2x^{2{\cdot}2}y^{3{\cdot}2} = 25x^4y^6$

(c) $\frac{a^3b^8}{a^2b^6} = ab^2$

9. ZERO POWER

Any number other than zero raised to the zero power is equal to 1. This applies to arithmetic numbers and to literal numbers, as well as to a polynomial enclosed in parentheses.

Example 1.4

(a) $\frac{(2)^3}{(2)^3}$

(b) $\frac{x^3}{x^3}$

Solution

(a) $\frac{2\cdot2\cdot2}{2\cdot2\cdot2} = (2)^{3-3} = (2)^0 = 1$

(b) $x^{3-3} = x^0 = 1$

10. NEGATIVE EXPONENTS

A number with a negative exponent is equal to 1 divided by the number with the sign of the exponent changed to positive.

Example 1.5

(a) $\frac{(2)^2}{(2)^4}$

(b) $\frac{(10)^3}{(10)^4}$

(c) $\frac{x^{-2}y^3}{x^3y^{-2}}$

(d) $\frac{a^{-3}}{b^{-2}}$

Solution

(a) $\frac{4}{16} = \frac{1}{4}$

(b) $10^{-1} = \frac{1}{10}$

(c) $\frac{y^3\cdot y^2}{x^2\cdot x^3} = \frac{y^5}{x^5}$

(d) $\frac{b^2}{a^3}$

Example 1.6

(a) $5\cdot5\cdot5$

(b) $a^3\cdot a^5$

(c) $(4)^3(4)^2$

(d) $(3x)(4x)$

(e) $x^3 \div x^{-2}$

(f) $(2)^2(3)^2(4)^2$

(g) $(ab^2)^3$

(h) $(-3x)^5$

(i) $(-2xy^2)^2$

(j) $\frac{x^3\cdot x^4}{x^2}$

(k) $\frac{(3)^7(3)^2\cdot3}{3^5}$

(l) $\frac{64a^3b^3}{4ab^2}$

(m) $\frac{5x^5y^5}{45x^6y^6}$

(n) $\frac{(-2x^2y^2)^3}{(2xy^2)^2}$

(o) $\frac{(-3ab^2)^3}{(12ab^2)^2}$

(p) $(5)^{-2}(5)^3$

(q) $a^3\cdot a^{-1}$

(r) $(10)^0$

(s) $3a^0$

(t) x^0y^2

(u) $(4-2)^0$

(v) $\frac{x^{-5}y^3}{xy}$

(w) $\frac{a^{-4}b^5c^{-6}}{ab^{-2}c^3}$

(x) $\frac{x^3y^{-2}}{xy^3}$

Solution

(a) 125

(b) a^8

(c) $(4)^5 = 1024$

(d) $12x^2$

(e) x^5

(f) $4\cdot9\cdot16 = 576$

(g) a^3b^6

(h) $-243x^5$

(i) $4x^2y^4$

(j) x^5

(k) $(3)^5 = 243$

(l) $16a^2b$

(m) $\frac{1}{9xy}$

(n) $-2x^4y^2$

(o) $\frac{-27ab^2}{144}$

(p) 5

(q) a^2

(r) 1

(s) 3

(t) y^2

(u) 1

(v) $\frac{y^2}{x^6}$

(w) $\frac{b^7}{a^5c^9}$

(x) $\frac{x^2}{y^5}$

11. MULTIPLYING A MONOMIAL AND A POLYNOMIAL

When a polynomial is multiplied by a monomial, each term of the polynomial must be multiplied by the monomial.

Example 1.7

$$-4x(2x^2 - 3x - 4)$$

Solution

$$\begin{aligned}&(-4x)(2x^2) - (-4x)(3x) - (-4x)(4)\\&= -8x^3 + 12x^2 + 16x\end{aligned}$$

12. MULTIPLYING BINOMIALS OR TRINOMIALS

In multiplying two binomials, each term of the multiplicand must be multiplied by each term of the multiplier.

Example 1.8

Multiply the following numbers.

(a)
$$\begin{array}{r}2x - 3y\\ 3x + 2y\\ \hline\end{array}$$

(b)
$$\begin{array}{r}x^2 + 4x - 3\\ 3x - 2\\ \hline\end{array}$$

Solution

(a)
$$\begin{array}{rrr}2x & - 3y & \\ 3x & + 2y & \\ \hline 6x^2 & - 9xy & \\ & + 4xy & - 6y^2\\ \hline 6x^2 & - 5xy & - 6y^2\end{array}$$

(b)
$$\begin{array}{rrrr} & x^2 & + 4x & - 3\\ & & 3x & - 2\\ \hline 3x^3 & + 12x^2 & - 9x & \\ & - 2x^2 & - 8x & + 6\\ \hline 3x^3 & + 10x^2 & - 17x & + 6\end{array}$$

In both solutions, the factors are arranged in a manner similar to multiplication in arithmetic with the multiplier placed under the multiplicand. Multiplication is performed from right to left.

Using Ex. 1.8., the steps in the operation are as follows.

step1: The left term in the multiplier is multiplied by the left term in the multiplicand, and the product is placed under the first column: $(3x)(2x) = 6x^2$.

step 2: The left term in the multiplier is multiplied by the next term (from left to right) in the multiplicand, and the product is placed under the second column: $(3x)(-3y) = -9xy$.

step 3: The second term in the multiplier is multiplied by the first term in the multiplicand, and the product is placed under $-9xy$ because it is a similar term: $(2y)(2x) = +4xy$.

step 4: The second term in the multiplier is multiplied by the second term in the multiplicand, and the product is placed in the third column: $(2y)(-3y) = -6y^2$.

step 5: Similar terms are combined to give the product: $6x^2 - 5xy - 6y^2$.

Example 1.9

(a) $3xy(x^2 - xz + z^2)$

(b) $2ab^2(a^2 + 2ab - 3b^2)$

(c) $-3x^2y(xy - 2xy^2 - 3y^3)$

(d) $(x + 3)(x + 4)$

(e) $(x - 2)(x^2 + 4x - 8)$

Solution

(a) $3x^3y - 3x^2yz + 3xyz^2$

(b) $2a^3b^2 + 4a^2b^3 - 6ab^4$

(c) $-3x^3y^2 + 6x^3y^3 + 9x^2y^4$

(d) $x^2 + 7x + 12$

(e) $x^3 + 2x^2 - 16x + 16$

13. DIVISION OF A POLYNOMIAL BY A MONOMIAL

To divide a polynomial by a monomial, each term in the dividend must be divided by the divisor.

Example 1.10

Divide the following numbers.

$$\frac{6ax^3y^3 - 8a^2x^2y^2 + 4ax^3y}{2axy}$$

Solution

$$\begin{aligned}\frac{6ax^3y^3 - 8a^2x^2y^2 + 4ax^3y}{2axy} &= \frac{6ax^3y^3}{2axy} - \frac{8a^2x^2y^2}{2axy} + \frac{4ax^3y}{2axy} \\ &= 3x^2y^2 - 4axy + 2x^2\end{aligned}$$

14. DIVISION OF A POLYNOMIAL BY A POLYNOMIAL

To perform division of a polynomial by a polynomial, the terms in the dividend and the divisor should be arranged in order of decreasing power. It is performed much like division in arithmetic.

Example 1.11

Divide the following number.

$$(15x^2 - x - 6) \div (5x + 3)$$

Solution

$$\begin{array}{r} 3x - 2 \\ 5x + 3\,\overline{)\,15x^2 - x - 6} \\ 15x^2 + 9x \\ -10x - 6 \\ -10x - 6 \end{array}$$

The steps in the procedure are as follows.

step 1: Divide the first term in the dividend by the first term in the divisor, $15x^2 \div 5x = 3x$, and place $3x$ over the dividend.

step 2: Multiply the divisor by the first term in the quotient: $3x(5x + 3) = 15x^2 + 9x$. Place these two terms under similar terms in the dividend and subtract: $(15x^2 - x) - (15x^2 + 9x) = -10x$.

step 3: Bring down the next term in the dividend to form the new dividend: $-10x - 6$.

step 4: Divide the first term in the new dividend by the first term in the divisor, $-10x \div 5x = -2$, and place -2 as the second term in the quotient.

step 5: Multiply the divisor by the second term in the quotient: $(-2)(5x+3) = -10x-6$. Place these two terms under similar terms in the dividend and subtract: $(-10x-6)-(-10x-6) = 0$.

If there had been a remainder of lower order than the first term in the divisor, it would be written over the divisor, as in arithmetic, to represent the remainder. Because there was no remainder, $(5x + 3)(3x - 2)$ are factors of $15x^2 - x - 6$.

Example 1.12

Perform the indicated divisions.

(a) $\dfrac{36a^3 - 24a^2 - 12a}{2a}$

(b) $\dfrac{21a^3b^2 - 14a^2b^3 + 7ab}{7ab}$

(c) $\dfrac{15a^4b^3c^2 + 20a^3b^2c^3 - 25a^2b^2c^3}{5a^2b^2c}$

(d) $\dfrac{9x^5y^4z^3 - 18x^3y^2z^2 + 3x^4y^3z^2}{3x^3y^2z}$

Solution

(a) $18a^2 - 12a - 6$

(b) $3a^2b - 2ab^2 + 1$

(c) $3a^2bc + 4ac^2 - 5c^2$

(d) $3x^2y^2z^2 - 6z + xyz$

15. FACTORING

Factoring is the reverse of multiplication. The product of $8 \cdot 9$ is 72. The factors of 72 are 8 and 9, as well as 12 and 6. The prime factors of 72 are $2{\cdot}2{\cdot}2{\cdot}3{\cdot}3$.

16. FACTORING A POLYNOMIAL CONTAINING A COMMON MONOMIAL

The first step in factoring a polynomial is to factor a common monomial, if one exists. The other factor is found by dividing each term of the polynomial by the common factor and writing the result as the product of the common factor and the quotient.

Example 1.13

Factor $20x^2 + 12x$.

Solution

The greatest monomial factor in the binomial is $4x$.

$$20x^2 + 12x = 4x(5x + 3)$$

Example 1.14

Factor $xy^2z^3 - x^2y^3z^4$.

Solution

$$xy^2z^3 - x^2y^3z^4 = xy^2z^3(1 - xyz)$$

17. FACTORING A TRINOMIAL THAT IS A PERFECT SQUARE

The number 4 is a *perfect square* because both factors of 4 are the same. Likewise, 9, 16, and 25 are perfect squares.

A trinomial is also a perfect square if both factors are the same. The product of $(x+2)(x+2)$ is x^2+4x+4, which is a perfect square. It can be said that $x^2+4x+4=(x+2)^2$.

To factor a trinomial that is a perfect square, it is necessary to recognize a trinomial that is a perfect square. Examine the following equation.

$$x^2+4x+4=(x+2)(x+2)=(x+2)^2$$

In the factor $(x+2)$,

(1) x is the square root of the first term in the trinomial.

(2) 2 is the square root of the third term in the trinomial.

(3) The product of these two terms is $2x$, which is one half the middle term of the trinomial.

(4) The sign between the two terms is the same as the sign of the middle term of the trinomial.

For a trinomial to be a perfect square, the following must be true.

(1) The first and third terms must be perfect squares.

(2) The middle term must be twice the product of the square roots of the first and third terms.

(3) The sign of the middle term can be plus or minus, but the sign of the second term of the factors must be the same as the sign of the middle term of the trinomial.

Examine the following trinomial and its factors.

$$x^2-14x+49=(x-7)^2$$

(1) Are the first and third terms perfect squares? Yes.

(2) Is the square root of the first term of the trinomial equal to the first term of the factors? Yes: $\sqrt{x^2}=x$.

(3) Is the square root of the third term of the trinomial equal to the second term of the factors? Yes: $\sqrt{49}=7$.

(4) Is the middle term of the trinomial twice the product of the terms of the factors? Yes: $14x=(2)(7x)$.

(5) Is the sign of the second term in the factors the same as the sign of the middle term of the trinomial? Yes.

Therefore, the trinomial $x^2-14x+49$ is a perfect square.

Example 1.15

(a) $25a^2-20a+4$

(b) $9-24x+16x^2$

(c) $4a^2-4a+1$

(d) x^2-6x+9

Solution

(a) $(5a-2)^2$

(b) $(3-4x)^2$

(c) $(2a-1)^2$

(d) $(x-3)^2$

18. FACTORING THE DIFFERENCE BETWEEN TWO SQUARES

The product of $(a+b)(a-b)$ is a^2-b^2.

$$\begin{array}{l} a+b \\ a-b \\ \hline a^2+ab \\ \quad -ab-b^2 \\ \hline a^2-b^2 \end{array}$$

Notice that the factors are the same except for the algebraic sign. Also notice that the product (a^2-b^2) is equal to the square of the first term of the factors minus the square of the second term. There is no middle term in the product because the sum of $+ab$ and $-ab$ is zero.

The factors of the difference between two squares are the product of the square root of the first term plus the square root of the second term times the square root of the first term minus the square root of the second term.

Example 1.16

(a) x^2-25

(b) $9x^2-16y^2$

(c) $36x^2-81y^2$

(d) $1-64a^2$

Solution

(a) $(x+5)(x-5)$

(b) $(3x+4y)(3x-4y)$

(c) $(9)(4x^2-9y^2)=(9)(2x+3y)(2x-3y)$

(d) $(1+8a)(1-8a)$

19. FACTORING A TRINOMIAL OF THE FORM $Ax^2 + Bx + C$

The factors of a perfect square, such as $a^2+2ab+b^2$, or the difference between two squares, such as a^2-b^2, are apparent as soon as the type of polynomial is identified. However, some polynomials, such as $2x^2-5x-12$, do not conform to either type.

Because factoring is the opposite of multiplication, start with the assumption that the factors of $2x^2-5x-12 = (2x+3)(x-4)$ and multiply the two factors.

$$\begin{array}{r} 2x+3 \\ x-4 \\ \hline 2x^2+3x \\ -8x-12 \\ \hline 2x^2-5x-12 \end{array}$$

The first term in the trinomial $2x^2-5x-12$ is found by multiplying the first terms of the factors: $x(2x) = 2x^2$, and the third term of the trinomial is found by multiplying the second terms of the factors: $(-4)(3) = -12$. The middle term of the trinomial is the algebraic sum of the cross products: $x(3)+(-4)(2x) = -5x$.

Now factor $2x^2-x-15$.

The factors are not immediately apparent, so set up two pairs of parentheses to contain the factors and insert trial numbers.

$$2x^2-x-15 = (\quad)(\quad)$$

The product of the first terms of the factors must equal $2x^2$. The only possibility is $x(2x)$. But which of these two terms will be placed in the first set of parentheses and which in the second?

$$2x^2-x-15 = (x\quad)(2x\quad)$$

Because the middle and third terms of the trinomial are negative, one of the signs in the parentheses must be positive and one negative. (If both signs were negative, the sign of the third term in the trinomial would be positive.) But which will be positive and which negative?

$$2x^2-x-15 = (x-\quad)(2x+\quad)$$

The product of the second terms of the factors must be -15, so insert 5 in the first parentheses and 3 in the second. The product of these two terms is -15, but the algebraic sum of the cross products gives $-10x$, so try

$$2x^2-x-15 = (x-3)(2x+5)$$

The product of these two factors is $(2x^2-x-15)$, so the trinomial is factored.

With practice, various combinations of numbers within the parentheses can be tried by performing mental multiplication and addition.

Example 1.17

Factor the following polynomials.

(a) x^2+6x+5

(b) x^2-4x+3

(c) $a^2-9a+14$

(d) $5a^2+11a+6$

(e) $9x^2-13x+4$

Solution

(a) $(x+5)(x+1)$

(b) $(x-3)(x-1)$

(c) $(a-7)(a-2)$

(d) $(5a+6)(a+1)$

(e) $(9x-4)(x-1)$

Example 1.18

Factor the following polynomials.

(a) $8x^4-4x^3$

(b) a^2-16

(c) x^2+6x+9

(d) a^4-16

(e) $5a^2-30a+45$

(f) $16x^2-24x+9$

(g) $36a^2-81b^2$

(h) x^2-5x+6

(i) $a^2+4a-12$

(j) x^3-10x^2+9x

Solution

(a) $4x^3(2x-1)$

(b) $(a+4)(a-4)$

(c) $(x+3)^2$

(d) $(a^2+4)(a+2)(a-2)$

(e) $(5)(a-3)^2$

(f) $(4x-3)^2$

(g) $(9)(2a+3b)(2a-3b)$

(h) $(x-3)(x-2)$

(i) $(a+6)(a-2)$

(j) $x(x-9)(x-1)$

20. EQUATIONS

Probably the most important concept in algebra is the *equation*. It is a means of solving problems in science, engineering, and everyday life.

An equation is a statement that two quantities are equal. The equal sign (=) separates the two quantities. The terms on the left of the equal sign are known as the *left member*; the terms on the right of the equal sign are known as the *right member*.

21. CONDITIONAL EQUATIONS

The statement "$2x + 4 = 10$" is true if, and only if, x is equal to 3. In other words, the equation is true on the condition that x is equal to 3. Therefore, 3 satisfies the equation.

It is important to be able to determine an unknown quantity in an equation.

22. ROOT OF AN EQUATION

In the equation $2x + 4 = 10$, 3 is the solution of the equation, or the *root* of the equation, and 3 is the only root of the equation.

23. SOLVING AN EQUATION

Solving an equation simply means finding the value of the literal number in the equation, or finding the root of the equation. The root of the equation $2x + 4 = 10$ can be found by inspection, but some equations are not so easily solved, so certain principles of algebra are necessary. These principles, or truths, are known as *axioms*.

24. AXIOMS

axiom 1: The same quantity may be added to both sides of an equation without altering the truth of the statement.

axiom 2: The same quantity may be subtracted from both sides of an equation without altering the truth of the statement.

axiom 3: Both sides of the equation may be multiplied by the same quantity, other than zero, without altering the truth of the statement.

axiom 4: Both sides of an equation may be divided by the same quantity, other than zero, without altering the truth of the statement.

axiom 5: The square root (or any root) of each side of an equation may be taken without altering the truth of the statement.

Example 1.19

Using axiom 1, solve the following equation.

$$x - 3 = 12$$

Solution

$$x - 3 = 12$$

Adding:

$$x - 3 + 3 = 12 + 3$$
$$x = 15$$

Example 1.20

Using axiom 2, solve the following equation.

$$x + 6 = 12$$

Solution

$$x + 6 = 12$$

Subtracting:

$$x + 6 - 6 = 12 - 6$$
$$x = 6$$

Example 1.21

Using axioms 3 and 4, solve the following equation.

$$4 = \frac{12}{x}$$

Solution

$$4 = \frac{12}{x}$$

Multiplying:

$$x(4) = \frac{x(12)}{x}$$
$$4x = 12$$

Dividing:

$$\frac{4x}{4} = \frac{12}{4}$$
$$x = 3$$

Example 1.22

Using axiom 5, solve the following equation.

$$x^2 = 16$$

Solution

$$x^2 = 16$$

Taking the square root:

$$\sqrt{x^2} = \sqrt{16}$$
$$x = \pm 4$$

25. TRANSPOSING

Note in Ex. 1.19 (axiom 1) that when 3 was added to both sides of the eqation, -3 disappeared from the left side, leaving only x on that side. Also, in Ex. 1.20 (axiom 2) when 6 was subtracted from both sides of the equation, $+6$ disappeared from the left side, leaving only x. Putting the unknown quantity x on the left side with no other quantity can be accomplished by a shortcut method known as *transposing*. It is not an axiom but gives the same results as were obtained using axiom 1 and axiom 2.

Transposing means that any term may be moved from one side of an equation to the other side if its sign is changed.

Example 1.23

Solve the following equation.

$$3x - 2 = x + 6$$

Solution

$$3x - 2 = x + 6$$

Transposing: $3x - x = 6 + 2$

$$2x = 8$$

Dividing: $x = 4$

Example 1.24

Solve the following equation.

$$4x + 2 - 3x - 6 = 5x + 4$$

Solution

$$4x + 2 - 3x - 6 = 5x + 4$$

Transposing: $4x - 3x - 5x = 4 - 2 + 6$

$$-4x = 8$$

Dividing by -4: $x = -2$

26. PARENTHESES

If a quantity within parentheses, brackets, or braces is preceded by a plus sign (+), the parentheses may be removed without changing the sign of the terms within the parentheses.

If a quantity within parentheses is preceded by a minus sign ($-$), the parentheses may be removed if the sign of each term within the parentheses is changed.

Example 1.25

Solve the following equation.

$$\begin{aligned} 7x - (2x - 4) - (2)(4x - 2) + (3x - 3) \\ = 8 - (x - 5) - (3)(x + 4) \end{aligned}$$

Solution

Removing parentheses:

$$\begin{aligned} 7x - 2x + 4 - 8x + 4 + 3x - 3 \\ = 8 - x + 5 - 3x - 12 \end{aligned}$$

Transposing: $7x - 2x - 8x + 3x + x + 3x$

$$= 8 + 5 - 12 - 4 - 4 + 3$$

Combining: $4x = -4$

Dividing: $x = -1$

27. FRACTIONAL EQUATIONS

To solve fractional equations, the first step is to get rid of the denominators. This is done by multiplying both sides of the equation by the lowest common denominator (LCD). Then, after the equation is cleared of fractions, the equation is solved using the axioms.

Example 1.26

Solve the following equation.

$$\frac{3x}{4} + 5 = 2x - \frac{1}{3}$$

Solution

To solve equations with fractions, it is convenient to write any term without a denominator with a denominator of 1.

$$\frac{3x}{4} + \frac{5}{1} = \frac{2x}{1} - \frac{1}{3}$$

The LCD is 12.

Multiplying by the LCD:

$$\begin{aligned} \left(\frac{12}{1}\right)\left(\frac{3x}{4}\right) + \left(\frac{12}{1}\right)\left(\frac{5}{1}\right) \\ = \left(\frac{12}{1}\right)\left(\frac{2x}{1}\right) - \left(\frac{12}{1}\right)\left(\frac{1}{3}\right) \end{aligned}$$

$$9x + 60 = 24x - 4$$

Transposing: $-15x = -64$

$$x = \frac{64}{15}$$

Example 1.27

Solve the following equations.

(a) $3x + 5 = 14$

(b) $5x - 8 = x + 16$

(c) $4x + (3)(2x - 5) = 7 + (5)(3x - 7)$

(d) $(2x - 3)(x + 4) - 23 = (x + 7)(2x)$

(e) $(2)(x+3)(x-4) = 6 + 2x(x-5)$

(f) $\dfrac{x-5}{x-3} - \dfrac{x+6}{x^2+x-12} = \dfrac{x-2}{x+4}$

Solution

(a)

$$\begin{aligned} 3x+5 &= 14 \\ 3x &= 14-5 \\ 3x &= 9 \\ x &= 3 \end{aligned}$$

(b)

$$\begin{aligned} 5x-8 &= x+16 \\ 5x-x &= 16+8 \\ 4x &= 24 \\ x &= 6 \end{aligned}$$

(c)

$$\begin{aligned} 4x+(3)(2x-5) &= 7+(5)(3x-7) \\ 4x+6x-15 &= 7+15x-35 \\ 4x+6x-15x &= 7-35+15 \\ -5x &= -13 \\ x &= \frac{13}{5} \end{aligned}$$

(d)

$$\begin{aligned} (2x-3)(x+4)-23 &= (x+7)(2x) \\ 2x^2+5x-12-23 &= 2x^2+14x \\ 5x-14x &= 35 \\ -9x &= 35 \\ x &= -\frac{35}{9} \end{aligned}$$

(e)

$$\begin{aligned} (2)(x+3)(x-4) &= 6+2x(x-5) \\ (2)(x^2-x-12) &= 6+2x^2-10x \\ 2x^2-2x-24 &= 6+2x^2-10x \\ -2x+10x &= 6+24 \\ 8x &= 30 \\ x &= \frac{30}{8} = \frac{15}{4} \end{aligned}$$

(f)

$$\begin{aligned} \frac{x-5}{x-3} - \frac{x+6}{x^2+x-12} &= \frac{x-2}{x+4} \\ (x+4)(x-5)-(x+6) &= (x-3)(x-2) \\ x^2-x-20-x-6 &= x^2-5x+6 \\ -x-x+5x &= 6+20+6 \\ 3x &= 32 \\ x &= \frac{32}{3} \end{aligned}$$

28. LITERAL EQUATIONS AND FORMULAS

Literal equations are equations in which some, or all, of the quantities are literal numbers.

Literal equations are solved in the same way as any other equation, by using the axioms mentioned previously.

For example, if the length and width of a room are known and the area is needed, the formula $A = LW$ is in proper form, but if the length and area are known and the width is needed, the formula can be written as $W = A/L$.

Example 1.28

From the formula for the area of a rectangle, $A = LW$, find the formula for the width of a rectangle.

Solution

$$A = LW$$

Transposing: $$LW = A$$

Dividing by L: $$W = \frac{A}{L}$$

Example 1.29

Solve the following formula for C.

$$F = \frac{9}{5}C + 32$$

Solution

$$\frac{F}{1} = \frac{9C}{5} + \frac{32}{1}$$

Multiplying by the LCD (5):

$$v5F = 9C + (5)(32)$$

Transposing: $$\begin{aligned} 9C &= 5F - (5)(32) \\ &= (5)(F-32) \end{aligned}$$

Dividing by 9: $$C = \frac{5(F-32)}{9}$$

Or, $$C = \frac{5}{9}(F-32)$$

29. QUADRATIC EQUATIONS

Equations that have been solved to this point have been first-degree equations known as *linear equations.*

A first-degree equation is an equation that contains only the first power of x. A *linear equation* is a first-degree equation of one or more variables, such as $2x + 3 = 9$ or $3x - 2y = 8$.

The general form of the linear equation is

$$Ax + By + C = 0 \qquad \textit{1.1}$$

A represents the coefficient of x, B represents the coefficient of y, and C represents any number that does not contain x. The C term is called the *constant.* In the equation $2x + 3 = 9$, the coefficient of y is zero (0) and the constant is 3.

A quadratic equation contains a term in the second degree x but no higher, such as $2x^2 - 3x + 4 = 0$.

The general form of the quadratic equation is

$$Ax^2 + Bx + C = 0 \qquad 1.2$$

When B is zero, the equation contains no term in x and is known as a *pure quadratic*, such as $3x^2 - 12 = 0$.

The root of a quadratic equation is any number that satisfies the equation. There are two roots for a quadratic equation.

30. SOLVING A PURE QUADRATIC EQUATION

In solving a pure quadratic equation, the term including x^2 is isolated on the left side of the equation and the other term is on the right. If the coefficient of x^2 is not 1, both sides of the equation are divided by this coefficient according to axiom 4, Sec. 24. The value of x is found by taking the square root of each side of the equation, according to axiom 5, Sec. 24.

Example 1.30

Solve the following equation.

$$3x^2 - 12 = 0$$

Solution

$$\begin{aligned} 3x^2 - 12 &= 0 \\ 3x^2 &= 12 \\ x^2 &= 4 \\ \sqrt{x^2} &= \sqrt{4} \\ x &= +2 \text{ or } -2 \end{aligned}$$

31. SOLVING A QUADRATIC EQUATION BY FACTORING

Factoring the left side of the equation $x^2 - 4x + 3 = 0$ gives

$$(x - 3)(x - 1) = 0$$

For this statement to be true—that is, for the product of two factors to be zero—one of the two factors must be zero or both must be zero. Consider two numbers, a and b, whose product is equal to zero.

$$ab = 0$$

For this statement to be true, either a must be zero, b must be zero, or both must be zero. Setting $x = 3$ in the equation $(x - 3)(x - 1) = 0$ gives

$$(3 - 3)(3 - 1) = 0$$

This is a true statement. Therefore, the roots of the equation $x^2 - 4x + 3 = 0$ are 3 and 1. For proof, substitute these numbers in the equation.

$$\begin{aligned} (3)^2 - (4)(3) + 3 &= 0 \\ 9 - 12 + 3 &= 0 \end{aligned}$$

and

$$\begin{aligned} (1)^2 - (4)(1) + 3 &= 0 \\ 1 - 4 + 3 &= 0 \end{aligned}$$

To simplify the operation, let each of the factors equal zero and solve for x.

$$\begin{aligned} (x - 3) &= 0 \\ x &= 3 \\ (x - 1) &= 0 \\ x &= 1 \end{aligned}$$

Example 1.31

Solve the following equation.

$$2x^2 - x - 15 = 0$$

Solution

$$\begin{aligned} 2x^2 - x - 15 &= 0 \\ (x - 3)(2x + 5) &= 0 \\ x - 3 &= 0 \\ x &= 3 \\ 2x + 5 &= 0 \\ x &= -\frac{5}{2} \end{aligned}$$

Example 1.32

Solve the following equation.

$$2x^2 = 3x$$

Solution

$$\begin{aligned} 2x^2 &= 3x \\ 2x^2 - 3x &= 0 \\ x(2x - 3) &= 0 \\ x &= 0 \\ 2x - 3 &= 0 \\ 2x &= 3 \\ x &= \frac{3}{2} \end{aligned}$$

32. SOLVING A QUADRATIC EQUATION BY COMPLETING THE SQUARE

The left side of the equation $x^2+4x+4=16$ is a perfect square because the factors are $(x+2)(x+2)=(x+2)^2$. The right side of the equation is also a perfect square: $\sqrt{16}=\pm 4$. The values of x can be found by taking the square root of each side of the equation.

$$\sqrt{(x+2)^2}=\sqrt{16}$$
$$x+2=\pm 4$$
$$x=+2 \text{ or } -6$$

Quadratic equations that are not perfect squares can be solved by a method known as *completing the square.* The method involves using the five axioms found in Sec. 24 to make both sides of the equation perfect squares and then solving. The method is lengthy and will not be explained here.

33. SOLVING A QUADRATIC EQUATION BY FORMULA

In the general form of the quadratic equation $Ax^2+Bx+C=0$, A is the coefficient of x^2, B is the coefficient of x, and C is the constant. As mentioned, if $B=0$, there will be no middle term Bx. In the equation $3x^2-5x-8=0$, $A=3$, $B=-5$, and $C=-8$.

The quadratic formula is

$$x=\frac{-B+\sqrt{B^2-4AC}}{2A} \quad \text{1.3(a)}$$

$$x=\frac{-B-\sqrt{B^2-4AC}}{2A} \quad \text{1.3(b)}$$

It is usually written as

$$x=\frac{-B\pm\sqrt{B^2-4AC}}{2A} \quad \text{1.3(c)}$$

Example 1.33

Solve the following equation by formula.

$$4x^2-5x-6=0$$

Solution

$4x^2-5x-6=0{:}A=4,\ B=-5,\ C=-6$

$$x=\frac{-(-5)+\sqrt{(-5)^2-(4)(4)(-6)}}{(2)(4)}$$
$$=\frac{5+\sqrt{25+96}}{8}$$
$$=\frac{5+\sqrt{121}}{8}$$
$$x=2$$

or

$$x=\frac{-(-5)-\sqrt{(-5)^2-(4)(4)(-6)}}{(2)(4)}$$
$$=\frac{5-\sqrt{25+96}}{8}$$
$$=\frac{5-\sqrt{121}}{8}$$
$$x=-\frac{3}{4}$$

In using the formula, the equation must be arranged in the form $Ax^2+Bx+C=0$ with A being positive. If the equation is not in this form, it must be rearranged.

Example 1.34

Arrange the equations in the form $Ax^2+Bx+C=0$ for solution by formula and write the values of A, B, and C.

(a) $3x^2=4x+2$

(b) $-4x^2+2x=6$

(c) $x^2=4-x$

(d) $x^2-7=0$

Solution

(a) $3x^2-4x-2=0{:}A=3,\ B=-4,\ C=-2$

(b) $4x^2-2x+6=0{:}A=4,\ B=-2,\ C=6$

(c) $x^2+x-4=0{:}A=1,\ B=1,\ C=-4$

(d) $x^2-7=0{:}A=1,\ B=0,\ C=-7$

PRACTICE PROBLEMS

1. Identify each of the following as a monomial, binomial, trinomial, or polynomial.

(a) $x^2+2xy+y^2$

(b) $4a^2b^2c^2$

(c) $3x^2+5xy^2$

(d) $2x^3+3x^2+3x+8$

2. Write the numerical coefficient of each of the following terms.

(a) $3x^2y$

(b) $6x^3$

(c) $4abc$

(d) $x^2y^2z^2$

3. Combine like terms into one algebraic expression.

Example:

$3x^3 + 5xy + 2x^3 + x^2 + 2xy + 3x^2 = 5x^3 + 4x^2 + 7xy$

(a) $2x^2 + xy + x^2 + 3xy + 4x^2$

(b) $x^2 + x^2y + xy^2 + 2x^2 + x^2y^2$

(c) $3ab^2 + 2ab - 4ab^2 - 2ab + 5$

(d) $x^2 + y^2 + 2 - y^2 - y^2 + 4x^2 - 4$

(e) $x^3 + 2x^2 + 4 - 2x^3 - 5x^2 - x^3$

(f) $3x^3 + x^2y - 6x^2y + 2x + 4 - 6x^3$

(g) $x^2 + 3xy - x^3 + 3x^2 - 5xy + x^3$

(h) $x^2y^2 - 2xy^2 + 2x^2y^2 + 3xy^2 - 4x^2y + x^2y^2$

(i) $a^2b^2c - ab^2c + 2a^2b^2c + 2ab^2c + 4ab^2c$

(j) $2a - 3 + 11a + 5 - 4a - 9$

4. Find the indicated products and quotients.

(a) $(2)^4$

(b) $(x^3)(x^4)$

(c) $(-3)^4$

(d) $-(-4)^2$

(e) $(2a)(3a)$

(f) $(-4x^2)(2x)$

(g) $(x^2y^3)(xy)$

(h) $(3xy^2)(-4x^2y)$

(i) $(2)^3(3^2)(4)^2$

(j) $(-4x^3)(-2x^2)$

(k) $\dfrac{x^8}{x^5}$

(l) $\dfrac{a^2}{a^{-3}}$

(m) $\dfrac{(x^3)(x^4)}{x^5}$

(n) $\dfrac{(2)^2(2)^3(2)}{(2)^5}$

(o) $\dfrac{15x^4y^3}{5x^3y^2}$

(p) $\dfrac{-64a^3b^2}{4ab^2}$

(q) $(x^2y)^3$

(r) $(-2ab^2c^3)^2$

(s) $\dfrac{(4xy^2)^4}{(-8x^2y)^2}$

(t) $\dfrac{a^3(b^2)^4}{(ab^3)^2}$

(u) $\dfrac{(3)^4}{(3)^4}$

(v) $\dfrac{x^4}{x^4}$

(w) $\dfrac{(2)^2}{(2)^5}$

(x) $\dfrac{a^{-2}b^3}{a^3b^{-2}}$

(y) $(4-2)^{-2}$

(z) $(a+b)^0$

5. Multiply the following terms.

(a) $2xy(x^2 - yz + z^2)$

(b) $3a^2b(a^2 - 2ab + 2b^2)$

(c) $5x^2(6x^2 - 3x + 4)$

(d) $-3xy(2x^2 - xy + 2y^2)$

(e) $\begin{array}{r} x+3 \\ \underline{x+2} \end{array}$

(f) $\begin{array}{r} a-4 \\ \underline{a+2} \end{array}$

(g) $\begin{array}{r} a^2+2a+6 \\ \underline{a-3} \end{array}$

(h) $\begin{array}{r} x^2+3x-2 \\ \underline{3x-2} \end{array}$

(i) $\begin{array}{r} a^3-3a^2+2a-8 \\ \underline{5a+4} \end{array}$

(j) $\begin{array}{r} x^2-2x-3 \\ \underline{x^2+\ x+4} \end{array}$

6. Divide the following terms.

(a) $\dfrac{8x^4 + 6x^3 + 4x^2 + 2x + 2}{2}$

(b) $\dfrac{24a^4 - 16a^3 - 12a}{4a}$

(c) $\dfrac{14a^2b^3 - 21a^2b^2 - 28ab}{7ab}$

(d) $\dfrac{18x^4y^3z^2 - 6x^3yz^2 + 3x^2y^2z}{3x^2yz}$

(e) $\dfrac{24a^5bc^3 - 12a^3bc^2 - 18a^4bc^4}{6a^3bc^2}$

(f) $\dfrac{25xy^3z^5 - 20x^3y^2z^2 - 15x^4y^3z^3}{5xy^2z^2}$

(g) $x+3\overline{)x^2+8x+15}$

(h) $x+2\overline{)2x^2+7x+6}$

(i) $x+5\overline{)x^2+8x+15}$

(j) $x+4\overline{)3x^2+10x-6}$

7. Factor the following terms.

(a) $12x^3 - 3x^2$

(b) $x^3y - 3x^2y$

(c) $a^2 + 2ab + b^2$

(d) $a^2 - b^2$

(e) $4a^2 - 8a + 4$

(f) $x^2 - 16$

(g) $x^3 + 2x^2y + xy^2$

(h) $25a^2 - 49$

(i) $3a^2 - 18a + 27$

(j) $a^4 - 81$

(k) $x^2 - 4x - 12$

(l) $a^2 - 4a - 5$

(m) $y^2 - 5y + 6$

(n) $x^2 + 7x + 12$

(o) $2x^2 - 5x - 12$

8. Solve the following equations.

Example:

$10x-(3)(x+3)(2x-5) = 3-(2)(4x-7)(x+1)-x(3-2x)$

$10x - 6x^2 - 3x + 45 = 3 - 8x^2 + 6x + 14 - 3x + 2x^2$

$10x - 6x^2 - 3x + 8x^2 - 2x^2 - 6x + 3x = 3 + 14 - 45;$

$4x = -28;\ x = -7$

(a) $3x + 4 = 6$

(b) $5a + 6 = a - 10$

(c) $15 - 2x = 1 - 5x$

(d) $9x - 12 = 7x - 11$

(e) $7x + 7 - x = 2x - 8$

(f) $5a - 7 - 4a - 8 + 8a - 15 = 0$

(g) $6 + 2x - 3 - 5x = x - 5 - 2x + 7$

(h) $12x - (4x - 6) = 3x - (9x - 27)$

(i) $5x - (x + 3)(x - 4) + 3 = 7 - x(x - 7)$

9. Solve the following equations.

Examples:

(1)

$$\frac{x}{3} + \frac{x}{4} = \frac{7}{2}$$
$$\text{LCD} = 12$$
$$\frac{12x}{3} + \frac{12x}{4} = \frac{(12)(7)}{2}$$
$$4x + 3x = 42;\ 7x = 42;\ x = 6$$

(2)

$$\frac{2x + 2}{3} - \frac{3x - 1}{4} = 1$$
$$\text{LCD} = 12$$
$$\frac{(12)(2x + 2)}{3} - \frac{(12)(3x - 1)}{4} = (12)(1)$$
$$8x + 8 - 9x + 3 = 12;\ -x = 1;\ x = -1$$

(3)

$$\frac{x - 4}{x - 2} - \frac{x}{x + 2} = \frac{8 - x}{x^2 - 4}$$
$$\text{LCD} = x^2 - 4$$
$$(x + 2)(x - 4) - x(x - 2) = 8 - x$$
$$x^2 - 2x - 8 - x^2 + 2x = 8 - x;\ x = 16$$

(a) $x + 3 = \frac{3x}{4} - \frac{x}{2}$

(b) $\frac{3x}{2} - x = \frac{x}{3} + 12$

(c) $\frac{3}{x + 4} - \frac{4}{x - 4} = \frac{x}{x^2 - 16}$

(d) $\frac{2}{x - 1} - \frac{4}{x - 3} = \frac{6}{x^2 - 4x + 3}$

10. Solve each of the following formulas for the letter indicated at the right.

(a) $C = \pi D$ (D)

(b) $C = 2\pi r$ (r)

(c) $A = \pi r^2$ (r)

(d) $A = \frac{\pi D^2}{4}$ (D)

(e) $V = \pi r^2 h$ (h)

(f) $V = \pi r^2 h$ (r)

(g) $A = \frac{\pi}{4} D^2$ (D)

(h) $A - P = Prt$ (P)

(i) $I = \frac{E}{R + r}$ (R)

(j) $\frac{a}{a + b} = \frac{c}{c + d}$ (b)

11. Write each of the following equations in the form $Ax^2 + Bx + C = 0$.

(a) $7x^2 = 4x - 3$

(b) $x^2 = 4 - x$

12. Write the value of A, B, and C in each of the following equations.

(a) $2x^2 + 2x + 5 = 0$

(b) $5x^2 - 9x = 0$

13. Solve the following equations by factoring.

(a) $3x^2 + 7x + 2 = 0$

(b) $4x^2 = 5x + 6$

(c) $3x^2 = 5x$

14. Solve the following equation by formula.

$$3x^2 + 7x + 2 = 0$$

SOLUTIONS

1. (a) trinomial (b) monomial
(c) binomial (d) polynomial

2. (a) 3 (b) 6 (c) 4 (d) 1

3. (a) $7x^2 + 4xy$
(b) $3x^2 + x^2y^2 + x^2y + xy^2$
(c) $-ab^2 + 5$
(d) $5x^2 - y^2 - 2$
(e) $-2x^3 - 3x^2 + 4$
(f) $-3x^3 - 5x^2y + 2x + 4$
(g) $4x^2 - 2xy$
(h) $4x^2y^2 + xy^2 - 4x^2y$
(i) $3a^2b^2c + 5ab^2c$
(j) $9a - 7$

4. (a) 16 (b) x^7
(c) 81 (d) -16
(e) $6a^2$ (f) $-8x^3$
(g) x^3y^4 (h) $-12x^3y^3$
(i) 1152 (j) $8x^5$
(k) x^3 (l) a^5
(m) x^2 (n) 2
(o) $3xy$ (p) $-16a^2$
(q) x^6y^3 (r) $4a^2b^4c^6$
(s) $4y^6$ (t) ab^2
(u) 1 (v) 1
(w) $\frac{1}{8}$ (x) $\frac{b^5}{a^5}$
(y) $\frac{1}{4}$ (z) 1

5. (a) $2x^3y - 2xy^2z + 2xyz^2$
(b) $3a^4b - 6a^3b^2 + 6a^2b^3$
(c) $30x^4 - 15x^3 + 20x^2$
(d) $-6x^3y + 3x^2y^2 - 6xy^3$
(e)
$$\begin{array}{r} x + 3 \\ \times\ x + 2 \\ \hline x^2 + 3x \\ 2x + 6 \\ \boxed{x^2 + 5x + 6} \end{array}$$
(f)
$$\begin{array}{r} a - 4 \\ \times\ a + 2 \\ \hline \boxed{a^2 - 2a - 8} \end{array}$$
(g)
$$\begin{array}{r} 2a - 8 \\ \hline a^2 - 2a - 8 \\ a^2 + 2a + 6 \\ \times\ a - 3 \\ \hline a^3 + 2a^2 + 6a \\ -3a^2 - 6a - 18 \\ \boxed{a^3 - a^2 - 18} \end{array}$$
(h)
$$\begin{array}{r} x^2 + 3x - 2 \\ \times\ 3x - 2 \\ \hline 3x^3 + 9x^2 - 6x \\ -2x^2 - 6x + 4 \\ \boxed{3x^3 + 7x^2 - 12x + 4} \end{array}$$
(i)
$$\begin{array}{r} a^3 - 3a^2 + 2a - 8 \\ \times\ 5a + 4 \\ \hline 5a^4 - 15a^3 + 10a^2 - 40a \\ 4a^3 - 12a^2 + 8a - 32 \\ \boxed{5a^4 - 11a^3 - 2a^2 - 32a - 32} \end{array}$$
(j)
$$\begin{array}{r} x^2 - 2x - 3 \\ \times\ x^2 + x + 4 \\ \hline x^4 - 2x^3 - 3x^2 \\ x^4 - 2x^3 - 2x^2 - 3x \\ 4x^2 - 8x - 12 \\ \boxed{x^4 - x^3 - x^2 - 11x - 12} \end{array}$$

6. (a) $4x^4 + 3x^3 + 2x^2 + x + 1$
(b) $6a^3 - 4a^2 - 3$
(c) $2ab^2 - 3ab - 4$
(d) $6x^2y^2z - 2xz + y$
(e) $4a^2c - 2 - 3ac^2$
(f) $5yz^3 - 4x^2 - 3x^3yz$
(g)
$$\begin{array}{r} \boxed{x + 5} \\ x + 3\,\overline{)\,x^2 + 8x + 15} \\ \underline{x^2 + 3x} \\ 5x + 15 \\ \underline{5x + 15} \end{array}$$
(h)
$$\begin{array}{r} \boxed{2x + 3} \\ x + 2\,\overline{)\,2x^2 + 7x + 6} \\ \underline{2x^2 + 4x} \\ 3x + 6 \\ \underline{3x + 6} \end{array}$$

(i)

$$\begin{array}{r l} & \boxed{x+3} \\ x+5\,\big) & x^2+8x+15 \\ & \underline{x^2+5x} \\ & 3x+15 \\ & \underline{3x+15} \end{array}$$

(j)

$$\begin{array}{r l} & \boxed{3x\;-\;2} \\ x+4\,\big) & 3x^2+10x-6 \\ & \underline{3x^2+12x} \\ & -2x\;-\;6 \\ & -\;2x-8 \\ & \underline{2} \end{array}$$

7. (a) $3x^2(4x-1)$

(b) $x^2y(x-3)$

(c) $(a+b)^2$

(d) $(a+b)(a-b)$

(e) $(4)(a-1)^2$

(f) $(x+4)(x-4)$

(g) $x(x+y)^2$

(h) $(5a+7)(5a-7)$

(i) $(3)(a-3)^2$

(j) $(a^2+9)(a^2-9)=(a^2+9)(a+3)(a-3)$

(k) $(x+2)(x-6)$

(l) $(a+1)(a-5)$

(m) $(y-2)(y-3)$

(n) $(x+3)(x+4)$

(o) $(2x+3)(x-4)$

8. (a) $3x+4=6;\ 3x=2;\ x=\boxed{\frac{2}{3}}$

(b) $5a+6=a-10;\ 4a=-16;\ a=\boxed{-4}$

(c) $15-2x=1-5x;\ 3x=-14;\ x=\boxed{-\frac{14}{3}}$

(d) $9x-12=7x-11;\ 2x=1;\ x=\boxed{\frac{1}{2}}$

(e) $7x+7-x=2x-8;\ 4x=-15;\ x=\boxed{-\frac{15}{4}}$

(f) $5a-7-4a-8+8a-15=0;$

$$9a=30;\ a=\boxed{\frac{10}{3}}$$

(g) $6+2x-3-5x=x-5-2x+7;$

$$-2x=-1;\ x=\boxed{\frac{1}{2}}$$

(h) $12x-(4x-6)=3x-(9x-27);$

$$12x-4x+6=3x-9x+27;$$

$$14x=21;\ x=\boxed{\frac{3}{2}}$$

(i) $5x-(x+3)(x-4)+3=7-x(x-7);$

$$5x-x^2+x+12+3=7-x^2+7x;$$

$$x=\boxed{8}$$

9. (a) $x+3=\frac{3x}{4}-\frac{x}{2};\ 4x+12=3x-2x;$

$$3x=12;\ x=\boxed{-4}$$

(b) $\frac{3x}{2}-x=\frac{x}{3}+12;\ 9x-6x=2x+72;$

$$x=\boxed{72}$$

(c) $\frac{3}{x+4}-\frac{4}{x-4}=\frac{x}{x^2-16};$

$$(3)(x-4)-(4)(x+4)=x;$$

$$3x-12-4x-16=x;$$

$$-2x=28;$$

$$x=\boxed{-14}$$

(d) $\frac{2}{x-1}-\frac{4}{x-3}=\frac{6}{x^2-4x+3};$

$$(2)(x-3)-(4)(x-1)=6;\ 2x-6-4x+4=6;$$

$$-2x=8;\ \boxed{x=-4}$$

10. (a) $D=\frac{C}{\pi}$ (b) $r=\frac{C}{2\pi}$

(c) $r=\sqrt{\frac{A}{\pi}}$ (d) $D=\sqrt{\frac{4A}{\pi}}$

(e) $h=\frac{V}{\pi r^2}$ (f) $r=\sqrt{\frac{V}{\pi h}}$

(g) $D=\sqrt{\frac{A}{\frac{\pi}{4}}}$ (h) $P=A-Prt=\frac{A}{1+rt}$

(i) $R=\frac{E}{I}-r$ (j) $b=\frac{ad}{c}$

11. (a) $7x^2 - 4x + 3 = 0$ (b) $x^2 + x - 4 = 0$

12. (a) $A = 2,\ B = 2,\ C = 5$

(b) $A = 5,\ B = -9,\ C = 0$

13. (a) $(3x + 1)(x + 2) = 0$

$$x = -\frac{1}{3}$$

$$x = \boxed{-2}$$

(b)

$$4x^2 - 5x - 6 = 0$$

$$(4x + 3)(x - 2) = 0$$

$$x = -\frac{3}{4}$$

$$x = \boxed{2}$$

(c)

$$3x^2 - 5x = 0$$

$$x(3x - 5) = 0$$

$$x = 0$$

$$x = \boxed{\frac{5}{3}}$$

14. $A = 3;\ B = 7;\ C = 2$

$$x = \frac{-7 \pm \sqrt{(7)^2 - (4)(3)(2)}}{(2)(3)}$$

$$= \frac{-7 \pm \sqrt{49 - 24}}{6}$$

$$= \frac{-7 \pm \sqrt{25}}{6}$$

$$x = \frac{-7 + 5}{6} = \boxed{-\frac{1}{3}}$$

$$x = \frac{-7 - 5}{6} = \boxed{-2}$$

2 Basic Geometry

1. DEFINITION

Geometry is a branch of mathematics that deals with the measurement, properties, and relationships of points, lines, angles, surfaces, and solids. The word "geometry" is derived from the words *geo*, meaning earth, and *metro*, meaning measure.

2. HISTORY

The Egyptians first used geometry in about 2500 B.C. because the seasonal overflowing of the Nile made it necessary to reestablish boundaries so that taxes could be levied and collected. In about 500 B.C. the Greeks began to develop information received from the Egyptians into the branch of mathematics we now know as geometry. By the 4th century A.D. they had developed arithmetic and geometry into separate branches of mathematical science.

3. POINTS AND LINES

Points and lines are undefined elements of geometry, yet everyone has some understanding of these terms. A *point* is understood to have no length, width, or thickness, and it indicates a location. A point is usually shown on paper as a small dot and is named with a capital letter such as A.

A *line* is considered to have length but not width or thickness. A line connecting two points is said to be a *straight line* if it does not curve. A straight line is usually designated by two points that it connects, such as AB. A curved line is a line of which no part is straight.

4. PARALLEL LINES

Two lines that lie in the same plane and do not intersect are called *parallel lines.*

5. ANGLE

There are many definitions of an *angle.* In geometry, an angle may be defined as the space between two lines diverging from a common point called the *vertex.* In trigonometry, an angle may be defined as the amount of rotation to bring one line into coincidence with another. In surveying, an angle may be defined as the difference in direction of two intersecting lines.

6. MEASURE OF ANGLES

The most common measure of an angle is the *degree.* It is defined as $1/360$ of a complete angle or turn. A circle can be divided into 360 equal arcs. If radii connect each end of these small arcs, the angle formed by the two radii measures 1 degree.

For closer measurement, the degree is divided into 60 equal parts, with one part measuring 1 minute. And for even closer measurement, the 1 minute angle is divided into 60 equal parts, with each part measuring 1 second.

The symbols used for degrees, minutes, and seconds are: degrees (°), minutes (′), and seconds (″). As an example, 36 degrees, 24 minutes, and 52 seconds is written as 36°24′52″.

A protractor can be used to measure angles on paper; a transit and theodolite measure angles in the field.

Figure 2.1 *Angle Measurements*

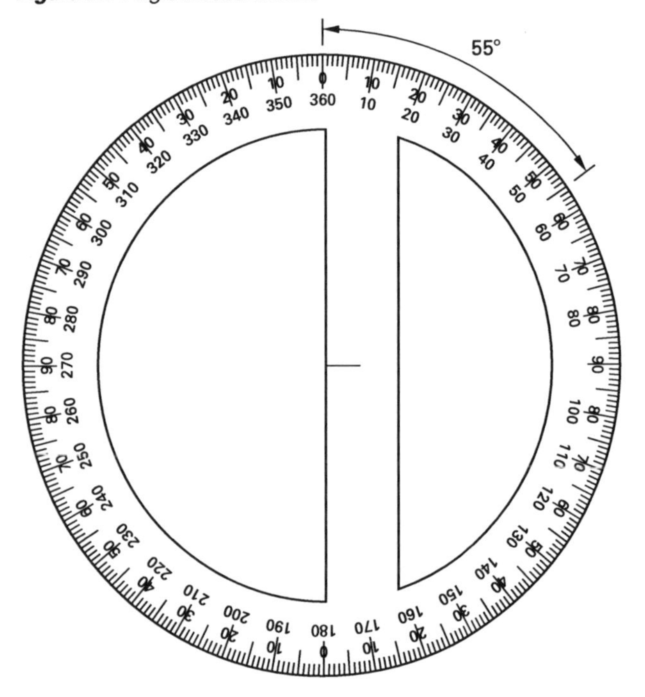

7. ACUTE ANGLE

An *acute angle* is an angle of less than 90°.

8. RIGHT ANGLE

A *right angle* is an angle of 90°.

9. OBTUSE ANGLE

An *obtuse angle* is an angle of more than 90° and less than 180°.

10. STRAIGHT ANGLE

A *straight angle* is an angle of 180°.

Figure 2.2 *Angles*

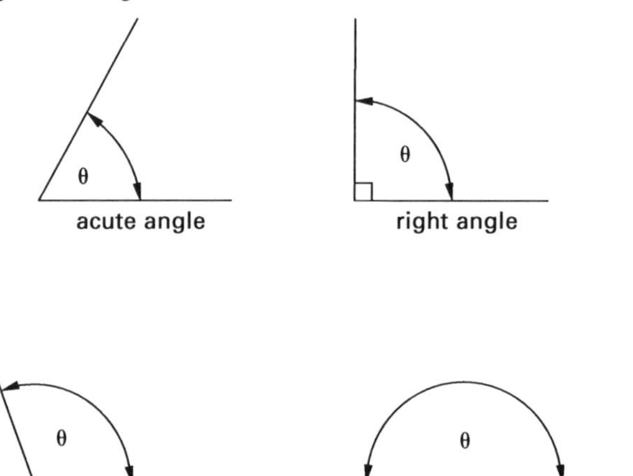

11. COMPLEMENTARY ANGLES

Two angles are said to be *complementary* if their sum is 90°.

Figure 2.3 *Complementary Angles*

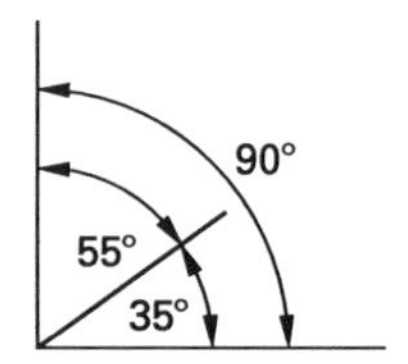

12. SUPPLEMENTARY ANGLES

Two angles are said to be *supplementary* if their sum is 180°.

Figure 2.4 *Supplementary Angles*

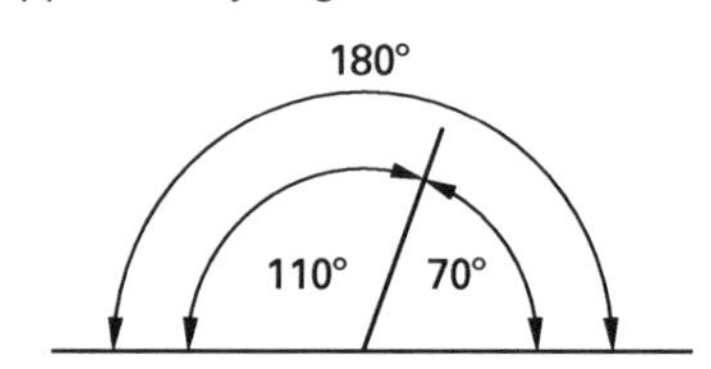

13. TRANSVERSAL

A line that cuts two or more lines is called a *transversal.*

14. ALTERNATE INTERIOR ANGLES

When two parallel lines are cut by a transversal, the *alternate interior angles* are equal.

Figure 2.5 *Alternate Interior Angles*

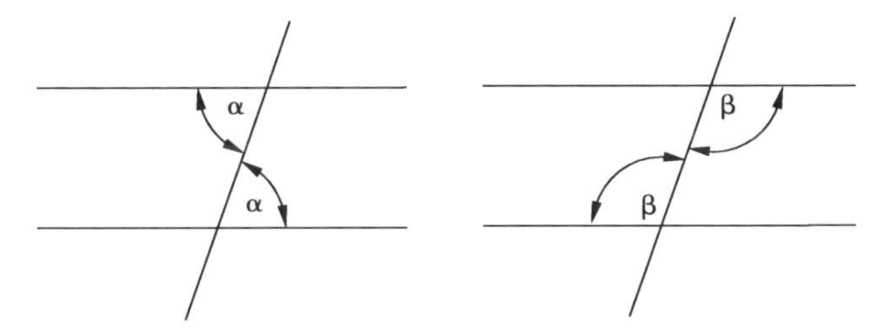

15. ALTERNATE EXTERIOR ANGLES

When two parallel lines are cut by a transversal, the *alternate exterior angles* are equal.

Figure 2.6 *Alternate Exterior Angles*

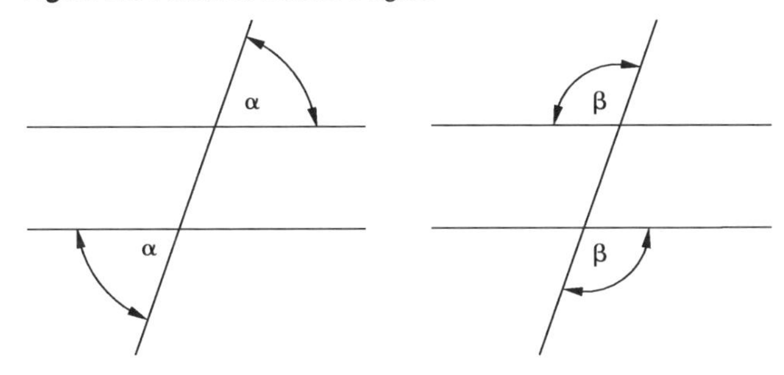

16. ADDING AND SUBTRACTING ANGLES

The surveying technician is often called on to add the measurements of two angles or to find the difference in the measurements of two angles. The procedures often involve "borrowing" 1 degree and converting it to 60 minutes and "borrowing" 1 minute and converting it to 60 seconds.

Example 2.1

Add 24°40′ and 16°30′.

Solution

When the number of minutes in the sum is 60 or more, 60′ are subtracted from the minutes column and 1° is added to the degree column. Likewise, when the number of seconds in the sum is 60 or more, 60″ are subtracted from the seconds column and 1′ is added to the minutes column.

$$\begin{array}{r} 24^\circ 40' \\ +\,16^\circ 30' \\ \hline 40^\circ 70' = 41^\circ 10' \end{array}$$

Example 2.2

Add 32°46′32″ and 14°22′44″.

Solution

$$\begin{array}{l} 32^\circ 46' 32'' \\ +\,14^\circ 22' 44'' \\ \hline 46^\circ 68' 76'' = 46^\circ 69' 16'' = 47^\circ 09' 16'' \end{array}$$

Example 2.3

Find the difference between 60°12′ and 40°32′.

Solution

Before finding the difference, convert 60°12′ to its equivalent 59°72′. "Borrow" 1° from 60° leaving 59° and add its equivalent, 60′ to 12′, making 72′.

$$\begin{array}{r} 60^\circ 12' \\ -40^\circ 32' \\ \hline \end{array} \qquad \begin{array}{r} 59^\circ 72' \\ -40^\circ 32' \\ \hline 19^\circ 40' \end{array}$$

Example 2.4

Find the difference between 96°08′14″ and 52°33′50″.

Solution

One degree was borrowed from the degree column, leaving 95° and making the minutes column 68′. One minute was borrowed from the minutes column, leaving 67′ and making the seconds column 74″.

When a number of angle measurements are added, as is common in surveying, each column (degrees, minutes, and seconds) is added separately and recorded. If the sum of either the minutes column or the seconds column, or both, is 60 or more, the same procedure is followed.

$$\begin{array}{r} 96^\circ 08' 14'' \\ -52^\circ 33' 50'' \\ \hline \end{array} \qquad \begin{array}{r} 95^\circ 67\ 74'' \\ -52^\circ 33' 50'' \\ \hline 43^\circ 34' 24'' \end{array}$$

Example 2.5

Add the following angle measurements.

$$\begin{array}{r} 93^\circ 18' 22'' \\ 65^\circ 13' 8'' \\ 218^\circ 19' 30'' \\ 67^\circ 05' 20'' \\ 96^\circ 04' 50' \end{array}$$

Solution

$$\begin{array}{rrr} 93^\circ & 18' & 22'' \\ 65^\circ & 13' & 08'' \\ 218^\circ & 19' & 30'' \\ 67^\circ & 05' & 20'' \\ 96^\circ & 04' & 50'' \\ \hline 539^\circ & 59' & 130'' \\ + & 02' & -120'' \\ \hline 539^\circ & 61' & 10'' \\ +1^\circ & -60' & \\ \hline 540^\circ & 01' & 10'' \end{array}$$

17. AVERAGE OF SEVERAL MEASUREMENTS OF AN ANGLE

Surveyors often measure angles by repetition. An angle of 36°30′30″ might have read on the first reading as 36°30′, but after turning the angle six times, the accumulated angle may have been read 219°03′00″. Dividing this by six gives 36°30′30″, which is closer to the true measurement.

Example 2.6

An angle was doubled, and the accumulated reading was 84°26′. What was the angle?

Solution

$$84^\circ 26' \div 2 = 42^\circ 13'$$

Example 2.7

An angle that was doubled read 314°13′. What was the closest value for the single angle?

Solution

$$314^\circ 13' \div 2 = 157^\circ 06' 30''$$

Example 2.8

An angle reads 318°03′ after having been turned six times. What was the average for the single angle?

Solution

$$318^\circ 03' = 318^\circ 00' 180'' \div 6 = 53^\circ 00' 30''$$

18. CHANGING DEGREES AND MINUTES TO DEGREES AND DECIMALS OF A DEGREE

In some situations and in some tables of trigonometric functions, angles are expressed in degrees and decimals of a degree. To change degrees and minutes to degrees and decimals of a degree, first express the minutes as a common fraction with a denominator of 60, and then convert the common fraction to a decimal fraction. Add this fraction to the degrees.

Example 2.9

Change 73°15′ to degrees and decimals of a degree.

Solution

$$73^\circ 15' = 73\tfrac{15}{60}^\circ = 73.25^\circ$$

19. CHANGING DEGREES, MINUTES, AND SECONDS TO DEGREES AND DECIMALS OF A DEGREE

To change degrees, minutes, and seconds to degrees and decimals of a degree, first convert degrees, minutes, and seconds to degrees, minutes, and decimals of a minute; then convert degrees, minutes, and decimals of a minute to degrees and decimals of a degree.

Example 2.10

Change 46°24′36″ to degrees and decimals of a degree.

Solution

$$46^\circ 24' 36'' = 46^\circ 24\tfrac{36'}{60} = 46^\circ 24.6' = 46\tfrac{24.6^\circ}{60} = 46.41^\circ$$

20. CHANGING DEGREES AND DECIMALS OF A DEGREE TO DEGREES, MINUTES, AND SECONDS

To change degrees and decimals of a degree to degrees, minutes, and seconds, multiply the decimal fraction by 60 and add the product (in minutes and decimals of a minute) to the degrees. Then multiply the decimal fraction in minutes by 60 and add the product (in seconds) to the degrees and minutes. The decimal fraction will be left as such.

Example 2.11

Change 36.12345° to degrees, minutes, and seconds.

Solution

$$(0.12345^\circ)(60) = 7.407'$$
$$(0.407')(60) = 24.42''$$
$$36.12345^\circ = 36^\circ 07' 24.42''$$

21. POLYGON

A closed figure bounded by straight lines lying in the same plane is known as a *polygon*.

The sum of the interior angles of a closed polygon is equal to

$$(n-2)(180^\circ)$$

n is the number of sides. Thus, the sum of the interior angles of a triangle is 180°, of a rectangle 360°, of a five-sided figure 540°, and so on.

22. TRIANGLE

A polygon of three sides is known as a *triangle.*

23. RIGHT TRIANGLE

A *right triangle* is a triangle that has one right angle (90°).

24. ISOSCELES TRIANGLE

An *isosceles triangle* is a triangle that has two equal sides and two equal angles.

25. EQUILATERAL TRIANGLE

An *equilateral triangle* is a triangle that has three equal sides and three equal angles.

26. OBLIQUE TRIANGLE

An *oblique triangle* is a triangle that has no right angle and no two sides equal.

Figure 2.7 *Triangles*

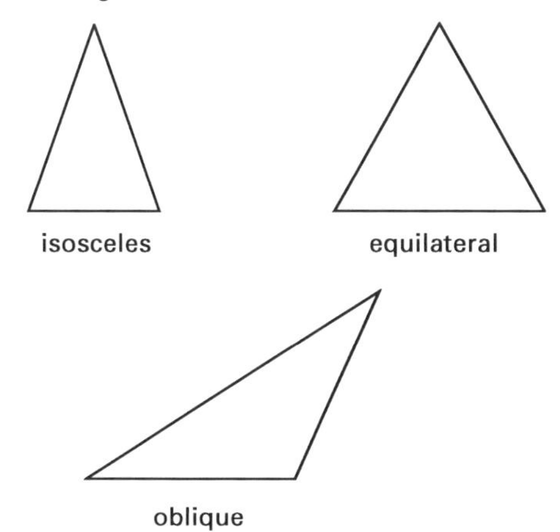

27. CONGRUENT TRIANGLES

Two triangles are *congruent* if their corresponding sides and corresponding angles are equal.

28. SIMILAR TRIANGLES

Two triangles are *similar* if their corresponding angles are equal and their corresponding sides are proportional.

Figure 2.8 *Similar Triangles*

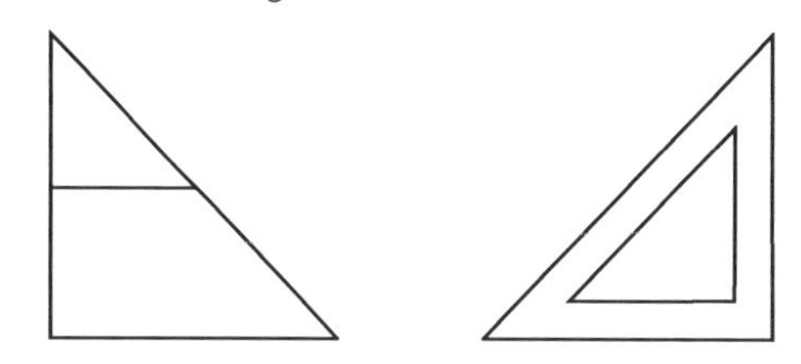

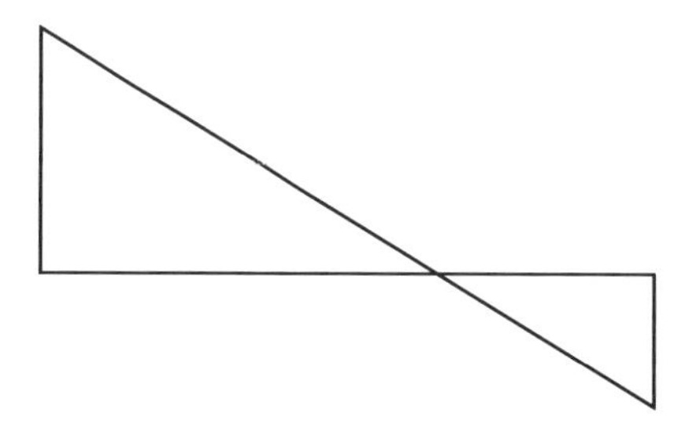

29. RECTANGLE

A *rectangle* is a four-sided polygon whose angles are right angles. A *square* is a rectangle whose sides are equal.

Figure 2.9 *Rectangles*

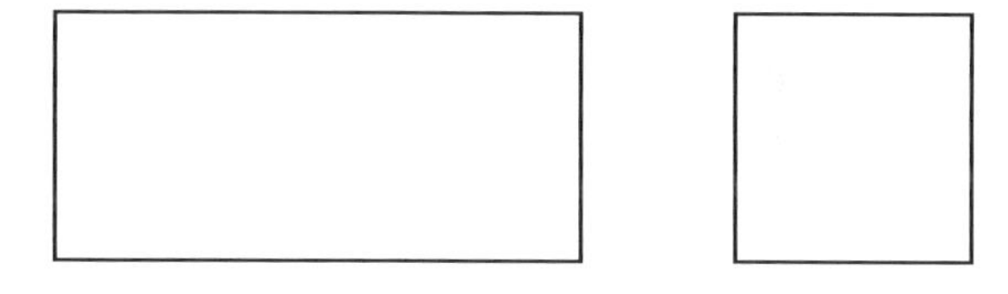

30. TRAPEZOID

A *trapezoid* is a four-sided polygon that has two parallel sides and two nonparallel sides.

Figure 2.10 *Trapezoids*

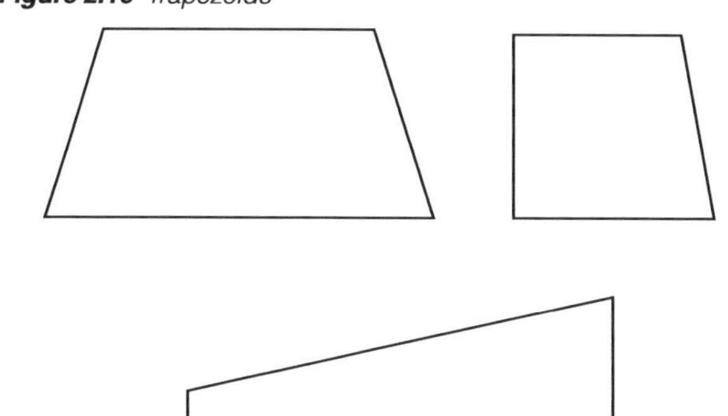

31. CIRCLE

A *circle* is a closed plane curve, all points on which are equidistant from a point within called the *center.*

32. RADIUS

The distance from the center of the circle to any point on the circle is called the *radius* of the circle.

33. DIAMETER

The distance across the circle through the center is called the *diameter.* One diameter is two radii.

34. CHORD

A straight line between points on a circle is called a *chord.* The length of the chord is designated LC.

35. SECANT

A *secant* of a circle is a line that intersects the circle at two points.

36. TANGENT

A *tangent* of a circle is a line that touches the circle at only one point.

37. ARC

Any part of a circle is called an *arc.*

Figure 2.11 *Elements of a Circle*

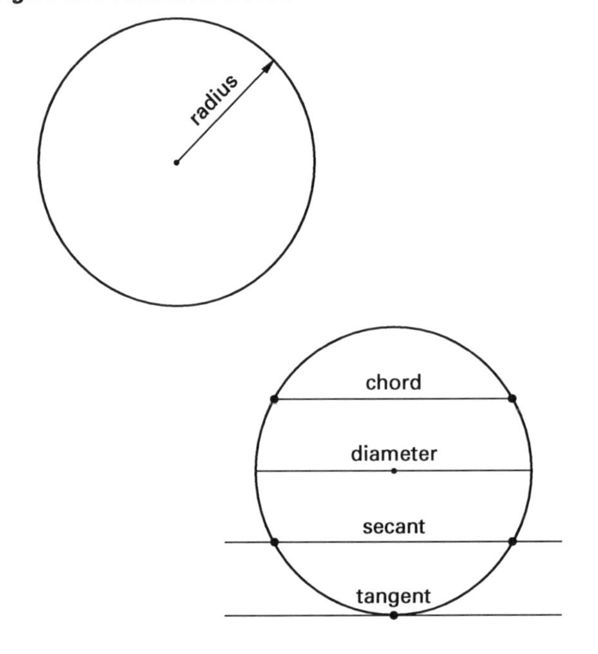

38. SEMICIRCLE

An arc equal to one-half the circumference of a circle is a *semicircle.*

Figure 2.12 *Semicircle and Central Angle*

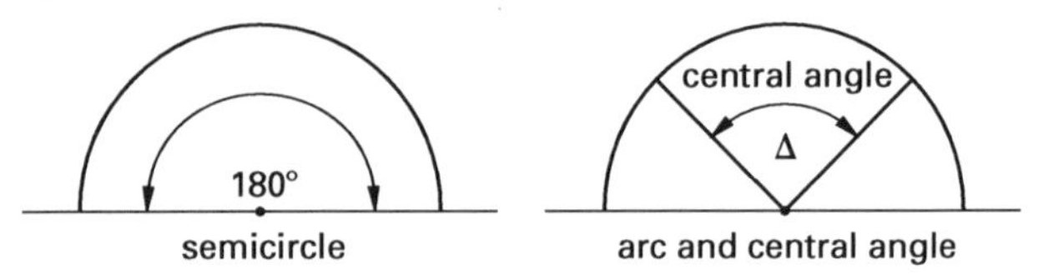

39. CENTRAL ANGLE

A *central angle* is an angle formed by two radii. The Greek letter Δ (delta) is often used to denote a central angle. A central angle has the same number of degrees as the arc it intercepts. A 60° central angle intercepts a 60° arc, and so on. Thus, a central angle is measured by its intercepted arc.

40. SECTOR

A figure bounded by an arc of a circle and two radii of the circle is called a *sector* of the circle.

41. SEGMENT

A figure bounded by a chord and an arc of a circle is called a *segment* of a circle.

42. CONCENTRIC CIRCLES

Two circles of different radius but with the same center are called *concentric circles.*

Figure 2.13 *Concentric Circles, Sector, and Segment of Circles*

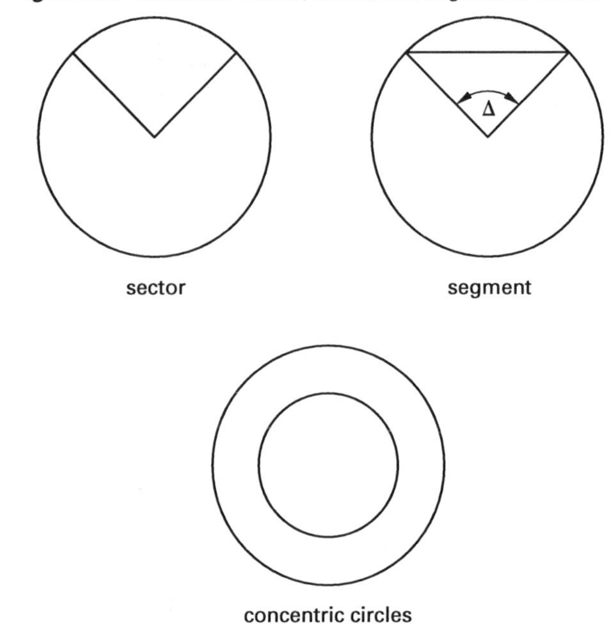

43. RADIUS PERPENDICULAR TO TANGENT

The radius of a circle is perpendicular to a tangent to the circle at the point of *tangency*.

44. RADIUS AS PERPENDICULAR BISECTOR OF A CHORD

The perpendicular *bisector* of a chord passes through the center of the circle.

Figure 2.14 *Tangents and Chords*

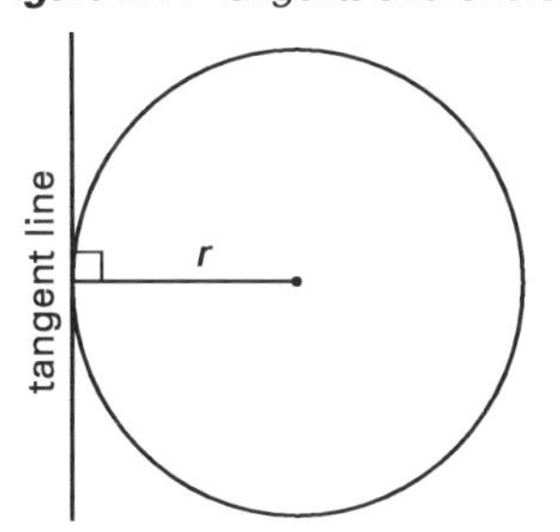

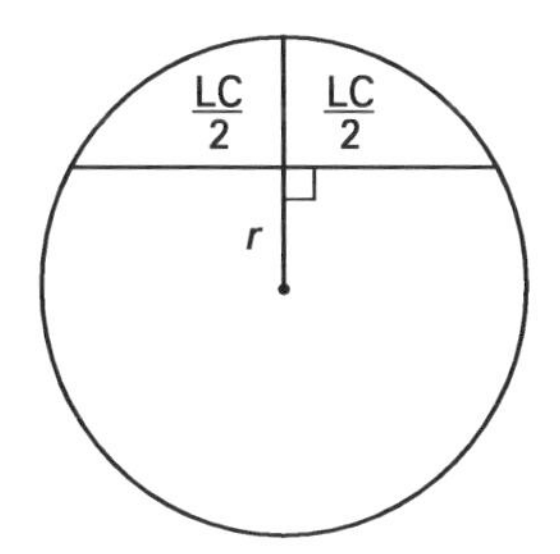

45. TANGENTS TO CIRCLE FROM OUTSIDE POINT

Tangents to a circle from an outside point are equal.

Figure 2.15 *Tangents to a Circle*

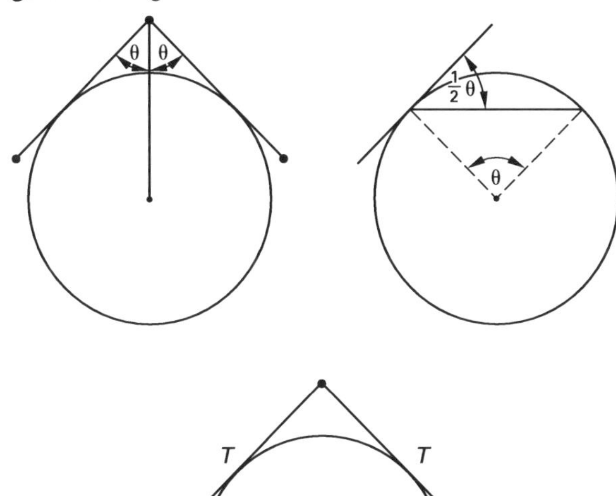

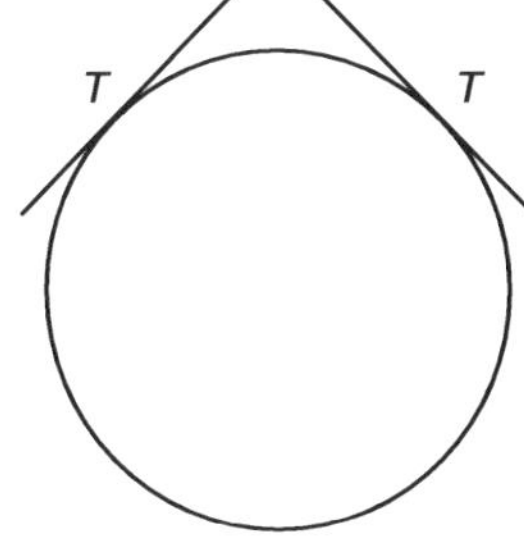

46. LINE FROM CENTER OF CIRCLE TO OUTSIDE POINT

A line from the center of a circle to an outside point bisects the angle between the tangents from the point to the circle.

47. ANGLE FORMED BY TANGENT AND CHORD

The angle formed by a tangent and a chord is equal to one-half its intercepted arc.

48. ANGLE FORMED BY TWO CHORDS

The angle formed by two chords is equal to one-half its intercepted arc.

49. SOLID GEOMETRY

Figures shown thus far are plane figures and are included in the study of *plane geometry*. *Solid geometry* is the study of figures of three dimensions such as cubes, cones, pyramids, and spheres.

50. POLYHEDRON

A *polyhedron* is any solid formed by plane surfaces.

51. PRISM

A *prism* is a polyhedron with parallel edges and parallel bases.

52. RIGHT PRISM

A prism with edges perpendicular to the bases is known as a *right prism*. A cylinder is a right prism with circular bases.

Figure 2.16 *Right Prisms*

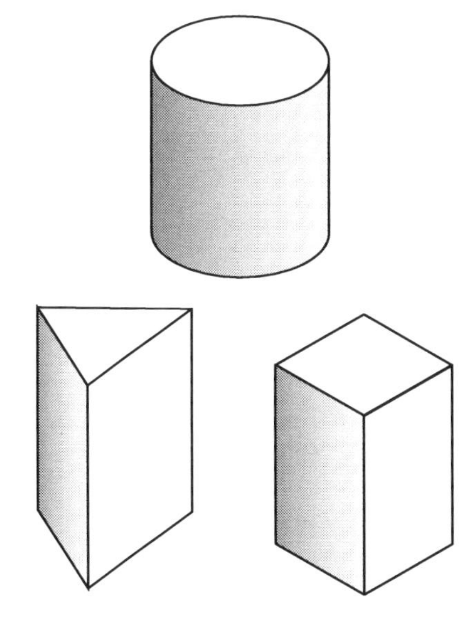

53. PYRAMID

A *pyramid* is a polyhedron having for its base a polygon and for its faces, triangles with a common vertex.

The altitude of a pyramid is the perpendicular distance from the vertex to the base.

A *right pyramid* is a pyramid in which the base is a regular polygon, and a line from the vertex to the center of the polygon is perpendicular to the polygon.

The slant height of a right pyramid is the altitude of one of the lateral faces.

54. FRUSTUM OF A PYRAMID

A *frustum of a pyramid* is that part left after cutting off the top part with a plane parallel to the base.

55. CONE

A *cone* is a polyhedron with a circular base and with sides that taper evenly up to a vertex.

The altitude of a cone is the perpendicular distance from the vertex to the base.

A *right circular cone* is a cone in which a line from the vertex to the center of the base is perpendicular to the base.

The slant height of a cone is the distance from the vertex to the base measured along the surface.

56. FRUSTUM OF A CONE

The *frustum of a right circular cone* is the part left after cutting off the top part with a plane parallel to the base.

Figure 2.17 *Frustums*

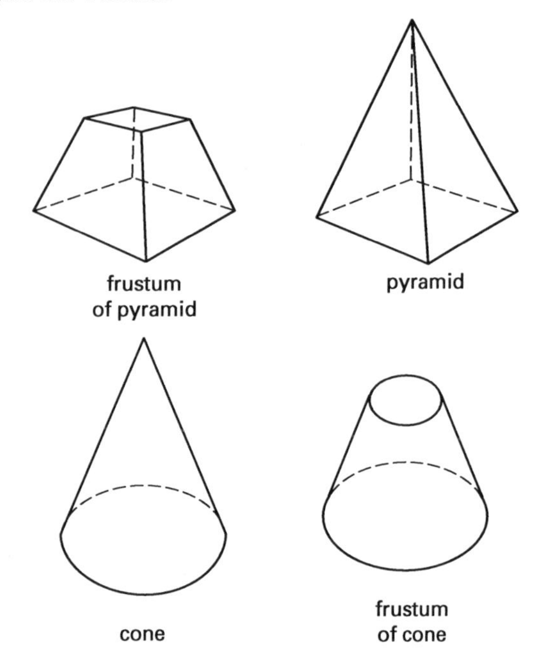

57. CONSTRUCTION OF GEOMETRIC FIGURES

Geometric figures may be constructed by using a compass and straight-edge.

To bisect a given angle A, construct an arc intersecting the sides of the angle; then, at these intersections, construct arcs of equal radii. A line from the vertex A to the intersection of these two arcs D bisects the angle A.

Figure 2.18 *Bisecting an Angle*

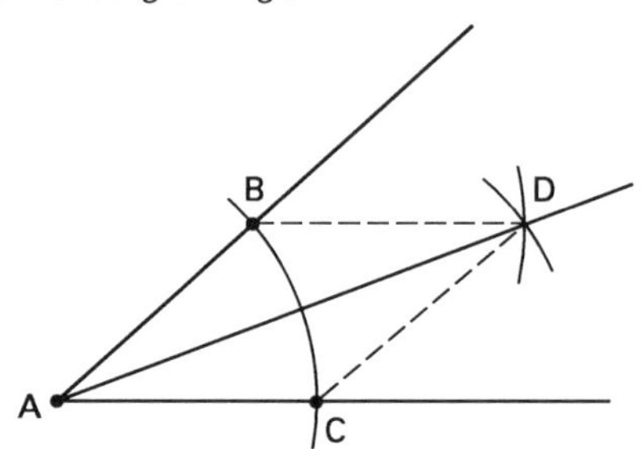

To construct a line perpendicular to a given line, w, through a given point, P, on the line, construct an arc intersecting the line at points A and B. Then use both A and B as a radius point to construct arcs with a radius more than half of AB. A line through the intersection of these arcs and P is perpendicular to line w.

Figure 2.19 *Constructing a Perpendicular*

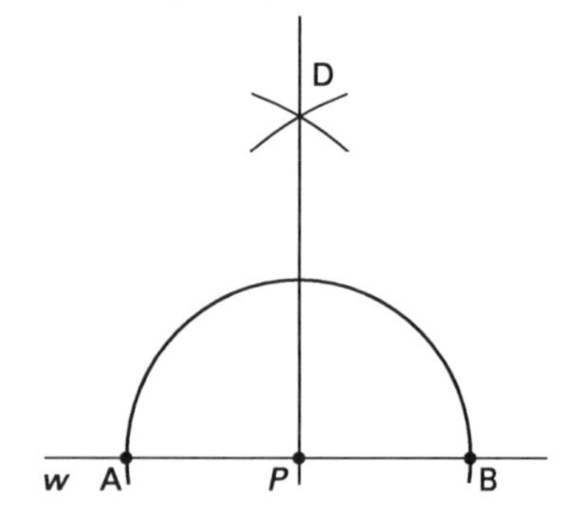

To construct the perpendicular bisector of a line AB, use A as a radius point and construct an arc with a radius more than half of AB. Then, use B as a radius point and construct an arc with the same radius. A line through the intersections of these arcs is the perpendicular bisector of line AB.

Figure 2.20 Constructing a Perpendicular Bisector

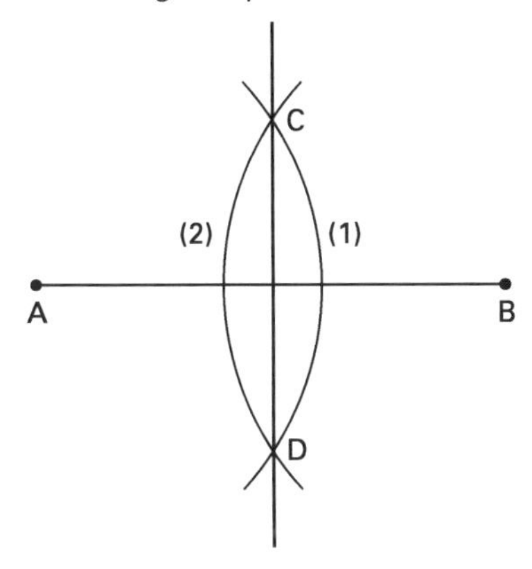

To circumscribe a circle about a triangle, construct the perpendicular bisectors of two sides of the triangle. Their intersection is at the center of the circle. The radius is the distance from the circle's center to any vertex of the triangle.

Figure 2.21 Circumscribing a Circle about a Triangle

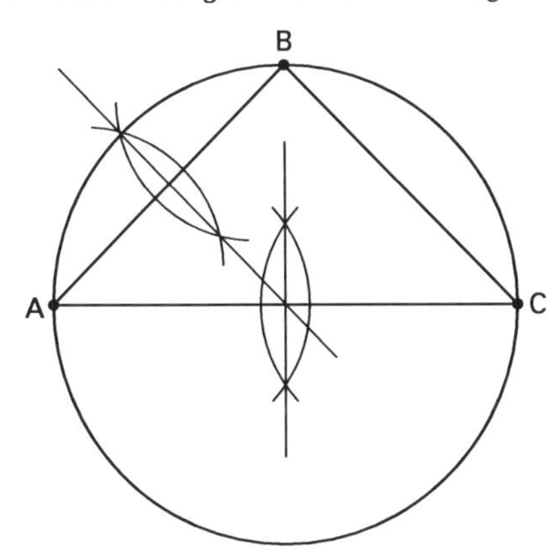

To locate the center of the circle, select three points on the circle and connect them with two chords; then, erect perpendicular bisectors to the chords. Their intersection is at the center of the circle.

Figure 2.22 Locating a Circle's Center

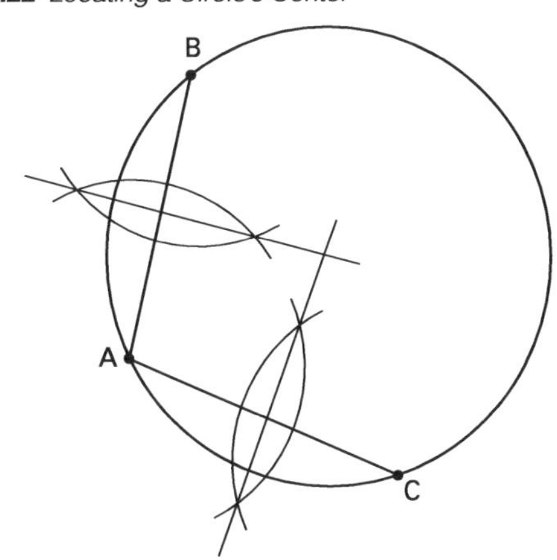

To inscribe a circle in a triangle, construct the bisectors of two of the angles of the triangle. Their intersection is at the center of the circle.

Figure 2.23 Inscribing a Circle in a Triangle

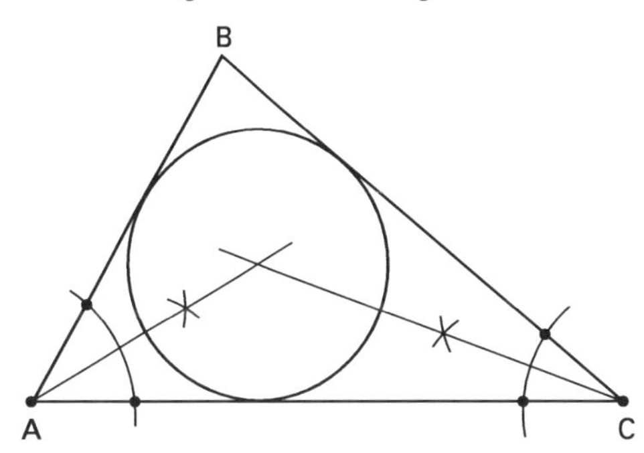

PRACTICE PROBLEMS

1. Add the following angles.

Example:

$$\begin{array}{l} 21°41'12'' \\ \underline{11°32'54''} \\ 32°73'66'' = 33°14'06'' \end{array}$$

(a) $\begin{array}{r} 46°27' \\ \underline{+22°24'} \end{array}$ (b) $\begin{array}{r} 56°24' \\ \underline{+33°26'} \end{array}$ (c) $\begin{array}{r} 35°52' \\ \underline{+47°39'} \end{array}$

(d) $\begin{array}{r} 21°46'52'' \\ \underline{+40°25'26''} \end{array}$ (e) $\begin{array}{r} 46°19'22'' \\ \underline{+35°51'40''} \end{array}$ (f) $\begin{array}{r} 13°49'58'' \\ \underline{+12°21'32''} \end{array}$

2. Find the average of the following angles that were doubled in the field with the accumulated value shown.

Example:

$$2\overline{)\,311°17'20''} = 2\overline{)\,310°76'80''}$$
$$155°38'40''$$

(a) 237°27′17″

(b) 329°47′16″

3. Find the average of the following angle that was repeated six times in the field with the accumulated value shown.

$$390°13'24''$$

4. Change the following to degrees and decimals of a degree.

Example:

$$36°14'52'' = 36°14\tfrac{52'}{60} = 36°14.8667' = 36\tfrac{14.8667°}{60} = 36.2478°$$

(a) 16°24′30″

(b) 24°30′

(c) 36°45′

(d) 68°44′05″

(e) 69°11′

(f) 118°55′11″

(g) 127°17′23″

(h) 173°32′56″

(i) 186°08′34″

(j) 223°37′48″

5. Change the following to degrees, minutes, and seconds. Show each step.

Example:

$$142.276843° = 142° + (60)(0.276843)' = 142°16.61058' = 142°16' + (60)(0.61058)'' = 142°16'37''$$

(a) 68.176°

(b) 96.564722°

(c) 145.882222°

(d) 221.347778°

(e) 303.107778°

6. The interior angles of polygons of 5, 6, and 7 sides were measured. Find the sum of the angles for each and indicate the error of measurement.

pt	angle	pt	angle	pt	angle
(a) A	83°23′	(b) A	96°34′	(c) A	98°08′05″
B	105°27′	B	111°42′	B	149°16′12″
C	158°31′	C	183°12′	C	134°12′55″
D	53°19′	D	88°57′	D	93°20′10″
E	139°18′	E	139°21′	E	152°39′47″
		F	100°18′	F	174°32′50″
				G	97°51′11″
error		error		error	

7. Write the missing word in each of the following sentences.

(a) Two lines that lie in the same plane and do not intersect are called ____ lines.

(b) An ____ angle is an angle of less than 90°.

(c) A ____ angle is an angle of 90°.

(d) An ____ angle is an angle of more than 90° and less than 180°.

(e) A ____ angle is an angle of 180°.

(f) Two angles are said to be ____ if their sum is 90°.

(g) Two angles are said to be ____ if their sum is 180°.

(h) A line that cuts two or more lines is known as a ____ .

(i) When two parallel lines are cut by a transversal, the alternate ____ angles are equal and the alternate ____ angles are equal.

(j) A polygon of three sides is known as a ____ .

(k) A ____ triangle is a triangle that has one right angle.

(l) An ____ triangle is a triangle that has two equal sides and two equal angles.

(m) An ____ triangle is a triangle that has three equal sides and three equal angles.

(n) An ____ triangle is a triangle that has no right angle and no two sides equal.

(o) Two triangles are ____ if their corresponding sides and corresponding angles are equal.

(p) Two triangles are ____ if their corresponding angles are equal and their corresponding sides are proportional.

(q) A ____ is a four-sided polygon that has two parallel sides and two nonparallel sides.

(r) A ____ is a closed plane curve, all points on which are equidistant from a point within called the center.

(s) The distance from the center of the circle to any point on the circle is called the ____

(t) The distance across a circle through the center is called the ____ .

(u) A straight line between two points on a circle is called a ____ .

(v) A ____ of a circle is a line that intersects the circle at two points.

(w) A ____ is an angle formed by two radii.

(x) A ____ of a circle is a line that touches the circle at only one point.

(y) A figure bounded by an arc of a circle and two radii of a circle is called a ____ of a circle.

(z) A figure bounded by a chord and an arc of a circle is called a ____ of a circle.

(aa) Two circles of different radii but having the same center are called ____ circles.

(bb) A ____ is a polyhedron with parallel edges and parallel bases.

(cc) A prism with edges perpendicular to the base is known as a ____ prism.

SOLUTIONS

1. (a)
$$\begin{array}{r} 46°27' \\ +24°24' \\ \hline 68°51' \end{array}$$

(b)
$$\begin{array}{r} 56°24' \\ +33°26' \\ \hline 89°50' \end{array}$$

(c)
$$\begin{array}{r} 35°52' \\ +47°39' \\ \hline 82°91' \\ = 83°31' \end{array}$$

(d)
$$\begin{array}{r} 21°46'52'' \\ +40°25'26'' \\ \hline 61°71'78'' \\ = 62°12'18'' \end{array}$$

(e)
$$\begin{array}{r} 46°19'22'' \\ +35°51'40'' \\ \hline 81°70'62'' \\ = 82°11'02'' \end{array}$$

(f)
$$\begin{array}{r} 13°49'58'' \\ +12°21'32'' \\ \hline 25°70'90'' \\ = 26°11'30'' \end{array}$$

2. (a) $2\overline{)\,237°27'17''} = 2\overline{)\,236°86'77''}$

$$\begin{array}{l} 118°43'38.5'' \\ = 118°43'38'' \end{array}$$

(b) $2\overline{)\,329°47'16''} = 2\overline{)\,328°106'76''}$

$$\begin{array}{l} 164°53'38'' \\ = 164°53'38'' \end{array}$$

3. (a) $6\overline{)\,390°13'24''} = 2\overline{)\,390°12'84''}$

$$\begin{array}{l} 65°02'14'' \\ = 65°02'14'' \end{array}$$

4. (a) 16.4083° (b) 24.50° (c) 36.75°
(d) 68.7347° (e) 69.18° (f) 118.9197°
(g) 127.2897° (h) 173.5489° (i) 186.1428°
(j) 223.6300°

5. (a) 68°10′34″ (b) 96°33′53″ (c) 145°52′56″
(d) 221°20′52″ (e) 303°06′28″

6.

	pt	angle		pt	angle		pt	angle
(a)	A	83°23′	(b)	A	96°34′	(c)	A	98°08′05″
	B	105°27′		B	111°42′		B	149°16′12″
	C	158°31′		C	183°12′		C	134°12′55″
	D	53°19′		D	88°57′		D	93°20′10″
	E	139°18′		E	139°21′		E	152°39′47″
		538°118′		F	100°18′		F	174°32′50″
		539°58′			717°184′		G	97°51′11″
		error = 02′			720°04′			897°178′190″
					error = 04′			900°01′10″
								error = 01′10″

7. (a) parallel
(b) acute
(c) right
(d) obtuse
(e) straight
(f) complementary
(g) supplementary
(h) transversal
(i) interior, exterior
(j) triangle
(k) right
(l) isosceles
(m) equilateral
(n) oblique
(o) congruent
(p) similar
(q) trapezoid
(r) circle
(s) radius
(t) diameter
(u) chord
(v) secant
(w) central
(x) tangent
(y) sector
(z) segment
(aa) concentric
(bb) prism
(cc) right

3 Dimensional Equations

1. MEASUREMENT

A *measurement* consists of a number that expresses quantity and a unit of measure. The surveyor and the surveying technician are intricately involved in measurements and in converting measurements expressed in one unit of measure to an equivalent in another unit of measure.

In converting values from one unit of measure to another, it is just as important to find the correct unit of measure as it is to find the correct quantity.

2. DEFINITION OF DIMENSIONAL EQUATION

A *dimensional equation* is one that contains units of measure but does not contain the corresponding numerical values. For example, to express in cubic yards the volume of a dump truck bed with dimensions of 6 ft by 8 ft by 4 ft, arithmetically multiply (6)(8)(4) to find the volume in cubic feet (192 ft^3). Since there are 27 ft^3 in a cubic yard, divide 192 by 27 to find that the bed has a volume of 7 yd^3. This operation is written as

$$\frac{(6\text{ ft})(8\text{ ft})(4\text{ ft})}{27\ \frac{\text{ft}^3}{\text{yd}^3}} = 7\text{ yd}^3$$

The dimensional equation that corresponds to this is

$$\frac{(\text{ft})(\text{ft})(\text{ft})}{\frac{\text{ft}^3}{\text{yd}^3}} = \frac{\frac{\text{ft}^3}{1}}{\frac{\text{ft}^3}{\text{yd}^3}} = \left(\frac{\text{ft}^3}{1}\right)\left(\frac{\text{yd}^3}{\text{ft}^3}\right) = \text{yd}^3$$

Including the numbers in the equation,

$$\frac{\left(\frac{6}{1}\text{ ft}\right)\left(\frac{8}{1}\text{ ft}\right)\left(\frac{4}{1}\text{ ft}\right)}{\frac{27\text{ ft}^3}{1\text{ yd}^3}} = \left(\frac{192\text{ ft}^3}{1}\right)\left(\frac{1}{27}\frac{\text{yd}^3}{\text{ft}^3}\right)$$
$$= 7\text{ yd}^3$$

Example 3.1

Write a dimensional equation for finding the area in acres of a rectangular tract of land 300 ft by 200 ft. Include the measured quantities in the equation.

Solution

$$\begin{aligned} A &= \frac{\left(\frac{300}{1}\text{ ft}\right)\left(\frac{200}{1}\text{ ft}\right)}{\frac{43{,}560}{1}\frac{\text{ft}^2}{\text{ac}}} \\ &= \left(\frac{60{,}000}{1}\text{ ft}^2\right)\left(\frac{1}{43{,}560}\frac{\text{ac}}{\text{ft}^2}\right) \\ &= 1.4\text{ ac} \end{aligned}$$

Example 3.2

Write a dimensional equation for finding the velocity in feet per second of a vehicle traveling 36 mi/hr. Include the measured quantities in the equation.

Solution

$$\begin{aligned} \text{v} &= \frac{\left(\frac{36}{1}\frac{\text{mi}}{\text{hr}}\right)\left(\frac{5280}{1}\frac{\text{ft}}{\text{mi}}\right)}{\frac{3600}{1}\frac{\text{sec}}{\text{hr}}} \\ &= \left(\frac{36}{1}\frac{\text{mi}}{\text{hr}}\right)\left(\frac{5280}{1}\frac{\text{ft}}{\text{mi}}\right)\left(\frac{1}{3600}\frac{\text{hr}}{\text{sec}}\right) \\ &= 53\text{ ft/sec} \end{aligned}$$

Example 3.3

Write a dimensional equation for finding the mass of water, in tons, in a fully rectangular tank that is 10 ft long, 8 ft wide, and 6 ft deep. Include the measured quantities in the equation.

Solution[1]

$$m = \frac{\left(\frac{10}{1}\text{ ft}\right)\left(\frac{8}{1}\text{ ft}\right)\left(\frac{6}{1}\text{ ft}\right)\left(\frac{62.5}{1}\frac{\text{lbm}}{\text{ft}^3}\right)}{\frac{2000\text{ lbm}}{1\text{ ton}}}$$

$$= \frac{\left(\frac{480}{1}\text{ ft}^3\right)\left(\frac{62.5\text{ lbm}}{1\text{ ft}^3}\right)}{\frac{2000\text{ lbm}}{1\text{ ton}}}$$

$$= \left(\frac{480}{1}\text{ ft}^3\right)\left(\frac{62.5\text{ lbm}}{1\text{ ft}^3}\right)\left(\frac{1\text{ ton}}{2000\text{ lbm}}\right)$$

$$= 15\text{ tons}$$

3. FORM FOR PROBLEM SOLVING

The use of common conversion factors often makes it unnecessary to set up a dimensional equation for solving problems involving several measured quantities. However, setting up a single equation that includes the numbers expressing quantity but not units of measure (similar to the dimensional equation) is advantageous. It allows cancellation and is easily followed by someone whose task is to check its accuracy. Each number that expresses quantity should be shown in the equation. For example, to find the area of a 10 in circle do not simply write $A = (\pi/4)D^2 = 78.5\text{ in}^2$. Write out $A = (\pi/4)(10\text{ in})^2 = 78.5\text{ in}^2$. For solutions that involve unfamiliar formulas, it is good practice to write the formula and then substitute the measured quantities.

Example 3.4

What is the cost of concrete, delivered to a site at $36 per cubic yard, for a parking lot 100 ft long, 54 ft wide, and 4 in thick?

Solution

(Note: Where length and width are measured in feet and thickness in inches, use a common fraction of a foot as the thickness measurement. Four inches is *exactly* one-third of a foot but *approximately* 0.33 of a foot.)

$$\text{cost} = \frac{(100\text{ ft})(54\text{ ft})(\$36)}{(27\text{ ft})(3\text{ ft})} = \$2400$$

Example 3.5

What is the cost of filling a rectangular tank, 100 ft long, 40 ft wide, and 10 ft deep, with water at $0.06 per 1000 gallons? (There are 7.5 gallons per cubic foot.)

Solution

$$\text{cost} = \frac{(100\text{ ft})(40\text{ ft})(10\text{ ft})\left(7.5\ \frac{\text{gal}}{\text{ft}^3}\right)(\$0.06)}{1000}$$

$$= \$18$$

[1]This book uses the unit lbm to distinguish pounds-mass from pounds-force (lbf).

PRACTICE PROBLEMS

1. Write a dimensional equation to convert the given quantities to an equivalent quantity in the unit of measure indicated.

Examples:

(1) Convert 48 in to feet.

$$\frac{48\text{ in}}{12\ \frac{\text{in}}{\text{ft}}} = (48\text{ in})\left(\frac{1\text{ ft}}{12\text{ in}}\right) = 4\text{ ft}$$

(2) Convert 3 ft to inches.

$$(3\text{ ft})\left(12\ \frac{\text{in}}{\text{ft}}\right) = 36\text{ in}$$

(3) Convert 72 ft^2 to square yards.

$$\frac{72\text{ ft}^2}{9\ \frac{\text{ft}^2}{\text{yd}^2}} = (72\text{ ft}^2)\left(\frac{1\text{ yd}^2}{9\text{ ft}^2}\right) = 8\text{ yd}^2$$

(a) Convert 588 ft to yards.

(b) Convert 121 yd to feet.

(c) Convert 2 mi to feet.

(d) Convert 4 ft^2 to square inches.

(e) Convert 432 in^2 to square feet.

(f) Convert 5 yd^2 to square feet.

(g) Convert 81 ft^2 to square yards.

(h) Convert 2 ac to square feet.

(i) Convert 21,780 ft^2 to acres.

(j) Convert 3 ft^3 to cubic inches.

(k) Convert 3456 in^3 to cubic feet.

(l) Convert 5 yd^3 to cubic feet.

(m) Convert 135 ft^3 to cubic yards.

(n) Convert 3 gal to cubic inches.

(o) Convert 693 in^3 to gallons.

(p) Convert 4 gal of water to pounds.

(q) Convert 25 lbm of water to gallons.

(r) Convert 87,120 ft^2 to acres.

(s) Convert 1320 ft to miles.

(t) Convert 7 yd^2 to square feet.

2. Find the required quantities by including the given quantities within a dimensional equation.

Examples:

(1) Find the number of square yards in a driveway 20 ft wide and 54 ft long.

$$\begin{aligned} A &= \frac{(20\ \text{ft})\,(54\ \text{ft})}{9\ \dfrac{\text{ft}^2}{\text{yd}^2}} \\ &= (20\ \text{ft})\,(54\ \text{ft})\left(\frac{1\ \text{yd}^2}{9\ \text{ft}^2}\right) \\ &= 120\ \text{yd}^2 \end{aligned}$$

(2) Find the cost of a concrete sidewalk 4 ft wide, 81 ft long, and 4 in thick at $30 per cubic yard.

$$\begin{aligned} \text{cost} &= \frac{(4\ \text{ft})(81\ \text{ft})\left(\dfrac{4\ \text{in}}{12\ \dfrac{\text{in}}{\text{ft}}}\right)}{\left(27\ \dfrac{\text{ft}^3}{\text{yd}^3}\right)\left(\dfrac{\$30}{1\ \text{yd}^3}\right)} \\ &= (4\ \text{ft})\,(81\ \text{ft})\left(\frac{1}{3}\ \text{ft}\right)\left(\frac{1\ \text{ft}}{12\ \text{in}}\right)\left(\frac{1\ \text{yd}^3}{27\ \text{ft}^3}\right)\left(\frac{\$30}{1\ \text{yd}^3}\right) \\ &= \$120 \end{aligned}$$

(a) How many acres [illegible] angular plot 545 ft long and 40[illegible]

(b) What [illegible] ons, of the water in a tan[illegible] gal?

(c) [illegible] y, in feet per second, of a vehicle [illegible] miles per hour.

(d) W[illegible] is the weight of water, in tons, in a full cylindrical tank of 10 ft diameter and 10 ft height?

(e) What is the cost of excavation of a ditch of rectangular cross section 3 ft wide, 4 ft deep, and 324 ft long at $0.30 per cubic yard?

3. Solve each of the following problems by writing an equation in the form of a dimensional equation.

Example: Find the mass of water in a full rectangular tank that is 8 ft long, 5 ft wide, and 5 ft deep.

$$\begin{aligned} m &= \frac{(8\ \text{ft})(5\ \text{ft})(5\ \text{ft})\left(62.5\ \dfrac{\text{lbm}}{\text{ft}^3}\right)}{2000\ \dfrac{\text{lbm}}{\text{ton}}} \\ &= 6.25\ \text{tons} \end{aligned}$$

(a) A 6 ft by 3 ft concrete box culvert, 54 ft long, is to be constructed. Walls, footing, and deck are 6 in thick. How many cubic yards of concrete are required? (Disregard wing walls.) Note: Culvert dimensions refer to waterway openings. The horizontal dimension is 6 ft, and the vertical dimension is 3 ft.

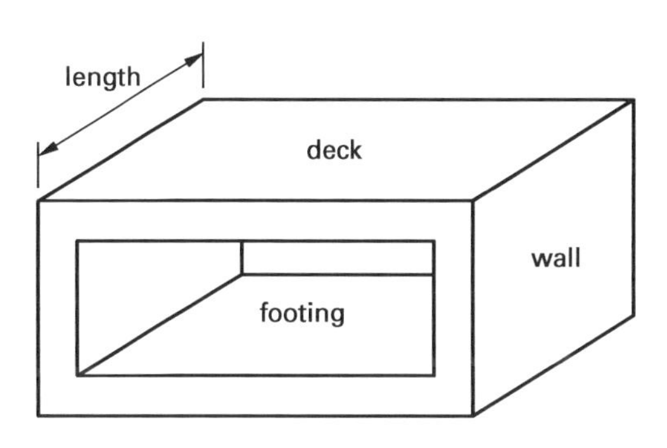

(b) The cross-sectional view of a concrete curb and gutter to be used in street paving is shown. How many lineal feet of curb and gutter can be poured with 1 yd^3 of concrete?

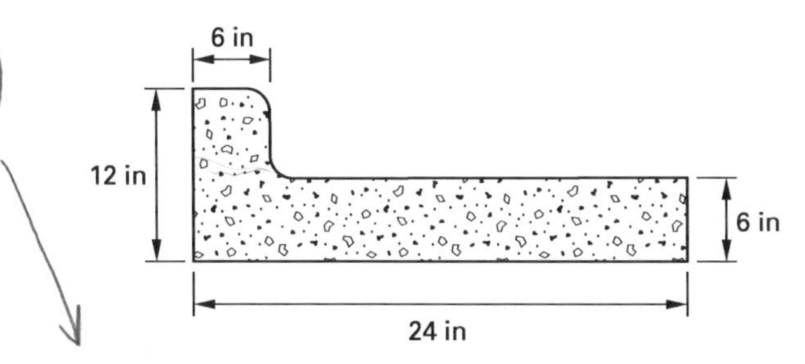

(c) A contractor is to be paid for sprinkling water in units of 1000 gallons. The empty weight of his water truck is 11,808 lbm. Loaded with water, the truck weighs 28,468 lbm. How many thousand gallons of water does the truck hold?

(d) A canal is to be excavated to a trapezoidal cross section, 30 ft at the top and 6 ft at the bottom with a 5 ft depth. What will be the cost of excavation at $0.50 per cubic yard if the length is 540 ft?

(e) A drainage ditch has a 4 ft flat bottom, 12 ft top width, and average depth of 2 ft through 162 ft of level ground. How many cubic yards of earth were excavated?

(f) How high must a cylindrical tank, 10 ft in diameter, be in order to have a capacity of 3000 gal? (Calculate to the nearest tenth of a foot.)

(g) A parking space is 100 ft by 81 ft. What is the cost of paving this area at $9.00 per square yard?

(h) A building lot has an area of 3840 ft^2. How deep is the lot if it is 32 ft wide?

(i) An electric power line is to be built from one city to another. One city is 16 mi due north and 12 mi due west of the other. (a) What length of wire is needed to connect the two cities? (b) If the wire weighs 50 lbm per 100 ft, what weight of wire is needed?

(j) A triangular piece of land has one side 320 yd long running north and south and another 1/4 mi long at right angles. A second piece of land, rectangular in shape and 250 yd on one side, has the same acreage as the triangular piece. Which piece of land would require the most fence to enclose?

(k) A 24 in shaft was drilled 54 ft deep and filled with concrete as part of a bridge pier. How many cubic yards of concrete were poured?

(l) A swimming pool 100 ft long, 50 ft wide, 2 ft deep at the shallow end, and 10 ft deep at the deep end is to be filled with water. What is the cost of the water at $0.20 per 1000 gallons?

(m) A cylindrical piece of cheese, 16 in in diameter and 8 in high, weighs 24 lbm. If a 30° sector is cut from it, (a) what is the cost of the sector at $1.00 per pound, and (b) how many cubic inches of cheese are in the sector?

(n) A rectangular concrete tank, 11 ft long, 6 ft wide, and 4 ft 6 in high (outside) is 3/4 full of water. The walls and floor of the tank are 6 in thick. How many gallons of water are in the tank?

(o) A piece of property to be purchased for highway right-of-way is bounded by an arc of a circle and a chord of that circle. The radius of the circle is 500 ft, and the central angle formed by radii to the ends of the chord is 90°. Find the area of the segment.

(p) A cylindrical water tank contains 60,000 gal of water when the water is 5 ft deep. What is the diameter of the tank?

(q) A lot 150 ft in depth and 100 ft wide is to be leveled for building construction. The fill at the front is 1.4 ft and at the rear is 2.2 ft. How many cubic yards of dirt will be required to make the fill? (Disregard shrinkage of soil.)

SOLUTIONS

1. (a)

$$\frac{588\ \text{ft}}{3\ \frac{\text{ft}}{\text{yd}}} = \boxed{196\ \text{yd}}$$

(b)

$$(121\ \text{yd})\left(3\ \frac{\text{ft}}{\text{yd}}\right) = \boxed{363\ \text{ft}}$$

(c)

$$(2\ \text{mi})\left(5280\ \frac{\text{ft}}{\text{mi}}\right) = \boxed{10{,}560\ \text{ft}}$$

(d)

$$(4\ \text{ft}^2)\left(144\ \frac{\text{in}^2}{\text{ft}^2}\right) = \boxed{576\ \text{in}^2}$$

(e)

$$\frac{432\ \text{in}^2}{144\ \frac{\text{in}^2}{\text{ft}^2}} = \boxed{3\ \text{ft}^2}$$

(f)

$$(5\ \text{yd}^2)\left(9\ \frac{\text{ft}^2}{\text{yd}^2}\right) = \boxed{45\ \text{ft}^2}$$

(g)

$$\frac{81\ \text{ft}^2}{9\ \frac{\text{ft}^2}{\text{yd}^2}} = \boxed{9\ \text{yd}^2}$$

(h)

$$(2\ \text{ac})\left(43{,}560\ \frac{\text{ft}^2}{\text{ac}}\right) = \boxed{87{,}120\ \text{ft}^2}$$

(i)

$$\frac{21{,}780\ \text{ft}^2}{43{,}560\ \frac{\text{ft}^2}{\text{ac}}} = \boxed{0.5\ \text{ac}}$$

(j)

$$(3\ \text{ft}^3)\left(1728\ \frac{\text{in}^3}{\text{ft}^3}\right) = \boxed{5184\ \text{in}^3}$$

(k)

$$\frac{3456\ \text{in}^3}{1728\ \frac{\text{in}^3}{\text{ft}^3}} = \boxed{2\ \text{ft}^3}$$

(l)

$$(5\ \text{yd}^3)\left(27\ \frac{\text{ft}^3}{\text{yd}^3}\right) = \boxed{135\ \text{ft}^3}$$

(m) $$\frac{135\ \text{ft}^3}{27\ \dfrac{\text{ft}^3}{\text{yd}^3}} = \boxed{5\ \text{yd}^3}$$

(n) $$(3\ \text{gal})\left(231\ \frac{\text{in}^3}{\text{gal}}\right) = \boxed{693\ \text{in}^3}$$

(o) $$\frac{693\ \text{in}^3}{231\ \dfrac{\text{in}^3}{\text{gal}}} = \boxed{3\ \text{gal}}$$

(p) $$(4\ \text{gal})\left(8\tfrac{1}{3}\ \frac{\text{lbm}}{\text{gal}}\right) = \boxed{33\tfrac{1}{3}\ \text{lbm}}$$

(q) $$\frac{25\ \text{lbm}}{8\tfrac{1}{3}\ \dfrac{\text{lbm}}{\text{gal}}} = \boxed{3\ \text{gal}}$$

(r) $$\frac{87{,}120\ \text{ft}^2}{43{,}560\ \dfrac{\text{ft}^2}{\text{ac}}} = \boxed{2\ \text{ac}}$$

(s) $$\frac{1320\ \text{ft}}{5280\ \dfrac{\text{ft}}{\text{mi}}} = \boxed{\frac{1}{4}\ \text{mi}}$$

(t) $$(7\ \text{yd}^2)\left(9\ \frac{\text{ft}^2}{\text{yd}^2}\right) = \boxed{63\ \text{ft}^2}$$

2. (a) The area is

$$A = \frac{(545\ \text{ft})(400\ \text{ft})}{43{,}560\ \dfrac{\text{ft}^2}{\text{ac}}} = \boxed{5\ \text{ac}}$$

(b) The weight is

$$W = \frac{(2000\ \text{gal})\left(8.33\ \dfrac{\text{lb}}{\text{gal}}\right)}{2000\ \dfrac{\text{lbm}}{\text{ton}}} = \boxed{8.33\ \text{ton}}$$

(c) The velocity is

$$\text{v} = \frac{\left(72\ \dfrac{\text{mi}}{\text{hr}}\right)\left(5280\ \dfrac{\text{ft}}{\text{mi}}\right)}{3600\ \dfrac{\text{sec}}{\text{hr}}} = \boxed{106\ \text{ft/sec}}$$

(d) The weight is

$$W = \frac{(0.785)(10\ \text{ft})^2(10\ \text{ft})\left(62.5\ \dfrac{\text{lbm}}{\text{ft}^3}\right)}{2000\ \dfrac{\text{lbm}}{\text{ton}}} = \boxed{24.5\ \text{ton}}$$

(e) The cost is

$$\frac{(3\ \text{ft})(4\ \text{ft})(324\ \text{ft})\left(\dfrac{\$0.30}{\text{yd}^3}\right)}{27\ \dfrac{\text{ft}^3}{\text{yd}^3}} = \boxed{\$43.20}$$

3. (a) The volume is

$$V = \frac{\big((7\ \text{ft})(4\ \text{ft}) - (6\ \text{ft})(3\ \text{ft})\big)(54\ \text{ft})}{27\ \dfrac{\text{ft}^3}{\text{yd}^3}} = \boxed{20\ \text{yd}^3}$$

(b) The length is

$$L = \frac{27\ \dfrac{\text{ft}^3}{\text{yd}^3}}{(1.0\ \text{ft})(0.5\ \text{ft}) + (1.5\ \text{ft})(0.5\ \text{ft})} = \boxed{21.6\ \text{ft}}$$

(c) The number of gallons is

$$\frac{28{,}468\ \text{lbm} - 11{,}808\ \text{lbm}}{\left(8.33\ \dfrac{\text{lbm}}{\text{gal}}\right)(1000)} = \boxed{2000\ \text{gal}}$$

(d) The cost is

$$\text{cost} = \frac{\left(\dfrac{30\ \text{ft} + 6\ \text{ft}}{2}\right)(5\ \text{ft})(540\ \text{ft})\left(\dfrac{\$0.50}{1\ \text{yd}^3}\right)}{27\ \dfrac{\text{ft}^3}{\text{yd}^3}} = \boxed{\$900.00}$$

(e) The volume is

$$V = \frac{\left(\dfrac{12\ \text{ft} + 4\ \text{ft}}{2}\right)(2\ \text{ft})(162\ \text{ft})}{27\ \dfrac{\text{ft}^3}{\text{yd}^3}} = \boxed{96\ \text{yd}^3}$$

(f) The height is

$$h = \frac{3000 \text{ gal}}{\left(7.5 \, \frac{\text{gal}}{\text{ft}^3}\right)\left(\frac{\pi}{4}\right)(10 \text{ ft})^2} = \boxed{5.1 \text{ ft}}$$

(g) The cost is

$$\text{cost} = \frac{(100 \text{ ft})(81 \text{ ft})\left(\frac{\$9.00}{1 \text{ yd}^2}\right)}{9 \, \frac{\text{ft}^2}{\text{yd}^2}} = \boxed{\$8100.00}$$

(h) The depth is

$$d = \frac{3840 \text{ ft}^2}{32 \text{ ft}} = \boxed{120 \text{ ft}}$$

(i) (a) The length is

$$L = \sqrt{(16 \text{ mi})^2 + (12 \text{ mi})^2} = \boxed{20 \text{ mi}}$$

(b) The weight is

$$W = \frac{(20 \text{ mi})\left(5280 \, \frac{\text{ft}}{\text{mi}}\right)(50 \text{ lbm})}{(100 \text{ ft})\left(2000 \, \frac{\text{lbm}}{\text{ton}}\right)} = \boxed{26.4 \text{ ton}}$$

(j) The area of the triangle is

$$\left(\frac{1}{2}\right)(320 \text{ yd})\left(3 \, \frac{\text{ft}}{\text{yd}}\right)(1320 \text{ ft}) = 633{,}600 \text{ ft}^2$$

The perimeter of the triangle is

$$\sqrt{(960)^2 + (1320 \text{ ft})^2} + 960 \text{ ft} + 1320 \text{ ft} = 3912 \text{ ft}$$

The side of the rectangle is

$$\frac{633{,}600 \text{ ft}^2}{(250 \text{ yd})\left(3 \, \frac{\text{ft}}{\text{yd}}\right)} = 845 \text{ ft}$$

The perimeter of the rectangle is

$$(2)\left(3 \, \frac{\text{ft}}{\text{yd}}\right)(250 \text{ yd}) + (2)(845 \text{ ft}) = 3190 \text{ ft}$$

The triangular piece would require the most fence.

(k) The volume is

$$V = \frac{(3.14)(1 \text{ ft})^2(54 \text{ ft})}{27 \, \frac{\text{ft}^3}{\text{yd}^3}} = \boxed{6.3 \text{ yd}^3}$$

(l) The cost is

$$\text{cost} = \frac{\left(\frac{2 \text{ ft} + 10 \text{ ft}}{2}\right)(50 \text{ ft})(100 \text{ ft}) \times \left(7.5 \, \frac{\text{gal}}{\text{ft}^3}\right)\left(\frac{\$0.20}{1000 \text{ gal}}\right)}{1000 \, \frac{\text{gal}}{1000 \text{ gal}}}$$

$$= \boxed{\$45.00}$$

(m) (a) The cost is

$$\left(\frac{30^\circ}{360^\circ}\right)(24 \text{ lbm})\left(\frac{\$1.00}{1 \text{ lbm}}\right) = \boxed{\$2.00}$$

(b) The volume is

$$V = \left(\frac{1}{12}\right)\pi(16 \text{ in})^2(8 \text{ in}) = \boxed{134 \text{ in}^3}$$

(n) The volume is

$$V = \left(\frac{3}{4}\right)(10 \text{ ft})(5 \text{ ft})(4 \text{ ft})\left(7.5 \, \frac{\text{gal}}{\text{ft}^3}\right)$$

$$= \boxed{1125 \text{ gal}}$$

(o) The area is

$$A = \left(\frac{90^\circ}{360^\circ}\right)\left(\frac{\pi}{4}\right)(1000 \text{ ft})^2 - \left(\frac{1}{2}\right)(500 \text{ ft})(500 \text{ ft})$$

$$= \boxed{71{,}350 \text{ ft}^2}$$

(p) The diameter is

$$D = 2\sqrt{\frac{60{,}000 \text{ gal}}{\left(7.5 \, \frac{\text{gal}}{\text{ft}^3}\right)(5 \text{ ft})\pi}} = \boxed{45 \text{ ft}}$$

(q) The volume is

$$V = \frac{\left(\frac{1.4 \text{ ft} + 2.2 \text{ ft}}{2}\right)(150 \text{ ft})(100 \text{ ft})}{27 \, \frac{\text{ft}^3}{\text{yd}^3}} = \boxed{1000 \text{ yd}^3}$$

4 Systems of Units

1. THE ENGLISH SYSTEM

American colonists from England brought the system of weights and measures in use at the time in England, called the *English system*.

The English system was originally based on standards determined by parts of the body, such as the foot, the hand, and the thumb. Several hundred years ago, the English king proclaimed the length of the English inch to be the length of three barley corn grains laid end to end. Later, of course, more sophisticated methods were used for standardization, but the system had no uniform conversion factors.

During the Colonial period, the English system was well standardized (12 inches per foot, 3 feet per yard, $5^1/_2$ yards per rod, and 16 ounces per pound), but such was not the case in the rest of Europe. There was such a wide variety of weights and measures in use that commerce was difficult.

2. THE METRIC SYSTEM

The variety of weights and measurements in Europe prompted the National Assembly of France to enact a decree in 1790 that directed the French Academy of Sciences to find standards for all weights and measures. The French Academy was supposed to work with the Royal Society of London, but the English did not participate so the French proceeded alone. The result was the *metric system*, which used the base ten in coverting units of measure.

The metric system spread rapidly in the 19th century. In 1872, France called an international meeting that was attended by 26 nations, including the United States, to further refine the system.

The meeting resulted in the establishment of the *International Bureau of Weights and Measures*, and in 1960 an extensive revision and simplification resulted in the *International System of Units*, which is in use in most countries today.

3. THE SI SYSTEM

The International System of Units—officially abbreviated SI—uses the base ten in expressing multiples and submultiples just as the metric system has always done. The six base units of measurement are as follows.

Table 4.1 *Six Base Units of Measurement*

length	meter	m
time	second	s
mass	kilogram	kg
temperature	kelvin	K
electric current	ampere	A
luminous intensity	candela	cd

The *meter* is defined as the distance traveled by light in a vacuum in 1/299 792 458 of a second. The SI unit of measure is the square meter, m^2. Land is measured by the *hectare* (10,000 square meters). The SI unit of volume is the cubic meter, m^3. Fluid volume is measured by the liter (0.001 cubic meter).

The *second* is defined as the duration of 9,192,631,770 oscillations between the two hyperfine levels of the ground state of a cesium-133 atom.

The standard for the *kilogram* is a cylinder of platinum-iridium alloy kept by the International Bureau of Weights and Standards (Bureau International des Poids et Mesures) in Sèvres, France. A duplicate is in the custody of the National Bureau of Standards, Washington, D.C.

To make a conversion in the metric system (SI system), the decimal is moved to the right or left just as in working with decimal fractions. To facilitate use of the system, names are given to the various powers of ten.

Table 4.2 *Names of Powers of Ten*

milli	one thousandth of	0.001
centi	one hundredth of	0.01
deci	one tenth of	0.1
deka	ten times	10
hecto	one hundred times	100
kilo	one thousand times	1000

Thus, centimeter means one hundredth of a meter, and kilometer means one thousand meters.

To express meters in centimeters, move the decimal two places to the right.

To express meters in kilometers, move the decimal three places to the left.

Prefixes and symbols for all SI units are as shown in Table 4.3.

Table 4.3 *Prefixes and Symbols for SI Units*

multiples and submultiples		prefixes	symbols
1,000,000,000,000 =	10^{12}	tera	T
1,000,000,000 =	10^{9}	giga	G
1,000,000 =	10^{6}	mega	M
1000 =	10^{3}	kilo	k
100 =	10^{2}	hecto	h
10 =	10	deka	da
0.1 =	10^{-1}	deci	d
0.01 =	10^{-2}	centi	c
0.001 =	10^{-3}	milli	m
0.000 001 =	10^{-6}	micro	μ
0.000 000 001 =	10^{-9}	nano	n
0.000 000 000 001 =	10^{-12}	pico	p
0.000 000 000 000 001 =	10^{-15}	femto	f
0.000 000 000 000 000 000 001 =	10^{-18}	atto	a

Table 4.4 *The English System of Weights and Measures*

linear measure	square measure
1 ft = 12 in	$1\ \text{ft}^2 = 144\ \text{in}^2$
1 yd = 3 ft	$1\ \text{yd}^2 = 9\ \text{ft}^2$
$16\frac{1}{2}$ ft = 1 rod	$1\ \text{ac} = 43{,}560\ \text{ft}^2$
5280 ft = 1 mi	$1\ \text{mi}^2 = 640\ \text{ac}$
cubic measure	**weight**
$1728\ \text{in}^3 = 1\ \text{ft}^3$	1 gal of water = 8.33 lbm
$27\ \text{ft}^3 = 1\ \text{yd}^3$	$1\ \text{ft}^3$ of water = 62.5 lbm
$231\ \text{in}^3 = 1\ \text{gal}$	1 kip = 1000 lbm
$1\ \text{ft}^3 = 7.5\ \text{gal}$	1 ton = 2000 lbm

Table 4.5 *Measures of the U.S. System of Rectangular Surveys*

1 m =	39.37 in exactly
1 m =	39.37/12
=	3.2808333 U.S. survey ft
1 U.S. survey ft =	12/39.37
=	0.3048006 m
1 international ft =	0.3048 m
1 Gunter's chain =	100 links
=	66 ft
1 Gunter's link =	7.92 in
1 Gunter's chain =	4 rods
=	4 poles
1 mi =	80 chains
$10\ \text{chains}^2$ =	$435{,}600\ \text{ft}^2$
=	1 ac

4. CONVERSION OF INCHES TO DECIMALS OF A FOOT

Engineering plans usually show dimensions of structures in feet and inches, while elevations are established in feet and decimals of a foot. The surveying technician's job is to make the necessary conversions to establish finished elevations. Construction stakes are usually set to the nearest hundredth of a foot for concrete, asphalt, pipe flow-lines, and so on. For earthwork, stakes are set to the nearest tenth of a foot.

The key to conversion is shown in Table 4.6. The values of 1 in and 1/8 in are important parts of the key.

Table 4.6 *Conversion of Inches to Decimals of a Foot*

$$1\ \text{in} = 0.08\ \text{ft}$$
$$\left(1\ \text{in} = \tfrac{1}{12}\ \text{ft} = 1 \div 12 = 0.083\ldots\text{ft}\right)$$
$$\tfrac{1}{8}\ \text{in} = 0.01\ \text{ft}$$
$$\left(\tfrac{1}{8}\ \text{in} = 1\ \text{in} \div 8 \div 12 = 0.01\ldots\text{ft}\right)$$
$$2\ \text{in} = 0.17\ \text{ft}$$
$$(1\ \text{in} + 1\ \text{in} = 0.166\ldots\text{ft})$$
$$3\ \text{in} = 0.25\ \text{ft}$$
$$\left(\tfrac{3}{12}\ \text{ft} = \tfrac{1}{4}\ \text{ft} = 0.250\ \text{ft}\right)$$
$$4\ \text{in} = 0.33\ \text{ft}$$
$$\left(\tfrac{4}{12}\ \text{ft} = \tfrac{1}{3}\ \text{ft} = 0.333\ldots\text{ft}\right)$$
$$5\ \text{in} = 0.42\ \text{ft}$$
$$(4\ \text{in} + 1\ \text{in} = 0.333 + 0.083 = 0.416\ \text{ft})$$
$$6\ \text{in} = 0.50\ \text{ft}$$
$$\left(\tfrac{6}{12}\ \text{ft} = \tfrac{1}{2}\ \text{ft} = 0.500\ \text{ft}\right)$$
$$7\ \text{in} = 0.58\ \text{ft}$$
$$(6\ \text{in} + 1\ \text{in} = 0.500 + 0.083 = 0.583\ \text{ft})$$
$$8\ \text{in} = 0.67\ \text{ft}$$
$$\left(\tfrac{8}{12}\ \text{ft} = \tfrac{2}{3}\ \text{ft} = 0.666\ldots\text{ft}\right)$$
$$9\ \text{in} = 0.75\ \text{ft}$$
$$\left(\tfrac{9}{12}\ \text{ft} = \tfrac{3}{4}\ \text{ft} = 0.750\ \text{ft}\right)$$
$$10\ \text{in} = 0.83\ \text{ft}$$
$$(9\ \text{in} + 1\ \text{in} = 0.750 + 0.083 = 0.833\ \text{ft})$$
$$11\ \text{in} = 0.92\ \text{ft}$$
$$(10\ \text{in} + 1\ \text{in} = 0.833 + 0.083 = 0.916\ \text{ft})$$
$$12\ \text{in} = 1.00\ \text{ft}$$

Conversions can be made mentally by using the following steps.

step 1: First memorize:

$$6\ \text{in} = 0.50\ \text{ft}$$
$$3\ \text{in} = 0.25\ \text{ft}$$
$$9\ \text{in} = 0.75\ \text{ft}$$

step 2: Next memorize:

$$4 \text{ in} = 0.33 \text{ ft}$$
$$8 \text{ in} = 0.67 \text{ ft}$$

step 3: Then memorize:

$$1 \text{ in} = 0.08 \text{ ft}$$
$$\tfrac{1}{8} \text{ in} = 0.01 \text{ ft}$$

step 4: In converting measurements expressed in feet, inches, and fractions of an inch to feet and decimals of a foot, convert the inches and fractions of an inch separately to decimals of a foot, then add the three parts. (In some cases subtraction can be used.)

Example 4.1

Convert the following measurements to feet and decimals of a foot.

(a) 1 ft 4 in

(b) 2 ft $8\frac{7}{8}$ in

(c) 5 ft $11\frac{1}{2}$ in

(d) 7 ft $5\frac{3}{4}$ in

(e) 11 ft $9\frac{1}{8}$ in

Solution

(a)
$$\begin{aligned} 1 \text{ ft} &= 1.00 \text{ ft} \\ 4 \text{ in} &= 0.33 \text{ ft} \\ \hline 1 \text{ ft } 4 \text{ in} &= 1.33 \text{ ft} \end{aligned}$$

(b)
$$\begin{aligned} 2 \text{ ft} &= 2.00 \text{ ft} \\ 8 \text{ in } \tfrac{7}{8} \text{ in} &= 0.74 \text{ ft} \\ \hline 2 \text{ ft } 8\tfrac{7}{8} \text{ in} &= 2.74 \text{ ft} \end{aligned}$$

(c)
$$\begin{aligned} 6 \text{ ft } 0 \text{ in} &= 6.00 \text{ ft} \\ -0 \text{ ft } \tfrac{1}{2} \text{ in} &= -0.04 \text{ ft} \\ \hline 5 \text{ ft } 11\tfrac{1}{2} \text{ in} &= 5.96 \text{ ft} \end{aligned}$$

(d)
$$\begin{aligned} 7 \text{ ft} &= 7.00 \text{ ft} \\ 5 \text{ in} &= 0.42 \text{ ft} \\ \tfrac{3}{4} \text{ in} &= 0.06 \text{ ft} \\ \hline 7 \text{ ft } 5\tfrac{3}{4} \text{ in} &= 7.48 \text{ ft} \end{aligned}$$

(e)
$$\begin{aligned} 11 \text{ ft} &= 11.00 \text{ ft} \\ 9 \text{ in} &= 0.75 \text{ ft} \\ \tfrac{1}{8} \text{ in} &= 0.01 \text{ ft} \\ \hline 11 \text{ ft } 9\tfrac{1}{8} \text{ in} &= 11.76 \text{ ft} \end{aligned}$$

5. CONVERSION OF DECIMALS OF A FOOT TO INCHES

In converting measurements expressed in feet and decimals of a foot to feet, inches, and fractions of an inch, mentally recall the "decimal-to-foot value" for the full inch that is nearest to and less than the given measurement. Then convert the remainder, which will be in hundredths of a foot, to a fraction that is expressed in eighths of an inch. Or, recall the "decimal of a foot value" for the full inch that is nearest to and more than the given measurement and subtract the given measurement from it. Remember that conversions are made only to the nearest 1/8 inch in this procedure.

Example 4.2

(a) 3.72 ft

(b) 3.79 ft

(c) 5.65 ft

(d) 6.34 ft

Solution

(a)
$$\begin{aligned} 3.00 \text{ ft} & &= 3 \text{ ft } 0 \text{ in} \\ 0.72 \text{ ft} &= 0.75 \text{ ft} - 0.03 \text{ ft} &= 0 \text{ ft } 8\tfrac{5}{8} \text{ in} \\ \hline 3.72 \text{ ft} & &= 3 \text{ ft } 8\tfrac{5}{8} \text{ in} \end{aligned}$$

(b)
$$\begin{aligned} 3.00 \text{ ft} & &= 3 \text{ ft } 0 \text{ in} \\ 0.79 \text{ ft} &= 0.75 \text{ ft} + 0.04 \text{ ft} &= 0 \text{ ft } 9\tfrac{1}{2} \text{ in} \\ \hline 3.79 \text{ ft} & &= 3 \text{ ft } 9\tfrac{1}{2} \text{ in} \end{aligned}$$

(c)
$$\begin{aligned} 5.00 \text{ ft} & &= 5 \text{ ft } 0 \text{ in} \\ 0.65 \text{ ft} &= 0.67 \text{ ft} - 0.02 \text{ ft} &= 0 \text{ ft } 7\tfrac{3}{4} \text{ in} \\ \hline 5.65 \text{ ft} & &= 5 \text{ ft } 7\tfrac{3}{4} \text{ in} \end{aligned}$$

(d)
$$\begin{aligned} 6.00 \text{ ft} & &= 6 \text{ ft } 0 \text{ in} \\ 0.34 \text{ ft} &= 0.33 \text{ ft} + 0.01 \text{ ft} &= 0 \text{ ft } 4\tfrac{1}{8} \text{ in} \\ \hline 6.34 \text{ ft} & &= 6 \text{ ft } 4\tfrac{1}{8} \text{ in} \end{aligned}$$

PRACTICE PROBLEMS

1. Write the missing word or number in each of the following sentences.

(a) 4 yd = ____ in

(b) 288 in^2 = ____ ft^2

(c) 54 ft^3 = ____ yd^3

(d) 3 acres = ____ ft^2

(e) 0.5 ft^2 = ____ in^2

(f) 2 gal = ____ in^3

(g) 15 gal = ____ ft^3

(h) 3 gal = ____ lbm (water)

(i) 250 lbm = ____ ft^3 (water)

(j) 3000 lbm = _____ ton

(k) kilo means _____ times

(l) centi means _____ of

(m) deci means _____ of

(n) milli means _____ of

(o) hecto means _____ times

(p) deka means _____ times

2. Convert the following feet and inches to feet and decimals of a foot (to two decimals only).

Example: 3 ft $9\frac{3}{8}$ in = 3.78 ft

(a) 1 ft $10\frac{1}{8}$ in

(b) 2 ft $6\frac{1}{2}$ in

(c) 2 ft $6\frac{3}{4}$ in

(d) 3 ft $7\frac{5}{8}$ in

(e) 4 ft $3\frac{3}{8}$ in

(f) 4 ft $6\frac{1}{8}$ in

(g) 4 ft $9\frac{3}{4}$ in

(h) 5 ft $0\frac{1}{4}$ in

(i) 5 ft $4\frac{3}{4}$ in

(j) 5 ft $10\frac{5}{8}$ in

(k) 6 ft $3\frac{1}{2}$ in

(l) 6 ft $4\frac{7}{8}$ in

(m) 7 ft $2\frac{1}{2}$ in

(n) 7 ft $2\frac{7}{8}$ in

(o) 8 ft $7\frac{1}{8}$ in

(p) 8 ft $8\frac{5}{8}$ in

(q) 9 ft $4\frac{1}{8}$ in

(r) 9 ft $8\frac{1}{2}$ in

(s) 10 ft $5\frac{1}{4}$ in

(t) 10 ft $11\frac{1}{4}$ in

3. Convert the following feet and decimals to feet and inches.

(a) 0.36 ft

(b) 1.35 ft

(c) 2.69 ft

(d) 2.94 ft

(e) 3.52 ft

(f) 3.87 ft

(g) 4.76 ft

(h) 4.79 ft

(i) 4.83 ft

(j) 5.06 ft

(k) 5.60 ft

(l) 6.08 ft

(m) 6.16 ft

(n) 6.25 ft

(o) 6.67 ft

(p) 7.81 ft

(q) 8.21 ft

(r) 8.72 ft

(s) 9.23 ft

(t) 9.27 ft

SOLUTIONS

1. (a) 144 in (b) 2 ft^2
(c) 2 yd^3 (d) 130,680 ft^2
(e) 72 in^2 (f) 462 in^3
(g) 2 ft^3 (h) 25 lbm (water)
(i) 4 ft^3(water) (j) 1.5 ton
(k) 1000 (l) 0.01
(m) 0.1 (n) 0.001
(o) 100 (p) 10

2. (a) 1.84 ft (b) 2.54 ft (c) 2.56 ft
(d) 3.64 ft (e) 4.28 ft (f) 4.51 ft
(g) 4.81 ft (h) 5.02 ft (i) 5.40 ft
(j) 5.88 ft (k) 6.29 ft (l) 6.41 ft
(m) 7.21 ft (n) 7.24 ft (o) 8.59 ft
(p) 8.72 ft (q) 9.34 ft (r) 9.71 ft
(s) 10.44 ft (t) 10.94 ft

3. (a) 0 ft $4\frac{3}{8}$ in (b) 1 ft $4\frac{1}{4}$ in
(c) 2 ft $8\frac{1}{4}$ in (d) 2 ft $11\frac{1}{4}$ in
(e) 3 ft $6\frac{1}{4}$ in (f) 3 ft $10\frac{1}{2}$ in
(g) 4 ft $9\frac{1}{8}$ in (h) 4 ft $9\frac{1}{2}$
(i) 4 ft 10 in (j) 5 ft $0\frac{3}{4}$ in
(k) 5 ft $7\frac{1}{4}$ in (l) 6 ft 1 in
(m) 6 ft $1\frac{7}{8}$ in (n) 6 ft 3 in
(o) 6 ft 8 in (p) 7 ft $9\frac{3}{4}$ in
(q) 8 ft $2\frac{1}{2}$ in (r) 8 ft $8\frac{5}{8}$ in
(s) 9 ft $2\frac{3}{4}$ in (t) 9 ft $3\frac{1}{4}$ in

5 Perimeter and Circumference

1. DEFINITION

The sum of the lengths of the sides of a polygon is called the *perimeter* of the polygon.

Example 5.1

Find the perimeter of the right triangle shown.

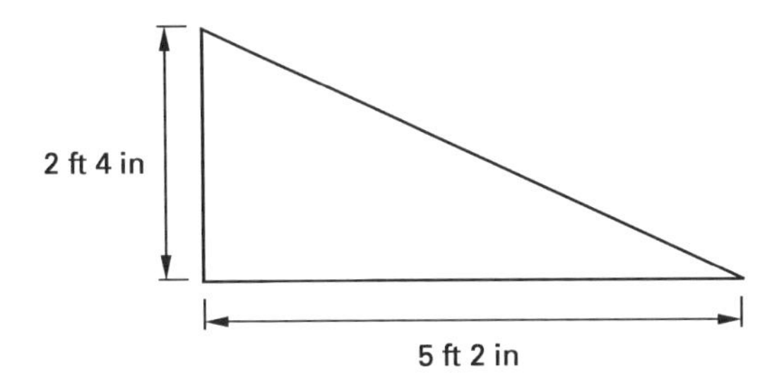

Solution

$$\begin{aligned} \text{side} &= \sqrt{(5.17\text{ ft})^2 + (2.33\text{ ft})^2} \\ &= 5.67\text{ ft} = 5\text{ ft } 8\text{ in} \\ \text{perimeter} &= 5\text{ ft } 2\text{ in} + 2\text{ ft } 4\text{ in} + 5\text{ ft } 8\text{ in} \\ &= 13\text{ ft } 2\text{ in} \end{aligned}$$

Example 5.2

Find the perimeter of the isosceles trapezoid shown.

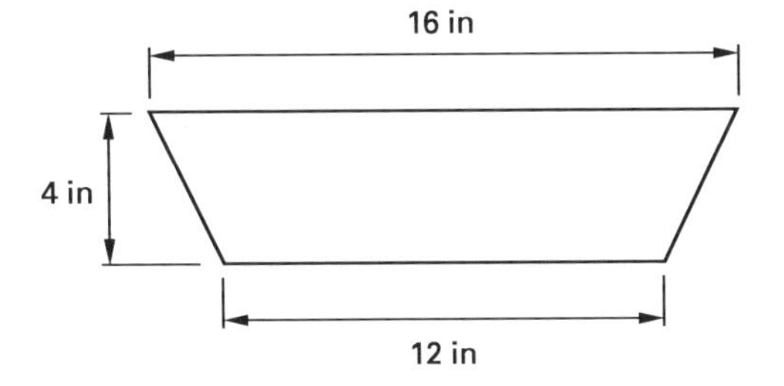

Solution

$$\text{side} = \sqrt{(4\text{ in})^2 + (2\text{ in})^2} = 4.5\text{ in}$$
$$\text{perimeter} = 12\text{ in} + 4.5\text{ in} + 16\text{ in} + 4.5\text{ in} = 37\text{ in}$$

Example 5.3

Find the perimeter of a right triangle with a base of 9 in and an altitude of 12 in.

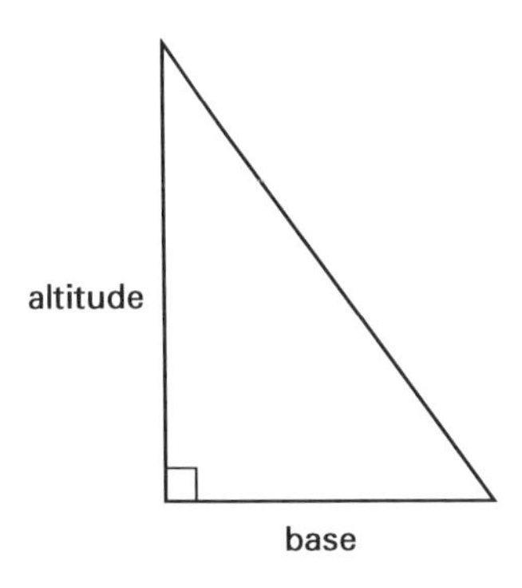

Solution

$$\text{side} = \sqrt{(9\text{ in})^2 + (12\text{ in})^2} = 15\text{ in}$$
$$\text{perimeter} = 9\text{ in} + 12\text{ in} + 15\text{ in} = 36\text{ in}$$

2. CIRCUMFERENCE OF A CIRCLE

The *circumference* of a circle is the distance around the circle. It contains 360°. Regardless of the size, the circumference of any circle is always approximately 3.14 times the length of the diameter. The exact ratio of the circumference to the diameter is the number π. It is an irrational number, but its value is usually considered to be 3.1416 or 3.14, depending on the accuracy desired, based on the accuracy of the measurement of the diameter. Thus, the circumference of any circle is

$$C = \pi D \qquad 5.1$$

C is the circumference and D is the diameter of the circle. Because the diameter is twice the radius,

$$C = 2\pi R \qquad 5.2$$

C is the circumference and R is the radius of the circle.

Example 5.4

Find the circumference of a 10 ft diameter circle.

Solution

$$C = \pi D = \pi(10 \text{ ft}) = 31.42 \text{ ft}$$
$$= 31 \text{ ft } 5 \text{ in}$$

Example 5.5

Find the circumference of a circle that has a radius of 21 ft 9 in.

Solution

$$C = 2\pi R = 2\pi(21.75 \text{ ft}) = 136.66 \text{ ft}$$
$$= 136 \text{ ft } 8 \text{ in}$$

Example 5.6

Find the outside circumference of a concrete pipe with an inside diameter of 36 in and a wall thickness of 3 in.

Solution

$$C = \pi D = \pi(42 \text{ in}) = 132 \text{ in}$$
$$= 11 \text{ ft } 0 \text{ in}$$

Example 5.7

Find the diameter of a tank that measures 62 ft 10 in around.

Solution

$$C = \pi D$$
$$D = \frac{C}{\pi} = \frac{62.83 \text{ ft}}{\pi} = 20 \text{ ft } 0 \text{ in}$$

3. LENGTH OF AN ARC OF A CIRCLE

The *length of an arc* of a circle is proportional to its central angle. A central angle of 90° (one-fourth of 360°) subtends an arc that is one-fourth the circumference in length. Thus, an arc whose central angle is 45° is (45°/360°)C in length.

Example 5.8

Find the length of an arc of a 100 ft diameter circle that has a central angle of 36°.

Solution

$$\text{arc} = \frac{(36°)(100 \text{ ft})\pi}{360°} = \frac{100\pi}{10} = 31 \text{ ft}$$

PRACTICE PROBLEMS

1. Find the perimeter of the floor of a room 18 ft by 22 ft.

2. Find the perimeter of a right triangle with a base of 12 in and an altitude of 5 in.

3. Find the perimeter of an isosceles triangle with a base of 12 in and an altitude of 8 in. (Note: For an isosceles triangle, a line from the vertex perpendicular to the base bisects the base.)

4. Find the circumference of a 10 in circle.

5. Find the circumference of a circle with a radius of 7 in.

6. Find the length of an arc of a circle of 24 in radius that has a central angle of 60°.

7. What is the diameter of a cylindrical tank that measures 47.10 ft around the outside?

8. Find the length of an arc of a circle of 16 in diameter that has a central angle of 45°.

9. Find the diameter of a tree that measures 3 ft $1\frac{3}{4}$ in around.

10. Find the perimeter of each of the following figures. Show computations for the length of any side needed that is not dimensioned.

(a)

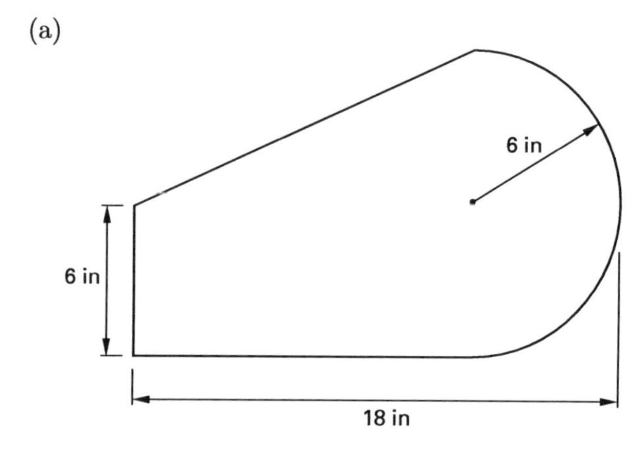

(b)

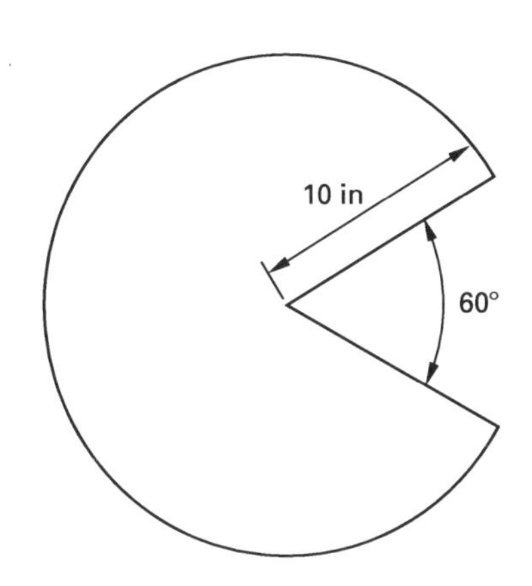

11. Find the length of the following steel reinforcing bars.

(a)

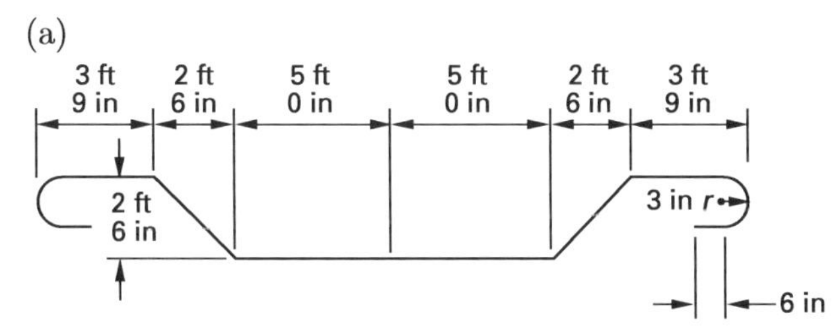

not to scale

(b)

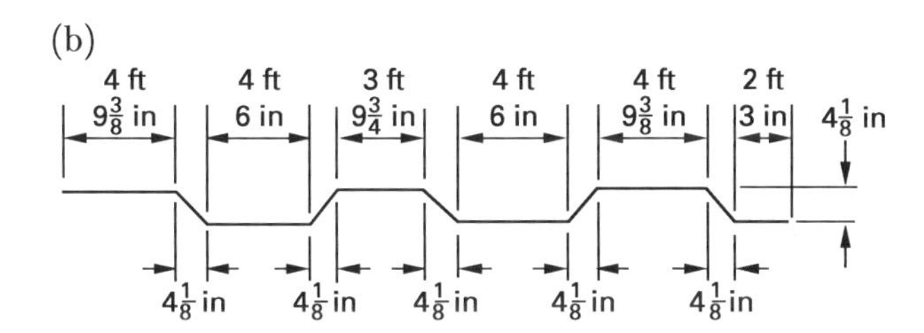

SOLUTIONS

1. $(2)(22\text{ ft}) + (2)(18\text{ ft}) = \boxed{80\text{ ft}}$

2.

$$\text{side} = \sqrt{(5\text{ in})^2 + (12\text{ in})^2} = \boxed{13\text{ in}}$$

$$\text{perimeter} = 5\text{ in} + 12\text{ in} + 13\text{ in} = \boxed{30\text{ in}}$$

3.

$$\text{side} = \sqrt{(8\text{ in})^2 + (6\text{ in})^2} = \boxed{10\text{ in}}$$

$$\text{perimeter} = 10\text{ in} + 10\text{ in} + 12\text{ in} = \boxed{32\text{ in}}$$

4. $\text{circumference} = \pi(10\text{ in}) = \boxed{31.4\text{ in}}$

5. $\text{circumference} = (2)\pi(7\text{ in}) = \boxed{44\text{ in}}$

6. $\text{arc} = \left(\frac{1}{6}\right)(2)\pi(24\text{ in}) = \boxed{25\text{ in}}$

7. $\text{diameter} = \frac{47.10\text{ ft}}{\pi} = \boxed{15.0\text{ ft}}$

8. $\text{arc} = \left(\frac{1}{8}\right)\pi(16\text{ in}) = \boxed{6.3\text{ in}}$

9. $\text{diameter} = \frac{3.14\text{ ft}}{\pi} = \boxed{1\text{ ft}}$

10. (a) $\text{arc} = \pi(6\text{ in}) = \boxed{19\text{ in}}$

$$\text{side} = \sqrt{(6\text{ in})^2 + (12\text{ in})^2} = \boxed{13\text{ in}}$$

$$\text{perimeter} = 19\text{ in} + 13\text{ in} + 6\text{ in} + 12\text{ in} = \boxed{50\text{ in}}$$

(b)

$$\text{arc} = (2)\left(\frac{5}{6}\right)\pi(10\text{ in}) = \boxed{52\text{ in}}$$

$$\text{side} = 10\text{ in} + 10\text{ in} = 20\text{ in}$$

$$\text{perimeter} = 52\text{ in} + 10\text{ in} + 10\text{ in} = \boxed{72}$$

11. (a) $\text{side} = \sqrt{(2.5\ \text{ft})^2 + (2.5\ \text{ft})^2} = \boxed{3.54\ \text{ft}}$

$$\text{arc} = \left(\frac{1}{2}\right)(2)\pi(0.25\ \text{ft}) = \boxed{0.78\ \text{ft}}$$

$$\begin{aligned}\text{length} &= (2)(0.50\ \text{ft} + 0.78\ \text{ft} + 3.50\ \text{ft} \\ &\quad + 3.54\ \text{ft} + 5.00\ \text{ft}) \\ &= \boxed{26.64\ \text{ft}}\end{aligned}$$

(b) $\text{side} = \sqrt{(0.34\ \text{ft})^2 + (0.34\ \text{ft})^2} = \boxed{0.48\ \text{ft}}$

$$\begin{aligned}\text{length} &= 4.78\ \text{ft} + 0.48\ \text{ft} + 4.50\ \text{ft} + 0.48\ \text{ft} \\ &\quad + 3.81\ \text{ft} + 0.48\ \text{ft} + 4.50\ \text{ft} + 0.48\ \text{ft} \\ &\quad + 4.78\ \text{ft} + 0.48\ \text{ft} + 2.25\ \text{ft} \\ &= \boxed{27.02\ \text{ft}}\end{aligned}$$

6 Area

Nomenclature

A area
b base
D diameter
h altitude
r radius
s half the perimeter

1. DEFINITION

Area is defined as the surface within a set of lines. Thus, the area of a triangle is the surface within the three sides; the area of a circle is the surface within the circumference.

Area is measured in square units: square inches (in^2), square feet (ft^2), square miles (mi^2), and so on.

A square inch is a square, each side of which is 1 in in length. The rectangle shown in Fig. 6.1 has an area of 6 in^2. It could be exactly covered by six squares, each 1 in on a side.

Figure 6.1 *Square Inch*

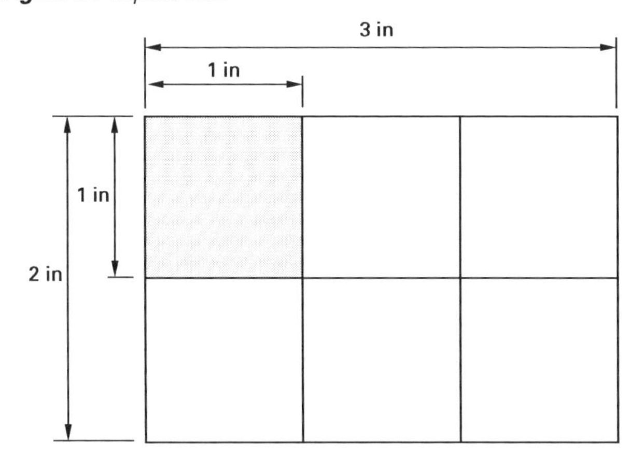

2. AREA OF A RECTANGLE

The area of a rectangle is equal to the product of the length and the width.

Example 6.1

Find the area of the floor of a room that is 20.25 ft long and 16.33 ft wide.

Solution

$$\begin{aligned}\text{area} &= (\text{length})(\text{width}) \\ &= (20.25\ \text{ft})(16.33\ \text{ft}) \\ &= 330.7\ \text{ft}^2\end{aligned}$$

Example 6.2

Find the area of the walls of a room that is 8.0 ft high if the length of the room is 20.0 ft and the width is 15.0 ft.

Solution

$$\begin{aligned}\text{area of two walls} &= (2)(8\ \text{ft})(20\ \text{ft}) = 320\ \text{ft}^2 \\ \text{area of two walls} &= (2)(8\ \text{ft})(15\ \text{ft}) = \underline{240\ \text{ft}^2} \\ \text{total area} &= 560\ \text{ft}^2\end{aligned}$$

3. AREA OF A TRIANGLE

The area of a triangle is expressed in terms of its base and its altitude. Any side of a triangle can be called the *base*. The *vertex* of a triangle is the vertex opposite the base. The *altitude* of a triangle is the perpendicular distance from the vertex to the base.

The area of any triangle is equal to one-half the product of the base and the altitude.

Figure 6.2 *Area of a Triangle*

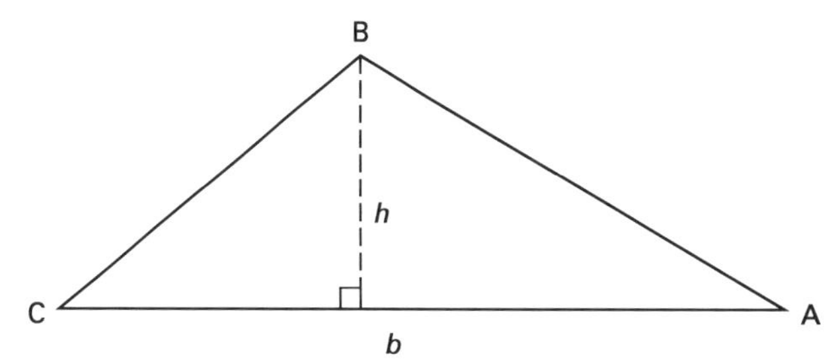

In the triangle ABC,

$$A = \tfrac{1}{2}bh \qquad 6.1$$

A = area
b = base
h = altitude

Example 6.3

Find the area of a triangle with a base of 12 in and an altitude of 4 in.

Solution

$$A = \frac{bh}{2} = \frac{(12 \text{ in})(4 \text{ in})}{2} = 24 \text{ in}^2$$

4. AREA OF A RIGHT TRIANGLE

The area of a right triangle is equal to one-half the product of the base and the altitude.

In Fig. 6.3, a rectangle has a side b and a side a. The area of the rectangle is ab. If the rectangle is cut into two equal triangles as shown by the dashed line, then a represents the altitude of a right triangle and b represents the base of the right triangle; the area of each triangle is $\frac{1}{2}ab$.

Figure 6.3 *Area of a Right Triangle*

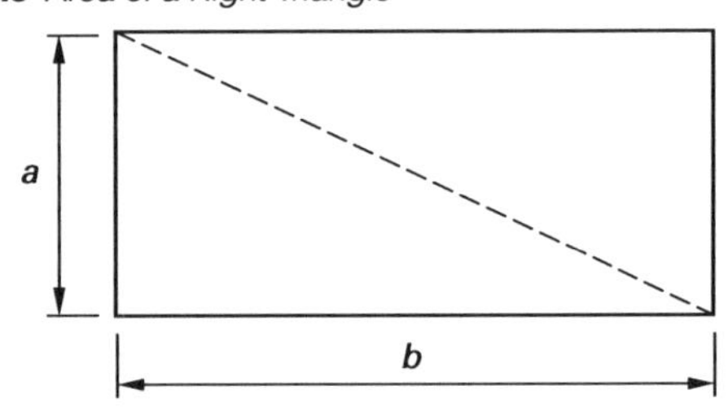

Example 6.4

Find the area of the right triangle shown.

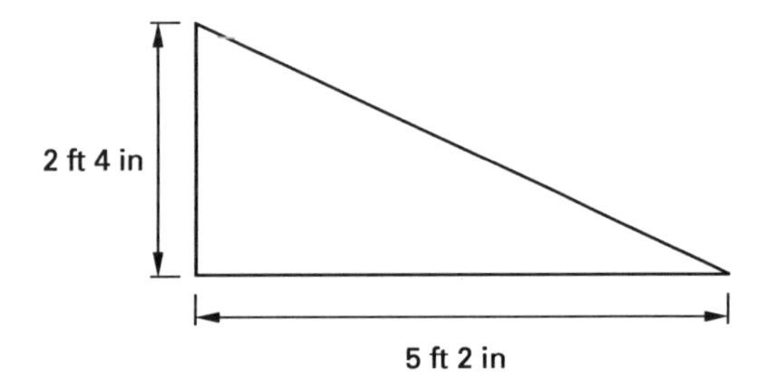

Solution

$$\text{base} = 5 \text{ ft } 2 \text{ in} = 5.16 \text{ ft}$$

$$\text{altitude} = 2 \text{ ft } 4 \text{ in} = 2.33 \text{ ft}$$

$$\text{area} = \frac{ab}{2} = \frac{(2.33 \text{ ft})(5.16 \text{ ft})}{2} = 6.01 \text{ ft}^2$$

5. AREA OF A TRIANGLE WITH KNOWN SIDES

If the lengths of the three sides of a triangle are known, the area of the triangle can be found from Eq. 6.2.

$$A = \sqrt{s(s-a)(s-b)(s-c)} \qquad 6.2$$

A = area
s = half the perimeter
a, b, and c = lengths of each of the sides

Example 6.5

Find the area of a triangle with sides 32 ft, 46 ft, and 68 ft.

Solution

$$A = \sqrt{s(s-a)(s-b)(s-c)}$$
$$= \sqrt{(73 \text{ ft})(73 \text{ ft} - 32 \text{ ft})(73 \text{ ft} - 46 \text{ ft})(73 \text{ ft} - 68 \text{ ft})}$$
$$= 636 \text{ ft}^2$$

6. AREA OF A TRAPEZOID

The area of a trapezoid is equal to the average width times the altitude. This may be expressed in another way: The area of a trapezoid is equal to one-half the sum of the bases times the altitude.

Figure 6.4 *Area of a Trapezoid*

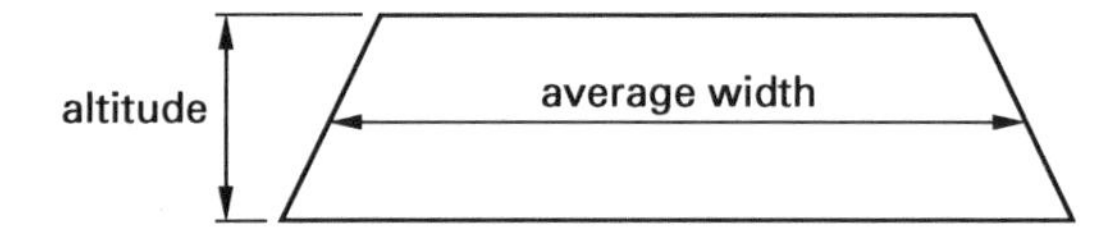

Example 6.6

Find the area of the trapezoid shown.

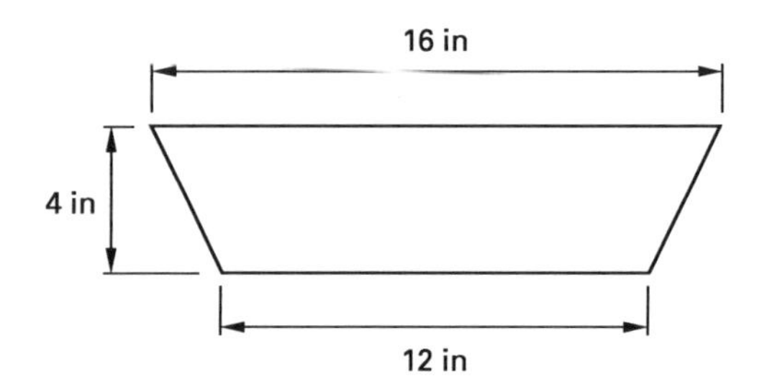

Solution

$$\begin{aligned} \text{area} &= \tfrac{1}{2}\,(\text{sum of bases})\,(\text{altitude}) \\ &= \left(\frac{16\text{ in} + 12\text{ in}}{2}\right)(4\text{ in}) \\ &= 56\text{ in}^2 \end{aligned}$$

Example 6.7

A swimming pool is 4 ft deep at one end, 8 ft deep at the other end, and 100 ft long. Find the area of the trapezoidal section through the long axis.

Solution

$$A = \frac{(4\text{ ft} + 8\text{ ft})(100\text{ ft})}{2} = 600\text{ ft}^2$$

Example 6.8

A drainage ditch has a trapezoidal cross section with a bottom width of 6 ft, a top width of 24 ft, and a depth of 4 ft. Find the area of the cross section.

Solution

$$A = \frac{(6\text{ ft} + 24\text{ ft})(4\text{ ft})}{2} = 60\text{ ft}^2$$

7. AREA OF A CIRCLE

The relation between the circumference of a circle and its radius is that the circumference is always 2π times the radius.

The area of a circle is always π times the square of its radius, or

$$A = \pi r^2 \qquad 6.3$$

A = the area of any circle

r = the radius of that circle

Since the radius of a circle is equal to half the diameter, the area of a circle can be expressed as

$$A = \pi\left(\frac{D}{2}\right)^2 = \left(\frac{\pi}{4}\right)D^2 \qquad 6.4$$

Example 6.9

Find the area of a 12.0 in circle.

Solution

$$\begin{aligned} A &= \pi r^2 = \pi(6\text{ in})^2 = \pi(36\text{ in}^2) \\ &= 113\text{ in}^2 \end{aligned}$$

Example 6.10

Find the area of a 10.0 in circle.

Solution

$$\begin{aligned} A &= \left(\frac{\pi}{4}\right)D^2 = \frac{\pi(10.0\text{ in})^2}{4} = \frac{\pi(100\text{ in}^2)}{4} \\ &= 78.5\text{ in}^2 \end{aligned}$$

Example 6.11

Find the area of an 11.0 in circle.

Solution

$$\begin{aligned} A &= \left(\frac{\pi}{4}\right)D^2 = \left(\frac{\pi}{4}\right)(11.0\text{ in})^2 \\ &= 95.0\text{ in}^2 \end{aligned}$$

8. AREA OF A SECTOR OF A CIRCLE

The area of a sector of a circle is a fractional part of the area of the circle. The central angle of the sector is a measure of the fraction. A sector whose central angle is 90° is one-fourth of a circle because 90° is one-fourth of 360°.

Figure 6.5 *Area of a Sector of a Circle*

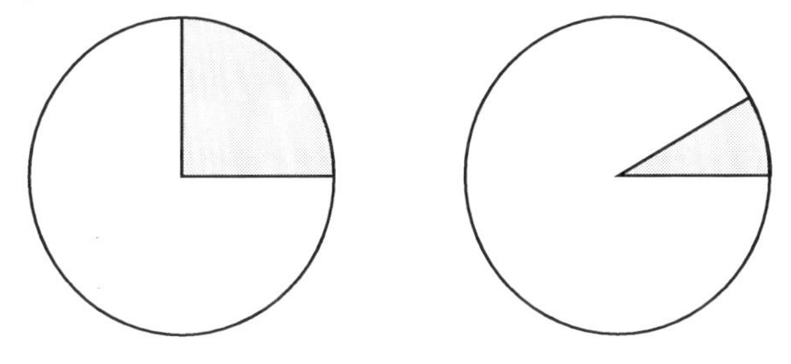

Example 6.12

Find the area of a 30° sector in a 6 in circle.

Solution

$$\begin{aligned} A &= \left(\frac{30^\circ}{360^\circ}\right)\left(\frac{\pi}{4}\right)D^2 \\ &= \left(\frac{30^\circ}{360^\circ}\right)\left(\frac{\pi}{4}\right)(36\text{ in}^2) \\ &= 2.4\text{ in}^2 \end{aligned}$$

9. AREA OF A SEGMENT OF A CIRCLE

A segment of a circle is bounded by an arc and a straight line that connects the ends of the arc. The area of a segment is found by subtracting the area of the triangle formed by the chord and the two radii to its end points from the area of the sector formed by the two radii and the arc.

Figure 6.6 Area of a Segment of a Circle

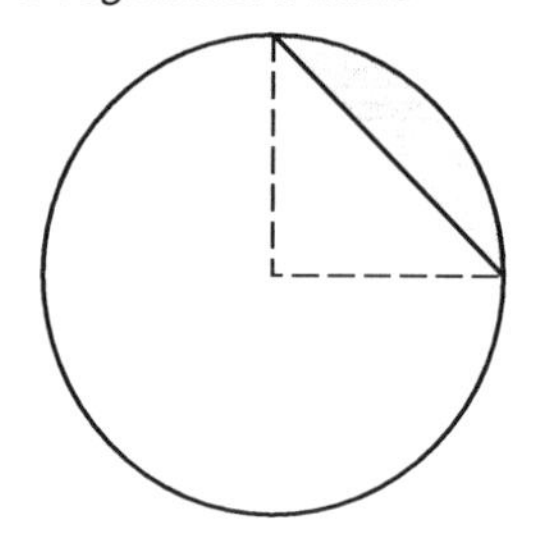

Example 6.13

Find the area of the segment whose arc subtends an angle of 90° in a circle with an 8 in radius.

Solution

$$\begin{aligned} A &= \left(\frac{90^\circ}{360^\circ}\right)\pi\,(8\text{ in})^2 - \frac{(8\text{ in})\,(8\text{ in})}{2} \\ &= 50\text{ in}^2 - 32\text{ in}^2 \\ &= 18\text{ in}^2 \end{aligned}$$

10. COMPOSITE AREAS

Irregularly shaped areas can sometimes be divided into components that consist of geometric figures, the areas of which can be found. Total area can be found by adding the areas of the components. In some cases it may be appropriate to subtract the areas of geometric figures in order to find the net area desired.

Example 6.14

Find the area of the following figure.

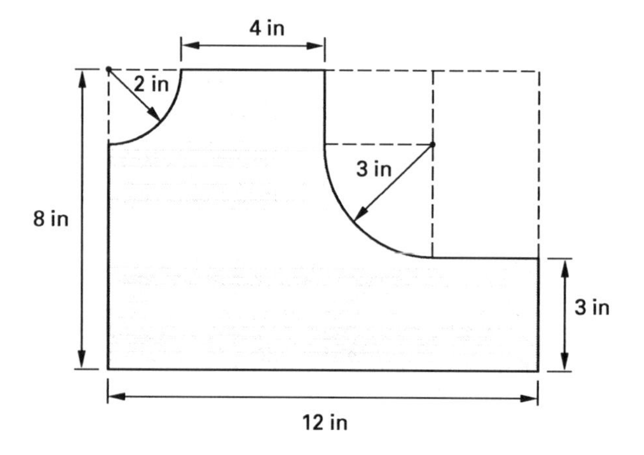

Solution

$$\begin{aligned} \text{area} &= (12\text{ in})(8\text{ in}) - \frac{\pi(2\text{ in})^2}{4} - (2\text{ in})(3\text{ in}) \\ &\quad - \frac{\pi(3\text{ in})^2}{4} - (3\text{ in})(5\text{ in}) \\ &= 65\text{ in}^2 \end{aligned}$$

PRACTICE PROBLEMS

1. Find the area of a right triangle with a base of 12 in and an altitude of 8 in.

2. Find the number of square feet of wallboard needed to cover the walls and ceiling of a room 24 ft long, 16 ft wide, and 8 ft high. Find the number of 4 ft by 8 ft sheets needed.

3. Find the cross-sectional area of a ditch of trapezoidal cross section with a top width of 28 ft, a bottom width of 4 ft, and a depth of 6 ft.

4. Find the cross-sectional area of a highway fill of trapezoidal cross section with a top width of 44 ft, a base width of 92 ft, and a height of 8 ft.

5. Find the area of a circle that has a 20 ft radius.

6. Find the area of a 10 ft diameter circle.

7. Find the area of a 60° sector of a 6 in circle.

8. Find the area of the segment whose arc subtends an angle of 90° in a 12 ft circle (12 ft diameter).

9. Find the area of a triangle with sides of 18 ft, 12 ft, and 10 ft.

10. Divide the following figures into component parts, then find the total area by either adding the areas of the component parts or by subtracting areas from a larger area that includes the area shown.

Example:

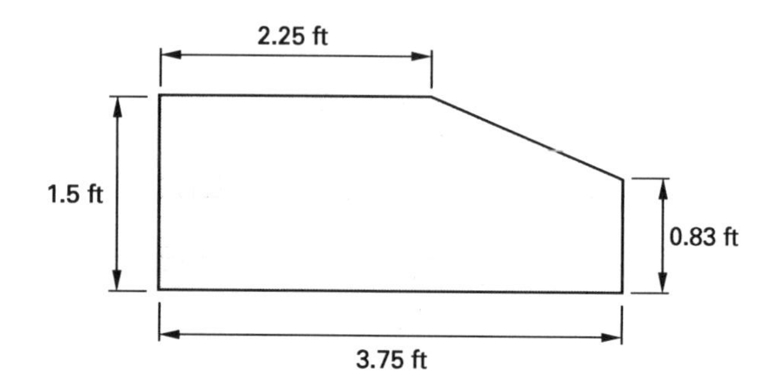

Solution:

$$A = (3.75\ \text{ft})(1.5\ \text{ft}) - \left(\frac{1}{2}\right)(1.50\ \text{ft})(0.67\ \text{ft})$$
$$= 5.1\ \text{ft}^2$$

(a)

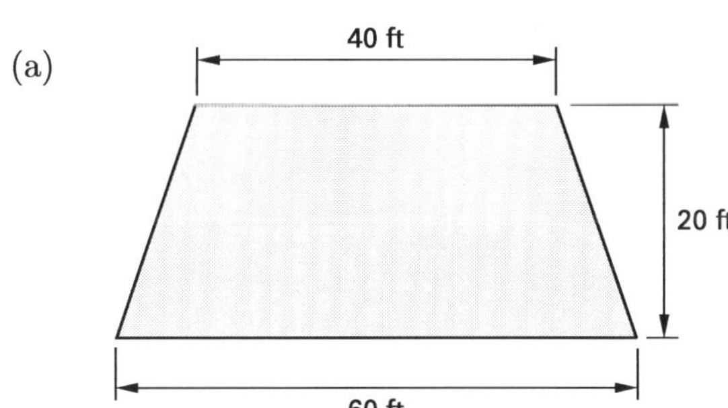

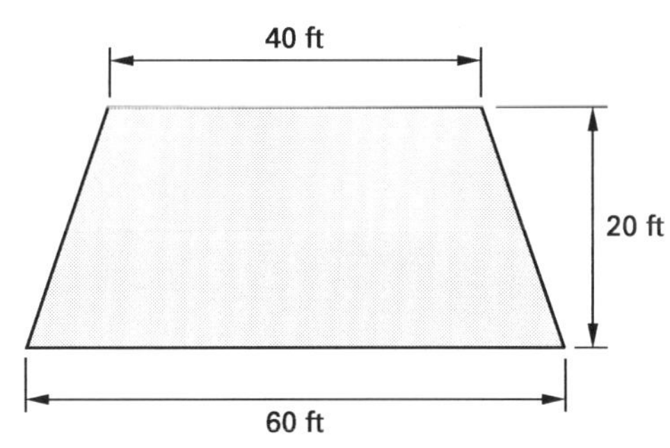

(b)

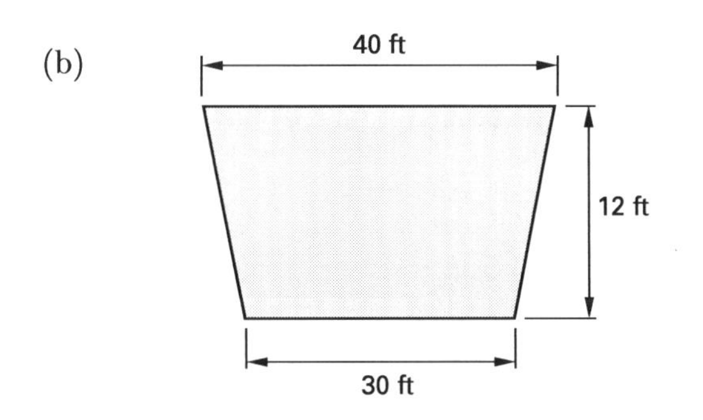

(c)

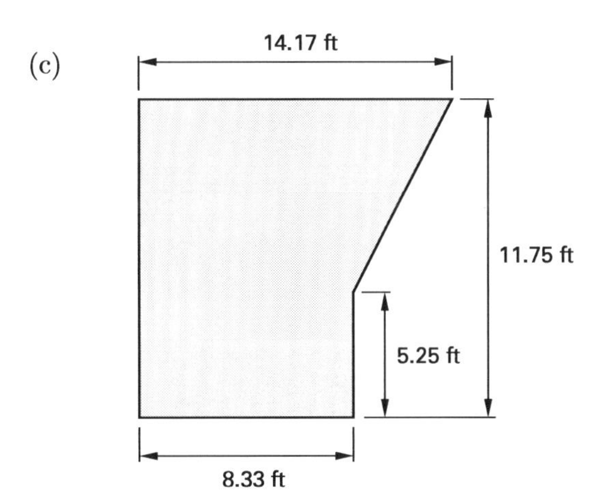

(d)

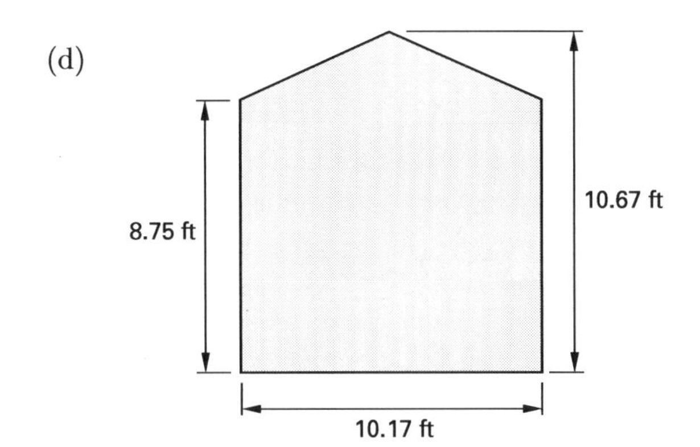

(e)

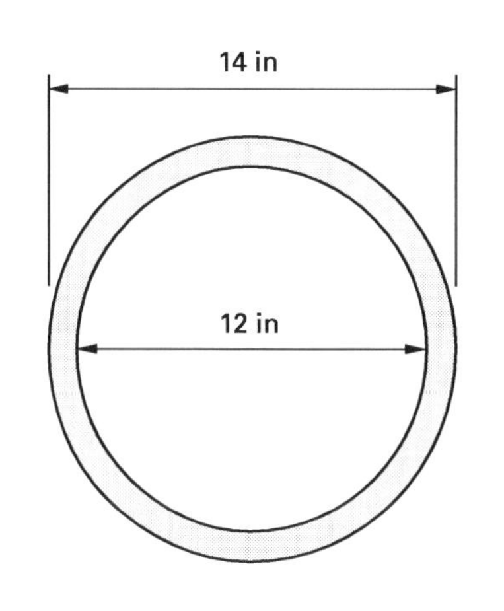

(f)

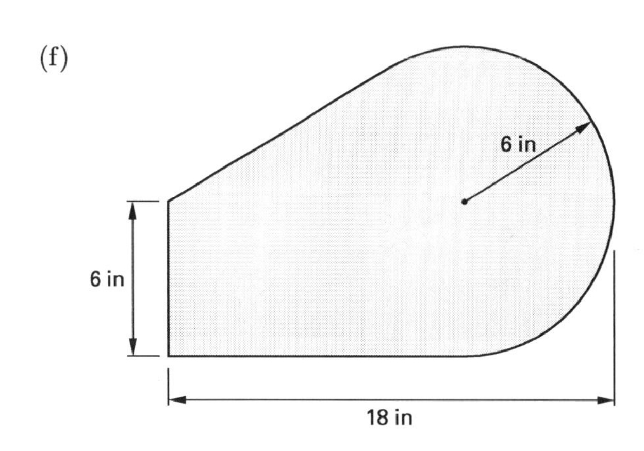

(g)

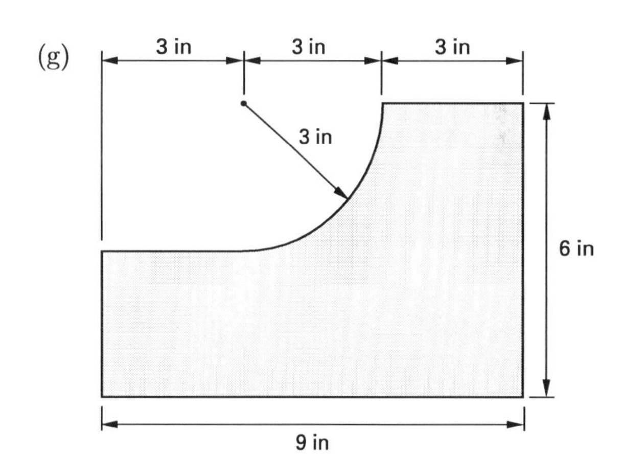

(h)

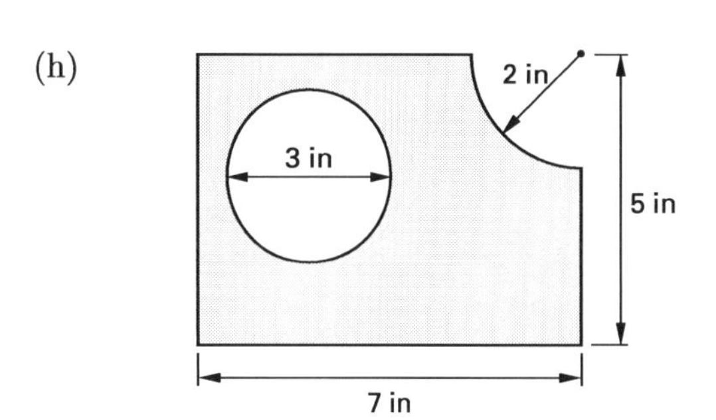

(i)

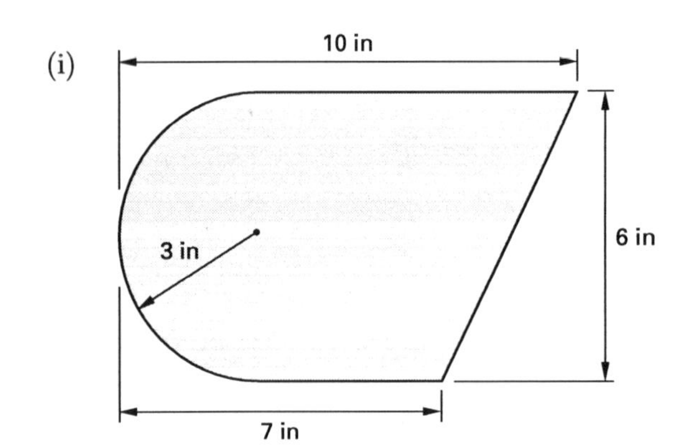

SOLUTIONS

1. The area is

$$A = \left(\frac{1}{2}\right)(12 \text{ in})(8 \text{ in}) = \boxed{48 \text{ in}^2}$$

2. The area is

$$A = (2)(24 \text{ ft})(8 \text{ ft}) + (2)(16 \text{ ft})(8 \text{ ft}) + (24 \text{ ft})(16 \text{ ft}) = 1024 \text{ ft}^2$$

$$\text{no. of sheets} = \frac{1024 \text{ ft}^2}{32 \dfrac{\text{ft}^2}{\text{sheet}}} = \boxed{32 \text{ sheets}}$$

3. The area is

$$A = \left(\frac{28 \text{ ft} + 4 \text{ ft}}{2}\right)(6 \text{ ft}) = \boxed{96 \text{ ft}^2}$$

4. The area is

$$A = \left(\frac{44 \text{ ft} + 92 \text{ ft}}{2}\right)(8 \text{ ft}) = \boxed{544 \text{ ft}^2}$$

5. The area is

$$A = \pi(20 \text{ ft})^2 = \boxed{1256 \text{ ft}^2}$$

6. The area is

$$A = \left(\frac{\pi}{4}\right)(10 \text{ ft})^2 = \boxed{78.5 \text{ ft}^2}$$

7. The area is

$$A = \left(\frac{1}{6}\right)\left(\frac{\pi}{4}\right)(6 \text{ in})^2 = \boxed{4.7 \text{ in}^2}$$

8. The area is

$$A = \left(\frac{1}{4}\right)\left(\frac{\pi}{4}\right)(12 \text{ ft})^2 - \left(\frac{1}{2}\right)(6 \text{ ft})(6 \text{ ft}) = \boxed{10.3 \text{ ft}^2}$$

9. The area is

$$A = \sqrt{\begin{array}{c}(20 \text{ ft})(20 \text{ ft} - 18 \text{ ft}) \\ \times (20 \text{ ft} - 12 \text{ ft})(20 \text{ ft} - 10 \text{ ft})\end{array}} = \boxed{57 \text{ ft}^2}$$

10. The area is

(a) $A = \left(\frac{40 \text{ ft} + 60 \text{ ft}}{2}\right)(20 \text{ ft}) = \boxed{1000 \text{ ft}^2}$

(b) $A = \left(\frac{30 \text{ ft} + 40 \text{ ft}}{2}\right)(12) = \boxed{420 \text{ ft}^2}$

(c) $A = (8.33 \text{ ft})(11.75 \text{ ft}) + \left(\frac{1}{2}\right)(5.84 \text{ ft})(6.50 \text{ ft}) = \boxed{116.9 \text{ ft}^2}$

(d) $A = (10.17 \text{ ft})(8.75 \text{ ft}) + \left(\frac{1}{2}\right)(1.92 \text{ ft})(10.17 \text{ ft}) = \boxed{99 \text{ ft}^2}$

(e) $A = \left(\frac{\pi}{4}\right)\left((14 \text{ in})^2 - (12 \text{ in})^2\right) = \boxed{41 \text{ in}^2}$

(f) $A = (6 \text{ in})(12 \text{ in}) + \left(\frac{1}{2}\right)(6 \text{ in})(12 \text{ in}) + \left(\frac{1}{2}\right)\pi(6 \text{ in})^2 = \boxed{165 \text{ in}^2}$

(g) $A = (9 \text{ in})(6 \text{ in}) - (3 \text{ in})(3 \text{ in}) - \left(\frac{1}{4}\right)\pi(3 \text{ in})^2 = \boxed{38 \text{ in}^2}$

(h) $A = (7 \text{ in})(5 \text{ in}) - \left(\frac{\pi}{4}\right)(3 \text{ in})^2 - \left(\frac{1}{4}\right)\pi(2 \text{ in})^2 = \boxed{25 \text{ in}^2}$

(i) $A = (6 \text{ in})(4 \text{ in}) + \left(\frac{1}{2}\right)\pi(3 \text{ in})^2 + \left(\frac{1}{2}\right)(3 \text{ in})(6 \text{ in}) = \boxed{47 \text{ in}^2}$

7 Volume

Nomenclature

A area
h altitude
r radius
V volume

Subscripts

i inside
o outside

1. DEFINITION

Volume is defined as the amount of substance occupying a certain space. It is measured in cubic units. The block shown in Fig. 7.1 has a volume of 6 cubic inches (6 in^3). One cubic inch is a cube that measures 1 in on each edge.

Figure 7.1 *Volume of a Block*

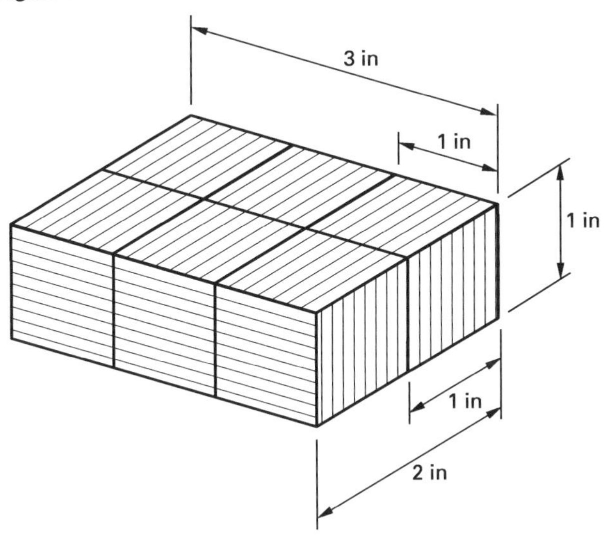

2. VOLUME OF RIGHT PRISMS AND CYLINDERS

The volume of a right prism or cylinder is the product of the area of the base and the altitude. Expressed as a formula,

$$V = Ah \quad 7.1$$

V is volume in cubic units, A is area in square units, and h is altitude in linear units.

Example 7.1

Find the volume of a rectangular prism with a base of 8 in by 6 in and an altitude of 10 in.

Solution

$$V = Ah = (8 \text{ in})(6 \text{ in})(10 \text{ in}) = 480 \text{ in}^3$$

Example 7.2

Find the volume of a triangular prism with a triangular base that has sides 3 in, 4 in, and 5 in, and with an 8 in altitude.

Solution

$$V = Ah = \left(\frac{1}{2}\right)(3 \text{ in})(4 \text{ in})(8 \text{ in}) = 48 \text{ in}^3$$

Example 7.3

Find the number of cubic yards of dirt in 500 ft of a highway fill of trapezoidal cross section with a bottom base of 112 ft, a top base of 40 ft, and a height of 12 ft.

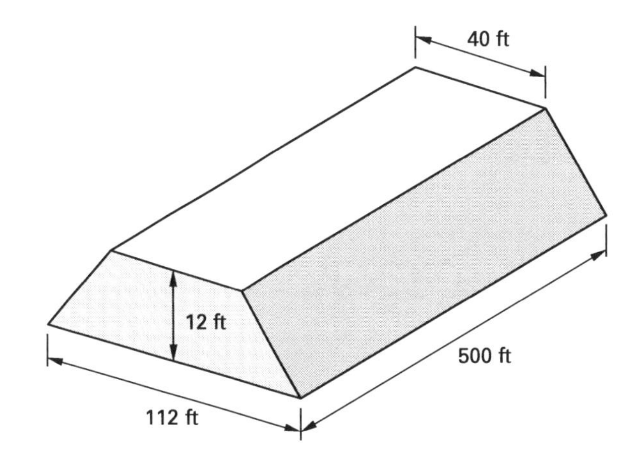

Solution

$$V = \frac{(112 \text{ ft} + 40 \text{ ft})(12 \text{ ft})(500 \text{ ft})}{(2)\left(27 \frac{\text{ft}^3}{\text{yd}^3}\right)} = 16{,}889 \text{ yd}^3$$

Example 7.4

Find the volume of the shell of a hollow cylinder that has an outside diameter of 8 in, an inside diameter of 6 in, and a height of 5 in.

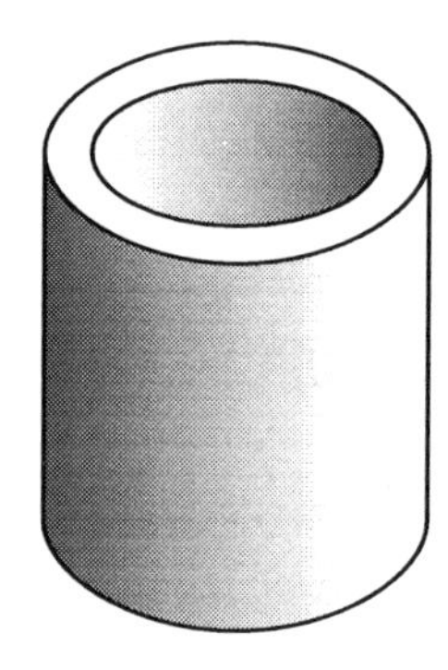

Solution

$$\begin{aligned} V &= Ah \\ &= (\text{outside area} - \text{inside area})h \\ &= \left(\frac{\pi}{4}\right)(D_o^2 - D_i^2)h \\ &= \left(\frac{\pi}{4}\right)((8 \text{ in})^2 - (6 \text{ in})^2)(5 \text{ in}) \\ &= 110 \text{ in}^3 \end{aligned}$$

3. VOLUME OF A CONE

The volume of a right circular cone is equal to one-third the product of the area of its base and its altitude.

$$V = \tfrac{1}{3}\pi r^2 h \qquad 7.2$$

Example 7.5

Find the volume of a cone that is 6 in high with a base of 4 in in diameter.

Solution

$$\begin{aligned} V &= \tfrac{1}{3}\pi r^2 h = \tfrac{1}{3}(\pi)(2 \text{ in})^2(6 \text{ in}) \\ &= 25 \text{ in}^3 \end{aligned}$$

4. VOLUME OF A PYRAMID

The volume of a pyramid is equal to one-third the product of the area of its base and its altitude.

$$V = \tfrac{1}{3}Ah \qquad 7.3$$

5. VOLUME OF A SPHERE

The volume of a sphere is equal to $4/3\pi r^3$.

$$V = \tfrac{4}{3}\pi r^3 \qquad 7.4$$

PRACTICE PROBLEMS

(Note: When dimensions are predominantly in feet but one dimension is in inches, convert inches to feet by using a common fraction: 3 in = 1/4 ft, 4 in = 1/3 ft, and 6 in = 1/2 ft. The denominators can be used for cancellation.)

1. Find the volume of a rectangular right prism with a base of 3 ft by 4 ft and an altitude of 6 ft.

2. Find the volume of a triangular right prism with base sides of 9 in, 12 in, and 15 in, and an altitude of 10 in.

3. Find the volume of a cylinder with a base with a diameter of 10 in and an altitude of 8 in.

4. Find the number of cubic feet of concrete (to the nearest tenth) in a pipe of 8 in inside diameter, 2 in wall thickness, and 30 in length.

5. Find the necessary height of a cylindrical tank 6 ft in diameter if its volume is to be 226 ft^3 (to the nearest tenth).

6. Find the volume of a right prism with an altitude of 10 in and a base which enscribes an isosceles triangle with a side length of 8 in and an altitude of 3 in.

7. Find the number of cubic yards of dirt in 810 ft of a highway fill of trapezoidal cross section with a base at the bottom of 120 ft, a base at the top of 80 ft, and an 8 ft height of fill.

8. How many cubic yards of concrete are needed to pour a parking area 30 ft long, 27 ft wide, and 4 in thick?

SOLUTIONS

1. The volume is

$$V = LWh = (3\ \text{ft})(4\ \text{ft})(6\ \text{ft}) = \boxed{72\ \text{ft}^3}$$

2. The volume is

$$\left(\frac{1}{2}\right)(9\ \text{in})(12\ \text{in})(10\ \text{in}) = \boxed{540\ \text{in}^3}$$

3. The volume is

$$\left(\frac{\pi}{4}\right)(10\ \text{in})^2(8\ \text{in}) = \boxed{628\ \text{in}^3}$$

4. The volume is

$$\frac{\left(\frac{\pi}{4}\right)\left((12\ \text{in})^2 - (8\ \text{in})^2\right)(30\ \text{in})}{1728\ \frac{\text{in}^3}{\text{ft}^3}} = \boxed{1.1\ \text{ft}^3}$$

5. The height is

$$h = \frac{V}{D} = \frac{226\ \text{ft}^3}{\left(\frac{\pi}{4}\right)(6\ \text{ft})^2} = \boxed{8\ \text{ft}}$$

6. The volume is

$$\left(\frac{1}{2}\right)(8\ \text{in})(3\ \text{in})(10\ \text{in}) = \boxed{120\ \text{in}^3}$$

7. The volume is

$$\frac{\left(\frac{120\ \text{ft} + 80\ \text{ft}}{2}\right)(8\ \text{ft})(810\ \text{ft})}{27\ \frac{\text{ft}^3}{\text{yd}^3}} = \boxed{24{,}000\ \text{yd}^3}$$

8. The volume is

$$\frac{(30\ \text{ft})(27\ \text{ft})(1\ \text{ft})}{\left(27\ \frac{\text{ft}^3}{\text{yd}^3}\right)(3)} = \boxed{10\ \text{yd}^3}$$

8 Trigonometry

1. DEFINITION OF AN ANGLE

In trigonometry, an *angle* is considered to be the measure of the rotation of a *ray* (line) from one position to another in a counterclockwise direction.

2. STANDARD POSITION OF AN ANGLE

An angle is in *standard position* when its vertex is at the origin and the initial side coincides with the positive x-axis of a rectangular coordinate system.

Figure 8.1 *Angle Terminology*

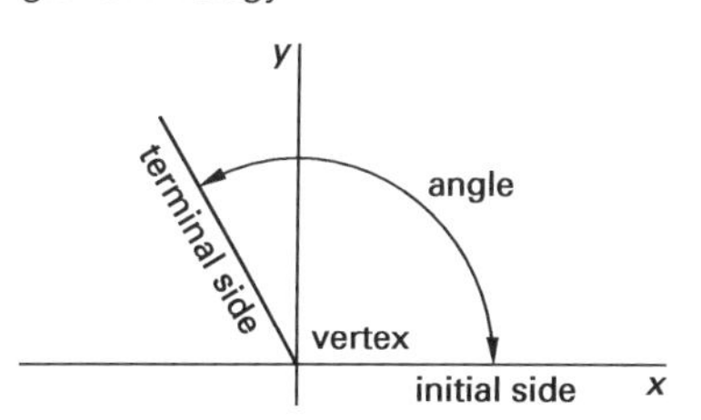

3. QUADRANTS

The coordinate axes divide the plane into four *quadrants* designated I, II, III, and IV, as shown in Fig. 8.2. An angle is in one of the quadrants when its *terminal side* is in that quadrant.

Figure 8.2 *Angle Quadrants*

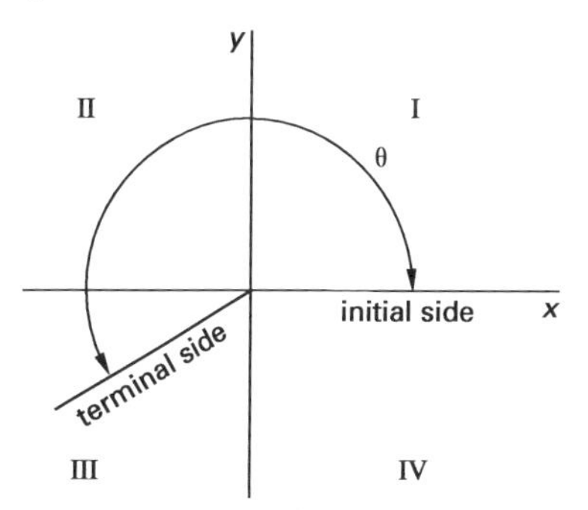

4. TRIGONOMETRIC FUNCTIONS OF ANY ANGLE

For any angle in standard position, the six trigonometric functions are given by Eqs. 8.1 through 8.6. Note that the abbreviations for sine, cosine, tangent, cosecant, secant, and cotangent are used in equations.

$$\sin\theta = \frac{y}{r} \qquad 8.1$$

$$\cos\theta = \frac{x}{r} \qquad 8.2$$

$$\tan\theta = \frac{y}{x} \quad 8.3$$

$$\csc\theta = \frac{r}{y} \quad 8.4$$

$$\sec\theta = \frac{r}{x} \quad 8.5$$

$$\cot\theta = \frac{x}{y} \quad 8.6$$

Figure 8.3 *Functions of an Angle*

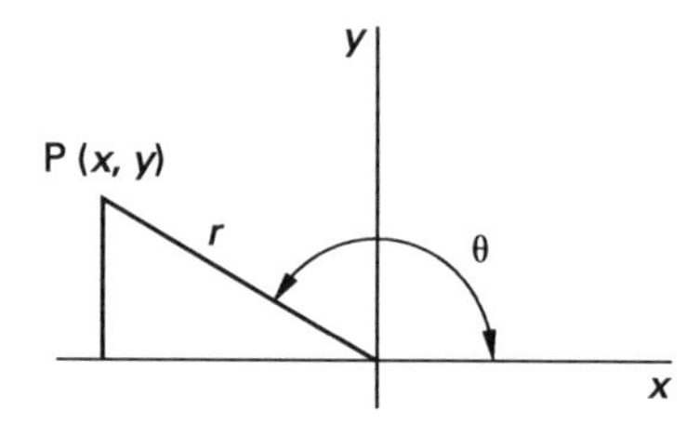

Because x, y, and r represent measured lengths, the six trigonometric functions are actually ratios of two numbers—that is, they are ratios of the length of one side of a triangle to the length of another side. A *ratio* is a comparison of two numbers. The ratio 2 to 3 can be written $2/3$, which is a fraction. Therefore, a trigonometric function is a number.

Figure 8.4 *Functions Related to Sides*

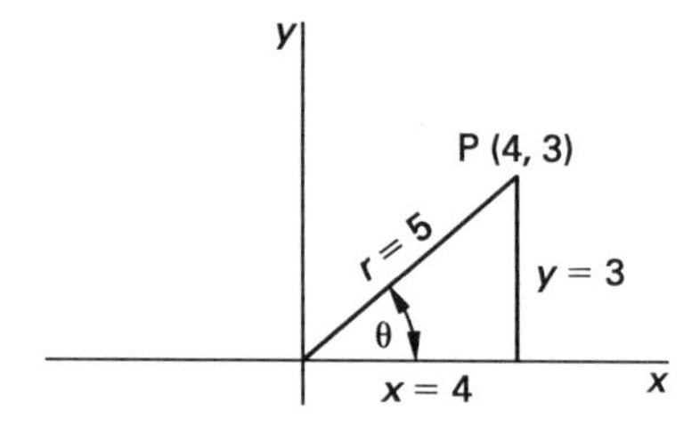

In Fig. 8.4,

$$\sin\theta = \frac{y}{r} = \frac{3}{5} = 0.60$$

$$\cos\theta = \frac{x}{r} = \frac{4}{5} = 0.80$$

$$\tan\theta = \frac{y}{x} = \frac{3}{4} = 0.75$$

The value of a trigonometric function of an angle θ depends on the value of θ. As θ changes, sin θ changes so that sin *0* is a *function* of θ.

5. RECIPROCAL OF A NUMBER

The *reciprocal* of a number is 1 divided by the number. Thus, the reciprocal of 3 is $1/3$, and the reciprocal of $2/3$ is $3/2$.

6. RECIPROCAL OF A TRIGONOMETRIC FUNCTION

Trigonometric functions are ratios of numbers and can be treated as such. Therefore, the reciprocal of $\sin\theta = 1\sin\theta$. It follows that if $\sin\theta = y/r$, the reciprocal is r/y.

$$\sin\theta = \frac{1}{\csc\theta} \quad 8.7$$

$$\cos\theta = \frac{1}{\sec\theta} \quad 8.8$$

$$\tan\theta = \frac{1}{\cot\theta} \quad 8.9$$

$$\csc\theta = \frac{1}{\sin\theta} \quad 8.10$$

$$\sec\theta = \frac{1}{\cos\theta} \quad 8.11$$

$$\cot\theta = \frac{1}{\tan\theta} \quad 8.12$$

Therefore, $\sin\theta$ and $\csc\theta$ are reciprocals, $\cos\theta$ and $\sec\theta$ are reciprocals, and $\tan\theta$ and $\cot\theta$ are reciprocals.

From the relation between a function and its reciprocal,

$$(\sin\theta)(\csc\theta) = \left(\frac{y}{r}\right)\left(\frac{r}{y}\right) = 1 \quad 8.13$$

$$(\cos\theta)(\sec\theta) = \left(\frac{x}{r}\right)\left(\frac{r}{x}\right) = 1 \quad 8.14$$

$$(\tan\theta)(\cot\theta) = \left(\frac{y}{x}\right)\left(\frac{x}{y}\right) = 1 \quad 8.15$$

7. ALGEBRAIC SIGN OF TRIGONOMETRIC FUNCTIONS

An angle in standard position is considered to be in the quadrant in which its terminal side lies; therefore, the values of x and y have algebraic signs. The value of r is always considered to be positive. It follows then, that functions expressed as ratios of positive and negative numbers will have positive and negative values. The algrebraic sign of any function can be determined by memorizing the terms shown in Fig. 8.5.

Figure 8.5 *Signs of the Natural Functions*

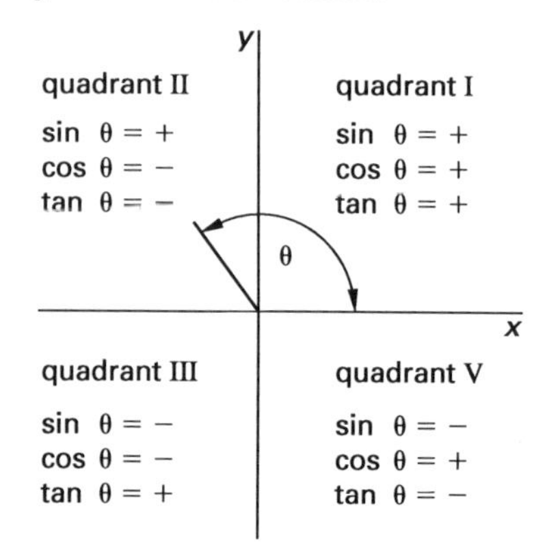

8. VALUES OF TRIGONOMETRIC FUNCTIONS OF QUADRANTAL ANGLES

The *quadrantal angles* are the angles that are common to two quadrants. They are 0°, 90°, 180°, 270°, and 360°. In Fig. 8.6, where $r = 5$, points P_1, P_2, P_3, and P_4 will have the coordinates shown.

Figure 8.6 *Quadrantal Angles*

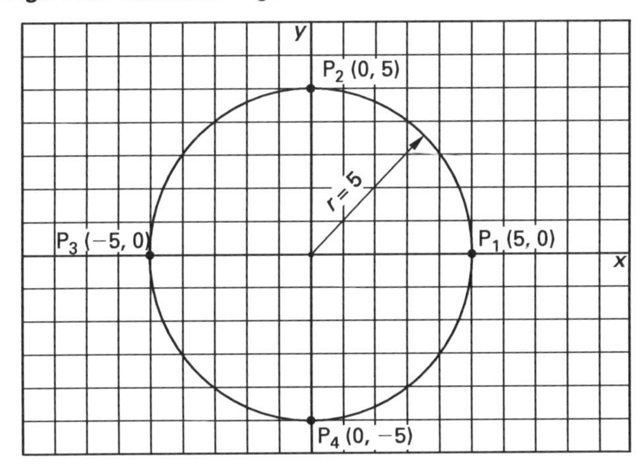

If the terminal side of θ_1 (0° or 360°) passes through point $P_1\,(5,0)$, θ_2 (90°) passes through point $P_2\,(0,5)$, θ_3 (180°) passes through point $P_3\,(-5,0)$, and θ_4 (270°) passes through $P_4\,(0,-5)$, then

$$\sin 0^\circ = \frac{y}{r} = \frac{0}{5} = 0$$
$$\sin 90^\circ = \frac{y}{r} = \frac{5}{5} = +1$$
$$\sin 180^\circ = \frac{y}{r} = \frac{0}{5} = 0$$
$$\sin 270^\circ = \frac{y}{r} = \frac{-5}{5} = -1$$
$$\sin 360^\circ = \frac{y}{r} = \frac{0}{5} = 0$$

As θ increases from 0° to 90°, $\sin\theta$ increases from 0 to +1; as θ increases from 90° to 180°, $\sin\theta$ decreases from +1 to 0; as θ increases from 180° to 270°, $\sin\theta$ decreases from 0 to −1; and as θ increases from 270° to 360°, $\sin\theta$ increases from −1 to 0.

An important fact that can be considered in the solution of right angles is that, except for the quadrantal angles, $\sin\theta$ always has a numerical value of less than 1.

Considering the cosine function for the angles in Fig. 8.6,

$$\cos 0^\circ = \frac{x}{r} = \frac{5}{5} = +1$$
$$\cos 90^\circ = \frac{x}{r} = \frac{0}{5} = 0$$
$$\cos 180^\circ = \frac{x}{r} = \frac{-5}{5} = -1$$
$$\cos 270^\circ = \frac{x}{r} = \frac{0}{5} = 0$$
$$\cos 360^\circ = \frac{x}{r} = \frac{5}{5} = +1$$

Except for the quadrantal angles, $\cos\theta$ also has a numerical value less than 1.

In Fig. 8.6, the tangent function varies as follows.

$$\tan 0^\circ = \frac{y}{x} = \frac{0}{5} = 0$$
$$\tan 90^\circ = \frac{y}{x} = \frac{5}{0} = \infty$$
$$\tan 180^\circ = \frac{y}{x} = \frac{0}{-5} = 0$$
$$\tan 270^\circ = \frac{y}{x} = \frac{-5}{0} = \infty$$
$$\tan 360^\circ = \frac{y}{x} = \frac{0}{5} = 0$$

The symbol ∞ is sometimes interpreted to mean infinity; however, $\tan 90^\circ$ does not exist because a number cannot be divided by 0.

Table 8.1 *Summary of Functions of Quadrantal Angles*

angle	sin	cos	tan	cot	sec	csc
0°	0	1	0	∞	1	∞
90°	1	0	∞	0	∞	1
180°	0	−1	0	∞	−1	∞
270°	−1	0	∞	0	∞	−1
360°	0	1	0	∞	1	∞

9. TRIGONOMETRIC FUNCTIONS OF AN ACUTE ANGLE

All angles in quadrant I are *acute angles* and, therefore, positive. In dealing with acute angles only, it is more convenient to consider them as part of a right triangle and express the trigonometric functions in terms of the sides of a triangle that are given the names *opposite*, *adjacent*, and *hypotenuse* as they are shown in Fig. 8.7.

$$\sin\theta = \frac{\text{opposite}}{\text{hypotenuse}} \qquad 8.16$$

$$\cos\theta = \frac{\text{adjacent}}{\text{hypotenuse}} \qquad 8.17$$

$$\tan\theta = \frac{\text{opposite}}{\text{adjacent}} \qquad 8.18$$

Figure 8.7 Sides of a Triangle

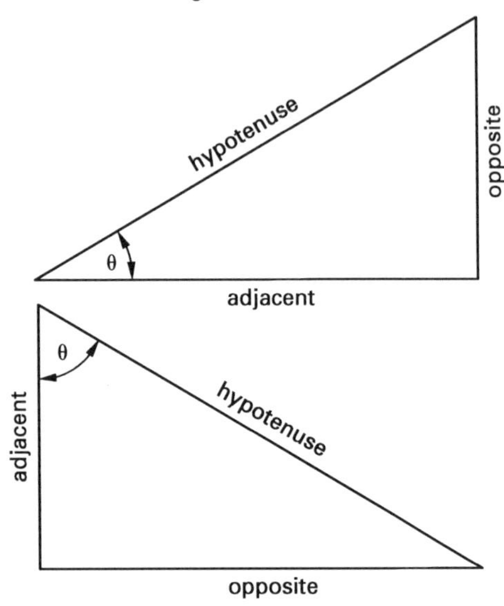

10. COFUNCTIONS

Any function of an acute angle is equal to the *cofunction* of its *complementary angle*. The sine and cosine are cofunctions, as are the tangent and the cotangent.

$$\sin 30° = \cos 60° \qquad 8.19$$

$$\tan 20° = \cot 70° \qquad 8.20$$

11. TRIGONOMETRIC FUNCTIONS OF 30°, 45°, AND 60°

Consider an equilateral triangle with sides of length 2 ft. The bisector of any of the 60° angles will bisect the opposite side and form a right triangle with acute angles of 30° and 60° as shown in Fig. 8.8.

Figure 8.8 A 30-60-90 Triangle

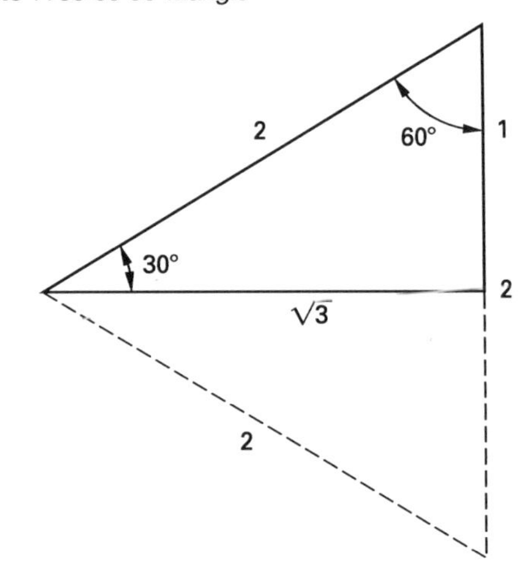

In Fig. 8.8,

$$\sin 30° = \frac{1}{2} = \cos 60° = 0.500$$

$$\sin 60° = \frac{\sqrt{3}}{2} = \cos 30° = 0.866\ldots$$

$$\tan 30° = \frac{1}{\sqrt{3}} = \cot 60° = 0.577\ldots$$

Consider an isosceles right triangle with two of the sides equal to 1 as shown in Fig. 8.9. The angles opposite the sides will be 45°. Then,

$$\sin 45° = \frac{1}{\sqrt{2}} = \frac{\sqrt{2}}{2} = 0.707\ldots$$

$$\cos 45° = \frac{1}{\sqrt{2}} = \frac{\sqrt{2}}{2} = 0.707\ldots$$

$$\tan 45° = \frac{1}{1} = 1 = 1.000$$

Figure 8.9 A 45-45-90 Triangle

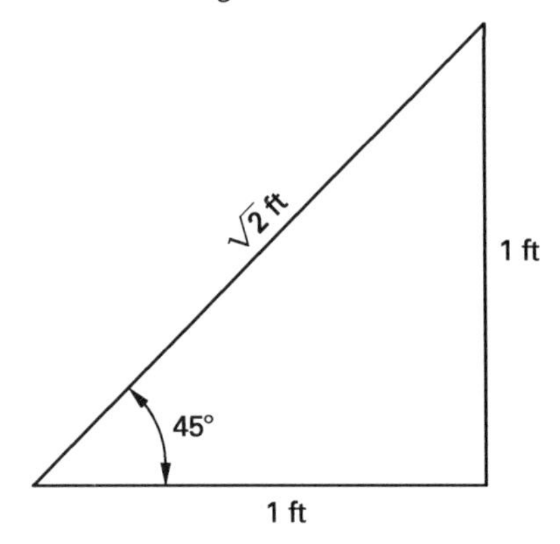

12. TABLE OF VALUES OF TRIGONOMETRIC FUNCTIONS

In Secs. 7 and 10, certain trigonometric functions of angles of 0°, 30°, 45°, 60°, 90°, 180°, 270°, and 360° were computed. Many years ago, mathematicians compiled tables showing values of trigonometric functions expressed in decimal form. Some tables show values for each degree of an angle, some for each degree and minute of an angle, and some for each degree, minute, and second of an angle. Values of functions are computed to four decimal places in some tables, and up to ten decimal places in others.

The use of handheld calculators has essentially eliminated the need for tables of trigonometric functions. However, the technique of interpolation is still often required in surveying work.

13. INTERPOLATION

Interpolation has been defined as finding an intermediate term in a sequence. Tables often give the sine and cosine functions for each minute of an angle, but to determine the sine and cosine functions of an angle measured in degrees, minutes, and seconds, interpolation is necessary. The process can be best explained with examples.

Example 8.1

Find $\sin 52°15'24''$ if $\sin 52°15' = 0.7906896$ and $\sin 52°16' = 0.7908676$.

Solution

The angle $52°15'24''$ is $24/60 = 0.4$ of the way from $52°15'00''$ to $52°16'00''$. The difference in the sine function of these two angles is $0.7908676 - 0.7906896 = 0.0001780$, remembering that as the angle increases, the sine increases. Because the angle $52°15'24''$ is 0.4 of the way between the other two angles, the sine of $52°15'24''$ is also 0.4 of the way between the sines of the other two angles.

$$\begin{aligned}\sin 52°15'24'' &= 0.7906896 + (0.4)(0.0001780)\\ &= 0.7907608\end{aligned}$$

Example 8.2

Find $\cos 37°25'48''$ given that $\cos 37°25' = 0.7942379$ and $\cos 37°26' = 0.7940611$.

Solution

As θ increases $\cos\theta$ decreases, so the cosine of $37°25'48''$ will be less than the cosine of $37°25'00''$.

$$\begin{aligned}\frac{48}{60} &= 0.8\\ 0.7942379 - 0.7940611 &= 0.0001768\\ \cos 37°25'48'' &= 0.7942379\\ &\quad - (0.8)(0.0001768)\\ &= 0.7940965\end{aligned}$$

Example 8.3

Find θ if $\cos\theta = 0.8047643$.

Solution

The cosine of $36°24'00''$, 0.8048938, and the cosine of $36°25'00''$, 0.8047211, are just greater than and just less than 0.8047643. So θ is more than $36°24'00''$ and less than $36°25'00''$. The cosine of θ is $1295/1727 = 0.75$ of the way between 0.8048938 and 0.8047211, so θ is 0.75 of the way between $36°24'00''$ and $36°25'00''$.

$$\begin{aligned}\arccos 0.8047643 &= 36°24'00'' + (0.75)(60'')\\ &= 36°24'45''\end{aligned}$$

Example 8.4

Find $\tan 44°17'06''$.

Solution

$$\begin{aligned}\tan 44°17'00'' &= 0.9752914\\ \tan 44°18'00'' &= 0.9758591\\ \tan 44°17'06'' &= 0.9752914\\ &\quad + \left(\frac{6}{60}\right)(0.9758591 - 0.9752914)\\ &= 0.9753482\end{aligned}$$

Example 8.5

Find θ if $\cot\theta = 1.0967405$.

Solution

$$\begin{aligned}\cot 42°21' &= 1.0970609\\ 1.0970609 - 1.0967405 &= 0.0003204\\ \cot 42°22' &= 1.0964201\\ 1.0970609 - 1.0964201 &= 0.0006408\\ \text{arccot}\, 1.0967405 &= 42°21'00''\\ &\quad + \left(\frac{0.0003204}{0.0006408}\right)(60'')\\ &= 42°21'30''\end{aligned}$$

14. BEARING OF A LINE

The *bearing of a line* is the acute horizontal angle between the meridian (north line) and the line.

15. ANGLE OF ELEVATION AND ANGLE OF DEPRESSION

The *angle of elevation* is the vertical angle between the horizontal and a line rotated upward from the horizontal. It is positive. The *angle of depression* is the vertical angle between the horizontal and a line rotated downward. It is negative.

16. SOLUTION OF RIGHT TRIANGLES

Solving a right triangle means finding the value of its three angles and the length of each of its sides. To solve a right triangle, two of these values must be known. Either two sides must be known, or one side and an acute angle must be known.

If an acute angle of a right triangle is known, the other acute angle is the complement of it, since the sum of the interior angles of a triangle equals 180°.

Example 8.6

Solve the right triangle ABC having angle $A = 23°30'$ and side $a = 400$ ft. (Note: It is customary to use capital letters to name the vertices of a triangle and the corresponding lower case letter to name the side opposite each of the vertices.)

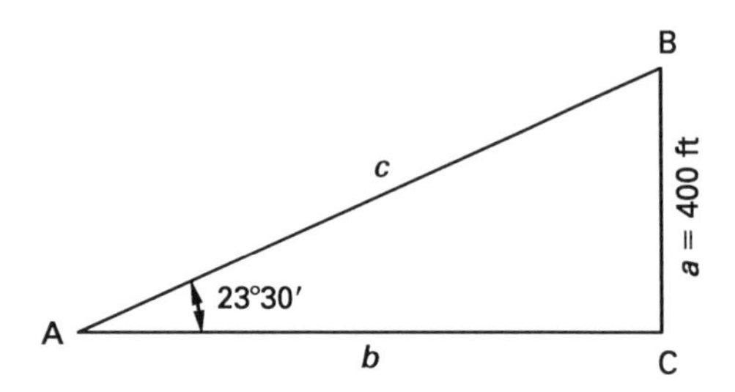

Solution

$$B = 90° - 23°30' = 66°30'$$

In solving for c, the side opposite the known angle A is the known side a, and side c is the hypotenuse. So, select the sine function for use in the solution.

$$\sin A = \frac{\text{opposite}}{\text{hypotenuse}}$$

$$\sin 23°30' = \frac{400}{c}$$

To remove c from the denominator, multiply both sides of the equation by c.

$$c \sin 23°30' = \frac{400c}{c}$$

$$c \sin 23°30' = 400$$

To isolate c, divide both sides of the equation by $\sin 23°30'$.

$$\frac{c \sin 23°30'}{\sin 23°30'} = \frac{400}{\sin 23°30'}$$

$$c = \frac{400}{\sin 23°30'}$$

From trigonometric tables or a calculator,

$$\sin 23°30' = 0.3987$$

$$c = \frac{400}{0.3987} = 1000 \text{ ft}$$

b is the side adjacent to the known angle A, so select the tangent function.

$$\tan A = \frac{\text{opposite}}{\text{adjacent}}$$

$$\tan 23°30' = \frac{400}{b}$$

Multiplying both sides of the equation by b and dividing both sides by $\tan 23°30'$,

$$b = \frac{400}{\tan 23°30'}$$

$$b = \frac{400}{0.4348} = 920 \text{ ft}$$

Example 8.7

Solve the right triangle EFG having angle $E = 40°$ and the hypotenuse, side $g = 200$ ft.

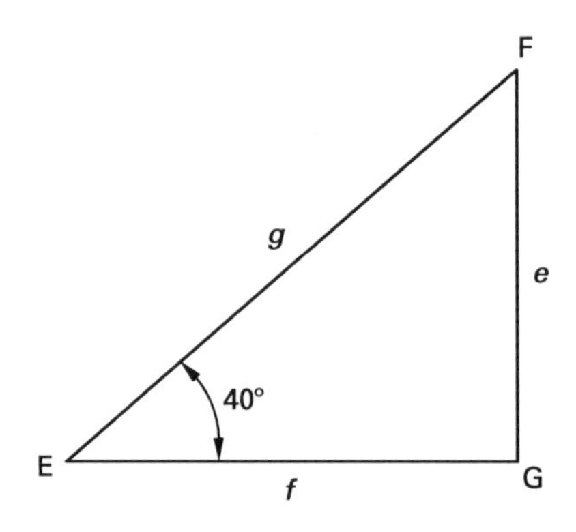

Solution

$$F = 90° - 40° = 50°$$

To solve for e, the sine function is selected.

$$\sin E = \frac{\text{opposite}}{\text{hypotenuse}}$$

$$\sin 40° = \frac{e}{200 \text{ ft}}$$

To isolate e, multiply both sides of the equation by 200 ft.

$$(200 \text{ ft}) \sin 40° = \left(\frac{200 \text{ ft}}{200 \text{ ft}}\right) e$$

$$e = 200 \sin 40°$$

From trigonometric tables or a calculator,

$$\sin 40° = 0.6428$$

$$e = (200 \text{ ft})(0.6428) = 130 \text{ ft}$$

To solve for f, select the cosine function.

$$\cos E = \frac{\text{adjacent}}{\text{hypotenuse}}$$

$$\cos 40° = \frac{f}{200 \text{ ft}}$$

Multiplying both sides by 200 ft,

$$f = (200\ \text{ft})(\cos 40°) = (200\ \text{ft})(0.7660) = 150\ \text{ft}$$

17. ALTERNATE SOLUTION METHODS FOR RIGHT ANGLES

Since there are many different relationships between the sides and angles of right triangles, it is not surprising that triangle problems may have more than a single correct solution method. Some solutions are simpler than others. Practice is required to be able to select the simplest solution procedure.

Additional examples are provided to illustrate various solution procedures.

Example 8.8

Solve the right triangle ABC having angle $A = 23°30'$ and side $a = 400$ ft. (Refer to the illustration in Ex. 8.6).

Solution

$$B = 90° - 23°30' = 66°30'$$

The known side is opposite the known angle. To find the hypotenuse, select the sine function.

$$c = \frac{400\ \text{ft}}{\sin 23°30'} = 1000\ \text{ft}$$

To solve for b, select the tangent function.

$$b = \frac{400\ \text{ft}}{\tan 23°30'} = 920\ \text{ft}$$

Example 8.9

Solve the right triangle EFG in Ex. 8.7 having angle $E = 40°$ and side $g = 200$ ft.

Solution

$$F = 90° - 40° = 50°$$

The hypotenuse, which is the longest side, is known; therefore, to find the other sides, the hypotenuse must be multiplied by the sine or cosine function.

To solve for e, use the sine function.

$$e = (200\ \text{ft})(\sin 40°) = 130\ \text{ft}$$

To solve for f, use the cosine function.

$$f = (200\ \text{ft})(\cos 40°) = 150\ \text{ft}$$

Example 8.10

Solve the right triangle ABC.

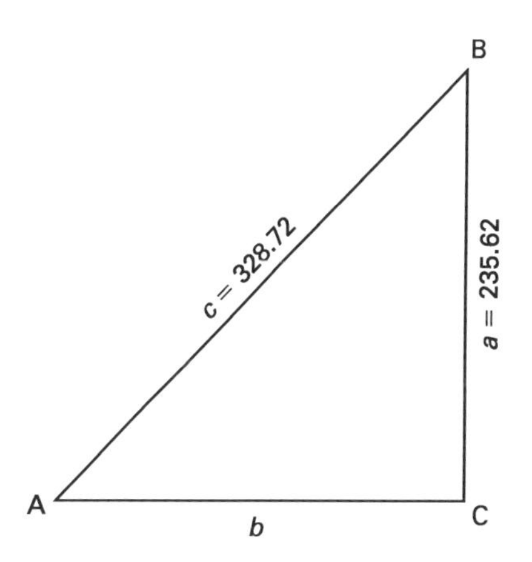

Solution

Find the solution from the inverse sine function on a calculator.

$$\sin A = \frac{235.62}{328.72} = 0.7167802$$

$$A = \arcsin 0.7167802 = 45°47'21''$$

$$B = 90°00'00'' - 45°47'21'' = 44°12'39''$$

$$b = \sqrt{(328.72)^2 - (235.62)^2} = 229.22$$

Example 8.11

To measure the width of a river, surveyors establish points A and B on the west bank and find the distance between them to be 90.0 ft. They set up a transit on point B and establish point C on the east bank so that BC is at right angles to AB. They then measure the angle at A between AB and AC and find it to be 68°20′. How wide is the river?

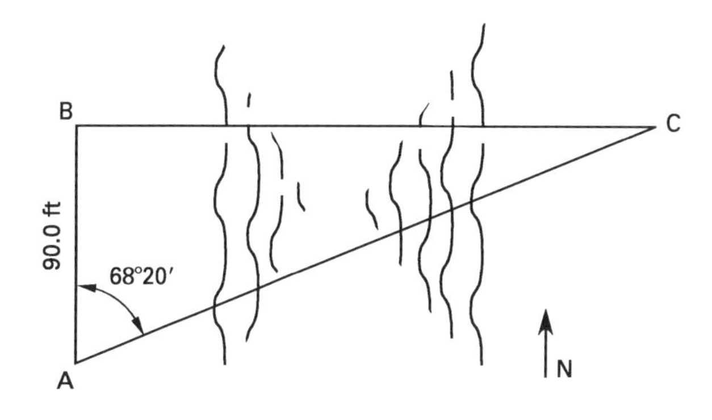

Solution

Select the tangent function.

$$\begin{aligned} BC &= (90.0\text{ ft})(\tan 68°20') \\ &= (90.0\text{ ft})(2.517) = 226.5\text{ ft} \end{aligned}$$

Example 8.12

San Angelo is due west of Waco, and Arlington is 100 mi due north of Waco. The bearing of Arlington from San Angelo is N 66°30′ E. (a) How far is San Angelo from Arlington? (b) How far is San Angelo from Waco?

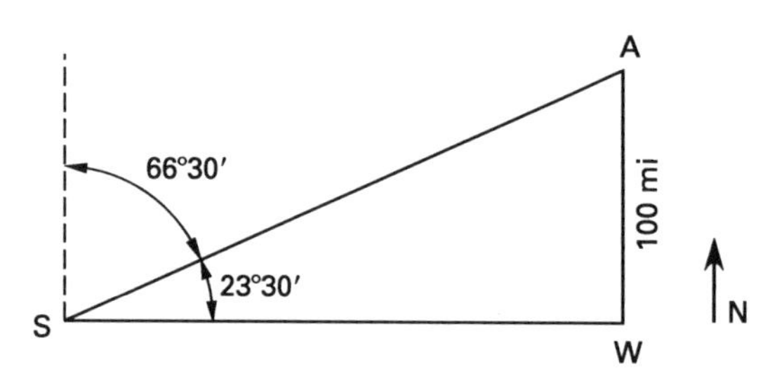

Solution

The triangle formed by the three cities is a right triangle. The angle at S is the complement of the bearing angle. In solving for SA, select the sine function.

$$SA = \frac{100\text{ mi}}{\sin 23°30'} = 250\text{ mi}$$

In solving for SW, select the tangent function.

$$SW = \frac{100\text{ mi}}{\tan 23°30'} = 230\text{ mi}$$

18. RELATED ANGLES

Trigonometric tables give values for acute angles only. Larger angles must be expressed in terms of an acute angle that has the same values for the functions. These angles are known as *related angles* or *reference angles.*

Figure 8.10 *Related Angles*

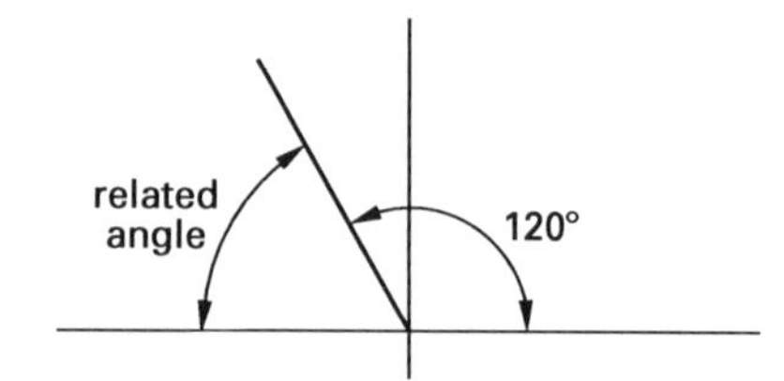

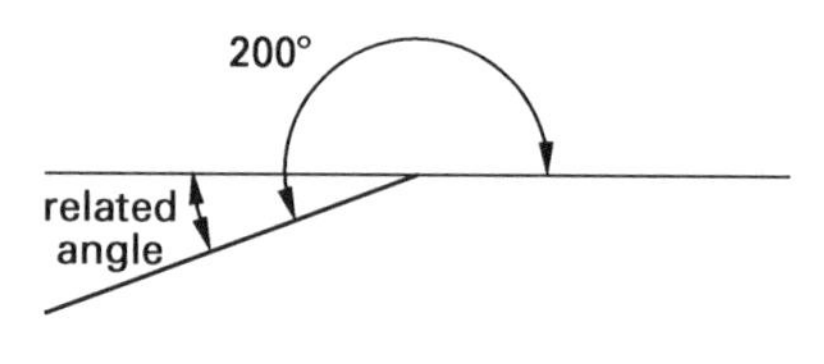

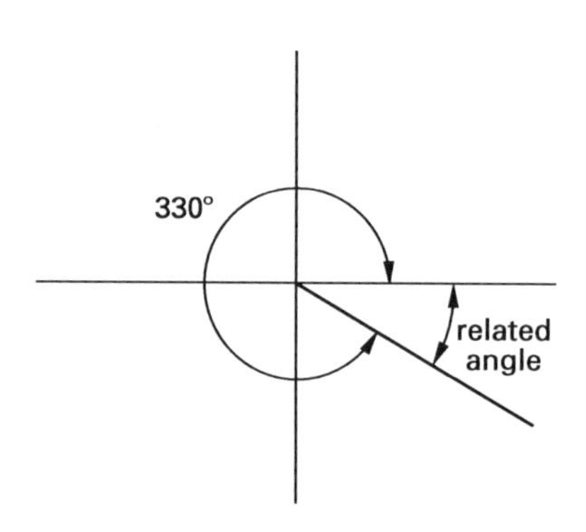

The related angle of θ is the positive acute angle between the x-axis and the terminal side of the angle. For example, in Fig. 8.10, the related angle of 120° is 60°, the related angle of 200° is 20°, and the related angle of 330° is 30°.

Regardless of which quadrant an angle lies in, the numerical value of any one of the six trigonometric functions is the same as that of the related angle. But the algebraic sign of any function depends on the quadrant in which the angle lies.

As illustrated in Fig. 8.2 and described in Sec. 3, an angle is a third-quadrant angle if its terminal side lies in the third quadrant. Likewise, an angle is in one of the other quadrants if its terminal side lies in that quadrant. Figure 8.5 gives the signs of the natural functions for each quadrant.

As solving oblique triangles involves first- and second-quadrant angles only, it is important to remember that the sine function will always be positive and the cosine function will be negative for second-quadrant angles (90° to 180°).

Example 8.13

Find the sine, cosine, and tangent of 150°.

Solution

The related angle is $180° - 150° = 30°$. The angle is less than 180° and more than 90°; therefore, it is a second-quadrant angle. The sign of the sine function is positive, the sign of the cosine function is negative, and the sign of the tangent function is negative.

$$\sin 150° = +0.500$$
$$\cos 150° = -0.866\ldots$$
$$\tan 150° = -0.577\ldots$$

Example 8.14

Find the sine, cosine, and tangent of 315°.

Solution

The related angle is $360° = 45°$.

$$\sin 315° = -0.707\ldots$$
$$\cos 315° = +0.707\ldots$$
$$\tan 315° = -1.000$$

19. SINE CURVE

The variations in the value of $\sin\theta$ can be shown by plotting θ as the abscissa and $\sin\theta$ as the ordinate on a system of coordinate axes. This is done in Fig. 8.11 as a *sine curve.*

Figure 8.11 *Sine Curve*

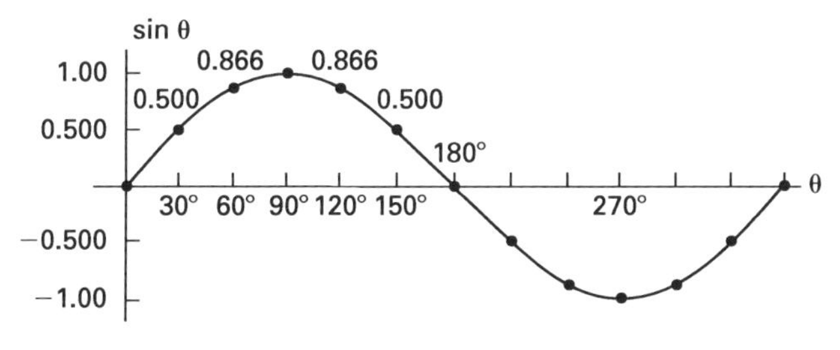

20. COSINE CURVE

The *cosine* curve has the same shape as the sine curve, but is offset 90°.

Figure 8.12 *Cosine Curve*

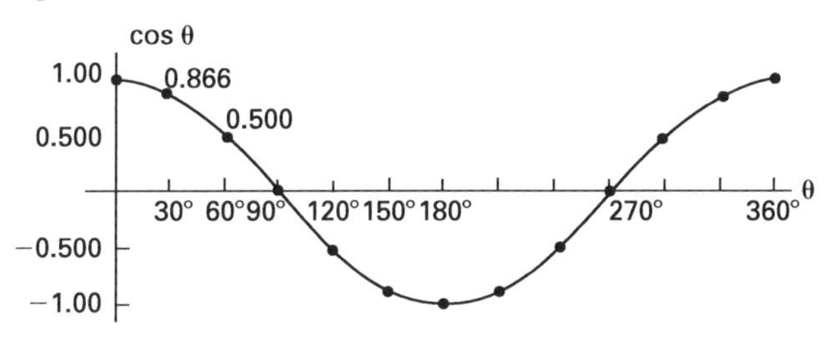

21. OBLIQUE TRIANGLES

An *oblique triangle* is a triangle that does not contain a right angle. All of the angles in an oblique triangle may be acute, or there may be one obtuse angle and two acute angles.

As with right triangles, the three angles in an oblique triangle are identified with capital letters, and most often the letters A, B, and C are used. The sides of the triangle are often identified with small (lower case) letters, with side a opposite angle A, side b opposite angle B, and side c opposite angle C. However, in surveys, angles can also be identified with letters other than A, B, and C and sometimes are identified with numbers. Sides can be identified with two capital letters.

The three angles and three sides of any triangle make up the six parts of the triangle. An oblique triangle can be solved if three of its parts, at least one of which is a side, are known. However, the solution is not as simple as the solution of a right triangle. Oblique triangles can be solved by forming two right triangles within the oblique triangle, but the task is made easier by the use of formulas. The most important formulas are the *law of sines* and the *law of cosines.* The choice of which of the two laws to use to solve a particular triangle depends on which three parts of the triangle are known.

If one side and two angles of a triangle are known, the triangle can be solved by the law of sines. This case is represented by the abbreviation SAA (side, angle, angle).

Example 8.15

What abbreviation represents the known triangle parameters?

Given: $a = 32.16$, $B = 64°20'$, $C = 50°20'$

Solution

If one side and two angles are known, the triangle can also be solved by the law of sines. This case is represented by the abbreviation SAA (side, angle, angle).

Example 8.16

What abbreviation represents the known triangle parameters?

Given: $a = 251.5$, $b = 647.3$, $A = 22°20'$

Solution

If two sides and the angle included between the two sides are known, the triangle cannot be solved by the law of sines alone. However, it can be solved by the law of cosines and the law of sines together. This case is represented by the abbreviation SAS (side, angle, side). After the third side is found by the law of cosines, the second angle can be found by the law of sines.

If three sides of a triangle are known, the triangle can be solved by the law of cosines. This case is represented by the abbreviation SSS (side, side, side).

22. LAW OF SINES

In any triangle, the sides are proportional to the sines of the opposite angles. Equations 8.21 and 8.22 show the *law of sines.*

$$\frac{a}{\sin A} = \frac{b}{\sin B} = \frac{c}{\sin C} \qquad 8.21$$

$$\frac{sinA}{a} = \frac{\sin B}{b} = \frac{\sin C}{c} \qquad 8.22$$

23. SAA CASE

In solving the SAA case, the law of sines can be expressed as Eqs. 8.23 through 8.25.

$$a = (\sin A)\left(\frac{b}{\sin B}\right) = (\sin A)\left(\frac{c}{\sin C}\right) \qquad 8.23$$

$$b = (\sin B)\left(\frac{a}{\sin A}\right) = (\sin B)\left(\frac{c}{\sin C}\right) \qquad 8.24$$

$$c = (\sin C)\left(\frac{a}{\sin A}\right) = (\sin C)\left(\frac{b}{\sin B}\right) \qquad 8.25$$

In the case with two angles known, the third angle can be readily found, so there will always be a known side opposite a known angle.

Example 8.17

Solve the triangle ABC with $C = 83°$, $B = 61°$, and $c = 150$ ft.

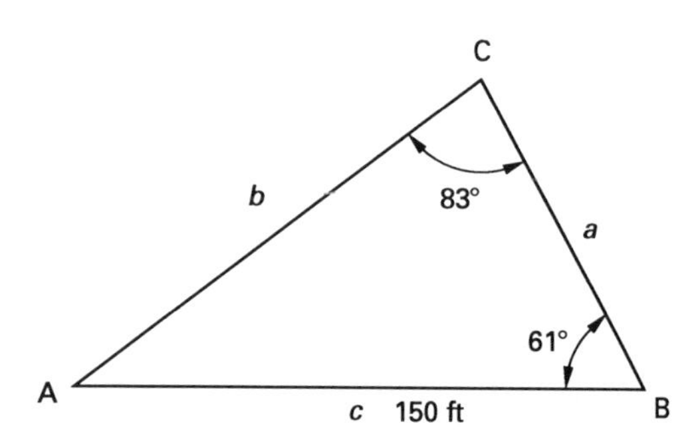

Solution

$$A = 180° - (83° + 61°) = 36°$$

$$a = (\sin A)\left(\frac{c}{\sin C}\right) = (\sin 36°)\left(\frac{150 \text{ ft}}{\sin 83°}\right) = 89 \text{ ft}$$

$$b = (\sin B)\left(\frac{c}{\sin C}\right) = (\sin 61°)\left(\frac{150 \text{ ft}}{\sin 83°}\right) = 132 \text{ ft}$$

Example 8.18

Solve the triangle EFG shown.

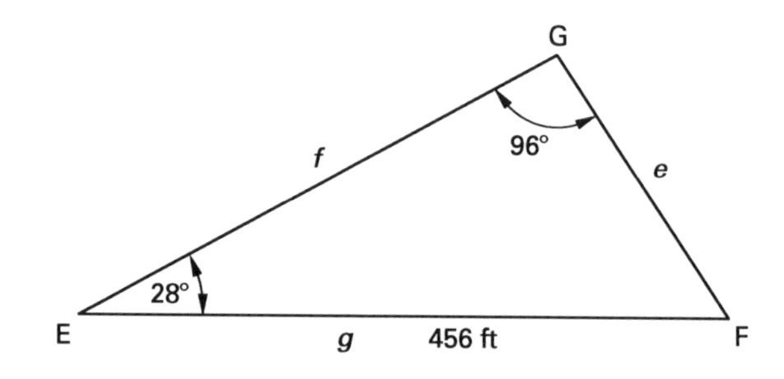

Solution

$$F = 180° - (28° + 96°) = 56°$$

$$e = (\sin 28°)\left(\frac{456 \text{ ft}}{\sin 96°}\right) = 215 \text{ ft}$$

$$f = (\sin 56°)\left(\frac{456 \text{ ft}}{\sin 96°}\right) = 380 \text{ ft}$$

24. SSA CASE

If two sides and an angle opposite one of them are known, the law of sines can be expressed as Eq. 8.26.

$$\sin A = \frac{a \sin B}{b} \qquad 8.26$$

Examplc 8.19

Solve the triangle ABC with $A = 36°$, $a = 50$ in, and $b = 70$ in.

In using the law of sines to solve triangles in which two sides and the angle opposite one of them are given, it is possible to construct two different triangles from the given information. For example, triangles ABC and AB′C can both be constructed from the given information.

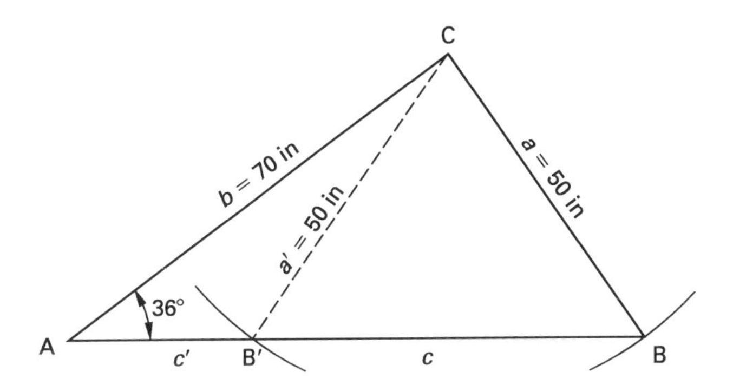

Solution for triangle ABC

$$\sin B = \frac{(70 \text{ in})(\sin 36^\circ)}{50 \text{ in}} = 0.8229 \text{ in}$$
$$B = 55^\circ$$
$$C = 180^\circ - (36^\circ + 55^\circ) = 89^\circ$$
$$c = (\sin 89^\circ)\left(\frac{50 \text{ in}}{\sin 36^\circ}\right) = 85 \text{ in}$$

Solution for triangle AB′C

Angle B could be the related angle to 55°. Then,

$$B' = 125^\circ$$
$$C = 180^\circ - (36^\circ + 125^\circ) = 19^\circ$$
$$c' = (\sin 19^\circ)\left(\frac{50 \text{ in}}{\sin 36^\circ}\right) = 28 \text{ in}$$

25. LAW OF COSINES

Equations 8.27 through 8.29 are alternate forms of the law of cosines. Referring to Fig. 8.13,

$$a^2 = b^2 + c^2 - 2bc\cos A \qquad 8.27$$
$$b^2 = c^2 + a^2 - 2ca\cos B \qquad 8.28$$
$$c^2 = a^2 + b^2 - 2ab\cos C \qquad 8.29$$

Figure 8.13 *A General Triangle*

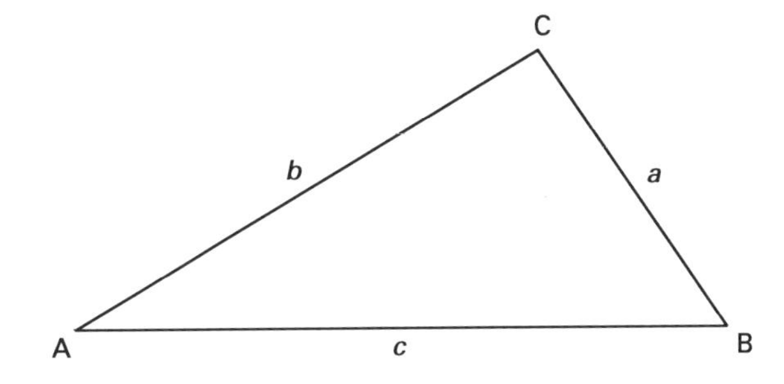

Remember that the cosine of an angle greater than 90° and less than 180° has a negative algebraic sign. In substituting the cosines of angles in that range, the negative sign must be included, meaning that the value of $2bc\cos A$ in Eq. 8.27 will be added to the value of $b^2 + c^2$.

$$\cos A = \frac{b^2 + c^2 - a^2}{2bc} \qquad 8.30$$
$$\cos B = \frac{a^2 + c^2 - b^2}{2ac} \qquad 8.31$$
$$\cos C = \frac{a^2 + b^2 - c^2}{2ab} \qquad 8.32$$

26. SAS CASE

The law of cosines can be used when two sides and the included angle of a triangle are known. In this case, the law of cosines can be expressed as Eqs. 8.33 through 8.35.

$$a = \sqrt{b^2 + c^2 - 2bc\cos A} \qquad 8.33$$
$$b = \sqrt{a^2 + c^2 - 2ac\cos B} \qquad 8.34$$
$$c = \sqrt{a^2 + b^2 - 2ab\cos C} \qquad 8.35$$

Example 8.20

Solve the triangle shown.

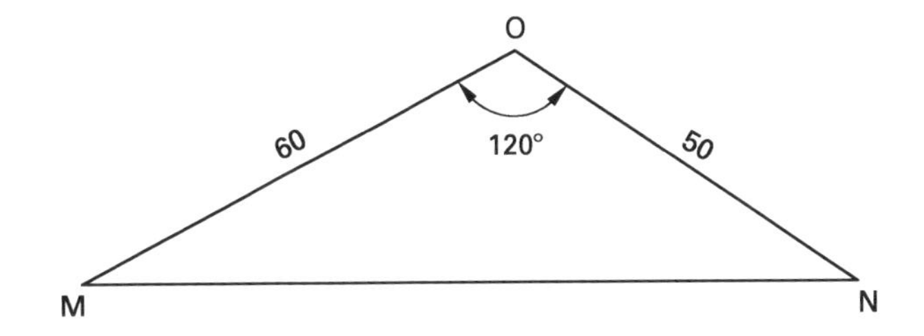

Solution

$$\begin{aligned} \text{MN} &= \sqrt{(60)^2 + (50)^2 - (2)(60)(50)(\cos 120^\circ)} \\ &= \sqrt{3600 + 2500 - (6000)(-0.500)} = 95 \text{ ft} \end{aligned}$$
$$\sin M = \frac{(50)(\sin 120^\circ)}{95} = 0.45580 \quad \text{[law of sines]}$$
$$M = 27^\circ$$
$$\sin N = \frac{(60)(\sin 120^\circ)}{95} = 0.54696 \quad \text{[law of sines]}$$
$$N = 33^\circ$$

27. SSS CASE

When three sides of a triangle are known (SSS), the law of cosines can be expressed as

$$\cos A = \frac{b^2 + c^2 - a^2}{2bc} \qquad 8.36$$

$$\cos B = \frac{a^2 + c^2 - b^2}{2ac} \quad \textit{8.37}$$

$$\cos C = \frac{a^2 + b^2 - c^2}{2ab} \quad \textit{8.38}$$

Example 8.21

Solve triangle ABC with sides $a = 3.0$, $b = 5.0$, and $c = 6.0$.

Solution

$$\cos A = \frac{(5.0)^2 + (6.0)^2 - (3.0)^2}{(2)(5.0)(6.0)} = \frac{25 + 36 - 9}{60}$$

$$A = 30°$$

$$\cos B = \frac{(3.0)^2 + (6.0)^2 - (5.0)^2}{(2)(3.0)(6.0)} = \frac{9 + 36 - 25}{36}$$

$$B = 56°$$

$$\cos C = \frac{(3.0)^2 + (5.0)^2 - (6.0)^2}{(2)(3.0)(5.0)} = \frac{9 + 25 - 36}{30} = -0.067$$

$$C = 94°$$

The negative value of $\cos C$ indicates that the angle is greater than 90° and the related angle is 86°.

28. OBLIQUE TRIANGLES USED IN SURVEYING

Since features rarely fall in a perfect right triangle, it is frequently necessary to use oblique triangles in surveying.

Example 8.22

Austin is 100 mi S 23° W from Waco. The bearing of Houston from Waco is S 40° E, and the bearing of Austin from Houston is N 76° W. What is the distance from Waco to Houston? What is the distance from Houston to Austin?

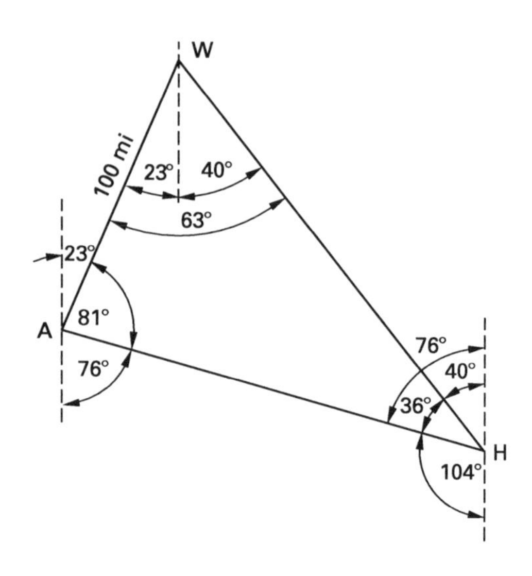

Solution

Draw a meridian (north line) and place point W (Waco) on it. From W draw a line that makes an angle of about 23° with the meridian in a southwesterly direction. Using any scale, place point A (Austin) on this line. Also from W draw a line in a southeasterly direction making an angle of about 40° with the meridian. From point A, draw a line in a southeasterly direction making an angle of about 76° with a meridian through A. This line intersects the southeasterly line from W at point H (Houston).

$$\text{WH} = \frac{(100 \text{ mi})(\sin 81°)}{\sin 36°} = 170 \text{ mi}$$

$$\text{HA} = \frac{(100 \text{ mi})(\sin 63°)}{\sin 36°} = 150 \text{ mi}$$

29. SELECTION OF LAW TO BE USED

To select the proper law for the solution of triangles, remember the following.

- *law of sines:* one side and two angles (SAA)

  two sides and the angle opposite one (SSA)

- *law of cosines:* two sides and the included angle (SAS)

  three sides (SSS)

30. RADIAN MEASURE

A *radian* is an angle that, when situated as a central angle of a circle, is subtended by an arc whose length is equal to the radius of the circle.

Figure 8.14 Radians

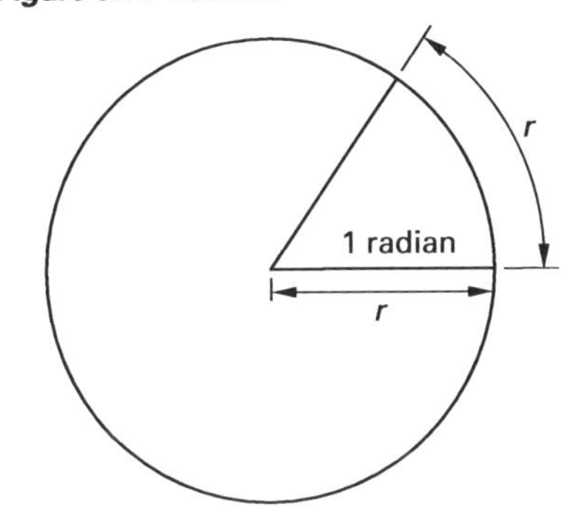

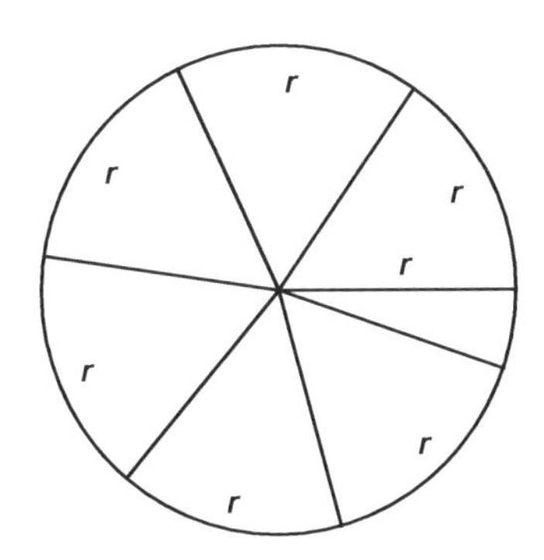

The circumference of a circle is 2π times the length of the radius, r. Therefore, the number of arcs of length r that can be applied to the circumference of a circle is 2π.

$$2\pi \text{ rad} = 360° \qquad 8.39$$

$$\pi \text{ rad} = 180° \qquad 8.40$$

Dividing both sides of Eq. 8.40 by π gives

$$1 \text{ rad} = \frac{180°}{\pi} \qquad 8.41$$

Dividing both sides of Eq. 8.40 by 180° gives

$$1° = \frac{\pi}{180°} \text{ rad} \qquad 8.42$$

rule: To convert degrees to radians, multiply the number of degrees by $\pi/180°$.

For example,

$$30° \equiv \frac{30°\pi}{180°} = \frac{\pi}{6} \text{ rad}$$

rule: To convert radians to degrees, multiply the number of radians by $180°/\pi$.

For example,

$$2 \text{ rad} = \frac{(2)(180°)}{\pi} = \frac{360°}{\pi}$$

31. LENGTH OF AN ARC OF A CIRCLE

If S equals the length of any arc of a circle, the relationship between the arc length and radius where θ is expressed in radians is

$$S = r\theta \qquad 8.43$$

Example 8.23

What is the length of the arc subtended by a central angle of 30° in a circle of 10 in radius?

Solution

$$S = r\theta = \frac{(10 \text{ in})(30°)\pi}{180°} = 5.2 \text{ in}$$

Example 8.24

What is the radius of the circle on which an arc of 100 ft subtends an angle of 1°?

$$r = \frac{S}{\theta} = \frac{100 \text{ ft}}{\frac{(1)\pi}{180°}} = \frac{18{,}000 \text{ ft}}{\pi} = 5729.58 \text{ ft}$$

32. AREA OF A SECTOR OF A CIRCLE

By geometry, the area of a sector of a circle is equal to one half its arc times the radius of the circle. If S equals the length of the arc, the area = $\frac{1}{2}Sr$.

The length of an arc of a circle equals $r\theta$, where θ is in radians. Therefore, where θ is in radians,

$$\text{area} = \tfrac{1}{2}r^2\theta \qquad 8.44$$

33. AREA OF A SEGMENT OF A CIRCLE

The area of a segment of a circle is found by subtracting the area of the triangle formed by the chord and the radii to the chord's end points from the area of the sector formed by the two radii and the arc of the segment. The area of the triangle equals $\frac{1}{2}ab(\sin C)$, where a and b are radii of the circle and C is the central angle. Therefore, where θ is the central angle in radians,

$$A = \tfrac{1}{2}r^2\theta - \tfrac{1}{2}r^2\sin\theta = \tfrac{1}{2}r^2(\theta - \sin\theta) \qquad 8.45$$

Example 8.25

Find the area of the segment with $r = 8$ in and central angle = 45°.

Solution

$$A = \left(\frac{1}{2}\right)(64\ \text{in})(0.7854\ \text{in} - 0.7071\ \text{in}) = 2.5\ \text{in}^2$$

PRACTICE PROBLEMS

1. Each of the following points lies on the terminal side of an angle θ in standard position. Plot the point, draw the terminal side through the point, and measure the angle with a protractor.

(a) $(12, 6)$

(b) $(-12, 4)$

(c) $(-10, -10)$

(d) $(7, -9)$

2. Write the sin, cos, tan, cot, sec, and csc of angles θ as a common fraction. Show the algebraic sign.

(a)

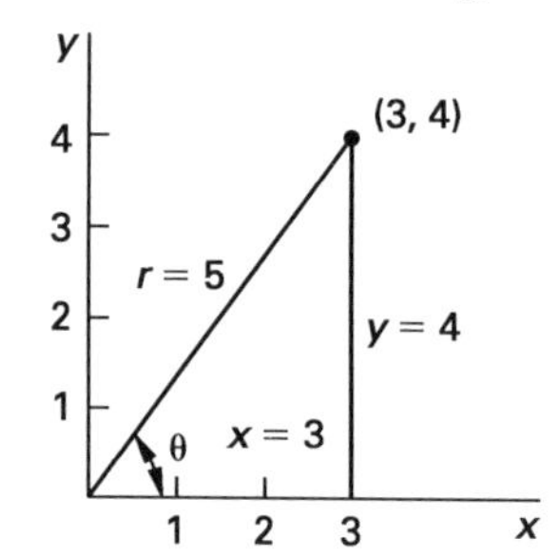

(b)

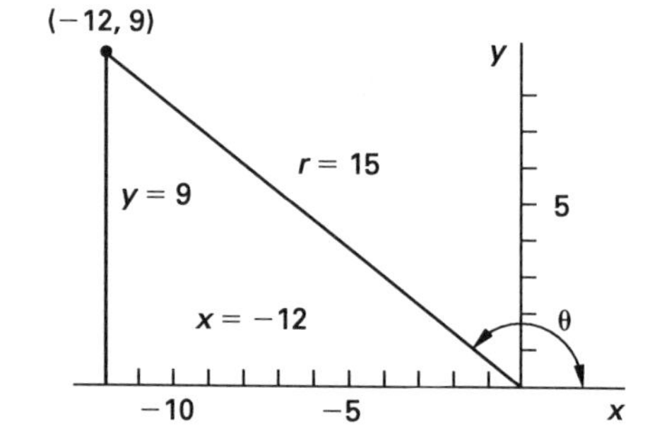

(c)

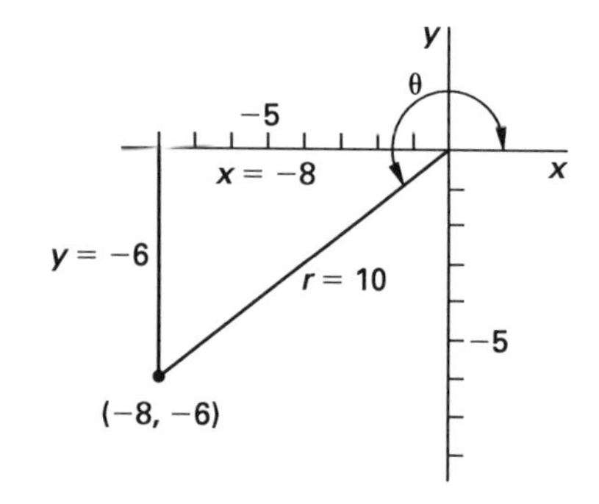

(d)

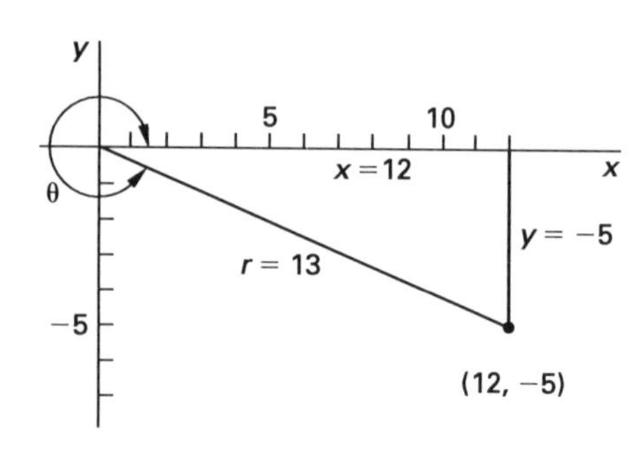

3. Write the reciprocal of each of the following numbers.

Example: $\frac{2}{3}$ $\qquad \frac{1}{\frac{2}{3}} = \frac{3}{2}$

(a) 3 $\qquad$ (b) $\frac{3}{4}$

(c) x $\qquad$ (d) y

(e) r $\qquad$ (f) $\frac{1}{x}$

(g) $\frac{1}{y}$ $\qquad$ (h) $\frac{x}{r}$

(i) $\frac{y}{r}$ $\qquad$ (j) $\frac{y}{x}$

4. Show that $\sin\theta$ and $\csc\theta$, $\cos\theta$ and $\sec\theta$, and $\tan\theta$ and $\cot\theta$ are reciprocals.

Example: $\sin\theta$ $\qquad \frac{1}{\sin\theta} = \frac{1}{\frac{y}{r}} = \frac{r}{y} = \csc\theta$

(a) $\cos\theta$

(b) $\tan\theta$

(c) $\cot\theta$

5. Using a calculator and the reciprocal, find the following numbers.

(a) $\sec 60°$

(b) $\csc 30°$

(c) $\cot 45°$

6. Using the following graph, find the sine, cosine, and tangent functions of the angles indicated.

Example:

$$\theta = 150°$$

$$\sin 150° = \frac{5.0}{10} = 0.50$$

$$\cos 150° = \frac{-8.7}{10} = -0.87$$

$$\tan 150° = \frac{5.0}{-8.7} = -0.57$$

(a) $\theta = 30°$

(b) $\theta = 135°$

(c) $\theta = 300°$

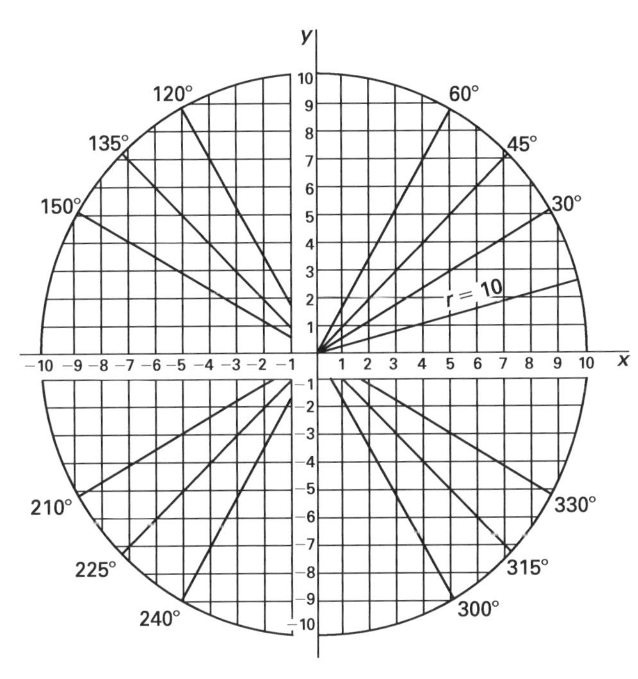

7. Indicate the algebraic sign of the sine, cosine, and tangent functions of each of the following angles.

Example: 30°: sin+
cos+
tan+

(a) 185° (b) 225°
(c) 350° (d) 190°
(e) 265° (f) 100°
(g) 89° (h) 175°
(i) 95° (j) 275°
(k) 300° (l) 110°
(m) 85° (n) 290°

8. Using the information provided, identify the quadrant in which the terminal side of the angle θ lies.

Examples: $\sin\theta = +,\ \cos\theta = -$:II

(a) $\sin\theta = -,\ \cos\theta = +$

(b) $\tan\theta = -,\ \sin\theta = -$

(c) $\tan\theta = +,\ \sin\theta = -$

(d) $\sin\theta = +,\ \cos\theta = +$

(e) $\tan\theta = +,\ \cos\theta = -$

(f) $\sin\theta = +,\ \tan\theta = +$

(g) $\sin\theta = -,\ \tan\theta = -$

9. Indicate the value of the sin, cos, and tan functions of the following quadrantal angles.

Example: 0° sin = 0; cos = +1; tan = 0

(a) 90°

(b) 180°

(c) 270°

(d) 360°

10. Using the figures, write sin, cos, and tan functions of each acute angle. Express as a common fraction and as a decimal fraction to two significant digits.

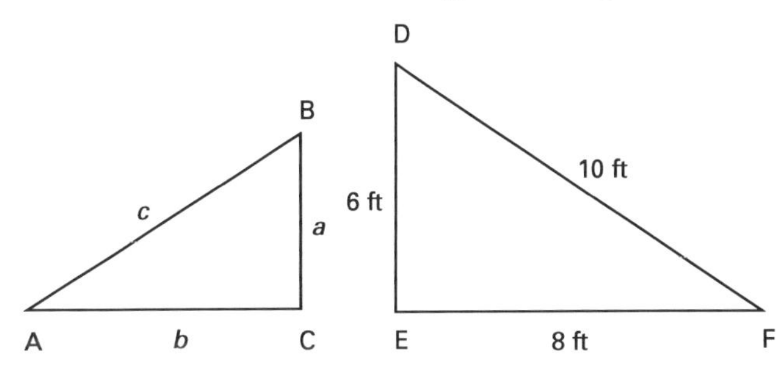

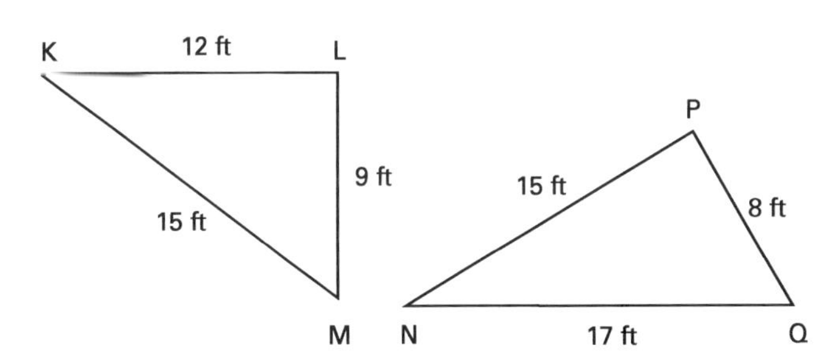

Examples: (1) $\sin A = \dfrac{a}{c}$

(2) $\sin K = \dfrac{9}{15} = 0.60$

(a) $\cos A$ (b) $\tan A$
(c) $\sin B$ (d) $\cos B$
(e) $\tan B$ (f) $\cos K$
(g) $\tan K$ (h) $\sin M$
(i) $\cos M$ (j) $\tan M$
(k) $\sin D$ (l) $\cos D$
(m) $\tan D$ (n) $\sin F$
(o) $\cos F$ (p) $\tan F$
(q) $\sin N$ (r) $\cos N$
(s) $\tan N$ (t) $\sin Q$
(u) $\cos Q$ (v) $\tan Q$

11. Find the trigonometric function for the angle indicated or find the angle for the function indicated.

Example: $\sin 53°13'36'' = 0.8010101$

(a) $\sin 37°29'16''$

(b) $\cos 52°42'51''$

(c) $\cos^{-1} 0.7918605$

(d) $\tan 43°17'28''$

12. For triangle ABC, select the trigonometric function to be used to solve the part of the triangle indicated.

Examples: (1)$A = 36°52'$, $a = 600$ ft, b: tan, c: sin

(2)$a = 300$ ft, $b = 400$ ft, A: tan, B: tan

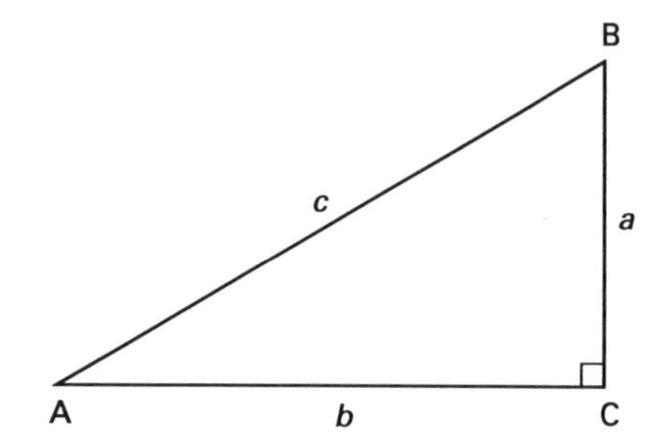

(a) $B = 51°40', a = 650$ ft
b: ____ c: ____

(b) $A = 46°44', b = 156$ ft
a: ____ c: ____

(c) $B = 53°21', c = 300$ ft
a: ____ b: ____

(d) $A = 38°19', c = 700$ ft
a: ____ b: ____

(e) $a = 600$ ft, $c = 1000$ ft
A: ____ B: ____

(f) $b = 400$ ft, $c = 500$ ft
A: ____ B: ____

(g) $B = 55°10', b = 378$ ft
a: ____ c: ____

(h) $A = 33°40', a = 250$ ft
b: ____ c: ____

13. Completely solve the right triangle ABC.

Example: $A = 36°52'$, $a = 600$ ft

$$B = 90° - 36°52' = 53°08'$$

$$b = \frac{600 \text{ ft}}{\tan 36°52'} = 800 \text{ ft}$$

$$c = \frac{600 \text{ ft}}{\sin 36°52'} = 1000 \text{ ft}$$

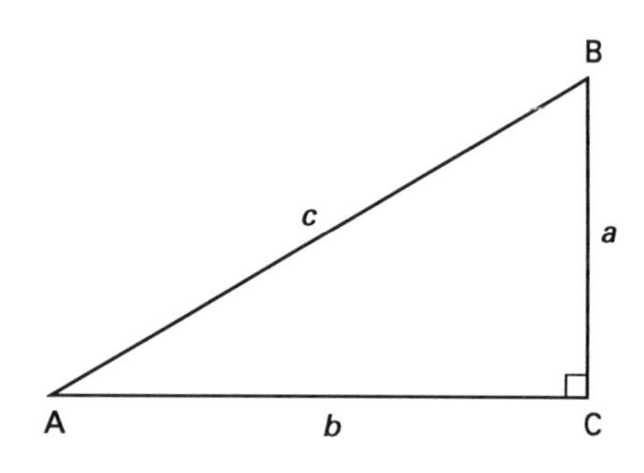

(a) $A = 28°41', b = 540$ ft

(b) $B = 55°13', a = 371$ ft

(c) $B = 61°29', b = 466$ ft

(d) $A = 33°15', c = 263$ ft

(e) $B = 58°55', c = 562$ ft

(f) $a = 300$ ft, $b = 400$ ft

14. Find the height to the nearest inch of a person who casts a shadow 10.0 ft long when the angle of elevation of the sun is 31°18′.

15. From a point 160 ft from the foot of a flagpole, the angle of elevation to the top of the flagpole is 50°40′. Find the height of the flagpole.

16. From the top of a tower 500 ft high, the angle of depression to a road intersection is 30°. How far from the tower is the road intersection?

17. Temple is due south of Fort Worth and 115 mi due east of Brady. The bearing of Brady from Fort Worth is S 45° W. How far is Brady from Fort Worth?

18. A 20 ft ladder reaches from the ground to the roof of a building. The angle of elevation of the ladder is 60°. How high is the roof?

19. Completely solve the right triangle ABC, as in Prob. 13.

(a) $B = 41°12'38'', a = 625.18$ ft

(b) $A = 66°22'37'', a = 492.72$ ft

(c) $B = 38°04'48'', c = 585.20$ ft

(d) $A = 22°13'50'', a = 376.26$ ft

(e) $B = 75°35'41'', b = 237.68$ ft

(f) $a = 427.82', b = 396.95$ ft

(g) $b = 445.64', c = 616.38$ ft

20. Write the related (or reference) angle of each angle.

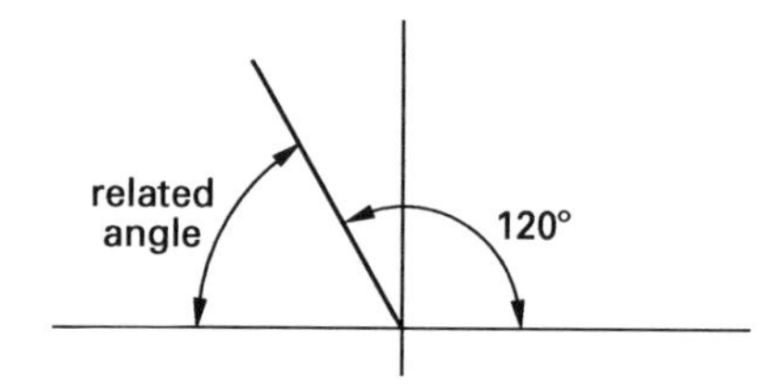

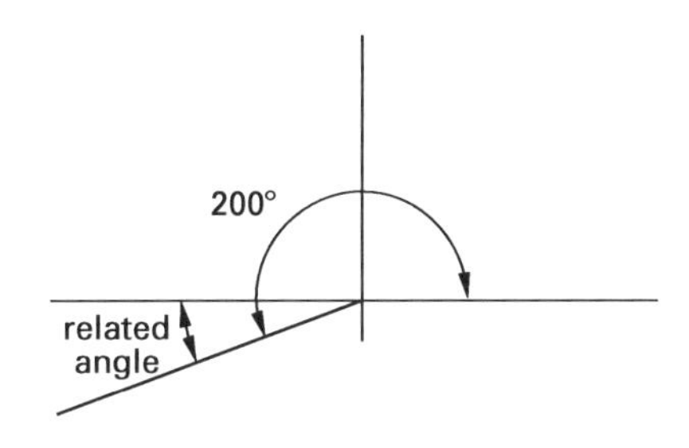

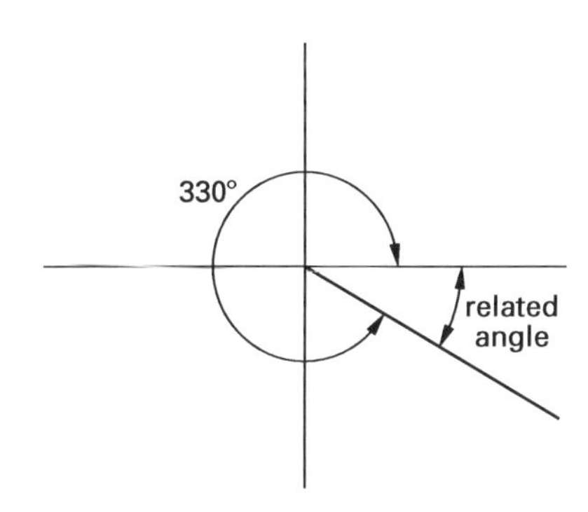

Example: 150°, 30°

(a) 260° (b) 210°
(c) 160° (d) 220°
(e) 190° (f) 350°
(g) 330° (h) 170°
(i) 200° (j) 300°
(k) 125° (l) 155°
(m) 110° (n) 140°
(o) 135° (p) 100°
(q) 120° (r) 175°
(s) 95° (t) 185°
(u) 280°

21. Solve for the missing sides and angles for the following oblique triangles (SAA).

Example: triangle MNO: $M = 38°48'45''$, $O = 82°23'56''$, MN = 298.34 ft

$$N = 180° - (38°48'45'' + 82°23'56'') = 58°47'19''$$

$$\text{NO} = \frac{(298.34 \text{ ft})(\sin 38°48'45'')}{\sin 82°23'56''} = 188.65 \text{ ft}$$

$$\text{OM} = \frac{(298.34 \text{ ft})(\sin 58°47'19'')}{\sin 82°23'56''} = 257.42 \text{ ft}$$

(a) triangle 1-2-3: $1 = 34°18'24'', 3 = 62°12'55''$ 1-2 = 1347.77 ft

(b) triangle PQR: $P = 118°34'24'', Q = 23°06'54''$, QR = 526.30 ft

(c) triangle 1-2-3: $1 = 82°46'58'', 2 = 58°54'20''$, 2-3 = 345.43 ft

22. Solve for the missing sides and angles for the following oblique triangles (SSA).

Example: triangle ABC: $A = 76°00'09''$, $a = 256.07$ ft, $b = 172.28$ ft

$$B\text{: } \sin B = \frac{(172.28 \text{ ft})(\sin 76°00'09'')}{256.07} \quad B = 40°45'13''$$

$$C = 180° - (76°00'09'' + 40°45'13'') = 63°14'38''$$

$$c = \frac{(\sin 63°14'38'')(256.07 \text{ ft})}{\sin 76°00'09''} = 235.65 \text{ ft}$$

(a) triangle EFG: $E = 25°40'55''$, $e = 646.13$ ft, $f = 1296.20$ ft

(b) triangle ABC: $A = 51°20'14''$, $a = 445.23$ ft, $b = 526.17$ ft (two solutions)

23. Solve for the missing sides and the angles for the following oblique triangles (SAS).

Example:

triangle ABC: $A = 58°33'47''$, $b = 204.38$ ft, $c = 152.15$ ft

$$a = \sqrt{\begin{array}{l}(204.38)^2 + (152.15)^2 \\ \quad - (2)(204.38)(152.15)\cos 58°33'47''\end{array}} = 180.23 \text{ ft}$$

$$B\text{: } \sin B = \frac{(204.38 \text{ ft})(\sin 58°33'47'')}{180.23} = 75°21'43''$$

$$C = 180° - (75°21'43'' + 58°33'47'') = 46°04'30''$$

(a) triangle EFG: $G = 95°12'50''$, FG = 146.25 ft, GE = 122.31 ft

(b) triangle KLM: $L = 35°19'16''$, KL = 595.45 ft, LM = 851.78 ft (Hint: The largest angle must be opposite the longest side. The sine function may represent a related angle.)

(c) triangle NOP: $N = 46°07'01''$, NO = 138.38 ft, PN = 165.12 ft

24. Solve for the missing sides and angles for the following oblique triangles (SSS).

Example: triangle ABC: $a = 48.79$ ft, $b = 62.45$ ft, $c = 30.13$ ft

$$A\text{: } \cos A = \frac{(62.45)^2 + (30.13)^2 - (48.79)^2}{(2)(62.45)(30.13)} = 49°49'59''$$

$$B\text{: } \cos B = \frac{(48.79^2) + (30.13)^2 - (62.45)^2}{(2)(48.79)(30.13)} = 102°00'32''$$

$$C\text{: } \cos C = \frac{(48.79)^2 + (62.45)^2 - (30.13)^2}{(2)(48.79)(62.45)} = 28°09'29''$$

check 180°00′00″

(a) triangle EFG: EF = 125.83 ft, FG = 171.25 ft, GE = 155.13 ft

(b) triangle MNO: MN = 298.34 ft, NO = 188.65 ft, OM = 257.42 ft

25. Given the parts of an oblique triangle ABC indicated, select the law applicable for the solution.

Example: a, c, A; law of sines

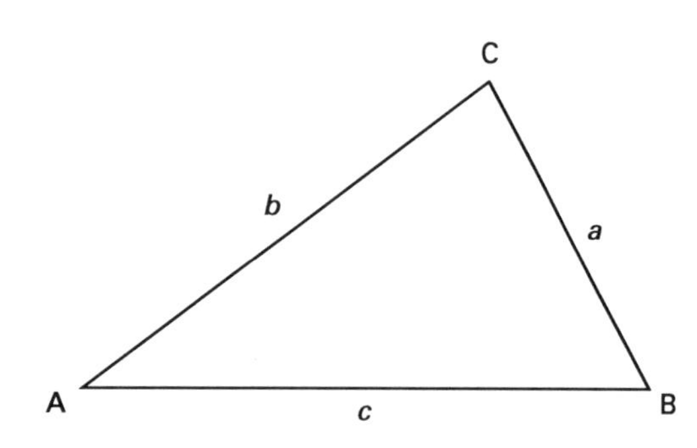

(a) a, b, A

(b) b, c, A

(c) b, A, C

(d) A, B, a

(e) a, c, B

(f) b, c, B

(g) A, a, b

(h) a, b, c

(i) A, C, a

(j) c, b, a

26. Express the following angles in radians using π in each answer.

Example: $30° = \frac{\pi}{6}$ rad

(a) 45° (b) 120° (c) 90° (d) 15° (e) 270°

27. Express the following angles in degrees.

Example: π rad = 180°

(a) $\frac{\pi}{3}$ rad (b) $\frac{\pi}{2}$ rad (c) $\frac{\pi}{4}$ rad (d) $\frac{\pi}{12}$ rad (e) $\frac{3\pi}{2}$ rad

28. What is the length of the arc subtended by a central angle of 60° in a circle with a 300 ft radius?

29. What is the radius of the circle on which an arc of 300 ft subtends an angle of 12°?

30. Find the radius of a circle on which an arc of 25 in has a central angle of 2.4 rad.

31. Find the length along the equator of an arc subtended by a central angle of 1° if the diameter of the earth at the equator is 7927 mi.

32. The end of a 25 in pedulum swings through a 3.4 in arc. What is the size of the angle through which the pendulum swings?

33. Find the area of a sector of a circle with a radius of 6 in and a central angle of 60°.

34. Find the area of a segment of a circle with a 100 ft radius and a 30° central angle.

35. Using a calculator, find the value of the sine function for each angle indicated, plot on the coordinate system marked sine curve, and connect the points with a curved line. Repeat the procedure for the cosine curve.

(c) $\sin 300° = \dfrac{-8.7}{10} = -0.87$

$\cos 300° = \dfrac{5.0}{10} = 0.50$

$\tan 300° = \dfrac{-8.7}{5.0} = -1.7$

7.

		sin	cos	tan
(a)	185°	−	−	+
(b)	225°	−	−	+
(c)	350°	−	+	−
(d)	190°	−	−	+
(e)	265°	−	−	+
(f)	100°	+	−	−
(g)	89°	+	+	+
(h)	175°	+	−	−
(i)	95°	+	−	−
(j)	275°	−	+	−
(k)	300°	−	+	−
(l)	110°	+	−	−
(m)	85°	+	+	+
(n)	290°	−	+	−

8. (a) IV
(b) IV
(c) III
(d) I
(e) III
(f) I
(g) IV

9. (a) sin = 1 cos = 0 tan = ∞
(b) sin = 0 cos = −1 tan = 0
(c) sin = −1 cos = 0 tan = ∞
(d) sin = 0 cos = +1 tan = 0

10. (a) $\cos A = \dfrac{b}{c}$

(b) $\tan A = \dfrac{a}{b}$

(c) $\sin B = \dfrac{b}{c}$

(d) $\cos B = \dfrac{a}{c}$

(e) $\tan B = \dfrac{b}{a}$

(f) $\cos K = \dfrac{12}{15} = 0.80$

(g) $\tan K = \dfrac{9}{12} = 0.75$

(h) $\sin M = \dfrac{12}{15} = 0.80$

(i) $\cos M = \dfrac{9}{15} = 0.60$

(j) $\tan M = \dfrac{12}{9} = 1.33$

(k) $\sin D = \dfrac{8}{10} = 0.80$

(l) $\cos D = \dfrac{6}{10} = 0.60$

(m) $\tan D = \dfrac{8}{6} = 1.33$

(n) $\sin F = \dfrac{6}{10} = 0.60$

(o) $\cos F = \dfrac{8}{10} = 0.80$

(p) $\tan F = \dfrac{6}{8} = 0.75$

(q) $\sin N = \dfrac{8}{17} = 0.47$

(r) $\cos N = \dfrac{15}{17} = 0.88$

(s) $\tan N = \dfrac{8}{15} = 0.53$

(t) $\sin Q = \dfrac{15}{17} = 0.88$

(u) $\cos Q = \dfrac{8}{17} = 0.47$

(v) $\tan Q = \dfrac{15}{8} = 1.88$

11. (a)
$$\begin{aligned}\sin 37°29'00'' &= 0.6085306\\ \sin 37°30'00'' &= 0.6087614\\ \sin 37°29'16'' &= 0.6085306\\ &\quad + \left(\frac{16}{60}\right)(0.0002308)\\ &= \boxed{0.6085922}\end{aligned}$$

(b)
$$\begin{aligned}\cos 52°42'00'' &= 0.6059884\\ \cos 52°43'00'' &= 0.6057570\\ \cos 52°42'51'' &= 0.6059884\\ &\quad - \left(\frac{51}{60}\right)(0.0002314)\\ &= \boxed{0.6057917}\end{aligned}$$

(c) $\cos 37°38'00'' = 0.7919345$

$\cos 37°39'00'' = 0.7917569$

$$\cos^{-1} 0.7918605 = 37°38'00'' + \left(\frac{740}{1776}\right)(60'') = \boxed{37°38'25''}$$

(d) $\tan 43°17'00'' = 0.9418033$

$\tan 43°18'00'' = 0.9423523$

$$\tan 43°17'28'' = 0.9418033 + \left(\frac{28}{60}\right)(0.0005490) = \boxed{0.9420595}$$

12. (a) $b : \tan \quad c : \cos$ (b) $a : \tan \quad c : \cos$

(c) $a : \cos \quad b : \sin$ (d) $a : \sin \quad b : \cos$

(e) $A : \sin \quad B : \cos$ (f) $A : \cos \quad B : \sin$

(g) $a : \tan \quad c : \sin$ (h) $b : \tan \quad c : \sin$

13. (a) $B = 90° - 28°41' = \boxed{61°19'}$

$$a = 540 \tan 28°41' = \boxed{295 \text{ ft}}$$

$$c = \frac{540}{\cos 28°41'} = \boxed{616 \text{ ft}}$$

(b) $A = 90° - 55°13' = 34°47'$

$b = (371 \text{ ft})(\tan 55°13') = 534 \text{ ft}$

$$c = \frac{371 \text{ ft}}{\cos 55°13'} = \boxed{650 \text{ ft}}$$

(c) $A = 90° - 61°29' = 28°31'$

$$a = \frac{466 \text{ ft}}{\tan 61°29'} = \boxed{253 \text{ ft}}$$

$$c = \frac{466 \text{ ft}}{\sin 61°29'} = \boxed{530 \text{ ft}}$$

(d) $B = 90° - 33°15' = \boxed{56°45'}$

$$a = (263 \text{ ft})(\sin 33°15') = \boxed{144 \text{ ft}}$$

$$b = (263 \text{ ft})(\cos 33°15') = \boxed{220 \text{ ft}}$$

(e) $A = 90° - 58°55' = \boxed{31°05'}$

$$a = (562 \text{ ft})(\cos 58°55') = \boxed{290 \text{ ft}}$$

$$b = (562 \text{ ft})(\sin 58°55') = \boxed{481 \text{ ft}}$$

$$A = \boxed{36°52'}$$

$$B = \boxed{53°08'}$$

14. $\tan 31°18' = \dfrac{h}{10 \text{ ft}}$

$$\text{height} = (10 \text{ ft})(\tan 31°18') = \boxed{6 \text{ ft, } 1 \text{ in}}$$

15. $\tan 50°40' = \dfrac{h}{160 \text{ ft}}$

$$\text{height} = (160 \text{ ft})(\tan 50°40') = \boxed{195 \text{ ft}}$$

16. $\tan 30° = \dfrac{500 \text{ ft}}{d}$

$$\text{distance} = \frac{500 \text{ ft}}{\tan 30°} = \boxed{866 \text{ ft}}$$

17. $\cos 45° = \dfrac{115 \text{ ft}}{d}$

$$\text{distance} = \frac{115 \text{ ft}}{\cos 45°} = \boxed{163 \text{ mi}}$$

18. $\sin 60° = \dfrac{h}{20 \text{ ft}}$

$$\text{height} = (20 \text{ ft})(\sin 60°) = \boxed{17 \text{ ft}}$$

19. (a) $A = 90° - 41°12'38'' = \boxed{48°47'22''}$

$$b = (625.18 \text{ ft})(\tan 41°12'38'') = \boxed{547.51 \text{ ft}}$$

$$c = \frac{625.18 \text{ ft}}{\cos 41°12'38''} = \boxed{831.03 \text{ ft}}$$

(b) $B = 90° - 66°22'37'' = \boxed{23°37'23''}$

$$b = \frac{492.72 \text{ ft}}{\tan 66°22'37''} = \boxed{215.50 \text{ ft}}$$

$$c = \frac{492.72 \text{ ft}}{\sin 66°22'37''} = \boxed{537.79 \text{ ft}}$$

(c) $A = 90° - 38°04'48'' = \boxed{51°55'12''}$

$a = (585.20 \text{ ft})(\cos 38°04'48'') = \boxed{460.64 \text{ ft}}$

$b = (585.20 \text{ ft})(\sin 38°04'48'') = \boxed{360.93 \text{ ft}}$

(d) $B = 90° - 22°13'50'' = \boxed{67°46'10''}$

$$b = \frac{376.26 \text{ ft}}{\tan 22°13'50''} = \boxed{920.59 \text{ ft}}$$

$$c = \frac{376.26 \text{ ft}}{\sin 22°13'50''} = \boxed{994.52 \text{ ft}}$$

(e) $A = 90° - 75°35'41'' = \boxed{14°24'19''}$

$$a = \frac{237.68 \text{ ft}}{\tan 75°35'41''} = \boxed{61.05 \text{ ft}}$$

$$c = \frac{237.68 \text{ ft}}{\sin 75°35'41''} = \boxed{245.40 \text{ ft}}$$

(f) $c = \sqrt{(427.82)^2 + (396.95)^2} = \boxed{583.61 \text{ ft}}$

$$\tan A = \frac{427.82}{396.95}; \; A = \boxed{47°08'37''}$$

$$\tan B = \frac{396.95}{427.82}; \; B = \boxed{42°51'23''}$$

(g) $a = \sqrt{(616.38)^2 - (445.64)^2} = \boxed{425.83 \text{ ft}}$

$$\cos A = \frac{445.64}{616.38}; \; A = \boxed{43°41'52''}$$

$$\sin B = \frac{445.64}{616.38}; \; B = \boxed{46°18'08''}$$

20. (a) 80° (b) 30° (c) 20° (d) 40°
(e) 10° (f) 10° (g) 30° (h) 10°
(i) 20° (j) 60° (k) 55° (l) 25°
(m) 70° (n) 40° (o) 45° (p) 80°
(q) 60° (r) 5° (s) 85° (t) 5°
(u) 80°

21. (a) Triangle 1–2–3: 1 = 34°18′24″;

$3 = \boxed{62°12'55''}$; $1\text{–}2 = \boxed{1347.77 \text{ ft}}$

$$2 = 180° - (34°18'24'' + 62°12'55'') = \boxed{83°28'41''}$$

$$2\text{–}3 = \frac{(1347.77 \text{ ft})(\sin 34°18'24'')}{\sin 62°12'55''} = \boxed{858.63 \text{ ft}}$$

$$3\text{–}1 = \frac{(1347.77 \text{ ft})(\sin 83°28'41'')}{\sin 62°12'55''} = \boxed{1513.55 \text{ ft}}$$

(b) Triangle PQR: $P = 118°34'24''$;

$Q = \boxed{23°06'54''}$; $\text{QR} = \boxed{526.30 \text{ ft}}$

$$R = 180° - (118°34'24'' + 23°06'54'') = \boxed{38°18'42''}$$

$$\text{PQ} = \frac{(526.30 \text{ ft})(\sin 38°18'42'')}{\sin 118°34'24''} = \boxed{371.52 \text{ ft}}$$

$$\text{RP} = \frac{(526.30 \text{ ft})(\sin 23°06'54'')}{\sin 118°34'24''} = \boxed{235.27 \text{ ft}}$$

(c) Triangle 1–2–3: 1 = 82°46′58″;

$2 = \boxed{58°54'20''}$; $2\text{–}3 = \boxed{345.43 \text{ ft}}$

$$3 = 180° - (82°46'58'' + 58°54'20'') = \boxed{38°18'42''}$$

$$1\text{–}2 = \frac{(345.43 \text{ ft})(\sin 38°18'42'')}{\sin 82°46'58''} = \boxed{215.86 \text{ ft}}$$

$$3\text{–}1 = \frac{(345.43 \text{ ft})(\sin 58°54'20'')}{\sin 82°46'58''} = \boxed{298.16 \text{ ft}}$$

22. (a) $\sin F = \frac{(1296.20 \text{ ft})(\sin 25°40'55'')}{646.13 \text{ ft}};$

$$F = \boxed{60°23'17''}$$

$$G = 180° - (25°40'55'' + 60°23'17'')$$

$$= \boxed{93°55'48''}$$

$$g = \frac{(646.13 \text{ ft})(\sin 93°55'48'')}{\sin 25°40'55''}$$

$$= \boxed{1487.42 \text{ ft}}$$

(b) *Solution 1:*

$$\sin B = \frac{(526.17 \text{ ft})(\sin 51°20'14'')}{445.23};$$

$$B = \boxed{67°20'13''}$$

$$C = 180° - (51°20'14'' + 67°20'13'')$$

$$= \boxed{61°19'33''}$$

$$c = \frac{(445.23 \text{ ft})(\sin 61°19'33'')}{\sin 51°20'14''}$$

$$= \boxed{500.27 \text{ ft}}$$

(b) *Solution 2:*

$$B = \boxed{112°39'47''}$$

$$C = 180° - (51°20'14'' + 112°39'47'')$$

$$= \boxed{15°59'59''}$$

$$c = \frac{(445.23 \text{ ft})(\sin 15°59'59'')}{\sin 51°20'14''}$$

$$= \boxed{157.16 \text{ ft}}$$

23. (a) $\text{EF} = \sqrt{\begin{array}{l}(146.25)^2 + (122.31)^2 \\ -(2)(146.25)(122.31)\cos 95°12'50''\end{array}}$

$$= \boxed{199.00 \text{ ft}}$$

$$\sin E = \frac{(146.25 \text{ ft})(\sin 95°12'50'')}{199.00}$$

$$E = \boxed{47°02'40''}$$

$$F = 180° - (95°12'50'' + 47°02'40'')$$

$$= \boxed{37°44'30''}$$

(b) $\text{MK} = \sqrt{\begin{array}{l}(595.45)^2 + (851.78)^2 \\ -(2)(595.45)(851.78)\cos 35°19'16''\end{array}}$

$$= \boxed{502.42 \text{ ft}}$$

$$\sin K = \frac{(851.78 \text{ ft})(\sin 35°19'16'')}{502.42}$$

$$K = \boxed{101°25'32''}$$

$$M = 180° - (35°19'16'' + 101°25'32'')$$

$$= \boxed{43°15'12''}$$

(c) $\text{OP} = \sqrt{\begin{array}{l}(138.38)^2 + (165.12)^2 \\ -(2)(138.38)(165.12)\cos 46°07'01''\end{array}}$

$$= \boxed{121.39}$$

$$\sin O = \frac{(165.12 \text{ ft})(\sin 46°07'01'')}{121.39}$$

$$O = \boxed{78°38'19''}$$

$$P = 180° - (46°07'01'' + 78°38'19'')$$

$$= \boxed{55°14'40''}$$

(Note: The largest angle must be opposite the longest side. Hint: The sine function may represent a related angle.)

24. (a) $\cos E = \dfrac{(125.83)^2 + (155.13)^2 - (171.25)^2}{(2)(125.83)(155.13)};$

$E = \boxed{74°17'18''}$

$\cos F = \dfrac{(125.83)^2 + (171.25)^2 - (155.13)^2}{(2)(125.83)(155.13)};$

$F = \boxed{60°41'40''}$

$\cos G = \dfrac{(171.25)^2 + (155.13)^2 - (125.83)^2}{(2)(171.25)(155.13)};$

$G = \boxed{45°01'02''}$

check: $E + F + G = 180°00'00''$

(b) $\cos M = \dfrac{(298.34)^2 + (257.42)^2 - (188.65)^2}{(2)(171.25)(155.13)};$

$M = \boxed{38°48'46''}$

$\cos N = \dfrac{(298.34)^2 + (188.65)^2 - (257.42)^2}{(2)(298.34)(188.65)};$

$N = \boxed{58°47'18''}$

$\cos O = \dfrac{(188.65)^2 + (257.42)^2 - (298.34)^2}{(2)(188.65)(257.42)};$

$O = \boxed{82°23'56''}$

check $180°00'00''$

25.
(a) law of sines (b) law of cosines
(c) law of sines (d) law of sines
(e) law of cosines (f) law of sines
(g) law of sines (h) law of cosines
(i) law of sines (j) law of cosines

26. (a) $\dfrac{\pi}{4}$ rad (b) $\dfrac{2\pi}{3}$ rad (c) $\dfrac{\pi}{2}$ rad
(d) $\dfrac{\pi}{12}$ rad (e) $\dfrac{3\pi}{2}$ rad

27. (a) 60° (b) 90° (c) 45°
(d) 15° (e) 270°

28. $S = r\theta = \dfrac{(300 \text{ ft})(60)\pi}{180} = \boxed{314 \text{ ft}}$

29. $r = \dfrac{S}{\theta} = \dfrac{300 \text{ ft}}{\dfrac{12\pi}{180°}} = \boxed{1432 \text{ ft}}$

30. $r = \dfrac{S}{\theta} = \dfrac{25 \text{ in}}{2.4} = \boxed{10.4 \text{ in}}$

31. $S = r\theta = \dfrac{(7927 \text{ mi})\pi}{(2)(180°)} = \boxed{69 \text{ mi}}$

32. $\theta = \dfrac{S}{r} = \dfrac{(3.4 \text{ in})(180°)}{(25 \text{ in})\pi} = \boxed{8°}$

33. $\text{area} = \frac{1}{2}r^2\theta = \left(\dfrac{1}{2}\right)(6 \text{ in})^2\left(\dfrac{\pi}{3}\right) = \boxed{19 \text{ in}^2}$

34. $\text{area} = \frac{1}{2}r^2(\theta - \sin\theta)$

$= \left(\dfrac{1}{2}\right)(100 \text{ ft})^2\left(\dfrac{\pi}{6} - 0.5\right)$

$= \boxed{118 \text{ ft}^2}$

35. (a)

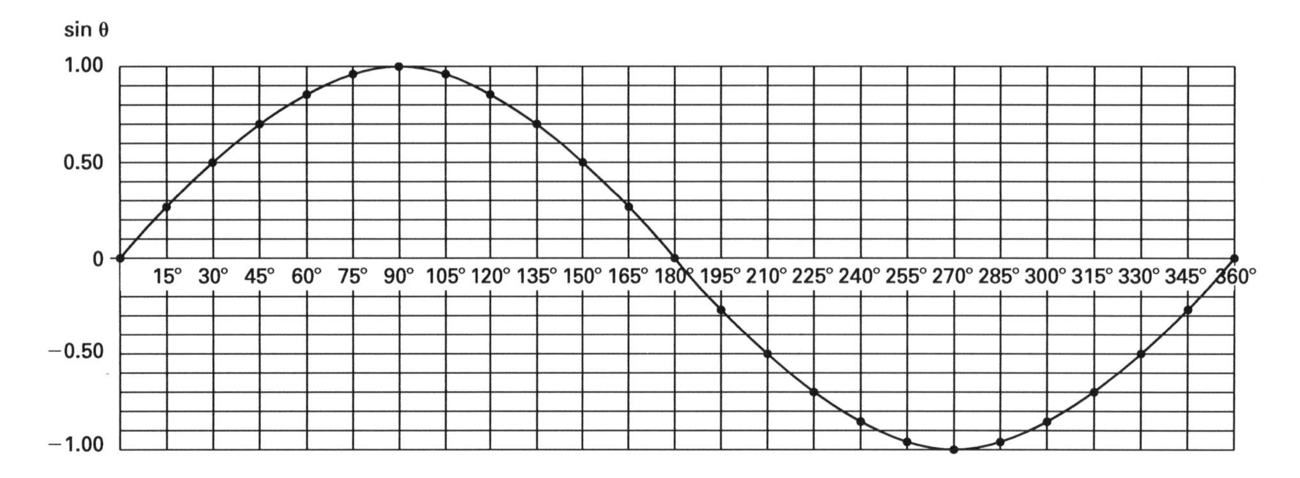
sin θ
1.00
0.50
0
−0.50
−1.00
15° 30° 45° 60° 75° 90° 105° 120° 135° 150° 165° 180° 195° 210° 225° 240° 255° 270° 285° 300° 315° 330° 345° 360°

(b)

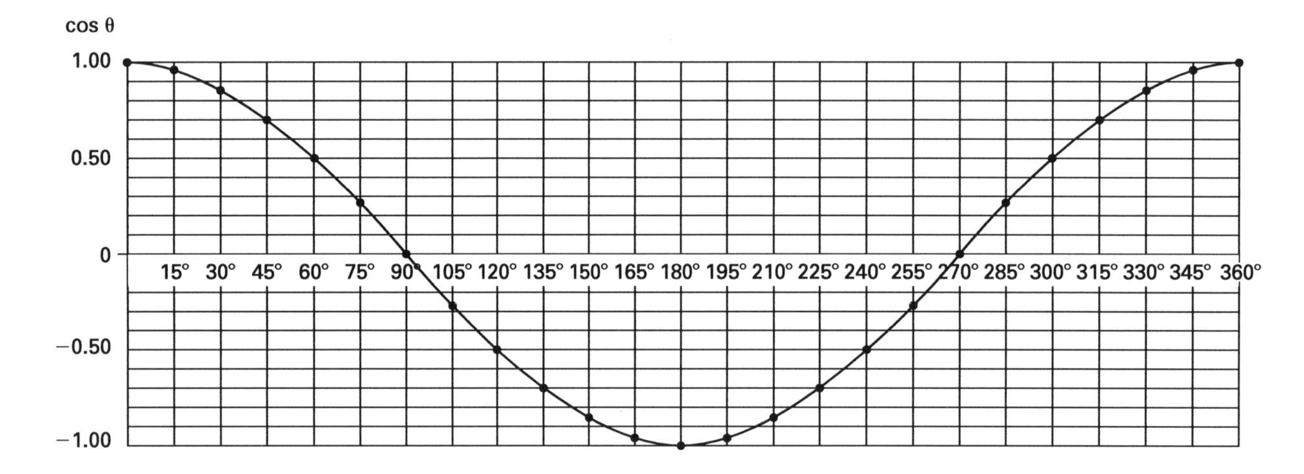
cos θ
1.00
0.50
0
−0.50
−1.00
15° 30° 45° 60° 75° 90° 105° 120° 135° 150° 165° 180° 195° 210° 225° 240° 255° 270° 285° 300° 315° 330° 345° 360°

9 Rectangular Coordinate System

1. DIRECTED LINE

Suppose a surveyor establishes a west-east line through a point on a monument called the *origin.* Moving east from the origin, the surveyor marks off points 1 ft apart, labeling them $+1$, $+2$, $+3$, ... Then, moving west from the origin, the surveyor marks off points 1 ft apart, labeling them $-1, -2, -3, \ldots$.

The surveyor then establishes a *directed line*, as shown in Fig. 9.1. From left to right (west to east), the direction is positive and numbers increase in value. From right to left (east to west), the direction is negative and numbers decrease in value.

Figure 9.1 *Directed Line*

A directed line, on paper, can be called an *axis.* A horizontal (west-east) line is called the *x-axis.* A point on the axis associated with a particular number is called the *graph* of the number, and the number is called the *coordinate* of the point. This coordinate is the directed distance from the origin to the point. Point A in Fig. 9.1 has the coordinate -4, which is the distance from 0 to A, not the distance from A to 0. Point B has the coordinate $+3$, which is the distance from 0 to B, not B to 0.

$$\text{distance AB} = 3 - (-4) = 7$$
$$\text{distance BA} = -4 - (+3) = -7$$

The word "distance" is used loosely in this text. A more appropriate term for distance on a directed line would be "the measure of travel in a specified direction." The actual length from A to B, or B to A, is 7. This actual length, which disregards the negative sign, is known as the *absolute value* or AB or BA. Symbolically, $|-7| = 7$. The enclosure indicates absolute value.

In general, if P_1 and P_2 are any two points on the x-axis with coordinates x_1 and x_2, then

$$P_1P_2 = x_2 - x_1 \qquad 9.1$$
$$P_2P_1 = x_1 - x_2 \qquad 9.2$$

The surveyor can also establish a south-north line through the same point on the monument and mark points on the line at 1 ft intervals from the origin in a northerly direction, labeling them $+1$, $+2$, $+3$, ... Points at 1 ft intervals on the line in a southerly direction are labeled $-1, -2, -3, \ldots$.

Figure 9.2 shows a vertical line through the origin representing this south-north line. It is a directed line that is positive from south to north and is called the *y-axis.*

Figure 9.2 *x-y Axis*

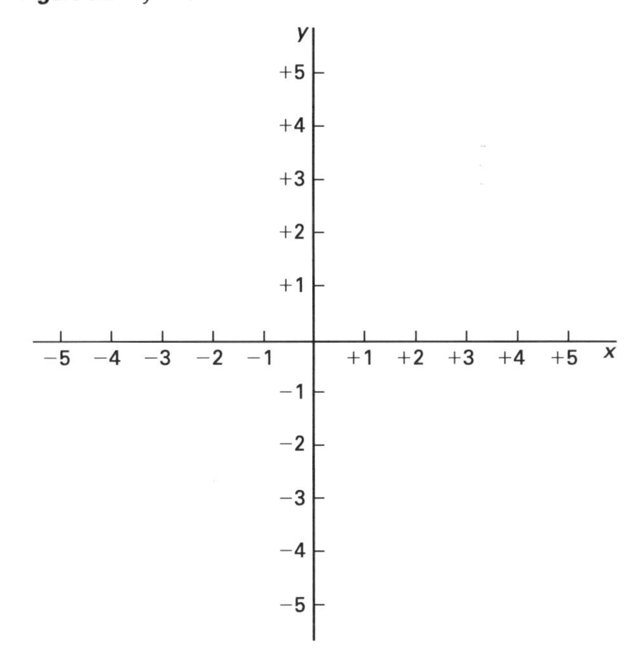

If P_1 and P_2 are any two points on the y-axis with coordinates y_1 and y_2, then

$$P_1P_2 = y_2 - y_1 \qquad 9.3$$
$$P_2P_1 = y_1 - y_2 \qquad 9.4$$

2. THE RECTANGULAR COORDINATE SYSTEM

The French mathematician René Descartes (17th century) devised the *rectangular coordinate system*, sometimes called the *Cartesian plane.* This system uses an ordered pair of coordinates to locate a point. The ordered pair of coordinates are the x-coordinate and the y-coordinate of a point, enclosed in parentheses with the x-coordinate always written first, followed by a comma and the y-coordinate.

The Cartesian plane consists of an x- (horizontal) and a y- (vertical) axis, which are directed lines as shown in Fig. 9.3.

Figure 9.3 *Abscissas and Ordinates*

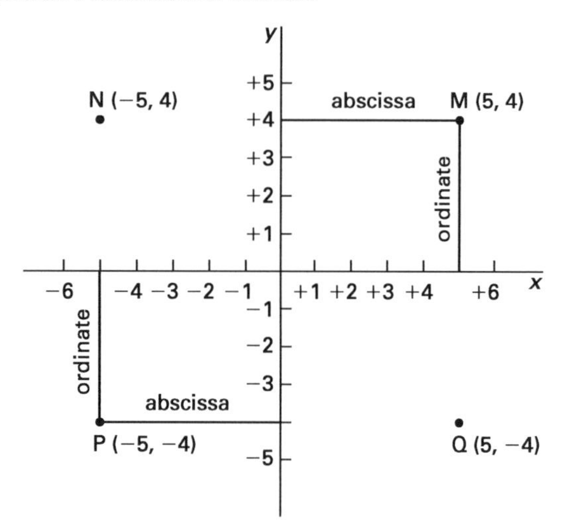

Point M has the coordinates $(5, 4)$. The horizontal distance 5, from the y-axis to M, is known as the *abscissa.* The vertical distance 4, from the x-axis to M, is known as the *ordinate.* The abscissa and ordinate are measured from the axis to point and not from point to the axis, which is in accordance with distances on a directed line. In Fig. 9.3, point N has an abscissa of -5 and an ordinate of $+4$; point P has an abscissa of -5 and an ordinate of -4; and point Q has an abscissa of $+5$ and an ordinate of -4.

The x- and y-axes divide the plane into four parts, numbered in a counterclockwise direction as shown in Fig. 9.4. Signs of the coordinates of points in each quadrant are also shown in Fig. 9.4. In quadrant I, x is positive and y is positive; in quadrant II, x is negative and y is positive; in quadrant III, x is negative and y is negative; and in quadrant IV, x is positive and y is negative.

Figure 9.4 *Signs of the Quadrants*

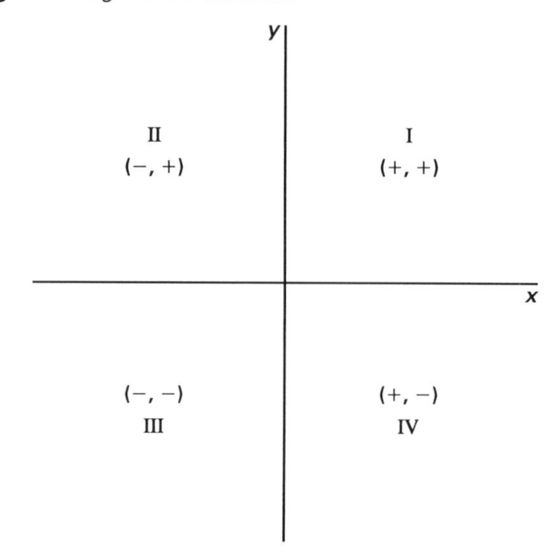

Example 9.1

Determine the coordinates of the points shown.

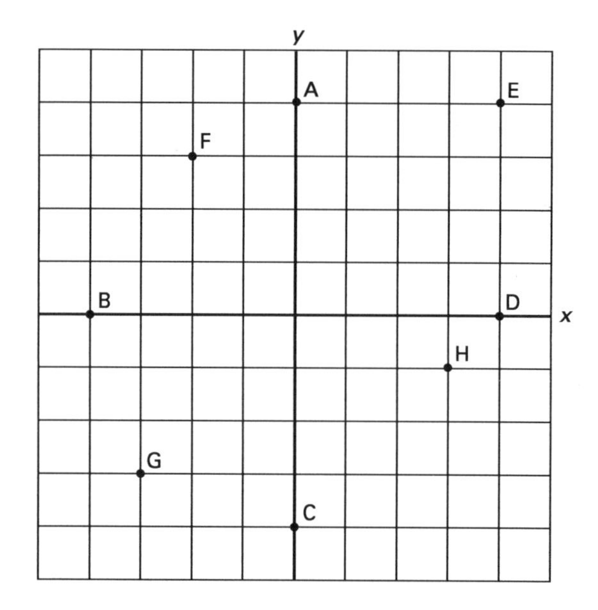

Solution

A: $(0, 4)$ B: $(-4, 0)$
C: $(0, -4)$ D: $(4, 0)$
E: $(4, 4)$ F: $(-2, 3)$
G: $(-3, -3)$ H: $(3, -1)$

3. DISTANCE FORMULA

A formula for finding the distance between any two points in a rectangular coordinate system can be derived from the *Pythagorean theorem*: in a right triangle, the square of the length of the hypotenuse equals the sum of the squares of the lengths of the other two sides.

If c represents the length of the hypotenuse and a and b are the lengths of the other two sides,

$$c^2 = a^2 + b^2 \tag{9.5}$$

Taking the square root of both sides of the equation,

$$c = \sqrt{a^2 + b^2} \tag{9.6}$$

In Fig. 9.5, P_1 and P_2 represent any two points in a rectangular coordinate system with the coordinates (x_1, y_1) and (x_2, y_2). If a horizontal line passes through P_1 and a vertical line passes through P_2, they will intersect at Q, forming a right triangle, with the line P_1P_2 being the hypotenuse. The x-coordinate of Q will be the same as the x-coordinate of P_2: x_2; and the y-coordinate of Q will be the same as the y-coordinate of P_1: y_1.

Figure 9.5 *Distance Between Points*

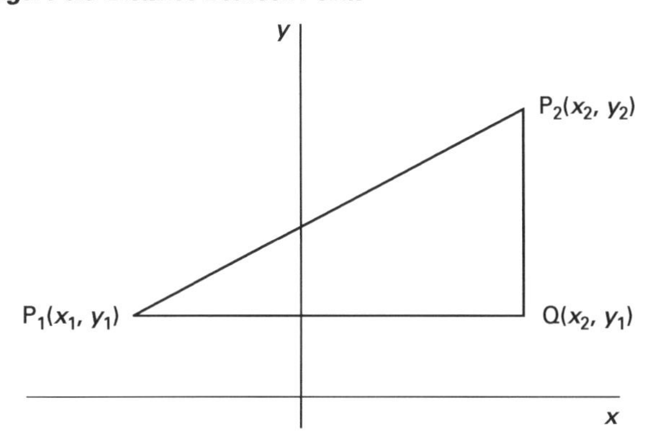

By the Pythagorean theorem,

$$(P_1P_2)^2 = (P_1Q)^2 + (QP_2)^2 \tag{9.7}$$

$$P_1P_2 = \sqrt{(P_1Q)^2 + (QP_2)^2} = \sqrt{(x_2 - x_1)^2 + (y_2 - y_1)^2} \tag{9.8}$$

It is also true that

$$P_1P_2 = \sqrt{(x_1 - x_2)^2 + (y_1 - y_2)^2} \tag{9.9}$$

Example 9.2

Find the distance between $P_1(-3, -1)$ and $P_2(5, 5)$.

Solution

$$\begin{aligned} P_1P_2 &= \sqrt{(x_2 - x_1)^2 + (y_2 - y_1)^2} \\ &= \sqrt{(5 - (-3))^2 + (5 - (-1))^2} \\ &= \sqrt{(5 + 3)^2 + (5 + 1)^2} \\ &= 10 \end{aligned}$$

The distance r from the origin to any point $P(x, y)$ is called the *radius*. This distance is always positive.

From the Pythagorean theorem,

$$r = \sqrt{x^2 + y^2} \tag{9.10}$$

Figure 9.6 *Radius*

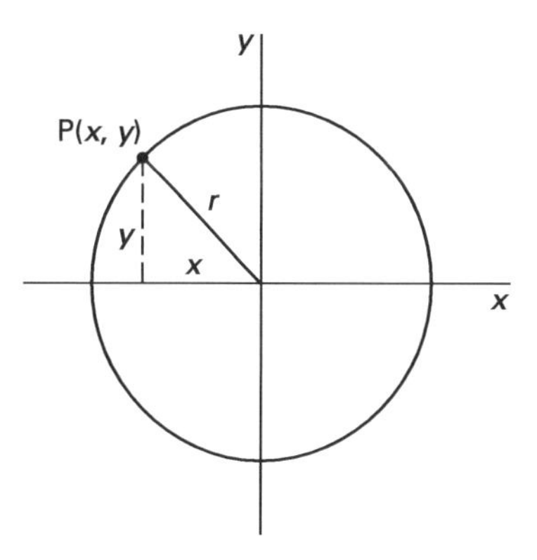

Example 9.3

Find the distance from the origin to the point $P(-3, 4)$.

Solution

$$r = \sqrt{(-3)^2 + (4)^2} = 5$$

4. MIDPOINT OF A LINE

In Fig. 9.7, point M is the midpoint of the line PQ.

The x-coordinate of M is the distance from the y-axis to M. This is equal to the distance from the y-axis to L, which is the average of the x-coordinates of P and Q.

$$\frac{6 + (-2)}{2} = 2$$

Figure 9.7 *Midpoint of a Line*

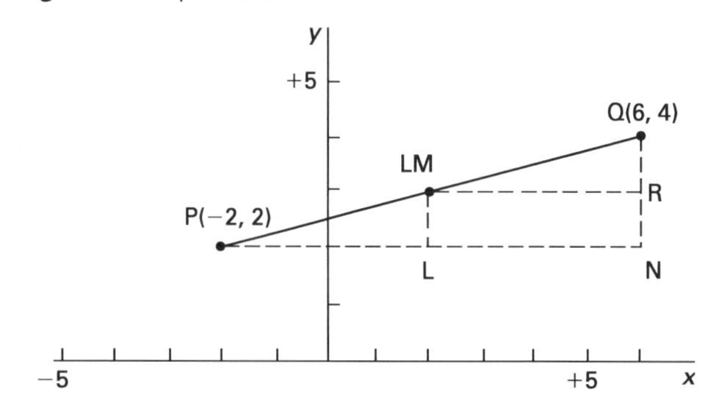

The y-coordinate of M is the distance from the x-axis to M. This is equal to the distance from the x-axis to R, which is the average of the y-coordinates of P and Q.

$$\frac{4+2}{2} = 3$$

From this it can be seen that the midpoint M of a line P_1P_2 has the coordinates

$$M\left(\frac{x_1+x_2}{2}, \frac{y_1+y_2}{2}\right) \qquad 9.11$$

Example 9.4

Find the midpoint M of line with endpoints $P_1(-2,3)$ and $P_2(-8,-3)$.

Solution

$$M\left(\frac{x_1+x_2}{2}, \frac{y_1+y_2}{2}\right) = M\left(\frac{-2-8}{2}, \frac{3-3}{2}\right) = M(-5,-0)$$

PRACTICE PROBLEMS

1. (a) In the rectangular coordinate system, plot the points and connect with lines in the order PQRSTP.

P(12, 16)

Q(−14, 18)

R(−12, 1)

S(−14, −17)

T(14, −14)

(b) Find the length of each line segment (PQ, QR, etc.) and the perimeter.

2. Find the radii for circles centered at the origin (0, 0) and with the points on the circles given.

(a) (3, 4)

(b) (−3, 3)

(c) (−5, −12)

(d) (5, −5)

SOLUTIONS

1. (a)

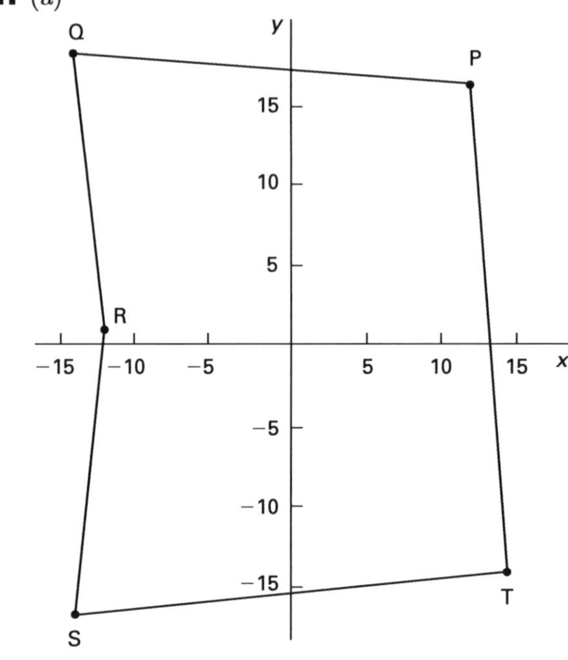

(b) $PQ = \sqrt{(-14-12)^2 + (18-16)^2} = \boxed{26}$

$QR = \sqrt{(-12-(-14))^2 + (1-18)^2} = \boxed{17}$

$RS = \sqrt{(-14-(-12))^2 + (-17-1)^2} = \boxed{18}$

$ST = \sqrt{(14-(-14))^2 + (-14-(-17))^2} = \boxed{28}$

$TP = \sqrt{(12-14)^2 + (16-(-14))^2} = \boxed{30}$

$\text{perimeter} = 26+17+18+28+30 = \boxed{119}$

2. (a)

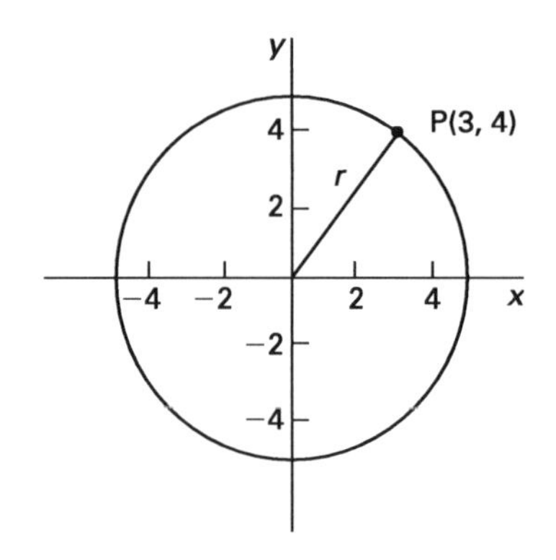

$$r = \sqrt{(3)^2 + (4)^2} = \boxed{5}$$

(b)

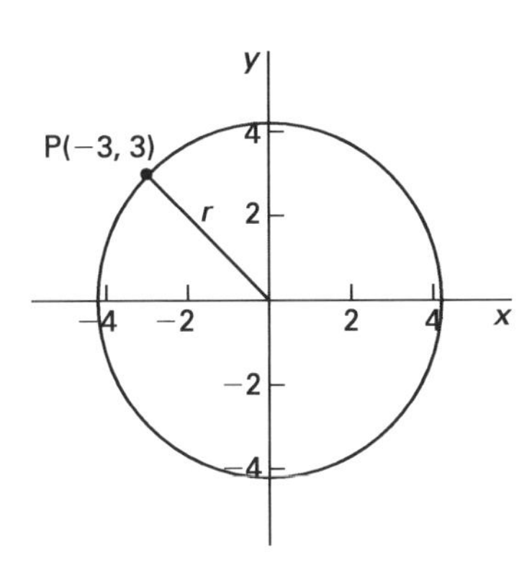

$$r = \sqrt{(-3)^2 + (3)^2} = \boxed{4.2}$$

(c)

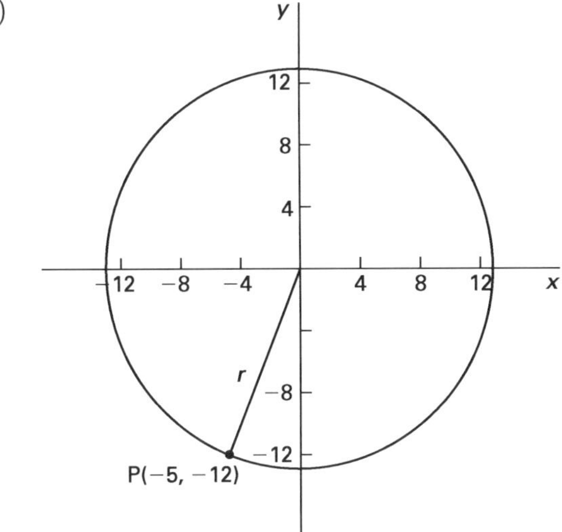

$$r = \sqrt{(-5)^2 + (-12)^2} = \boxed{13}$$

(d)

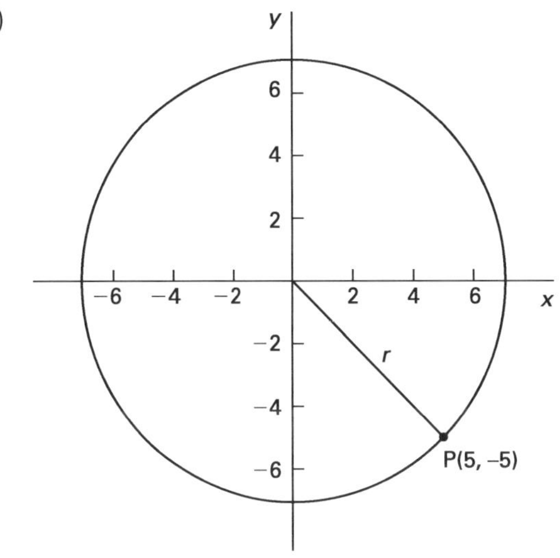

$$r = \sqrt{(5)^2 + (-5)^2} = \boxed{7.1}$$

10 Analytical Geometry

Nomenclature

m slope of a line

Symbols

α inclination of a line

1. FIRST-DEGREE EQUATIONS

The statement "four times a number minus three is equal to five" can be expressed in algebraic terms as $4x - 3 = 5$. The letter x represents the unknown number. The number 2 satisfies the equation, and it is called the *root* of the equation.

The statement "the sum of two numbers is eight" can be written in algebraic terms as $x + y = 8$. In this equation there are two unknowns, represented by x and y, and more than one pair of numbers will make the statement true. (If $x = 1$, $y = 7$; if $x = 2$, $y = 6$; if $x = 3$, $y = 5$; etc.) If these pairs of numbers are always expressed in the order of x first and y second, they are called *ordered pairs* and are written symbolically as $(1,7)$, $(2,6)$, $(3,5)$, and so on. Because x and y are of the first power, $x + y = 8$ is an *equation of the first degree*. The equation $x + y = 8$ has two *unknowns*, or two *variables*. The equation $4x - 3 = 5$ is an equation of the first degree with one unknown.

2. GRAPHS OF FIRST-DEGREE EQUATIONS WITH TWO VARIABLES

Consider each ordered pair that satisfies a first-degree equation with two variables (the roots of the equation, or the *solution set* of the equation) to be coordinates of a point. If several of these points are plotted on a rectangular coordinate system and connected with a line, the system will show the *graph of the equation*.

Consider the following equation.

$$2x - 3y = 12$$

To solve the equation—that is, to find ordered pairs that satisfy it, first rearrange the equation.

$$-3y = -2x + 12$$
$$y = \tfrac{2}{3}x - 4$$

Next, give various values to x and solve for the corresponding values of y (for instance, when $x = 0$, $y = -4$),

$$x = -3,\ 0,\ 3,\ 6,\ 9$$
$$y = -6,\ -4,\ -2,\ 0,\ 2$$

Plotting these ordered pairs as coordinates of a point creates the graph of the equation $2x - 3y = 12$ shown in Fig. 10.1. When the points are connected, they lie in a straight line. An infinite number of values could be given to x and y.

Figure 10.1 *Straight Line*

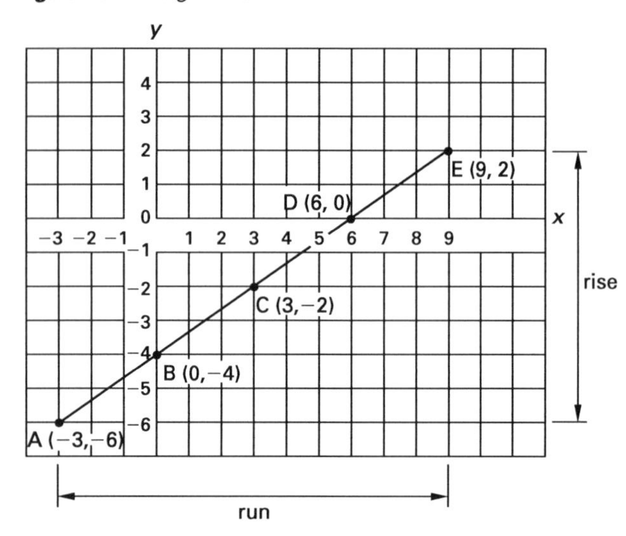

Example 10.1

Find three roots of each equation. Give your answers as (x, y).

(a) $2x + y = 10$ (b) $3x + 2y = 0$

(c) $x + y = 5$ (d) $x - y = 0$

Solution

(a) $(3,4); (2,6); (5,0)$ (b) $(2,-3); (4,-6); (0,0)$

(c) $(1,4); (2,3); (0,5)$ (d) $(-4,-4); (0,0); (2,2)$

Example 10.2

Find the coordinates of the points at which each of the following equations intersects the x- and y-axes.

(a) $x - y + 5 = 0$ (b) $3x - y + 6 = 0$

(c) $2x - 5y = 10$ (d) $2x + 3y + 6 = 0$

Solution

Let $y = 0$ and solve for x. Then let $x = 0$ and solve for y.

(a) $(0,5);\ (-5,0)$ (b) $(0,6);\ (-2,0)$

(c) $(0,-2);\ (5,0)$ (d) $(0,-2);\ (-3,0)$

3. SLOPE OF A LINE

Slope is the ratio of the change in the vertical distance to the change in the horizontal distance. If m represents the slope for the line connecting points (x_1, y_1) and (x_2, y_2), then

$$m = \frac{y_2 - y_1}{x_2 - x_1} \qquad 10.1$$

To describe the steepness or *slope* of a line (such as the graph of the equation $2x - 3y = 12$ shown in Fig. 10.1), choose two points on it (such as A(−3, −6) and E(9, 2)).

$$\text{slope} = \frac{\text{rise}}{\text{run}} = \frac{\text{ordinate of E} - \text{ordinate of A}}{\text{abscissa of E} - \text{abscissa of A}}$$
$$= \frac{2-(-6)}{9-(-3)} = \frac{8}{12} = 2/3$$

Alternatively,

$$\text{slope} = \frac{\text{ordinate of A} - \text{ordinate of E}}{\text{abscissa of A} - \text{abscissa of E}}$$
$$= \frac{-6-2}{-3-9} = \frac{8}{12} = 2/3$$

Example 10.3

Determine the slope of the line through each pair of points.

(a) $(1,2);\ (3,6)$ (b) $(-4,3);\ (3,-4)$

(c) $(-2,-3);\ (2,5)$ (d) $(-4,3);\ (4,3)$

Solution

(a) $m = \dfrac{6-2}{3-1} = +2$

(b) $m = \dfrac{-4-3}{3-(-4)} = -1$

(c) $m = \dfrac{5-(-3)}{2-(-2)} = +2$

(d) $m = \dfrac{3-3}{4-(-4)} = 0$

In plotting the lines it can be seen that when a line rises from left to right, its slope is positive; and when a line falls from left to right, its slope is negative.

4. LINEAR EQUATIONS

The slope of the line represented by the linear equation $2x - 3y = 12$ is $+2/3$. The graph of the equation shown in Fig. 10.1 rises from left to right. Assume the equation is written in the following form.

$$y = \frac{2}{3}x - \frac{12}{3}$$

The coefficient of x is $+2/3$, which is the slope of the line. The numerator 2 is the coefficient of x, and the denominator 3 is the coefficient of y when the equation is written in the following form.

$$2x - 3y = 12$$

Also notice that when the coefficient of x is $+2$ and the coefficient of y is -3, the slope is positive.

Equation 10.2 represents the *general form* of a linear equation, where A is the positive coefficient of x, B is the coefficient of y, and C is a constant.

$$Ax + By + C = 0 \quad \textit{10.2}$$

The slope is then

$$m = -\frac{A}{B} \quad \textit{10.3}$$

Example 10.4

Rearrange the following equations to the form $Ax + By + C = 0$ with A positive. Determine the slope m of each.

(a) $3x + 4y = 8$

(b) $2x = 6y$

(c) $2y + 3x - 6 = 0$

(d) $-3x + 4y + 10 = 0$

Solution

(a) $3x + 4y - 8 = 0$

$$m = -\frac{A}{B} = -\frac{+3}{+4} = -3/4$$

(b) $2x - 6y = 0$

$$m = -\frac{+2}{-6} = 1/3$$

(c) $3x + 2y - 6 = 0$

$$m = -\frac{+3}{+2} = -3/2$$

(d) $3x - 4y - 10 = 0$

$$m = -\frac{+3}{-4} = 3/4$$

5. EQUATIONS OF HORIZONTAL AND VERTICAL LINES

Consider the line that contains the points $(-4, 5)$ and $(4, 5)$. The slope of the line is

$$m = \frac{5-5}{4+4} = \frac{0}{8} = 0$$

This is a horizontal line that has the equation $y = 5$ (Fig. 10.2). In considering the general equation of a line, $Ax + By + C = 0$, the coefficient of x for a horizontal line is 0. So the equation of a horizontal line is

$$m = \frac{5-(-5)}{4-4} = \frac{10}{0}$$

$$By = C$$

The line that contains the points $(4, 5)$ and $(4, -5)$ has the slope

$$m = \frac{5-(-5)}{4-4} = \frac{10}{0}$$

This number is infinite, so the line has an infinite slope. The equation of the line is $x = 4$. Therefore, the general equation of the vertical line is

$$Ax + 0y + C = 0$$

Figure 10.2 *Horizontal and Vertical Lines*

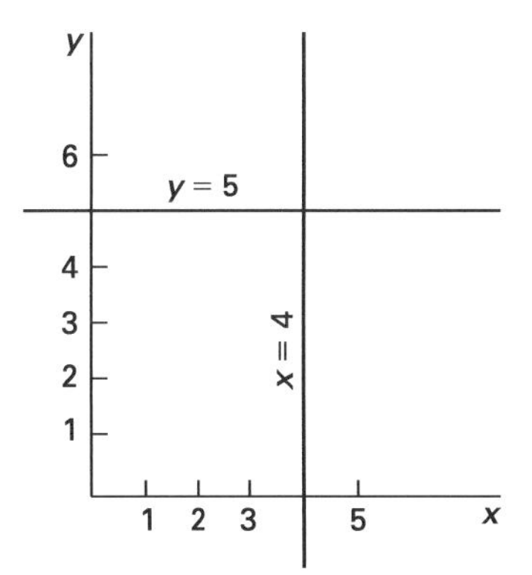

6. *x*- AND *y*-INTERCEPTS

If $x = 0$ in the equation $2x - 3y = 12$, then $y = -4$. The point $(0, -4)$ is the point where the graph of the equation crosses the y-axis (Fig. 10.1). The distance from the x-axis to this point is known as the *y-intercept* and is given the symbol b.

If $y = 0$ in the equation $2x - 3y = 12$, then $x = 6$. The point $(6, 0)$ is where the graph of the equation crosses the x-axis. The distance from the y-axis to this point is known as the *x-intercept* and is given the symbol a.

Therefore, the coordinates of the point where any line crosses the x-axis are $(a, 0)$ and the coordinates of the point where any line crosses the y-axis are $(0, b)$.

If the equation $2x - 3y = 12$ is written as $y = {}^2\!/\!_3x - 4$, the constant term -4 is the y-coordinate of the point where the line crosses the y-axis. It is, in fact, the y-intercept. In finding this equivalent equation of $2x - 3y = 12$, both sides of the equation were divided by -3, which is the coefficient of y, so that the constant 12 was divided by -3 to obtain the quotient -4.

For any equation $Ax + By + C = 0$, the y-intercept is

$$b = -\frac{C}{B} \quad \textit{10.4}$$

If the equation $2x - 3y = 12$ is written as $x = {}^3\!/\!_2y + 6$, the constant term 6 is the x-coordinate of the point where the line crosses the x-axis; it is the x-intercept. For any equation $Ax + By + C = 0$, the x-intercept is

$$a = -\frac{C}{A} \tag{10.5}$$

In summary, for any linear equation $Ax + By + C = 0$,

$$m = \text{slope} = -\frac{A}{B} \tag{10.6}$$

$$a = x\text{-intercept} = -\frac{C}{A} \tag{10.7}$$

$$b = y\text{-intercept} = -\frac{C}{B} \tag{10.8}$$

Example 10.5

Find the slope, x-intercept, and y-intercept of the line $3x + 2y - 6 = 0$.

Solution

$$m = -\frac{A}{B} = -\frac{3}{2} = -1.5$$

$$a = -\frac{C}{A} = -\frac{-6}{3} = 2$$

$$b = -\frac{C}{B} = -\frac{-6}{2} = 3$$

7. PARALLEL LINES

Two different lines having the same slope are called *parallel lines.*

Figure 10.3 Parallel Lines

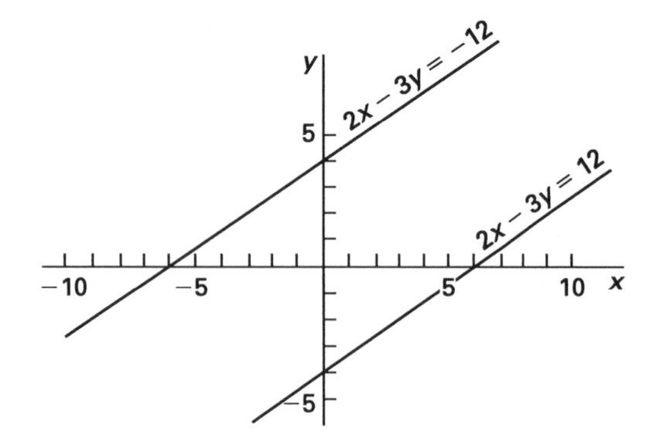

In Fig. 10.3, the lines $2x - 3y = -12$ and $2x - 3y = 12$ have the same slope, $+2/3$. Writing the equations in equivalent form shows that the slope for each line is the same; only the y-intercepts differ.

$$y = \frac{2}{3}x + 4$$

$$y = \frac{2}{3}x - 4$$

8. PERPENDICULAR LINES

Two lines intersecting at right angles are called *perpendicular lines.* For two lines to be perpendicular, the slope of one must be the negative reciprocal of the other, or

$$m_1 = -\frac{1}{m_2} \tag{10.9}$$

In Fig. 10.4, the lines $2x - 3y = 12$ and $3x + 2y = 8$ are perpendicular lines, as can be seen when written in the form $y = mx + b$.

$$y = \frac{2}{3}x - 4$$

$$y = -\frac{3}{2}x + 4$$

Figure 10.4 Perpendicular Lines

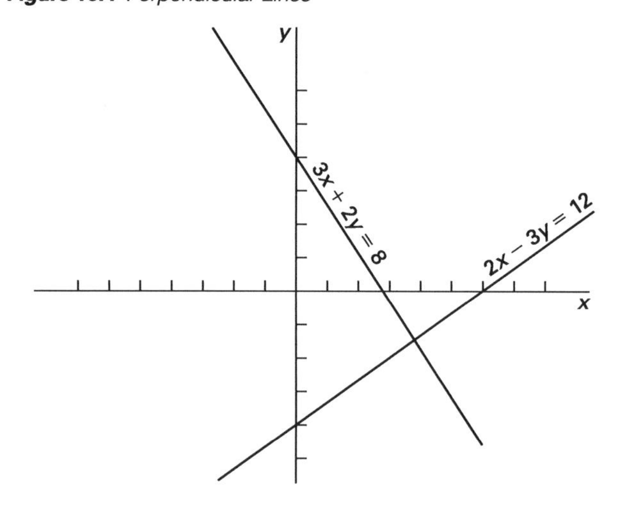

9. PERPENDICULAR DISTANCE FROM A POINT TO A LINE

A formula for finding the perpendicular distance from a point of known coordinates (x, y) to a line of known equation can be found from Eq. 10.10.

$$D = \frac{|Ax + By + C|}{\sqrt{A^2 + B^2}} \tag{10.10}$$

Example 10.6

Find the perpendicular distance D from the point $P(-2,4)$ to the line $4x - 3y - 16 = 0$.

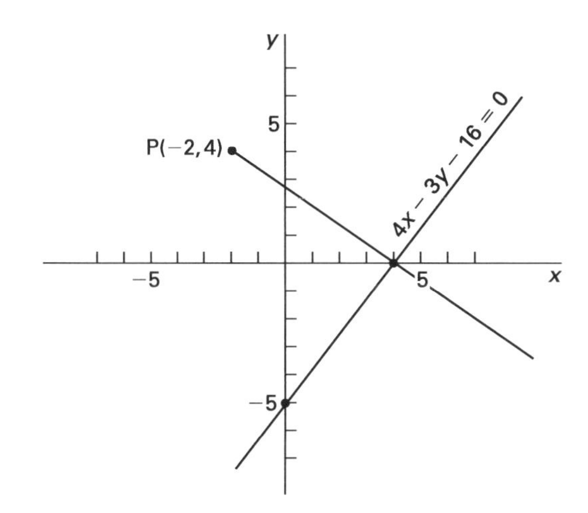

Solution

$$D = \frac{|(4)(-2) - (3)(4) - 16|}{\sqrt{(4)^2 + (-3)^2}} = \frac{|(-36)|}{5} = \frac{36}{5}$$
$$= 7.2$$

10. WRITING THE EQUATION OF A LINE

An equation may be written for a straight line if sufficient information is known, for example, if any of the following is true.

- one point on the line and the slope are known
- two points on the line are known
- the x-intercept and y-intercept are known
- the slope of the line and the y-intercept are known
- the slope of the line and the x-intercept are known

11. POINT-SLOPE FORM OF THE EQUATION OF A LINE

If $P(x,y)$ is any point on a line with slope m through point $P_1(x_1,y_1)$, then the *point-slope form* of the line is

$$\frac{y - y_1}{x - x_1} = m \quad \text{10.11}$$

$$y - y_1 = m(x - x_1) \quad \text{10.12}$$

Example 10.7

Write the equation of the line through the given point with the given slope m in the form $Ax + By + C = 0$.

(a) $(4,-2)$; $m = 2$

(b) $(3,-2)$; $m = -\frac{3}{2}$

(c) $(-4,5)$; $m = 0$

Solution

(a)

$$\begin{aligned} y - y_1 &= m(x - x_1) \\ y - (-2) &= (2)(x - 4) \\ y + 2 &= 2x - 8 \\ -2x + y + 10 &= 0 \\ 2x - y - 10 &= 0 \end{aligned}$$

(b) The slope m can be written $-(3/2)$, $(-3)/2$, or $3/(-2)$. For ease in performing the algebraic operation, the numerator should carry the negative sign.

$$\begin{aligned} y - y_1 &= m(x - x_1) \\ y + 2 &= \frac{(-3)(x - 3)}{2} \\ 2y + 4 &= -3x + 9 \\ 3x + 2y - 5 &= 0 \end{aligned}$$

(c)

$$\begin{aligned} y - y_1 &= m(x - x_1) \\ y - 5 &= (0)(x + 4) \\ y &= 5 \end{aligned}$$

This is a linear equation. The graph is a line parallel to the x-axis, 5 units above.

12. TWO-POINT FORM OF THE EQUATION OF A LINE

If $P(x,y)$ is any point on a line that passes through the points $P_1(x_1,y_1)$ and $P_2(x_2,y_2)$, then the *two-point form* of the line is

$$y - y_1 = m(x - x_1) \quad \text{10.13}$$

$$y - y_1 = \left(\frac{y_2 - y_1}{x_2 - x_1}\right)(x - x_1) \quad \text{10.14}$$

$$\frac{y - y_1}{x - x_1} = \frac{y_2 - y_1}{x_2 - x_1} \quad \text{10.15}$$

In writing the two-point form, either point may be designated as point 1.

Example 10.8

Write the equation of the line through the following two points.

(a) $(1,4)$; $(3,-2)$ (b) $(-2,2)$; $(1,-3)$
(c) $(1,-3)$; $(-2,1)$ (d) $(3,4)$; $(1,4)$

Solution

(a)

$$\begin{aligned} \frac{y - y_1}{x - x_1} &= \frac{y_2 - y_1}{x_2 - x_1} \\ \frac{y - 4}{x - 1} &= \frac{-2 - 4}{3 - 1} = \frac{-6}{2} = -3 \\ y - 4 &= (-3)(x - 1) \\ y - 4 &= -3x + 3 \\ 3x + y - 7 &= 0 \end{aligned}$$

(b)

$$\frac{y-y_1}{x-x_1} = \frac{y_2-y_1}{x_2-x_1}$$

$$\frac{y-2}{x+2} = \frac{-3-2}{1+2} = -5/3$$

$$(-5)(x+2) = (3)(y-2)$$

$$-5x-10 = 3y-6$$

$$-5x-3y-4 = 0$$

$$5x+3y+4 = 0$$

(c)

$$\frac{y-y_1}{x-x_1} = \frac{y_2-y_1}{x_2-x_1}$$

$$\frac{y+3}{x-1} = \frac{1+3}{-2-1} = 4/-3$$

$$(4)(x-1) = (-3)(y+3)$$

$$4x-4 = -3y-9$$

$$4x+3y+5 = 0$$

(d)

$$\frac{y-y_1}{x-x_1} = \frac{y_2-y_1}{x_2-x_1}$$

$$\frac{y-4}{x-3} = \frac{4-4}{1-3} = \frac{0}{-2} = 0$$

$$(-2)(y-4) = (0)(x-3)$$

$$-2y+8 = 0$$

$$2y-8 = 0$$

$$y = 4$$

13. INTERCEPT FORM OF THE EQUATION OF A LINE

A line with x-intercept a and y-intercept b (where both a and b are not zero) has the equation

$$\frac{x}{a} + \frac{y}{b} = 1 \qquad 10.16$$

Example 10.9

Write the equation of the line with x-intercept 3 and y-intercept -4.

Solution

$$\frac{x}{3} + \frac{y}{-4} = 1$$

$$-4x+3y = -12$$

$$4x-3y = 12$$

14. SLOPE-INTERCEPT FORM OF THE EQUATION OF A LINE

If the slope of a line and its y-intercept are known, the *slope-intercept form* of the line is

$$y = mx + b \qquad 10.17$$

Example 10.10

Write the equation of the line of slope 2 and y-intercept -3.

Solution

$$y = 2x-3$$

$$2x-y = 3$$

Example 10.11

Write the equation of the line through $(3, -1)$ perpendicular to the line $2x+3y=6$.

Solution

The slope of $2x+3y$ is $m_1 = -2/3$.

The slope of the perpendicular line is $m_2 = 3/2$.

The equation of the perpendicular line is

$$y - y_1 = m_2(x-x_1)$$

$$y-(-1) = \left(\frac{3}{2}\right)(x-3)$$

$$3x-2y = 11$$

15. SYSTEMS OF LINEAR EQUATIONS

If the graphs of two linear equations lie in the same xy plane, then one of three conditions must be true:

- The two lines are parallel and will never intersect.
- The two lines coincide.
- The two lines will intersect at a point.

Figure 10.5 *System of Linear Equations*

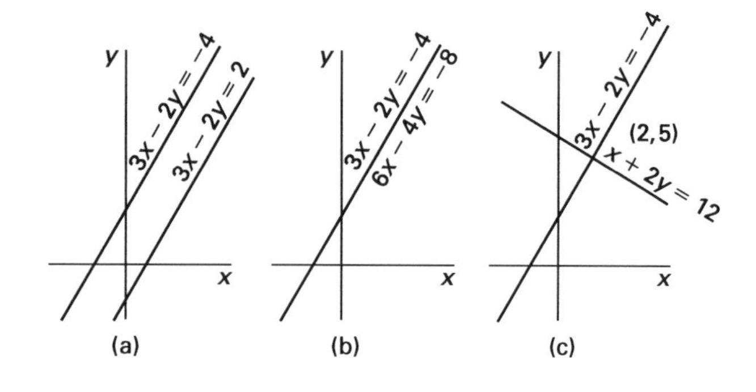

When a point must be located on two straight lines, the line equations form a system of simultaneous equations. Any ordered pair that satisfies both equations is called a *solution* or a *root* of the system.

When two lines intersect at a point, there can be only one root, and this root can be found by solving the two equations simultaneously. The ordered pair found to be the root of a system will be the coordinates of the point of intersection of the two graphs of the equations.

In Fig. 10.5(a), the two lines are parallel and will not intersect. Therefore, the two equations cannot be solved simultaneously. The slopes of the two lines will indicate whether or not they are parallel.

In Fig. 10.5(b), the two lines intersect everywhere because they have the same solution set. It can be seen that the equation $6x - 4y = -8$ is equivalent to the equation $3x - 2y = -4$. If both sides of the equation are divided by 2, the result will be $3x - 2y = -4$. Thus, both equations have the same graph.

In Fig. 10.5(c), the two lines will intersect at a point. This point can be found by solving the equations simultaneously.

16. SOLVING SYSTEMS OF SIMULTANEOUS EQUATIONS

Several methods can be used to solve a system of equations. One method is known as the *method of reduction.*

Consider the equations $x + y = 9$ and $x - y = 3$. If the two equations are added,

$$\begin{array}{r} x + y = 9 \\ \underline{x - y = 3} \\ 2x \quad = 12 \\ x = 6 \end{array}$$

The set of equations has been reduced to an equation of one variable. Substituting the value of x in either equation and solving for y,

$$y = 3$$

The same results can be obtained by subtracting one equation from the other.

$$\begin{array}{r} x + y = 9 \\ \underline{x - y = 3} \\ 2y = 6 \\ y = 3 \end{array}$$

Substituting,

$$x = 6$$

In this example, the coefficient of x and the coefficient of y are the same, but this will not always be the case. Consider the equations $3x + 2y = 4$ and $2x - 3y = 7$. Adding or subtracting the two equations will not eliminate one of the unknowns as it did in the first example. However, one or both of the two equations can be converted into an equivalent equation that will make it possible to do so.

$$3x + 2y = 4$$
$$2x - 3y = 7$$

Multiplying the first equation by 3 and the second equation by 2 and adding will reduce the system of equations to a single variable equation.

$$\begin{array}{r} 9x + 6y = 12 \\ \underline{4x - 6y = 14} \\ 13x \quad = 26 \\ x = 2 \\ y = -1 \quad \text{[obtained by substitution]} \end{array}$$

If the graphs of the two equations are plotted, the two lines will intersect at $(2, -1)$.

Figure 10.6 *Circle Centered at (0, 0)*

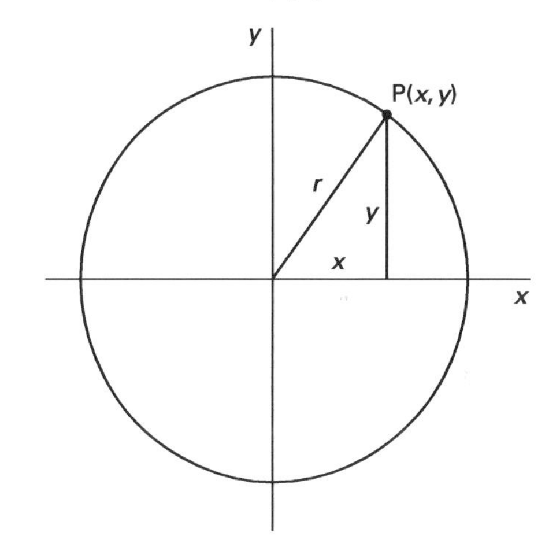

17. EQUATION OF A CIRCLE

A *circle* is a curve, all points on which are equidistant from a point called the *center.* The distance of all points from the center is known as the *radius.*

If the center of the circle is at the origin as in Fig. 10.6, the equation of the circle is

$$x^2 + y^2 = r^2 \qquad 10.18$$

If P is any point on the circle, its coordinates must satisfy the equation $x^2 + y^2 = r^2$.

If the center of the circle is at point $Q(h, k)$, as in Fig. 10.7 the equation becomes

$$(x - h)^2 + (y - k)^2 = r^2 \qquad 10.19$$

Figure 10.7 *Circle Centered at (h, k)*

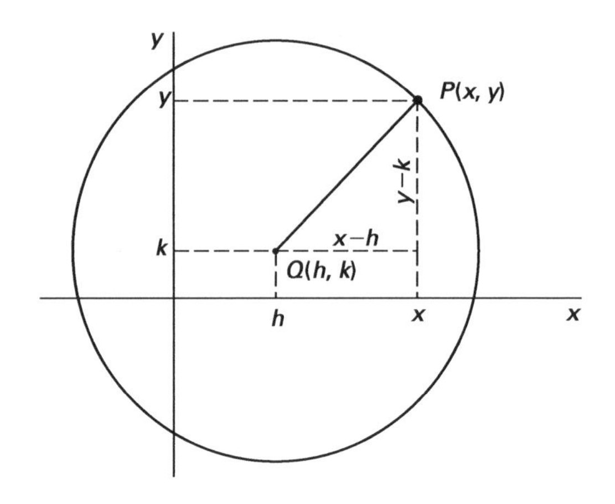

The *general form* for this equation is

$$x^2 + y^2 + Dx + Ey + F = 0 \qquad 10.20$$

Example 10.12

Find the equation of the circle with center $(2, -1)$ and radius 3.

Solution

$$(x-2)^2 + (y+1)^2 = (3)^2$$
$$x^2 + y^2 - 4x + 2y - 4 = 0$$

Example 10.13

Write the equation $x^2 + 10x + y^2 - 6y + 18 = 0$ in the form $(x-h)^2 + (y-k)^2 = r^2$.

Solution

Complete the square.

$$x^2 + 10x + y^2 - 6y = -18$$
$$x^2 + 10x + 25 + y^2 - 6y + 9 = -18 + 25 + 9$$
$$(x+5)^2 + (y-3)^2 = 16$$

The center of the circle is at $(-5, 3)$, and the radius is 4.

18. LINEAR-QUADRATIC SYSTEMS

The intersections of a circle and a straight line can be found by solving the system of the linear equation and the quadratic equation. This is illustrated in Ex. 10.14.

Example 10.14

Find the intersections of the graphs of the following system.

$$x^2 + y^2 = 100$$
$$2x - y = 8$$

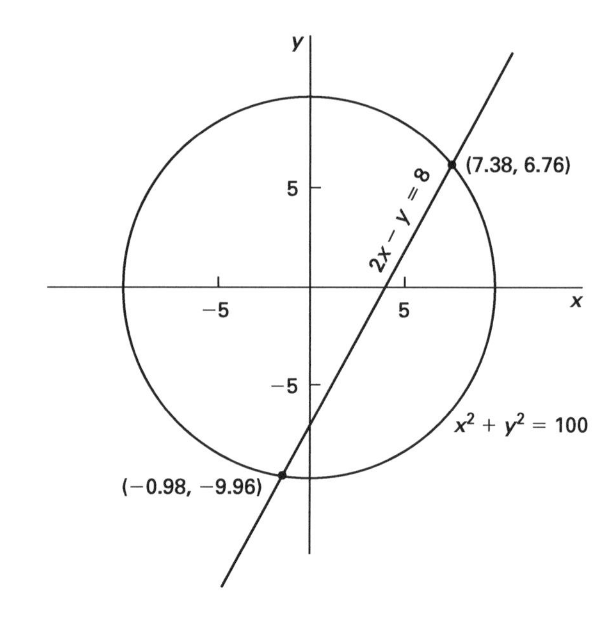

Solution

Transform the linear equation by isolating one of the variables.

$$y = 2x - 8$$

Substitute this value of y into the quadratic equation.

$$x^2 + (2x-8)^2 = 100$$
$$x^2 + 4x^2 - 32x + 64 = 100$$
$$5x^2 - 32x - 36 = 0$$

Use the quadratic formulas to solve for x.

$$x = \frac{-B \pm \sqrt{B^2 - 4AC}}{2A}$$
$$x = \frac{-(-32) + \sqrt{-(32)^2 - (4)(5)(-36)}}{(2)(5)}$$
$$= \frac{-(-32) - \sqrt{-(32)^2 - (4)(5)(-36)}}{(2)(5)}$$
$$= 7.38 \text{ and } -0.98$$

Substituting the values of x in the equation $y = 2x - 8$ gives

$$y = (2)(7.38) - 8 = 6.76$$

and

$$y = (2)(-0.98) - 8 = -9.96$$

The intersections are $(7.38, 6.76)$ and $(-0.98, -9.96)$.

19. INCLINATION OF A LINE

The *inclination of a line* not parallel to the x-axis is the angle measured counterclockwise from the positive direction of the x-axis. (The inclination of a line parallel to the x-axis is zero.) The symbol α denotes inclination.

Figure 10.8 Inclination of a Line

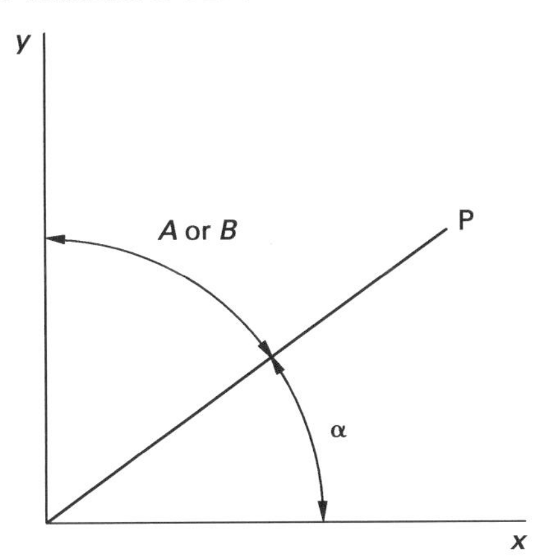

In Fig. 10.8, the inclination of line OP is α. Considering the trigonometric ratios,

$$m = \frac{y}{x} = \tan\alpha \qquad 10.21$$

The *azimuth from north A* (*bearing angle B*) of a line calculated from the slope is the complement of inclination, and it can be calculated from the slope.

$$m = \cot A \qquad 10.22(a)$$

$$m = \cot B \qquad 10.22(b)$$

20. THE ACUTE ANGLE BETWEEN TWO LINES

If the equations of two intersecting lines are known, the acute angle between them can be found by using the *law of tangents.*

$$\tan\theta = \tan(\alpha_2 - \alpha_1) \qquad 10.23(a)$$

$$= \frac{\tan\alpha_2 - \tan\alpha_1}{1 + \tan\alpha_1 \tan\alpha_2} \qquad 10.23(b)$$

$$= \left|\frac{m_2 - m_1}{1 + m_1 m_2}\right| \quad [\text{for } m_1 m_2 \neq -1] \qquad 10.23(c)$$

Figure 10.9 Angle Between Two Lines

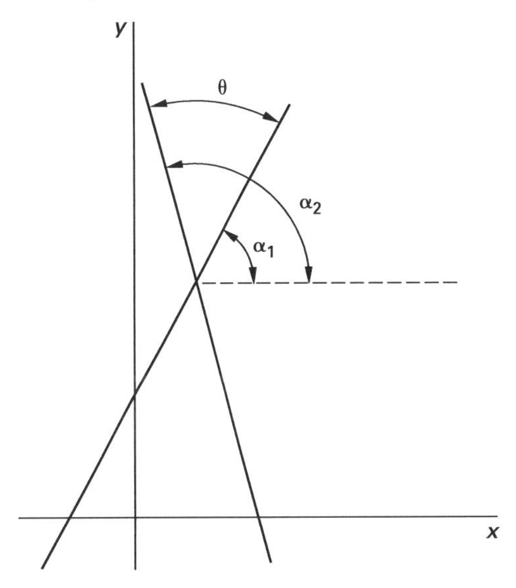

Example 10.15

Find the acute angle θ between the two lines.

$$2x + 3y - 12 = 0$$

$$3x - 4y - 12 = 0$$

Solution

$$m_1 = -\frac{A}{B} = -\frac{2}{3}$$

$$m_2 = -\frac{A}{B} = +\frac{3}{4}$$

$$\tan\theta = \frac{\frac{3}{4} + \frac{2}{3}}{1 - \left(\frac{2}{3}\right)\left(\frac{3}{4}\right)} = 2.833$$

$$\theta = 7°30'$$

21. TRANSLATION OF AXES

Solving simultaneous equations in which the coefficients of x and y are large numbers can be simplified by reducing the value of the coefficients. This can be done without changing the values of the equations by translating the axes.

In Fig. 10.10, let P be any point with coordinates (x, y) with respect to the axes OX and OY. Establish the new axes, O′X′ and O′Y′, respectively, parallel to the old axes, so that the new origin O′ has the coordinates (h, k) with respect to the old axes. The coordinates of the point P will then be $(x'y')$ with respect to the new axes.

$$x = x' + h \qquad 10.24$$

$$x' = x - h \quad (10.25)$$

$$y = y' + k \quad (10.26)$$

$$y' = y - k \quad (10.27)$$

Figure 10.10 *Transformation of Axes*

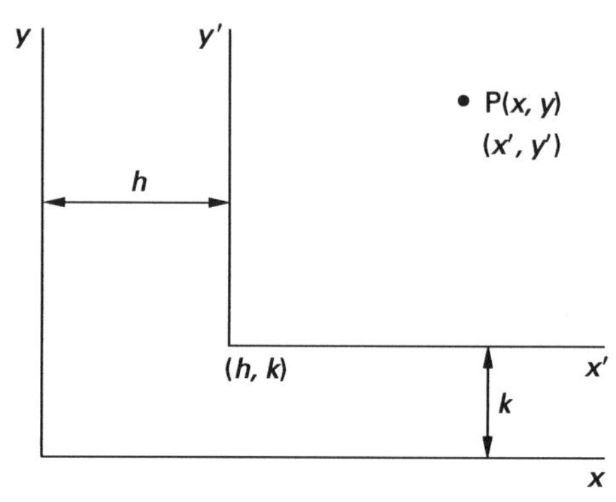

Example 10.16

Point P has the coordinates $(5,3)$. Find the coordinates of P from the origin $O'(3,1)$.

Solution

$$x' = x - h = 5 - 3 = 2$$

$$y' = y - k = 3 - 1 = 2$$

The coordinates of P are $(2,2)$.

PRACTICE PROBLEMS

1. Find two sets of roots for each equation.

Example: $2x + y = 7$ $\quad (2,3), (3,1)$

(a) $2x - 3y = 5$

(b) $x - 2y = 4$

(c) $3x + 2y = 6$

2. Find the coordinates of the points at which the graph of each equation intersects the two axes.

Example: $3x - 2y = 12$ $\quad (0,-6), (4,0)$

(a) $x - y = 0$

(b) $2x + 3y = 18$

(c) $x + y = 4$

3. Determine the slope of the line through each pair of points.

Example: $(-2,4); (4,-3)$ $\quad m = \frac{-3-4}{4+2} = -7/6$

(a) $(3,2); (6,8)$

(b) $(1,3); (4,5)$

(c) $(0,-2); (-3,5)$

(d) $(6,-4); (2,-3)$

(e) $(-3,4); (3,4)$

(f) $(1,-5); (-1,3)$

4. Rearrange the equation in the form $Ax+By+C=0$ with A positive, and determine the slope m of each.

Example: $3y - 2x - 4 = 0$ $\quad 2x - 3y + 4 = 0$

$$m = -\frac{A}{B} = -\frac{+2}{-3} = 2/3$$

(a) $3x + 4y = 6$

(b) $y = -2x + 5$

(c) $-4x + 2y + 8 = 0$

(d) $y = -5x$

5. Find the slope of each line. Express as a common fraction showing the algebraic sign.

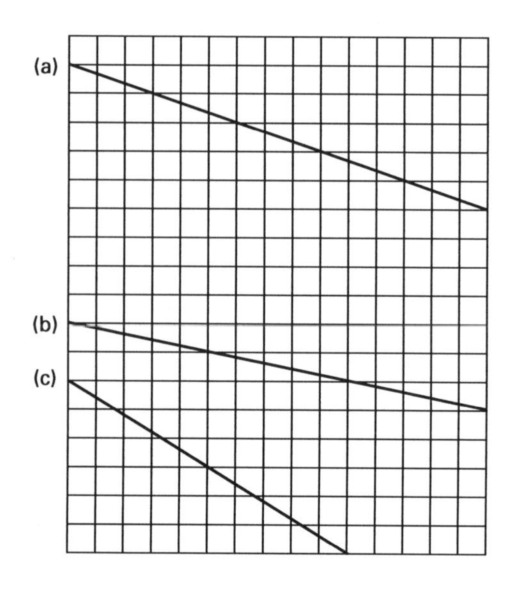

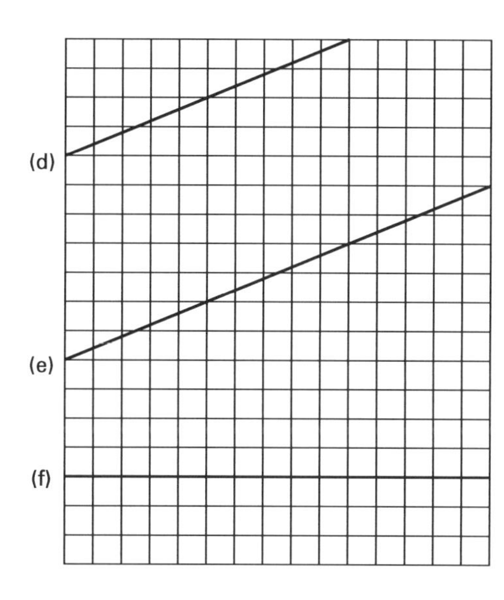

6. Graph each equation, plotting at least three points, and write the equation along the line.

(a) $3x - 2y = 0$

(b) $2x - 3y + 30 = 0$

(c) $3x + 2y + 6 = 0$

(d) $x + y = 0$

(e) $y + 11 = 0$

7. Write each equation in the form $y = mx + b$.

Example: $3x + 2y + 6 = 0 \qquad y = -\frac{3}{2}x - 3$

(a) $4x + 5y + 10 = 0$

(b) $2x - y - 10 = 0$

(c) $3x + 4y + 8 = 0$

(d) $x - y = 0$

(e) $2x - 3y - 12 = 0$

8. Write each equation in the form $Ax + By + C = 0$.

Example: $y = -\frac{2}{3}x - 2 \qquad 2x + 3y + 6 = 0$

(a) $y = \frac{2}{5}x - 2$

(b) $y = 2x - 10$

(c) $y = \frac{3}{2}x + \dfrac{5}{2}$

(d) $y = -x + 5$

(e) $y = -\frac{2}{3}x + 4$

9. Write each equation in the form $y = mx + b$ and plot the graph. Write the equation in the form $Ax + By + C = 0$ along each graph.

Example: $2x - 3y + 30 = 0 \qquad y = \frac{2}{3}x + 10$

(a) $x + 5y - 60 = 0$

(b) $x - y = 0$

(c) $3x + 4y + 24 = 0$

(d) $x + y + 4 = 0$

(e) $2x - 5y - 50 = 0$

10. Write each equation in the form $Ax + By + C = 0$ and indicate the slope, the x-intercept, and the y-intercept of each. (Hint: $m = -A/B$, x-intercept $= -C/A$, and y-intercept $= -C/B$.)

Example: $4y - 3x = 10 \qquad 3x - 4y + 10 = 0$

$m = 3/4 \qquad x_{y=0} = -10/3 \qquad y_{x=0} = 5/2$

(a) $x - y + 5 = 0$

(b) $-2x + 3y = 12$

(c) $x - y = 0$

(d) $6y = 4x + 2$

(e) $2x - 4y = 2$

11. Indicate which lines are parallel and which lines are perpendicular.

Example:

$$3x + 4y = 8$$
$$m = -\frac{3}{4}$$

(a) $2x - 3y - 3 = 0$

(b) $10x - 6y = 18$

(c) $y = -\frac{3}{2}x + 5$

(d) $5x - 3y = 9$

(e) $6x + 8y + 12$

(f) $4x - 3y = 7$

(g) $3x + 5y = -10$

(h) $y = -\frac{3}{4}x - 3$

12. Find the perpendicular distance from point P to the line indicated.

Example:

$$\text{P}(-8, -1);\ 2x - 3y - 6 = 0$$
$$D = \frac{|(2)(-8) + (-3)(-1) + (-6)|}{\sqrt{(2)^2 + (-3)^2}}$$
$$= \frac{|-16 + 3 - 6|}{\sqrt{13}} = 5.3$$

(a) $P(3,3)$; $3x - 2y + 4 = 0$
(b) $P(-10,8)$; $x - y + 5 = 0$
(c) $P(9,6)$; $5x - 2y + 10 = 0$
(d) $P(-8,-6)$; $3x + 2y = 0$
(e) $P(12,-6)$; $3x - 2y - 6 = 0$

13. Write the equation of the line through the given point with the given slope.

Example:

$(1,4)$; $m = -\frac{1}{2}$

$y - 4 = \frac{(-1)(x-1)}{2} \quad 2y - 8 = -x + 1 \quad x + 2y - 9 = 0$

(a) $(3,1)$; $m = -2$
(b) $(-4,3)$; $m = \frac{2}{3}$
(c) $(-2,5)$; $m = 0$
(d) $(2,-3)$; $m = -\frac{2}{3}$

14. Write the equation of the line through the two given points in the form $Ax + By + C = 0$.

Example:

$$(-4,3);\ (0,-2)$$
$$\frac{y-3}{x+4} = \frac{-2-3}{0+4}$$
$$\frac{y-3}{x+4} = \frac{-5}{4}$$
$$(-5)(x+4) = (4)(y-3)$$
$$-5x - 20 = 4y - 12$$
$$5x + 4y + 8 = 0$$

(a) $(-3,2)$; $(1,4)$
(b) $(2,-3)$; $(5,-2)$
(c) $(3,4)$; $(-3,4)$
(d) $(-2,-6)$; $(3,-4)$

15. Write the equation of the lines with the given x- and y-intercepts in the form $Ax + By + C = 0$.

Example:

$$a = -3;\ b = 4$$
$$\frac{x}{-3} + \frac{y}{4} = 1$$
$$4x - 3y = -12$$
$$4x - 3y + 12 = 0$$

(a) $a = -4$; $b = 3$
(b) $a = 1$; $b = -4$

16. Write the equation of the line that has a slope of $3/4$ and a y-intercept of -3.

17. Write the equation of the line whose y-intercept is 4 and that is perpendicular to the line $4x + 3y + 9 = 0$.

18. Write the equation of the line through the point $(0,8)$ and parallel to the line whose equation is $y = -3x + 4$.

19. Write the equations of two lines through the point $(5,5)$, one parallel and one perpendicular to the line $2x + y - 4 = 0$.

20. Write the equation of the line through the point $(-2,1)$ and parallel to the line through the points $(1,4)$ and $(2,-3)$.

21. Solve the following systems of simultaneous equations by addition or subtraction.

(a) $x + y = 8$; $x - y = 4$
(b) $x + 2y = 6$; $x + 2y = 4$
(c) $2x + 5y = -8$; $2x + 3y = 5$
(d) $5x - 4y = -15$; $2x - 12y = 7$
(e) $7x - 2y = 3$; $2x + 3y = 9$
(f) $7x - 2y = -11$; $8x + 3y = -39$
(g) $5x - 7y = 3$; $-3x + 6y = 4$
(h) $5x - 4y = -17$; $2x - 12y = 14$
(i) $9x + 10y = 9$; $6x - 25y = -13$

22. Graph the following systems of equations and find the intersection of the two lines in each system if they intersect.

(a) $4x + 3y = 24$; $4x - 3y = -48$
(b) $3x + 2y = 24$; $x - 2y = -11$
(c) $5x + 11y = -55$; $5x + 11y = -11$
(d) $2x + 4y = -48$; $x + 2y = -24$

23. Find the coordinates of the point of intersection of the diagonal lines CA and EB.

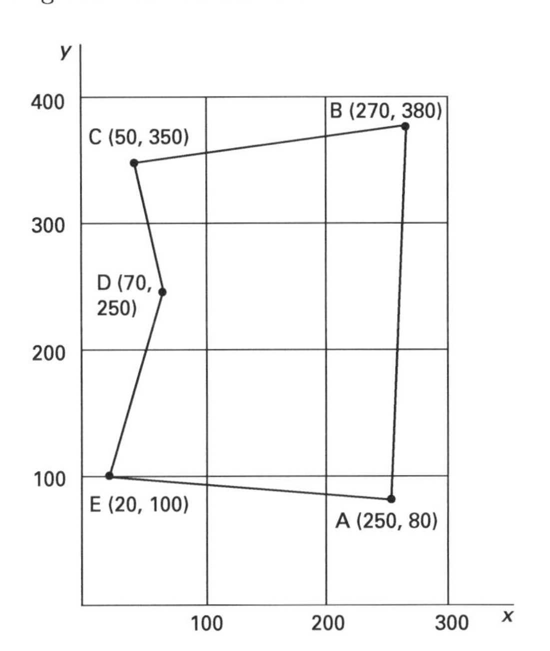

24. The figure ABCDEA represents a tract of land that is to be subdivided by a line from D parallel to EA. Find the coordinates of the point of intersecion of DH and AB. (Designate the point of intersection as H.)

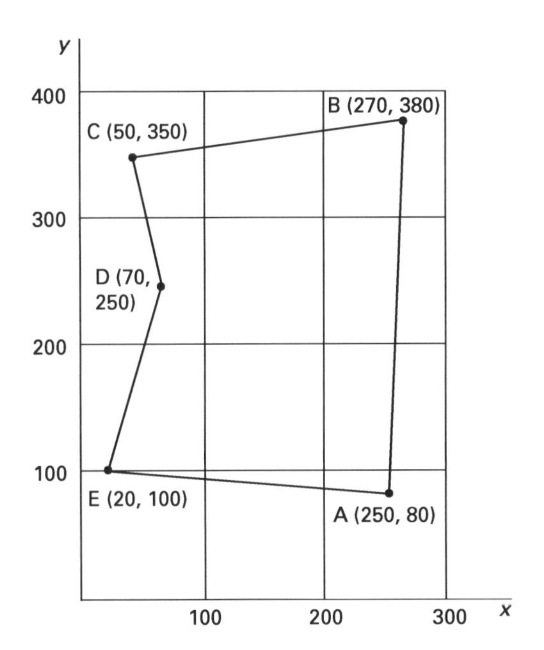

25. Find the acute angle θ between the two lines.

(a) $4x + 3y = 24$
$4x - 3y = -48$

(b) $3x + 2y = 24$
$x - 2y = -11$

(c) $2x - y = 5$
$4x + y = 2$

(d) $2x + y = -2$
$6x - 5y = 18$

(e) $x - y = 5$
$x + 2y = 2$

26. Find the new coordinates if the axes are translated to a new origin located at $(4, 3)$.

(a) $(4, 6)$

(b) $(-7, 3)$

(c) $(-3, -2)$

(d) $(0, 0)$

(e) $(8, -2)$

(f) (7, 7)

SOLUTIONS

1. (a) $(7, 3)$, $(4, 1)$ (b) $(6, 1)$, $(8, 2)$
(c) $(4, -3)$, $(6, -6)$

2. (a) $(0, 0)$ (b) $(9, 0)$, $(0, 6)$
(c) $(0, 4)$, $(4, 0)$

3. (a) $m = \frac{8-2}{6-3} = \boxed{2}$

(b) $m = \frac{5-3}{4-1} = \boxed{2/3}$

(c) $m = \frac{5+2}{-3} = \boxed{-7/3}$

(d) $m = \frac{-3+4}{2-6} = \boxed{-1/4}$

(e) $m = \frac{4-4}{3+3} = \boxed{0}$

(f) $m = \frac{3+5}{-1-1} = \boxed{-4}$

4. (a) $3x + 4y = 6$

$\boxed{3x + 4y - 6 = 0}$

$m = -\frac{+3}{+4} = \boxed{-3/4}$

(b) $y = -2x + 5$

$\boxed{2x + y - 5 = 0}$

$m = -\frac{+2}{+1} = \boxed{-2}$

(c) $-4x + 2y + 8 = 0$

$\boxed{4x - 2y - 8 = 0}$

$m = -\frac{+4}{-2} = \boxed{2}$

(d) $y = -5x$

$\boxed{5x + y = 0}$

$m = -\frac{+5}{+1} = \boxed{-5}$

5. (a) $-1/3$ (b) $-1/5$ (c) $-3/5$
(d) $2/5$ (e) $2/5$ (f) 0

6.

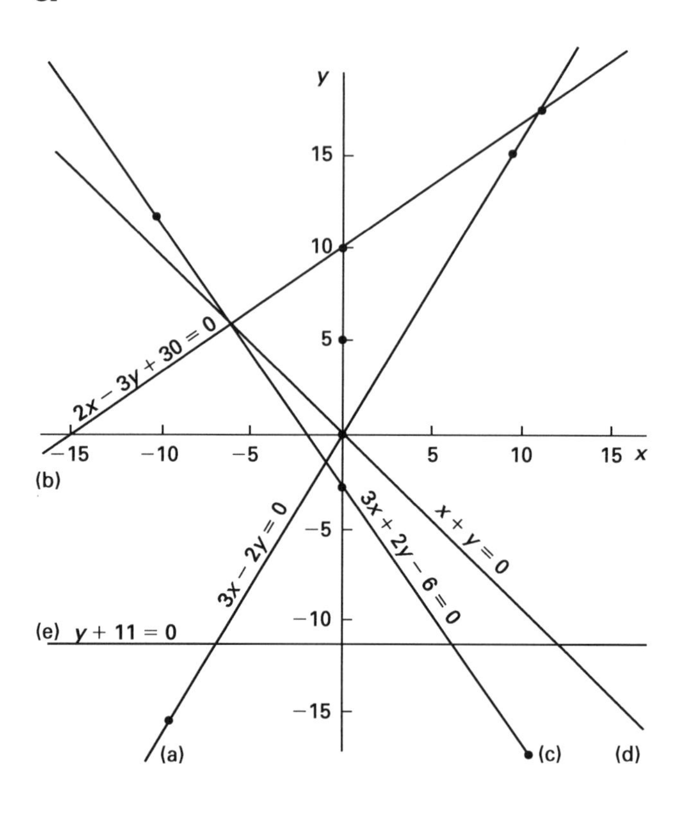

7. (a) $4x + 5y + 10 = 0$

$5y = -4x - 10$

$\boxed{y = -\frac{4}{5}x - 2}$

(b) $2x - y - 10 = 0$

$-y = -2x + 10$

$\boxed{y = 2x - 10}$

(c) $3x + 4y + 8 = 0$

$4y = -3x - 8$

$\boxed{y = -\frac{3}{4}x - 2}$

(d) $x - y = 0$

$-y = -x$

$\boxed{y = x}$

(e) $2x - 3y - 12 = 0$

$-3y = -2x + 12$

$\boxed{y = \frac{2}{3}x - 4}$

8. (a) $y = \frac{2}{5}x - 2$

$5y = 2x - 10$

$\boxed{2x - 5y - 10 = 0}$

(b) $y = 2x - 10$

$-2x + y + 10 = 0$

$\boxed{2x - y - 10 = 0}$

(c) $y = -\frac{3}{2}x + \dfrac{5}{2}$

$2y = -3x + 5$

$\boxed{3x + 2y - 5 = 0}$

(d) $y = -x + 5$

$\boxed{x + y - 5 = 0}$

(e) $y = -\frac{2}{3}x + 4$

$3y = -2x + 12$

$\boxed{2x + 3y - 12 = 0}$

9. (a) $x + 5y - 60 = 0$

$\boxed{y = -\frac{1}{5}x + 12}$

(b) $x - y = 0$

$\boxed{y = x}$

(c) $3x + 4y + 24 = 0$

$\boxed{y = -\frac{3}{4}x - 6}$

(d) $x + y + 4 = 0$

$\boxed{y = -x - 4}$

(e) $2x - 5y - 50 = 0$

$\boxed{y = \frac{2}{5}x - 10}$

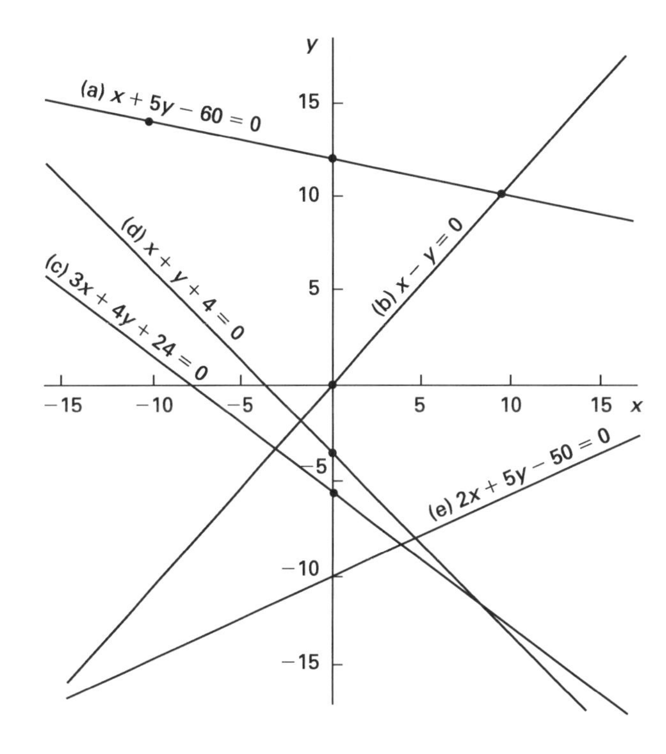

10.

		$Ax + By + C = 0$	m	x-int	y-int
(a)	$x - y + 5 = 0$	$x - y + 5 = 0$	1	-5	5
(b)	$-2x + 3y = 12$	$2x - 3y + 12 = 0$	$\frac{2}{3}$	-6	4
(c)	$x - y = 0$	$x - y = 0$	1	0	0
(d)	$6y = 4x + 2$	$4x - 6y + 2 = 0$	$\frac{2}{3}$	$-\frac{1}{2}$	$\frac{1}{3}$
(e)	$2x - 4y = 2$	$2x - 4y - 2 = 0$	$\frac{1}{2}$	1	$-\frac{1}{2}$

11.

		slope	parallel to problem no.	perpendicular to problem no.
(a)	$2x - 3y - 3 = 0$	$\frac{2}{3}$	–	c
(b)	$10x - 6y = 18$	$\frac{5}{3}$	d	g
(c)	$y = -\frac{3}{2}x + 5$	$-\frac{3}{2}$	–	a
(d)	$5x - 3y = 9$	$\frac{5}{3}$	b	g

		slope	parallel to problem no.	perpendicular to problem no.
(e)	$6x + 8y = 12$	$-\frac{3}{4}$	h	f
(f)	$4x - 3y = 7$	$\frac{4}{3}$	–	e & h
(g)	$3x + 5y = -10$	$-\frac{3}{5}$	–	b & d
(h)	$y = -\frac{3}{4}x - 3$	$-\frac{3}{4}$	e	f

12. (a) $D = \dfrac{(3)(3) - (2)(3) + 4}{\sqrt{(3)^2 + (2)^2}} = \boxed{1.9}$

(b) $D = \dfrac{(1)(-10) + (-1)(8) + 5}{\sqrt{(1)^2 + (1)^2}} = \boxed{9.2}$

(c) $D = \dfrac{(5)(9) - (2)(6) + 10}{\sqrt{(5)^2 + (-2)^2}} = \boxed{8.0}$

(d) $D = \dfrac{(3)(-8) + (2)(-6)}{\sqrt{(3)^2 + (2)^2}} = \boxed{10.0}$

(e) $D = \dfrac{(3)(12) + (-2)(-6) - 6}{\sqrt{(3)^2 + (-2)^2}} = \boxed{11.6}$

13. (a) $y - 1 = (-2)(x - 3)$

$y - 1 = -2x + 6$

$\boxed{2x + y - 7 = 0}$

(b) $y - 3 = \dfrac{(2)(x + 4)}{3}$

$3y - 9 = 2x + 8$

$\boxed{2x - 3y + 17 = 0}$

(c) $y - 5 = 0$

$\boxed{y = 5}$

(d) $y + 3 = \dfrac{(-2)(x - 2)}{3}$

$3y + 9 = -2x + 4$

$2x + 3y + 5 = 0$

$\boxed{2x + 3y + 5 = 0}$

14. (a) $\dfrac{y - 2}{x + 3} = \dfrac{4 - 2}{1 + 3} = \dfrac{1}{2}$

$x + 3 = (2)(y - 2)$

$x + 3 = 2y - 4$

$\boxed{x - 2y + 7 = 0}$

(b) $\dfrac{y + 3}{x - 2} = \dfrac{-2 + 3}{5 - 2} = \dfrac{1}{3}$

$x - 2 = (3)(y + 3)$

$x - 2 = 3y + 9$

$\boxed{x - 3y - 11 = 0}$

(c) $\dfrac{y - 4}{x - 3} = \dfrac{4 - 4}{-3 - 3} = 0$

$\boxed{y = 4}$

(d) $\dfrac{y + 6}{x + 2} = \dfrac{-4 + 6}{3 + 2} = \dfrac{2}{5}$

$(2)(x + 2) = (5)(y + 6)$

$2x + 4 = 5y + 30$

$\boxed{2x - 5y - 26 = 0}$

15. (a) $\dfrac{x}{-4} + \dfrac{y}{3} = 1$

$3x - 4y = -12$

$\boxed{3x - 4y + 12 = 0}$

(b) $\dfrac{x}{1} + \dfrac{y}{-4} = 1$

$-4x + y = -4$

$\boxed{4x - y - 4 = 0}$

16. $y = \dfrac{3x}{4} - 3$

$4y = 3x - 12$

$\boxed{3x - 4y - 12 = 0}$

17. For the perpendicular line,

$m = \dfrac{3}{4}$

$y = \dfrac{3x}{4} + 4$

$4y = 3x + 16$

$\boxed{3x - 4y + 16 = 0}$

18. For the parallel line,

$m = -3$

$y - 8 = (-3)(x - 0)$

$y - 8 = -3x$

$\boxed{3x + y - 8 = 0}$

19. $m = -2$

$$y - 5 = (-2)(x - 5)$$
$$y - 5 = -2x + 10$$
$$\boxed{2x + y - 15 = 0}$$
$$y - 5 = \frac{(1)(x-5)}{2}$$
$$2y - 10 = x - 5$$
$$\boxed{x - 2y + 5 = 0}$$

20. $m = \frac{4+3}{1-2} = -7$

$$y - 1 = (-7)(x + 2)$$
$$y - 1 = -7x - 14$$
$$\boxed{7x + y + 13 = 0}$$

21. (a)

$$x + y = 8$$
$$x - y = 4$$
$$2x = 12$$
$$\boxed{x = 6}$$
$$\boxed{y = 2}$$

(b)

$$x + 2y = 6$$
$$x + 2y = 4$$

$\boxed{\text{parallel}}$

(c)

$$2x + 5y = -8$$
$$2x + 3y = 5$$
$$2y = -13$$
$$\boxed{y = -6\tfrac{1}{2}}$$
$$\boxed{x = 12\tfrac{1}{4}}$$

(d)

$$5x - 4y = -15$$
$$2x - 12y = 7$$
$$10x - 8y = -30$$
$$10x - 60y = 35$$
$$52y = -65$$
$$\boxed{y = -1.25}$$
$$\boxed{x = -4}$$

(e)

$$7x - 2y = 3$$
$$2x + 3y = 9$$
$$21x - 6y = 9$$
$$4x + 6y = 18$$
$$25x - 27$$
$$\boxed{x = 1.08}$$
$$\boxed{y = 2.28}$$

(f)

$$7x - 2y = -11$$
$$8x + 3y = -39$$
$$21x - 6y = -33$$
$$16x + 6y = -78$$
$$37x = -111$$
$$\boxed{x = -3}$$
$$\boxed{y = -5}$$

(g)

$$5x - 7y = 3$$
$$-3x + 6y = 4$$
$$15x - 21y = 9$$
$$-15x + 30y = 20$$
$$9y = 29$$
$$\boxed{y = 29/9 = 3.22}$$
$$\boxed{x = 46/9 = 5.11}$$

(h)

$$5x - 4y = -17$$
$$2x - 12y = 14$$
$$-15x + 12y = 51$$
$$2x - 12y = 14$$
$$-13x = 65$$
$$\boxed{x = -65/13 = -5}$$
$$\boxed{y = 2}$$

(i)

$$9x + 10y = 9$$
$$6x - 25y = -13$$
$$54x + 60y = 54$$
$$54x - 225y = -117$$
$$285y = 171$$
$$\boxed{y = 0.6}$$
$$\boxed{x = 0.333}$$

22. (a) $4x + 3y = 24$

$4x + 3y = -48$

(b) $3x + 2y = 24$

$x - 2y = -11$

(c) $5x + 11y = -55$

$5x + 11y = -11$

(d) $2x + 4y = 48$

$x + 2y = -24$

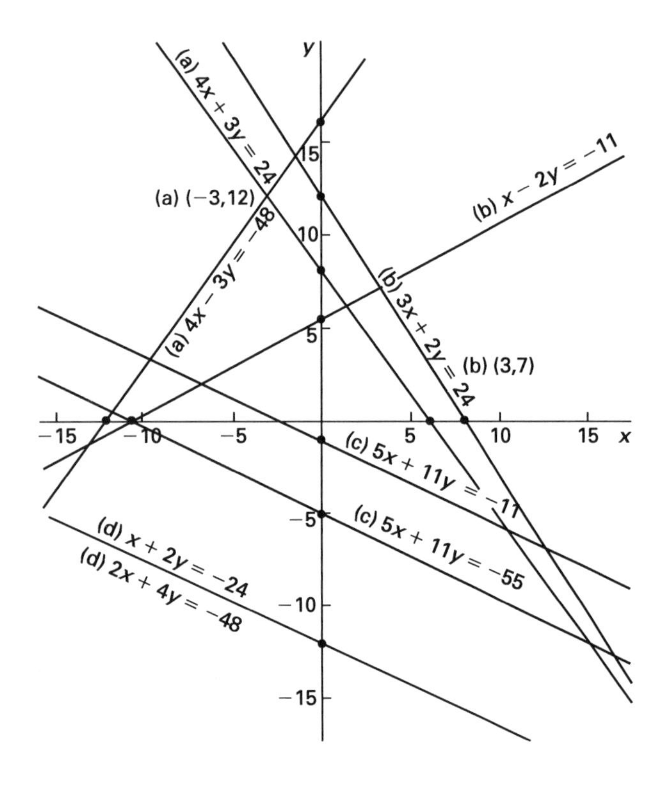

23. The equation of line CA can be developed as follows.

$$\frac{y - 350}{x - 50} = \frac{80 - 350}{250 - 50} = \frac{-270}{200} = \frac{-27}{20}$$

$$(-27)(x - 50) = (20)(y - 350)$$

$$-27x + 1350 = 20y - 7000$$

$$-27x - 20y = -8350$$

$$27x + 20y = 8350$$

The equation of line EB can be developed as follows.

$$\frac{y - 100}{x - 20} = \frac{380 - 100}{270 - 20} = \frac{28}{25}$$

$$(28)(x - 20) = (25)(y - 100)$$

$$28x - 560 = 25y - 2500$$

$$28x - 25y = -1940$$

The solution of the system equations is

$$\begin{aligned} 27x + 20y &= 8350 \\ 28x - 25y &= -1940 \\ \hline 675x + 500y &= 208{,}750 \\ 560x - 500y &= -38{,}800 \\ \hline 1235x &= 169{,}950 \\ x &= 137.61 \\ 27x &= 3715.47 \\ 3715.47 + 20y &= 8350 \\ 20y &= 4634.53 \\ y &= 231.73 \end{aligned}$$

The coordinates are (138, 232).

24. The slope of EA is

$$m = \frac{80 - 100}{250 - 20} = \frac{-20}{230} = \frac{-2}{23}$$

The equation of line DH can be developed as follows.

$$y - 250 = \frac{(-2)(x - 70)}{23}$$

$$(-2)(x - 70) = (23)(y - 250)$$

$$-2x + 140 = 23y - 5750$$

$$-2x - 23y = -5890$$

$$2x + 23y = 5890$$

The equation of line AB can be developed as follows.

$$\frac{y - 80}{x - 250} = \frac{380 - 80}{270 - 250} = 15$$

$$15x - 3750 = y - 80$$

$$15x - y = 3670$$

The solution of the system of equations is

$$\begin{aligned} 2x + 23y &= 5890 \\ 15x - y &= 3670 \\ \hline 30x + 345y &= 88{,}350 \\ 30x - 2y &= 7340 \\ \hline 347y &= 81{,}010 \\ y &= 233.46 \\ 15x - 233.46 &= 3670 \\ x &= 260.21 \end{aligned}$$

The coordinates are (260, 233).

25. (a)

$$m_1 = \frac{-4}{3}$$

$$m_2 = \frac{4}{3}$$

$$\tan\theta = \frac{\frac{4}{3} + \frac{4}{3}}{1 + \left(\frac{-4}{3}\right)\left(\frac{4}{3}\right)} = \frac{\frac{8}{3}}{\frac{9}{9} - \frac{16}{9}}$$

$$= \frac{\frac{8}{3}}{\frac{-7}{9}} = \frac{24}{7}$$

$$\theta = \boxed{73°44'}$$

(b)

$$m_1 = \frac{-3}{2}$$

$$m_2 = \frac{1}{2}$$

$$\tan\theta = \frac{\frac{1}{2} + \frac{3}{2}}{1 + \left(\frac{-3}{4}\right)} = \frac{\frac{4}{2}}{\frac{1}{4}} = 8$$

$$\theta = \boxed{82°52'}$$

(c)

$$m_1 = 2$$

$$m_2 = -4$$

$$\tan\theta = \frac{-4 - 2}{1 + (-8)} = \frac{-6}{-7}$$

$$\theta = \boxed{40°36'}$$

(d)

$$m_1 = -2$$

$$m_2 = \frac{6}{5}$$

$$\tan\theta = \frac{\frac{6}{5} + \frac{10}{5}}{1 + \left(\frac{-12}{5}\right)} = \frac{\frac{16}{5}}{\frac{-7}{5}} = \frac{16}{-7}$$

$$\theta = \boxed{66°22'}$$

(e)

$$m_1 = 1$$

$$m_2 = -\frac{1}{2}$$

$$\tan\theta = \frac{\frac{-1}{2} - \frac{2}{2}}{1 + \left(-\frac{1}{2}\right)} = \frac{\frac{-3}{2}}{\frac{1}{2}} = -3$$

$$\theta = \boxed{71°34'}$$

26. (a) $\boxed{(0, 3)}$ (b) $\boxed{(-11, 0)}$ (c) $\boxed{(-7, -5)}$

(d) $\boxed{(-4, -3)}$ (e) $\boxed{(4, -5)}$ (f) $\boxed{(3, 4)}$

Topic II: Field Data Acquisition

11 Taping

1. LINEAR MEASUREMENT

The distance between two points can be determined by pacing, taping, electronic distance measurement (EDM), tacheometry (stadia), using an odometer, or scaling on a map. Of these methods, only taping will be discussed in this chapter.

Taping consists of aligning the tape, pulling the tape tight, using the plumb bob on unlevel ground, marking tape lengths, and reading the tape.

2. GUNTER'S CHAIN

The *Gunter's chain* was once used extensively in surveying the public lands of the United States, but that is no longer the case. However, the term *chaining* is still used to mean taping, and surveyors skilled at taping are still called *chainpersons*.

The Gunter's chain is 66 ft long and consists of 100 links, each link being 7.92 in in length. One chain = 1/80 mi, and 10 square chains = $(10)(66)^2 = 43{,}560\ \text{ft}^2 = 1\ \text{ac}$.

Knowledge of the Gunter's chain is important to surveyors when retracing old surveys in which the Gunter's chain was used.

3. STEEL TAPES

Steel tapes are available in widths of from 3/8 in to 5/16 in with thicknesses varying from 0.016 in to 0.025 in. Lengths of steel tapes can be 50 ft, 100 ft, 200 ft, 300 ft, and 500 ft. The 100 ft tape is most common for surveying. Tapes are also made in 30 m, 50 m, and 100 m lengths.

4. INVAR TAPES

Invar tapes are made of a steel alloy containing 35% nickel. The expansion and contraction of the Invar tape because of changes in temperature is only about 3% of the change of a steel tape.

These tapes are used when measurements of extreme accuracy are required, such as measuring base lines or calibrating steel tapes. Invar tapes are, however, too fragile for normal use.

5. MARKING PINS

Marking pins, also known as *chaining pins*, are made of 3/16 in steel and are used to mark tape lengths. They usually are 12 in to 18 in long, are sharpened on one end, and have a ring about 2 1/2 in in diameter at the other end. A set of marking pins contains 11 pins.

6. TYPES OF STEEL TAPES

Some 100 ft tapes measure 100 ft from the outer edges of the end loops. Most tapes, however, are in excess of 100 ft from end loop to end loop and have graduations for every foot from 0 to 100 ft.

An *add tape* has an extra graduated foot beyond the zero mark. The extra foot is usually graduated in tenths of a foot, but it is sometimes graduated in tenths and hundredths of a foot.

A *cut tape*, or a *subtract tape*, does not have the extra graduated foot, but the last foot at each end is graduated in tenths of a foot or in tenths and hundredths of a foot.

Figure 11.1 shows the distance between point A and point B measured by both an add tape and a cut tape. The distance using add tape is 26 ft + 0.18 ft = 26.18 ft. The distance using the cut tape is 27 ft − 0.82 ft = 26.18 ft.

7. HORIZONTAL TAPING

In surveying, the distance between two points is the horizontal distance, regardless of the slope.

8. TAPING WITH TAPE SUPPORTED THROUGHOUT ITS LENGTH

The rear chainperson wraps the leather thong at the end of the tape tightly around his right hand near the knuckles and faces at right angles to the line of measurement. He kneels with his left knee near the pin (or other mark) and braces his right arm against his right leg near the knee with the heel of his right hand firmly against the ground. To bring the end mark of the tape exactly on the pin, he shifts his weight to the left knee, or right foot, as desired, keeping the heel of his right hand firmly braced on the ground. In this position, he is off the line of sight, and his eyes are directly over the end mark of the tape and the pin. (Left-handed people will use the opposite positions.)

The forward chainperson wraps the leather thong around her left hand, faces at right angles to the line of measurement, and kneels on her right knee. She increases or decreases the pull on the tape by shifting her body weight. With her right hand, she sticks the pin at the zero mark on a call from the rear chainperson, which indicates the 100 ft mark is on the pin. In this position, she is also off the line of sight.

9. TAPING ON SLOPE WITH TAPE SUPPORTED AT ENDS ONLY

Taping downhill, the rear chainperson proceeds as in taping on level ground.

The forward chainperson wraps the leather thong around her left hand and takes a position facing at right angles to the line of sight as she did in taping on level ground, but in this procedure she remains standing. With her right hand, she makes one loop of the plumb bob string around the tape. With her forefinger under the tape and her thumb on the top of the tape, she can roll the string to the proper mark with her thumb. The last two fingers of her right hand grasp the loose end of the string. To get the proper length of string, she rests the plumb bob on the ground and feeds the string with her right hand.

She holds the tape as nearly horizontal as possible. Her feet should be placed well apart and her left elbow should be braced against her body. To apply tension, her left knee is bent so that the weight of her body pushes against the arm holding the tape. The plumb bob is steadied by lowering it to the ground. When the plumb bob is just slightly above ground and steady, the tape is horizontal, and when the chainperson feels the proper tension, she lets the plumb bob drop and then marks the point with a pin.

Taping uphill, the rear chainperson holds the plumb bob over the pin or point on the ground, and the forward chainperson proceeds as when taping on level ground. The rear chainperson holds the tape and the plumb bob as the foward chainperson does when taping downhill.

10. STATIONING WITH PINS AND RANGE POLE ON LEVEL GROUND

As stated earlier, a set of marking pins consists of 11 pins. A *station* is 100 ft (one tape) in length. In route surveying, stationing is carried along continuously from a starting point designated as sta 0+00.

The rear chainperson stations himself at the beginning point with one pin in hand or in the ground if the beginning point was not previously marked.

The forward chainperson takes 10 pins. With the zero end of the tape and the range pole, she advances in the direction of the stationing. She counts her paces from the beginning point so that, if she does not hear the rear chainperson call, she knows when she has advanced approximately one station.

Figure 11.1 *Add and Cut Tapes*

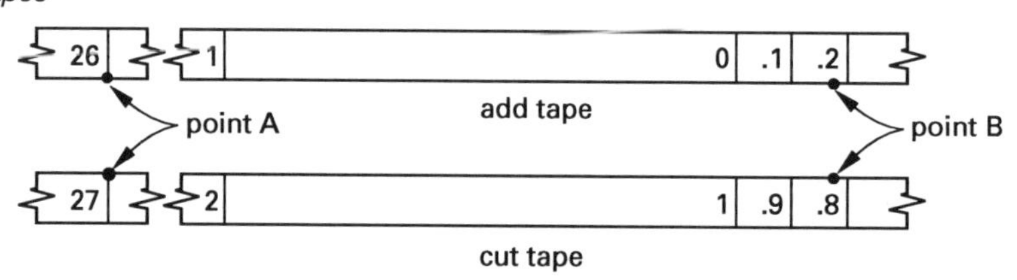

The rear chainperson watches the tape pass his beginning station. When the end is about 6 ft from his position, he calls, "chain," to the foward chainperson. He grabs the leather thong on the end of the tape as it nears him and proceeds as explained in Sec. 8.

On hearing "chain," the foward chainperson immediately turns and faces the rear. She observes the rear chainperson grab the leather thong. With the tape in her left hand and the range pole in her right, she puts tension on the tape, flips it to straighten it, and, holding the range pole vertically, places it near and slightly to the rear of the zero mark. She immediately drops the tape and, with legs spread fairly wide apart, takes the range between the forefinger and thumb of each hand and observes the transitperson (or rear chainperson) for alignment. She keeps the range pole vertical and her legs apart so that, on long sights, the transitperson will have a clear view of the range pole between her legs. When she recieves an OK from the transitperson, she presses the point of the range pole in the ground, removes it, and places a chaining pin in the hole left by the range pole. She then wraps the thong around her left hand, flips the tape for alignment, and pulls the edge of the tape over to the pin.

On observing the forward chainperson reach this point, the rear chainperson checks the 100 ft mark to see that it is on the pin and calls out his station number, such as "eight." Besides keeping up with the station number, this call "eight" is also saying "I am on my mark" to the forward chainperson.

On hearing the rear chainperson call "eight," the forward chainperson quickly and carefully sticks her pin at the zero mark and calls her station, "nine." The call "nine," besides keeping up with the station number, says to the rear chainperson, "I have marked my point. Drop the tape and start walking forward."

The rear chainperson should never hang on the tape as he moves forward, but he should keep the end of the tape in view.

The system whereby both chainpersons call out the station numbers is a double-check on counting the pins. It is also a simple way to communicate when the forward pin should be set.

In taping long distances, both chainpersons should choose distance objects on their line to walk toward.

Chaining pins should be stuck at an angle of 45° with the ground and at right angles to the line of measurement.

11. STATIONING WHEN DISTANCE IS MORE THAN TEN LENGTHS

When the forward chainperson has set her last pin in the ground, she should have just heard the rear chainperson call "nine." She should have replied, "ten." Her last pin in the ground indicates that she has taped 10 stations, or 1000 ft. She waits at this last pin until the rear chainperson comes forward and hands her his pins. Both chainpersons count the pins to be certain there are ten in hand and one in the ground. As taping is resumed, the situation is the same as it was in the beginning. One pin is in the ground in front of the rear chainperson, and ten pins are in the hand of the forward chainperson.

12. STATIONING AT END OF LINE OR WHEN PLUS IS DESIRED AT POINT ON LINE

Using the add tape. The rear chainperson moves to the forward station and holds a foot mark on the pin.

If the forward chainperson needs more tape, she calls, "Give me a foot." The rear chainperson slides the next larger foot mark to the pin.

If the forward chainperson has too much tape, she calls, "Take a foot." The rear chainperson slides the next smaller foot mark to the pin.

The forward chainperson calls, "What are you holding?"

The rear chainperson calls, "Holding 46," for example.

The forward chainperson then calls "Reading 46.32," for example.

The forward chainperson then calls "Station?"

The rear chainperson counts the pins in his possession but does not count the pin in the ground at the last full station. The station number is the same as the number of pins in his hand if the station is less than ten. If it is more than ten, the station number is the same as the number of pins in his hand plus ten for each exchange of ten pins. He calls out the station number.

The forward chainperson checks the rear chainperson's count. The difference between ten and the number of pins in her hand, plus ten for each exchange of ten pins, is the station number.

The forward chainperson calls out the full station number and plus. Both chainpersons record it.

Using the cut tape. The procedure is the same for the rear chainperson in placing a foot mark on the pin.

The foreward chainperson calls, "What are you holding?"

The rear chainperson calls, "Holding 47."

The foreward chainperson then calls, "Cut 68."

The rear chainperson calls, "46.32."

The forward chainperson repeats, "46.32. Station?"

Both chainpersons then check the number of pins in hand as described in the procedure for using an add tape. They call the station number.

13. BREAKING TAPE

Where the slope is so great that a 100 ft length of the tape cannot be held horizontally without plumbing above the shoulders, a procedure known as *breaking tape* can be used as illustrated in Fig. 11.2.

The forward chainperson pulls the tape forward a full length as usual.

She puts the tape approximately on line and walks back along the tape to a point where the tape can be held horizontal below the shoulder level.

She then picks up a foot mark ending in 0 or 5 (70, for example) and measures a partial tape length (30 ft, for example), using the plumb bob as described in Sec. 9, and marking the point with a pin.

Figure 11.2 *Breaking Tape*

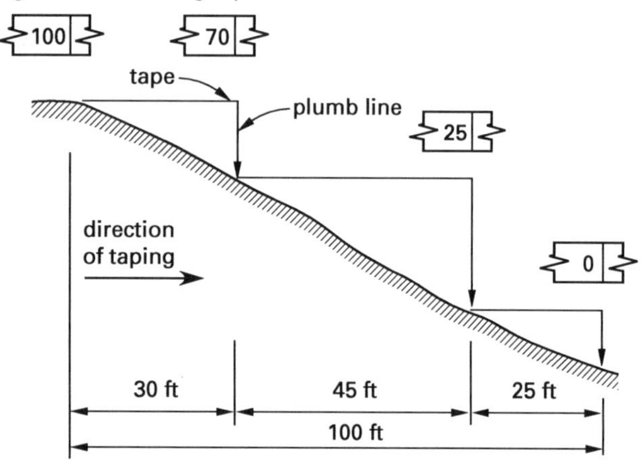

After the forward chainperson has placed the pin, she waits for the rear chainperson to come forward and then tells him what foot mark she was holding, such as, "Holding 70."

The rear chainperson repeats, "Holding 70." He hands the forward chainperson a pin to replace the one used to mark the intermediate point.

The chainpersons continue the procedure at as many intermediate points as necessary. The rear chainperson always picks up the intermediate pin. When he moves forward to an intermediate point, he hands the forward chainperson a pin. He does *not* hand a pin to the forward chainperson when he moves forward to the zero mark.

14. TAPING AT AN OCCUPIED STATION

When taping at a station that is occupied by an instrument, chainpersons must be extremely careful not to hit the leg of the instrument. If a plumb bob is needed at the point, the plumb bob string hanging from the instrument can be used. In some cases, it may be necessary to use the point on top of the instrument on the vertical axis as a measuring point.

15. CARE OF THE TAPE

The tape will not be broken by pulling on it unless there is a kink (loop) in it. Chainpersons should always be alert to "kinking." The tape is easily broken if it is pulled when there is a kink in it.

If the tape has been used in wet grass or mud, it should be cleaned and oiled lightly by pulling it through an oily rag.

16. SLOPE MEASUREMENTS

On fairly level ground where the slope is uniform, it is sometimes easier to determine the slope and make corrections for changing the slope measurement to horizontal measurement rather than to break tape every few feet. To determine horizontal distance, the correction will be subtracted from the slope distance.

Figure 11.3 *Measurements on Slope*

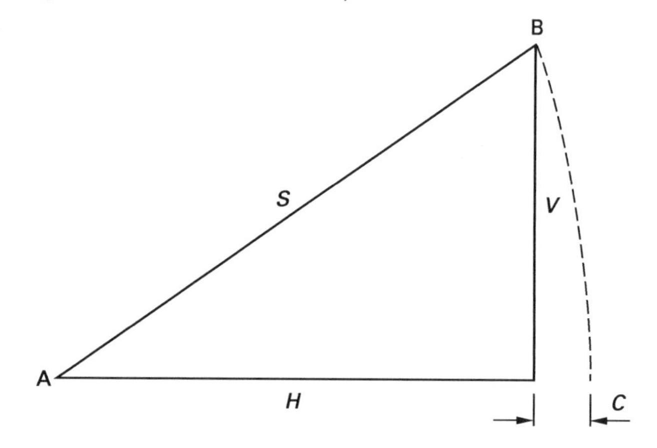

In Fig. 11.3, H is the horizontal distance from A to B. S is the slope distance from A to B. V is the difference in elevation from A to B, and C is the correction.

Thus,

$$V^2 = S^2 - H^2 = (S - H)(S + H) \qquad 11.1$$

Where the slope difference is small, $S + H$ is approximately $2S$. Therefore,

$$V^2 \approx (2S)(S - H) \qquad 11.2$$

Because $S - H =$ correction C,

$$C \approx \frac{V^2}{2S} \qquad 11.3$$

Where $S = 100$ ft (one tape length),

$$C \approx \frac{V^2}{200 \text{ ft}} \qquad 11.4$$

The approximate value will be within 0.007 ft of the actual value when the difference in elevation per 100 ft of slope distance is not more than 15 ft. For steeper slopes, more exact formulas may be used.

Example 11.1

The difference in elevation between two points is 4.0 ft and the slope distance is 100.00 ft. What is the horizontal distance?

Solution

The correction is

17. TENSION

Tapes are not guaranteed by their manufacturers to be of exact length. The National Bureau of Standards will, for a fee, compare any tape with a standard tape or distance, and it will certify the exact length of the tape under certain conditions.

In the United States, steel tapes are standardized for use at 68°F. The standard pull for a 100 ft tape with the tape supported throughout its length is usually, depending on the cross-sectional area of the tape, 10 lbf. When tapes are standardized, they are usually standardized for use under two conditions: supported throughout, and supported only at the ends. When supported only at the ends, the pull is usually 30 lbf, but this can be varied on request. (See Fig. 11.4.)

When the tape is pulled with more or less than the standard amount of tension (10 lbf when supported), the actual distance is more or less than 100.00 ft. However, variations in pull when the tape is supported do not affect the distance greatly. (If a pull of 20 lbf is exerted on a tape that was standardized for a pull of 10 lbf, the increased length is 0.006 ft.)

By using spring-balance handles, chainpersons can get the feel of a 10 lbf pull so that, in ordinary taping with the tape supported, the error caused by variation in tension is neglible.

18. CORRECTION FOR SAG

When the tape is supported only at the ends, it sags and takes the form of a catenary. The correction for sag can be determined by formula or can be offset by increased tension. For a medium-weight tape standardized for a pull of 10 lbf, a pull of 30 lbf will offset the difference in length caused by sag. If the tape is supported at 25 ft intervals, the pull only need be 14 lbf.

Chainpersons should use the spring-balance handle to familiarize themselves with various "pulls."

19. EFFECT OF TEMPERATURE ON TAPING

Steel tapes are standardized for 68°F in the United States. For a change in temperature of 15°F, a steel tape will undergo a change in length of about 0.01 ft, introducing an error of about 0.5 ft per mile.

The coefficient of thermal expansion for steel is approximately 0.00000645 per unit length per degree Fahrenheit.

For a 100 ft tape where T is the temperature (in °F) at time of measurement, the correction in length C due to change in temperature is

$$C = (0.00000645)(T - 68^\circ)(100) \qquad 11.5$$

For example, assume that a line was measured to be 675.48 ft at 30°F. The change in the recorded length due to temperature change is

$$C = (0.00000645)(30^\circ\text{F} - 68^\circ)(675.48 \text{ ft}) = -0.17 \text{ ft}$$

The corrected length is

$$L = 675.48 \text{ ft} - 0.17 \text{ ft} = 675.31 \text{ ft}$$

If the same line were measured when the temperature was 106°F, the change would be

$$C = (0.00000645)(106^\circ\text{F} - 68^\circ\text{F})(675.48 \text{ ft}) = +0.17 \text{ ft}$$

The corrected length would be

$$L = 675.48 \text{ ft} + 0.17 \text{ ft} = 675.65 \text{ ft}$$

20. EFFECT OF IMPROPER ALIGNMENT

Improper alignment is probably the least important error in taping. Many transitpersons and chainpersons spend time aligning that is not justified by the effect of improper alignment. The linear error when one end of the tape is off line can be computed in the same way slope correction is computed. For example, for a 100 ft tape with one end off line 1.0 ft,

$$C = \frac{V^2}{200 \text{ ft}} = \frac{(1.0)^2}{200 \text{ ft}} = 0.005 \text{ ft}$$

When the error in alignment is 0.5 ft, the linear error is 0.001 ft per tape length, or about 0.05 ft per mile.

Field Data Acquisition

Figure 11.4 *National Bureau of Standards Certificate*

UNITED STATES DEPARTMENT OF COMMERCE
WASHINGTON

National Bureau of Standards

Certificate

100-Foot

Steel Tape

Maker: Keuffel & Esser Co. Submitted by NBS No. 10565

This tape has been compared with the standards of the United States, and the intervals indicated have the following lengths at 68° Fahrenheit (20° centigrade) under the conditions given below:

Supported on a horizontal flat surface

Tension (pounds)	Interval (feet)	Length (feet)
10	0 to 100	100.002
5 1/2	0 to 100	100.000

Supported at the 0, 50, and 100-foot points

Tension (pounds)	Interval (feet)	Length (feet)
30	0 to 100	100.000

Supported at the 0 and 100-foot points

Tension (pounds)	Interval (feet)	Length (feet)
38 1/2	0 to 100	100.000

See Note 3(a) on the reverse side of this certificate.

the Director
National Bureau of Standards

Test No. 2.4/1426
Date: June 13, 1955

Lewis V. Judson

Lewis V. Judson

Chief Length Section
Optics and Metrology Division

21. INCORRECT LENGTH OF TAPE

A standardized tape can be used to check other tapes. If a 100 ft tape is known to be of incorrect length, the correction factor to be used for measurements made with the correct tape 100 ft is

$$C = \text{actual length} - 100.00 \text{ ft}$$

For example, the correction for a tape found to be 100.02 ft long after comparison with a standardized tape is

$$C = 100.02 \text{ ft} - 100.00 \text{ ft} = +0.02 \text{ ft per } 100 \text{ ft}$$

If a line is measured to be 662.35 ft with this tape, the corrected length would be

$$662.35 \text{ ft} + (6.6235 \text{ ft})(+0.02) = 662.48 \text{ ft}$$

For a tape found to be 99.98 ft long after comparison with the standardized tape, the correction would be

$$C = 99.98 \text{ ft} - 100.00 \text{ ft} = -0.02 \text{ ft per } 100 \text{ ft}$$

If a line is measured to be 662.35 ft with this tape, the corrected length would be

$$662.35 \text{ ft} + (6.6235 \text{ ft})(-0.02) = 662.22 \text{ ft}$$

A line measured with a tape that is longer than 100 ft is actually longer than the measurement shown. A line measured with a tape that is shorter than 100 ft is acutally shorter than the measurement shown. A rule to remember is, "for a tape too long, add; for a tape too short, subtract." This rule can also be applied to temperature correction.

22. COMBINED CORRECTIONS

Corrections for incorrect length of tape, temperature, and slope can be combined.

Example 11.2

A tape that is 100.03 ft long was used to measure a line that was recorded as 1238.22 ft when the temperature was 18°F. The difference in elevation from beginning to end was 12.1 ft. What is the corrected length?

Solution

The tape correction is

$$(12.3822)(+0.03) = +0.37$$

The temperature correction is

$$(0.00000645)(18^\circ - 68^\circ)(1238.22) = -0.40$$

The slope correction is

$$\frac{(12.1)^2}{(2)(1238.22)} = -0.06$$

The total correction is -0.09.

The corrected length is

$$1238.22 - 0.09 = 1238.13 \text{ ft}$$

PRACTICE PROBLEMS

1. Compared to steel tape, the expansion and contraction of an Invar tape caused by temperature change is

(A) slightly more
(B) slightly less
(C) much less
(D) about the same

2. How many pins are in a set of marking pins?

(A) 10
(B) 11
(C) 12
(D) 20

3. An add tape has

(A) the last foot of each end graduated in tenths or hundredths of a foot
(B) an extra graduated foot beyond the zero mark
(C) both (A) and (B)
(D) neither (A) nor (B)

4. A Gunter's chain

(A) is 66 ft long
(B) has 66 links
(C) is 100 ft long
(D) both (A) an (B)

5. A square, one chain on a side, equals

(A) 4356 ft^2
(B) 43,560 ft^2
(C) 1 ac
(D) none of the above

6. The area of a rectangle 6 chains by 5 chains equals

(A) 3 ac
(B) 30 ac
(C) 300 ac
(D) none of the above

7. In marking 100 ft tape lengths, at what angle should the pin be set?

(A) 30° to the ground
(B) 45° to the ground
(C) 60° to the ground
(D) 90° to the ground

Field Data Acquisition

8. In taping from A to B, a distance of 1646.32 ft, how many pins should the rear chainperson have in his hand when he holds the tape on the last full station?

(A) 4
(B) 5
(C) 6
(D) 7

9. In taping as in Prob. 8, how many pins should the foreward chainperson have in her hand when she reaches point B?

(A) 3
(B) 4
(C) 5
(D) 6

10. A distance of 1000.00 ft was measured when the temperature was 68°F with a tape calibrated to be 100.000 ft in length. The difference in elevation from sta 0+00 to sta 10+00 was 11 ft. What is the corrected distance?

(A) 999.82 ft
(B) 999.94 ft
(C) 1000.06 ft
(D) 1000.18 ft

11. The coefficient of thermal expansion for steel per unit length per degree Fahrenheit is

(A) 0.00000645
(B) 0.00000465
(C) 0.00000654
(D) 0.0000645

12. A distance of 787.35 ft was measured on level ground with a tape that was 100.000 ft in length when the temperature was 98°F. What is the corrected length?

(A) 787.20 ft
(B) 787.50 ft
(C) 787.55 ft
(D) none of the above

13. A line was measured on level ground when the temperature was 68°F and was found to be 582.32 ft in length. Later, the tape was calibrated and found to be 100.03 ft in length. What is the corrected length of the line?

(A) 582.15 ft
(B) 582.49 ft
(C) 582.65 ft
(D) none of the above

14. For a change in temperature of 15°F, a 100 ft steel tape will undergo a change in length of about

(A) 0.005 ft
(B) 0.01 ft
(C) 0.02 ft
(D) 0.04 ft

15. At 28°F, a tape that measured 100.000 ft in length at 68°F will measure

(A) 99.97 ft
(B) 100.01 ft
(C) 100.59 ft
(D) none of the above

16. A tape that was 100.02 ft in length at 68°F was used to measure a line recorded as 1196.44 ft when the temperature was 28°F. The difference in elevation from the beginning to the end of the line was 10 ft. What is the corrected length?

(A) 1196.27 ft
(B) 1196.33 ft
(C) 1196.55 ft
(D) 1196.61 ft

17. A tape that was 99.97 ft in length at 68°F was used to measure a line recorded as 713.19 ft when the temperature was 98°F. The difference in elevation from the beginning to the end of the line was 8 ft. What is the corrected length?

(A) 712.99 ft
(B) 713.05 ft
(C) 713.07 ft
(D) 713.14 ft

SOLUTIONS

1. Compared to steel tape, the expansion and contraction of an Invar tape caused by temperature change is *much less*.

The answer is (C).

2. There are eleven pins in a set of marking pins.

The answer is (B).

3. An add tape has an extra graduated foot beyond the zero mark.

The answer is (B).

4. A Gunter's chain is 66 ft long.

The answer is (A).

5. A square, one chain on a side, equals 4356 ft^2.

The answer is (A).

6. The area of a rectangle 6 chains by 5 chains equals 3 ac.

The answer is (A).

7. In marking 100 ft tape lengths, the pin should be set at an angle 45° to the ground.

The answer is (B).

8. When the rear chainperson is at the last full station (sta 16+00), he should have pins for sta 10+00 through sta 15+00 in hand. Thus, he should have six pins in hand.

The answer is (C).

9. When the forward chainperson reaches point B, she should have four pins in hand since she should have started with ten pins in hand at sta 10+00 and placed six of those pins at sta 11+00 through sta 16+00.

The answer is (B).

10. Using Eq. 11.3, calculate the slope correction.

$$C \approx \frac{V^2}{2S} = \frac{11^2}{(2)(1000.00 \text{ ft})} = 0.06 \text{ ft}$$

The corrected length is

$$L = 1000.00 \text{ ft} - 0.06 \text{ ft} = 999.94 \text{ ft}$$

The answer is (B).

11. The coefficient of thermal expansion for steel is 0.00000645 per unit length per degree Fahrenheit.

The answer is (A).

12. Using Eq. 1.5, calculate the temperature correction.

$$\begin{aligned} C &= (0.00000645)(T - 68°)(\text{distance}) \\ &= (0.00000645)(98° - 68°)(787.35 \text{ ft}) = +0.15 \text{ ft} \end{aligned}$$

The corrected length is

$$L = 787.35 \text{ ft} + 0.15 \text{ ft} = 787.50 \text{ ft}$$

The answer is (B).

13. The correction for the incorrect length is

$$C = 100.03 \text{ ft} - 100.00 \text{ ft} = +0.03 \text{ ft}/100 \text{ ft}$$

The corrected length is

$$L = 582.32 \text{ ft} + (5.8232)(0.03 \text{ ft}) = 582.49 \text{ ft}$$

The answer is (B).

14. For a change in temperature of 15°F, a 100 ft steel tape will undergo a change in length of approximately 0.01 ft.

The answer is (B).

15. At 28°F, a tape that measured 100.000 ft in length at 68°F will measure 99.97 ft.

The answer is (A).

16. The correction for the temperature is

$$\begin{aligned} C &= (0.00000645)(T - 68°)(\text{distance}) \\ &= (0.00000645)(28° - 68°)(1196.44 \text{ ft}) = -0.31 \text{ ft} \end{aligned}$$

The correction for the incorrect length is

$$C = (11.9644)(+0.02) = +0.24 \text{ ft}$$

From Eq. 11.3, the correction for the slope is

$$C \approx \frac{V^2}{2S} = \frac{10^2}{(2)(1196.44 \text{ ft})} = -0.04 \text{ ft}$$

The corrected length is

$$L = 1196.44 \text{ ft} - 0.31 \text{ ft} + 0.24 \text{ ft} - 0.04 \text{ ft} = 1196.33 \text{ ft}$$

The answer is (B).

17. The correction for the temperature is

$$\begin{aligned} C &= (0.00000645)(T - 68°)(\text{distance}) \\ &= (0.00000645)(98° - 68°)(713.19 \text{ ft}) = +0.14 \text{ ft} \end{aligned}$$

The correction for the incorrect length is

$$C = (7.1319)(-0.03 \text{ ft}) = -0.21 \text{ ft}$$

From Eq. 11.3, the correction for the slope is

$$C = \frac{V^2}{2S} = \frac{8^2}{(2)(713.19 \text{ ft})} = -0.04 \text{ ft}$$

The corrected length is

$$L = 713.19 \text{ ft} + 0.14 \text{ ft} - 0.21 \text{ ft} - 0.04 \text{ ft} = 73.08 \text{ ft}$$

The answer is (C).

12 Electronic Distance Measurement

Nomenclature

c	system constant	–	–
d	phase difference distance	ft	m
D	distance	ft	m
elev	elevation	ft	m
f	frequency	1/sec	Hz
h.i.	height of instrument above hub	ft	m
HI	height of instrument above mean sea level or similar datum	ft	m
n	number of wavelengths	–	–
s	scale factor	–	–
v	velocity of light	ft/sec	m/s

Symbols

λ	wavelength	ft	m
Δ	difference	ft	m

Subscripts

I	instrument
H	horizontal
R	rod
S	slope

1. ELECTRONIC DISTANCE MEASUREMENT

For over two millennia, surveyors have used taping for measurement of distance. The advent of *electronic distance measurement* (EDM) in the 20th century not only represented a significant change in the method of distance measurement, it also marked the beginning of a movement toward the use of electronics in all aspects of surveying that has revolutionized the profession.

SI units are frequently used for EDM measurements and satellite geodesy (discussed later in this book), and calibration tables and other reference data often use these units.

Probably the first use of electronics for distance measurement was the fathometer. That device, invented in 1919 by Reginald Fessenden, measured the distance from a vessel to the ocean floor by observing the time it took a sound wave to be reflected and returned.

The earliest electronic instruments for distance measurement on land were developed in Sweden in the early 1950s. Typical of these was the Geodimeter. That device transmitted a modulated light beam to a mirror at the end of the line being measured and then made phase comparisons between the transmitted and reflected pulses. The device measured the elapsed time and calculated the distance based on the velocity of light. The first generation of electro-optical instruments had maximum ranges of 2–3 mi (3–5 km) during the day and 15–20 mi (25–30 km) at night under optimum conditions. The units were quite heavy and bulky, as well as expensive.

Shortly thereafter, during the late 1950s, EDM instruments that used microwave signals were developed in South Africa. Typical of these was the Tellurometer. That device consisted of two identical instruments, one set up at each end of the line being measured. The sending unit would transmit a series of microwaves that were received by the other unit and sent back to allow measurement of the elapsed time. That process determined the distance based on the velocity of the microwaves. These early electromagnetic instruments could measure distances up to 50 mi (80 km) under favorable conditions. They were somewhat lighter than the early electro-optical instruments. They did not require intervisibility, and thus they could be used in weather conditions that would hinder electro-optical equipment.

Figure 12.1 *Early Use of a Tellurometer by Coast and Geodetic Survey Personnel*

Source: U.S. National Geodetic Survey

Due to their cost and size, early EDM instruments were used primarily for geodetic surveys performed by government agencies. But, beginning in the mid-1960s and continuing to the present, solid-state electronic components and mass production allowed the development of smaller, lighter instruments that used much less power and were considerably less expensive. This allowed the adoption of EDM technology as the standard mode for distance measuring by the land surveying profession.

2. PRINCIPLES OF OPERATION

Most modern EDM instruments are electro-optical, and use a laser or infrared light source. The units are portable, have low power requirements, and are easily operated. These devices generally have shorter ranges than the first-generation units, typically 2 mi (5 km), and are thus geared to the practicing land surveyor market. A majority of EDM devices manufactured today are incorporated into total stations to provide angulation and distance measurement in one instrument. Some of the latest developments in EDM technology are "reflectorless" units with typical ranges of 600–1000 ft that measure distance to objects without using retro-prisms.

The laser or infrared light used by most electro-optical EDM instruments is modulated into wavelengths of a certain frequency. Measurement of a line is accomplished by setting up the transmitter at one end and a reflector (usually a prism) at the other end. The reflected light is converted to an electrical signal by the EDM to allow phase comparison of the transmitted and reflected signals. The integral number of wavelengths included in the double path is determined by transmitting pulses on multiple frequencies and comparing the results. The phase difference is then used to determine the length of the line, which would be equivalent to one-half of the sum of the number of wavelengths in the double path distance plus the partial wavelength represented by the phase difference. This calculation is represented by Eq. 12.1. D is the length of the line, n is the integral number of wavelengths in the double path of the light, λ is the wavelength of the modulated beam, and d is the distance representing the phase difference.

$$D = \frac{1}{2}(n\lambda + d) \qquad 12.1$$

The measurement process is illustrated by Fig. 12.2.

For the measurement process, the wavelength is determined from the modulation frequency of the EDM using Eq. 12.2. f is the modulation frequency in hertz (cycles per second), and v is the velocity of the light in meters per second.

$$\lambda = \frac{\mathrm{v}}{f} \qquad 12.2$$

The reflectors usually consist of retrodirective prisms that cause the incident light to be reflected parallel to itself. This is achieved by having the reflective faces of the prism at right angles as illustrated in Fig. 12.3. As may be seen, the path of the light beam within the prism is the sum of distances a, b, and c, which is equal to twice the depth of the prism. Due to the refractive properties of the glass in the prism, the equivalent travel in air would be (1.57)(a + b + c). Therefore, the ideal prism would be mounted so that the front face of the

Figure 12.2 *Measurement Process for Electro-Optical EDM*

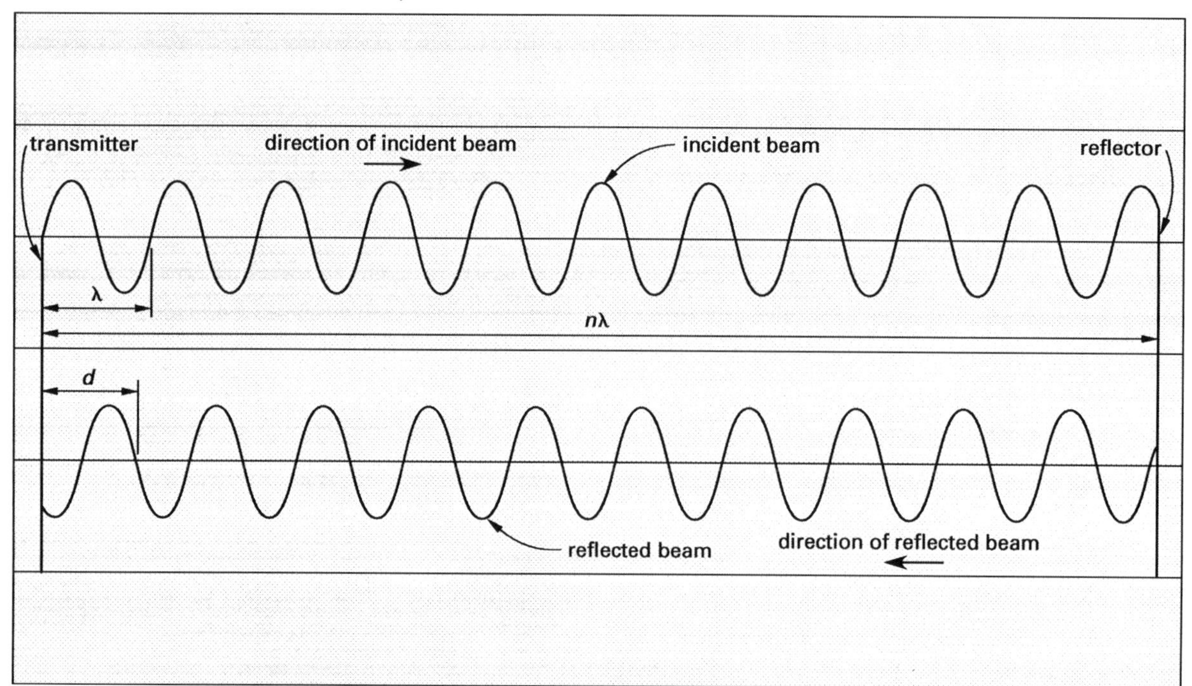

prism is set at a distance of 1.57 times twice the depth of the prism forward of the survey point. Yet, that distance is so far behind the prism that mounting it at that point would create an unbalanced prism. Therefore, most prisms manufactured today are created with a small *prism offset* that can be compensated for by an adjustment in the EDM instrument. This standard prism offset is typically either 30 mm or 0.

Figure 12.3 *Retrodirective Prism*

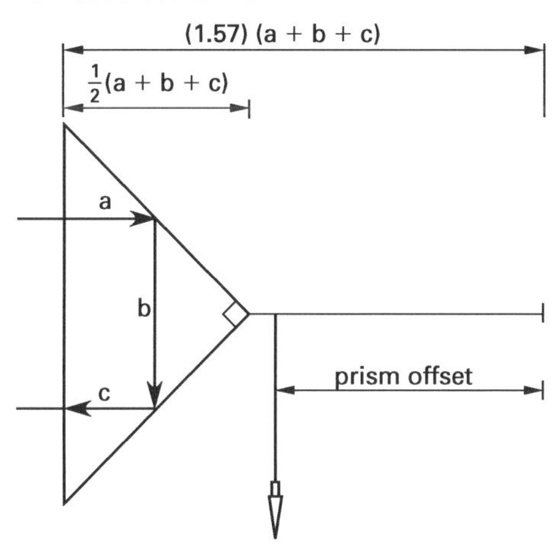

Example 12.1

If an EDM with a wavelength of 60 ft measures a distance resulting in a double path length of 50.25 cycles, what is the length of the line measured?

Solution

Using Eq. 12.1,

$$\begin{aligned} D &= \frac{1}{2}(n\lambda + d) = \left(\frac{1}{2}\right)\left((50)(60\text{ ft}) + (0.25)(60\text{ ft})\right) \\ &= \left(\frac{1}{2}\right)(3000\text{ ft} + 15\text{ ft}) \\ &= 1507.5\text{ ft} \end{aligned}$$

3. SOURCES OF ERROR

The accuracy of modern electro-optical EDM devices is usually stated as plus or minus the sum of a constant plus a certain number of parts per million (ppm) for the standard error. A typical high-end type EDM might have an accuracy of ±(2 mm + 2 ppm). This is equivalent to ±(0.007 ft + 2 ppm).

Achieving the stated standard accuracy depends upon proper observation procedures, calibration, and care in avoiding errors. A surveyor needs a thorough knowledge of the major causes of error when using an EDM instrument in order to avoid mistakes.

Atmospheric conditions that affect the speed of light are a major source of error in EDM. The speed of light for a given set of atmospheric conditions is defined by Eq. 12.3. v_0 is the velocity of light in a vacuum (186,282.05 mi/sec or 299 792.5 km/s). r is the refractive index for the defined atmospheric conditions.

$$v = \frac{v_0}{r} \qquad 12.3$$

The refractive index r varies with atmospheric pressure, air temperature and to a lesser degree, vapor pressure. Therefore, it is important that meteorological parameters be measured and a correction applied. Fortunately, most EDM instruments calculate the correction value automatically when the meteorological readings are entered. An important precaution is to avoid measurements along lines where the light beam passes close to the ground. In such areas there can be considerable variation in temperature.

Uncertainty in the offset correction for the prism is another frequently encountered error. As discussed earlier in this chapter, most prisms have an offset that requires a correction to measurements. Fortunately, prisms typically use a standard offset of 30 mm or 0 have the offset marked on the prism assembly. However, even with the correct offset, errors can appear under certain conditions. Where distances are measured on a slope, light rays do not strike the face of a vertical prism in a direction perpendicular to its front face. This changes the position of the effective center of the prism, causing significant errors with steep slope's distances. In such conditions, use a reflector that can be adjusted to an angle perpendicular to the slope. If such a reflector is not available, a correction to compensate for the incident angle may be applied to the measurements.

Frequency drift is another source of error. Periodic calibration of the EDM instrument against a known distance ensures accurate and consistent results. The next section of this chapter addresses such calibrations.

Another frequent source of error in EDM is *tribrach error*. If the optical plummet on the tribrach supporting either the EDM instrument or the prism is misadjusted, this could result in the instrument or the prism being off the point. In that situation, the line being measured is not the line between occupied points.

Example 12.2

If an EDM has a standard error of ±(0.007 ft + 2 ppm), what would be the expected error in measuring a 3000 ft line?

Solution

$$\begin{aligned} \text{standard error} &= \pm(0.007\text{ ft} + 2\text{ ppm}) \\ &\quad \times (\text{measured distance}) \\ &= \pm\left(0.007\text{ ft} + \left(\frac{2}{1{,}000{,}000}\right)(3000\text{ ft})\right) \\ &= \pm(0.007\text{ ft} + 0.006\text{ ft}) \\ &= 0.013\text{ ft} \end{aligned}$$

4. EDM CALIBRATION

Periodic calibration of EDM instruments against a known distance is essential for obtaining consistent and accurate measurements. A calibration is best performed on a precise baseline established specifically for that purpose. Most states have a number of such baselines, typically established in cooperative programs with the National Geodetic Survey, and data for such sites are available on that agency's internet site. A typical baseline configuration includes four marks set at 0, 150 m, 800 m, and 1500 m. It is recommended that each mark be occupied and that redundant measurements be made to each of the other marks. For a typical calibration, this process would result in data similar to Table 12.1. The table shows the published distances and the average observed distances corrected for atmospheric conditions and reduced to the horizontal.

Table 12.1 *Typical Calibration Data*

observation number	from (m)	to (m)	published distance (m)	observed distance (m)	difference (m)
1	0	150	150.0023	150.0070	−0.0047
2	150	0	150.0023	150.0047	−0.0024
3	0	800	800.0146	800.0160	−0.0014
4	800	0	800.0146	800.0171	−0.0025
5	0	1500	1491.8610	1491.8625	−0.0015
6	1500	0	1491.8610	1491.8697	−0.0087
7	150	800	650.0123	650.0028	0.0095
8	800	150	650.0123	650.0162	−0.0039
9	150	1500	1341.8587	1341.8618	−0.0031
10	1500	150	1341.8587	1341.8606	−0.0019
11	800	1500	691.8467	691.8446	0.0021
12	1500	800	691.8467	691.8423	0.0044
total			10,251.1912	10,251.2053	−0.0141

The difference, Δ, between the published and observed distances can be attributed to either a scale correction or a constant correction, or both. Equation 12.4 expresses that relationship. D_p is the published distance, D_o is the observed distance, s is the scale factor, and c is the system constant.

$$\Delta = D_p - D_o = sD_o + c \tag{12.4}$$

If it is significant, the scale factor is applied by multiplying it by the observed distance and algebraically adding the results to the observed distance. If it is significant, the system constant is applied by algebraically adding it to the observed distance. The recommended method of determining the factors is to do a least-squares regression of the data using Eq. 12.4. Such an analysis may be accomplished by statistical software or spreadsheet programs, or the factors may be calculated by Eqs. 12.5 and 12.6.

$$s = \frac{n\sum(D_o\Delta) - (\sum D_o)(\sum \Delta)}{n\sum D_o^2 - (\sum D_o)^2} \tag{12.5}$$

$$c = \frac{(\sum D_o^2)(\sum \Delta) - (\sum D_o)(\sum D_o\Delta)}{n\sum D_o^2 - (\sum D_o)^2} \tag{12.6}$$

As previously stated, the scale factor is applied by multiplying it by the observed distance and algebraically adding the results to the observed distance, and the system constant is applied by algebraically adding it to the observed distance. If the values are statistically insignificant, they need not be applied. It is recommended that standard statistical methods, such as application of Student's t test, be used to test the significance of the determined values.

Example 12.3

Using the data provided in Table 12.1, determine the scale factor and system constant.

Solution

See the following table.

Table for Example 12.3

observation number	from (m)	to (m)	published distance (m)	observed distance (m)	difference (m)	$D_o\Delta$ (m^2)	D_o^2 (m^2)
1	0	150	150.0023	150.0070	−0.0047	−0.7050	22,500.6900
2	150	0	150.0023	150.0047	−0.0024	−0.3600	22,500.6900
3	0	800	800.0146	800.0160	−0.0014	−1.1200	640,023.3602
4	800	0	800.0146	800.0171	−0.0025	−2.0000	640,023.3602
5	0	1500	1491.8610	1491.8625	−0.0015	−2.2378	2,225,649.2433
6	1500	0	1491.8610	1491.8697	−0.0087	−12.9792	2,225,649.2433
7	150	800	650.0123	650.0028	0.0095	6.1751	422,515.9902
8	800	150	650.0123	650.0162	−0.0039	−2.5350	422,515.9902
9	150	1500	1341.8587	1341.8618	−0.0031	−4.1598	1,800,584.7708
10	1500	150	1341.8587	1341.8606	−0.0019	−2.5495	1,800,584.7708
11	800	1500	691.8467	691.8446	0.0021	1.4529	478,651.8563
12	1500	800	691.8467	691.8423	0.0044	3.0441	478,651.8563
total			10,251.1912	10,251.2053	−0.0141	−17.9745	11,179,851.8215

Solution (continued)

$$s = \frac{n\sum(D_o\Delta) - (\sum D_o)(\sum \Delta)}{n\sum D_o^2 - (\sum D_o)^2}$$

$$= \frac{(12)(-17.9745\ \text{m}^2) - (10\,251.2053\ \text{m})(-0.0141\ \text{m})}{(12)(11\,179\,887.7703\ \text{m}^2) - (10\,251.2053\ \text{m})^2}$$

$$= \frac{-71.152\ \text{m}^2}{29\,071\,443.1\ \text{m}^2}$$

$$= -0.000\,002\,447$$

$$c = \frac{(\sum D_o^2)(\sum \Delta) - (\sum D_o)(\sum D_o\Delta)}{n\sum D_o^2 - (\sum D_o)^2}$$

$$= \frac{(11\,179\,887.7703\ \text{m}^2)(-0.0141\ \text{m}) - (10\,251.2053\ \text{m})(-17.9745\ \text{m}^2)}{(12)(11\,179\,887.7703\ \text{m}^2) - (10\,251.2053\ \text{m})^2}$$

$$= \frac{26\,623.8721\ \text{m}^3}{29\,071\,443.1\ \text{m}^2}$$

$$= 0.000\,915\,8\ \text{m}$$

Example 12.4

For a measured distance of 3000.00 m, what would be the corrected distance given the values for the scale factor, $s = -0.000\,002\,447$, and system constant, $c = 0.000\,915\,8$ m?

Solution

Using Eq. 12.4,

$$D_p - D_o = sD_o + c$$

$$D_p = D_o + sD_o + c$$

$$= 3000.00\ \text{m} + (-0.000\,002\,447)(3000.00\ \text{m}) + 0.000\,915\,8\ \text{m}$$

$$= 2999.99\ \text{m}$$

5. SLOPE DISTANCE REDUCTION

Unlike taping, where the normal procedure is to make horizontal distance measurements, electronic measurements are typically made using slope distances. Figure 12.4 illustrates the need to reduce slope distance to horizontal distance. Many total stations automatically apply slope distance reduction, so care should be taken to apply a correction but not apply it twice. Equations 12.7 and 12.8 are used to reduce slope distances to horizontal distances using zenith or vertical angles. D_H is the horizontal distance, D_S is the slope distance, θ_Z is the zenith angle, and θ_V is the vertical angle.

$$D_H = D_S \sin\theta_Z \qquad 12.7$$

$$D_H = D_S \cos\theta_V \qquad 12.8$$

Figure 12.4 *Slope Distance Correction*

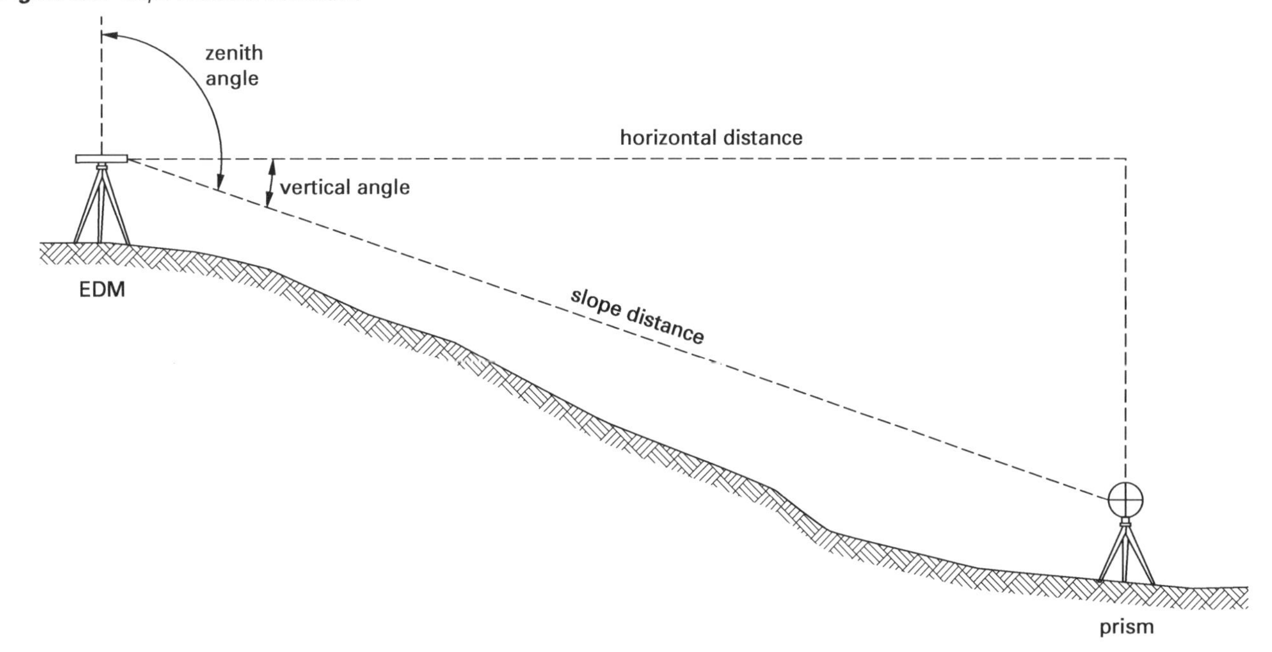

The reduction of slope distance to horizontal may also be accomplished by finding elevations at both ends of the measured line, instead of vertical angles, and using Eq. 12.9. The elevation at the rod is represented by elev_R, the elevation at the instrument by elev_I, the height of instrument at the rod by h.i._R, and the height of instrument at the instrument by h.i._I.

$$D_H = \sqrt{D_S^2 - \left((\text{elev}_R + \text{h.i.}_R) - (\text{elev}_I + \text{h.i.}_I)\right)^2} \qquad 12.9$$

Example 12.5

For a measured slope distance of 662.5 ft and a zenith angle of 92°20 min, what is the corresponding horizontal distance?

Solution

Using Eq. 12.7,

$$\begin{aligned} D_H &= D_S \sin\theta_Z = (662.5 \text{ ft})(\sin 92°20') \\ &= (662.5 \text{ ft})(0.99917) \\ &= 661.9 \text{ ft} \end{aligned}$$

Example 12.6

An EDM instrument is set up with a height of instrument of 4.00 ft over a point with an elevation of 205.25 ft. The reflector is set at a height of 6.00 ft over a point with an elevation of 159.15 ft. If the measured slope distance between the points is 1260.50 ft, what is the horizontal distance?

Solution

Using Eq. 12.9,

$$\begin{aligned} D_H &= \sqrt{D_S^2 - \left((\text{elev}_R + \text{h.i.}_R) - (\text{elev}_I + \text{h.i.}_I)\right)^2} \\ &= \sqrt{1260.50 \text{ ft}^2 - \begin{pmatrix} (159.15 \text{ ft} + 6.00 \text{ ft}) \\ - (205.25 \text{ ft} + 4.00 \text{ ft}) \end{pmatrix}^2} \\ &= \sqrt{1{,}588{,}860.25 \text{ ft}^2 - 1944.81 \text{ ft}^2} \\ &= 1259.73 \text{ ft} \end{aligned}$$

PRACTICE PROBLEMS

1. If an EDM has a standard error of ±(5 mm + 5 ppm), what would be the expected error in measuring a 2000 ft line?

(A) ±0.03 ft
(B) ±0.04 ft
(C) ±0.05 ft
(D) ±0.06 ft

2. An EDM has a modulation frequency of 15 MHz that provides a wavelength of 65 ft. If it measures a distance and obtains a double path length of 62.15 cycles, what is the length of the line measured?

(A) 2019.9 ft
(B) 2115.7 ft
(C) 2330.8 ft
(D) 2411.3 ft

3. For a measured distance of 2500.50 ft, what would be the corrected distance given a scale factor of −0.000005 and a system constant of −0.001 m as determined by a calibration?

(A) 2500.46 ft
(B) 2500.48 ft
(C) 2500.52 ft
(D) 2500.54 ft

4. For a measured slope distance of 3364.5 ft and a zenith angle of 84°40′, what is the corresponding horizontal distance?

(A) 3349.9 ft
(B) 3355.7 ft
(C) 3361.2 ft
(D) 3402.3 ft

5. An EDM instrument is set up with a height of instrument of 4.25 ft over a point with an elevation of 100.50 ft. The reflector is set at a height of 5.00 ft over a point with an elevation of 150.25 ft. If the measured slope distance between the points is 1500.15 ft, what is the horizontal distance?

(A) 1499.12 ft
(B) 1499.22 ft
(C) 1499.30 ft
(D) 1500.00 ft

6. If the vertical angle from point 1 to point 2 is 2°30′ and the slope distance as measured by an EDM instrument is 5604.32 ft, what is the horizontal distance?

(A) 5317.21 ft
(B) 5598.99 ft
(C) 5602.84 ft
(D) 5687.14 ft

SOLUTIONS

1.

$$\begin{aligned}\text{standard error} &= \pm(5\text{ mm} + 5\text{ ppm})\\ &= \pm\left(5\text{ mm} + \left(\frac{5}{1{,}000{,}000}\right)(2000\text{ ft})\right)\\ &= \pm\left((5\text{ mm})\left(0.00328\ \frac{\text{ft}}{\text{mm}}\right) + 0.01\text{ ft}\right)\\ &= \boxed{\pm 0.03\text{ ft}}\end{aligned}$$

The answer is (A).

2.

$$\begin{aligned}D &= \tfrac{1}{2}(n\lambda + d)\\ &= \left(\tfrac{1}{2}\right)\left((62)(65\text{ ft}) + (0.15)(65\text{ ft})\right)\\ &= \left(\tfrac{1}{2}\right)(4030\text{ ft} + 9.75\text{ ft})\\ &= \boxed{2019.9\text{ ft}}\end{aligned}$$

The answer is (A).

3.

$$\begin{aligned}D_p - D_o &= sD_o + c\\ D_p &= D_o + sD_o + c\\ &= 2500.50\text{ ft} + (-0.000005)(2500.50\text{ ft})\\ &\quad + (-0.001\text{ m})\left(3.2808333\ \frac{\text{ft}}{\text{m}}\right)\\ &= \boxed{2500.48\text{ ft}}\end{aligned}$$

The answer is (B).

4.

$$\begin{aligned}D_H &= D_S \sin\theta_Z\\ &= (3364.5\text{ ft})(\sin 84^\circ 40')\\ &= (3364.5\text{ ft})(0.99567)\\ &= \boxed{3349.9\text{ ft}}\end{aligned}$$

The answer is (A).

5.

$$\begin{aligned}D_H &= \sqrt{\begin{array}{l}D_S^2 - \big((\text{elev}_R + \text{h.i.}_R)\\ \quad - (\text{elev}_I + \text{h.i.}_I)\big)^2\end{array}}\\ &= \sqrt{\begin{array}{l}1500.15\text{ ft}^2\\ \quad - \left(\begin{array}{l}(150.25\text{ ft} + 5.00\text{ ft})\\ \quad - (100.50\text{ ft} + 4.25\text{ ft})\end{array}\right)^2\end{array}}\\ &= \sqrt{2{,}250{,}450.02\text{ ft}^2 - 2550.25\text{ ft}^2}\\ &= \boxed{1499.30\text{ ft}}\end{aligned}$$

The answer is (C).

6.

$$\begin{aligned}D_H &= D_S \cos\theta_V = (5604.32\text{ ft})(\cos 2^\circ 30')\\ &= (5604.32\text{ ft})(0.99905)\\ &= \boxed{5598.99\text{ ft}}\end{aligned}$$

The answer is (B).

13 Leveling

Nomenclature

C	departure of the earth from a horizontal line	ft	m
d	length of smallest division on rod scale	ft	m
D	distance that rod is out of plumb	ft	m
E	error caused by out-of-plumb rod	ft	m
F	distance	10^3 ft	–
h	error caused by refraction	ft	m
H	horizontal distance	ft	m
HI	height of instrument from datum	ft	m
L	length of rod	ft	m
M	distance	mi	–
n	number of divisions on vernier	–	–
v	length of a vernier division	ft	m
V	difference in elevation between horizontal and slope distances	ft	m

1. DEFINITIONS

Understanding leveling requires a vocabulary of terms used in the study of the earth's surface. The following terms are important to know.

- *vertical line:* a line from any point on the earth to the center of the earth
- *plumb line:* a vertical line, usually established by a pointed metal bob hanging on a string or cord
- *level surface:* Because the earth is round, a level surface is actually a curved surface. Although a lake appears to have a flat surface, it follows the curvature of the earth. A level surface is a curved surface that, at any point, is perpendicular to a plumb line.
- *horizontal line:* a line perpendicular to the vertical
- *datum:* any level surface to which elevations are referred. Mean sea level is usually used for a datum.
- *elevation:* the vertical distance from a datum to a point on the earth
- *leveling:* the process of finding the difference in elevation of points on the earth
- *spirit level:* a device for establishing a horizontal line by centering a bubble in a slightly curved glass tube (vial) filled with alcohol or another liquid
- *bench mark:* a marked point of known elevation from which other elevations may be established
- *turning point:* a temporary point on which an elevation has been established and which is held while an engineer's level is moved to a new location
- *height of instrument* (HI)*:* the vertical distance from the datum to the line of sight of the level

Figure 13.1 *Leveling Terms*

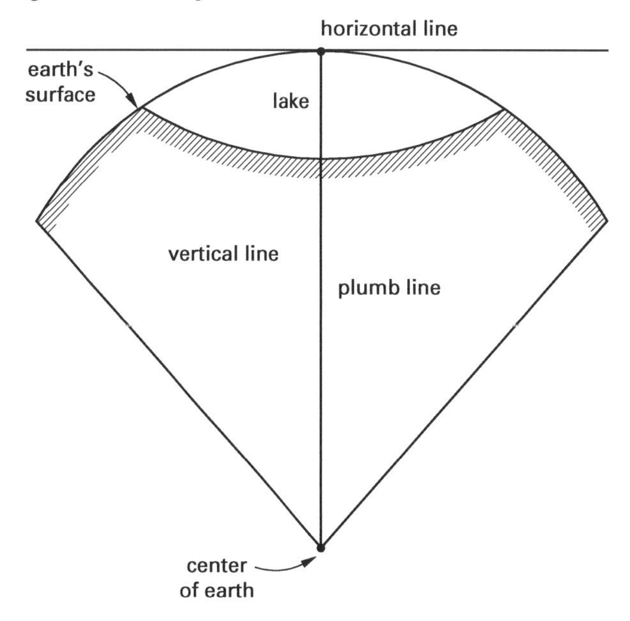

2. DIFFERENTIAL LEVELING

As the term implies, *differential leveling* is the process of finding the difference in elevation between two points. A surveyor's level and level rod are used in differential leveling. A rod is merely a piece of wood, fiberglass, or aluminum that is marked off from one end to the other in meters or in feet, tenths of a foot, and hundredths of a foot. It is held in vertical position. The level is a telescope with crosshairs attached to a spirit level. By keeping the level bubble centered in the vial, the horizontal crosshair in the telescope can be kept on the same elevation while the telescope turns in any direction in the horizontal plane. It establishes a horizontal plane in space from which measurements can be made with the rod. As the observer focuses on the level rod, a measurement can be made from the horizontal plane to the point on which the rod rests merely by reading the measured markings on the rod where the horizontal crosshair is imposed on it.

Figure 13.2 *Differential Leveling*

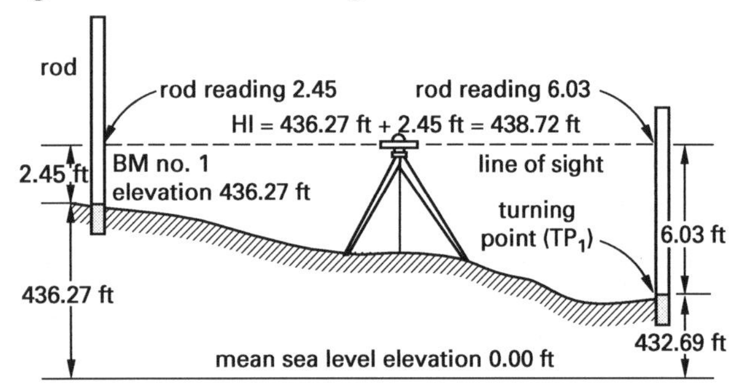

Figure 13.2 illustrates how the level and rod can be used to find the difference in elevation between two points.

Bench mark 1 (BM no. 1) is a semipermanent object, the elevation of which is 436.27 ft above mean sea level. Turning point 1 (TP_1) is a temporary object (in this case, the top of a stake), the elevation of which is to be determined.

The level is set up so that both objects can be seen through the telescope.

The rod is first placed on BM no. 1, and a reading of the rod is made and recorded with the bubble centered. This reading is known as a *backsight* (BS) and is added to the elevation of BM no. 1 to find the HI. Backsight is commonly called a *plus sight* because it is always added to the known elevation.

The rod is then placed on the TP_1, and a reading of the rod is made and recorded with the bubble centered. This reading is known as a *foresight* (FS) and is subtracted from the HI to find the elevation of TP_1. It is commonly called a *minus sight* because it is always subtracted from a known HI.

The elevations of a continuing line of objects can be determined by moving the level along the line. While the level is being moved forward, the rodperson must hold the turning point so that the levelperson may make a backsight reading on it from the new location of the level.

Figure 13.3 shows a profile view of differential levels between BM no. 3 and BM no. 4 and field notes recorded at the time the levels were run. It can be seen that leveling is a series of vertical measurements that alternate in sequence from a plus sight to a minus sight.

Field notes show columns for plus readings (+), for minus readings (−), for HIs, and for elevations of bench marks and turning points.

Notice that a plus reading is shown on the same horizontal line as BM no. 3, but a minus reading is not shown on that line. Only one reading was taken on BM no. 3. Also notice that a plus reading and a minus reading are shown on the same horizontal line as each turning point. The minus reading is subtracted from HI on the line above it to determine the elevation of the turning point, which is shown on the same horizontal line.

After the elevation of the turning point has been determined, the level is moved forward. Then a backsight (+) is read on the same turning point. This plus reading is added to the elevation just determined to new HI. Notice that a minus reading is shown on the same horizontal line as BM no. 4, but a plus reading is not shown on that line. BM no. 4 is the end of the level line; only one reading was made on it.

At the bottom of the field notes, the plus column and the minus column are totalled and the smaller total is subtracted from the larger total. This difference should be the same as the difference between the beginning and ending elevations. If it is not, a mistake has been made in arithmetic.

The "rod" column is not used for differential leveling but it is used for other types of leveling.

3. THE PHILADELPHIA ROD

The *Philadelphia rod* is commonly used in leveling. It consists of two sliding parts and can be extended from 7 ft to 13 ft in length. When it is 7 ft long, it is known as a "low rod." When it is 13 ft long, it is known as a "high rod."

The graduations on the face of the rod are in hundredths of a foot. They measure continuously from zero at the bottom to 13 ft at the top. Each full foot is marked with a red number (white in Fig. 13.4); each tenth of a foot is marked with black numbers (1 to 9) between two red numbers.

The hundredth graduations alternate from black to white. At each tenth mark, the black graduation is extended and slashed so that the top edge is read for 0.10, 0.20, 0.30, and so on. The black graduation at the halfway distance between each tenth mark (0.05) is also extended and slashed. The bottom edge of the black graduation is read.

Figure 13.3 *Continuous Differential Leveling*

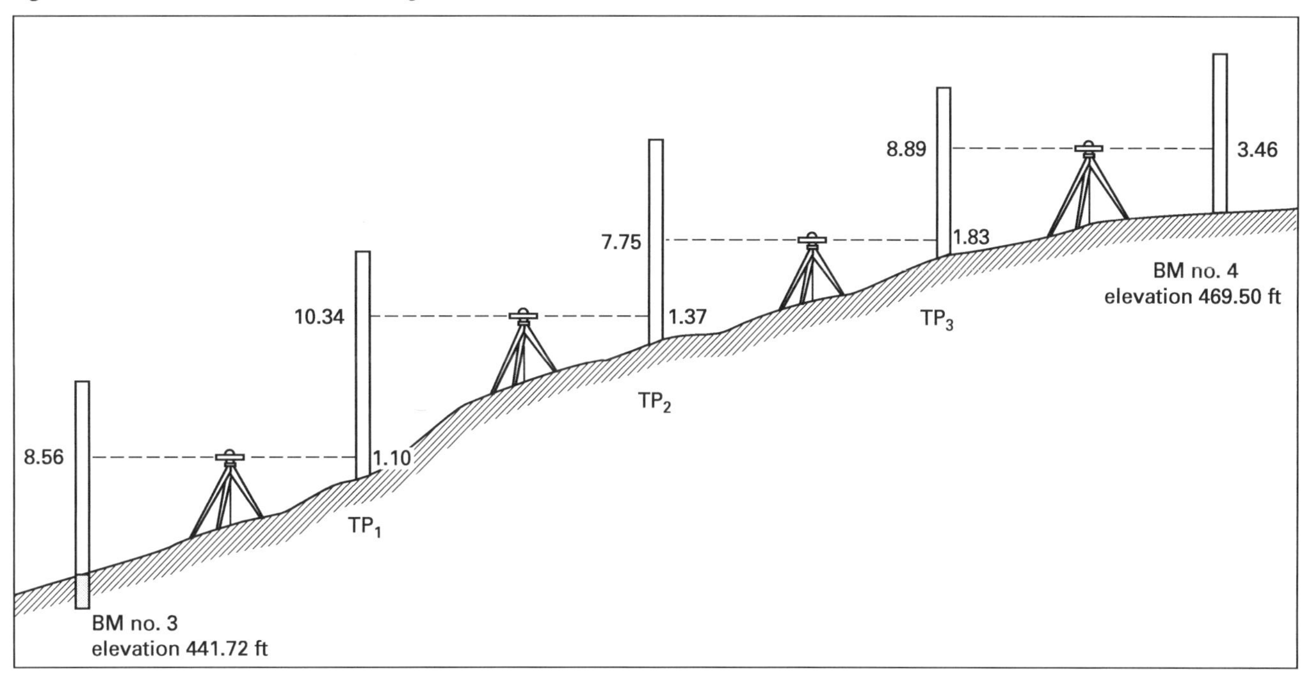

differential levels					
sta	+	HI	–	rod	elev
BM no. 3	8.56	450.28			441.72
TP	10.34	459.52	1.10		449.18
TP	7.75	465.90	1.37		458.15
TP	8.89	472.96	1.83		464.07
BM no. 4			3.46		469.50
	35.54		7.76		469.50
	7.76				441.72
	27.78	----	check	----	27.78

hwy 345	3-28-72 Jones ⊼
bolthead in concrete mon. – Lt. sta. 10+00	Smith ⏀
spike in 12 in elm 100 ft – Lt. sta. 25+50	

4. USING BLACK NUMBERS TO READ THE PHILADELPHIA ROD

In Fig. 13.4(a), the top edge of the black 1 is in line with the bottom edge of a black hundredth graduation; the reading, then, is 5.13. All the black numbers are accurately and consistently marked on the rod so that whenever the top edge of any black number aligns with the crosshair, the reading ends in 0.03. Whenever the bottom edge of any black number aligns with the crosshair, the reading ends in 0.07.

In Fig. 13.4(b) the top edge of the base of the black 2 is in line with the top edge of a black hundredth graduation, so that the reading is 5.18. The bottom edge of the top part of the black 2 is in line with the top edge of a black hundredth graduation, so that, if the middle crosshair were on that line, the reading would be 5.22.

The black numbers are easier to read than the black hundredth graduations, so the numbers can be used from bottom to top to read 0.07, 0.08, 0.02, and 0.03 at any black number. The 0.05 reading can be identified by the extended black graduation, which is slashed. Therefore, almost any reading can be made by using the black numbers, except near the red numbers. The red numbers are not made so that they can be read in a useful way other than for the full foot.

Figure 13.4 *Graduations on a Philadelphia Rod*

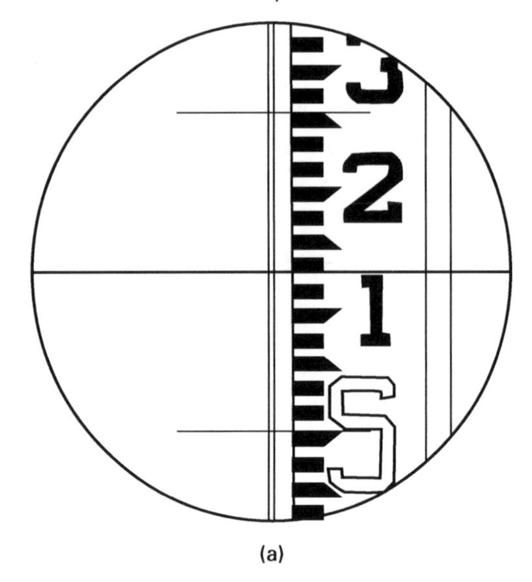

(a)

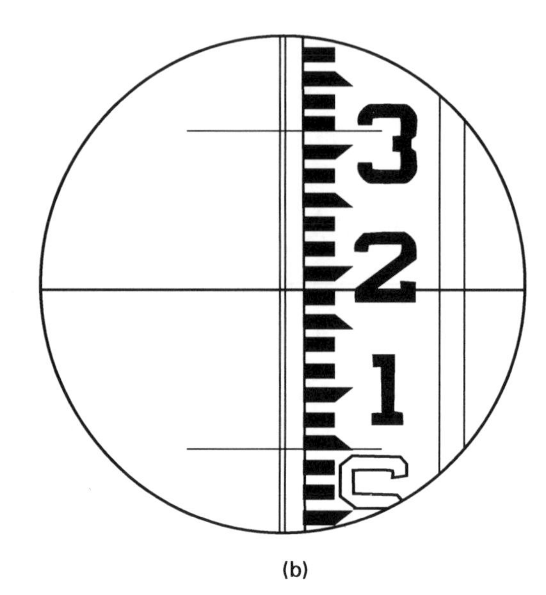

(b)

5. TARGETS

For long sights or for readings to the thousandth of a foot, a target attached to a vernier is used.

For readings less than 7 ft, the target is moved up or down the rod at the direction of the levelperson until it coincides with the middle horizontal crosshair. The rodperson then clamps it at that position. The reading is made by the rodperson using the vernier.

For readings greater than 7 ft, the target is clamped at 7.000 ft, and the top section of the rod is moved up or down at the direction of the levelperson. The reading is made by use of the vernier on the back of the rod.

6. VERNIERS

A *vernier* can be used to find a fractional part of the smallest division of a scale. Using a vernier on a Philadelphia rod, readings can be made to the thousandth of a foot.

It can be seen in Fig. 13.5(a) that ten spaces on the vernier cover nine of the smallest divisions on the scale, which on a Philadelphia rod are hundredths of a foot. For any vernier,

$$(n-1)d = nv \qquad 13.1$$

n is the number of divisions on the vernier, $n-1$ is the number of the smallest divisions on the scale, d is the length of the smallest scale division, and v is the length of a vernier division.

For a Philadelphia rod, $n = 10$, $d = 0.01$ ft, and $v = 0.09/10 = 0.009$ ft or $(10-1)(0.01) = (10)(0.009)$.

In Fig. 13.5(b), the vernier has moved so that the first division past the zero on the vernier is in line with the 0.31 mark on the scale. The vernier has moved a distance of 0.001, so the reading is 0.301.

In Fig. 13.5(c), the eighth division on the vernier is in line with a mark on the main scale, so the reading is 0.308.

In summary, to determine the rod reading, read the red number directly, determine the tenths reading by observing the zero mark on the vernier, and determine the thousandths reading by determining which mark on the vernier is in line with a hundredth graduation on the rod.

Figure 13.5 *Verniers*

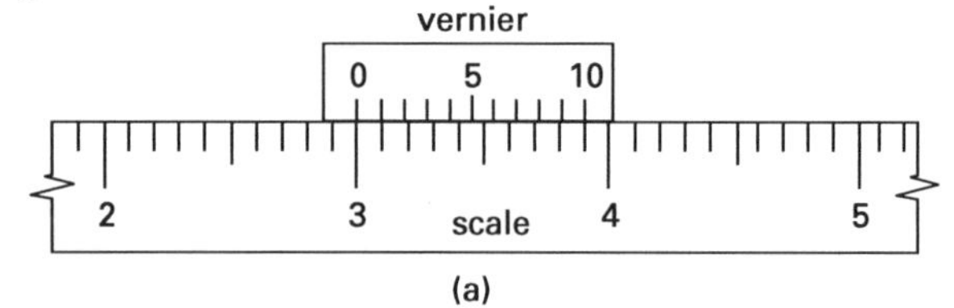

(a)

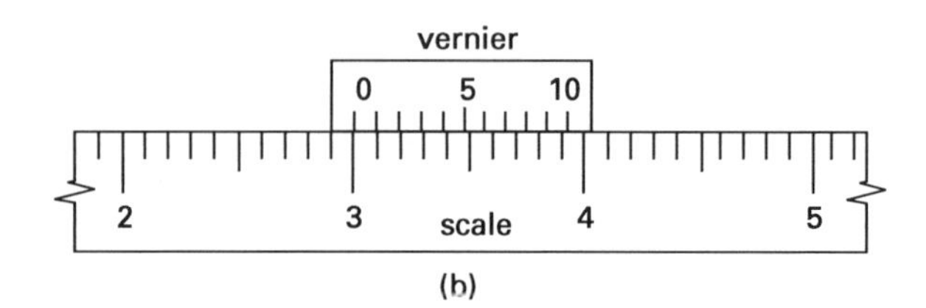

(b)

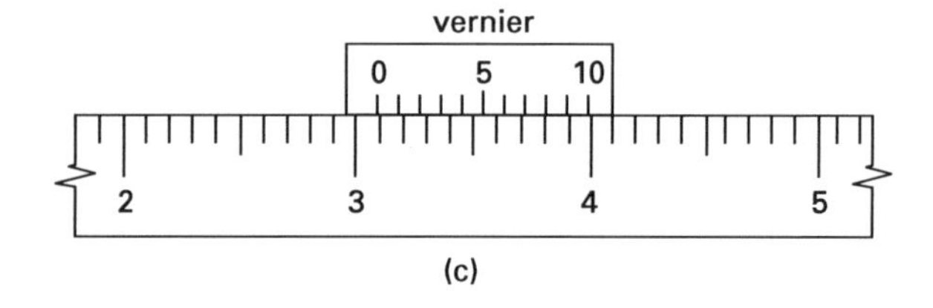

(c)

7. LEAST COUNT OF A VERNIER

The *least count of a vernier* is the smallest reading that can be determined without interpolation. For any vernier, the least count is d/n, where d is the length of smallest scale division and n is the number of divisions on the vernier. It can easily be remembered as

$$\text{least count} = \frac{\text{value of smallest division on the scale}}{\text{number of divisions on the vernier}} \quad 13.2$$

8. EFFECT OF CURVATURE OF THE EARTH

By definition, a level surface is a curved surface and a horizontal line is a straight line (see Fig. 13.1).

If a level sight were made on a level rod 1 mi away from any point on the earth, the reading, if one could be made, would be greater by 0.667 ft because of the curvature of the earth. The departure of the earth from the horizontal line varies as the square of the distance from the level to the rod. Two formulas can be used to find this distance.

$$C = 0.667M^2 \quad 13.3$$

$$C = 0.024F^2 \quad 13.4$$

C is the departure in feet, M is the distance in miles, and F is the distance in thousands of feet.

At a distance of 100 ft, $C = 0.00024$ ft. At a distance of 300 ft, $C = 0.0022$ ft.

Where sights are held to 300 ft and read to hundredths of a foot, the effect on elevations that are expressed to hundredths of a foot is very small when it is considered that the effect of the curvature is offset somewhat by refraction.

9. REFRACTION

Light passing through the atmosphere is bent so that in reading a rod, the reading is less. This offsets the effect of the earth's curvature by about 14%. A formula for the combined effect of curvature and refraction is

$$h = 0.574M^2 = 0.0206F^2 \quad 13.5$$

h is the error in feet, M is in miles, and F is in thousands of feet.

For a sight of 200 ft, $h = 0.0008$ ft, and for a sight of 300 ft, $h = 0.0019$ ft.

10. WAVING THE ROD

It is extremely important that the rod be plumb when a reading is taken. The levelperson can bring the rod into plumb in one direction by observing the vertical crosshair and signaling the rodperson to plumb the rod, but the levelperson cannot tell whether the rod is leaning toward or away from the rodperson. For low rods, the rodperson can hold the rod lightly between finger tips just in front of his nose and balance the rod. For high rods, however, this is more difficult.

If the bench mark or turning point is not a flat surface (it should not be), the plumb position can be found by a method known as *waving the rod*. The rodperson moves the rod slowly toward and then away from the level while the levelperson observes the horizontal crosshair on the rod. The rod is plumb at the lowest reading. The error caused by the rod being out of plumb can be approximated by adapting Eq. 11.3 from Ch. 11, Sec. 16.

$$E = \frac{D^2}{2L} \quad 13.6$$

E is the error caused by the rod being out of plumb, D is the distance in feet that the top of the rod is out of plumb, and L is the length of the rod.

Example 13.1

A rod reading of 12.000 was made on a 12 ft rod when it was 9 in out of plumb. What is the error, E?

Solution

$$D = \frac{9 \text{ in}}{12 \frac{\text{in}}{\text{ft}}} = 0.75 \text{ ft}$$

The error E is

$$E = \frac{(0.75 \text{ ft})^2}{(2)(12 \text{ ft})} = 0.0234 \text{ ft}$$

If the rod reading were 10.500 under the same conditions, the error E would be

$$E = \frac{(10.500)(0.75 \text{ ft})^2}{(12)(2)(12 \text{ ft}^2)} = 0.021 \text{ ft}$$

11. PARALLAX

Parallax is the apparent change in the position of the crosshair as viewed through the telescope. Because the reticle (the ring that holds the crosshairs in the telescope) is stationary, the distance between it and the eyepiece must be adjusted to suit the eye of each individual observer. The eyepiece is adjusted by turning it slowly until the crosshair is as black as possible. After the eyepiece is adjusted, the object viewed should be brought into sharp focus by means of the focusing knob for the objective lens. If the crosshairs seem to move across the object when the viewer moves his eye slightly, parallax exists. It is eliminated by carefully adjusting the eyepiece and the objective lens. If parallax is not eliminated, it can affect the accuracy of the rod readings.

12. BALANCING SIGHTS

The most common cause of errors in leveling is imperfect adjustment of the level. Centering the level bubble establishes a horizontal plane for the observer. If the level is not properly adjusted, however, the line of sight may not be parallel to the axis of the level vial, causing the rod reading to be greater or less than the true reading. The error can be offset by *balancing sights*—that is, by making the horizontal length of plus sights and minus sights approximately equal for each setup of the level. Leveling uphill or downhill makes this impossible, but if the total length of plus sights equals the total length of minus sights for a line of levels, the result will be the same. Distances can be determined by means of the stadia hairs in modern levels.

It is extremely important to make sure that the level bubble is exactly centered at the instant of a rod reading. The bubble should be centered, the telescope should be focused on the rod, a check on the bubble should be made, and a final reading should be made without touching the level. All this can be done in a few seconds.

13. RECIPROCAL LEVELING

Running a line of levels across a river or other obstacle where the horizontal distance is more than the desired maximum can be performed using *reciprocal leveling.*

The level is set up on the bank of the river, and turning point A is established nearby on the same side of the river. Turning point B is established on the other side of the river. A reading is taken on turning point A, and several readings are made on turning point B by unleveling, releveling, and then averaging the readings. The level is then set up on the side of the river opposite point A and near point B. Readings are made on B and A in the same manner as before. The difference in elevation between A and B is determined from the average readings.

14. DOUBLE-RODDED LEVELS

When saving time is important, using two rods and two sets of turning points on the same line of levels provides a good check on the difference in elevations. The notekeeper will, in effect, have two sets of notes.

15. THREE-WIRE LEVELING

Reading the two stadia hairs and the middle crosshair at each turning point and bench mark provides an excellent check for a line of levels. The difference in the middle crosshair reading and the upper stadia hair reading should be very near the difference in the middle crosshair reading and the lower stadia hair reading. If there is a discrepancy, one of the readings can be disregarded; otherwise, differences in elevations can be determined by averaging the three readings.

16. PROFILE LEVELING

In planning highways, canals, pipelines, and so on, a vertical section of the earth is needed to determine the vertical location of the centerline of the project. This vertical section is known as a *profile.* It is plotted on paper from field notes. *Profile leveling* is similar to differential leveling except that many minus readings are taken in addition to the usual plus and minus readings taken on bench marks and turning points.

At each setup of the level, readings are taken on the ground along the centerline at each full station and at each break on the ground. (A *break* on the ground is a point on the ground where the slope changes.) These readings are all minus readings. They are measurements made to determine the elevation at each point on the profile, and they are subtracted from the HI at each level setup. For clarity, these ground readings are recorded in the rod column. Bench mark and turning point readings are recorded as they are in differential leveling. To determine elevations of ground points, all readings taken at one level setup are subtracted from the HI at that level setup.

After elevations at each ground point are determined, they are plotted on specially ruled paper known as *profile paper* or *profile sheets.*

Figure 13.7 shows the profile plotted from the field notes in Fig. 13.6.

Figure 13.6 Profile Leveling Field Notes

profile levels						FM ROAD 123
sta	+	HI	−	rod	elev	
BM no. 4	4.87	483.13			478.26	R.R. spike in 12 in oak 75 ft Lt. sta. 33+50
32+00				11.5	471.6	
33+00				9.4	473.7	
+75				10.1	473.0	
34+00				8.2	474.9	
35+00				3.0	480.1	
+15				1.9	481.2	
+70				2.3	480.8	
36+00				5.2	477.0	
+50				6.8	476.3	
37+00				5.9	477.2	
38+00				13.3	469.8	
TP	4.54	476.95	10.72		472.41	T/stake 38+00 Lt
38+60				13.2	463.8	
39+00				12.0	465.0	
40+00				3.9	473.1	
41+00				1.2	475.8	
42+00				0.8	476.2	
+70				0.7	476.3	
+80				1.5	475.5	
43+00				0.4	476.6	
BM no. 5			0.17		476.78	R.R. spike in 16 in elm 100 ft rt. sta. 42+50
	9.41		10.89			
	10.89 − 9.41 = 1.48		478.26 − 476.78 = 1.48			

Figure 13.7 Profile Plotted from Field Notes (Fig. 13.6)

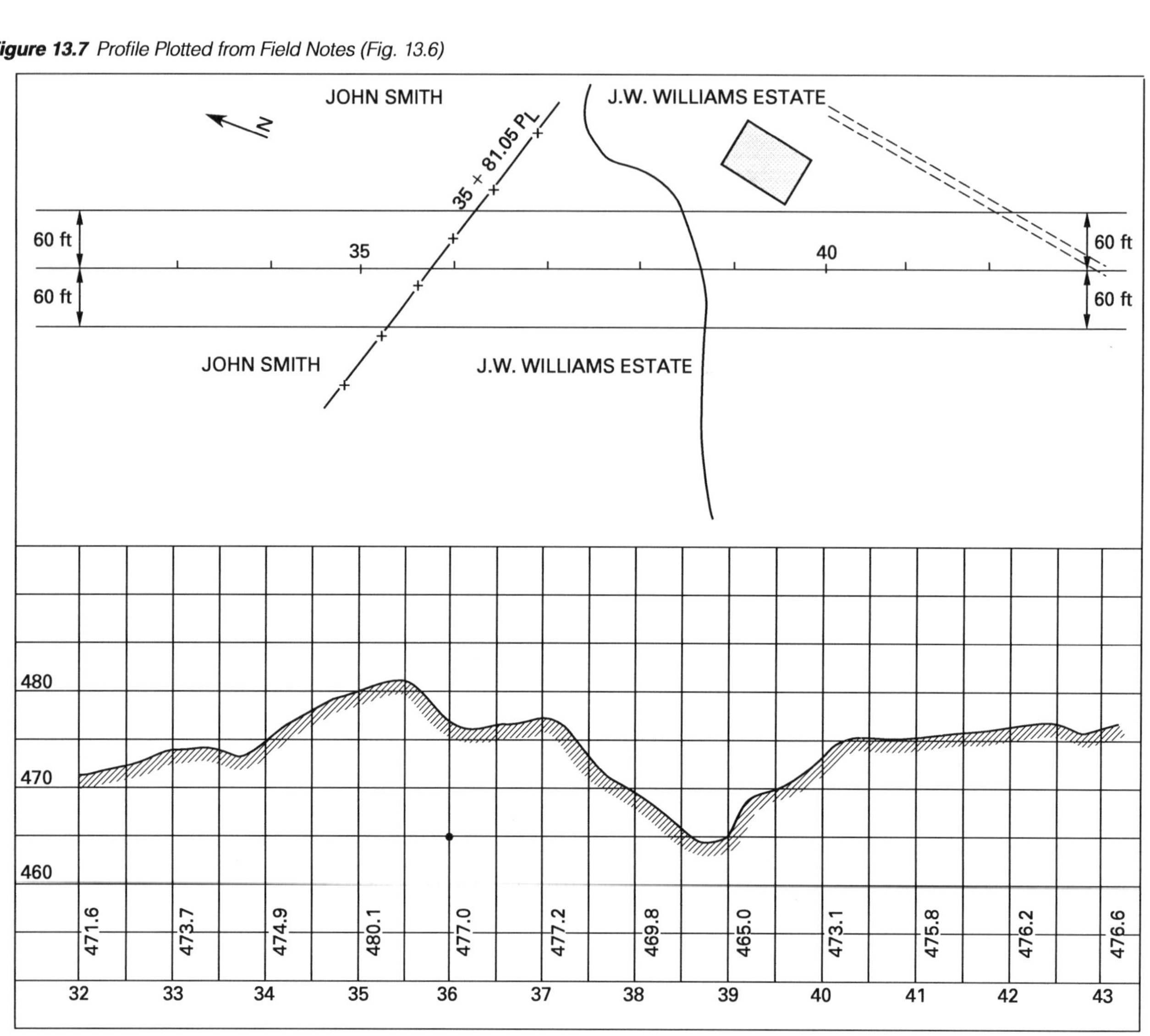

PRACTICE PROBLEMS

1. A level surface is

(A) a flat surface perpendicular to a horizontal line

(B) a horizontal surface perpendicular to a plumb line

(C) a curved surface perpendicular, at any point, to a plumb line

(D) all of the above

2. A plumb line is

(A) a horizontal line established by a spirit level

(B) a line established by a plumb bob

(C) a vertical line

(D) both (B) and (C)

3. A datum is

(A) a horizontal plane

(B) any level surface to which elevations are referred

(C) the field notes kept by a survey party

(D) none of the above

4. The term *backsight* in leveling means

(A) a sight toward the beginning bench mark

(B) a rod reading on a turning point

(C) a rod reading on a point whose elevation is known

(D) all of the above

5. The term *foresight* in leveling means

(A) a sight on a point whose elevation is to be determined
(B) a minus sight
(C) both (A) and (B)
(D) neither (A) nor (B)

6. A plus sight is taken on

(A) a point whose elevation is known
(B) a point whose elevation is to be determined
(C) the starting bench mark
(D) both (A) and (C)

7. The height of instrument, as used in leveling, is the

(A) distance from the ground to the axis of the telescope
(B) elevation of the line of sight above the datum plane
(C) height of the line of sight above a bench mark
(D) all of the above

8. A point observed through a level telescope appears to be higher than it actually is because of

(A) parallax
(B) refraction
(C) curvature of the earth
(D) all of the above

9. Balancing distances to backsights and foresights eliminates

(A) errors caused by curvature of the earth
(B) errors caused by refraction
(C) parallax
(D) both (A) and (B)

10. A rod reading of 10.00 was made on a 12 ft rod when it was 10 in out of plumb. What is the error caused by the rod's being out of plumb?

(A) 0.01 ft
(B) 0.02 ft
(C) 0.04 ft
(D) 0.20 ft

11. If a level sight were made on a level rod 1000 ft away, what would be the correction, C, for curvature of the earth? (Note: $C = 0.024F^2$.)

(A) −2.4 ft
(B) −0.024 ft
(C) +0.0024 ft
(D) +0.24 ft

12. What would be the combined error, h, for curvature and refraction for a 500 ft sight? (Note: $h = 0.0206F^2$.)

(A) 0.004 ft
(B) 0.005 ft
(C) 0.01 ft
(D) 0.02 ft

13. Record the rod reading for each of the following figures.

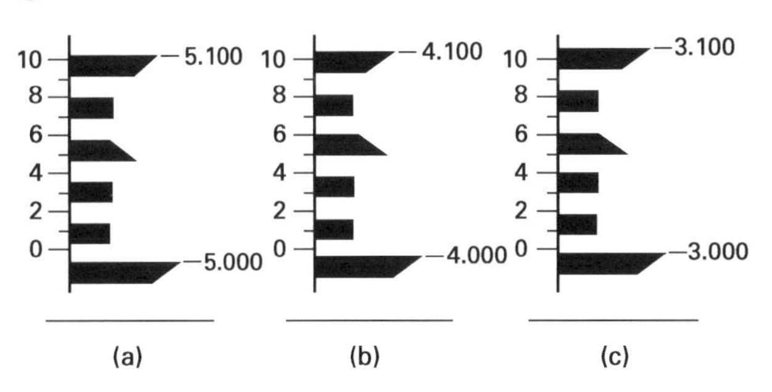

14. Complete the following level notes and show the arithmetic check.

differential levels					
sta	+	HI	−	rod	elev
BM no. 1	9.45				431.71
TP	11.88		1.08		
TP	10.99		0.62		
TP	8.33		1.30		
TP	11.21		2.47		
BM no. 2	10.54		3.90		
TP	5.60		6.77		
TP	4.18		8.32		
TP	3.90		9.74		
BM no. 3			5.12		

15. Complete the following level notes.

differential levels					
sta	+	HI	−	rod	elev
BM no. 1		455.58			447.23
TP		465.14	0.88		
TP	9.56	473.67			464.11
TP	7.16	480.08			472.92
TP		488.76	1.65		478.43
BM no. 2	9.22		2.70		486.06
TP	4.04	491.41	7.91		487.37
TP	3.51	487.82	7.10		
TP		482.04			479.21
BM no. 3			5.17		476.87

16. Complete the following field notes, show the arithmetic check, and plot the profile on the blank profile sheet on the following page.

profile levels					
sta	+	HI	−	rod	elev
BM no. 1	0.17				476.78
32+00				0.4	
33+00				0.8	
34+00				1.2	
35+00				3.9	
36+00				12.0	
+40				1.2	
TP	10.72		4.54		
37+00				13.3	
38+00				5.9	
+50				6.8	
39+00				5.2	
+30				2.3	
+85				1.9	
40+00				3.0	
41+00				8.2	
+25				10.1	
42v00				9.4	
43+00				11.5	
BM no. 2			4.87		

Problem 16: Blank Profile Sheet

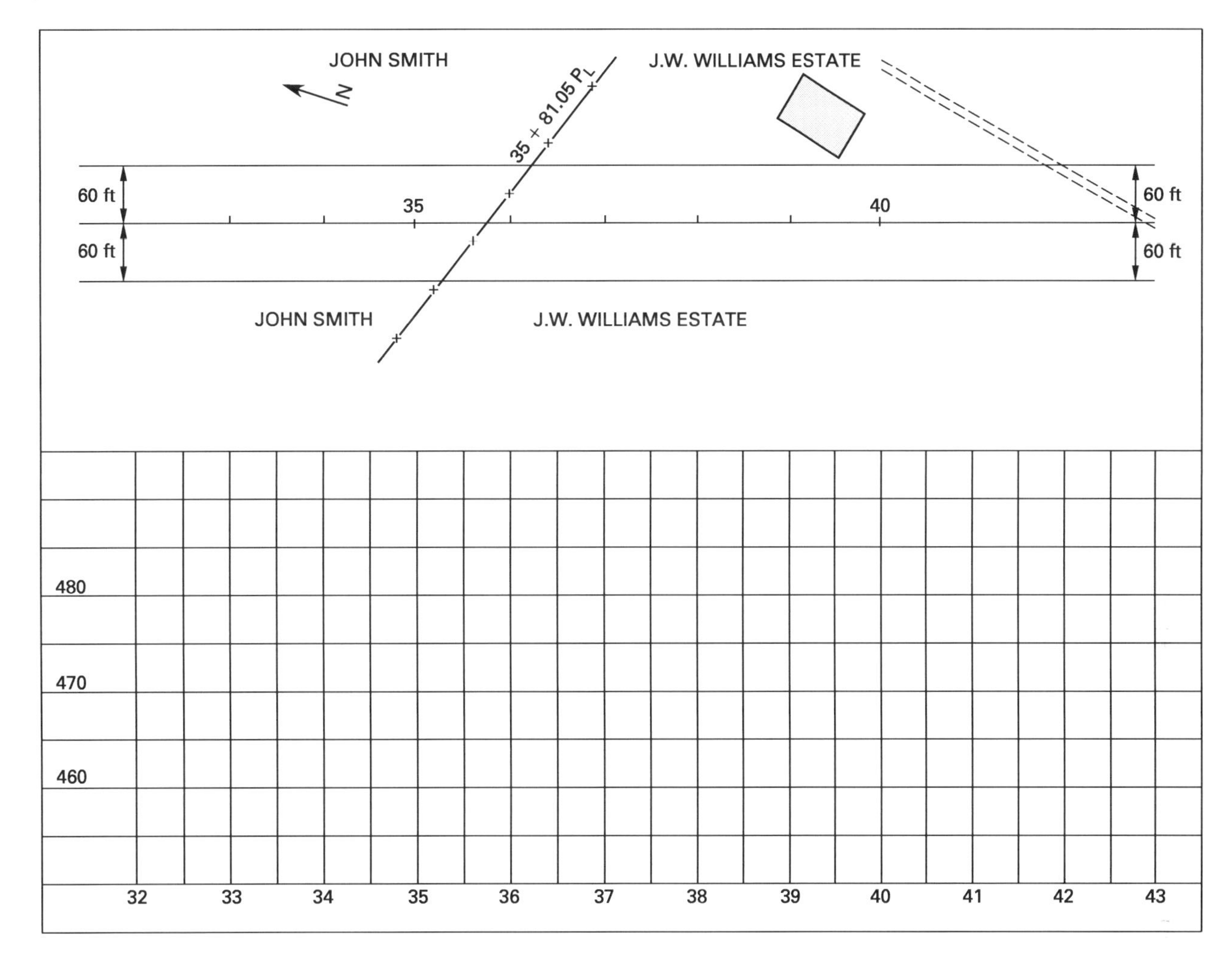

Field Data Acquisition

SOLUTIONS

1. A level surface is a curved surface perpendicular, at any point, to a plumb line.

The answer is (C)

2. A plumb line is both a vertical line and a line established by a plumb bob.

The answer is (D)

3. A datum is any level surface to which elevations are referred.

The answer is (B)

4. Backsight in leveling can mean a sight toward the beginning bench mark, a rod reading on a turning point, or a rod reading on a point whose elevation is known.

The answer is (D)

5. Foresight in leveling can mean a sight on a point whose elevation is to be determined, or a minus sight.

The answer is (C)

6. A plus sight is taken on a point whose elevation is known, or a point whose elevation is to be determined.

The answer is (D)

7. The height of instrument, as used in leveling, is the elevation of the line of sight above the datum plane.

The answer is (B)

8. A point observed through a level telescope appears to be higher than it actually is because of refraction.

The answer is (B)

9. Balancing distances to backsights and foresights eliminates errors caused by curvature of the earth and errors caused by refraction.

The answer is (D)

10. Using Eq. 13.6,

$$E = \frac{D^2}{2L}\left(\frac{\text{rod reading}}{\text{rod length}}\right) = \left(\frac{\left(\frac{10 \text{ in}}{12 \frac{\text{in}}{\text{ft}}}\right)^2}{(2)(12.00 \text{ ft})}\right)\left(\frac{10.00 \text{ ft}}{12.00 \text{ ft}}\right)$$

$$= 0.02 \text{ ft}$$

The error caused by the rod's being out of plumb is 0.02 ft.

The answer is (B)

11. Using Eq. 13.4,

$$C = 0.024F^2 = (0.024)(1^2) = -0.024 \text{ ft}$$

The refraction correction is always negative. Therefore, the correction for the curvature would be −0.024 ft.

The answer is (B)

12. Using Eq. 13.5,

$$h = 0.0206F^2 = (0.0206)(0.5^2) = 0.005 \text{ ft}$$

The error is positive, although the correction is negative. Therefore, the combined error would be 0.005 ft.

The answer is (B)

13.

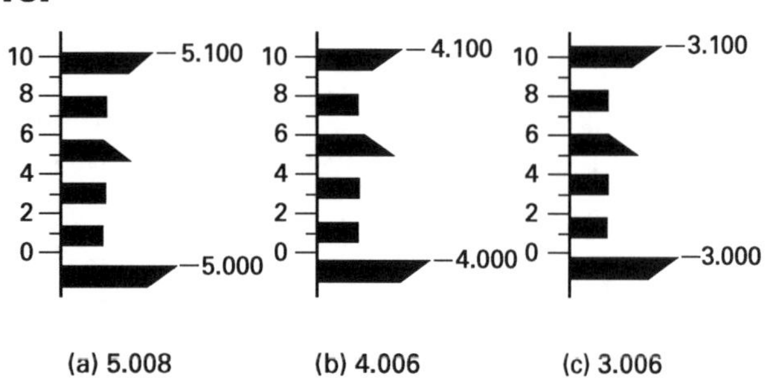

14.

differential levels					
sta	+	HI	−	rod	elev
BM no. 1	9.45	**441.16**			431.71
TP	11.88	**451.96**	1.08		**440.08**
TP	10.99	**462.33**	0.62		**451.34**
TP	8.33	**469.36**	1.30		**461.03**
TP	11.21	**478.10**	2.47		**466.89**
BM no. 2	10.54	**484.74**	3.90		**474.20**
TP	5.60	**483.57**	6.77		**477.97**
TP	4.18	**479.43**	8.32		**475.25**
TP	3.90	**473.59**	9.74		**469.69**
BM no. 3			5.12		**468.47**
	76.08		**39.32**		
	39.32				**468.47**
	36.76				**431.71**
					36.76

15.

differential levels					
sta	+	HI	–	rod	elev
BM no. 1	**8.35**	455.58			447.23
TP	**10.44**	465.14	0.88		**454.70**
TP	9.56	473.67	**1.03**		464.11
TP	7.16	480.08	0.75		472.92
TP	**10.33**	488.76	1.65		478.43
BM no. 2	9.22	**495.28**	2.70		486.06
TP	4.04	491.41	7.91		484.37
TP	3.51	487.82	7.10		**484.31**
TP	2.83	482.04	**8.61**		479.21
BM no. 3			5.17		476.87
	65.44		**35.80**		
	35.80				**476.87**
	29.64				**447.23**
					29.64

16. See the field notes and profile sheet on the following page.

profile levels					
sta	+	HI	–	rod	elev
BM no. 1	0.17	**476.95**			476.78
32+00				0.4	**476.6**
33+00				0.8	**476.2**
34+00				1.2	**475.8**
35+00				3.9	**473.1**
36+00				12.0	**465.0**
+40				1.2	**475.8**
TP	10.72	**483.13**	4.54		**472.41**
37+00				13.3	**469.8**
38+00				5.9	**477.2**
+50				6.8	**476.3**
39+00				5.2	**477.9**
+30				2.3	**480.8**
+85				1.9	**481.2**
40+00				3.0	**480.1**
41+00				8.2	**474.9**
+25				10.1	**473.0**
42+00				9.4	**473.7**
43+00				11.5	**471.6**
BM no. 2			4.87		**478.26**
	10.89		**9.41**		**476.78**
	9.41				**1.48**
	1.48				

Field Data Acquisition

Solution 16: Profile Sheet

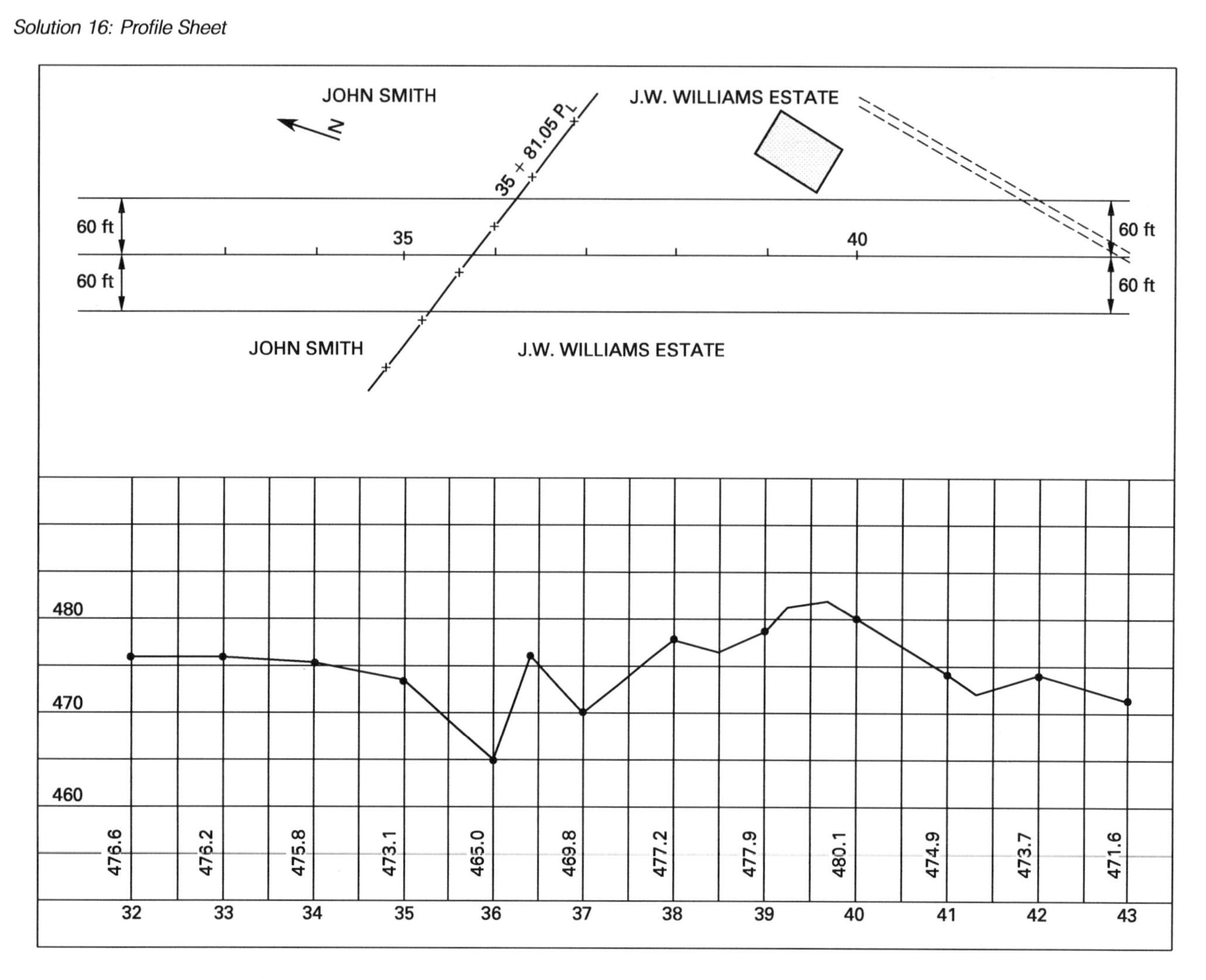

14 Compass Survey

Nomenclature

C	correction of a slope measurement
d	length of smallest division on rod scale
F	per 1000 ft
h	error in feet caused by refraction
H	horizontal distance
M	per mile
n	number of divisions on vernier
S	slope distance
T	temperature
v	length of a vernier division
V	difference in elevation between horizontal and slope distances

1. MAGNETIC NEEDLE

A *magnetic needle* is a slender, magnetized steel rod that, when freely suspended at its center of gravity, points to magnetic north.

2. MAGNETIC DIP

In the northern hemisphere, the magnetic needle dips toward the north magnetic pole. In the southern hemisphere, the needle dips toward the south magnetic pole. To counteract the dip so that the needle will be horizontal, a counterweight is attached to the south end of the needle in the northern hemisphere and to the north end in the southern hemisphere. This weight is usually a short piece of fine brass wire.

3. THE MAGNETIC COMPASS

The *magnetic compass* consists of a magnetic needle mounted on a pivot at the center of a graduated circle in a metal box covered with a glass plate. It is constructed so that the angle between a line of sight and the magnetic meridian can be measured. The line of sight, with the horizontal circle, can be rotated in the horizontal plane while the needle continues to point to magnetic north. The point of the needle marks the angle made by the magnetic meridian and the line of sight.

Figure 14.1 Typical Compass

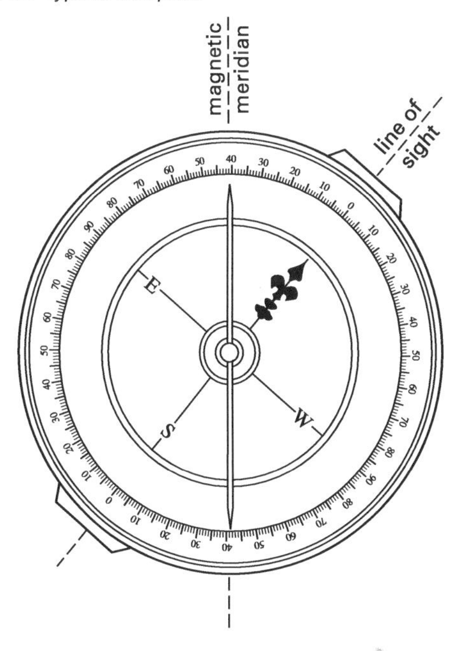

4. THE SURVEYOR'S COMPASS

The horizontal circle in the *surveyor's compass* is usually graduated in half degrees. (Figure 14.1 does not show these graduations.) The letters E and W on the compass are reversed so that direct readings of bearings can be made. In the figure, the bearing of the line of sight is N 40°E. When the letters are reversed, the north end of the needle lies in the northeast quadrant of the horizontal circle.

5. MAGNETIC DECLINATION

The magnetic poles do not coincide with the axis of the earth. The horizontal angle between the magnetic meridian and the true (geodetic) meridian is known as *declination*. In some areas, the needle points east of true north, and in some areas it points west of true north. Zero declination is found along a line between areas. This line is known as the *agonic line*. It passes, generally, through Florida and the Great Lakes, but it is constantly changing its location. East of this line, declination is west (−); west of this line, declination is east (+). In the United States, declination varies from 0° to 23°.

6. VARIATIONS IN DECLINATION

Declination in any one point varies daily, annually, and secularly (over a long period of time such as a century). Declination for a particular location for a particular year can be obtained from the United States Geological Survey.

7. IMPORTANCE OF COMPASS SURVEYING

Compass surveying is as obsolete as the Gunter's chain, but it is important for the modern surveyor to understand it when retracing old lines. The modern surveyor needs to be able to convert magnetic bearings to true bearings. In order to do so, the surveyor must know whether the declination is east or west and what the declination is, or was, on a certain day. Typical problems are given as follows.

Example 14.1

Convert the following magnetic bearings to true bearings.

(a) N 68°20′ E, decl 8°00′ W

(b) S 12°30′ W, decl 3°45′ E

(c) S 20°30′ E, decl 6°30′ W

(d) N 3°15′ W, decl 4°20′ E

Solution

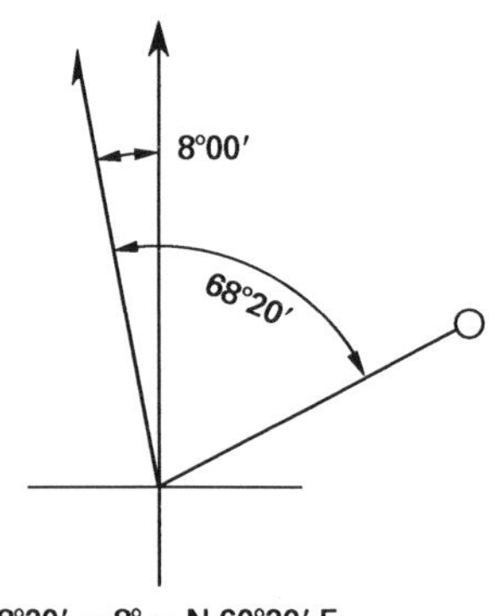

68°20′ − 8° = N 60°20′ E
(a)

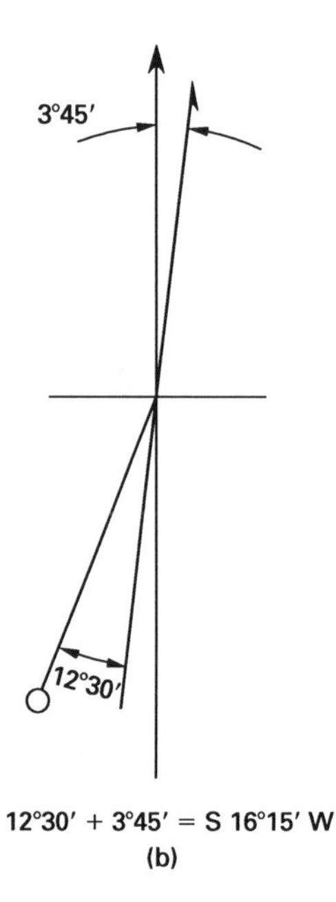

12°30′ + 3°45′ = S 16°15′ W
(b)

Magnetic bearings can also be converted to true bearings by use of azimuths.

Example 14.2

The magnetic bearing of a line was S 85°15′ W at a location where the declination was 8°30′ E. Find the true bearing.

Solution

$$\text{magnetic bearing} = \text{S}\,85°15'\,\text{W}$$

$$\text{magnetic azimuth} = 265°15'$$

$$\text{declination} = +8°30'$$

$$\text{true azimuth} = 273°45'$$

$$\text{true bearing } \beta = \text{N}\,86°15'\,\text{W}$$

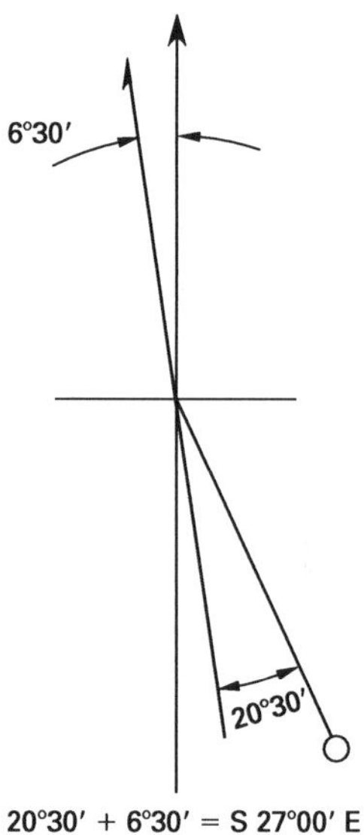

20°30′ + 6°30′ = S 27°00′ E
(c)

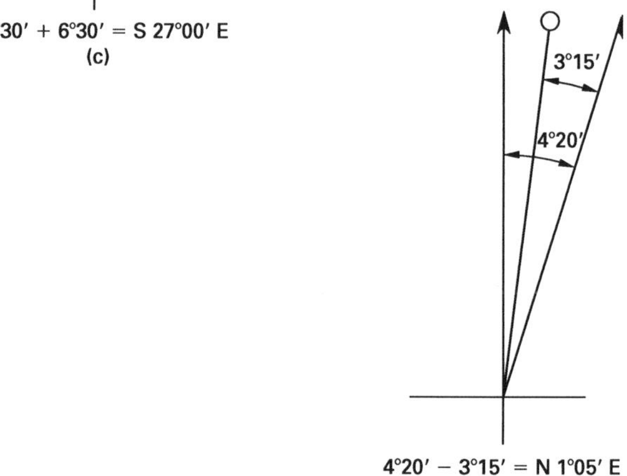

4°20′ − 3°15′ = N 1°05′ E
(d)

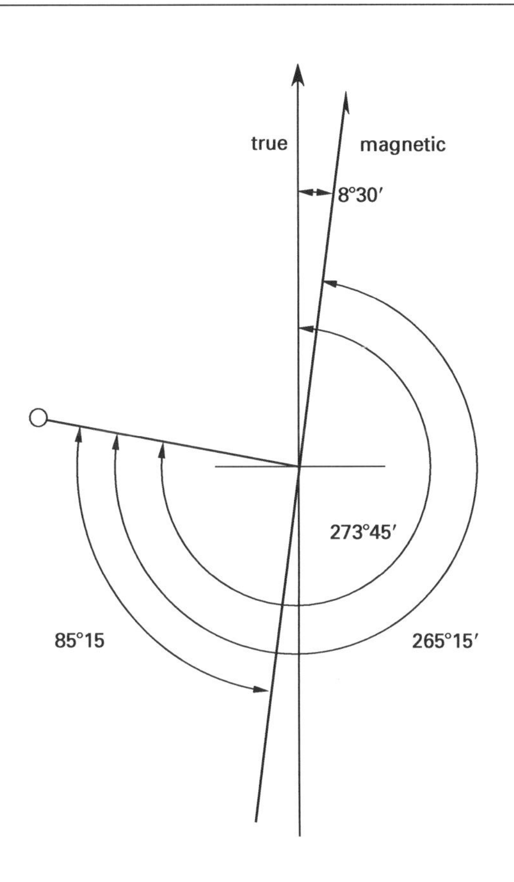

PRACTICE PROBLEMS

For these problems, convert the magnetic bearings to true (geodetic) bearings. Draw sketches.

1. bearing: S 56°10′ W
declination: 7°30′ W

2. bearing: N 48°30′ W
declination: 4°20′ E

3. bearing: S 68°10′ E
declination: 5°40′ W

For Probs. 4 and 5, convert the bearings to azimuths and determine the true bearings. Draw sketches showing all angles.

4. magnetic bearing: S 89°50′ E
declination: 6°30′ W

5. magnetic bearing: N 88°15′ W
declination: 3°20′ W

SOLUTIONS

1.

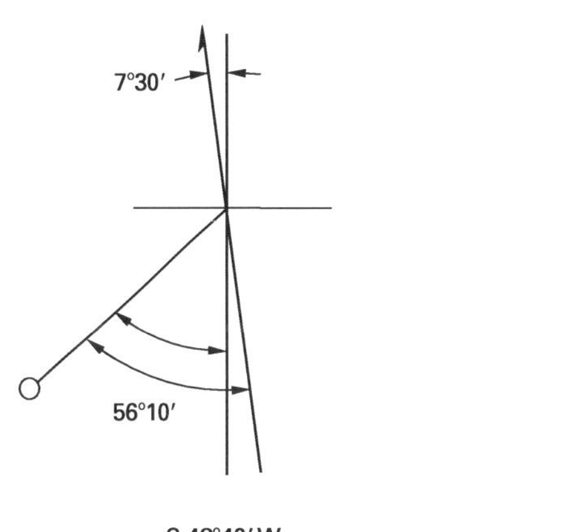

true brg S 48°40′ W

2.

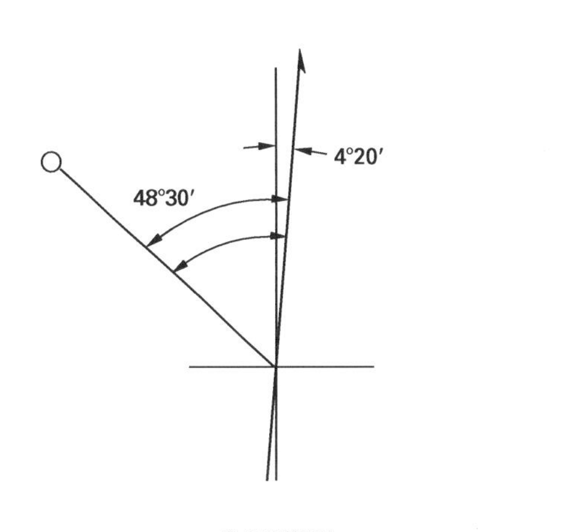

true brg N 44°10′ W

3.

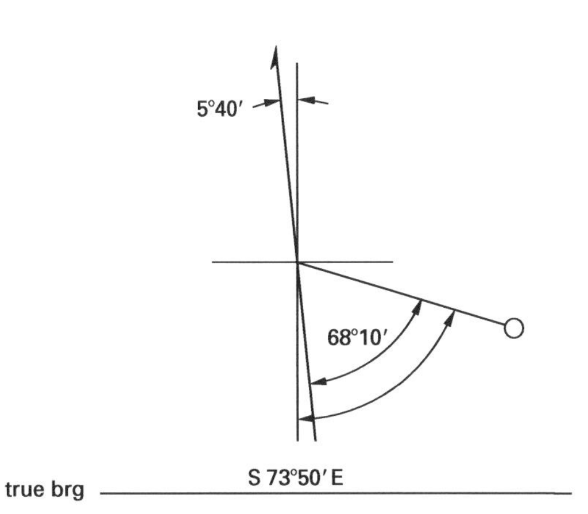

true brg S 73°50′ E

Field Data Acquisition

4.

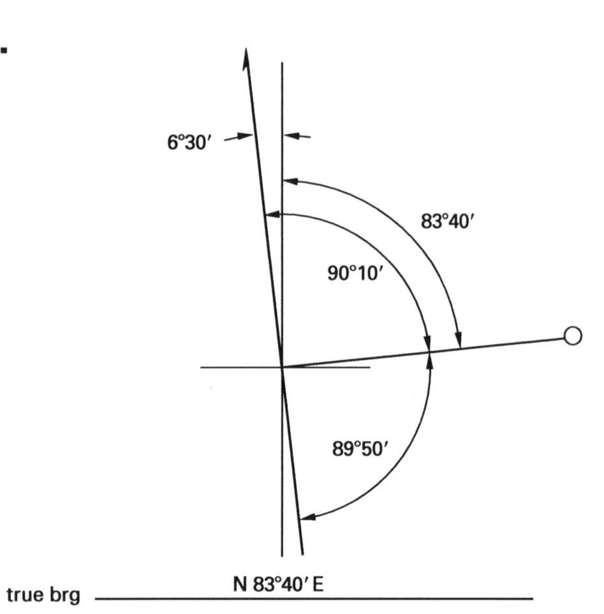

true brg N 83°40′ E

5.

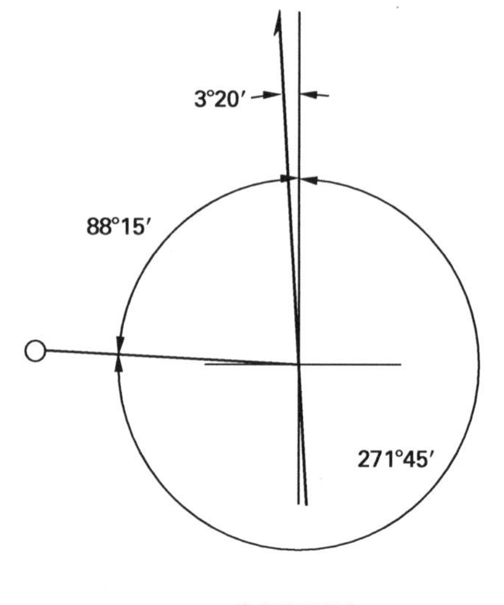

true brg S 88°25′ W

Topic III: Plane Survey Calculations

Chapter

15 Traverse

1. INTRODUCTION

A *traverse* is a series of lines connecting successive instrument stations of a survey. The relative position of the stations is determined by the direction and length of the lines. Typically, an angular measurement is taken at each point where the traverse changes direction, and a distance measurement is made along each connecting line.

2. OPEN TRAVERSE

An *open traverse* is a series of lines that do not return to the starting point. It is used in route surveying for the location of highways, pipelines, canals, and so on. To check the accuracy of an open traverse, the traverse must start and end at points of known position.

Figure 15.1 *Open Traverse*

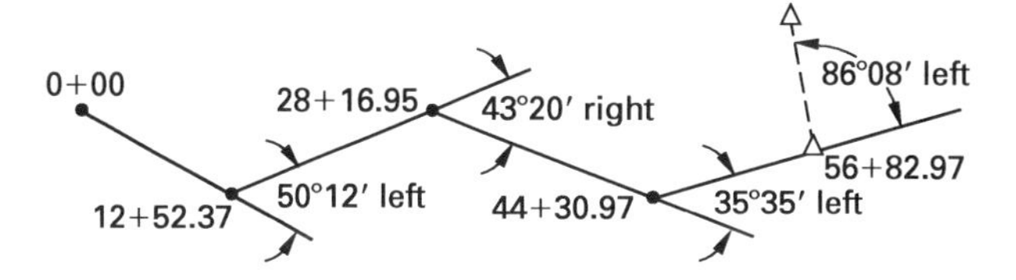

3. CLOSED TRAVERSE

A *closed traverse*, also called a *loop traverse*, starts and ends at the same point. Because it is a closed polygon, the interior angles and the lengths of the sides may be checked for accuracy and mathematically adjusted.

Figure 15.2 *Closed Traverse*

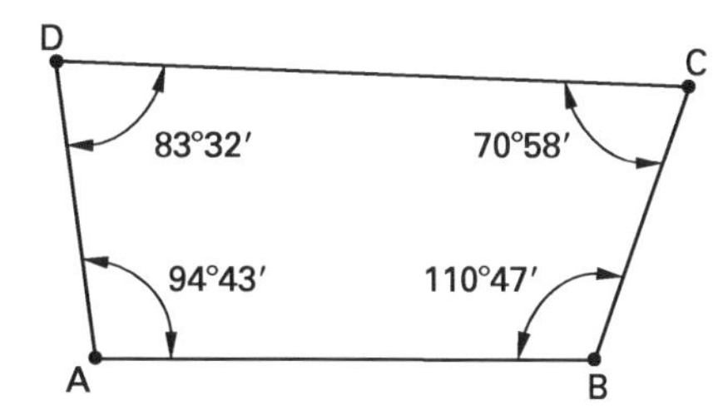

Plane Survey Calculations

4. HORIZONTAL ANGLES

Angles measured for open traverses are usually deflection angles as shown in Fig. 15.1. Angles measured for closed traverses are usually interior angles as shown in Fig. 15.2, but they can be deflection angles as shown in Fig. 15.9.

Interior angles can be turned clockwise (right) or counterclockwise (left), but usually are turned by the method known as *angles to the right.*

5. DEFLECTION ANGLES

A *deflection angle* is an angle between a line and the extension of the preceding line, as shown in Fig. 15.1. It may be turned either right or left from the extension, but the direction of turning must be recorded with the angular measurement. In an open traverse, the straight lines between the points of change in direction are known as *tangents*, and the points of change in direction are known as *points of intersection* (PI).

6. ANGLES TO THE RIGHT

The interior angles of the closed traverse in Fig. 15.2 are known as *angles to the right.* They are measured from the backsight station to the foresight station in a clockwise direction. With the instrument at station A, a backsight was made on station D, the telescope was turned in a clockwise direction to station B, and the angle 94°43′ was read and recorded. With the instrument at station B, the backsight was on station A and foresight was on station C. At each point, the angle was measured in a clockwise direction, or to the right.

The stations in the traverse ABCDA run in a counterclockwise direction, alphabetically, which is appropriate for the angles-to-the-right method. Most theodolites measure angles only to the right, which makes the counterclockwise lettering of the stations necessary. But it is not improper to letter traverse points in a clockwise direction.

7. DIRECTION OF SIDES

The *direction of the sides* of the traverse may be determined if the direction of one of the sides is known. If the direction of none of the sides is known, the direction of one side may be determined by measuring the angle at the intersection of this line and a line outside the traverse that has a known direction, or by making an observation on the sun or the stars. Otherwise, the direction of one of the lines must be assumed.

8. ANGLE CLOSURE

The sum of the interior angles of a polygon depends on the number of sides of the polygon. For a triangle, the sum is 180°; for a four-sided polygon, the sum is 360°. For any polygon with n sides, the sum is

$$\text{sum of angles} = (n - 2)(180°) \qquad 15.1$$

After the interior angles of a traverse have been measured, they should be adjusted so that their sum agrees with Eq. 15.1. The error may be distributed evenly at each angle, or it may be distributed arbitrarily in accordance with the surveyor's knowledge of the conditions of the survey. The error should be within the limits allowed in specifications for the survey. In land surveying, these specifications are usually based on the value of the land. Surveys of metropolitan areas are performed at much more rigid standards than are surveys of arid ranch lands.

The interior angles of an arbitrary five-sided traverse are balanced in the following example.

angles as measured	
A	96°03′30″
B	95°19′30″
C	65°13′00″
D	216°19′30″
E	67°06′00″
	540°01′30″

balanced angles $(n - 2)(180°) = 540°$	
A	96°03′
B	95°19′
C	65°13′
D	216°19′
E	67°06′
	540°00′

The angles in this traverse were measured with a one-minute vernier, and the balanced angles reflect this accuracy. However, the total discrepancy of 90 sec could be distributed evenly over the five angles by subtracting 18 sec from each recorded angle.

A	96°03′12″
B	95°19′12″
C	65°12′42″
D	216°19′12″
E	67°05′42″
	540°00′00″

9. METHODS OF DESIGNATING DIRECTION

The *direction of a line* is expressed as the angle between a meridian and the line. The *meridian* may be a *true meridian* (a great circle of the earth passing through the poles), a *magnetic meridian* (the direction of which is defined by a compass needle), or a *grid meridian* (established for a plane coordinate system).

The direction of a line may be expressed as its *bearing* or its *azimuth*. *True bearing* or *true azimuth* is measured from true north, referred to as *geodetic north*.

Old land surveys were usually referenced to the magnetic meridian. The direction of a line was determined by reading the angle between the line and the compass needle. This method has long since been discarded for most surveys; modern surveys refer to geodetic north or grid north. But since old surveys must be retraced, an understanding of the magnetic meridian is essential to the land surveyor.

10. BEARING

The *bearing of a line* is the horizontal acute angle between the meridian and the line. Because the bearing of a line cannot exceed 90°, the full horizontal circle is divided into four *quadrants:* northeast, southeast, southwest, and northwest. An angle of 40°, measured between the meridian and a line in each of the four quadrants, is shown in Fig. 15.3

Figure 15.3 *Bearing Quadrants*

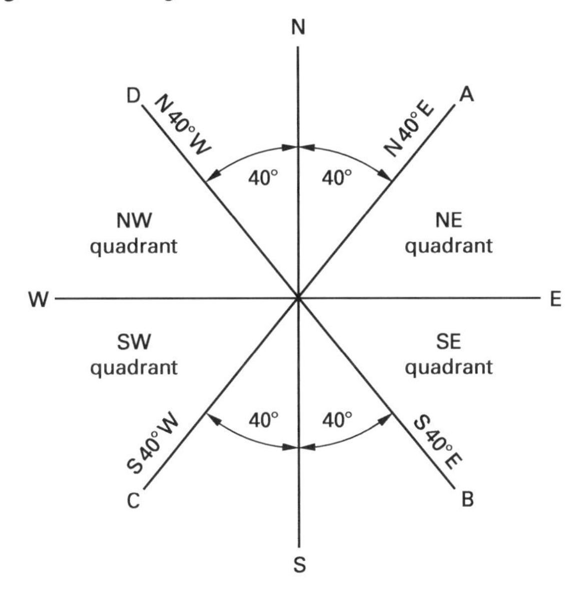

Angles are measured from either the north or south, but are never measured from the east or west. The quadrants are designated in the bearing of a line by preceding the angle with north (N) or south (S) and following the angle with east (E) or west (W).

11. BACK BEARING

If the bearing of a line AB is N 65°E, the back bearing of AB is S 65°W. In other words, someone standing at A looking at B is looking northeast; if the person stands at B and looks at A, the person will be looking southwest. The angle with the meridian is the same. The prefix is changed from N to S, and the suffix is changed from E to W. This is illustrated in Fig. 15.4.

When two parallel lines are cut by a transversal, the alternate interior angles are equal. In Fig. 15.4, the meridians through A and B are parallel lines. The line AB is the transversal, and the alternate interior angles are the bearing angles at A and B. The interior angle at A is the bearing angle of AB, and the interior angle at B is the bearing angle of BA. This also demonstrates that the bearing angle of AB is equal to the back bearing angle. Only the prefix and the suffix differ.

Figure 15.4 *Back Bearing*

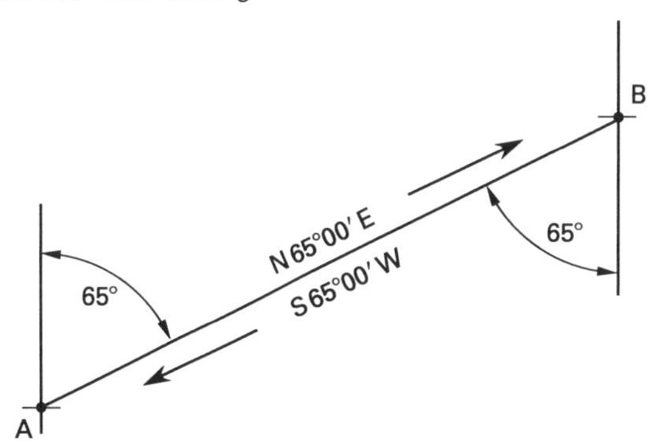

12. COMPUTATION OF BEARINGS OF A CLOSED TRAVERSE

Before the directions of the sides of a closed traverse are computed, it is essential that the interior angles of the traverse be adjusted so that their sum agrees with Eq. 15.1.

In Fig. 15.5, the bearing AB is known and the interior angles have been measured and adjusted. The bearings of the other sides must be computed.

Figure 15.5 *Interior Angles of a Closed Traverse*

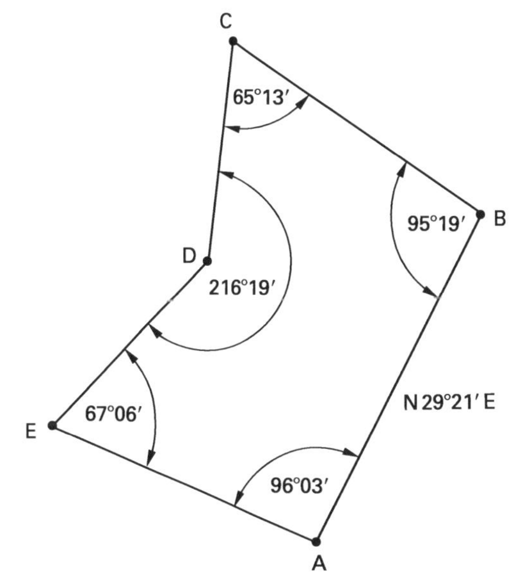

The first step in computing the bearing of BC is to draw a sketch around point B showing:

Figure 15.6 Angles About a Point

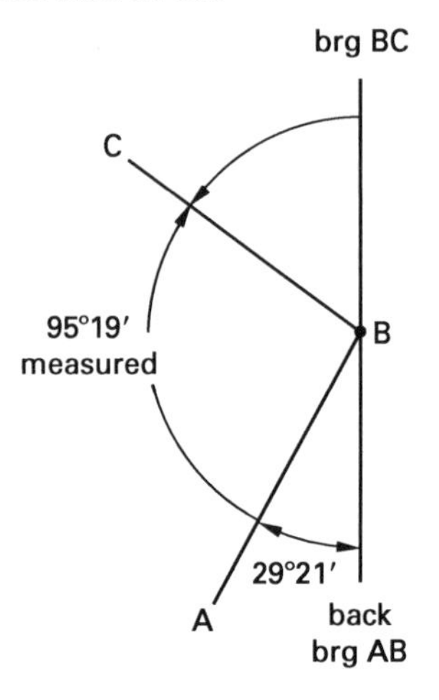

- the meridian through B
- the angle made by BA with the meridian. (The bearing of BA is the back bearing of AB, which is known.)
- the interior angle at B with its field measurement
- the bearing angle of BC, which is the angle to be computed and, by definition, is the angle between the meridian and the line BC

A straight angle is an angle that equals 180°. The meridian makes an angle of 180° at point B. Therefore, the bearing angle of BC is

$$180° - (29°21' + 95°19') = 55°20'$$

From Fig. 15.6, BC bears northwest. Therefore, the bearing of BC is

$$\text{N}\,55°20'\,\text{W}$$

Computations for bearing of CD, DE, and EA are shown in Fig. 15.7. In each case, a meridian is drawn through the next traverse point after a bearing has been computed. At each point, three angles are identified.

- the angle between the meridian and the preceding side
- the measured interior angle
- the angle between the meridian and the succeeding side

Accuracy of the computations can be checked by calculating the bearing of AB (the known bearing) using the computed bearing of EA. The computed bearing of AB must be equal to the given bearing.

Figure 15.7 Bearing Computations for Traverse in Figure 15.5

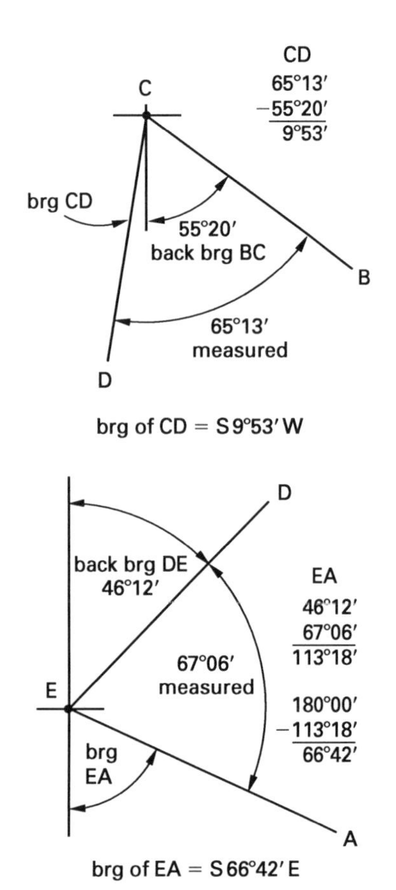

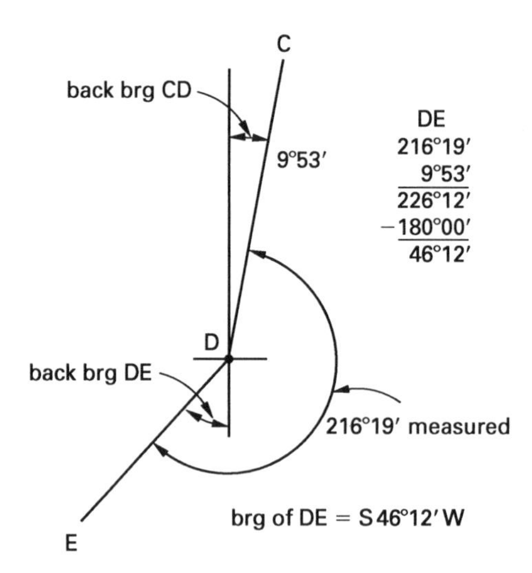

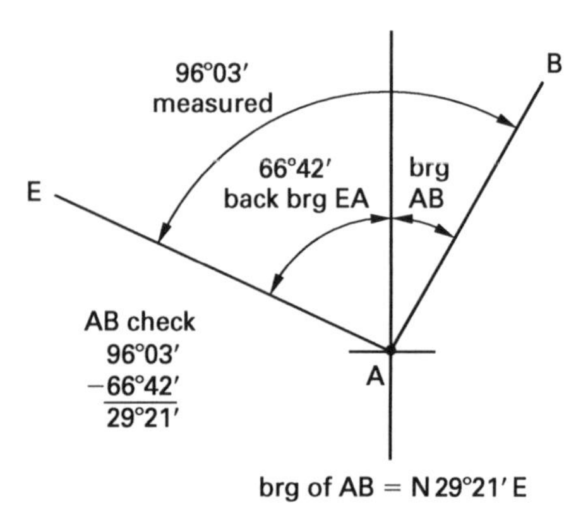

The importance of a sketch at each traverse point cannot be overemphasized. No set rule for computing bearings can be made, but a sketch that properly identifies the three angles mentioned makes a lengthy explanation unnecessary. The meridian through each traverse point always makes an angle of 180°, and the bearing angle is the angle between the meridian and the line (not between an east-west line and the line).

13. AZIMUTH

The *azimuth* of a line is the horizontal angle measured clockwise from the meridian. Azimuth is usually measured from the north.

Azimuths are not limited to 90°. Therefore, there is no need to divide the circle into quadrants. Azimuths vary from 0° to 360°. Fig. 15.8 shows azimuths of 40°, 140°, 220°, and 320°.

Figure 15.8 *Representative Azimuths*

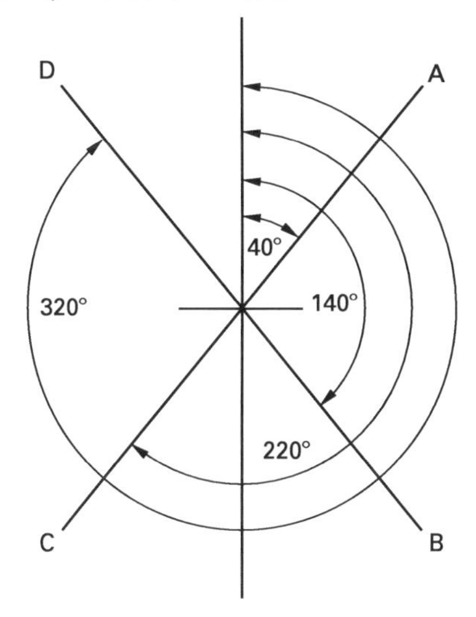

14. BACK AZIMUTH

As with bearing, the *back azimuth* of a line AB is the azimuth of line BA. The back azimuth of a line may be found by either adding 180° to the azimuth of the line or by subtracting 180° from the azimuth of the line. If the azimuth is less than 180°, add 180°.

Referring to Fig. 15.8,

- for azimuth 40°, the back azimuth = 220°
- for azimuth 140°, the back azimuth = 320°
- for azimuth 220°, the back azimuth = 40°
- for azimuth 320°, the back azimuth = 140°

15. CONVERTING BEARING TO AZIMUTH

Bearing is used to give the direction of a course in most land surveys. Azimuth is used in topographic surveys and some route surveys. Bearing may be converted to azimuth, and azimuth may be converted to bearing.

In converting bearing to azimuth in the northeast quadrant, the azimuth angle equals the bearing angle.

$$\text{N}\,76^\circ 30'\,\text{E} = 76^\circ 30'$$

In the southeast quadrant, the azimuth angle equals 180° minus the bearing angle.

$$\text{S}\,42^\circ 28'\,\text{E} = 180^\circ - 42^\circ 28' = 137^\circ 32'$$

In the southwest quadrant, 180° is added to the bearing angle.

$$\text{S}\,36^\circ 47'\,\text{W} = 180^\circ + 36^\circ 47' = 216^\circ 47'$$

In the northwest quadrant, the bearing angle is subtracted from 360°.

$$\text{N}\,62^\circ 56'\,\text{W} = 360^\circ - 62^\circ 56' = 297^\circ 04'$$

16. CONVERTING AZIMUTH TO BEARING

In the northeast quadrant, the bearing angle equals the azimuth. The prefix N and the suffix E must be added.

In the southeast quadrant, the azimuth is subtracted from 180° and the prefix S and suffix E are added.

$$168^\circ 40' = 180^\circ - 168^\circ 40' = \text{S}\,11^\circ 20'\,\text{E}$$

In the southwest quadrant, 180° is subtracted from the azimuth.

$$195^\circ 22' = 195^\circ 22' - 180^\circ = \text{S}\,15^\circ 22'\,\text{W}$$

In the northwest quadrant, the azimuth is subtracted from 360°.

$$314^\circ 35' = 360^\circ - 314^\circ 35' = \text{N}\,45^\circ 25'\,\text{W}$$

17. CLOSED DEFLECTION ANGLE TRAVERSE

In a *closed deflection angle traverse* (Fig. 15.9), the difference between the sum of the right deflection angles and the sum of the left deflection angles is 360°. Before bearings are computed, deflection angles must be adjusted.

If, during the adjustment of angles, it is found that the sum of the right deflection angles is greater than the sum of the left deflection angles, the sum of the right deflection angles must be reduced and the sum of the left deflection angles must be increased. The correction may be distributed arbitrarily or evenly.

Bearings of the sides of the traverse are computed in much the same manner as with the interior angle traverse, with a sketch drawn at each traverse point showing the angles involved in the computation.

Figure 15.9 *Closed Deflection Angle Traverse*

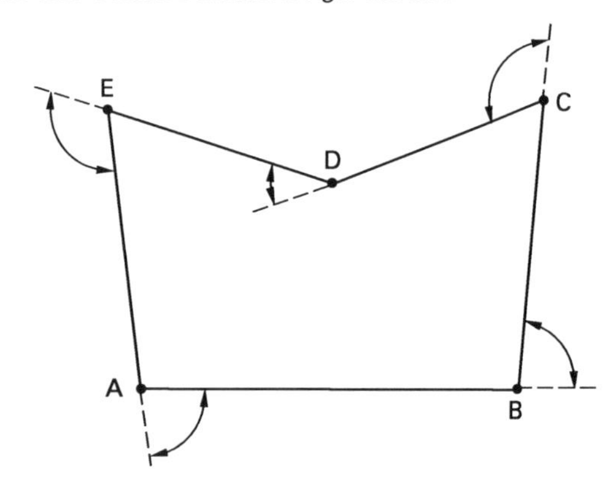

18. ANGLE-TO-THE-RIGHT TRAVERSE

Open or closed traverses can be run by the angle-to-the-right method. All angles are measured from the backsight to the foresight in a clockwise direction, as shown in Fig. 15.10.

Figure 15.10 *Angle-to-the-Right Traverse*

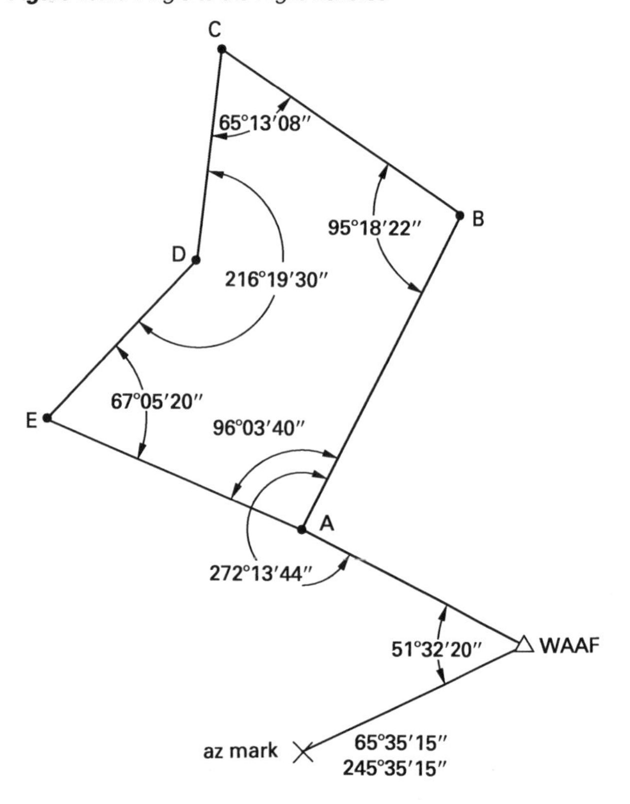

To determine the forward azimuth from a traverse point, the angle to the right is added to the back azimuth of the preceding line. In other words, the angle to the right is added to the azimuth of the preceding line ±180°. (It is sometimes necessary to subtract 360°.)

Traverse ABCDEA in Fig. 15.10 is tied to triangulation station WAAF for direction. National Geodetic Survey data show the azimuth to the azimuth mark from station WAAF to be 65°35′15″ (from the south). Converted to azimuth from the north, this azimuth is 180° + 65°35′15″ = 245°35′15″. Angles to the right, adjusted, are shown in Fig. 15.10. Computations for the azimuths of the traverse courses are shown in Table 15.1.

Table 15.1 *Computation of Azimuth*

245°35′15″	az WAAF-mark	124°39′41″	az CB
51°32′20″	angle right	65°13′08″	angle right
297°07′35″		189°52′49″	az CD
−180°		−180°	
117°07′35″	az A-WAAF	9°52′49″	az DC
272°13′44″	angle right	216°19′30″	angle right
389°21′19″		226°12′19″	az DE
−360°		−180°	
29°21′19″	az AB	46°12′19″	az ED
+180°		67°05′20″	angle right
209°21′19″	az BA	113°17′39″	az EA
95°18′22″	angle right	180°	
304°39′41″	az BC	293°17′39″	az AE
−180°		96°03′40″	angle right
124°39′41″	az CB	389°21′19″	
		−360°	
	check	29°21′19″	az AB

19. LATITUDES AND DEPARTURES

Latitudes and departures are similiar in concept to the projections of a line. The projection of a line can be compared to the shadow of a building. When the sun is nearly overhead, the shadow is short; when the sun is sinking in the west, the shadow becomes long. The height has not changed, but the length of its shadow has changed.

In Fig. 15.11, the line AB is projected on the y-axis of a rectangular coordinate system by dropping perpendiculars from A to the y-axis and from B to the y-axis. The interval between these two perpendiculars along the y-axis is the projection of AB on the y-axis. As AB changes its position relative to the y-axis, as shown in Fig. 15.11, the length of the projection becomes longer or shorter. As the position of AB nears the vertical, the projection nears the length of the line. As the projection of AB nears the horizontal, the projection of AB becomes very short.

In Fig. 15.12, AB is projected on the x-axis. As in Fig. 15.11, the length of the projection of AB on the x-axis changes as the position of AB changes relative to the x-axis.

Figure 15.11 *Projections on the y-Axis*

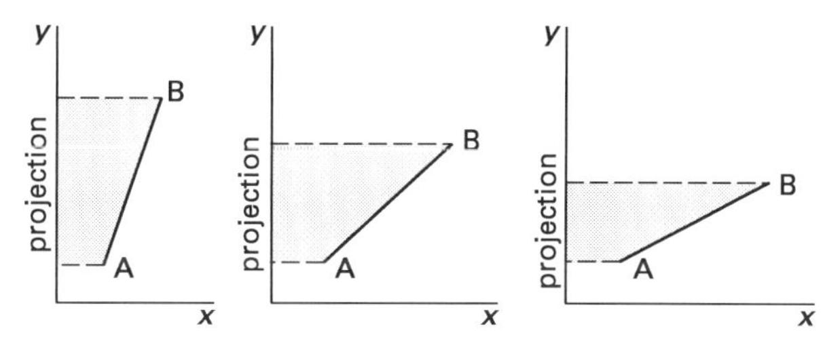

Figure 15.12 *Projections on the x-Axis*

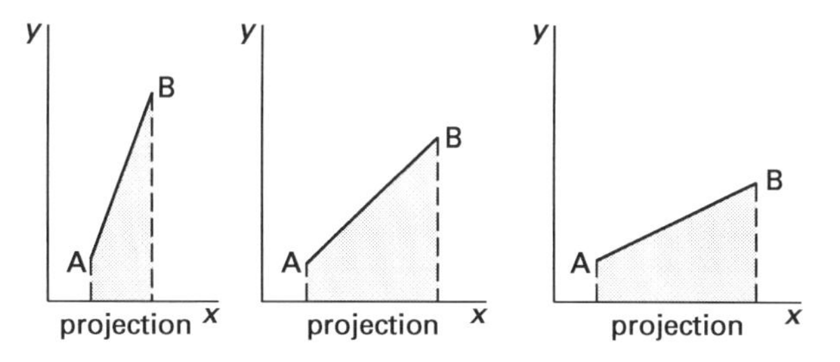

In surveying, the projection of a side of a traverse on the north-south (y-) axis of a rectangular coordinate system is known as its *latitude*. The projection on the east-west (x-) axis is known as its *departure*.

Figure 15.13 shows the projection of a line AB on the y- and x-axes. The latitude of line AB is the length of the right angle projection of AB on a meridian. The departure of line AB is the length of the right angle projection of line AB on a line perpendicular to the meridian, an east-west line.

Figure 15.13 *Latitude and Departure*

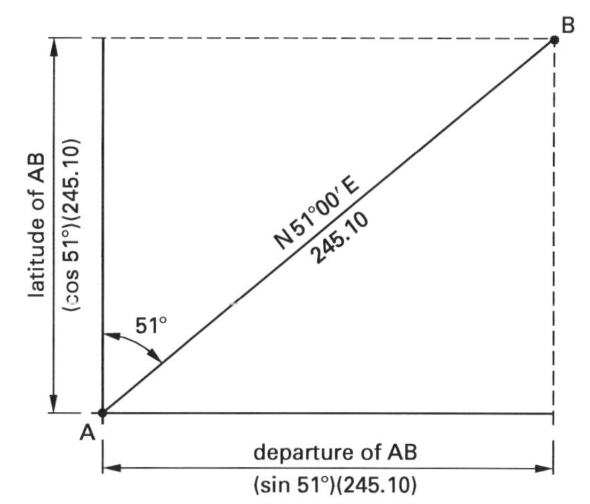

The *latitude of a course* is equal to the cosine of its bearing angle multiplied by its length. The *departure of a course* is equal to the sine of its bearing angle multiplied by its length.

$$\text{latitude} = (\cos \text{bearing})(\text{length}) \qquad 15.2$$

$$\text{departure} = (\sin \text{bearing})(\text{length}) \qquad 15.3$$

The latitude of the course AB in Fig. 15.13 is

$$\text{latitude} = (\cos 51°00')(245.10 \text{ ft}) = 154.25$$

The departure of the course AB is

$$\text{departure} = (\sin 51°00')(245.10 \text{ ft}) = 190.48$$

Figure 15.14 shows the latitude and departure for each of the courses in the traverse ABCDEA. Course AB has a north latitude and an east departure, BC has a south latitude and an east departure, CD has a south latitude and a west departure, DE has a south latitude and a west departure, and EA has a north latitude and a west departure.

Figure 15.14 *Latitudes and Departures of Traverse Legs*

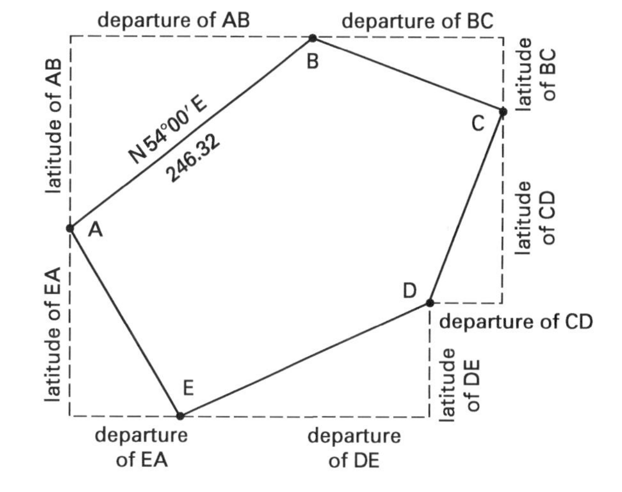

20. ERROR OF CLOSURE

If the traverse ABCDEA in Fig. 15.14 starts at point A and ends at point A, it is obvious that the distance traversed north equals the distance traversed south. Likewise, the distance traversed east equals the distance traversed west. In other words, the sum of the north latitudes must equal the sum of the south latitudes; and the sum of the east departures must equal the sum of the west departures.

However, because measurements are not exact due to human and instrument errors, these sums will not be equal and their differences may be used to determine the *linear error of closure* due to errors caused by angular and linear measurements. The actual error for each course cannot be determined, but the total error can be distributed over the entire length of the traverse so that north latitudes equal south latitudes and east departures equal west departures. This is called *balancing the traverse* or *closing the traverse.*

The *error of closure* of a traverse is a measure of the precision of a survey.

Figure 15.15 shows a traverse that does not close. Point A is the *point of beginning.* Point A′ is the *ending point,* found by plotting the latitude and departure of each course before the traverse is balanced. The distance from A′ to A is the error of closure. For the traverse to close, A′ would move in the direction of A′A. Balancing the traverse does just that—it makes A′ coincide with A.

Figure 15.15 *Error of Closure*

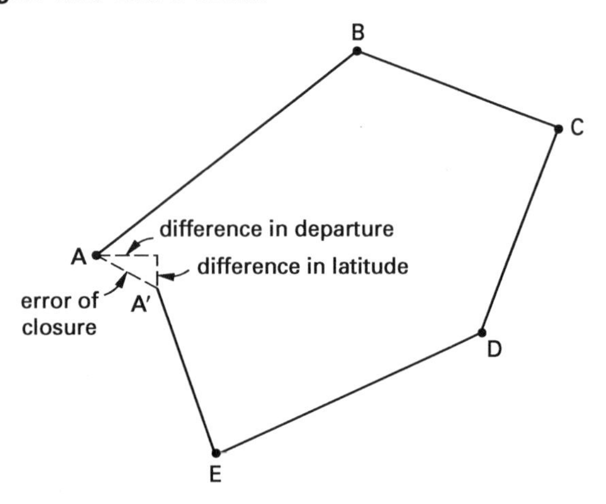

The error of closure is found by the Pythagorean theorem.

$$\text{error of closure} = \sqrt{\begin{array}{l}(\text{difference in latitude})^2 \\ \quad + (\text{difference in departure})^2\end{array}} \quad 15.4$$

The direction of the line A′A is found by determining the angle it makes with the meridian. The tangent of this bearing angle is equal to the difference in departures divided by the difference in latitudes.

$$\text{tangent bearing} \text{A}'\text{A} = \frac{\text{difference in departure}}{\text{difference in latitude}} \quad 15.5$$

21. BALANCING THE TRAVERSE

The traverse can be balanced by using one of several methods: the least squares adjustment, the compass rule, the transit rule, or the Crandall method.

The *least squares method* is adaptable to any traverse, whether angular accuracy is higher, equal to, or less than linear accuracy. It is difficult and seldom used.

The *compass rule,* also known as the *Bowditch rule,* is adaptable to traverses in which angular accuracy and linear accuracy are about the same. In this type of traverse, the compass rule will give very nearly the same results as the least squares adjustment. It is used in a great majority of traverse closures.

The *transit rule* is adaptable to traverses in which angular accuracy is much higher than linear accuracy. It is seldom used.

The *Crandall method* is also used where angular accuracy is much higher than linear accuracy. It is more accurate than the transit rule but requires more time. It, also, is seldom used.

22. THE COMPASS RULE

With the compass rule, the difference in the sums of the north and south latitudes is distributed over the latitudes of the traverse. A correction is made in the latitude of each side to bring the north and south latitudes into balance. The difference in the sums of the east and west departures is distributed in the same way.

The correction to be applied to the latitude of each side is a fraction of the total difference in the north and south latitudes. The fraction is the ratio of the length of each side to the perimeter of the entire traverse.

For example, if the difference in north and south latitudes of a traverse is 0.86, the length of side AB is 356.73, and the perimeter of the traverse is 2156.78, the correction in latitude for the side AB is

$$\text{correction AB} = \left(\frac{356.73}{2156.78}\right)(0.86) = 0.14$$

This can be expressed as a proportion.

$$\frac{\text{correction for latitude of AB}}{\text{difference in N-S latitudes}} = \frac{\text{length of AB}}{\text{traverse perimeter}} \quad 15.6$$

$$\text{correction for AB} = \left(\frac{\text{length AB}}{\text{perimeter}}\right)\left(\begin{array}{c}\text{difference} \\ \text{in latitude}\end{array}\right) \quad 15.7$$

This can be written as Eq. 15.8.

$$\text{correction for AB} = \left(\frac{\begin{array}{c}\text{difference} \\ \text{in latitude}\end{array}}{\text{perimeter}}\right)(\text{length AB}) \quad 15.8$$

Using Eq. 15.8 in the example,

$$\text{correction for AB} = \left(\frac{0.86}{2156.78}\right)(356.73) = 0.14 \text{ ft}$$

Equation 15.8 is more efficient because 0.86/2156.78 is a constant that can be applied to the other sides.

After the corrections are made in the latitude and departure for each side, the traverse is balanced.

23. RATIO OF ERROR

The *ratio of error*, or *precision*, of a traverse is the ratio of the error of closure to the perimeter. It is expressed with the numerator as one (1) and the denominator in round numbers. It is a measure of the precision of a traverse.

If the error of closure of a traverse is 0.76 and the perimeter is 5214.75, the ratio of error is

$$\frac{0.76}{5214.75}$$

Dividing the numerator and denominator by 0.76 (in order to have 1 in the numerator), and rounding off,

$$\text{ratio of error} = \frac{1}{6900}$$

This indicates an error of 1 ft per 6900 ft in distance.

24. SUMMARY OF COMPUTATIONS FOR BALANCING A TRAVERSE

The steps in balancing a traverse are summarized as follows.

step 1: Compute the angular error and adjust to make the sum of the angles agree with Eq. 15.1.

step 2: Compute the bearings for each course.

step 3: Compute the latitudes and departures.

step 4: Compute the error of closure.

step 5: Compute the ratio of error.

step 6: Compute the latitude and departure corrections for each course.

step 7: Adjust the latitudes and departures.

Example 15.1

Given the traverse ABCDEA shown with angles adjusted, balance the traverse using the compass rule and compute error of closure and ratio of error.

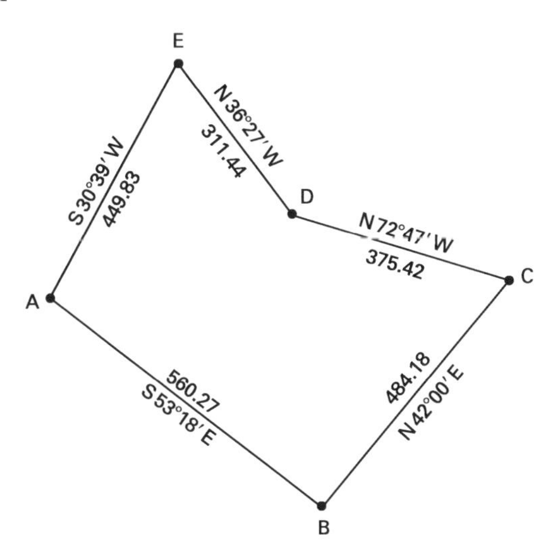

Solution

For each course, the cosine of the bearing angle times the distance has been recorded in one of the two latitude columns. Latitudes of courses with a north prefix are recorded under N; latitudes of courses with a south prefix are recorded under S.

Likewise, the sine of the bearing angle times the distance has been recorded in one of the two departure columns. Departures of courses with an east suffix are recorded under E; departures of courses with a west suffix are recorded under W.

Next, the N, S, E, and W columns have been added and the totals recorded. The smaller of the two latitude totals has been subtracted from the larger total, and the smaller of the departure totals has been subtracted from the larger.

The error of closure is the square root of the sum of the squares of these two numbers.

The ratio of error, or precision, is the ratio of the error of closure to the perimeter. Ratio of error, in this case, indicates that for every 5000 ft measured, an error of 1 ft was made.

To balance the traverse, the difference in the north and south latitude totals has been prorated to the latitudes of each course. A correction is made for each latitude in proportion to the ratio of the length of the course to the perimeter. Computations for corrections to latitudes and departures are shown in the following table.

course	latitude corrections	departure corrections
AB	$\left(\frac{0.37}{2181}\right)(560.27) = -0.10$	$\left(\frac{0.24}{2181}\right)(560.27) = -0.06$
BC	$\left(\frac{0.37}{2181}\right)(484.18) = +0.08$	$\left(\frac{0.24}{2181}\right)(484.18) = -0.05$
CD	$\left(\frac{0.37}{2181}\right)(375.42) = +0.06$	$\left(\frac{0.24}{2181}\right)(375.42) = +0.04$
DE	$\left(\frac{0.37}{2181}\right)(311.44) = +0.05$	$\left(\frac{0.24}{2181}\right)(311.44) = +0.04$
EA	$\left(\frac{0.37}{2181}\right)(449.83) = -0.08$	$\left(\frac{0.24}{2181}\right)(449.83) = +0.05$

It can be seen from both tables in this solution that the sum of the north latitudes is smaller than the sum of the south latitudes. Therefore, corrections for north latitudes are positive and are added to the computed latitudes. Corrections for south latitudes are negative and are subtracted from the computed latitudes.

The sum of the west departures is smaller than the sum of the east departures. Therefore, corrections for west departures were added, and corrections for east departures were subtracted. After corrections were made, north latitudes equal south latitudes, and east departures equal west departures. The traverse is balanced.

Note that the sum of the correction column equals the difference in the sums of the north and south latitudes.

Table for Example 15.2

				latitude			departure			latitude		departure	
point	bearing	distance	cosine sine	N	S	correction	E	W	correction	N	S	E	W
A			0.597625										
	S 53°18′ E	560.27	0.801776		334.83	−0.10	449.21		−0.06		334.73	449.15	
B			0.743145										
	N 42°00′ E	484.18	0.669131	359.81		+0.08	323.98		−0.05	359.89		323.93	
C			0.295986										
	N 72°47′ W	375.42	0.955192	111.12		+0.06		358.60	+0.04	111.18			358.64
D			0.804376										
	N 36°27′ W	311.44	0.594121	250.52		+0.05		185.03	+0.04	250.57			185.07
E			0.860298										
	S 30°39′ W	449.83	0.509792		386.99	−0.08		229.32	+0.05		386.91		229.37
A		2181.14		721.45	721.82		773.19	772.95		721.64	721.64	773.08	773.08
					721.45		772.95						
					0.37		0.24						

$$\text{error of closure} = \sqrt{(0.37)^2 + (0.24)^2} = 0.44$$

$$\text{ratio of error} = 0.44/2181 = 1/5000$$

Likewise, the departure corrections equal the difference in east and west departures.

25. COORDINATES

After latitudes and departures have been computed, coordinates of the traverse points are easily computed. Coordinates for one of the points may be known, or they can be assumed. If coordinates are assumed, they should be large enough that no coordinates will be negative.

Coordinates of a point are computed by adding a north latitude to or subtracting a south latitude from the y-coordinate of the preceding point, and by adding an east departure to or by subtracting a west departure from the x-coordinate of the preceding point.

Example 15.2

Latitudes and departures for the traverse ABCDEA have been computed and balanced as shown in the following table. Assume the coordinates of point A to be $y = 1000.00$, $x = 1000.00$. Compute the coordinates of points B, C, D, and E. Check the arithmetic by computing coordinates of point A from coordinates of point E.

Solution

	latitude		departure		coordinates	
point	north	south	east	west	y	x
A					1000.00	1000.00
		334.73	449.15		−334.73	+449.15
B					665.27	1449.15
	359.89		323.93		+359.89	+323.93
C					1025.16	1773.08
	111.18			358.64	+111.18	−358.64
D					1136.34	1414.44
	250.57			185.07	+250.57	−185.07
E					1386.91	1229.37
		386.91		229.37	−386.91	−229.37
A					1000.00	1000.00
	721.64	721.64	773.08	773.08		

In the solution, north latitudes are given a positive sign and south latitudes are given a negative sign; east departures are given a positive sign and west departures are given a negative sign. y-coordinates are associated with latitude; x-coordinates are associated with departure. The cosine function is associated with latitude; the sine function is associated with departure.

26. FINDING BEARING AND LENGTH OF A LINE FROM COORDINATES

It is often necessary to find the bearing and length of a line between two points of known coordinates. Figure 15.16 illustrates that the tangent of the bearing angle can be determined from the latitude and departure.

$$\text{tangent of bearing angle} = \frac{\text{departure}}{\text{latitude}} \tag{15.9}$$

$$\text{tangent of bearing angle} = \frac{\text{difference in } x}{\text{difference in } y} \tag{15.10}$$

$$\begin{aligned}\text{length} &= \sqrt{(\text{departure})^2 + (\text{latitude})^2} \\ &= \sqrt{\begin{array}{l}(\text{difference in } x)^2 \\ \quad + (\text{difference in } y)^2)\end{array}} \\ &= \frac{\text{latitude}}{\text{cosine bearing}} \\ &= \frac{\text{departure}}{\text{sine bearing}}\end{aligned} \tag{15.11}$$

Equation 15.2 is also true.

$$\begin{aligned}\text{cotangent of bearing angle} &= \frac{\text{latitude}}{\text{departure}} \\ &= \frac{\text{difference in } y}{\text{difference in } x}\end{aligned} \tag{15.12}$$

For large angles, Eq. 15.12 is recommended.

Figure 15.16 *Bearing and Length from Coordinates*

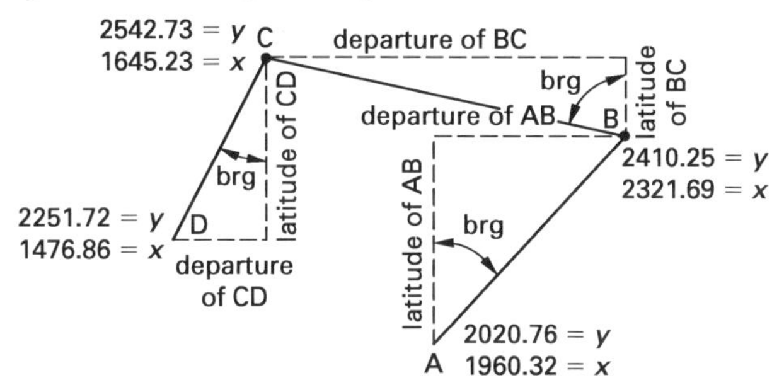

Example 15.3

Using the coordinates of points A, B, C, and D as shown in Fig. 15.16, find the bearing and lengths of AB, BC, and CD.

Solution

$$\tan \text{brg AB} = \frac{2321.69 - 1960.32}{2410.25 - 2020.76}$$

$$\text{brg AB} = \text{N}\,42^\circ 51'\,\text{E}$$

AB is northerly because the y-coordinate of B is greater than the y-coordinate of A. AB is easterly because the x-coordinate of B is greater than the x-coordinate of A.

$$\begin{aligned}\text{AB} &= \sqrt{(2321.69 - 1960.32)^2 + (2410.25 - 2020.76)^2} \\ &= 531.31 \text{ ft}\end{aligned}$$

$$\tan \text{brg BC} = \frac{2321.69 - 1645.23}{2542.73 - 2410.25}$$

$$\text{brg BC} = \text{N}\,78^\circ 55'\,\text{W}$$

$$\begin{aligned}\text{BC} &= \sqrt{(2321.69 - 1645.23)^2 + (2542.73 - 2410.25)^2} \\ &= 689.31 \text{ ft}\end{aligned}$$

$$\tan \text{brg CD} = \frac{1645.23 - 1476.86}{2542.73 - 2251.72}$$

$$\text{brg CD} = \text{S}\,30^\circ 03'\,\text{W}$$

$$\begin{aligned}\text{CD} &= \sqrt{(1645.23 - 1476.86)^2 + (2542.73 - 2251.72)^2} \\ &= 336.21 \text{ ft}\end{aligned}$$

27. COMPUTING TRAVERSES WHERE TRAVERSE POINTS ARE OBSTRUCTED

In land surveying, it is common to find boundary corners occupied by fence posts or other obstructions, making it impractical to retrace courses as they were originally run. In such cases, a traverse can be run very near the original one, and ties can be made to the original corners. The adjacent traverse can be closed, and the original survey can be computed. By making the points on the adjacent traverse very close to the corresponding points on the original survey, error in measurement of the ties is lessened. However, error of closure cannot be computed for the original survey.

The direction and distance from an adjacent point to the corresponding original corner can be determined by measurement, and from this information latitude and departure for this tie can be computed. If the coordinates of each point in the adjacent traverse are known, the coordinates of original corners can be determined by using these coordinates and the latitudes and departures of each of the ties to the original corners.

If the coordinates of the original corners are known, the bearing and length of each course of the original survey can be computed.

$$\text{tangent bearing} = \frac{\text{difference in } x}{\text{difference in } y} \tag{15.13}$$

$$\text{length} = \sqrt{(\text{difference in } x)^2 + (\text{difference in } y)^2} \tag{15.14}$$

Example 15.4

The traverse ABCDEA has been run inside the original survey represented by the traverse MNOPQM as shown. The traverse ABCDEA has been closed, and balanced latitudes and departures are shown. Ties to the original survey have been made from corresponding points on the adjacent traverse, and azimuth and distance for each are shown. Coordinates of the point A are assumed to be $x = 1000.00$, $y = 1000.00$.

Plane Survey Calculations

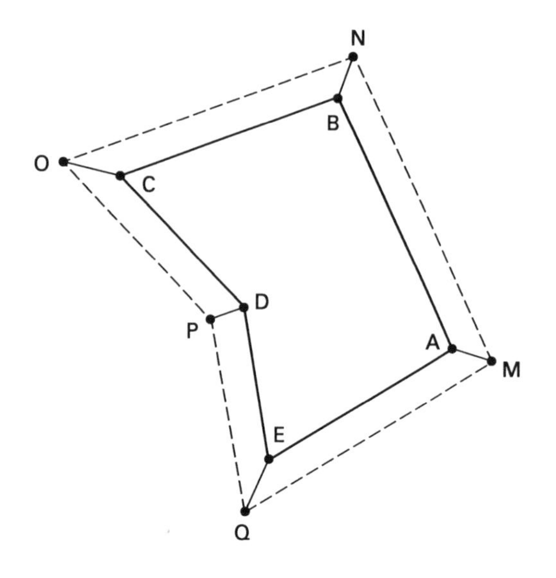

line	bearing	length	balanced north	balanced south	balanced east	balanced west
AB	N 25°45′ W	560.27	504.60			243.45
BC	S 69°33′ W	484.18		169.19		453.70
CD	S 45°14′ E	375.42		264.40	266.51	
DE	S 08°54′ E	311.44		307.70	48.16	
EA	N 58°11′ E	449.83	236.69		382.48	
			741.29	741.29	697.15	697.15

Corner Ties

line	azimuth	length
AM	106°20′	58.30
BN	33°52′	64.65
CO	271°59′	71.22
DP	256°11′	72.30
EQ	203°47′	77.35

Solution

The coordinates of points B, C, D, and E are computed as follows.

point	latitude north	latitude south	departure east	departure west	coordinates y	coordinates x
A					1000.00	1000.00
	504.60			243.45		
B					1504.60	756.55
		169.19		453.70		
C					1335.41	302.85
		264.40	266.51			
D					1071.01	569.36
		307.70	48.16			
E					763.31	617.52
	236.69		382.48			
A					1000.00	1000.00

After the coordinates of the traverse points of ABCDEA are computed, latitudes and departures of the ties to each survey corner are determined. Then the coordinates of each survey corner are computed. Directions of the ties have been recorded in azimuth, so these azimuths need to be converted to bearings. The tie from traverse point A to survey corner M is shown below.

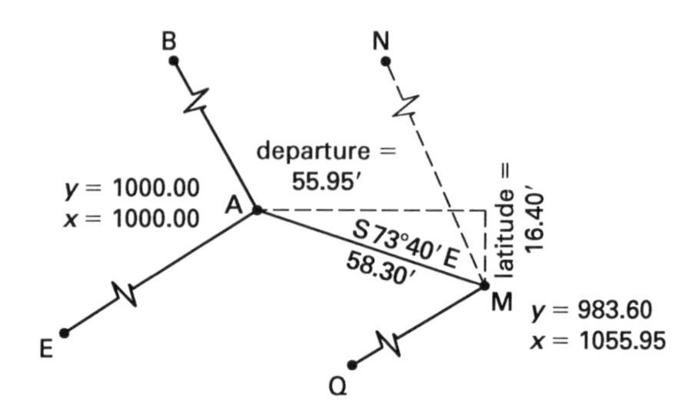

The coordinates of M are

$$\text{bearing AM} = 180° - 106°20' = \text{S}\,73°40'\,\text{E}$$

$$\text{latitude AM} = 58.30 \cos 73°40' = 16.40 \text{ ft}$$

$$\text{departure AM} = 58.30 \sin 73°40' = 55.95 \text{ ft}$$

$$y\text{-coordinate of M} = 1000.00 - 16.40 = 983.60$$

$$x\text{-coordinate of M} = 1000.00 + 55.95 = 1055.95$$

Computations of the coordinates of N, O, P, and Q are performed in a similar manner. Tabulations showing data used in these computations are shown in the following table.

With the coordinates of M and N known, the bearing and length of MN are computed.

$$\tan \text{brg MN} = \frac{1055.95 - 792.58}{1558.28 - 983.60}$$

$$\text{brg MN} = \text{N}\,24°37'\,\text{W}$$

$$\text{length MN} = \sqrt{(263.37)^2 + (574.68)^2} = 632.16 \text{ ft}$$

Tabulations of bearings and lengths of NO, OP, PQ, and QM are shown in the table on the following page.

Example 15.5

The tract of land represented by the illustration MNOPQM has been resurveyed. It is completely enclosed by a fence, so that corners could not be occupied. The traverse ABCDEA was run as close to the survey as possible. Point A is a monument of known position, and the azimuth from A to monument X is known to be 24°10′30″. The traverse was tied to this line for direction. All angles were turned to the right, beginning with the instrument at A with backsight on X, using the known azimuth. The coordinates of A are known to be y = 1276.28, x = 1533.45, and coordinates of the traverse points and the original survey corners were referred to the coordinates of A. The interior angle at each point of the traverse and the angle to the survey corner were turned from the same instrument setup in the order shown in the field notes. Horizontal distances were also measured in the sequence shown in the notes.

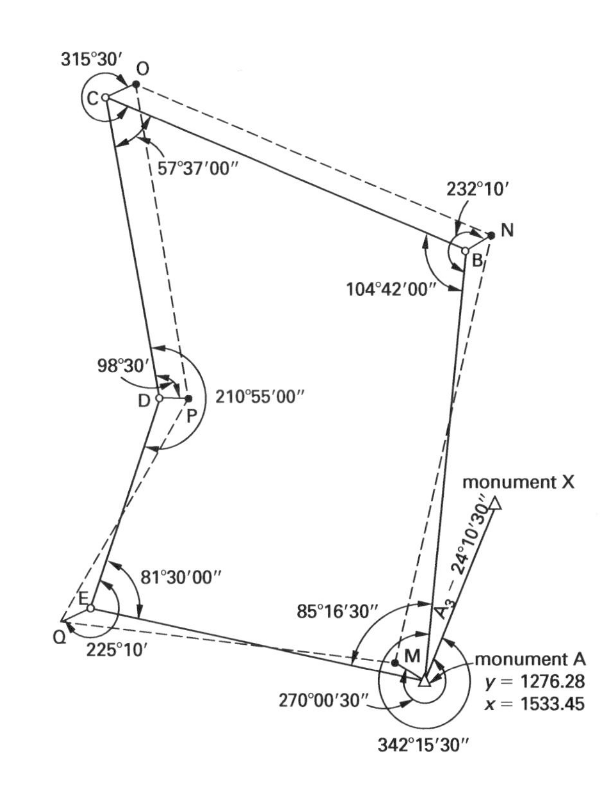

Table for Example 15.4

point	bearing	length	latitude north	latitude south	departure east	departure west	coordinates y	coordinates x
A							1000.00	1000.00
	S 73°40′ E	58.30		16.40	55.95			
M							983.60	1055.95
B							1504.60	756.55
	N 33°52 E	64.65	53.68		36.03			
N							1558.28	792.58
C							1335.41	302.85
	N 88°01′ W	71.22	2.46			71.18		
O							1337.87	231.67
D							1071.01	569.36
	S 76°11′ W	72.30		17.27		70.21		
P							1053.74	499.15
E							763.31	617.52
	S 23°47′ W	77.35		70.78		31.19		
Q							692.53	586.33
	computation of traverse MNOPQM							
M							983.60	1055.95
	N 24°37′ W	632.16	574.68			263.37		
N							1558.28	792.58
	S 68°33′ W	602.66		220.41		560.91		
O							1337.87	231.67
	S 43°16′ E	390.22		284.13	267.48			
P							1053.74	499.15
	S 13°34′ E	371.58		361.21	87.18			
Q							692.53	586.33
	N 58°12′ E	552.51	291.07		469.62			
M							983.60	1055.95
			865.75	865.75	824.28	824.28		

Field Notes

⊼ station	angle right	measured angle	adjusted angle	distance from-to	distance
A	X-M	270°00′30″		A-M	85.20
A	X-B	342°15′30″		A-B	986.10
A	E-B	85°16′30″	85°16′00″		
B	A-C	104°42′00″	104°42′00″	B-N	72.67
B	A-N	232°10′00″		B-C	930.01
C	B-D	57°37′00″	57°37′00″	C-O	81.31
C	B-O	315°30′00″		C-D	690.70
D	C-P	98°30′00″		D-P	70.54
D	C-E	210°55′00″	210°55′00″	D-E	510.22
E	D-A	81°30′00″	81°30′00″	E-Q	77.80
E	D-Q	225°10′00″		E-A	809.00
			540°00′00″		

Solution

The interior angles were adjusted and recorded in the field notes. Using the azimuth of A-X, azimuths of the traverse sides and corner ties were computed and converted to bearings. Computations are shown in the first two of the following tables.

Traverse closure and coordinate computations for the traverse ABCDEA are shown in the third table. Computations for coordinates of M, N, O, P, and Q and for bearings and lengths of MN, NO, OP, and PQ are shown in the last table.

Traverse Bearing Computations

point	angle right	azimuth	bearing
X			
		24°10′30″	
A	342°15′30″		
		6°26′00″	N 06°26′ E
B	104°42′00″		
		291°08′00″	N 68°52′ W
C	57°37′00″		
		168°45′00″	S 11°15′ E
D	210°55′00″		
		199°40′00″	S 19°40′ W
E	81°30′00″		
		101°10′00″	S 78°50′ E
A	85°16′00″		
		6°26′00″	check
B			

Corner Tie Bearing Computation

point	angle right	azimuth	bearing
X			
		24°10′30″	
A	270°00′30″		
		294°11′00″	N 65°49′ W
M			
A			
		6°26′00″	
B	232°10′00″		
		58°36′00″	N 58°36′ E
N			
B			
		291°08′00″	
C	315°30′00″		
		66°38′00″	N 66°38′ E
O			
C			
		168°45′00″	
D	98°30′00″		
		87°15′00″	N 87°15′ E
P			
D			
		199°40′00″	
E	225°10′00″		
		244°50′00″	S 64°50′ W
Q			

Traverse Computation

point	bearing	length	latitude N	latitude S	cor	departure E	departure W	cor	latitude N	latitude S	departure E	departure W	coordinates y	coordinates x
A													1276.28	1533.45
	N 06°26′ E	986.10	979.89		−0.16	110.49		+0.06	979.73		110.55			
B													2256.01	1644.00
	N 68°52′ W	930.01	335.31		−0.15		867.46	−0.06	335.16			867.40		
C													2591.17	776.60
	S 11°15′ E	690.70		677.43	+0.11	134.75		+0.04		677.54	134.79			
D													1913.63	911.39
	S 19°40′ W	510.22		480.46	+0.08		171.71	−0.03		480.54		171.68		
E													1433.09	739.71
	S 78°50′ E	809.00		156.67	+0.14	793.68		+0.06		156.81	793.74			
A													1276.28	1533.45
		3926.03	1315.20	1314.56		1038.92	1039.17		1314.89	1314.89	1039.08	1039.08		
			1314.56				1038.92							
			0.64				0.25							

$$\text{error of closure} = \sqrt{(0.64)^2 + (0.25)^2} = 0.69$$

$$\text{precision} = \frac{0.69}{3926.03} = \frac{1}{5700}$$

Summary of Example 15.5

point	bearing	length	latitude north	latitude south	departure east	departure west	coordinates y	coordinates x
			computations of coordinates					
A							1276.28	1533.45
	N 65°49′ W	85.20	34.90			77.72		
M							1311.18	1455.73
B							2256.01	1644.00
	N 58°36′ E	72.67	37.86		62.03			
N							2293.87	1706.03
C							2591.17	776.60
	N 66°38′ E	81.31	32.25		74.64			
O							2623.42	851.24
D							1913.63	911.39
	N 87°15′ E	70.54	3.38		70.46			
P							1917.01	981.85
E							1433.09	739.71
	S 64°50′ W	77.80		33.08		70.41		
Q							1400.01	669.30
			computation of MNOPQM					
M							1311.18	1455.73
	N 14°17′ E	1014.07	982.69		250.30			
N							2293.87	1706.03
	N 68°55′ W	916.12	329.55			854.79		
O							2623.42	851.24
	S 10°28′ E	718.38		706.41	130.61			
P							1917.01	981.85
	S 31°09′ W	604.13		517.00		312.55		
Q							1400.01	669.30
	S 83°33′ E	791.43		88.83	786.43			
M							1311.18	1455.73

Plane Survey Calculations

28. LATITUDES AND DEPARTURES USING AZIMUTH

In the preceding example, it was not necessary to convert azimuth to bearing; latitudes and departures can be found from azimuth with calculators that will give the algebraic sign of the latitudes and departures.

Example 15.6

Given the azimuths and lengths of the sides of the traverse ABCDEA as shown tabulated, compute latitudes and departures, balance the traverse, and compute the errors of closure and precision.

line	azimuth	length
AB	29°21′23″	560.06
BC	304°39′45″	484.14
CD	189°52′53″	375.48
DE	226°12′23″	311.53
EA	113°17′43″	449.79

Solution

line	azimuth	length	latitude	departure	balanced latitude	balanced departure
AB	29°21′23″	560.06	+488.14	+274.56	+488.11	+274.52
BC	304°39′45″	484.14	+275.35	−398.21	+275.33	−398.25
CD	189°52′53″	375.48	−369.91	−64.44	−369.93	−64.47
DE	226°12′23″	311.52	−215.59	−224.87	−215.61	−224.89
EA	113°17′43″	449.79	−177.88	+413.12	−177.90	+413.09
		2180.99	+0.11	+0.16	0.00	0.00

$$\text{error of closure} = \sqrt{(0.11)^2 + (0.16)^2} = 0.19$$

$$\text{precision} = \frac{0.19}{2180.99} = 1/11{,}500$$

29. ROUTE LOCATION BY DEFLECTION ANGLE TRAVERSE

Deflection angle traverses are suitable for highway locations because the deflection angle at the point of intersection of two tangents along the centerline of a highway is equal to the central angle of the circular arc that is inserted to connect two tangents. Straight sections along the centerline are known as *tangents*, and circular arcs are known as *simple curves*. Curves are not always circular arcs, but normally they are. Curves will be discussed in detail in Ch. 18.

The deflection angle traverse shown in Fig. 15.17 begins at point A (a point on the line XA, the azimuth of which is 346°06′) and ends at point E (on the line YE, the azimuth of which is 17°34′). The azimuths of these two lines are used to check the angular closure of the traverse. The azimuth of AB is found by adding the deflection angle at A to the azimuth XA. Azimuths of the other lines of the traverse are found by adding right deflection angles to the forward azimuth of the preceding line and subtracting left deflection angles from the forward azimuth of the preceding line.

Figure 15.17 *Deflection Angle Traverse and Calculations*

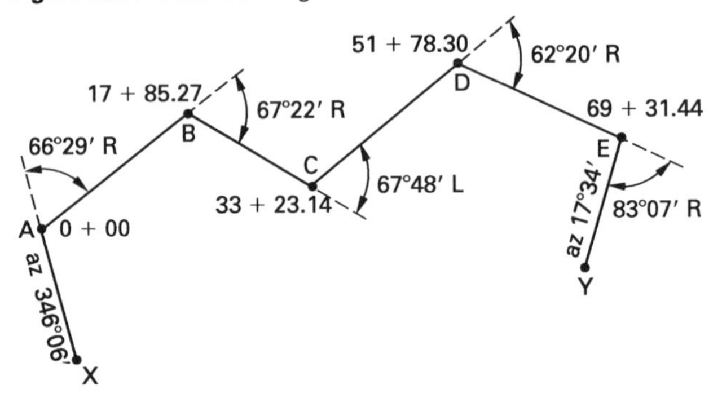

Table 15.2 *Calculations for Fig. 15.17*

line	azimuth	correction	adjusted azimuth	adjusted bearing
XA	346°06′	fixed	346°06′00″	
	+66°29′			
	412°35′			
	−360°00′			
AB	52°35′	−0′30″	52°34′30″	N 52°34′30″ E
	+67°22′			
BC	119°57′	−1′00″	119°56′00″	S 60°04′00″ E
	−67°48′			
CD	52°09′	−1′30″	52°07′30″	N 52°07′30″ E
	+62°20′			
DE	114°29′	−2′00″	114°27′00″	S 65°33′00″ E
	+83°07′			
EY	197°36′	fixed	197°34′00″	S 17°34′00″ W
	−180°00′			
YE	+17°36′			
	−17°34′			
	+02′ = closure error			

30. CONNECTING TRAVERSE

Figure 15.18 shows a *connecting traverse* between triangulation station WAAF and triangulation station PRICE. Traverse stations A, B, and C are established as part of the connecting traverse. x- and y-coordinates for these stations are to be computed.

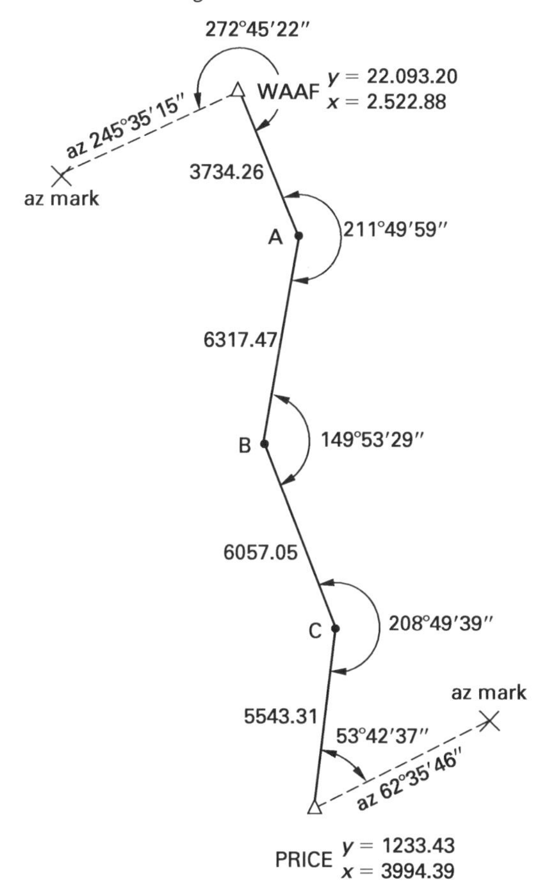

Figure 15.18 *Connecting Traverse*

Table 15.3 *Computations of Directions for Fig. 15.18*

azimuth		correction	adjusted azimuth
245°35′15″	WAAF-mk		245°35′15″
+272°45′22″	angle right	−07″	272°45′15″
518°20′37″			518°20′30″
−360°00′00″			360°00′00″
158°20′37″	WAAF-A		158°20′30″
+211°49′59″	angle right	−07″	211°49′52″
370°10′36″			370°10′22″
−180°00′00″			180°00′00″
190°10′36″	A-B		190°10′22″
+149°53′29″	angle right	−07″	149°53′22″
340°04′05″			340°03′44″
−180°00′00″			180°00′00″
160°04′05″	B-C		160°03′44″
+208°49′39″	angle right	−07″	208°49′32″
368°53′44″			368°53′16″
−180°00′00″			180°00′00″
188°53′44″	C-PRICE		188°53′16″
+ 53°42′37″	angle right	−07″	53°42′30″
+242°36′21″			242°35′46″
−180°00′00″			180°00′00″
62°36′21″	PRICE-mk	check	62°35′46″
− 62°35′46″	PRICE-mk		
clos = +35″			

Coordinates for triangulation stations WAAF and PRICE are known and shown in Fig. 15.18. Azimuth from each triangulation station to an azimuth mark is also known and shown in Fig. 15.18.

Computations for determining coordinates of stations A, B, and C are shown in Tables 15.3 through 15.5.

Table 15.3 shows computations for direction of traverse courses. The direction of the WAAF azimuth mark is fixed, as is the direction of the PRICE azimuth mark. Angular closure was found to be 0′35″, which was distributed evenly to angles at stations WAAF, A, B, C, and PRICE.

Table 15.4 shows traverse computations with the error of closure and precision.

Table 15.5 shows latitudes and departures balanced by the compass rule and the coordinates of stations A, B, and C.

Table 15.4 *Traverse Computations for Fig. 15.18*

line	azimuth	length	latitude	departure
WAAF-A	158°20′30″	3734.26	−3470.63	+1378.21
A-B	190°10′22″	6317.47	−6218.16	−1115.77
B-C	160°03′44″	6057.05	−5694.01	+2065.45
C-PRICE	188°53′16″	5543.31	−5476.75	−856.44
		21652.09	−20859.55	+1471.45
			+20859.77	−1471.51
			+0.22	−0.06

$$\text{error of closure} = \sqrt{(0.22)^2 + (0.06)^2} = 0.23$$

$$\text{precision} = 1/94{,}000$$

Table 15.5 Coordinate Computations for Fig. 15.18

station	balanced latitude	balanced departure	coordinates y	coordinates x
WAAF			22,093.20	2522.88
	−3470.67	+1378.22		
A			18,622.53	3901.10
	−6218.22	−1115.75		
B			12,404.31	2785.35
	−5694.07	+2065.47		
C			6710.24	4850.82
	−5476.81	−856.43		
PRICE			1233.43	3994.39
	−20,859.77	+1471.51		

31. ERRORS IN TRAVERSING

Errors made in linear and angular measurements in traversing are of two types: (a) *systematic errors* and (b) *accidental errors*, often called *random errors.*

Systematic errors are inherent in surveying instruments such as steel tapes, EDM instruments, theodolites, and engineer's levels. These errors can be computed and corrected. They can be prevented or minimized by using calibrated tapes, by repeating and averaging horizontal angular measurements, and by balancing backsight and foresight distances in leveling.

32. SYSTEMATIC ERRORS IN TAPING

A steel tape that is calibrated to be 100.01 ft long introduces a systematic positive error of 0.01 ft each time it is used; a tape calibrated to be 99.99 ft long introduces a systematic negative error of 0.01 ft each time it is used. Changes in length brought about by temperature changes can be computed and the algebraic sign of the correction determined, depending on whether the temperature was below or above the calibration temperature.

33. SYSTEMATIC ERRORS IN ANGULAR MEASUREMENT

Systematic errors in angular measurement occur because of the inherent inaccuracy of the transit or theodolite. The expected angular closure for the interior angles of a traverse is expressed by the formula

$$c = k\sqrt{n} \qquad 15.15$$

n is number of angles, and k is the least count of the instrument circle.

For a five-sided closed traverse measured with a 20″ theodolite, the permissible closure error is

$$20''\sqrt{5} = 45'' \qquad 15.16$$

For a closure greater than 45″, the surveyor of this traverse should examine the notes for an accidental error.

34. ACCIDENTAL (RANDOM) ERRORS

Accidental errors are not always easily found and cannot be corrected by known formulas. They may be large or small; they may be positive or negative. A 100 ft station may be inadvertently dropped or added in taping, numbers in angle measurements may be transposed in recording, and so on. Large accidental errors are sometimes referred to as *blunders.*

35. LOCATING ERRORS IN A TRAVERSE

After a traverse is plotted, large errors are often found by close inspection of the plot. If there is only one large error in angle measurement, the perpendicular bisector of the line of closure will point to the traverse station where the error was probably made.

In Fig. 15.19, the perpendicular bisector of the closure line A′A points toward station B as the possible point of angular error. If the perpendicular bisector points toward two traverse stations, it is possible that there are two accidental errors.

Figure 15.19 Locating an Error in a Traverse

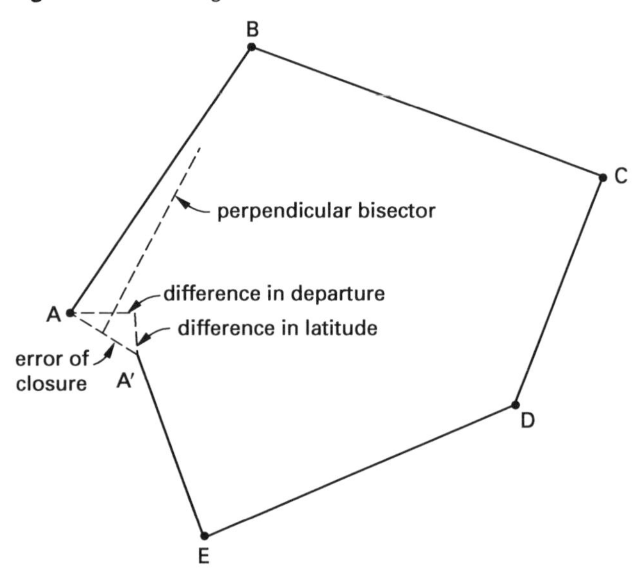

If there are no accidental errors in angle measurement, the line of closure may nearly parallel the side that contains an accidental error in distance. The length of the closure line may indicate the magnitude of the accidental error. In Fig. 15.19, the side BC would be suspect.

36. INTERSECTIONS OF TRAVERSE LINES

The method for finding the point of intersection of two traverse lines using analytic geometry was explained in

Ch. 10. It is more common, however, to determine the point of intersection by using trigonometry. There are three trigonometric methods for determining the point of intersection: (a) the bearing-bearing method, (b) the bearing-distance method, and (c) the distance-distance method. Selection of the method to use depends on the known information.

37. BEARING-BEARING METHOD OF DETERMINING INTERSECTIONS

If the coordinates of the end points of one side of a triangle are known and the bearings of the other two sides are known, the coordinates of the point of intersection of the other two sides can be found by using the *bearing-bearing* method.

Example 15.7

Find the coordinates of the point of intersection of the sides AC and BC of the triangle ABC, given the information shown.

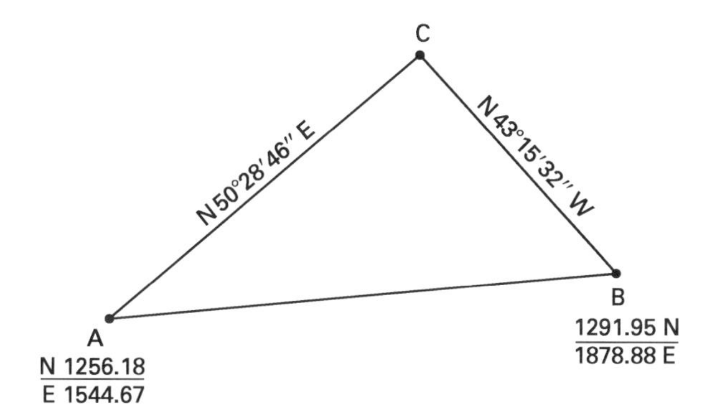

Solution

$$\tan \text{brg AB} = \frac{1878.88 - 1544.67}{1291.95 - 1256.18}$$

$$\text{brg AB} = \text{N}\,83°53'28''\,\text{E}$$

$$\text{dist AB} = \sqrt{\begin{array}{l}(1878.88 - 1544.67)^2 \\ \quad + (1291.95 - 1256.18)^2\end{array}} = 336.12$$

$$\text{angle } A = 33°24'42''$$

$$\text{angle } B = 52°51'00''$$

$$\text{angle } C = 93°44'18''$$

$$\text{AC} = \frac{336.12 \sin 52°51'00''}{\sin 93°44'18''} = 268.48$$

$$\text{BC} = \frac{336.12 \sin 33°24'42''}{\sin 93°44'18''} = 185.48$$

$$\text{north coord. of C} = 1256.18 + 268.48 \cos 50°28'46'' = 1427.03$$

$$\text{east coord. of C} = 1544.67 + 268.48 \sin 50°28'46'' = 1751.77$$

38. BEARING-DISTANCE METHOD OF DETERMINING INTERSECTIONS

If the coordinates of the end points of one side of a triangle are known and the bearing of one of the other sides and the distance of the third are known the coordinates of the point of intersection of the latter two sides can be found using the *bearing-distance* method.

Example 15.8

Find the coordinates of the point of intersection of the sides AC and BC of the triangle ABC, given the information shown.

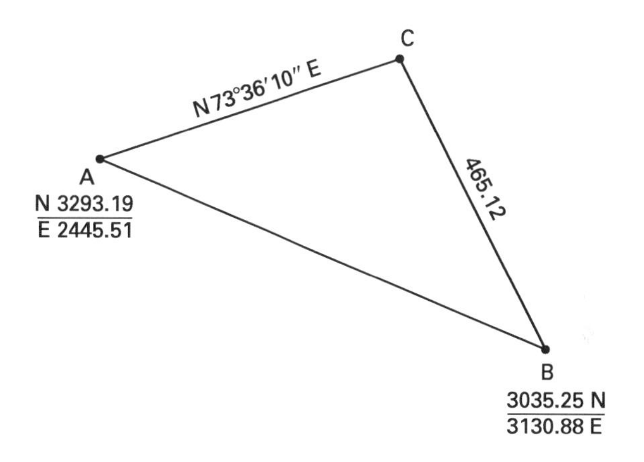

Solution

$$\tan \text{brg AB} = \frac{3130.88 - 2445.51}{3293.19 - 3035.25}$$

$$\text{brg AB} = \text{S}\,69°22'34''\,\text{E}$$

$$\text{dist AB} = \sqrt{\begin{array}{l}(3130.88 - 2445.51)^2 \\ \quad + (3293.19 - 3035.25)^2\end{array}} = 732.30$$

$$\text{angle } A = 37°01'16''$$

Note: When two sides of a triangle and the angle opposite one of them are known (the SSA case), it is possible to construct two triangles from the known data. In this case, if an arc with a radius of 465.12 ft is swung from point B, it will intersect a line from A with bearing N 73°36′10″ at two points. (See Ex. 8.19.)

$$\sin C' = \frac{732.30 \sin 37°01'16''}{465.12}$$

$$C' = 71°26'17''$$

$$\text{angle } C = 180° - 71°26'17'' = 108°33'43''$$

$$\text{angle } B = 34°25'01''$$

$$\text{north coord. of C} = 3293.19 + (436.62)(\cos 73°36'10'') = 3416.45$$

$$\text{east coord. of C} = 2445.51 + (436.62)(\sin 73°36'10'') = 2864.37$$

39. DISTANCE-DISTANCE METHOD OF DETERMINING INTERSECTIONS

If the coordinates of the end points of one side of a triangle are known and the distances of the other two sides are known, the coordinates of the point of intersection of the other two sides can be found by using the *distance-distance* method.

Example 15.9

Find the coordinates of the point of intersection of sides AC and BC given the information shown.

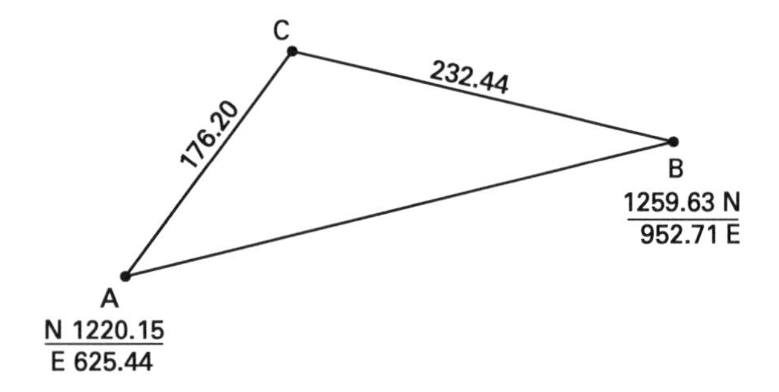

Solution

$$\tan \text{brg AB} = \frac{952.71 - 625.44}{1259.63 - 1220.15}$$

$$\text{brg AB} = \text{N}\,83°07'17''\text{E}$$

$$\begin{aligned}\text{dist AB} &= \sqrt{\begin{array}{l}(952.71 - 625.44)^2 \\ + (1259.63 - 1220.15)^2\end{array}} \\ &= 329.64\end{aligned}$$

$$\begin{aligned}\cos A &= \frac{\text{AC}^2 + \text{AB}^2 - \text{BC}^2}{(2)(\text{AC})(\text{AB})} \\ &= \frac{(176.20)^2 + (329.64)^2 - (232.44)^2}{(2)(176.20)(329.64)}\end{aligned}$$

$$\text{angle } A = 42°28'29''$$

$$\text{brg AC} = \text{N}\,40°38'48''\,\text{E}$$

$$\begin{aligned}\text{north coord. of C} &= 1220.15 + 176.20 \cos 40°38'48'' \\ &= 1353.84\end{aligned}$$

$$\begin{aligned}\text{east coord. of C} &= 625.44 + 176.20 \sin 40°38'48'' \\ &= 740.22\end{aligned}$$

PRACTICE PROBLEMS

1. Adjust the interior angles of the following closed traverses arbitrarily.

(a)

point	measured angle
A	67°06′30″
B	216°19′
C	65°12′30″
D	95°18′30″
E	96°02′

(b)

point	measured angle
A	92°38′
B	117°21′30″
C	129°13′
D	261°44′30″
E	55°28′30″
F	207°10′
G	70°52′30″
H	145°34′

2. Adjust the interior angles of the following closed traverse by applying the same correction to each angle.

point	measured angle
A	90°19′59″
B	106°19′55″
C	318°51′26″
D	48°29′12″
E	150°55′29″
F	195°44′47″
G	87°11′56″
H	193°08′23″
I	212°18′51″
J	47°27′37″
K	288°30′28″
L	60°43′21″

3. Compute the bearings of the sides of the traverse ABCDEA. Interior angles are as shown. Draw a sketch for each traverse station showing the two known angles and designate the unknown bearing angle with a question mark. Show computations at each station. Bearing of AB is N 15°22′ E.

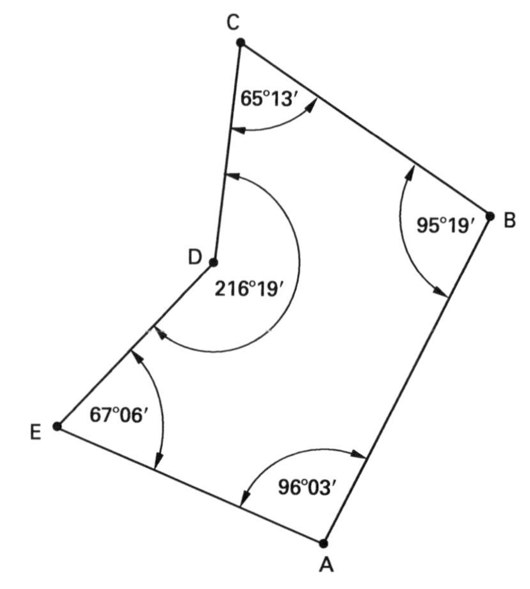

4. Compute the bearings of the sides of the traverse ABCDEA. Interior angles are as shown. Draw a sketch for each traverse station showing the two known angles and designate the unknown bearing angle with a question mark. Show computations at each station. Bearing of AB is N 65°04′ W.

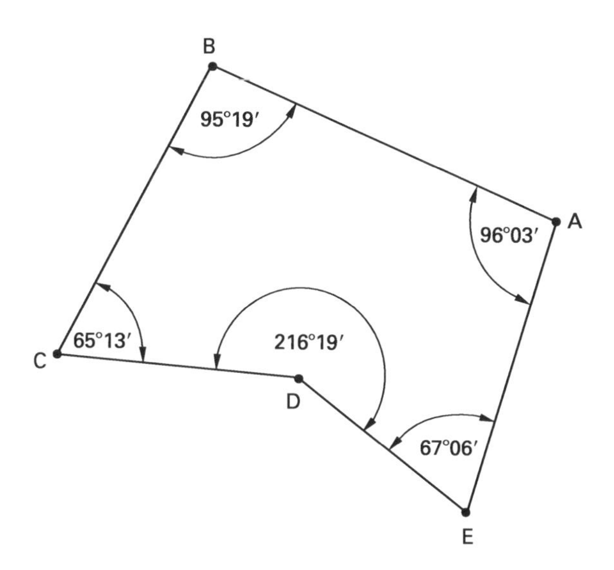

5. Use the figure showing the traverse ABCDEA to compute the interior angles of the traverse. Draw a meridian (north line) through each traverse station, identify the angles at each station that are used in the computations, and show the computations. Then record the interior angles on the figure.

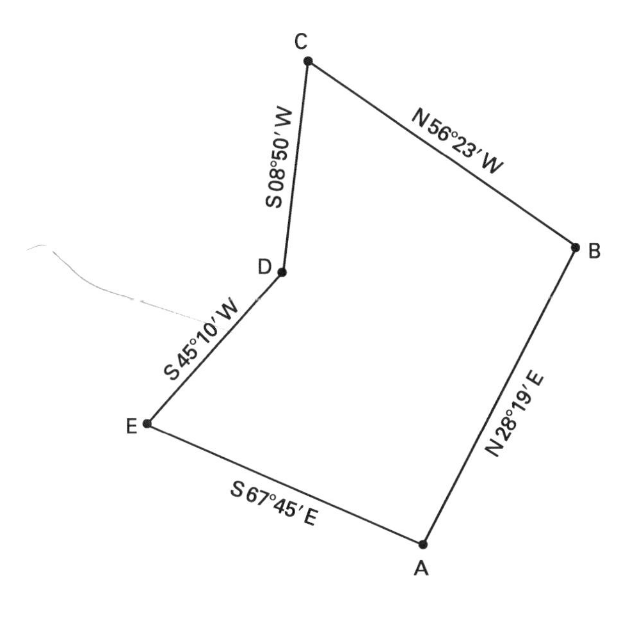

6. Using a protractor and a scale, plot the traverse ABCDEFGHA to a scale of 1 in = 100 ft.

AB: N 21°12′ E, 289.07
BC: S 78°41′ E, 366.63
CD: S 25°24′ W, 228.60
DE: S 35°23′ E, 232.44
EF: S 47°51′ W, 281.24
FG: N 78°30′ W, 316.68
GH: N 29°29′ E, 192.13
HA: N 25°57′ W, 174.36

7. On a Cartesian x-y plane, draw lines with azimuths: OA = 30°, OB = 60°, OC = 105°, OD = 135°, OE = 180°, OF = 210°, OG = 265°, and OH = 270°, and OI = 315°. Show the back azimuth of each line.

8. Traverse ABCDEA is tied to station WAAF for direction. Azimuth to the azimuth mark from station WAAF is 14°37′19″ (from the north). The angle to the right at station WAAF (az mk to A) is 21°14′46″. Angles to the right at other stations are shown. Compute the azimuth of each side of the travese.

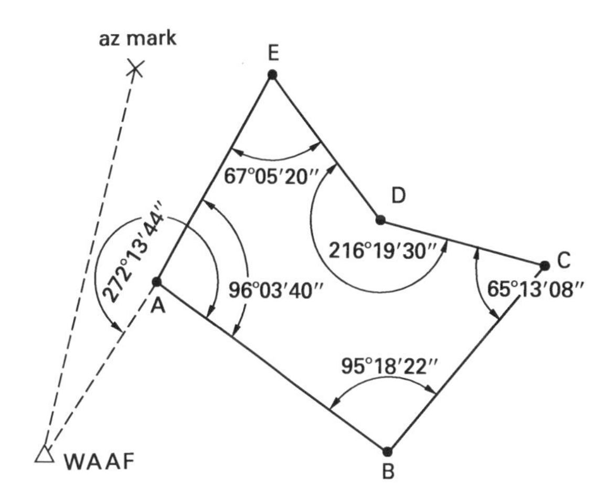

9. Convert the following bearings to azimuths.

(a) S 85°13′16″ W (b) N 74°24′01″ W
(c) S 08°19′19″ E (d) N 84°28′13″ E
(e) S 83°03′28″ E (f) N 07°26′33″ W
(g) N 27°57′45″ E (h) S 05°17′25″ W
(i) S 04°18′12″ E (j) N 57°08′02″ W

10. Convert the following azimuths to bearings.

(a) 291°37′06″ (b) 12°13′47″
(c) 106°12′46″ (d) 232°31′18″
(e) 93°04′02″ (f) 65°11′37″
(g) 337°15′11″ (h) 267°25′51″
(i) 102°27′38″ (j) 317°40′53″

11. Determine the latitude and departure of each side of the traverse ABCDA and record in the proper column. Total the north and south latitudes and the east and west departures.

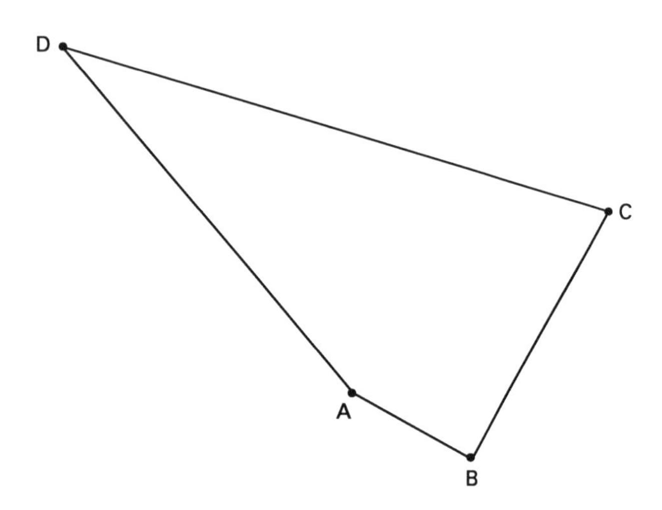

point	bearing	length
A		
	S 60°00′ E	50.0
B		
	N 30°00″ E	100.0
C		
	N 75°00′ W	200.0
D		
	S 41°30′ E	151.0
A		

12. Compute the error of closure and ratio of error, and balance the traverse by the compass rule. Show computations for corrections for each side of the traverse.

point	bearing	length
A		
	N 28°19′ E	560.27
B		
	N 56°23′ W	484.18
C		
	S 08°50′ W	375.42
D		
	S 45°10′ W	311.44
E		
	S 67°45′ E	449.83
A		

13. Balance the following traverse by the compass rule.

point	bearing	length
A		
	S 53°18′ E	560.27
B		
	N 42°00′ E	484.18
C		
	N 72°47′ W	375.42
D		
	N 36°27′ W	311.44
E		
	S 30°39′ W	449.83
A		

14. Balance the following traverse.

point	bearing	length
A		
	N 80°00′ W	1015.43
B		
	S 66°30′ W	545.52
C		
	S 12°00′ E	480.97
D		
	S 88°30′ E	750.26
E		
	N 69°00′ E	639.18
F		
	N 10°00′ E	306.28
A		

15. Given the balanced latitudes and departures of the traverse ABCDEA with the coordinates of station A being $y = 9012.34$ and $x = 1234.56$, determine the coordinates of stations B, C, D, and E.

			latitude		departure		coordinates	
pt	bearing	length	N	S	E	W	y	x
A							9012.34	1234.56
	N 28°19′ E	560.27	493.12		265.66			
B								
	N 56°23′ W	484.18	267.97			403.29		
C								
	S 08°50′ W	375.42		371.04		57.72		
D								
	S 45°10′ W	311.44		219.64		220.91		
E								
	S 67°45′ E	449.83		170.41	416.26			
A								

16. Given the open traverse ABCD with coordinates of A, B, C, and D as shown, compute the bearings and lengths of AB, BC, and CD.

point	coordinates y	coordinates x
A	9090.90	9090.90
B	9480.39	9452.27
C	9612.87	8890.70
D	9321.86	8775.81

17. The traverse ABCDEA has been run inside the tract MNOPQM because corners of the tract cannot be occupied. The traverse has been balanced, and coordinates of each traverse station have been computed. Ties to each corner of the tract have been made from the adjacent traverse station and are shown. Compute the bearing and length of each course of the trace MNOPQM.

			latitude		departure		coordinates	
pt	bearing	length	N	S	E	W	y	x
A							1000.00	1000.00
	N 28°19 E	560.27	493.12		265.66			
B							1493.12	1265.66
	N 56°23′ W	484.18	267.97			403.29		
C							1761.09	862.37
	S 08°50′ W	375.42		371.04		57.72		
D							1390.05	804.65
	S 45°10′ W	311.44		219.64		220.91		
E							1170.41	583.74
	S 67°45′ E	449.83		170.41	416.26			
A							1000.00	1000.00

corner ties

line	azimuth	length
AM	167°58′	59.20
BN	88°13′	65.13
CO	346°08′	70.85
DP	303°51′	72.45
EQ	258°31′	76.51

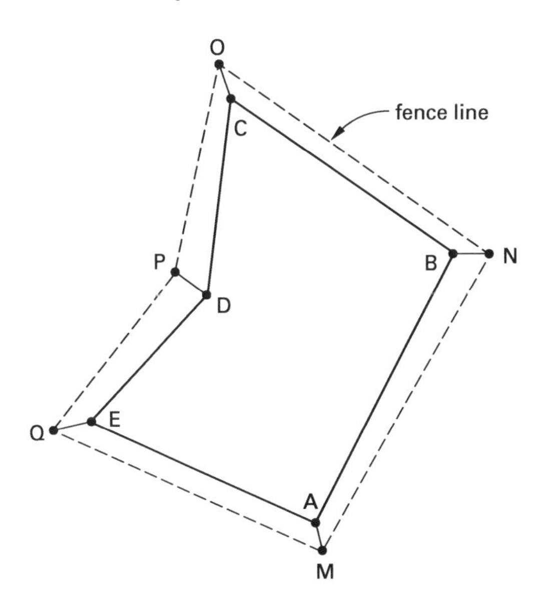

18. The traverse ABCDEA has been run and ties to the corners of the tract MNOPQM, which cannot be occupied, have been made. The traverse has been balanced and coordinates of each traverse station have been computed as shown. Compute the bearing and length for each course of the tract MNOPQM.

			latitude		departure		coordinates	
pt	bearing	length	N	S	E	W	y	x
A							1000.00	1000.00
	N 28°19′ E	560.27	493.12		265.66			
B							1493.12	1265.66
	N 56°23′ W	484.18	267.97			403.29		
C							1761.09	862.37
	S 08°50′ W	375.42		371.04		57.72		
D							1390.05	804.65
	S 45°10′ W	311.44		219.64		220.91		
E							1170.41	583.74
	S 67°45′ E	449.83		170.41	416.26			
A							1000.00	1000.00

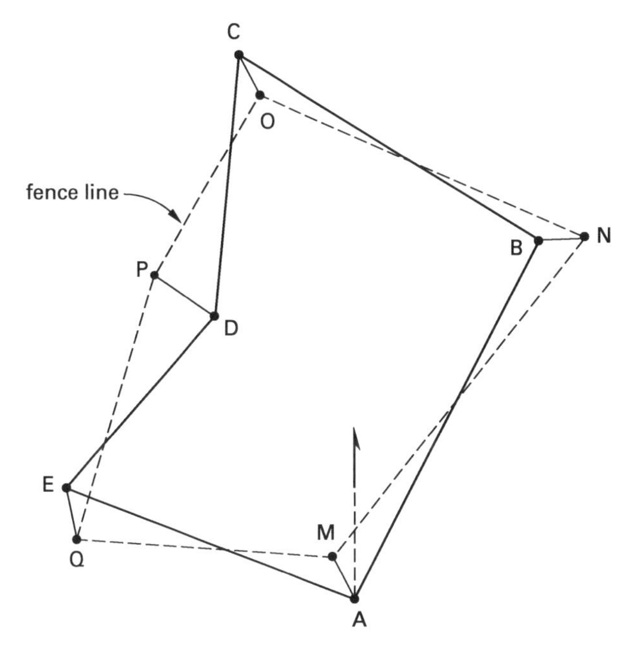

corner ties

line	azimuth	length
AM	331°12′	58.46
BN	87°33′	66.72
CO	152°13′	68.89
DP	304°31′	74.13
EQ	166°22′	66.52

19. Balance the following traverse, and find coordinates of points B, C, D, E, F, G, and H. Enter the azimuth directly into a calculator with north and east plus (+) and south and west minus (−).

line	corner ties azimuth	length
AB	21°12′	289.07
BC	101°19′	366.63
CD	205°24′	228.60
DE	144°37′	232.44
EF	227°51′	281.24
FG	281°30′	316.68
GH	29°29′	192.13
HA	334°03′	174.06

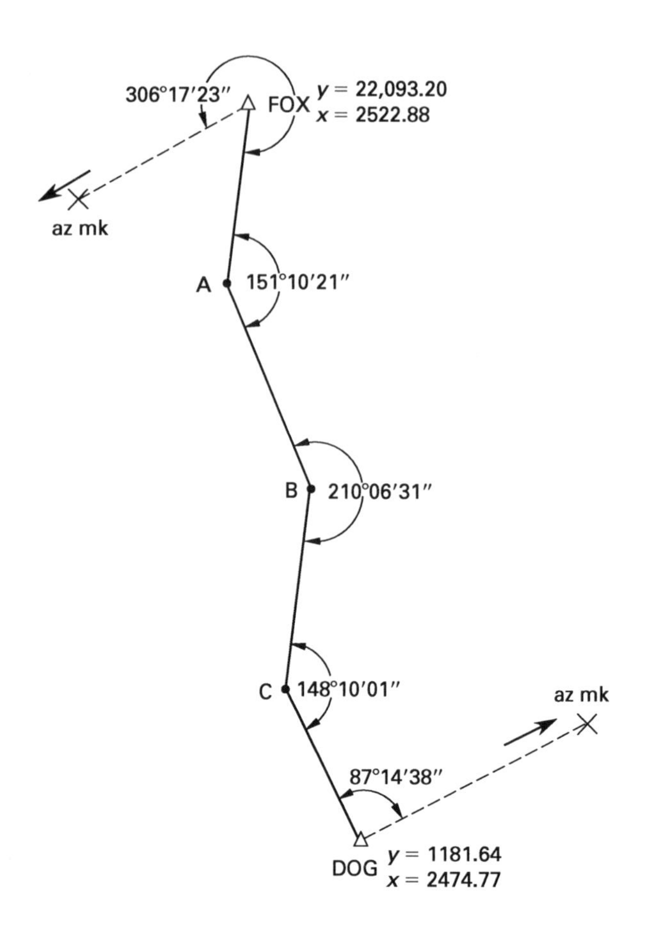

20. The open traverse ABCDE for a highway location begins at point A (which lies on line XA, the azimuth of which is known to be 21°44′) and ends at point E (which lies on line YE, the azimuth of which is known to be 350°14′). Using the azimuth of XA and YE as fixed, find the angular error in the traverse, correct arbitrarily, and convert azimuths to bearings.

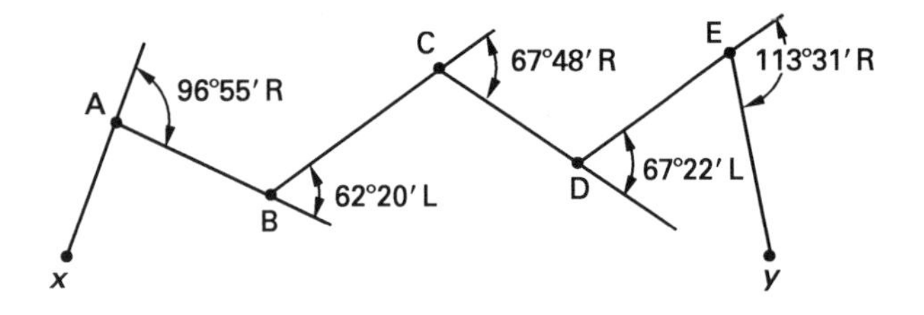

21. Traverse FOX-A-B-C-DOG connects triangulation stations FOX and DOG, coordinates of which are known and shown. The direction FOX-az mk and DOG-az mk are fixed. The traverse stations A, B, and C were set. Find the angular closure, correct azimuths, and compute the coordinates of A, B, and C. Az FOX-az mk = 246°45′46″. Az DOG-az mk = 69°45′15″.

line	length
FOX	
−A	5543.31
AB	6057.05
BC	6317.47
C−	3734.26
DOG	

22. Corner C of the traverse ABCDA is the center of a 24 in tree. Traverse ABC′DA was run as detailed in the figure with C′ being 18.6 ft off the center of the tree. Find bearing and length of BC and CD.

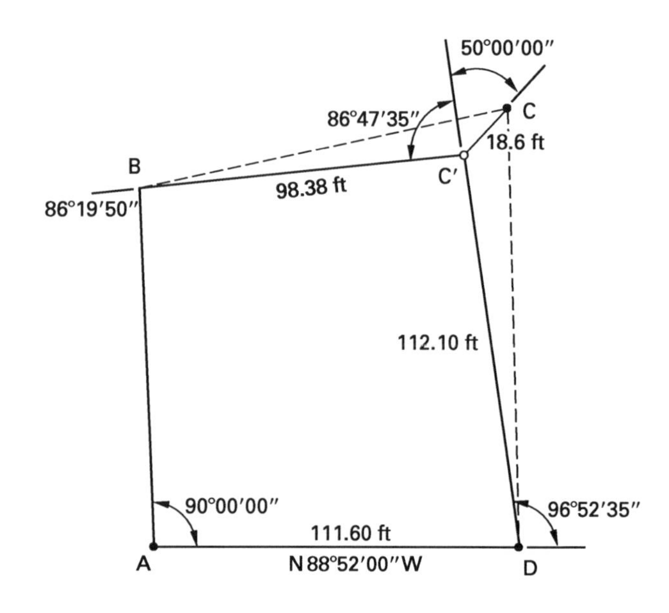

23. Given the figure shown with bearing of AB due south and length of lines as shown, find the bearing of each line and the area of the total figure.

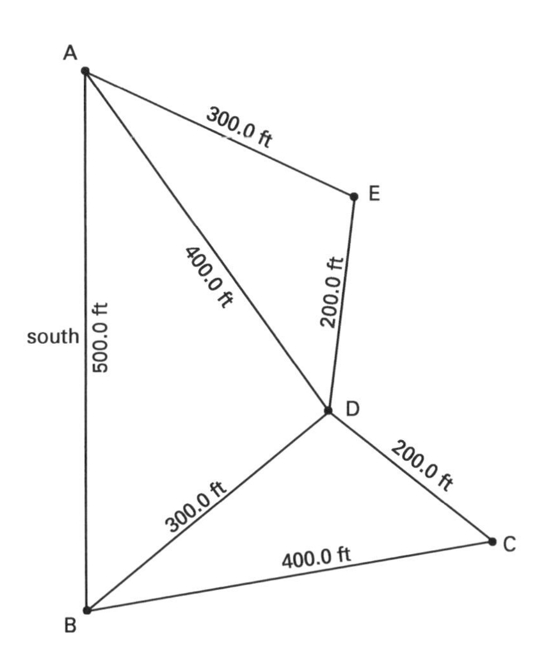

24. Given the tract ABCDA with sides CD and DA, and angles C and D as shown, and area of 43,560 ft^2, find sides AB and BC and angles A and B.

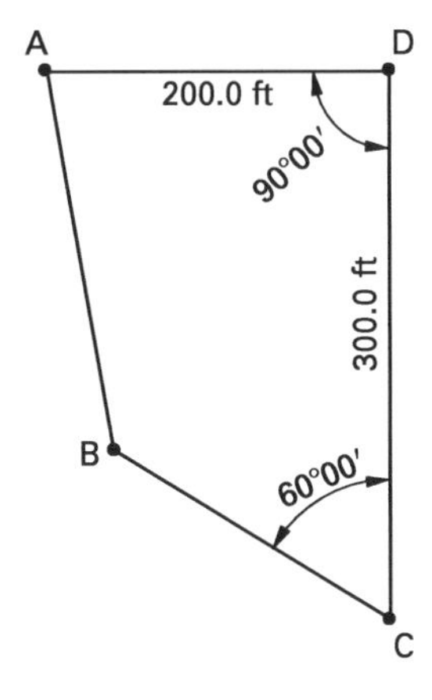

25. The trapezoidal tract ABCDA is to be divided into two equal areas by a line originating at the midpoint of the south line. Determine the necessary information to describe the property.

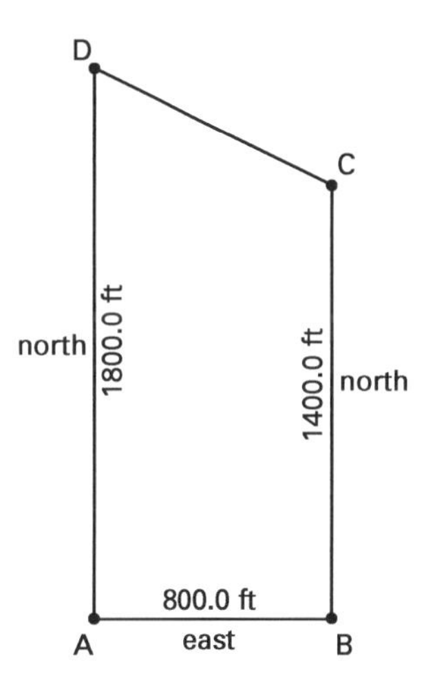

26. Find the coordinates of the point of intersection of the sides AC and BC of the triangle ABC given the information shown.

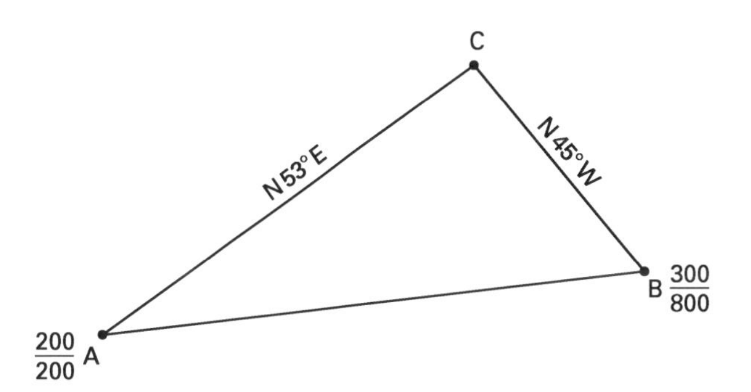

27. Find the coordinates of the point of intersection of the sides AC and BC of the triangle ABC given the information shown.

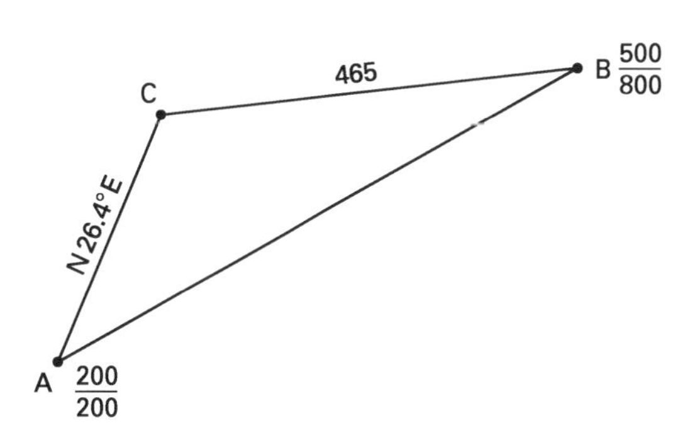

28. Find the coordinates of the point of intersection of the sides AC and BC of the triangle given the information shown.

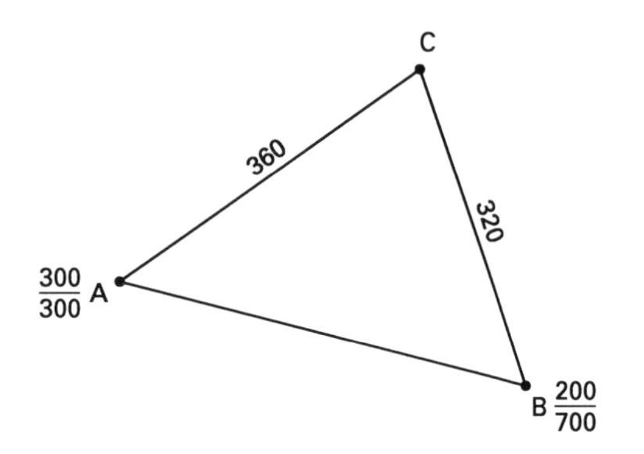

SOLUTIONS

1. (a)

point	measured angle	balanced angle
A	67°06′30″	67°07′
B	216°19′	216°19′
C	65°12′30″	65°13′
D	95°18′30″	95°19′
E	96°02′	96°02′
	539°58′30″	540°00′

(b)

point	measured angle	balanced angle
A	92°38′	92°38′
B	117°21′30″	117°21′
C	129°13′	129°13′
D	261°44′30″	261°44′
E	55°28′30″	55°28′
F	207°10′	207°10′
G	70°52′30″	70°52′
H	145°34′	145°34′
	1080°02′	1080°00′

2.

point	measured angle	balanced angle
A	90°19′59″	90°19′52″
B	106°19′55″	106°19′48″
C	318°51′26″	318°51′19″
D	48°29′12″	48°29′05″
E	150°55′29″	150°55′22″
F	195°44′47″	195°44′40″
G	87°11′56″	87°11′49″
H	193°08′23″	193°08′16″
I	212°18′51″	212°18′44″
J	47°27′37″	47°27′30″
K	288°30′28″	288°30′21″
L	60°43′21″	60°43′14″
	1800°01′24″	1800°00′00″

3.

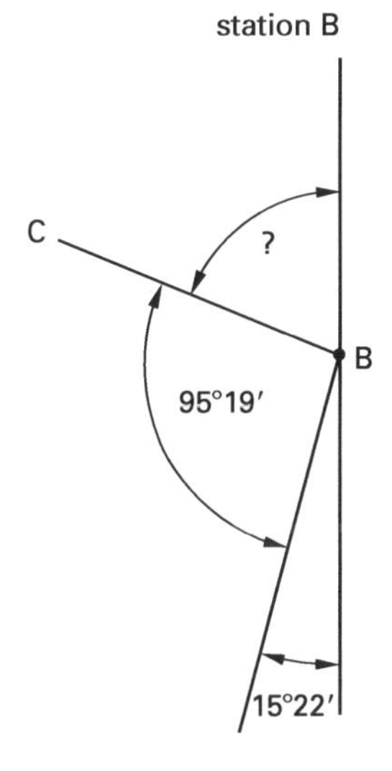

$$\begin{array}{r} 95°19' \\ \underline{+\ 15°22'} \\ 110°41' \\ 179°60' \\ \underline{-\ 110°41'} \\ 69°19' \end{array}$$

bearing BC: N 69°19′ W

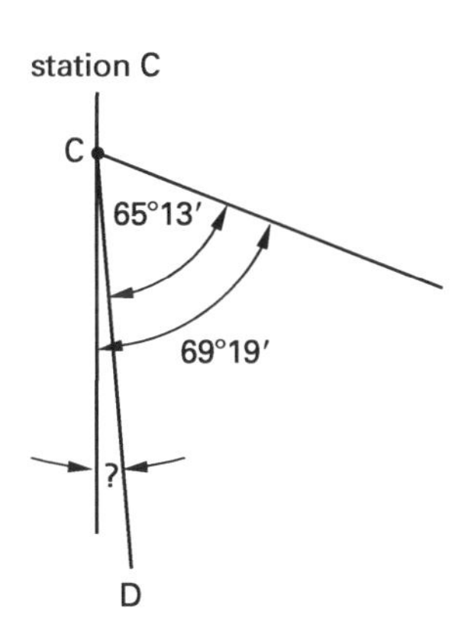

$$\begin{array}{r} 69°19' \\ \underline{-65°13'} \\ 4°06' \end{array}$$

bearing CD: S 04°06′ E

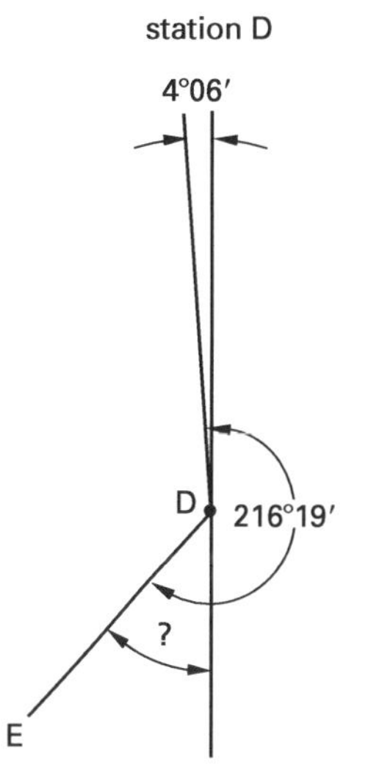

$$\begin{array}{r} 216°19' \\ \underline{-\ \ 4°06'} \\ 212°13' \\ \underline{-180°00'} \\ 32°13' \end{array}$$

bearing DE: S 32°13′ W

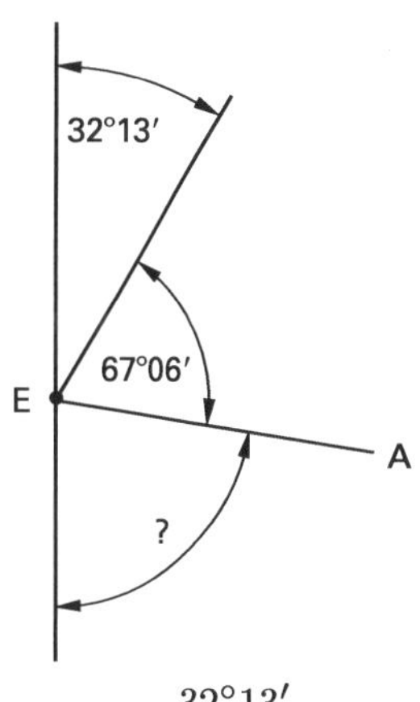

$$\begin{array}{r} 32°13' \\ \underline{+67°06'} \\ 99°19' \\ 179°60' \\ \underline{-\ 99°19'} \\ 80°41' \end{array}$$

Plane Survey Calculations

bearing EA: S 80°41′ E

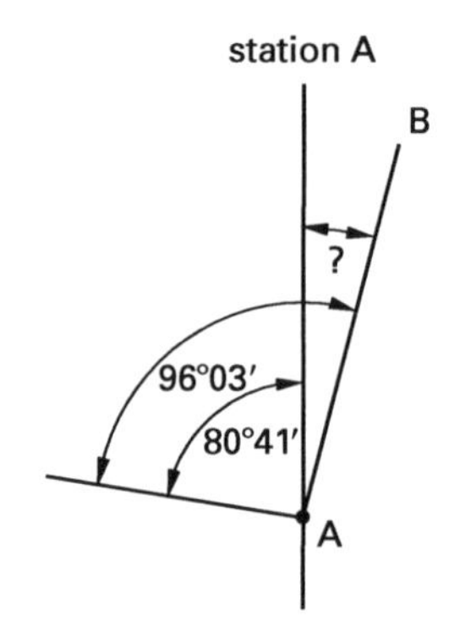

96°03′
−80°41′
15°22′

bearing AB: N 15°22′ E

4.

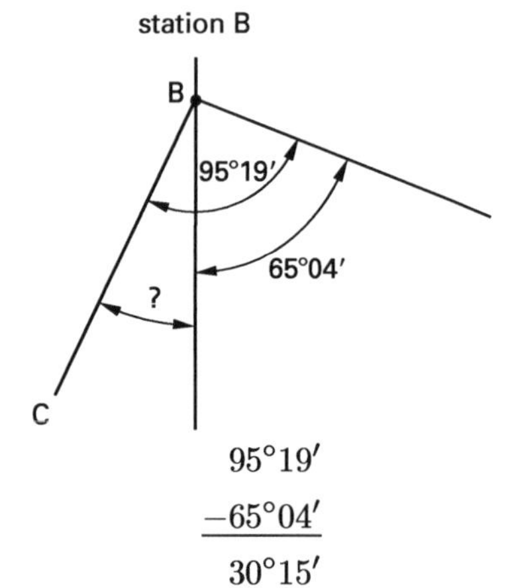

95°19′
−65°04′
30°15′

bearing BC: S 30°15′ W

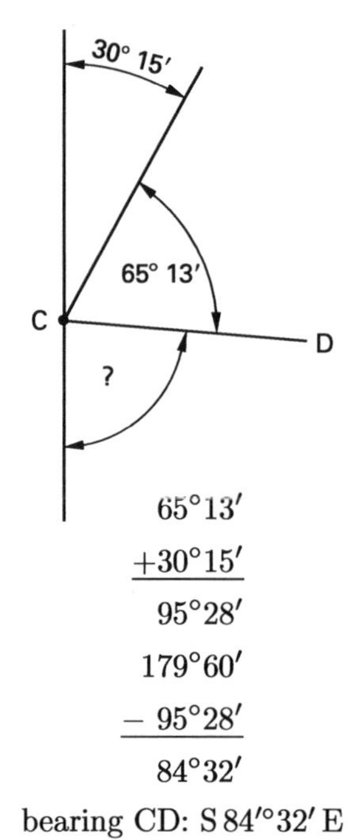

65°13′
+30°15′
95°28′

179°60′
− 95°28′
84°32′

bearing CD: S 84′°32′ E

station D

84°32′
216°19′
D
?
E

216°19′
− 84°32′
131°47′

179°60′
−131°47′
48°13′

bearing DE: S 48°13′ E

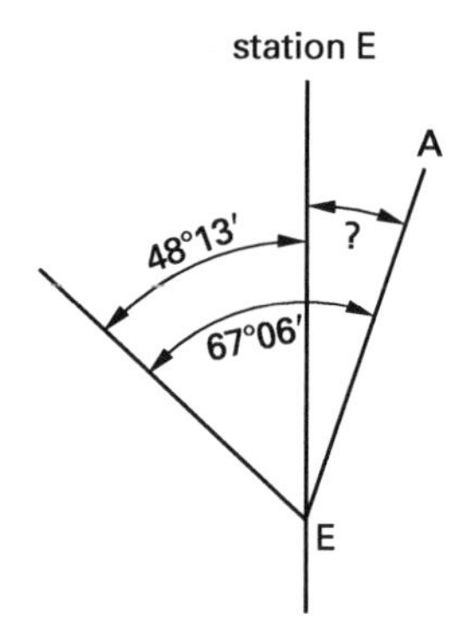

67°06′
−48°13′
18°53′

bearing EA: N 18°53′ E

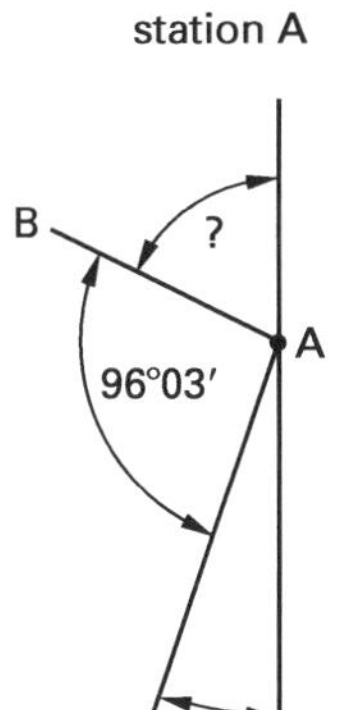

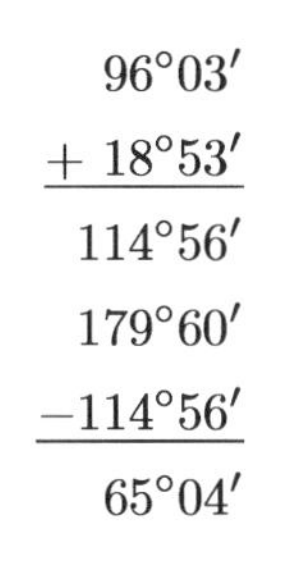

$$
\begin{array}{r}
96^\circ 03' \\
+\ 18^\circ 53' \\ \hline
114^\circ 56' \\
179^\circ 60' \\
-114^\circ 56' \\ \hline
65^\circ 04'
\end{array}
$$

bearing AB: N 65°04′ W

5.

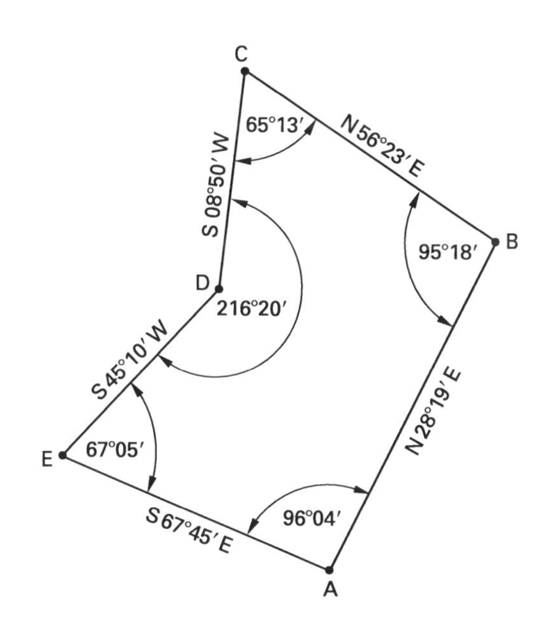

angle A	angle B	angle C	angle D	angle E
67°45′	179°60′	56°23′	179°60′	179°60′
+28°19′	−56°23′	+8°50′	−8°50′	−45°10′
	−28°19′		+45°10′	−67°45′
96°04′	95°18′	65°13′	216°20′	67°05′

6.

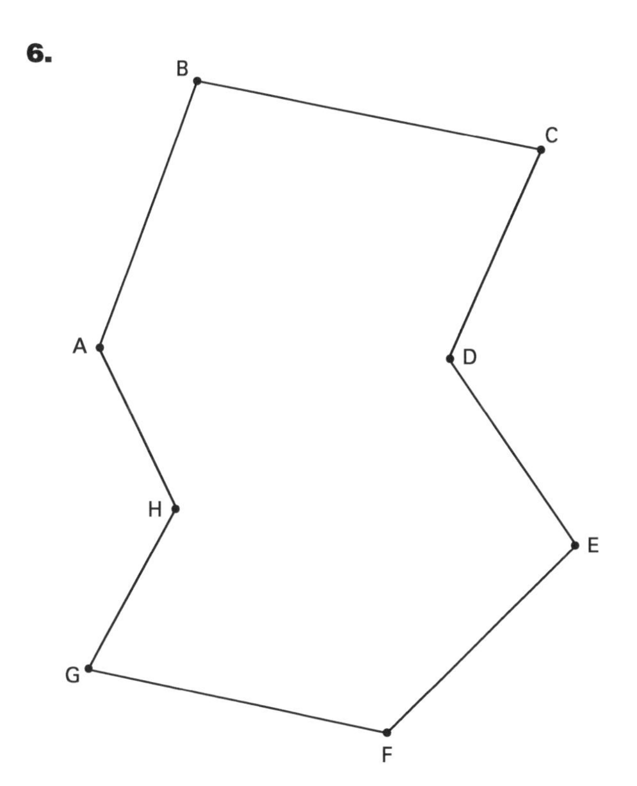

drawn at a scale of 1 in = 200 ft

7.

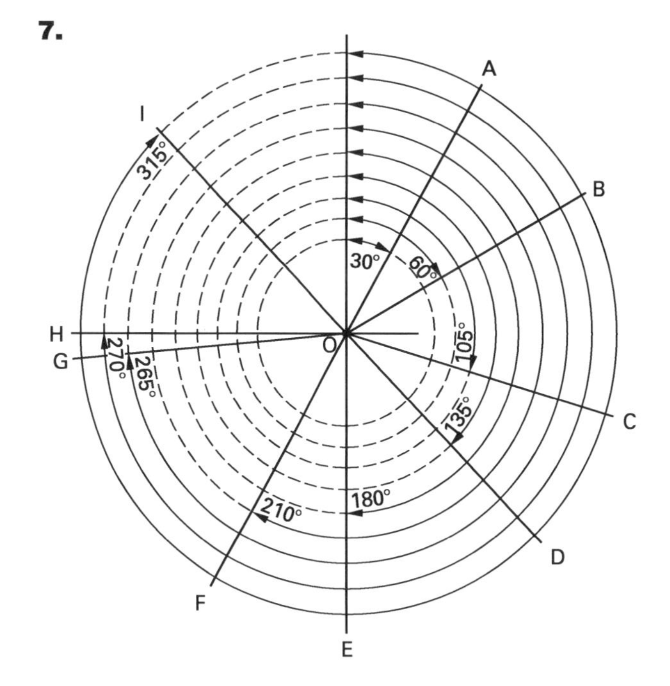

line	azimuth	back azimuth
OA	30°	210°
OB	60°	240°
OC	105°	285°
OD	135°	315°
OE	180°	0°
OF	210°	30°
OG	265°	85°
OH	270°	90°
OI	315°	135°

8.

	14°37′19″	
+	21°14′46″	
	35°52′05″	WAAF-A
+	180°00′00″	
	215°52′05″	
+	272°13′44″	
	488°05′49″	
−	360°00′00″	
	128°05′49″	AB
+	180°00′00″	
	308°05′49″	
+	95°18′22″	
	403°24′11″	
−	360°00′00″	
	43°24′11″	BC
+	180°00′00″	
	223°24′11″	
+	65°13′08″	
	288°37′19″	CD
−	180°00′00″	
	108°37′19″	
+	216°19′30″	
	324°56′49″	DE
−	180°00′00″	
	144°56′49″	
+	67°05′20″	
	212°02′09″	EA
−	180°00′00″	
	32°02′09″	
+	96°03′40″	
	128°05′49″	AB

9. (a) 265°13′16″ (b) 285°35′59″
(c) 171°40′41″ (d) 84°28′13″
(e) 96°56′32″ (f) 352°33′27″
(g) 27°57′45″ (h) 185°17′25″
(i) 175°41′48″ (j) 302°51′58″

10. (a) N 68°22′54″ W (b) N 12°13′47″ E
(c) S 73°47′14″ E (d) S 52°31′18″ W
(e) S 86°55′58″ E (f) N 65°11′37″ E
(g) N 22°44′49″ W (h) S 87°25′51″ W
(i) S 77°32′22″ E (j) N 42°19′07″ W

11.

					latitude		departure	
point	bearing	length	cos	sin	N	S	E	W
A								
	S 60°00 E	50.0	0.5000	0.8660		25.0	43.3	
B								
	N 30°00′ E	100.0	0.8660	0.5000	86.6		50.0	
C								
	N 75°00′ W	200.0	0.2588	0.9659	51.8			193.2
D								
	S 41°30′ E	151.0	0.7490	0.6626		113.1	100.1	
A								
					138.4	138.1	193.4	193.2

12.

point	bearing	length	latitude N	latitude S	cor	departure E	departure W	cor	latitude N	latitude S	departure E	departure W
A												
	N 28°19′ E	560.27	493.23		−0.11	265.76		−0.10	493.12		265.66	
B												
	N 56°23′ W	484.18	268.06		−0.09		403.21	+0.08	267.97			403.29
C												
	S 08°50′ W	375.42		370.97	+0.07		57.65	+0.07		371.04		57.72
D												
	S 45°10′ W	311.44		219.58	+0.06		220.86	+0.05		219.64		220.91
E												
	S 67°45′ E	449.83		170.33	+0.08	416.34		−0.08		170.41	416.26	
A		2181.14	761.29	760.88		682.10	681.72		761.09	761.09	681.92	681.92
			760.88			681.72						
			0.41			0.38						

error of closure $= \sqrt{(0.41)^2 + (0.38)^2} = 0.56$

ratio of error $= 1/3900$

course	latitude corrections
AB	$\left(\frac{0.41}{2181.14}\right)(560.27) = 0.11$
BC	$\left(\frac{0.41}{2181.14}\right)(484.18) = 0.09$
CD	$\left(\frac{0.41}{2181.14}\right)(375.42) = 0.07$
DE	$\left(\frac{0.41}{2181.14}\right)(311.44) = 0.06$
EA	$\left(\frac{0.41}{2181.14}\right)(449.83) = 0.08$

course	departure corrections
AB	$\left(\frac{0.38}{2181.14}\right)(560.27) = 0.10$
BC	$\left(\frac{0.38}{2181.14}\right)(484.18) = 0.08$
CD	$\left(\frac{0.38}{2181.14}\right)(375.42) = 0.07$
DE	$\left(\frac{0.38}{2181.14}\right)(311.44) = 0.05$
EA	$\left(\frac{0.38}{2181.14}\right)(449.83) = 0.08$

13.

point	bearing	length	latitude N	latitude S	cor	departure E	departure W	cor	latitude N	latitude S	departure E	departure W
A												
	S 53°18′ E	560.27		334.83	−0.10	449.21		−0.06		334.73	449.15	
B												
	N 42°00′ E	484.18	359.82		+0.08	323.98		−0.05	359.90		323.93	
C												
	N 72°47′ W	375.42	111.12		+0.06		358.60	+0.04	111.18			358.64
D												
	N 36°27′ W	311.44	250.51		+0.05		185.03	+0.04	250.56			185.07
E												
	S 30°39′ W	449.83		386.99	−0.08		229.32	+0.05		386.91		229.37
A			721.45	721.82		773.19	772.95		721.64	721.64	773.08	773.08
				721.45		772.95						
				0.37		0.24						

14.

point	bearing	length	latitude N	latitude S	cor	departure E	departure W	cor	latitude N	latitude S	departure E	departure W
A												
	M 80°00′ W	1015.43	176.33		+0.17		1000.00	−0.10	176.50			999.90
B												
	S 66°30′ W	545.52		217.53	−0.09		500.27	−0.05		217.44		500.22
C												
	S 12°00′ E	480.97		470.46	−0.08	100.00		+0.05		470.38	100.05	
D												
	S 88°30′ E	750.26		19.64	−0.12	750.00		+0.07		19.52	750.07	
E												
	N 69°00′ E	639.18	229.06		+0.10	596.73		+0.06	229.16		596.79	
F												
	N 10°00′ E	306.28	301.63		+0.05	53.18		+0.03	301.68		53.21	
A		3737.64	707.02	707.63		1499.91	1500.27		707.34	707.34	1500.12	1500.12

error of closure $= \sqrt{(0.61)^2 + (0.36)^2} = 0.71$

ratio of error $= 1/5300$

15.

point	bearing	length	latitude N	latitude S	departure E	departure W	coordinates y	coordinates x
A							9012.34	1234.56
	N 28°19′ E	560.27	493.12		265.66			
B							9505.46	1500.22
	N 56°23′ W	484.18	267.97			403.29		
C							9773.43	1096.93
	S 08°50′ W	375.42		371.04		57.72		
D							9402.39	1039.21
	S 45°10′ W	311.44		219.64		220.91		
E							9182.75	818.30
	S 67°45′ E	449.83		170.41	416.26			
A							9012.34	1234.56

16.

point	coordinates y	coordinates x	latitude N	latitude S	departure E	departure W	bearing	length
A	9090.90	9090.90						
			389.49		361.37		N 42°51′ E	531.31
B	9480.39	9452.27						
			132.48			561.57	N 76°44′ W	576.99
C	9612.87	8890.70						
				291.01		114.89	S 21°33′ W	312.87
D	9321.86	8775.81						

17.

point	bearing	length	latitude N	latitude S	departure E	departure W	coordinates y	coordinates x
A							1000.00	1000.00
	S 12°02′ E	59.20		57.90	12.34			
M							942.10	1012.34
B							1493.12	1265.66
	N 88°13′ E	65.13	2.03		65.10			
N							1495.15	1330.76
C							1761.09	862.37
	N 13°52′ W	70.85	68.79			16.98		
O							1829.88	845.39
D							1390.05	804.65
	N 56°09′ W	72.45	40.36			60.17		
P							1430.41	744.48
E							1170.41	583.74
	S 78°31′ W	76.51		15.23		74.98		
Q							1155.18	508.76
M							942.10	1012.34
	N 29°56′ E	638.17	553.05		318.42			
N							1495.15	1330.76
	N 55°24′ W	589.60	334.73			485.37		
O							1829.88	845.39
	S 14°11′ W	412.02		399.47		100.91		
P							1430.41	744.48
	S 40°35′ W	362.37		275.23		235.72		
Q							1155.18	508.76
	S 67°04′ E	546.81		213.08	503.58			
M							942.10	1012.34

18.

point	bearing	length	latitude N	latitude S	departure E	departure W	coordinates y	coordinates x
A							1000.00	1000.00
	N 28°48′ W	58.46	51.23			28.16		
M							1051.23	971.84
B							1493.12	1265.66
	N 87°33′ E	66.72	2.85		66.66			
N							1495.97	1332.32
C							1761.09	862.37
	S 27°47′ E	68.89		60.95	32.11			
O							1700.14	894.48
D							1390.05	804.65
	N 55°29′ W	74.13	42.01			61.08		
P							1432.06	743.57
E							1170.41	583.74
	S 13°38′ E	66.52		64.65	15.68			
Q							1105.76	599.42
M							1051.23	971.84
	N 39°02′ E	572.49	444.74		360.48			
N							1495.97	1332.32
	N 65°00′ W	483.10	204.17			437.84		
O							1700.14	894.48
	S 29°23′ W	307.64		268.08		150.91		
P							1432.06	743.57
	S 23°50′ W	356.72		326.30		144.15		
Q							1105.76	599.42
	S 81°40′ E	376.39		54.53	372.42			
M							1051.23	971.84

19.

point	line	azimuth	length	latitude	departure	balanced latitude	balanced departure	coordinates y	coordinates x
A								1000.00	1000.00
	AB	21°12′	289.07	269.51	104.53	269.55	104.51		
B								1269.55	1104.51
	BC	101°19′	366.63	−71.94	359.50	−71.89	359.48		
C								1197.66	1463.99
	CD	205°24′	228.60	−206.50	−98.05	−206.47	−98.06		
D								991.19	1365.93
	DE	144°37′	232.44	−189.51	134.59	−189.48	134.57		
E								801.71	1500.50
	EF	227°51′	281.24	−188.73	−208.51	−188.69	−208.53		
F								613.02	1291.97
	FG	281°30′	316.68	63.14	−310.32	63.18	−310.34		
G								676.20	981.63
	GH	29′°29′	192.13	167.25	94.56	167.27	94.55		
H								843.47	1076.18
	HA	334′°03′	174.06	156.51	−76.17	156.53	−76.18		
A								1000.00	1000.00
				−0.27	+0.13				

error of closure $= \sqrt{(0.27)^2 + (0.13)^2} = 0.30$

precision $= 1/6900$

20.

line	azimuth	correction	adjusted azimuth	adjusted bearing
XA	21°44′	fixed	21°44′	N 21°44′ E
	96°55′			
AB	118°39′	−0°00′24″	118°38′36″	S 61°21′24″ E
	−62°20′			
BC	56°19′	−0°00′48″	56°18′12″	N 56°18′12″ E
	67°48′			
CD	124°07′	−0°01′12″	124°05′48″	S 55°54′12″ E
	−67°22′			
DE	56°45′	−0°01′36″	56°43′24″	N 56°43′24″ E
	113°31′			
EY	170°16′	−0°02′00″	170°14′00″	S 9°46′00″ E
	180°			
YE	350°16′			
YE	350°14′	fixed		

21.

	azimuth	cor	adjusted azimuth
	246°45′46″		
	+306°17′23″	+07″	
	553°03′09″		
	−360°		
FOX-A	193°03′09″		193°03′16″
	−180°		
	13°03′09″		
	+151°10′21″	+07″	
AB	164°13′30″		164°13′44″
	−180°		
	344°13′30″		
	+210°06′31″	+07″	
BC	194°20′01″		194°20′22″
	−180°		
	14°20′01″		
	+148°10′01″	+07″	
C-DOG	162°30′02″		162°30′30″
	+180°		
	342°30′02″		
	+ 87°14′38″	+07″	
	429°44′40″		
	−360°		
	69°44′40″		
DOG-MK	−69°45′15″		69°45′15″
	35″		

							coordinates		
line	azimuth	length	latitude	departure	latitude	departure	y	x	point
FOX							22,093.20	2522.88	FOX
−A	193°03′16″	5543.31	−5400.05	−1252.11	−5400.11	−1252.12			
							16,693.09	1270.76	A
AB	164°13′44″	6057.05	−5829.03	1646.28	−5829.10	+1646.26			
							10,863.99	2917.02	B
BC	194°20′22″	6317.47	−6120.65	−1564.62	−6120.72	−1564.64			
							4743.27	1352.38	C
−C	162°30′30″	3734.26	−3561.59	1122.40	−3561.63	+1122.39			
							1181.64	2474.77	DOG
DOG		21,652.09	−20,911.32	−48.05					
			−20,911.56	48.11					
			+0.24	−0.06					

$$\sqrt{(0.24)^2 + (0.06)^2} = 0.25$$

22. Solve triangle BCC′ and triangle CDC′.

In ΔBCC′:

$$\text{BC} = \sqrt{\begin{array}{l}(98.38)^2 + (18.6)^2 \\ \quad - (2)(98.38)(18.6)(\cos 136°47'35'')\end{array}} = 112.66$$

$$\sin \text{B} = \frac{18.6 \sin 136°47'35''}{112.66}$$

$$\text{B} = 6°29'25''$$

$$\begin{array}{r}\text{bearing} \\ \text{BC}\end{array} = \text{N}\,80°58'25''\,\text{E}$$

In ΔCDC′:

$$\text{CD} = \sqrt{\begin{array}{l}(18.6)^2 + (112.10)^2 \\ \quad - (2)(18.6)(112.10)(\cos 130°00'00'')\end{array}} = 124.87$$

$$\sin \text{D} = \frac{18.6 \sin 130°00'00''}{124.87}$$

$$\text{D} = 6°33'07''$$

$$\begin{array}{r}\text{bearing} \\ \text{CD}\end{array} = \text{S}\,0°48'32''\,\text{W}$$

$$\begin{array}{r}\text{bearing} \\ \text{BC}\end{array} = \text{N}\,80°58'25''\,\text{E} \qquad \text{distance BC} = 112.66$$

$$\begin{array}{r}\text{bearing} \\ \text{CD}\end{array} = \text{S}\,0°48'32''\,\text{W} \qquad \text{distance CD} = 124.87$$

23. Solve ABD:

$$\cos \text{A} = \frac{(500)^2 + (400)^2 - (300)^2}{(2)(500)(400)} \qquad \text{A} = 36°52'$$

$$\cos \text{B} = \frac{(500)^2 + (300)^2 - (400)^2}{(2)(500)(400)} \qquad \text{B} = 53°08'$$

$$\text{D} = \frac{90°00'}{180°00'}$$

ΔABD is a right triangle (3-4-5).

Solve BCD:

$$\cos \text{B} = \frac{(300)^2 + (400)^2 - (200)^2}{(2)(300)(400)} \qquad \text{B} = 28°57'$$

$$\cos \text{C} = \frac{(400)^2 + (200)^2 - (300)^2}{(2)(400)(200)} \qquad \text{C} = 46°34'$$

$$\cos \text{D} = \frac{(300)^2 + (200)^2 - (400)^2}{(2)(300)(200)} \qquad \text{D} = \frac{104°29'}{180°00'}$$

Solve DEA:

$$\text{congruent to BCD} \qquad \begin{array}{l}\text{D} = 46°34' \\ \text{E} = 104°29' \\ \text{A} = \underline{28°57'} \\ \quad\;\; 180°00'\end{array}$$

Bearings:

$\text{AB} = \text{South}$

$\text{AD} = \text{S}\,36°52'\,\text{E}$

$\text{AE} = \text{S}\,65°50'\,\text{E}$

$\text{BD} = \text{N}\,53°08'\,\text{E}$

$\text{BC} = \text{N}\,82°05'\,\text{E}$

$\text{CD} = \text{N}\,51°21'\,\text{W}$

$\text{ED} = \text{S}\,09°42'\,\text{W}$

Area:

$$\text{ABD} = \sqrt{\begin{array}{l}(600)(600-500) \\ \quad \times (600-400)(600-300)\end{array}} = 60{,}000 \text{ ft}^2$$

$$\text{BCD} = \sqrt{\begin{array}{l}(450)(450-400) \\ \quad \times (450-300)(450-200)\end{array}} = 29{,}047 \text{ ft}^2$$

$$\text{DEA} = \text{congruent} \qquad = \underline{29{,}047 \text{ ft}^2}$$

$$2.711 \text{ ac}$$

24. Construct triangles ABC, ACD, and BCE.

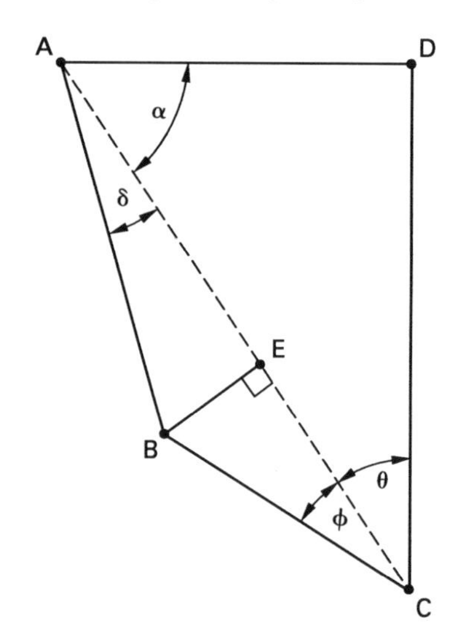

Solve ACD:

$$\theta = \tan^{-1}\left(\frac{200.0}{300.0}\right) = 33°41'$$

$$\alpha = 90° - 33°41' = 56°19'$$

$$\text{AC} = \sqrt{(200)^2 + (300)^2} = 360.6$$

Solve ABC:

$$\text{area ACD} = \left(\tfrac{1}{2}\right)(200.0)(300.0) = 30{,}000.0$$

$$\text{area ABC} = 43{,}560.0 - 30{,}000 = 13{,}560.0$$

$$13{,}560.0 = \left(\tfrac{1}{2}\right)(\text{BE})(360.6)$$

$$\text{BE} = (2)\left(\frac{13{,}560.0}{360.6}\right) = 75.2$$

$$\phi = 60° - 33°41' = 26°19'$$

$$BC = \frac{75.2}{\sin 26°19'} = 169.6$$

$$AB = \sqrt{\begin{matrix}(169.6)^2 + (360.6)^2 \\ - (2)(169.6)(360.6)(\cos 26°19')\end{matrix}} = 221.7$$

$$\delta = \sin^{-1}\left(\frac{169.6 \sin 33°41'}{221.7}\right) = 19°50'$$

$$A = 19°50' + 56°19' = 76°09'$$

$$B = 180° - (19°50' + 26°19') = 133°51'$$

Summary:

$$AB = 221.7 \text{ ft}$$
$$BC = 169.6 \text{ ft}$$
$$\text{angle } A = 76°09'$$
$$\text{angle } B = 133°51'$$

25. Construct line MN parallel to BC and DA and originating at midpoint M. Construct MO to represent the boundary between the two tracts when they have been partitioned. Construct right triangle CPN with CP perpendicular to MN.

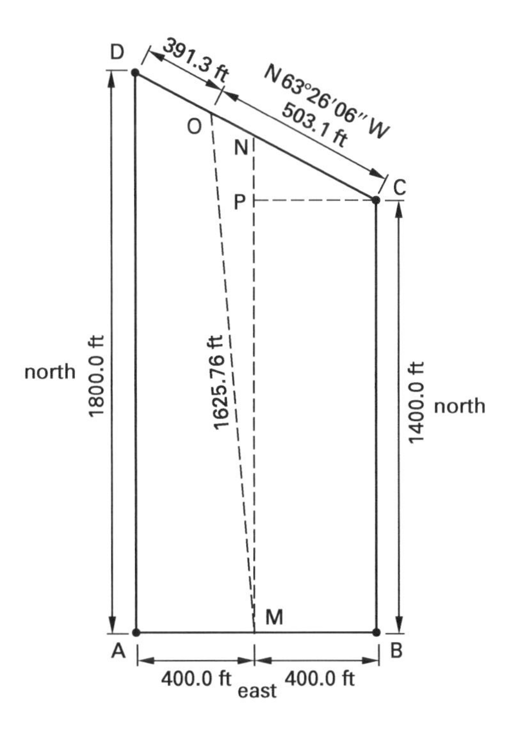

By proportion, MN = 1600.

$$\text{area ABCDA} = \left(\tfrac{1}{2}\right)(1400 + 1800)(800) = 1{,}280{,}000$$

$$\text{area MBCN} = \left(\tfrac{1}{2}\right)(1400 + 1600)(400) = 600{,}000$$

$$\text{area MNO} = 640{,}000 - 600{,}000 = 40{,}000$$

Solve triangle CPN:

$$\angle C = \tan^{-1}\left(\frac{200}{400}\right) = 26°33'54''$$

$$CN = \sqrt{(200)^2 + (400)^2} = 447.21$$

In ΔMNO:

$$\angle N = 90° + 26°33'54'' = 116°33'54''$$

$$\text{area MNO} = \left(\tfrac{1}{2}\right)(NO)(MN)(\sin 116°33'54'')$$

$$40{,}000 = \left(\tfrac{1}{2}\right)(NO)(1600)(\sin 116°33'54'')$$

$$NO = \frac{(2)(40{,}000)}{1600 \sin 116°33'54''} = 55.90$$

$$MO = \sqrt{\begin{matrix}(1600)^2 + (55.90)^2 \\ + (2)(1600)(55.90)(\cos 116°33'54'')\end{matrix}} = 1625.76$$

$$\sin M = \frac{55.90 \sin 116°33'54''}{1625.76}$$

$$M = 1°45'44''$$

$$CO = 447.21 + 55.90 = 503.11$$

$$OD = 447.21 - 55.90 = 391.31$$

$$\text{bearing MO} = \text{N } 1°45'44'' \text{ W}$$

$$\text{bearing CD} = \text{N } 63°26'06'' \text{ W}$$

26. *Bearing-Bearing Method:*

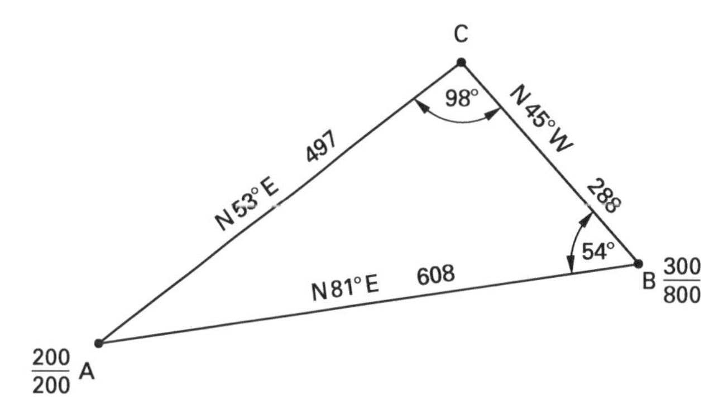

$$\text{tan bearing AB} = \frac{800 - 200}{300 - 200} = 6$$

$$\text{bearing AB} = \text{N}\,81^\circ\,\text{E}$$

$$\text{distance AB} = \sqrt{(800 - 200)^2 + (300 - 200)^2} = 608$$

$$\text{angle } A = 28^\circ$$

$$\text{angle } B = 54^\circ$$

$$\text{angle } C = 98^\circ$$

$$\text{AC} = \frac{608 \sin 54^\circ}{\sin 98^\circ} = 497$$

$$\text{BC} = \frac{608 \sin 28^\circ}{\sin 98^\circ} = 288$$

$$\text{north coord. of C} = 200 + 497 \cos 53^\circ = 499$$

$$\text{east coord. of C} = 200 + 497 \sin 53^\circ = 597$$

27. *Bearing-Distance Method:*

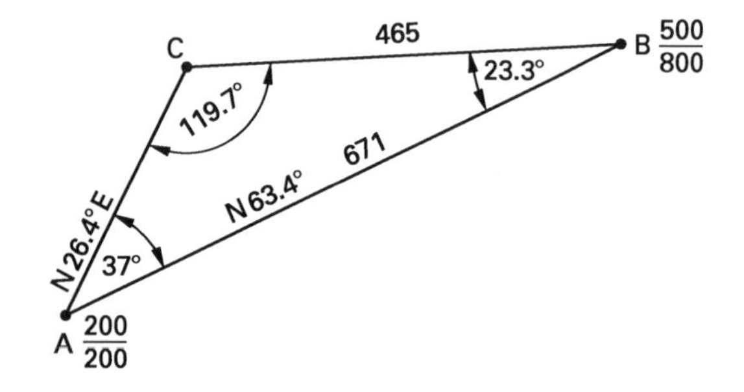

$$\text{tangent bearing AB} = \frac{800 - 200}{500 - 200} = 2$$

$$\text{bearing AB} = \text{N}\,63.4\,\text{E}$$

$$\text{distance AB} = \sqrt{600^2 + 300^2} = 671$$

$$\text{angle } A = 37.0^\circ$$

$$\sin \text{C}' = \frac{671 \sin 37.0^\circ}{465}$$

$$\text{C}' = 60.3^\circ$$

$$\text{C} = 180^\circ - 60.3^\circ = 119.7^\circ$$

$$\begin{aligned}\text{B} &= 180^\circ - \text{A} - \text{C} \\ &= 180^\circ - 37.0^\circ - 119.7^\circ \\ &= 23.3^\circ\end{aligned}$$

$$b = \frac{\sin 23.3^\circ (465)}{\sin 37^\circ} = 306$$

$$\text{north coord.} = 200 + 306 \cos 26.4^\circ = 474$$

$$\text{east coord.} = 200 + 306 \sin 26.4^\circ = 336$$

28. *Distance-Distance Method:*

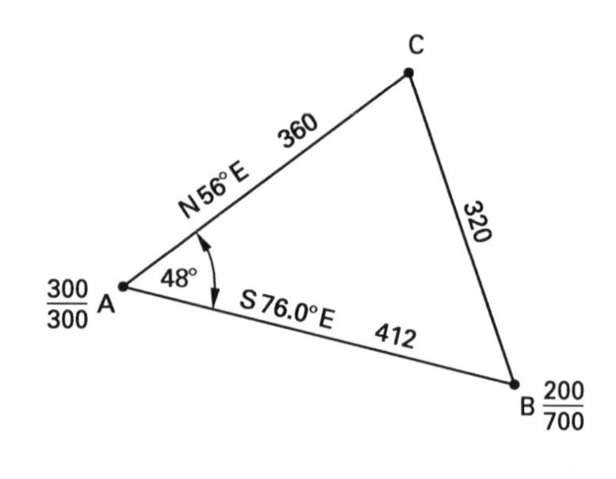

$$\text{tangent bearing AB} = \frac{700 - 300}{300 - 200} = 4$$

$$\text{bearing AB} = \text{S}\,76^\circ\,\text{E}$$

$$\text{distance AB} = \sqrt{(100)^2 + (400)^2} = 412$$

$$\begin{aligned}\cos \text{A} &= \frac{b^2 + c^2 - a^2}{2bc} \\ &= \frac{(360)^2 + (412)^2 - (320)^2}{(2)(360)(412)}\end{aligned}$$

$$A = 48^\circ$$

$$\text{bearing AC} = \text{N}\,56^\circ\,\text{E}$$

$$\text{north coord. of C} = 300 + 360 \cos 56^\circ = 501$$

$$\text{east coord. of C} = 300 + 360 \sin 56^\circ = 598$$

16 Area of a Traverse

1. METHODS FOR COMPUTATION OF AREA

The area within a traverse can be computed by the *double meridian distance* (DMD) *method*, by the coordinate method, and by use of geometric or trigonometric formulas. The most frequently used method for land area calculations is the DMD method.

2. DOUBLE MERIDIAN DISTANCES

The DMD method makes use of balanced latitudes and departures. It is widely used because it allows areas to be computed quickly.

The DMD method sets up a series of trapezoids and triangles, both inside and outside of the traverse. It calculates each of these areas and determines the area of the traverse from them.

The double meridian distance is simply twice the meridian distance (see Sec. 3). The DMD is used instead of the meridian distance (MD) to simplify the arithmetic. If the MD were used, division by two would be required several times. Using DMD, division by two is required only once.

The area of a trapezoid is one-half the sum of the bases times the altitude (i.e., the average of the bases times the altitude.) In the DMD method, the meridian distance for each course of the traverse serves as the average of the bases of a trapezoid. The DMDs of the courses are obtained from the departures of the courses. Thus, the only data needed are latitudes and departures.

3. MERIDIAN DISTANCES

The *meridian distance* (MD) of a course is the right angle distance from the midpoint of the course to a reference meridian. MDs are illustrated in Fig. 16.1. Since east and west departures are used, algebraic signs must be considered. To simplify the use of plus and minus values for departure, the entire traverse should be placed in the northeast quadrant. This can be done by taking the reference meridian through the most westerly point in the traverse.

Figure 16.1 *Meridian Distance*

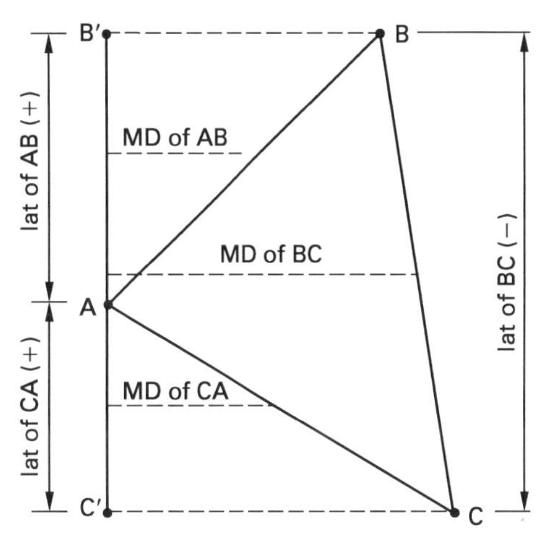

4. DETERMINING THE MOST WESTERLY POINT

The most westerly point can usually be determined by studying the east and west departures.

5. RULES FOR USING DMD CALCULATIONS

In Fig. 16.2, the MD of EA = 1/2 departure of EA. The DMD of EA = departure of EA.

The MD of AB = MD of EA + 1/2 departure EA + 1/2 departure AB. DMD of AB = DMD of EA + departure EA + departure AB.

Figure 16.2 DMD Rules

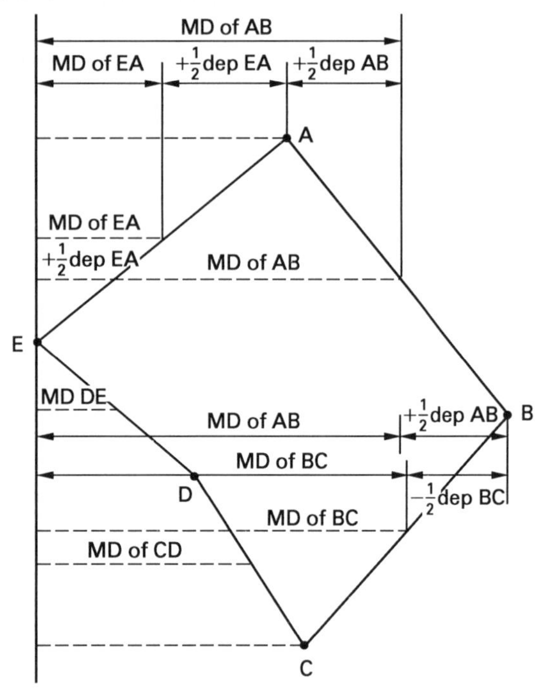

The MD of BC = MD of AB + ½ departure AB − ½ departure BC. The DMD of BC = DMD of AB + departure AB − departure BC.

From these examples and from further examination, the following rules can be derived for computing DMDs.

step 1: The DMD of the first course is equal to the departure of the first course.

step 2: The DMD of any course is equal to the DMD of the preceding course plus the departure of the preceding course plus the departure of the course itself.

step 3: The DMD of the last course is equal to the departure of the last course with opposite sign. ("Opposite sign" means that for a westerly course, the DMD will be positive.)

6. AREA BY DMD

The area of a traverse can be found by multiplying the DMD of each course by the latitude of that course, with north latitudes producing positive areas and south latitudes producing negative areas, adding the areas algebraically, and dividing by two. The algebraic sign of DMDs is always positive.

In Fig. 16.1,

$$\text{area ABC} = \text{area B}'\text{BCC}' - \text{area AB}'\text{B} - \text{area ACC}'$$

If MDs are positive, north latitudes are positive, and south latitudes are negative, then

$$\text{area of B}'\text{BCC}' = (\text{MDof BC})(\text{latitude of BC}) \quad [\text{negative}]$$

$$\text{area of AB}'\text{B} = (\text{MD of AB})(\text{latitudeofAB}) \quad [\text{positive}]$$

$$\text{area of ACC}' = (\text{MD of CA})(\text{latitudeofCA}) \quad [\text{positive}]$$

$$\text{area of ABC} = \text{algebraic sum of B}'\text{BCC}', \text{AB}'\text{B}, \text{ACC}'$$

The sign of the sum can be either plus or minus.

If DMDs are used, the area will be double the area determined by using MDs, and the area of the traverse can be determined by dividing the double area by two.

Example 16.1

The following table shows the tabulation of computations for the area of a traverse ABCDEA by the DMD method. Latitudes and departures shown are balanced. Find the DMD.

Table for Example 16.1

(All distances in feet.)

			latitude		departure			double	area
line	bearing	distance	north	south	east	west	DMD	plus	minus
AB	N 65°04′ W	560.27	236.11			507.97	995.41	235,026	
BC	S 30°14′ W	484.14		418.39		243.72	243.72		101,970
CD	S 84°33′ E	375.42		35.71	373.77		373.77		13,347
DE	S 48°13′ E	311.44		207.56	232.27		979.81		203,369
EA	N 18°53′ E	449.83	425.55		145.65		1357.73	577,782	
								812,808	318,686

Solution

The first step is to determine the most westerly traverse point. In lieu of a sketch showing the traverse, it is found as follows.

step 1: Because AB has a northwest direction, B is west of A. Looking at the departure column, it is 507.97 ft west of A.

step 2: C is 243.72 ft west of B, so B is not the most westerly point.

step 3: Courses, CD, DE, and EA have east departures; therefore, C is the most westerly point.

With C the most westerly point, the first DMD computed is for the course CD. Remembering that the DMD for the first course is the departure of the first course, and also remembering the definition of the DMD for any course,

$$\text{departure of CD} = +373.77 \text{ ft} = \text{DMD of CD}$$

$$\begin{aligned}\text{departure of CD} &= +373.77 \text{ ft}\\ \text{departure of DE} &= +232.27 \text{ ft}\\ &\quad + 979.81 \text{ ft} = \text{DMD of DE}\end{aligned}$$

$$\begin{aligned}\text{departure of DE} &= +232.27 \text{ ft}\\ \text{departure of EA} &= +145.16 \text{ ft}\\ &\quad + 1357.73 \text{ ft} = \text{DMD of EA}\end{aligned}$$

$$\begin{aligned}\text{departure of EA} &= +145.65 \text{ ft}\\ \text{departure of AB} &= -507.97 \text{ ft}\\ &\quad + 995.41 \text{ ft} = \text{DMD of AB}\end{aligned}$$

$$\begin{aligned}\text{departure of AB} &= -507.97 \text{ ft}\\ \text{departure of BC} &= -243.72 \text{ ft}\\ &\quad + 243.72 \text{ ft} = \text{DMD of BC}\end{aligned}$$

Note that the DMD of BC is the same as its departure except that it has a positive sign.

The area of the traverse ABCDEA is found from the double area sums in the table.

$$\text{area} = \left(\tfrac{1}{2}\right)(812{,}808 \text{ ft}^2 - 318{,}686 \text{ ft}^2) = 247{,}061 \text{ ft}^2$$

If necessary, this area can be converted to acres by dividing by 43,560.

$$\text{area} = \frac{247{,}061 \text{ ft}^2}{43{,}560 \ \frac{\text{ft}^2}{\text{ac}}} = 5.672 \text{ ac}$$

7. AREA BY COORDINATES

After the coordinates of the corners of a tract of land are determined, the area of the tract can be computed by the *coordinate method.*

The coordinate formula is derived by forming trapezoids and determining their areas just as is done in the DMD method. Meridian distances are not used; trapezoids are formed by the abscissas of the corners. Ordinates of the corners serve as the altitude of the trapezoids. Alternately, the trapezoids can be formed by the ordinates of the corners, and the abscissas serve as the altitudes. In Fig. 16.3, the abscissas are used to form the trapezoids.

Figure 16.3 *Coordinates of Traverse Points*

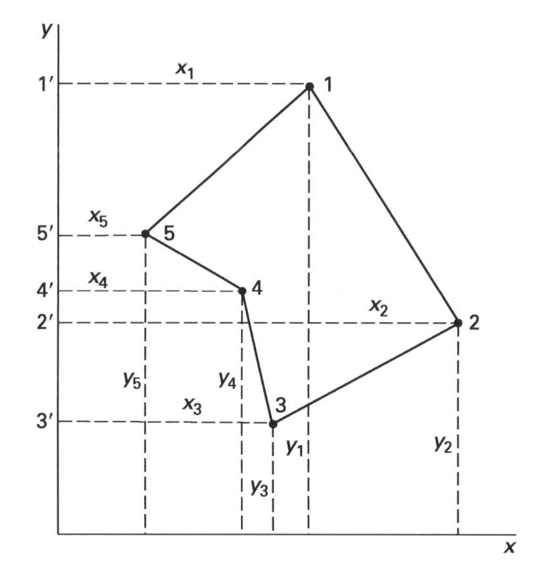

$$\begin{aligned}\text{area 1-2-3-4-5-1} &= \text{area } 1'\text{-1-2-}2' + \text{area } 2'\text{-2-3-}3'\\ &\quad - \text{area } 1'\text{-1-5-}5' - \text{area } 5'\text{-5-4-}4'\\ &\quad - \text{area } 4'\text{-4-3-}3'\\ &= \left(\tfrac{1}{2}\right)\big(x_1(y_5 - y_2) + x_2(y_1 - y_3)\\ &\quad + x_3(y_2 - y_4) + x_4(y_3 - y_5)\\ &\quad + x_5(y_4 - y_1)\big)\end{aligned}$$

The same results can be found by using the ordinates to form the trapezoids.

$$\begin{aligned}\text{area 1-2-3-4-5-1} &= \left(\tfrac{1}{2}\right)\big(y_1(x_5 - x_2) + y_2(x_1 - x_3)\\ &\quad + y_3(x_2 - x_4) + y_4(x_3 - x_5)\\ &\quad + y_5(x_4 - x_1)\big)\end{aligned}$$

Example 16.2

Given the traverse 1-2-3-4-5-1 in Fig. 16.3 with the coordinates as shown, find the area inside the traverse by the coordinate method.

	coordinates	
point	y (ft)	x (ft)
1	1000.00	1000.00
2	1236.11	492.03
3	817.72	248.31
4	782.01	622.08
5	574.45	854.35

Solution

	coordinates (ft)			double area (ft²)	
station	y	x	$y_1 - y_2$	plus	minus
1	1000.00	1000.00	(1000.00)(574.45 − 1236.11)		661,660
2	1236.11	492.03	(492.03)(1000.00 − 817.72)	89,687	
3	817.72	248.31	(248.31)(1236.11 − 782.01)	112,758	
4	782.01	622.08	(622.08)(817.72 − 574.45)	151,333	
5	574.45	854.35	(854.35)(782.01 − 1000.00)		186,240
				353,778	847,900
					353,778
					2) 494,122
					247,061

$$\text{area} = \frac{847{,}900 \text{ ft}^2 - 353{,}778 \text{ ft}^2}{(2)\left(43{,}560 \ \frac{\text{ft}^2}{\text{ac}}\right)} = 5.672 \text{ ac}$$

8. AREA BY TRIANGLES

When small traverses do not warrant computations of latitudes and departures, their areas can be determined by using formulas for the area of a triangle.

$$\text{area} = \tfrac{1}{2}ab\sin C \qquad 16.1$$

(a and b are any two sides, and C is the angle included between them.)

In Fig. 16.4(a), a tract of land has been divided into two triangles. Two sides and an included angle have been measured in each triangle. The areas of the triangles can be computed by using Eq. 16.1. Their sum is the area of the tract.

In Fig. 16.4(b), the property line has become covered with brush so that the four triangles have been formed from a central point. Angles at the central point have been measured for each triangle, and distances from the central point to each corner have also been measured. Areas can again be computed by using Eq. 16.1.

Also applicable in computing areas is Eq. 16.2.

$$A = \sqrt{(s)(s-a)(s-b)(s-c)} \qquad 16.2$$

s is one-half of the perimeter of the triangle, and a, b, and c are the sides of the triangle.

Figure 16.4 *Area by Triangles*

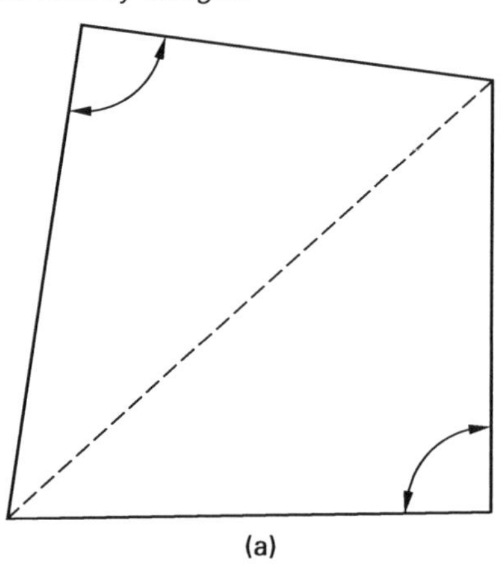

(a)

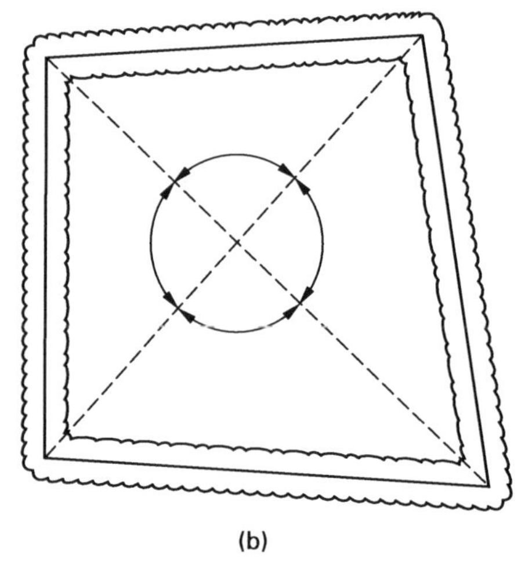

(b)

Example 16.3

Find the area of a triangle with sides 32 ft, 46 ft, and 68 ft.

Solution

$$s = \left(\tfrac{1}{2}\right)(32 \text{ ft} + 46 \text{ ft} + 68 \text{ ft}) = 73 \text{ ft}$$

$$\begin{aligned}\text{area} &= \sqrt{\begin{matrix}(73 \text{ ft})(73 \text{ ft} - 32 \text{ ft}) \\ \times (73 \text{ ft} - 46 \text{ ft})(73 \text{ ft} - 68 \text{ ft})\end{matrix}} \\ &= 636 \text{ ft}^2\end{aligned}$$

9. AREA ALONG AN IRREGULAR BOUNDARY

When a tract of land is bounded on one side by an irregular boundary such as a stream or lake, the traverse can be composed of straight lines so that closure can be computed. Points along the irregular side can be tied to one of the sides of the traverse by right angle offset measurements. The area between the irregular side and the traverse line is approximated by dividing the area into trapezoids and triangles formed by the ties to the breaks in the irregular side. This irregular area is then added to the traverse area. The irregular area can be computed by applying the trapezoidal rule or Simpson's one-third rule.

10. THE TRAPEZOIDAL RULE

In using the *trapezoidal rule*, it is assumed that the irregular boundary is made up of a series of straight lines. When the ties are taken close enough, a curved line connecting the ends of any two ties is very nearly a straight line, and no significant error is introduced.

The trapezoidal rule applies only to the part of the area where the ties are at regular intervals and form trapezoids. Triangles and trapezoids that do not have altitudes of the regular interval are computed separately and are added to the area found by applying the trapezoidal rule.

The rule is given by Eq. 16.3.

$$\text{area} = D\left(\frac{T_1}{2} + T_2 + T_3 + T_4 + \cdots + \frac{T_n}{2}\right) \qquad 16.3$$

D is the regular interval, and T is the tie distance.

Example 16.4

Using the trapezoidal rule, find the area in the following illustration.

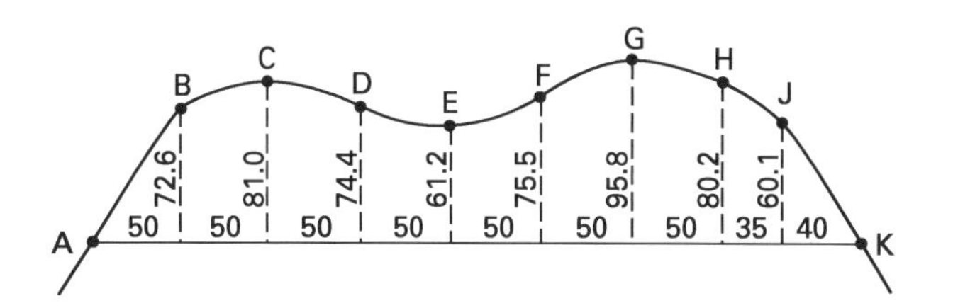

Solution

$$\begin{aligned}\text{area} &= (50)\left(\frac{72.6}{2}\right) \\ &\quad + (50)\left(\frac{72.6}{2} + 81.0 + 74.4 + 61.2 \right. \\ &\quad \left. +75.5 + 95.8 + \frac{80.2}{2}\right) \\ &\quad + (35)\left(\frac{80.2 + 60.1}{2}\right) \\ &\quad + (40)\left(\frac{60.1}{2}\right) \\ &= 28{,}687\end{aligned}$$

11. AREA OF A SEGMENT OF A CIRCLE

Land along highways, streets, and railroads often has a circular arc for a boundary. A traverse of straight lines can be run by using the *long chord* (LC) of the circular arc as one of the sides of the traverse. The area of the tract can be found by adding the area of the segment formed by the chord and the arc to the area within the traverse. It is usually practical to measure the chord length and the middle ordinate length. Using these two lengths and formulas derived for computing circular curves, the area of the segment can be found.

Figure 16.5 *Circular Segment Areas*

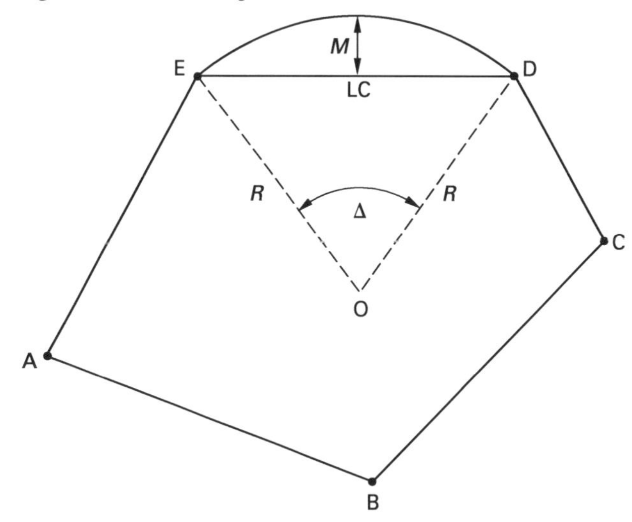

Δ is the *central angle*, M is the *middle ordinate*, LC is the *long chord*, and R is the *radius*.

$$\text{area ofsector} = \frac{\Delta^\circ \pi R^2}{360^\circ} \qquad 16.4$$

$$\text{area of triangle} = \frac{R(R\sin\Delta)}{2} = \frac{R^2\sin\Delta}{2} \qquad 16.5$$

The radius R is not known, but it can be determined if the long chord DE and the middle ordinate M are known. These lengths can be measured. Two formulas used to compute R are

$$\tan\frac{\Delta}{4} = \frac{2M}{\text{LC}} \qquad 16.6$$

$$R = \frac{\text{LC}}{2\sin\frac{\Delta}{2}} \qquad 16.7$$

If A = the area of the segment, A_s = area of the sector, and A_t = area of the triangle, then

$$A = A_s - A_t = \frac{\Delta\pi R^2}{360^\circ} - \frac{R^2\sin\Delta}{2} = R^2\left(\frac{\Delta\pi}{360^\circ} - \frac{\sin\Delta}{2}\right) \qquad 16.8$$

Example 16.5

The tract of land shown in Fig. 16.5 consists of the area within the traverse ABCDEA plus the area in the segment bounded by side DE and arc DE. The area of the segment is equal to the area of the sector DOE minus the area of the triangle DOE.

Find the area in the segment bounded by side DE and arc DE in Fig. 16.5 if the long chord is 325.48 ft and the middle ordinate is 42.16 ft.

$$\tan\frac{\Delta}{4} = \frac{(2)(42.16\text{ ft})}{325.48\text{ ft}}$$

$$\Delta = 58.0958^\circ$$

$$R = \frac{\text{LC}}{2\sin\frac{\Delta}{2}} = \frac{325.48\text{ ft}}{2\sin\frac{58.0958^\circ}{2}} = 335.17\text{ ft}$$

$$A = (335.17\text{ ft})^2\left(\frac{(58.0958^\circ)\pi}{360^\circ} - \frac{\sin 58.0958^\circ}{2}\right) = 9270\text{ ft}^2$$

12. SPECIAL FORMULA

The formula for the area of a triangle given in Sec. 8, $A = \frac{1}{2}ab\sin C$, can be used in deriving another useful formula for the area of a triangle.

Use the law of sines and substitute $b = a\sin B/\sin A$ in the formula $A = \frac{1}{2}\sin C$ to get

$$\text{area} = \frac{a^2\sin B\sin C}{2\sin A} \qquad 16.9$$

Also,

$$\text{area} = \frac{b^2\sin C\sin A}{2\sin B} = \frac{c^2\sin A\sin B}{2\sin C} \qquad 16.10$$

Example 16.6

A 3 ac triangular tract is to be cut off the northeast corner of a larger tract with bearings as shown. Find the lengths of the sides of the triangle.

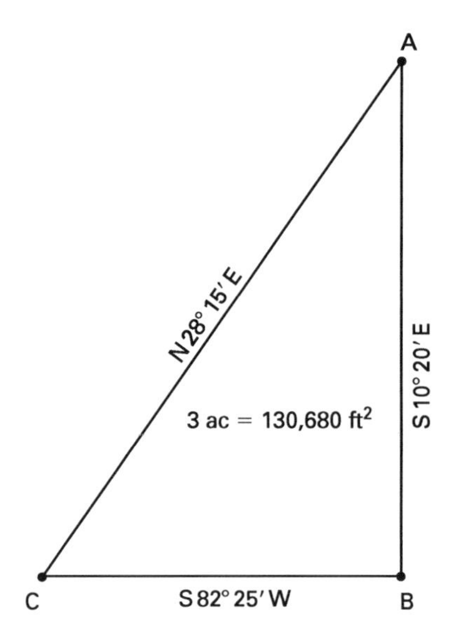

Solution

$A = 38°35'$, $B = 87°15'$, and $C = 54°10'$.

$$\text{area} = \frac{a^2\sin B\sin C}{2\sin A}$$

$$a^2 = \frac{(\text{area})(2\sin A)}{\sin B\sin C}$$

Also,

$$\text{area} = \frac{b^2\sin C\sin A}{2\sin B}$$

$$b^2 = \frac{(\text{area})(2\sin B)}{\sin C\sin A}$$

Also,

$$\text{area} = \frac{c^2\sin A\sin B}{2\sin C}$$

$$c^2 = \frac{(\text{area})(2\sin C)}{\sin A\sin B}$$

Therefore,

$$a = \sqrt{\frac{(\text{area})(2\sin A)}{\sin B \sin C}} = \sqrt{\frac{(130{,}680\ \text{ft}^2)(2\sin 38°35')}{(\sin 87°15')(\sin 54°10')}}$$
$$= 448.6\ \text{ft}$$

$$b = \sqrt{\frac{(\text{area})(2\sin B)}{\sin C \sin A}} = \sqrt{\frac{(130{,}680\ \text{ft}^2)(2\sin 87°15')}{(2\sin 54°10')(\sin 38°35')}}$$
$$= 718.5\ \text{ft}$$

$$c = \sqrt{\frac{(\text{area})(2\sin C)}{\sin A \sin B}} = \sqrt{\frac{(130{,}680\ \text{ft}^2)(2\sin 54°10')}{(\sin 38°35')(\sin 87°13')}}$$
$$= 583.2\ \text{ft}$$

In summary,

$$\text{AB} = 583.2\ \text{ft}$$
$$\text{BC} = 448.6\ \text{ft}$$
$$\text{CA} = 718.5\ \text{ft}$$

Example 16.7

Compute the area of the tract 1-2-3-4-5-6-1.

1-2	N 02°27′50″ W	761.49
2-3	N 87°35′37″ E	1076.62
3-4	S 02°24′23″ E	290.00
4-5	S 09°49′21″ W	826.10
5-6	S 30°30′21″ W	68.00
6-1	N 67°56′54″ W	949.09

Illustration for Example 16.7

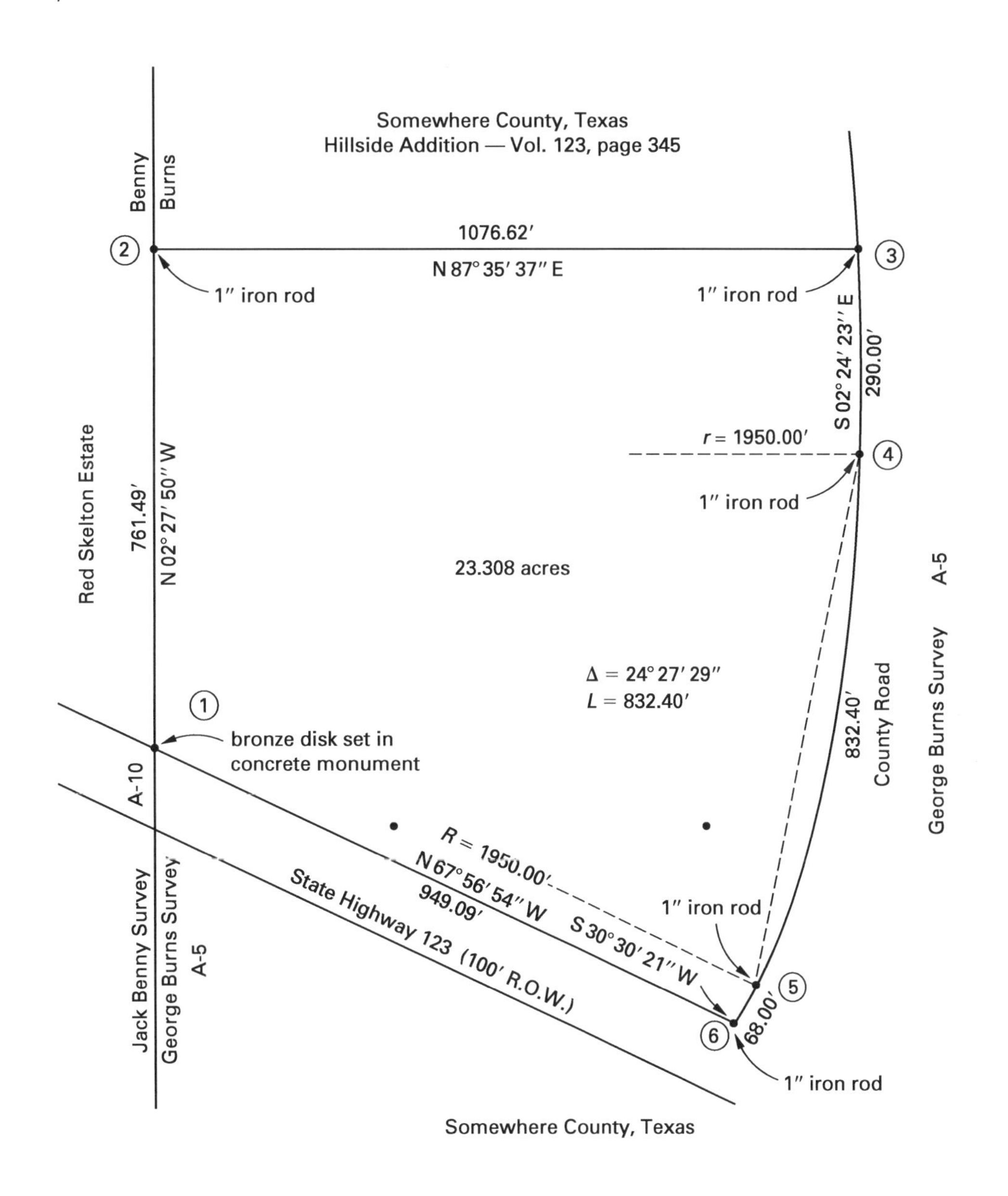

Solution

Area of Traverse 1-2-3-4-5-6-1

	bearing	distance	latitude	departure	DMD	area
1-2	N 02°27′50″ W	761.49	+760.79	−32.74	32.74	+24,908
2-3	N 87°35′37″ E	1076.62	+45.20	+1075.67	1075.67	+48,620
3-4	S 02°24′23″ E	290.00	−289.74	+12.18	2163.52	−626,858
4-5	S 09°49′21″ W	826.10	−813.99	−140.93	2034.77	−1,656,282
5-6	S 30°30′21″ W	68.00	−58.59	−34.52	1859.32	−108,938
6-1	N 67°56′54″ W	949.09	+356.33	−879.66	945.14	− 336,782
						−1,981,768

$$\frac{1{,}981{,}768\ \text{ft}^2}{2} = 990{,}884\ \text{ft}^2$$

$$\text{area 1-2-3-4-5-6-1} = 990{,}884\ \text{ft}^2$$

$$\begin{aligned}\text{area of segment} &= R^2\left(\frac{\Delta\pi}{360} - \frac{\sin\Delta}{2}\right) = (1950\ \text{ft})^2\left(\frac{(24°27'29'')\pi}{360} - \frac{\sin 24°27'29''}{2}\right)\\ &= 24{,}425\ \text{ft}^2\\ \text{total area} &= 990{,}884\ \text{ft}^2 + 24{,}425\ \text{ft}^2\\ &= 1{,}015{,}309\ \text{ft}^2\end{aligned}$$

PRACTICE PROBLEMS

(All dimensions and distances are in feet.)

1. Find the most westerly point in each traverse, and indicate which side would be the first for DMD computations.

(a)

line	departure east	departure west
AB	600	
BC	50	
CD		1000
DE		500
EF	100	
FA	750	

(b)

line	departure east	departure west
AB		500
BC	100	
CD	750	
DE	600	
EF	50	
FA		1000

(c)

line	departure east	departure west
AB	559.88	
BC	60.84	
CD		357.90
DE		294.87
EA	32.05	

(d)

line	departure east	departure west
AB		507.90
BC		243.75
CD	373.76	
DE	232.26	
EA	145.63	

(e)

line	departure
AB	+536.87
BC	+96.62
CD	+487.82
DE	−102.54
EF	−629.31
FG	−583.40
GA	+193.96

(f)

line	departure
AB	+629.31
BC	+102.54
CD	−487.82
DE	−96.62
EF	−536.87
FG	−193.94
GA	+583.40

2. Compute the area of the traverse ABCDEFA by the DMD method.

point	bearing	length	latitude N	latitude S	departure E	departure W
A						
	north	500.0	500		0	
B						
	N 45°00′ W	848.6	600			600
C						
	S 69°27′ W	854.4		300		800
D						
	S 11°19′ W	1019.8		1000		200
E						
	S 79°42′ E	1118.0		200	1100	
F						
	N 51°20′ E	640.3	400		500	
A						

3. Compute the area of the traverse ABCDEA by the DMD method.

point	bearing	length	latitude N	latitude S	departure E	departure W
A						
	N 28°19′ E	560.27	493.12		265.66	
B						
	N 56°23′ W	484.18	267.97			403.29
C						
	S 08°50′ W	375.42		371.04		57.72
D						
	S 45°10′ W	311.44		219.64		220.91
E						
	S 67°45′ E	449.83		170.41	416.26	
A						

4. Compute the area of the traverse ABCDEFGHA by the DMD method.

line	bearing	length	latitude	departure
AB	S 25°57′ E	174.36	−156.78	+76.30
BC	S 29°29′ W	192.13	−167.25	−94.56
CD	S 78°30′ E	316.68	−63.14	+310.32
DE	N 47°51′ E	281.24	+188.73	+208.51
EF	N 35°23′ W	232.44	+189.51	−134.59
FG	N 25°24′ E	228.60	+206.50	+98.05
GH	N 78°41′ W	366.63	+71.94	−359.50
HA	S 21°12′ W	289.07	−269.51	−104.53

5. Compute the area of the traverse ABCDEA by the coordinate method.

point	coordinates y	x
A	1000.00	1000.00
B	1493.12	1265.66
C	1761.09	862.37
D	1390.05	804.65
E	1170.41	583.74

6. Compute the area of the city lot shown using the formula $A = {}^1\!/\!_2 ab \sin C$. AB = 218.5 ft, BC = 199.8 ft, CD = 231.2 ft, and DA = 231.2 ft.

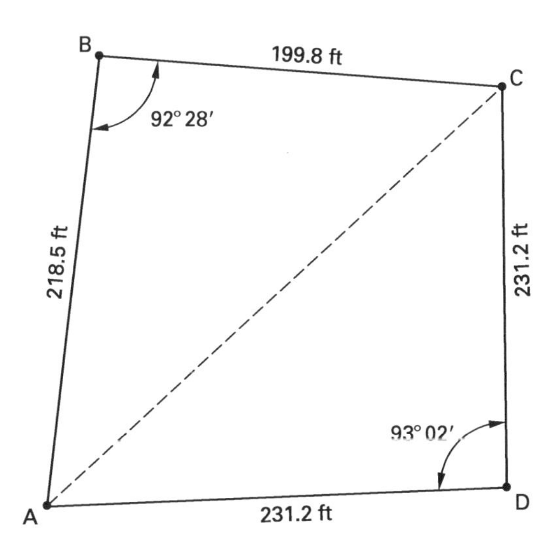

7. Compute the area of the city lot shown by using the formula ${}^1\!/\!_2 ab \sin C$. OE = 140.4 ft, OF = 131.8 ft, OG = 144.8 ft, and OH = 172.0 ft.

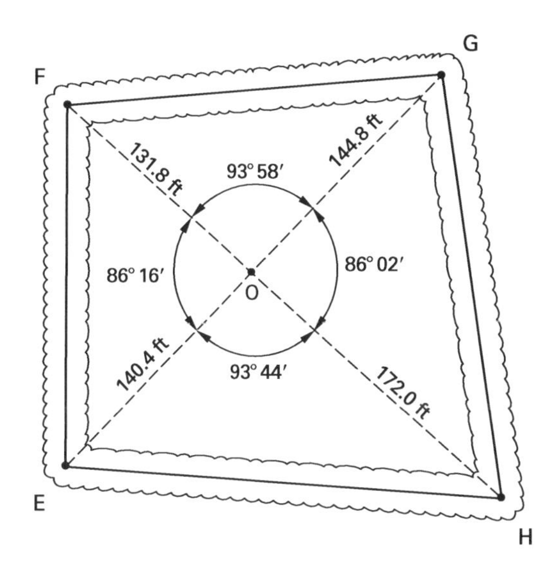

8. Find the area of the triangle with sides 12 ft, 14 ft, and 20 ft using the formula

$$A = \sqrt{s(s-a)(s-b)(s-c)}$$

9. Compute the area along the irregular boundary by using the trapezoidal rule.

Ties: B = 48.1, C= 52.6, D = 46.8, E = 39.9, F = 43.7, G = 58.0, H = 51.6, and J = 40.0.

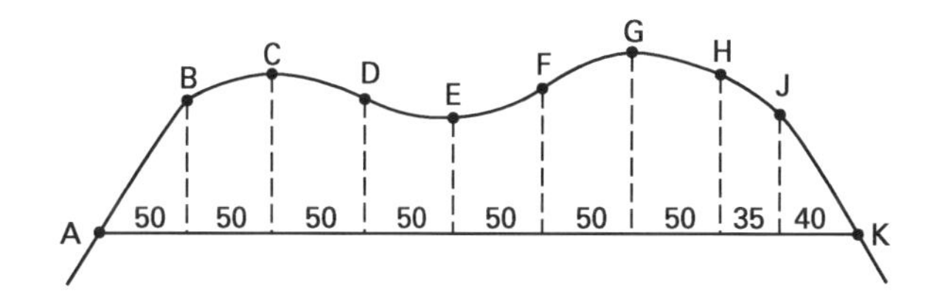

10. Compute the area of the segment with long chord = 491.67 ft and middle ordinate = 98.23 ft.

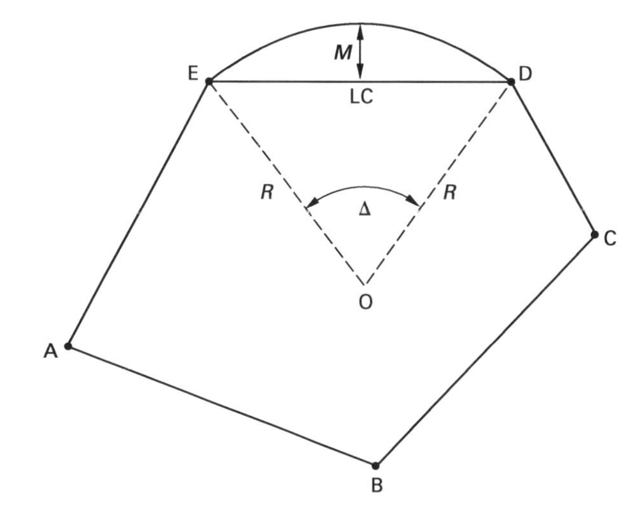

Plane Survey Calculations

SOLUTIONS

1. (a) most westerly point: E
begin DMD: EF

(b) most westerly point: B
begin DMD: BC

(c) most westerly point: E
begin DMD: EA

(d) most westerly point: C
begin DMD: CD

(e) most westerly point: G
begin DMD: GA

(f) most westerly point: G
begin DMD: GA

2.

			latitude		departure			area	
point	bearing	length	N	S	E	W	DMD	plus	minus
A									
	north	500.0	500		0		3200	1,600,000	
B									
	N 45°00′ W	848.6	600			600	2600	1,560,000	
C									
	S 69°27′ W	854.4		300		800	1200		360,000
D									
	S 11°19′ W	1019.8		1000		200	200		200,000
E									
	S 79°42′ E	1118.0		200	1100		1100		220,000
F									
	N 51°20′ E	640.3	400		500		2700	1,080,000	
A								4,240,000	780,000
								780,000	
								3,460,000	

$$\text{area} = \frac{3{,}460{,}000\ \text{ft}^2}{(2)\left(43{,}560\ \frac{\text{ft}^2}{\text{ac}}\right)} = \frac{1{,}730{,}000\ \text{ft}^2}{43{,}560\ \frac{\text{ft}^2}{\text{ac}}} = \boxed{39.715\ \text{ac}}$$

3.

			latitude		departure			area	
point	bearing	length	N	S	E	W	DMD	plus	minus
A									
	N 28°19′ E	560.27	493.12		265.66		1098.18	541,535	
B									
	N 56°23′ W	484.18	267.97			403.29	960.55	257,399	
C									
	S 08°50′ W	375.42		371.04		57.72	499.54		185,349
D									
	S 45°10′ W	311.44		219.64		220.91	220.91		48,521
E									
	S 67°45′ E	449.83		170.41	416.26		416.26		70,935
A									
								798,934	304,805
								304,805	
								494,129	

$$\text{area} = \frac{494{,}129\ \text{ft}^2}{(2)\left(43{,}560\ \frac{\text{ft}^2}{\text{ac}}\right)} = \frac{247{,}065\ \text{ft}^2}{43{,}560\ \frac{\text{ft}^2}{\text{ac}}} = \boxed{5.672\ \text{ac}}$$

4.

line	bearing	length	latitude	departure	DMD	area
AB	S 25°57′ E	174.36	−156.78	+76.30	112.82	−17,688
BC	S 29°29′ W	192.13	−167.25	−94.56	94.56	−15,815
CD	S 78°30′ E	316.68	−63.14	+310.32	310.32	−19,594
DE	N 47°51′ E	281.24	+188.73	+208.51	829.15	+156,485
EF	N 35°23′ W	232.44	+189.51	−234.59	903.07	+171,141
FG	N 25°24′ E	228.60	+206.50	+98.05	866.53	+178,938
GH	N 78°41′ W	366.63	+71.94	−359.50	605.08	+43,529
HA	S 21°12′ W	289.07	−269.51	−104.53	141.05	−38,014
						458,982

$$\text{area} = \frac{458{,}982\ \text{ft}^2}{(2)\left(43{,}560\ \frac{\text{ft}^2}{\text{ac}}\right)} = \frac{229{,}491\ \text{ft}^2}{43{,}560\ \frac{\text{ft}^2}{\text{ac}}} = \boxed{5.268\ \text{ac}}$$

5.

	y	x	x_n	$(y_{n-1} - y_{n+1})$	area
A	1000.00	1000.00	(1000.00)(1170.41 − 1493.12)	=	−322,710
B	1493.12	1265.66	(1265.66)(1000.00 − 1761.09)	=	−963,281
C	1761.09	862.37	(862.37)(1493.12 − 1390.05)	=	+88,884
D	1390.05	804.65	(804.65)(1761.09 − 1170.41)	=	+475,291
E	1170.41	583.74	(583.74)(1390.05 − 1000.00)	=	+227,688

$$\text{area} = \frac{494{,}128\ \text{ft}^2}{(2)\left(43{,}560\ \frac{\text{ft}^2}{\text{ac}}\right)} = \boxed{5.67\ \text{ac}}$$

6.

$$\begin{aligned} A &= \tfrac{1}{2}ab\sin C \\ &= \left(\tfrac{1}{2}\right)(218.5\ \text{ft})(199.8\ \text{ft})\sin 92°28' = 21{,}808\ \text{ft}^2 \\ &= \left(\tfrac{1}{2}\right)(231.2\ \text{ft})(231.2\ \text{ft})\sin 93°02' = \underline{26{,}689\ \text{ft}^2} \\ &\qquad \boxed{48{,}497\ \text{ft}^2} \end{aligned}$$

7.

$$\begin{aligned} A &= \tfrac{1}{2}ab\sin C \\ &= \left(\tfrac{1}{2}\right)(131.8\ \text{ft})(144.8\ \text{ft})\sin 93°58' = 9{,}519\ \text{ft}^2 \\ &= \left(\tfrac{1}{2}\right)(144.8\ \text{ft})(172.0\ \text{ft})\sin 86°02' = 12{,}423\ \text{ft}^2 \\ &= \left(\tfrac{1}{2}\right)(172.0\ \text{ft})(140.4\ \text{ft})\sin 93°44' = 12{,}049\ \text{ft}^2 \\ &= \left(\tfrac{1}{2}\right)(140.4\ \text{ft})(131.8\ \text{ft})\sin 86°16' = \underline{9{,}233\ \text{ft}^2} \\ &\qquad \boxed{43{,}224\ \text{ft}^2} \end{aligned}$$

8.

$$\begin{aligned} A &= \sqrt{s(s-a)(s-b)(s-c)} \\ &= \sqrt{\begin{array}{l}(23\ \text{ft})(23\ \text{ft} - 12\ \text{ft})(23\ \text{ft} - 14\ \text{ft}) \\ \quad \times (23\ \text{ft} - 20\ \text{ft})\end{array}} \\ &= \boxed{83\ \text{ft}^2} \end{aligned}$$

9.

$$\begin{aligned} A &= \left(\tfrac{1}{2}\right)(50)(48.1) \\ &\quad + (50)\left(\begin{array}{l}\frac{48.1}{2} + 52.6 + 46.8 + 39.9 \\ \quad + 43.7 + 58.0 + \frac{51.6}{2}\end{array}\right) \\ &\quad + (35)\left(\frac{51.6 + 40.0}{2}\right) \\ &\quad + \left(\tfrac{1}{2}\right)(40.0)(40.0) \\ &= \boxed{18{,}148\ \text{ft}^2} \end{aligned}$$

10.

$$\begin{aligned} \tan\frac{\Delta}{4} &= \frac{2M}{C} = \frac{(2)(98.23\ \text{ft})}{491.67\ \text{ft}} \\ \Delta &= 87.122° \\ R &= \frac{C}{2\sin\frac{\Delta}{2}} = \frac{491.67\ \text{ft}}{2\sin\left(\frac{87.122°}{2}\right)} \\ &= 356.73\ \text{ft} \\ \text{area} &= R^2\left(\frac{\pi\Delta}{360°} - \frac{\sin\Delta}{2}\right) \\ &= (356.73\ \text{ft})^2\left(\frac{\pi(87.122°)}{360°} - \frac{\sin 87.122°}{2}\right) \\ &= \boxed{33{,}203\ \text{ft}^2} \end{aligned}$$

17 Partitioning of Land

1. INTRODUCTION

An important part of the surveyor's work is the subdivision of tracts of land into two or more parts. Each partition of land is a separate problem, but there are basic techniques that can be used. These techniques will be illustrated in this chapter through the use of examples.

In solving the examples, latitudes, departures, and double meridian distances (DMDs) will be used where applicable. Unknown sides and angles of triangles will be solved by trigonometry. Formulas for area of triangles and trapezoids will be used to find lengths and bearings of sides.

2. LENGTH AND BEARING OF ONE SIDE UNKNOWN (THE CUTOFF LINE)

Due to incomplete field work or measurement blunders, it is sometimes desirable to do a preliminary closure on a traverse with a missing side.

Example 17.1

The bearing and length of the side DE of the traverse ABCDEA are missing and are to be computed.

line	bearing	length (ft)
AB	N 28°19′ E	560.27
BC	N 56°23′ W	484.18
CD	S 08°50′ W	375.42
DE		
EA	S 67°45′ E	449.83

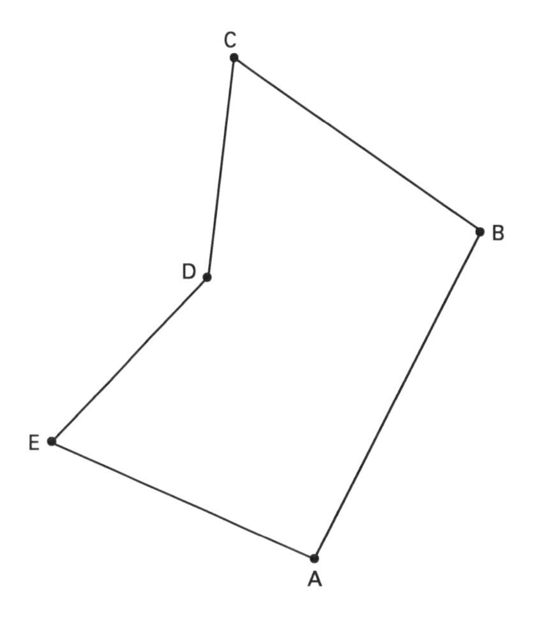

Solution

With the bearing and length of a side missing, the error of closure and precision cannot be computed. However, the traverse can be mathematically closed by giving a value to the latitude and departure of the side with missing bearing and length. These assigned values should be chosen to balance north and south latitudes and east and west departures of the traverse. This assumes that the latitudes and departures of the other sides are correct and any errors in them will be contained in the latitude and departure of the side with missing measurements.

line	bearing	length	latitude	departure
AB	N 28°19′ E	560.27	+493.23	+265.76
BC	N 56°23′ W	484.18	+268.06	−403.21
CD	S 08°50′ W	375.42	−370.97	−57.65
DE				
EA	S 67°45′ E	449.83	−170.33	+416.34
			+219.99	+221.24

$$\text{tangent of bearing DE} = \frac{\text{dep}}{\text{lat}} = \frac{221.24\ \text{ft}}{219.99\ \text{ft}}$$

$$\text{bearing DE} = \text{S}\,45°09'44''\,\text{W}$$

$$\begin{aligned}\text{length DE} &= \sqrt{\text{lat}^2 + \text{dep}^2} \\ &= \sqrt{(219.99\ \text{ft})^2 + (221.24\ \text{ft})^2} \\ &= 312.00\ \text{ft}\end{aligned}$$

In the solution, the sum of the latitude column is +219.99 ft, which represents the latitude of side DE with opposite sign. The sum of the departure column is +221.24 ft, which represents the departure of side DE with opposite sign.

3. LENGTHS OF TWO SIDES UNKNOWN

Example 17.2

The lengths for sides DE and EA of the traverse ABCDEA are unknown and are to be computed.

line	bearing	length (ft)
AB	N 65°04′ W	560.27
BC	S 30°14′ W	484.18
CD	S 84°33′ E	375.42
DE	S 48°13′ E	
EA	N 18°53′ E	

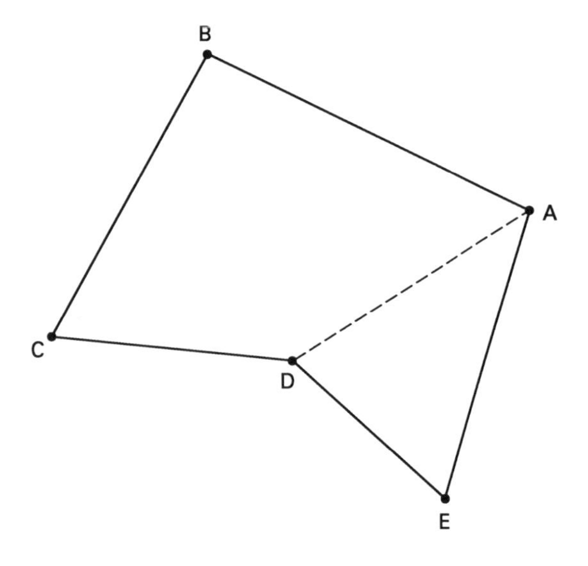

Solution

With measurements of two sides missing, the method of the *cutoff line* can be applied. The traverse ABCDA can be formed, excluding the sides with missing measurements. The side DA of this traverse will have an unknown bearing and length that can be computed by the method used in the preceding example. The side DA is the cutoff line. The cutoff line is usually used to isolate sides with missing measurements from sides with known measurements.

After the bearing and length of DA are computed, the side DA, together with sides DE and EA, can be considered to be the triangle ADE, which can be solved by the law of sines. When the triangle is solved, the solution is complete.

line	bearing	length	latitude	departure
AB	N 65°04′ W	560.27	+236.19	−508.05
BC	S 30°14′ W	484.18	−418.32	−243.80
CD	S 84°33′ E	375.42	−35.66	+373.72
DA				
			−217.79	−378.13

$$\text{tangent of bearing DA} = \frac{378.13\ \text{ft}}{217.79\ \text{ft}}$$

$$\text{bearing DA} = \text{N}\,60°03'34''\,\text{E}$$

$$\begin{aligned}\text{length DA} &= \sqrt{(378.13\ \text{ft})^2 + (217.79\ \text{ft})^2} \\ &= 436.37\ \text{ft}\end{aligned}$$

In triangle DEA,

$$\text{D} = 180° - (60°04' + 48°13') = 71°43'$$

$$\text{A} = 60°04' - 18°53' = 41°11'$$

$$\text{E} = 180° - (71°43' + 41°11') = 67°07'$$

$$\begin{aligned}\text{DE} &= \frac{(436.37\ \text{ft})(\sin 41°11')}{\sin 67°07'} \\ &= 311.88\ \text{ft}\end{aligned}$$

$$\begin{aligned}\text{EA} &= \frac{(436.37\ \text{ft})(\sin 71°43')}{\sin 67°07'} \\ &= 449.74\ \text{ft}\end{aligned}$$

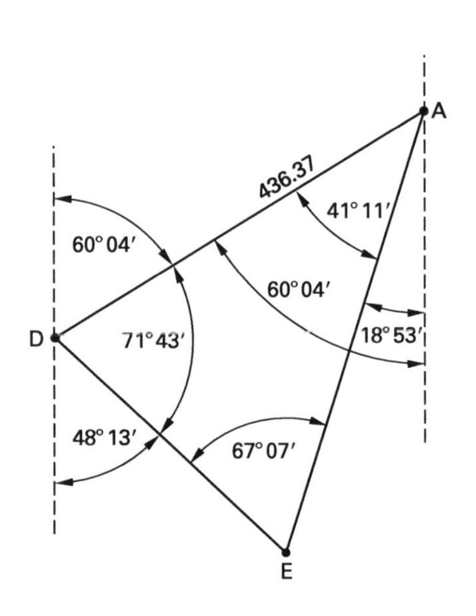

4. BEARING OF TWO SIDES UNKNOWN

Example 17.3

The bearing of the sides DE and EA of the traverse ABCDEA are unknown and are to be computed.

line	bearing	length (ft)
AB	N 65°04′ W	560.27
BC	S 30°14′ W	484.18
CD	S 84°33′ E	375.42
DE		311.88
EA		449.74

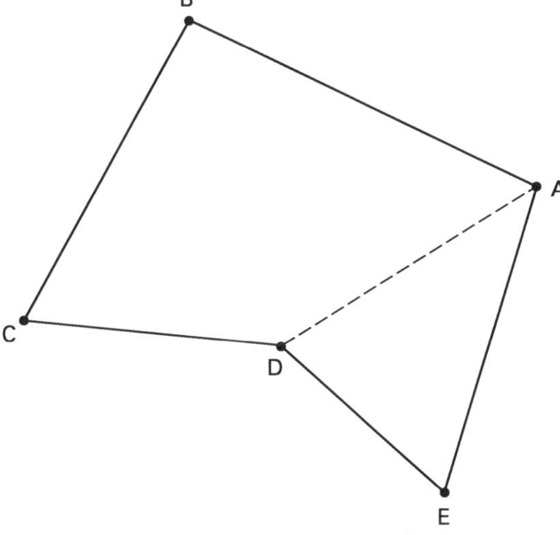

Solution

Bearings and lengths for sides AB, BC, and CD are the same as they are in the preceding example. Side DA is used as a cutoff line in traverse ABCDA again. The sides DA, DE, and EA form the triangle ADE again, but the known information is different from that in the preceding example. In this example, the lengths of all three sides of the triangle are known and the interior angles are to be determined in order to find bearings of the sides of the triangle. The law of cosines is appropriate for the solution.

$$\text{bearing DA} = \text{N } 60°04' \text{ E}$$

$$\text{length DA} = 436.37 \text{ ft}$$

In triangle DEA,

$$\cos \text{D} = \frac{(436.37 \text{ ft})^2 + (311.88 \text{ ft})^2 - (449.74 \text{ ft})^2}{(2)(436.37 \text{ ft})(311.88 \text{ ft})}$$

$$\angle \text{D} = 71°42'35'' \quad (71°43')$$

$$\cos \text{E} = \frac{(311.88 \text{ ft})^2 + (449.74 \text{ ft})^2 - (436.37 \text{ ft})^2}{(2)(311.88)(449.74)}$$

$$\angle \text{E} = 67°06'35'' \quad (67°07')$$

$$\angle \text{A} = 180° - (71°43' + 67°07') = 41°11'$$

$$\text{bearing DE} = 180° - (60°04' + 71°43') = \text{S } 48°13' \text{ E}$$

$$\text{bearing EA} = 60°04' - 41°11') = \text{N } 18°53' \text{ E}$$

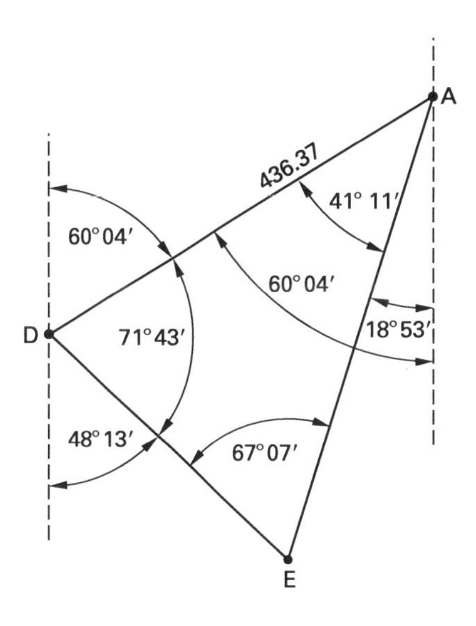

5. BEARING OF ONE SIDE AND LENGTH OF ANOTHER SIDE UNKNOWN

Example 17.4

The length of CD and the bearing of EA of the traverse ABCDEA are unknown and are to be computed.

line	bearing	length (ft)
AB	N 65°04′ W	560.27
BC	S 30°14′ W	484.18
CD	S 84°33′ E	
DE	S 48°13′ E	311.82
EA		449.87

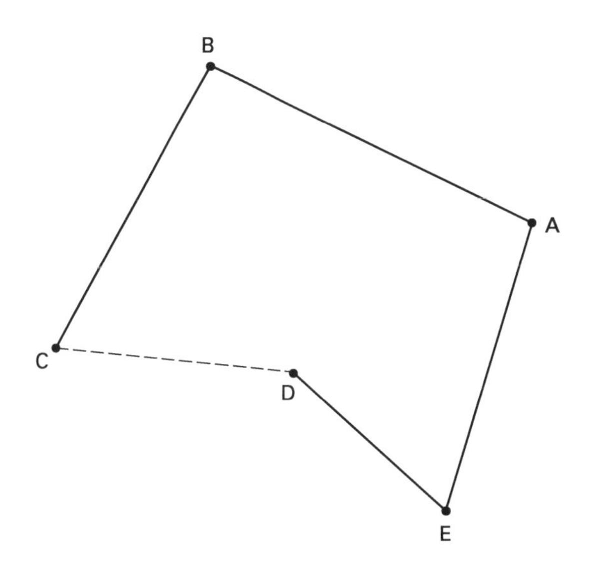

Solution

The sides AB, BC, and DE, which have no missing measurements, can be connected in sequence to form a traverse with the missing side x in the following illustration. Arranging the sides in this order does not change the latitudes and departures of the sides. (Designating the sides in the manner shown is done to avoid confusion.) The bearing and length of the cutoff line x can be computed as in previous examples.

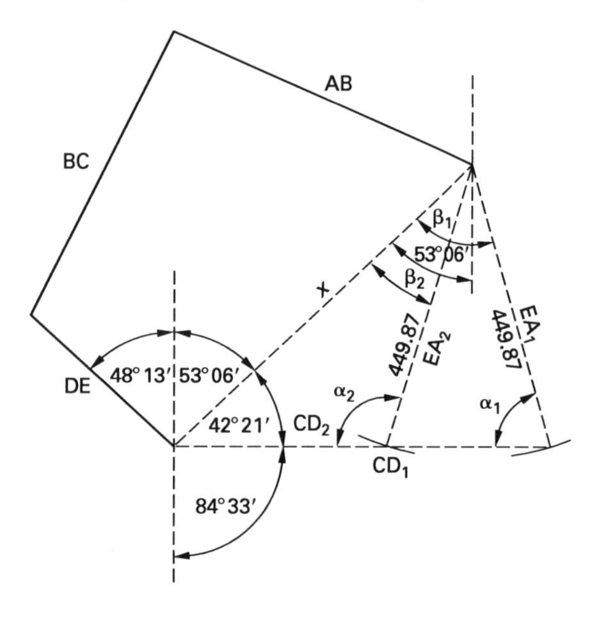

The line CD is given its correct bearing from the terminal of DE, but with an indefinite length. The length of EA is used as the radius of an arc, with center at the beginning point of AB, to intersect CD. In swinging the arc, it can be seen that it intersects CD at two points, giving two possible locations for EA and making two bearings. From the information given, the correct solution cannot be determined. Further information from the field would be necessary before determining the correct solution. (Not all problems of this nature will have two possible solutions.)

line	bearing	length	latitude	departure
AB	N 65°04′ W	560.27	+236.19	−508.05
BC	S 30°14′ W	484.18	−418.32	−243.80
DE	S 48°13′ E	311.82	−207.77	+232.51
			−389.90	−519.34

$$\text{tangent bearing } x = \frac{519.34 \text{ ft}}{389.90 \text{ ft}}$$

$$\text{bearing } x = \text{N } 53°06' \text{ E}$$

$$\text{length } x = \sqrt{(519.34 \text{ ft})^2 + (389.90 \text{ ft})^2} = 649.41 \text{ ft}$$

In the triangle bounded by sides x, CD_1, and EA_1,

$$\sin\alpha_1 = \frac{(649.41 \text{ ft})(\sin 42°21')}{449.87}$$

$$\alpha_1 = 76°31'$$

$$\beta_1 = 180° - (76°31' + 42°21') = 61°08'$$

$$CD_1 = \frac{(449.87 \text{ ft})(\sin 61°08')}{\sin 42°21'} = 584.82 \text{ ft}$$

$$\text{bearing } EA_1 = 61°08' - 53°06' = \text{N } 08°02' \text{ W}$$

In the triangle bounded by sides x, CD_2, and EA_2,

$$\sin\alpha_2 = \frac{(649.41 \text{ ft})(\sin 42°21')}{449.87}$$

$$\alpha_2 = (103°29')(\text{related angle to } 76°31')$$

$$\beta_2 = 180° - (103°29' + 42°21') = 34°10'$$

$$CD_2 = \frac{(449.87 \text{ ft})(\sin 34°10')}{\sin 42°21'} = 375.04 \text{ ft}$$

$$\text{bearing } EA_2 = 53°06' - 34°10' = \text{N } 18°56' \text{ E}$$

6. AREAS CUT OFF BY A LINE BETWEEN TWO POINTS ON THE PERIMETER

Example 17.5

The tract of land represented by the traverse ABCDEA is to be divided into two parts by a line from D to A.

line	bearing	length (ft)
AB	N 65°04′ W	560.27
BC	S 30°14′ W	484.18
CD	S 84°33′ E	375.42
DE	S 48°13′ E	311.88
EA	N 18°53′ E	449.74

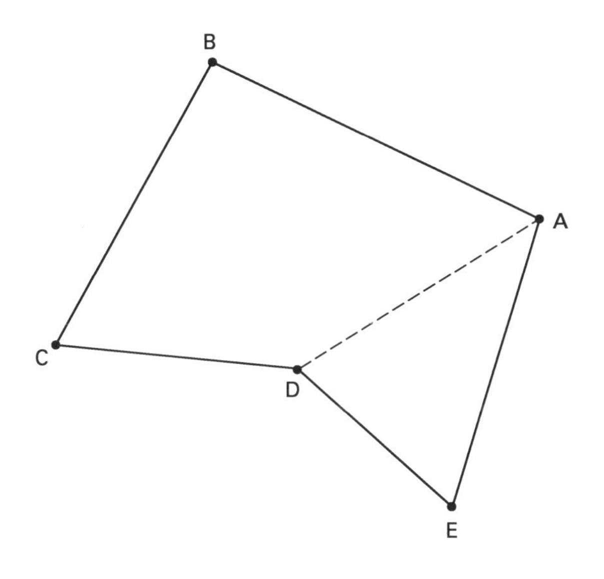

Solution

The area of the entire tract can be computed by DMD. The bearing and length for DA can be computed as in previous examples, and the areas of the two parts can be computed by DMD and their sum checked against the area of the entire tract.

			total area			
line	bearing	length	latitude	departure	DMD	area (ft^2)
AB	N 65°04′ W	560.27	+236.20	−508.04	995.64	+235,170
BC	S 30°14′ W	484.18	−418.30	−243.80	243.80	−101,982
CD	S 84°33′ E	375.42	−35.65	+373.73	373.73	−13,323
DE	S 48°13′ E	311.88	−207.80	+232.56	980.02	−203,648
EA	N 18°53′ E	449.74	+425.55	+145.55	1358.13	+577,952
						494,169

$$\text{area ABCDEA} = \frac{494{,}169 \text{ ft}^2}{2} = 247{,}085 \text{ ft}^2$$

$$\text{tangent bearing DA} = \frac{378.11 \text{ ft}}{217.75 \text{ ft}}$$

$$\text{bearing DA} = \text{N } 60°04' \text{ E} \quad (60°03'46'')$$

$$\text{length DA} = \sqrt{(217.75 \text{ ft})^2 + (378.11 \text{ ft})^2} = 436.33 \text{ ft}$$

			area ABCDA			
line	bearing	length	latitude	departure	DMD	area (ft^2)
AB	N 65°04′ W	560.27	+236.20	−508.04	995.64	+235,170
BC	S 30°14′ W	484.18	−418.30	−243.80	243.80	−101,982
CD	S 84°33′ E	375.42	−35.65	+373.73	373.73	−13,323
DA	N 60°04′ E	436.33	+217.75	+378.11	1125.57	+245,092
						364,957

$$\text{area ADEA} = \frac{364{,}957 \text{ ft}^2}{2} = 182{,}479 \text{ ft}^2$$

			area ADEA			
line	bearing	length	latitude	departure	DMD	area (ft^2)
AD	S 60°04′ W	436.33	−217.75	−378.11	378.11	−82,333
DE	S 48°13′ E	311.88	−207.80	+232.56	232.56	−48,326
EA	N 18°53′ E	449.74	+425.55	+145.55	610.67	+259,871
						129,212

$$\text{area ADEA} = \frac{129{,}212 \text{ ft}^2}{2} = 64{,}606 \text{ ft}^2$$

$$\text{total area check} = 182{,}479 \text{ ft}^2 + 64{,}606 \text{ ft}^2 = 247{,}085 \text{ ft}^2$$

7. AREAS CUT OFF BY A LINE IN A GIVEN DIRECTION FROM A POINT ON THE PERIMETER

Example 17.6

The tract of land represented by the traverse ABCDEA is to be divided into two parts by a line from point D parallel to BC.

line	bearing	length (ft)
AB	N 65°04′ W	560.27
BC	S 30°14′ W	484.18
CD	S 84°33′ E	375.42
DE	S 48°13′ E	311.88
EA	N 18°53′ E	449.74

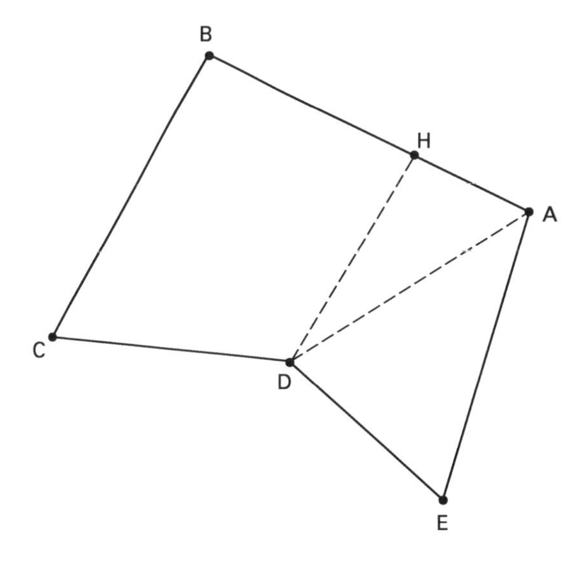

Solution

The line DH is drawn parallel to BC to represent the dividing line. The line DA is used as a cutoff line for the traverse ABCDA. The bearing and length for DA are computed as in previous examples. The triangle AHD is solved by the law of sines for sides AH and DH. HB is found by subtracting length AH from AB. Areas of HBCDH and AHDEA are computed by DMD and checked against the total area of ABCDEA.

From the preceding example,

$$\text{bearing DA} = \text{N } 60°04' \text{ E}$$
$$\text{length DA} = 436.33 \text{ ft}$$
$$\text{bearing DH} = \text{bearing CB} = \text{N } 30°14' \text{ E}$$
$$\text{bearing AH} = \text{bearing AB} = \text{N } 65°04' \text{ W}$$

In triangle AHD,

$$\angle A = 180° - (\text{bearing AB} + \text{bearing AD}) = 54°52'$$
$$\angle D = \text{bearing DA} - \text{bearing DH} = 29°50'$$
$$\angle H = 180° - (\angle A + \angle D) = 95°18'$$
$$\text{AH} = \frac{(436.33)(\sin 29°50')}{\sin 95°18'} = 218.00 \text{ ft}$$
$$\text{DH} = \frac{(436.33)(\sin 54°52')}{\sin 95°18'} = 358.37 \text{ ft}$$
$$\text{HB} = \text{AB} - \text{AH} = 560.27 \text{ ft} - 218.00 \text{ ft} = 342.27 \text{ ft}$$

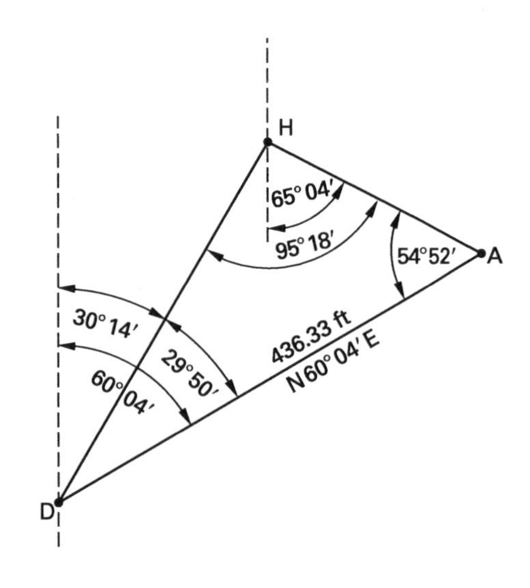

area HBCDH

line	bearing	length	latitude	departure	DMD	area
HB	N 65°04′ W	342.27	+144.29	−310.37	797.97	+115,139
BC	S 30°14′ W	484.18	−418.30	−243.80	243.80	−101,982
CD	S 84°33′ E	375.42	−35.65	+373.73	373.73	−13,323
DH*	S 30°14′ E	358.37	+309.66	+180.44	927.90	+287,334
						287,168

*closure forced

$$\text{area AHDEA} = \frac{287{,}168 \text{ ft}^2}{2} = 143{,}584 \text{ ft}^2$$

area AHDEA

line	bearing	length	latitude	departure	DMD	area
AH*	N 65°04′ W	218.00	+91.91	−197.67	558.55	+51,336
HD	S 30°14′ W	358.37	−309.66	−180.55	180.44	−55,875
DE	S 48°13′ E	311.88	−207.80	+232.56	232.56	−48,326
EA	N 18°53′ E	449.74	+425.55	+145.55	610.67	+259,871
						207,006

*closure forced

$$\text{area AHDEA} = \frac{207{,}006 \text{ ft}^2}{2} = 103{,}503 \text{ ft}^2$$
$$\text{total area check} = 143{,}584 \text{ ft}^2 + 103{,}503 \text{ ft}^2 = 247{,}087 \text{ ft}^2$$

8. DIVIDING TRACTS INTO TWO EQUAL PARTS BY A LINE FROM A POINT ON THE PERIMETER

Example 17.7

The tract represented by traverse ABCDEA is to be divided into two equal parts by a line from point D. The traverse ABCDEA is the same as that in the preceding example.

line	bearing	length (ft)
AB	N 65°04′ W	560.27
BC	S 30°14′ W	484.18
CD	S 84°33′ E	375.42
DE	S 48°13′ E	311.88
EA	N 18°53′ E	449.74

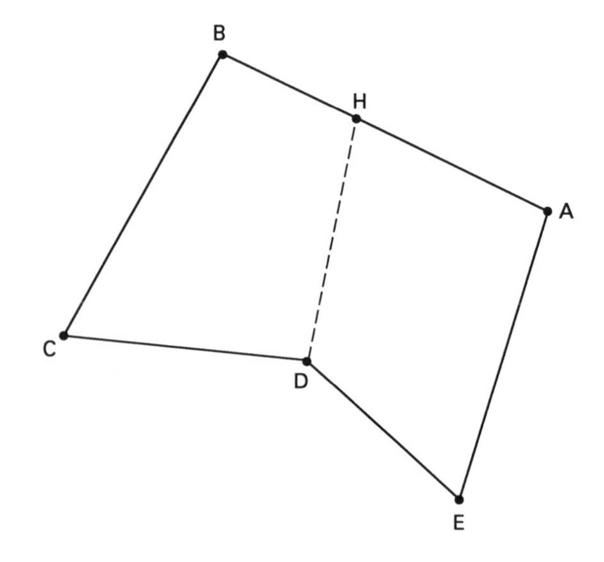

Solution

The line DH is drawn by inspection to divide the tract into two equal parts. In making computations for the solution of the problem, the line is considered to be in the exact location. The area of the entire tract is computed by DMD, and the area of the traverse HBCDH must be exactly one-half of the total area.

The area of the traverse ABCDA can be computed by DMD after the bearing and length of DA have been computed as in previous examples. Then the area of the triangle AHD can be found by subtracting area HBCDH from area ABCDA. Angle A of the triangle AHD can be found from bearings. Using the equation for the area of a triangle, the formula for triangle AHD is written $A = {}^1\!/\!_2(\text{AH})(\text{DA})\sin A$.

The triangle AHD is solved for DH by the law of cosines. Angle D is found by the law of sines. Bearing DH can now be found, as can length HB. Areas can be checked by DMD.

From Ex. 17.5,

$$\text{bearing DA} = \text{N}\,60°04'\,\text{E}$$

$$\text{length DA} = 436.33\ \text{ft}$$

$$\text{area ABCDEA} = 247{,}085\ \text{ft}^2$$

$$\text{area ABCDA} = 182{,}479\ \text{ft}^2$$

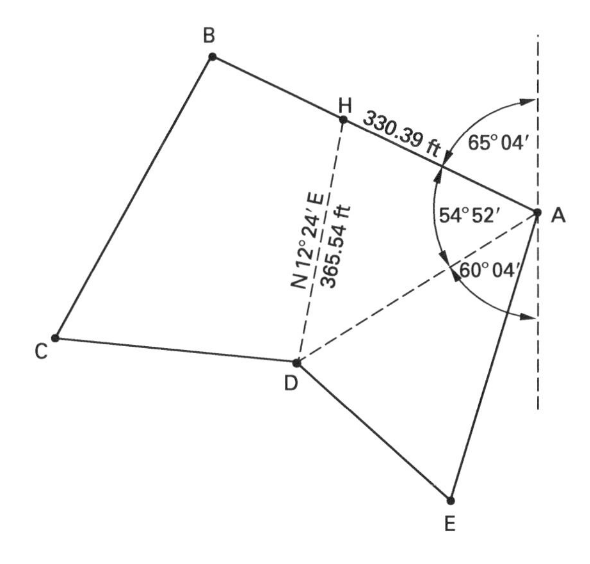

$$\text{area HBCDH} = \left(\tfrac{1}{2}\right)(\text{area ABCDEA}) = 123{,}542\ \text{ft}^2$$

$$\begin{aligned}\text{area AHD} &= \text{area ABCDA} - \text{area HBCDH}\\ &= 182{,}479\ \text{ft}^2 - 123{,}542\ \text{ft}^2\\ &= 58{,}937\ \text{ft}^2\end{aligned}$$

In triangle AHD,

$$\begin{aligned}\text{angle } A &= 180° - (65°04' + 60°04')\\ &= 54°52'\end{aligned}$$

$$\text{area AHD} = \frac{(\text{AH})(\text{AD})(\sin 54°52')}{2}$$

$$\begin{aligned}\text{AH} &= \frac{2A}{\text{AD}\sin 54°52'}\\ &= \frac{(2)(58{,}937\ \text{ft}^2)}{(436.33\ \text{ft})(\sin 54°52')}\\ &= 330.33\ \text{ft}\end{aligned}$$

In triangle AHD, using the law of cosines,

$$\begin{aligned}\text{DH} &= \sqrt{\begin{array}{l}(330.33\ \text{ft})^2 + (436.33\ \text{ft})^2\\ \quad - (2)(330.33\ \text{ft})(436.33\ \text{ft})(\cos 54°52')\end{array}}\\ &= 365.53\ \text{ft}\end{aligned}$$

In triangle AHD, using the law of sines,

$$\sin D = \frac{(330.33\ \text{ft})(\sin 54°52')}{365.53}$$

$$D = 47°39'05''\quad(47°39')$$

$$\text{bearing DH} = 60°04' - 47°39' = \text{N}\,12°25'\,\text{E}$$

Also,

$$\begin{aligned}\text{HB} &= \text{AB} - \text{AH} = 560.27\ \text{ft} - 330.33\ \text{ft}\\ &= 229.94\ \text{ft}\end{aligned}$$

			area AHDEA			
line	bearing	length	latitude	departure	DMD	area
AH	N 65°04′ W	330.33	+139.26	−299.54	456.66	+63,594
HD*	S 12°25′ W	365.54	−356.98	−78.60	78.60	−28,059
DE	S 48°13′ E	311.88	−207.81	+232.56	232.56	−48,328
EA	N 18°53′ E	449.7	+425.50	+145.54	610.66	+259,836
						247,043

*closure forced

$$\text{area AHDEA} = \frac{247{,}043\ \text{ft}^2}{2} = 123{,}522\ \text{ft}^2$$

9. DIVIDING AN IRREGULAR TRACT INTO TWO EQUAL PARTS

Frequently it is desirable to divide tracts of land that are not necessarily regular in shape into equal parts.

Example 17.8

The tract of land represented by the traverse ABCDEFA is to be divided into two equal parts by a line parallel to CD.

line	bearing	length (ft)
AB	N 80°00′ W	1015.43
BC	S 66°30′ W	545.22
CD	S 12°00′ E	480.97
DE	S 88°30′ E	750.26
EF	N 69°00′ E	639.18
FA	N 10°00′ E	306.78

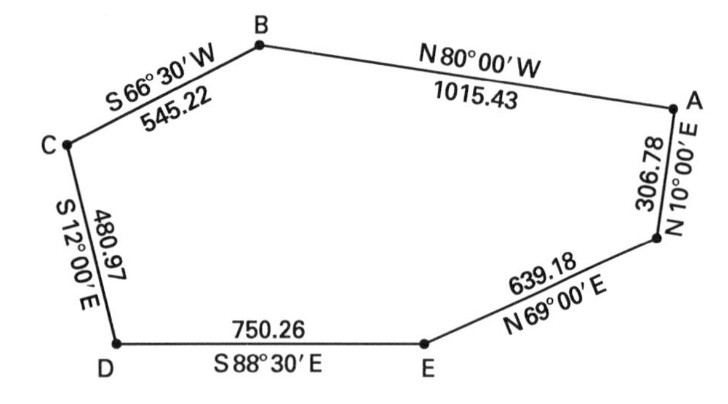

Solution

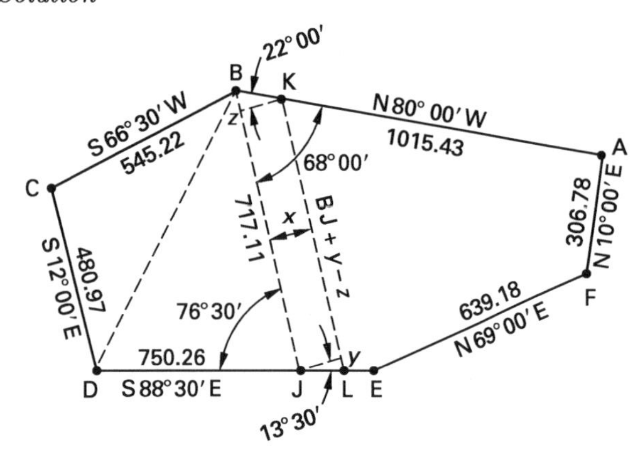

The problem can be solved in a manner similar to that used in the preceding example. A line KL can be drawn parallel to CD at the approximate location of the dividing line, using a scale drawing of the traverse. It can be seen by inspection that K will fall on AB and L will fall on DE.

From the point nearest to K, point B, the line BJ is drawn parallel to CD. (Point E could be used in place of point B.)

There are now four traverses formed:

(1) the original traverse ABCDEFA, for which the total area can be found by the DMD method

(2) the traverse KBCDLK, for which the area can be found by taking one-half of the total area

(3) the traverse BCDJB, for which the area can be found after first determining the bearing and length of BJ by use of the cutoff line DB as in previous examples

(4) the traverse KBJLK, a trapezoid, for which the area can be found by subtracting the area of BCDJB from the area of KBCDLK

The altitude x of a trapezoid can be found as the unknown quantity in the formula for the area of a trapezoid in which the area and the lengths of bases BJ and KL are known. KL can be expressed in terms of BJ and x to give a quadratic equation of the form $Ax^2+Bx+C=0$, which can be solved by the quadratic formula

$$x = \frac{-B \pm \sqrt{B^2 - 4AC}}{2A}$$

total area

line	bearing	length	latitude	departure	DMD	area
AB	N 80°00′ W	1015.43	+176.33	−1000.00	2000.00	+352,660
BC	S 66°30′ W	545.22	−217.41	−500.00	500.00	−108,705
CD	S 12°00′ E	480.97	−470.46	+100.00	100.00	−47,046
DE	S 88°30′ E	750.26	−19.64	+750.00	950.00	−18,658
EF	N 69°00′ E	639.18	+229.06	+596.73	2296.73	+526,089
FA	N 10°00′ E	306.78	+302.12	+53.27	2946.73	+890,266
			0.00	0.00		1,594,606

$$\text{total area} = \frac{1{,}594{,}606 \text{ ft}^2}{2} = 797{,}303 \text{ ft}^2$$

$$\text{area KBCDLK} = \frac{797{,}303 \text{ ft}^2}{2} = 398{,}652 \text{ ft}^2$$

To find the bearing and length of BJ, the cutoff line DB is found as the missing side of BCDB and the triangle BDJ is solved.

line	latitude	departure
BC	−217.41	−500.00
CD	−470.46	+100.00
DB	+687.87	+400.00

$$\text{tangent bearing DB} = \frac{400.00 \text{ ft}}{687.87 \text{ ft}}$$

$$\text{bearing DB} = \text{N } 30°11' \text{ E}$$

$$\text{length DB} = \sqrt{(687.87 \text{ ft})^2 + (400.00 \text{ ft})^2} = 795.72 \text{ ft}$$

In triangle DJB,

$$\text{DJ} = \frac{(795.72 \text{ ft})(\sin 42°11')}{\sin 76°30'} = 549.51 \text{ ft}$$

$$\text{JB} = \frac{(795.72 \text{ ft})(\sin 61°19')}{\sin 76°30'} = 717.91 \text{ ft}$$

$$\text{bearing JB} = \text{bearing DC} = \text{N } 12°00' \text{ W}$$

BCDJB area

line	bearing	length	latitude	departure	DMD	area
BC	S 66°30′ W	545.22	−217.41	−500.00	500.00	−108,705
CD	S 12°00′ E	480.97	−470.46	+100.00	100.00	−47,046
DJ	S 88°30′ E	549.51	−14.38	+549.32	749.32	−10
JB*	N 12°00′ W	717.91	+702.25	−149.32	1149.32	+ 80
			0.00	0.00		640,584

*closure forced

$$\text{area BCDJB} = \frac{640{,}584\ \text{ft}^2}{2} = 320{,}292\ \text{ft}^2$$

In the trapezoid BJLK,

$$y = x\tan 13°30'$$
$$z = x\tan 22°00'$$
$$\text{LK} = \text{JB} + y - z$$
$$= \text{JB} + x\tan 13°30' - x\tan 22°00'$$

$$\text{area KBCDLK} = 398{,}652\ \text{ft}^2$$
$$\text{area BCDJB} = 320{,}292\ \text{ft}^2$$
$$\text{area BJLKB} = 78{,}360\ \text{ft}^2$$
$$\text{area BJLKB} = \frac{(\text{JB} + \text{LK})x}{2}$$

$$78{,}360 = \frac{(\text{JB} + \text{JB} + x\tan 13°30' - x\tan 22°00')x}{2}$$
$$= \frac{(2\text{JB} + x\tan 13°30' - x\tan 22°00')x}{2}$$
$$= (\text{JB})x - \frac{(\tan 22°00' - \tan 13°30')x^2}{2}$$
$$= (717.91)x - (0.0820)x^2(0.0820)x^2$$
$$- 717.91x + 78{,}360$$
$$= 0$$

Substituting in the quadratic formula,

$$x = \frac{717.91 \pm \sqrt{(-717.91)^2 - (4)(0.0820)(78{,}360)}}{(2)(0.0820)}$$
$$= 110.55\ \text{ft}$$

Then,

$$\text{JL} = \frac{110.55\ \text{ft}}{\cos 13°30'} = 113.69\ \text{ft}$$

$$\text{KB} = \frac{110.55\ \text{ft}}{\cos 22°00'} = 119.23\ \text{ft}$$

$$\text{LE} = \text{DE} - \text{DJ} - \text{JL}$$
$$= 750.26\ \text{ft} - 549.51\ \text{ft} - 113.69\ \text{ft}$$
$$= 87.06\ \text{ft}$$

$$\text{AK} = \text{AB} - \text{KB} = 1015.43\ \text{ft} - 119.23\ \text{ft}$$
$$= 896.20\ \text{ft}$$

$$\text{LK} = \text{JB} + y - z$$
$$= 717.91\ \text{ft} + (110.55\ \text{ft})(\tan 13°30')$$
$$- (110.55\ \text{ft})(\tan 22°00')$$
$$= 699.79\ \text{ft}$$

$$\text{DL} = \text{DE} - \text{LE} = 750.26\ \text{ft} - 87.06\ \text{ft}$$
$$= 663.20\ \text{ft}$$

			total area			
line	bearing	length	latitude	departure	DMD	area
AK	N 80°00′ W	896.20	+155.62	−882.58	882.58	+137,347
KL*	S 12°00′ E	699.79	−684.52	+145.55	145.55	−99,632
LE	S 88°30′ E	87.06	−2.28	+87.03	378.13	−862
EF	N 69°00′ E	639.18	+229.06	+596.73	1061.89	+243,237
FA	N 10°00′ E	306.78	+302.12	+53.27	1711.89	+517,196
						797,286

*closure forced

$$\text{area AKLEFA} = \frac{797{,}286\ \text{ft}^2}{2} = 398{,}643\ \text{ft}^2$$
$$\text{error in area} = 398{,}652\ \text{ft}^2 - 398{,}643\ \text{ft}^2 = 9\ \text{ft}^2$$

10. CUTTING A GIVEN AREA FROM AN IRREGULAR TRACT

In the subdivision of land, it is often necessary to cut known areas from irregularly shaped parent tracts.

Example 17.9

Two acres are to be cut off of the westerly end of the tract represented by the traverse ABCDA. At what easterly distance from point D along line DC will a boundary line parallel to AD intersect DC?

line	bearing	length
AB	N 51°00′ E	647.81
BC	S 30°22′ E	449.76
CD	S 64°14′ W	596.15
DA	N 39°00′ W	308.17

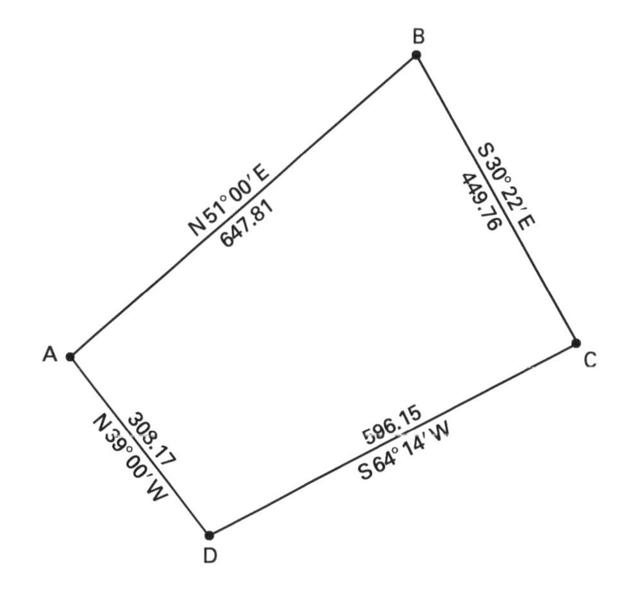

Solution

After plotting the traverse, it can be seen that the interior angle at A is a right angle.

Cutting two acres off the westerly end of the tract implies that exactly two acres are to be cut off and that the line cutting off the tract will be parallel to the westerly side DA.

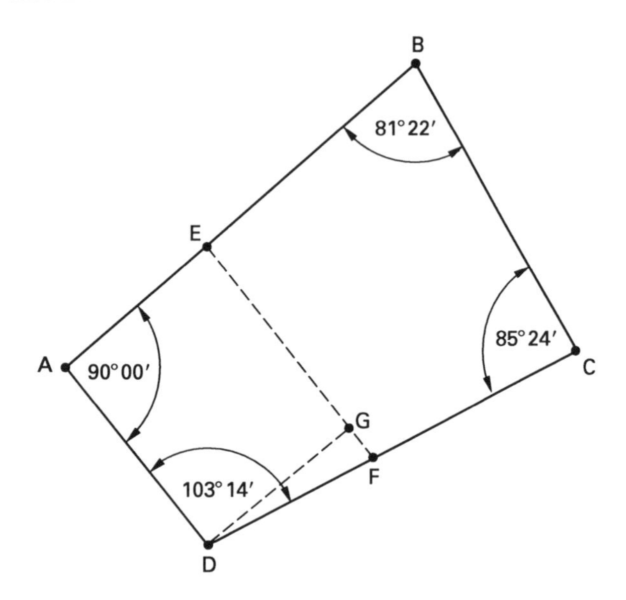

Computing the area of the entire tract by DMD gives an area of 223,440 ft^2 = 5.359 ac. The area to be cut off is 2 ac = 87,120 ft^2.

First draw EF parallel to DA at the approximate location of the dividing line, forming the trapezoid AEFD. Then drop a perpendicular to EF from D, which intersects EF at G. DG is parallel to AE. Using bearings,

$$\begin{aligned}\text{angle } FDG &= 39°00' + 64°14' - 90°00' \\ &= 13°14'\end{aligned}$$

Also,

$$\text{GF} = \text{EF} - 308.17$$

In triangle DGF,

$$\text{DG} = \frac{\text{GF}}{\tan 13°14'} = \frac{\text{EF} - 308.17 \text{ ft}}{\tan 13°14'}$$

In trapezoid AEFD,

$$\begin{aligned}\text{area} &= \frac{(\text{base} + \text{base})(\text{altitude})}{2} \\ 87{,}120 &= \frac{(\text{EF} + 308.17 \text{ ft})\text{DG}}{2} \\ 87{,}120 &= \frac{(\text{EF} + 308.17 \text{ ft})(\text{EF} - 308.17 \text{ ft})}{2\tan 13°14'} \\ 87{,}120 &= \frac{(\text{EF}^2 - 308.17 \text{ ft})^2}{2\tan 13°14'} \\ \text{EF}^2 &= (2\tan 13°14')(87{,}120 \text{ ft}^2) + (308.17 \text{ ft})^2 \\ \text{EF} &= 368.70 \text{ ft}\end{aligned}$$

Solving for DG,

$$\begin{aligned}87{,}120 \text{ ft}^2 &= \frac{(\text{EF} + 308.17 \text{ ft})\text{DG}}{2} \\ 87{,}120 \text{ ft}^2 &= \frac{(368.70 \text{ ft} + 308.17 \text{ ft})\text{DG}}{2} \\ \text{DG} &= \frac{(2)(87{,}120 \text{ ft}^2)}{368.70 \text{ ft} + 308.17 \text{ ft}} = 257.42 \text{ ft} \\ \text{DF} &= \frac{\text{DG}}{\cos 13°14'} = \frac{257.42 \text{ ft}}{\cos 13°14'} = 264.44 \text{ ft}\end{aligned}$$

Check:

$$\frac{(368.70 \text{ ft} + 308.17 \text{ ft})(257.42 \text{ ft})}{2} = 87{,}120 \text{ ft}^2$$

11. ANALYTIC GEOMETRY IN PARTING LAND

In many situations, analytic geometry is more applicable to the solution of problems encountered in surveying than is trigonometry. Both analytic geometry and surveying deal with the coordinates of points. The equations of analytic geometry can be used in surveying problems as written or with slight modification. (Analytic geometry for surveyors has been discussed in Ch. 10.)

Example 17.10

For the traverse ABCDEA shown in Ex. 17.5, find the intersection of the line connecting points A and C and the line connecting points B and D.

	coordinates	
point	y	x
A	1000.00	1000.00
B	1236.08	492.10
C	817.70	248.35
D	782.01	622.11
E	574.45	854.37

Solution

Using the two-point form of the equation of a line and designating A as point 1 and C as point 2,

$$\begin{aligned}y - y_1 &= \left(\frac{y_2 - y_1}{x_2 - x_1}\right)(x - x_1) \\ y - 1000.00 &= \left(\frac{817.70 - 1000.00}{248.35 - 1000.00}\right)(x - 1000.00) \\ y - 1000.00 &= (0.24253)(x - 1000.00) \\ y - 1000.00 &= 0.24253x - 2242.53\end{aligned}$$

$$0.24253x - y = -757.47$$

Designating B as point 1 and D as point 2,

$$y - y_1 = \left(\frac{y_2 - y_1}{x_2 - x_1}\right)(x - x_1)$$
$$y - y_1 = \left(\frac{782.01 - 1236.08}{622.08 - 492.03}\right)(x - x_1)$$
$$y - 1236.08 = (-3.49150)(x - 492.03)$$
$$y - 1236.08 = -3.49150x + 1717.92$$
$$3.49150x + y = 2954.00$$

Solving simultaneously,

$$\begin{aligned} 0.24253x - y &= -757.47 \\ 3.49150x + y &= 2954.00 \\ \hline 3.73403x &= 2196.53 \\ x &= 588.25 \end{aligned}$$

Substituting, $y = 900.14$.

The two lines intersect at (588.25, 900.14).

12. AREAS CUT OFF BY A LINE IN A GIVEN DIRECTION FROM A POINT ON THE PERIMETER USING ANALYTIC GEOMETRY

As with the previously covered partitioning processes, the subdivision of land sometimes requires the use of lines with a given bearing for partitioning.

Example 17.11

The tract of land represented by the traverse ABCDEA, Ex. 17.6, is to be divided into two parts by a line from point D parallel to side BC. Coordinates of traverse points are shown.

	coordinates	
point	y	x
A	1000.00	1000.00
B	1236.08	492.03
C	817.72	248.31
D	782.01	622.08
E	574.45	854.35

Solution

$$\text{slope DH} = \text{slope CB} = \frac{1236.08 - 817.72}{492.03 - 248.31} = 1.716560$$

(The slope of line DH can also be found as the cotangent of the bearing angle.)

The equation of line DH is

$$y - y_1 = m(x - x_1)$$
$$y - 782.01 = (1.716560)(x - 622.08)$$
$$y - 782.01 = 1.716560x - 1067.84$$
$$1.7167x - y = 285.83$$

The equation of line AB is

$$y - y_1 = \left(\frac{y_2 - y_1}{x_2 - x_1}\right)(x - x_1)$$
$$y - 1000.00 = \left(\frac{1236.08 - 1000.00}{492.03 - 1000.00}\right)(x - 1000.00)$$
$$y - 1000.00 = (-0.464752)(x - 1000.00)$$
$$y - 1000.00 = -0.4648x + 464.75$$
$$0.4648x + y = 1464.75$$

Solving simultaneously,

$$\begin{aligned} 1.7167x - y &= 285.83 \\ 0.4648x + y &= 1464.75 \\ \hline 2.1815x &= 1750.58 \\ x &= 802.47 \\ y &= 1091.76 \end{aligned}$$

The coordinates of point H are (802.47, 1091.76).

$$\text{length DH} = \sqrt{(802.47 - 622.08)^2 + (1091.76 - 782.01)^2} = 358.45$$

$$\text{length AH} = \sqrt{(1000.00 - 802.47)^2 + (1000.00 - 1091.76)^2} = 217.80$$

$$\text{length HB} = 560.27 - 217.80 = 342.47$$

PRACTICE PROBLEMS

(All distances and dimensions are in feet.)

1. The bearing and length of the side DE of the traverse ABCDEA are missing and are to be computed.

line	bearing	length
AB	N 28°19′ E	560.27
BC	N 56°23′ W	484.18
CD	S 08°50′ W	375.42
DE		
EA	S 67°45′ E	449.83

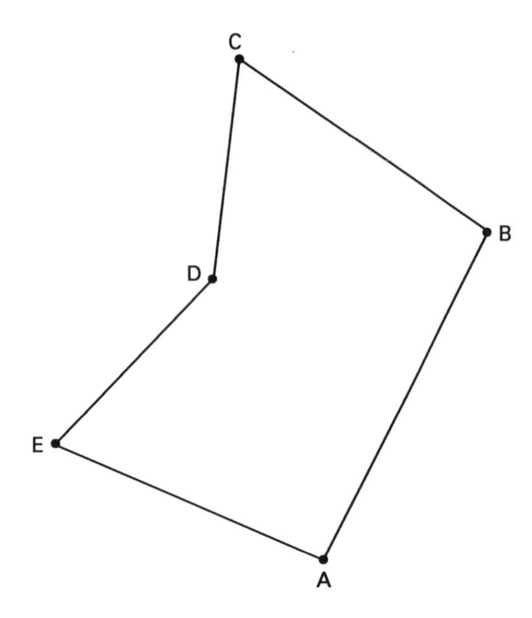

2. Compute the length of the sides DE and EA of the traverse ABCDEA. Use DA as a cutoff line. Draw a sketch of triangle DEA to a scale of 1 in = 200 ft. Show the size of angles in sketch.

line	bearing	length
AB	N 28°19′ E	560.27
BC	N 56°23′ W	484.18
CD	S 08°50′ W	375.42
DE	S 45°10′ W	
EA	S 67°45′ E	

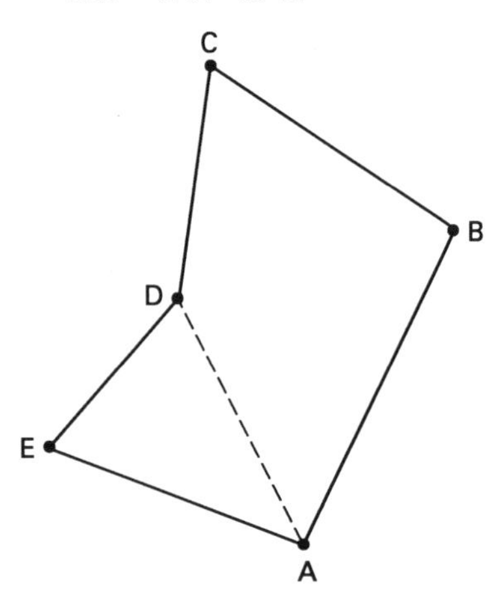

3. Compute the bearing of the sides DE and EA of the traverse ABCDEA. Use DA as a cutoff line. Draw a sketch of triangle DEA to a scale of 1 in = 200 ft. Show the size of angles in sketch.

line	bearing	length
AB	N 28°19′ E	560.27
BC	N 56°23′ W	484.18
CD	S 08°50′ W	375.42
DE		311.44
EA		449.83

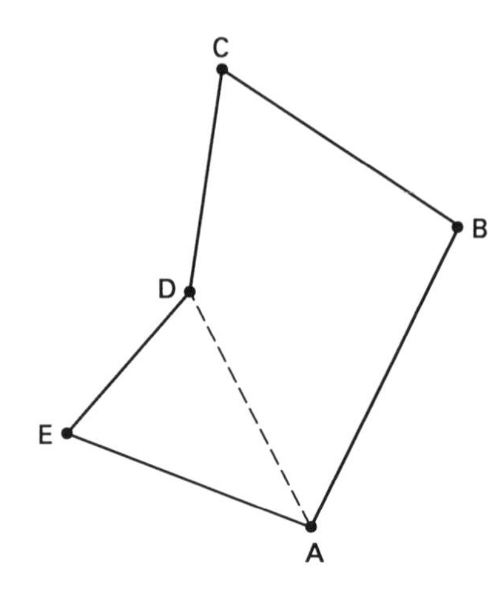

4. Compute the length of side CD and the bearing of side EA of the traverse ABCDEA. Use the following procedure.

step 1: Designate the cutoff line as XA.

step 2: From X draw CD in its given direction.

step 3: Using A as a compass point and length of side EA as the radius, draw an arc to intersect CD. A line from this intersection to A gives the direction of EA.

There are two intersections so there will be two solutions. AB: N 28°19′ E, 560.27; BC: N 56°23′ W, 484.18; CD: S 08°50′ W; DE: S 45°10′ W, 311.44; and EA: 449.83. Solve for both triangles.

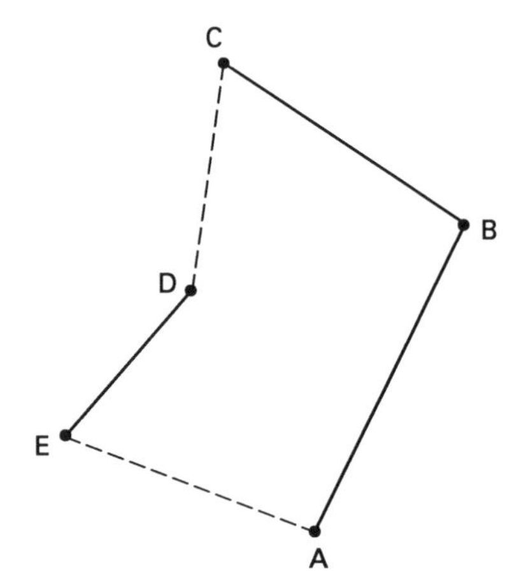

5. The tract of land represented by the traverse ABCDEA has been divided into two parts by a line from D to A. Balanced latitudes and departures are shown. Compute the area of the entire tract, the area of the tract ABCDA, and the area of the tract DEAD by the DMD method, and check the sum of the two areas against the total area.

line	bearing	length	latitude	departure
AB	N 28°19′ E	560.27	+493.12	+265.66
BC	N 56°23′ W	484.18	+267.97	−403.29
CD	S 08°50′ W	375.42	−371.04	−57.72
DE	S 45°10′ W	311.44	−219.64	−220.91
EA	S 67°45′ E	449.83	−170.41	+416.26

6. The tract of land represented by the traverse ABCDEA is to be divided into two parts by a line from D parallel to the side BC, which intersects the side AB at point H. The length AH must be computed so that the point H can be located by measuring from point A. To find AH, the triangle AHD is formed by using the cutoff line AD. The bearing of AD has been computed to be N 26°36′ W and the length of AD has been computed to be 436.23′. The area of the tract has been computed as shown. Find the lengths AH, DH, and HB, and compute the area of the two subdivided tracts by the DMD method, and check the sum of the two areas against the total area.

line	bearing	length	latitude	departure	DMD	area
AB	N 28°19′ E	560.27	+493.12	+265.66	1098.18	+541,535
BC	N 56°23′ W	484.18	+267.97	−403.29	960.55	+257,399
CD	S 08°50′ W	375.42	−371.04	−57.72	499.54	−185,349
DE	S 45°10′ W	311.44	−219.64	−220.91	220.91	−48,521
EA	S 67°45′ E	449.83	−170.41	+416.26	416.26	−70,935
						494,129

$$\text{area} = \frac{494{,}129\ \text{ft}^2}{(2)\left(43{,}560\ \frac{\text{ft}^2}{\text{ac}}\right)} = \frac{247{,}065\ \text{ft}^2}{43{,}560\ \frac{\text{ft}^2}{\text{ac}}} = 5.672\ \text{ac}$$

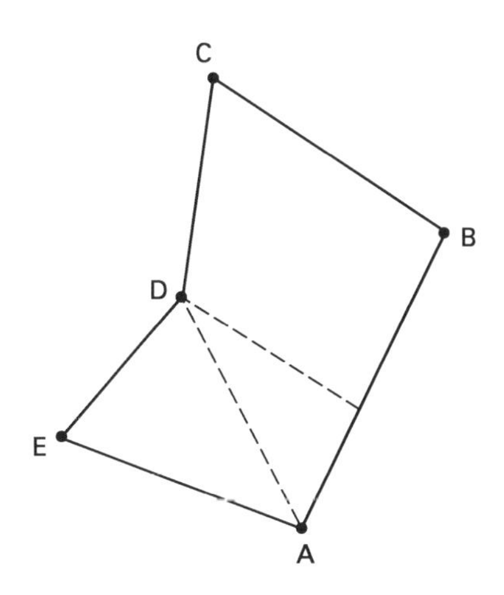

7. The tract of land represented by the traverse ABCDEA in Prob. 6 is to be divided into two equal parts by a line from point D. The dividing line intersects side AB at the point H. The tract has been surveyed and the following information has been found from computations.

(1) area ABCDEA = 247,065 ft^2

(2) area ABCDA = 182,300 ft^2

(3) bearing DA = S 26°36′ E

(4) length DA = 436.23 ft

Compute the length of AH, HB, and DH, and the bearing DH. Solve for the triangle AHD. Check areas by DMD.

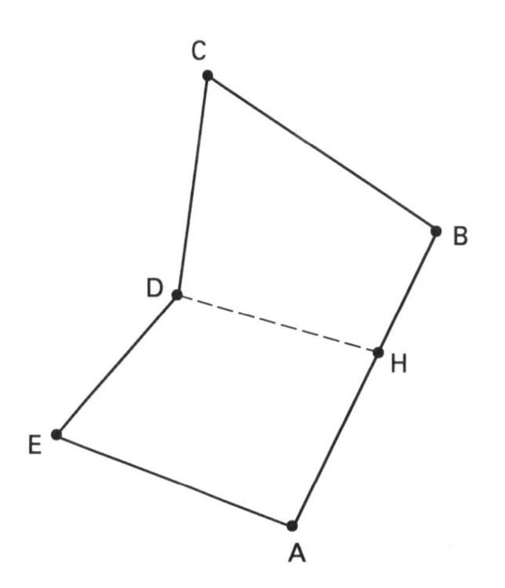

8. The tract of land represented by the traverse ABCDEFA is to be divided into two equal parts by a line parallel to AB. Designate dividing line as KL and the line parallel to AB as FJ. Find the length of LC, BL, FK, JL, and KL, and compute the area of ABLKFA by the DMD method.

line	bearing	length	latitude	departure	DMD	area
AB	S 08°30′ E	480.97	−475.69	+71.09	71.09	−33,817
BC	S 85°00′ E	750.26	−65.39	+747.41	889.59	−58,170
CD	N 72°30′ E	639.18	+192.21	+609.60	2246.60	+431,819
DE	N 13°30′ E	306.78	+298.30	+71.62	2927.82	+873,369
EF	N 76°30′ W	1015.43	+237.05	−987.37	2012.07	+476,961
FA	S 70°00′ W	545.22	−186.48	−512.35	512.35	− 95,543
						1,594, 619

$$\text{area} = \frac{1{,}594{,}619\ \text{ft}^2}{(2)\left(43{,}560\ \frac{\text{ft}^2}{\text{ac}}\right)} = \frac{797{,}310\ \text{ft}^2}{43{,}560\ \frac{\text{ft}^2}{\text{ac}}} = 18.304\ \text{ac}$$

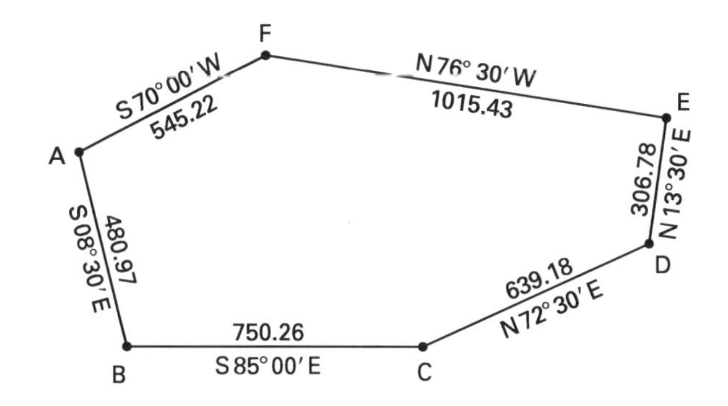

SOLUTIONS

1.

line	bearing	length	latitude	departure
AB	N 28°19′ E	560.27	+493.23	+265.76
BC	N 56°23′ W	484.18	+268.06	−403.21
CD	S 08°50′ W	375.42	−370.97	−57.65
DE				
EA	S 67°45′ E	449.83	−170.33	+416.34
			219.99	221.24

$$\text{bearing DE} = \tan^{-1}\frac{221.24\text{ ft}}{219.99\text{ ft}} = 45°10'$$

$$= \boxed{\text{S }45°10'\text{ W}}$$

$$\text{length DE} = \sqrt{(221.24\text{ ft})^2 + (219.99\text{ ft})^2}$$

$$= \boxed{312.00\text{ ft}}$$

2.

line	bearing	length	latitude	departure
AB	N 28°19′ E	560.27	+493.23	+265.76
BC	N 56°23′ W	484.18	+268.06	−403.21
CD	S 08°50′ W	375.42	−370.97	−57.65
DA				
			+390.32	−195.10

$$\text{bearing DA} = \tan^{-1}\frac{195.10\text{ ft}}{390.32\text{ ft}} = 26°33' = \text{S }26°33'\text{ E}$$

$$\text{length DA} = \sqrt{(195.10\text{ ft})^2 + (390.32\text{ ft})^2} = 436.36\text{ ft}$$

Solution of triangle DEA:

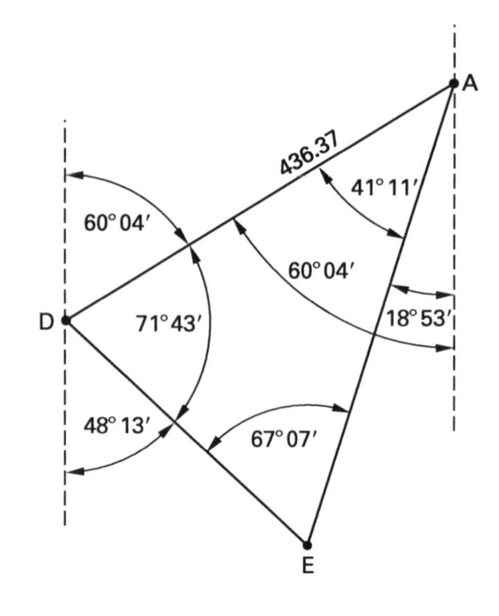

$$\text{length DE} = \frac{(436.36\text{ ft})(\sin 41°12')}{\sin 67°05'} = \boxed{311.06\text{ ft}}$$

$$\text{length EA} = \frac{(436.36\text{ ft})(\sin 71°43')}{\sin 67°05'} = \boxed{449.84\text{ ft}}$$

3.

line	bearing	length	latitude	departure
AB	N 28°19′ E	560.27	+493.23	+265.76
BC	N 56°23′ W	484.18	+268.06	−403.21
CD	S 08°50′ W	375.42	−370.97	−57.65
DA			−390.32	+195.10

$$\text{bearing DA} = \tan^{-1}\left(\frac{195.10\text{ ft}}{390.32\text{ ft}}\right) = 26°33'$$

$$= \text{S }26°33'\text{ E}$$

$$\text{length DA} = \sqrt{(195.10\text{ ft})^2 + (390.32\text{ ft})^2}$$

$$= 436.36\text{ ft}$$

Solution of triangle DEA:

$$\cos E = \frac{(311.44\text{ ft})^2 + (449.83\text{ ft})^2 - (436.36\text{ ft})^2}{(2)(311.44\text{ ft})(449.83\text{ ft})}$$

$$E = 67°07'$$

$$\cos D = \frac{(311.44\text{ ft})^2 + (436.36\text{ ft})^2 - (449.83\text{ ft})^2}{(2)(311.44\text{ ft})(436.36\text{ ft})}$$

$$D = 71°46'$$

$$\cos A = \frac{(436.36\text{ ft})^2 + (449.83\text{ ft})^2 - (311.44\text{ ft})^2}{(2)(436.36\text{ ft})(449.83\text{ ft})}$$

$$A = 41°07'$$

$$\text{bearing DE} = \boxed{\text{S }45°13'\text{ W}}$$

$$\text{bearing EA} = \boxed{\text{S }67°40'\text{ E}}$$

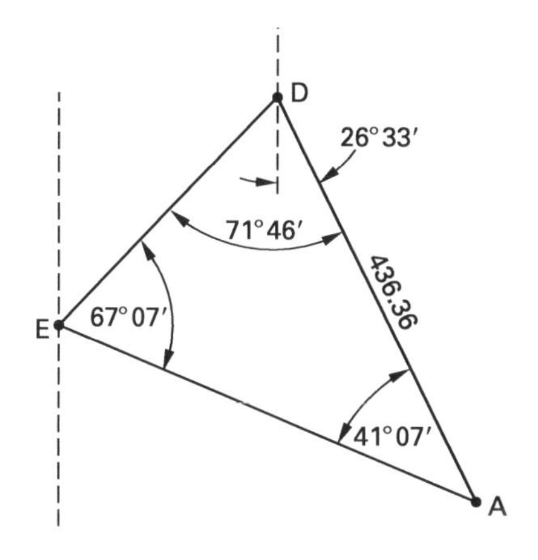

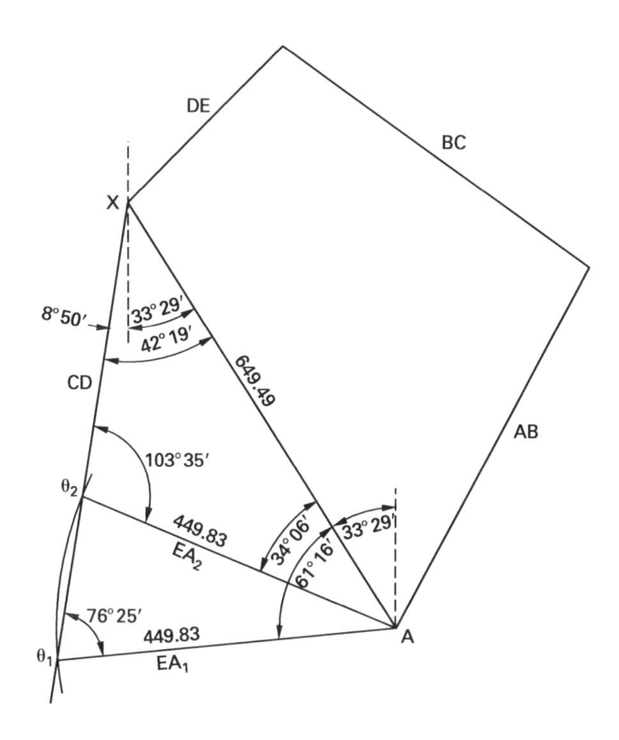

4. The computations for cutoff line are as follows.

line	bearing	length	latitude	departure
AB	N 28°19′ E	560.27	+493.23	+265.76
BC	N 56°23′ W	484.18	+268.06	−403.21
DE	S 45°10′ W	311.44	−219.58	−220.86
			+541.71	−358.31

$$\text{bearing XA} = \tan^{-1}\frac{358.31 \text{ ft}}{541.71 \text{ ft}} = \text{S } 33°29' \text{ E}$$

$$\begin{aligned}\text{length XA} &= \sqrt{(541.71 \text{ ft})^2 + (358.31 \text{ ft})^2}\\ &= 649.49 \text{ ft}\end{aligned}$$

Solution of triangle no. 1:

$$\begin{aligned}\sin\theta_1 &= \frac{(649.49 \text{ ft})(\sin 42°19')}{449.83 \text{ ft}}\\ \theta_1 &= 76°25'\\ \text{CD}_1 &= \frac{(449.83 \text{ ft})(\sin 61°16')}{\sin 42°19'} = 585.90 \text{ ft}\\ \text{bearing EA}_1 &= \boxed{\text{N } 85°15' \text{ E}}\\ \text{length CD}_1 &= \boxed{585.90 \text{ ft}}\end{aligned}$$

Solution of triangle no. 2:

$$\begin{aligned}\sin\theta_2 &= \frac{(649.49 \text{ ft})(\sin 42°19')}{449.83 \text{ ft}}\\ \theta_2 &= 103°35'\\ \text{CD}_2 &= \frac{(449.83 \text{ ft})(\sin 34°06')}{\sin 42°19'} = 374.60 \text{ ft}\\ \text{bearing EA}_2 &= \boxed{\text{S } 67°35' \text{ E}}\\ \text{length CD}_2 &= \boxed{374.60 \text{ ft}}\end{aligned}$$

5.

line	bearing	length	latitude	departure	DMD	area
AB	N 28°19′ E	560.27	+493.12	+265.66	1098.18	+541,535
BC	N 56°23′ W	484.18	+267.97	−403.29	960.55	+257,399
CD	S 08°50′ W	375.42	−371.04	−57.72	499.54	−185,349
DE	S 45°10′ W	311.44	−219.64	−220.91	220.91	−48,521
EA	S 67°45′ E	449.83	−170.41	+416.26	416.26	−70,935
						494,129

$$\text{area} = \frac{494{,}129 \text{ ft}^2}{(2)\left(43{,}560 \ \frac{\text{ft}^2}{\text{ac}}\right)} = \frac{247{,}065 \text{ ft}^2}{43{,}560 \ \frac{\text{ft}^2}{\text{ac}}} = 5.672 \text{ ac}$$

The computations for DA are as follows.

line	latitude	departure
DE	−219.64	−220.91
EA	−170.41	+416.26
	−390.05	+195.35

$$\begin{aligned}\text{bearing DA} &= \tan^{-1}\frac{195.35 \text{ ft}}{390.05 \text{ ft}} = 26°36'\\ &= \text{S } 26°36' \text{ E}\end{aligned}$$

$$\begin{aligned}\text{length DA} &= \sqrt{(390.05 \text{ ft})^2 + (195.35 \text{ ft})^2}\\ &= 436.23 \text{ ft}\end{aligned}$$

line	latitude	departure	DMD	area
AB	+493.12	+265.66	656.36	+323,664
BC	+267.97	−403.29	518.73	+139,004
CD	−371.04	−57.72	57.72	−21,416
DA	−390.05	+195.35	195.35	−76,196
				365,056

$$\text{area} = \frac{365{,}056\ \text{ft}^2}{(2)\left(43{,}560\ \frac{\text{ft}^2}{\text{ac}}\right)} = \frac{182{,}528\ \text{ft}^2}{43{,}560\ \frac{\text{ft}^2}{\text{ac}}} = 4.190\ \text{ac}$$

line	latitude	departure	DMD	area
DE	−219.64	−220.91	220.91	−48,521
EA	−170.41	+416.26	416.26	−70,935
AD	+390.05	−195.35	637.17	−248,528
				129,072

$$\text{area} = \frac{129{,}072\ \text{ft}^2}{(2)\left(43{,}560\ \frac{\text{ft}^2}{\text{ac}}\right)} = \frac{64{,}536\ \text{ft}^2}{43{,}560\ \frac{\text{ft}^2}{\text{ac}}} = \boxed{1.482\ \text{ac}}$$

total area check: $182{,}528\ \text{ft}^2 + 64{,}536\ \text{ft}^2 = 247{,}064\ \text{ft}^2$

6. Solution of triangle AHD:

$$\text{length DH} = \frac{(436.23\ \text{ft})(\sin 54°55')}{\sin 95°18'} = \boxed{358.51\ \text{ft}}$$

$$\text{length AH} = \frac{(436.23\ \text{ft})(\sin 29°47')}{\sin 95°18'} = \boxed{217.62\ \text{ft}}$$

$$\text{length HB} = 560.27 - 217.62 = \boxed{342.65\ \text{ft}}$$

bearing DH = S 56°23′ E

line	bearing	length	latitude	departure	DMD	area
HB	N 28°19′ E	342.65	+301.59	+162.48	759.48	+229,052
BC	N 56°23′ W	484.18	+267.98	−403.28	518.68	+138,996
CD	S 08°50′ W	375.42	−371.03	−57.70	57.70	−21,408
DH	S 56°23′ E	358.51	−198.54	+298.50	298.50	−59,264
						287,376

$$\text{area} = \frac{287{,}376\ \text{ft}^2}{(2)\left(43{,}560\ \frac{\text{ft}^2}{\text{ac}}\right)} = \frac{143{,}688\ \text{ft}^2}{43{,}560\ \frac{\text{ft}^2}{\text{ac}}} = 3.299\ \text{ac}$$

line	bearing	length	latitude	departure	DMD	area
AH	N 28°19′ E	217.62	+191.56	+103.20	935.78	+179,258
HD	N 56°23′ W	358.51	+198.44	−298.59	740.39	+146,923
DE	S 45°10′ W	311.44	−219.62	−220.90	220.90	−48,514
EA	S 67°45′ E	449.83	−170.38	+416.29	416.29	−70,927
						206,740

$$\text{area} = \frac{206{,}740\ \text{ft}^2}{(2)\left(43{,}560\ \frac{\text{ft}^2}{\text{ac}}\right)} = \frac{103{,}370\ \text{ft}^2}{43{,}560\ \frac{\text{ft}^2}{\text{ac}}} = \boxed{2.373\ \text{ac}}$$

total area check $= 143{,}688 + 103{,}370 = 247{,}058\ \text{ft}^2$

7. Solution of triangle AHD:

$$\begin{aligned}\text{area AHD} &= 182{,}300\ \text{ft}^2 - \left(\tfrac{1}{2}\right)(247{,}065\ \text{ft}^2)\\ &= 58{,}768\ \text{ft}^2\end{aligned}$$

$$\text{AH} = \frac{(2)(58{,}768\ \text{ft}^2)}{(436.23\ \text{ft})(\sin 54°55')} = 329.25610265$$

$$\begin{aligned}\text{DH} &= \sqrt{\begin{aligned}&(436.23\ \text{ft})^2 + (329.26\ \text{ft})^2\\ &\quad - (2)(436.23\ \text{ft})(329.26\ \text{ft})(\cos 54°55')\end{aligned}}\\ &= 365.51\ \text{ft}\end{aligned}$$

$$\sin \text{D} = \frac{(329.26\ \text{ft})(\sin 54°55')}{365.51\ \text{ft}}$$

$$\text{D} = 47°29'$$

$$\text{HB} = 560.27\ \text{ft} - 329.26\ \text{ft} = 231.01\ \text{ft}$$

$$\text{AH} = \boxed{329.26\ \text{ft}}$$

$$\text{DH} = \boxed{365.51\ \text{ft}}$$

bearing DH – S 74°05′ E

$$\text{HB} = \boxed{231.01\ \text{ft}}$$

line	bearing	length	latitude	departure	DMD	area
AH	N 28°19′ E	329.26	+289.86	+156.18	988.70	+286,585
HD*	N 74°05′ W	365.51	+100.19	−351.53	793.35	+79,486
DE	S 45°10′ W	311.44	−219.64	−220.91	220.91	−48,521
EA	S 67°45′ E	449.83	−170.41	+416.26	416.26	−70,935
						246,615

*closure forced

$$\text{area} = \frac{246{,}615\ \text{ft}^2}{(2)\left(43{,}560\ \frac{\text{ft}^2}{\text{ac}}\right)} = 2.830\ \text{ac}$$

8.

$$\text{area ABLKFA} = \frac{797{,}310\ \text{ft}^2}{2} = 398{,}655\ \text{ft}^2$$

The bearing and length of BF can be computed as follows.

line	latitude	departure
AB	−475.69	+71.09
BF	(+662.17)	(+441.25)
FA	−186.48	−512.34

$$\text{tangent bearing BF} = \frac{441.25 \text{ ft}}{662.17 \text{ ft}}$$

$$\text{bearing BF} = \text{N}\,33°41'\,\text{E}$$

$$\text{length} = \sqrt{(441.25 \text{ ft})^2 + (662.17 \text{ ft})^2} = 795.72 \text{ ft}$$

Solution of triangle BJF:

$$\text{BJ} = \frac{(795.72 \text{ ft})(\sin 42°11')}{\sin 76°30'} = 549.51 \text{ ft}$$

$$\text{JF} = \frac{(795.72 \text{ ft})(\sin 61°19')}{\sin 76°30'} = 717.91 \text{ ft}$$

line	bearing	length	latitude	departure	DMD	area
AB	S 08°30′ E	480.97	−475.69	+71.09	71.09	−33,817
BJ	S 85°00′ E	549.51	−47.89	+547.42	689.60	−33,025
JF*	N 08°30′ W	717.91	+710.06	−106.17	1130.85	+802,971
FA	S 70°00′ W	545.22	−186.48	−512.34	512.34	−95,541
						640,588

*closure forced

$$\text{area} = \frac{640{,}588 \text{ ft}^2}{2} = 320{,}294 \text{ ft}^2$$

$$y = x \tan 13°30'$$

$$z = x \tan 22°00'$$

$$\begin{aligned} \text{LK} &= \text{FJ} + y - z \\ &= \text{FJ} + x\tan 13°30' - x\tan 22°00' \end{aligned}$$

$$\begin{aligned} \text{area ABLKFA} &= 398{,}655 \text{ ft}^2 \\ \text{area ABJFA} &= 320{,}294 \text{ ft}^2 \\ \text{area FJLK} &= 78{,}361 \text{ ft}^2 \end{aligned}$$

$$\begin{aligned} 78{,}361 \text{ ft}^2 &= \frac{(\text{FJ} + \text{FJ} + x\tan 13°30' - x\tan 22°00')x}{2} \\ &= \frac{(2\text{FJ} + x\tan 13°30' - x\tan 22°00')x}{2} \\ &= \frac{(\text{FJ})x - (\tan 22°00' - \tan 13°30')x^2}{2} \\ &= 717.91x \quad 0.0819737x^2 \end{aligned}$$

$$0.819737x^2 - 717.91x + 78{,}361 = 0$$

$$x = \frac{717.91 \pm \sqrt{(-717.91)^2 - (4)(0.0819737)(78{,}361)}}{(2)(0.0819737)} = 110.55 \text{ ft}$$

$$\text{JL} = \frac{110.55 \text{ ft}}{\cos 13°30'} = 113.69 \text{ ft}$$

$$\text{FK} = \frac{110.55 \text{ ft}}{\cos 22°00'} = 119.23 \text{ ft}$$

$$\text{LC} = 750.26 \text{ ft} - 549.51 \text{ ft} - 113.69 \text{ ft} = \boxed{87.06 \text{ ft}}$$

$$\text{BL} = 549.51 \text{ ft} + 113.69 \text{ ft} = \boxed{663.20 \text{ ft}}$$

$$\begin{aligned} \text{KL} &= 717.91 \text{ ft} + (110.55 \text{ ft})(\tan 13°30') \\ &\quad - (1101.55 \text{ ft})(\tan 22°00') \\ &= \boxed{699.79 \text{ ft}} \end{aligned}$$

line	bearing	length	latitude	departure	DMD	area
AB	S 08°30′ E	480.97	−475.69	+71.09	71.09	−33,817
BL	S 85°00′ E	663.20	−57.80	+660.68	802.86	−46,405
LK	N 08°30′ W	699.79	+692.12	−103.66	1360.08	+941,339
KF	N 76°30′ W	119.23	+27.85	−115.96	1140.66	+31,767
FA	S 70°00′ W	545.22	−186.48	−512.35	512.35	−95,543
						797,341

$$\text{area} = \frac{797{,}341 \text{ ft}^2}{(2)\left(43{,}560 \dfrac{\text{ft}^2}{\text{ac}}\right)} = \frac{398{,}670 \text{ ft}^2}{43{,}560 \dfrac{\text{ft}^2}{\text{ac}}} = \boxed{9.152 \text{ ac}}$$

$$\begin{aligned} \text{total area check} &= 398{,}670 \text{ ft}^2 - 398.655 \text{ ft}^2 \\ &= 15 \text{ ft}^2 \\ &= 0.0003 \text{ ac} \quad \text{[error]} \end{aligned}$$

18 Horizontal Curves

1. SIMPLE CURVES

Highways consist of a series of straight sections joined by curved sections. The straight sections are known as *tangents*. The curves are most often circular arcs, known as *simple curves*, but may be spiral curves. Spiral curves are encountered more often on railroads.

Initial locations of highways usually consist of straight lines. Curves are later inserted to connect two intersecting tangents. Many curves of different radii, or degrees of curve, may be selected for any given intersection of tangents.

Figure 18.1 *Curves Connecting Tangent Lines*

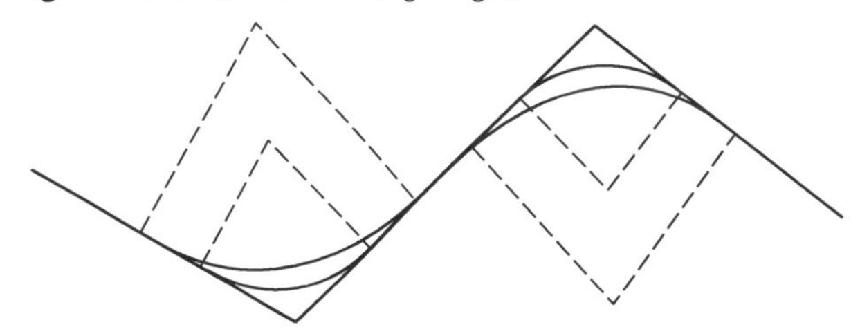

Figure 18.1 shows that a choice of circular arcs, or curves, can be made after the tangent locations have been made. The curves are usually classified as to their *degree of curve*, the angle subtended by a portion of the curve 100 ft long. In selecting the degree of curve, consideration is given to design speed, topographic features, economy, and other variables.

2. GEOMETRY

All the formulas for computations involving circular curves depend on certain principles of geometry and trigonometry.

3. INSCRIBED ANGLE

An *inscribed angle* is an angle that has its vertex on a circle and that has chords for its sides as shown in Fig. 18.2(a).

4. MEASURE OF AN INSCRIBED ANGLE

An inscribed angle is measured by one-half its intercepted arc as shown in Fig. 18.2(a).

5. MEASURE OF AN ANGLE FORMED BY A TANGENT AND A CHORD

An angle formed by a tangent and a chord is measured by one-half its intercepted arc as shown in Fig. 18.2(b).

Figure 18.2 Tangent and Arc Geometry

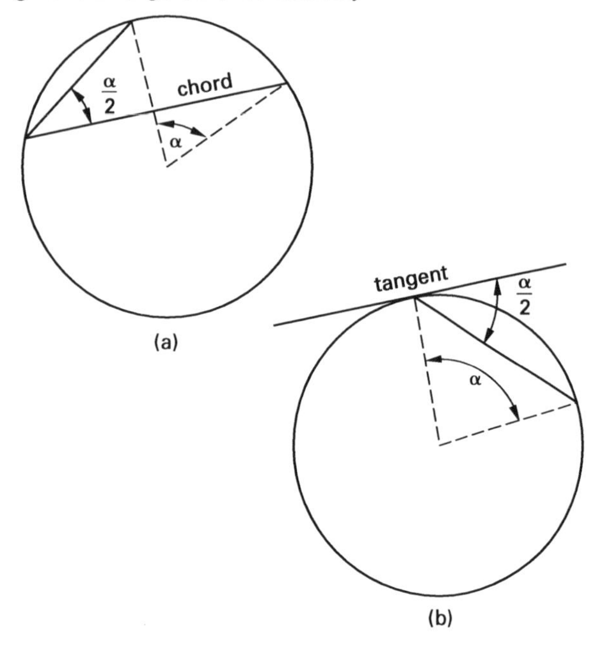

6. RADIUS IS PERPENDICULAR TO TANGENT

The radius of a circle is perpendicular to a tangent at the point of tangency as shown in Fig. 18.3(a).

7. RADIUS IS PERPENDICULAR TO BISECTOR OF A CHORD

The perpendicular bisector of a chord passes through the center of the circle as shown in Fig. 18.3(b).

Figure 18.3 Chord Geometry

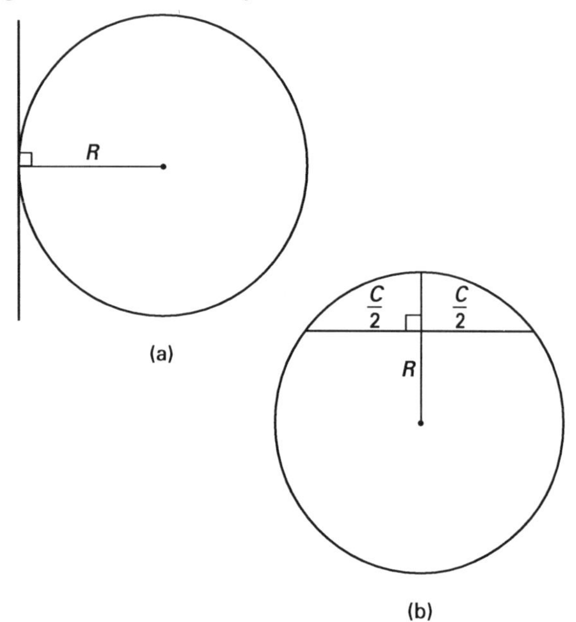

8. DEFINITIONS AND SYMBOLS

The following definitions of symbols are used in curve computations.

- C *(length of chord):* chord length
- Δ *(deflection angle):* central angle: angle at PI, or angle at center.
- D *(degree of curve):* central angle that subtends a 100 ft arc (arc basis)
- E *(external distance):* distance from PI to middle of curve
- L *(length of curve):* distance from PC to PT along the arc
- LC *(length of long chord):* distance from PC to PT; chord length for angle Δ
- M *(middle ordinate):* length of ordinate from middle of long chord to middle of curve
- PC *(point of curvature):* beginning of curve
- PI *(point of intersection):* point where two tangents intersect
- PT *(point of tangency):* end of curve
- R *(radius):* a straight line from the center of a circle to the circumference
- T *(tangent distance):* distance from PI to PC, or distance from PI to PT

Figure 18.4 Circular Arc

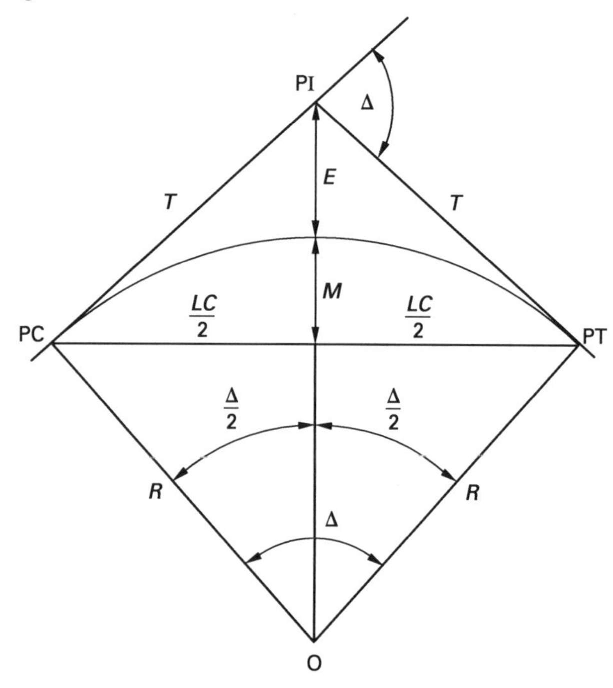

9. DEFLECTION ANGLE EQUALS CENTRAL ANGLE

In Fig. 18.4, the sum of the interior angles of the polygon O-PC-PI-PT is 360°. The angles at the PC and PT each equal 90°. The sum of the interior angle at the PI and the deflection angle is 180°. The sum of the interior angle at the PI and the central angle is 180°. Therefore, the deflection angle equals the central angle.

10. HORIZONTAL CURVE FORMULAS

In computing various components of a circular curve, certain formulas derived from trigonometry are useful and necessary. In addition to the usual six trigonometric functions, two others are used in curve computations: *versed sine* (vers) and *external secant* (exsec).

Most surveying textbooks include tables for these functions.

$$\operatorname{vers}\phi = 1 - \cos\phi \quad \text{18.1}$$

$$\operatorname{exsec}\phi = \sec\phi - 1 \quad \text{18.2}$$

Other formulas commonly encountered are

$$T = R\tan\frac{\Delta}{2} \quad \text{18.3}$$

$$L = \frac{2\pi R\Delta}{360°} \quad \text{18.4}$$

$$C = 2R\sin\frac{D}{2} \quad \text{18.5}$$

$$M = R\left(1 - \cos\frac{\Delta}{2}\right) = R\operatorname{vers}\frac{\Delta}{2} \quad \text{18.6}$$

$$E = R\left(\sec\frac{\Delta}{2} - 1\right) = R\operatorname{exsec}\frac{\Delta}{2} \quad \text{18.7}$$

11. DEGREE OF CURVE

There are two definitions of *degree of curve*. The *arc definition* is used for highways and streets. The *chord definition* is used for railroads. By the arc definition, degree of curve, D, is the central angle that subtends a 100 ft arc. By the chord definition, degree of curve, D, is the central angle that subtends a 100 ft chord.

Figure 18.5 *Arc and Chord Definitions*

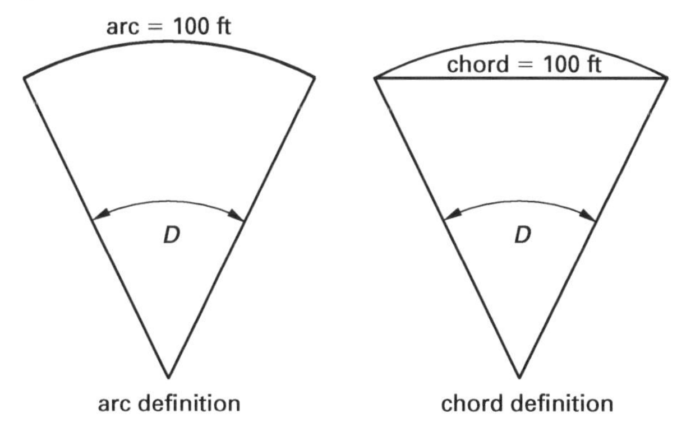

Using the arc definition,

$$R = \frac{(360°)(100\text{ ft})}{2\pi D} = \frac{5729.58}{D} \quad \left[\begin{array}{c}\text{for } D = 1°,\\ R = 5729.58\text{ ft}\end{array}\right] \quad \text{18.8}$$

Using the chord definition,

$$R = \frac{50}{\sin\frac{D}{2}} \quad \left[\begin{array}{c}\text{for } D = 1°,\\ R = 5729.65\text{ ft}\end{array}\right] \quad \text{18.9}$$

Values of R for various values of D are given in App. B.

When using the arc definition for curve computations with a 100 ft tape to lay out the curve in the field, measurements are actually chord lengths of 100 ft, and the arc length is somewhat greater. For curves up to 4°, the difference in arc length and chord length is negligible. For instance, the chord length for a 100 ft arc on a 4° curve is 99.980 ft. For curves of greater degree of curve, the actual chord length can be found in the tables in App. C. Chord lengths can be measured accordingly.

12. CURVE LAYOUT

Because of their long radii, most curves cannot be laid out by swinging an arc from the center of the circle. They must be laid out by a series of straight lines (chords). This is done by use of transit and tape.

13. DEFLECTION ANGLE METHOD

The *deflection angle method* is based on the fact that the angle between a tangent and a chord, or between two chords that form an inscribed angle, is one-half the intercepted arc (see Fig. 18.2). In Fig. 18.6, the angle formed by the tangent at the PC and a chord from the PC to a point 100 ft along the arc is equal to one-half the degree of curve. Likewise, the angle formed by this chord and a chord from the PC to a point 100 ft farther along the arc is also equal to one-half the degree of curve. These angles are known as *deflection angles.* The deflection angle from the PC to the PT is one-half the central angle Δ, which provides an important check in computing deflection angles.

Figure 18.6 *Laying Out a Curve*

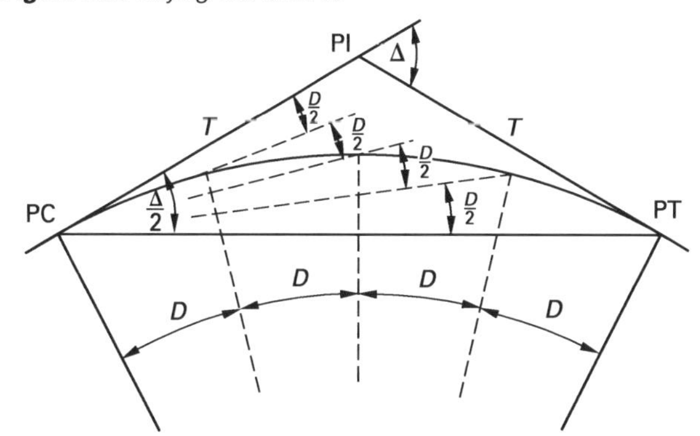

Plane Survey Calculations

In laying out the curve, the transit is set up at the PC, the PT, or some other point on the curve, and deflection angles are turned for 100 ft arcs along the curve as 100 ft arcs are marked off by a 100 ft tape. For degree of curve up to 4°, the difference in length between chord and arc is slight. For sharper curves, this discrepancy can be corrected by laying out a chord slightly less than 100 ft. This length can be found either in App. C or from Eq. 18.5.

On route surveys, stationing is carried continuously along tangent and curve. Thus, the PC and PT will seldom fall on a full station. The first full station will not likely be 100 ft from the PC, and the deflection angle to the first full station will not be $D/2$ but will be a fraction of $D/2$.

When stakes are required on closer intervals than full stations, such as 50 ft, the true length of the 50 ft chord (known as a *subchord*) will be less than 50 ft and this length can be found in App. C or from Eq. 18.5. (The central angle that subtends a 50 ft arc is $D/2$).

14. LENGTH OF CURVE

The *length of curve*, L (arc definition), is the distance along the arc from the PC to PT. As any two arcs are proportional to their central angles,

$$L = (100 \text{ ft})\left(\frac{\Delta}{D}\right) \tag{18.10}$$

15. FIELD PROCEDURE IN STAKING A SIMPLE CURVE

The following steps can be used when staking out a curve from the PC or PT.

step 1: Measure the deflection angle.

step 2: Select D by considering the design criteria.

step 3: Compute the tangent distance T from the PI to the PC.

Tangent distances for a 1° curve for various values of Δ can be found in App. A. Tangent distance for any degree of curve can be found by dividing the tangent distance for a 1° curve by D.

step 4: Measure the tangent distance T from the PI to the PT and set the tacked hub. Measure T from the PI to the PC and set the tacked hub.

step 5: Compute PC station by subtracting T from PI station.

step 6: Compute the length of curve L. Compute the PT station by adding L to the PC station.

step 7: Compute the deflection angles at the PC for each station, checking to see that the deflection angle to the PT is $\Delta/2$.

step 8: (a) Set up a transit at PC and take a foresight on PI with the telescope normal and with the A vernier set on 0°00′. Turn the deflection angle for each station as chainpeople mark off corresponding arc length. Make the check at the PT for angle and distance.

(b) Set up the transit at the PT and take a backsight reading on the PC with the telescope normal and with the A vernier set on 0°00′. Turn the deflection angle for each station as chainpeople mark off corresponding arc lengths, with chainpeople starting at the PC. Deflection angles are the same as those computed for the staking curve from the PC. Make the check by sighting on the PI with the deflection angle to the PT.

16. CIRCULAR CURVE COMPUTATIONS

Example 18.1

Compute the parameters of a simple curve.

$$\text{PI} = \text{sta } 25{+}05$$
$$\Delta = 20°$$
$$D = 2°$$

Solution

$$L = (100 \text{ ft})\left(\frac{\Delta}{D}\right) = (100 \text{ ft})\left(\frac{20°}{2°}\right) = 1000 \text{ ft}$$

From App. B,

$$R = 2864.79 \text{ ft}$$

From Eq. 18.8,

$$R = \frac{5729.58}{D} = \frac{5729.58}{2} = 2864.79 \text{ ft}$$
$$T = \frac{T \text{ for } 1° \text{ curve}}{D} = \frac{1010.28 \text{ ft}}{2°} = 505 \text{ ft}$$

From Eq. 18.3,

$$T = R\tan\frac{\Delta}{2} = (2864.79 \text{ ft})\tan\left(\frac{20°}{2}\right) = 505 \text{ ft}$$

From App. A,

$$E = \frac{88.39 \text{ ft}}{2} = 44.19 \text{ ft}$$

From Eq. 18.7,

$$E = R\left(\sec\frac{\Delta}{2} - 1\right) = (2864.79 \text{ ft})\left(\sec\left(\frac{20^\circ}{2}\right) - 1\right)$$
$$= 44.19 \text{ ft}$$

From Eq. 18.6,

$$M = R\left(1 - \cos\frac{\Delta}{2}\right) = (2864.79 \text{ ft})\left(1 - \cos\frac{20^\circ}{2}\right)$$
$$= 43.52 \text{ ft}$$

PI	=	sta	25+05
T	=		−5+05
PC	=	sta	20+00
L	=		+10+00
PT	=	sta	30+00

The deflection angles are

point	station			deflection angle
PC	20+00			0°00′
	21+00 =	$D/2$		1°00′
	22+00 =	deflection angle of station	21+00 + $D/2$ =	2°00′
	23+00 =		22+00 + $D/2$ =	3°00′
	24+00 =		23+00 + $D/2$ =	4°00′
	25+00 =		24+00 + $D/2$ =	5°00′
	26+00 =		25+00 + $D/2$ =	6°00′
	27+00 =		26+00 + $D/2$ =	7°00′
	28+00 =		27+00 + $D/2$ =	8°00′
	29+00 =		28+00 + $D/2$ =	9°00′
PT	30+00 =		29+00 + $D/2$ =	10°00′

(Check: deflection angle to the PT = $\Delta/2$ = 10°00′)

Example 18.2

Compute the parameters of a simple curve.

$$\text{PI} = \text{sta } 45+78.39$$
$$\Delta = 43^\circ 39'$$
$$D = 4^\circ 15'$$

Solution

From Eq. 18.10,

$$L = (100 \text{ ft})\left(\frac{\Delta}{\text{D}}\right)$$
$$= (100 \text{ ft})\left(\frac{43^\circ 39'}{4^\circ 15'}\right) = (100 \text{ ft})\left(\frac{43.65^\circ}{4.25^\circ}\right)$$
$$= 1027.06 \text{ ft}$$

From App. B,

$$R = 1348.14 \text{ ft}$$

From App. A,

$$T = \frac{T \text{ for } 1^\circ \text{ curve}}{D}$$
$$= \frac{2294.57 \text{ ft}}{4.25^\circ} = 539.90 \text{ ft}$$

From App. B,

$$R = 1348.14 \text{ ft}$$

PI	=	sta	45+78.39
T	=		−5+39.90
PC	=	sta	40+38.49
L	=		+10+27.06
PT	=	sta	50+65.55

The deflection angles are

point	station			deflection angle
PC	40+38.49			0°00′
	41+00	=	$\left(\frac{100 - 38.49}{100}\right)\left(\frac{D}{2}\right)$ =	1°18′
	42+00	=	1°18′ + $D/2$ =	3°26′
	43+00	=	3°26′ + $D/2$ =	5°33′
	44+00	=	5°33′ + $D/2$ =	7°41′
	45+00	=	7°41′ + $D/2$ =	9°48′
	46+00	=	9°48′ + $D/2$ =	11°56′
	47+00	=	11°56′ + $D/2$ =	14°03′
	48+00	=	14°03′ + $D/2$ =	16°11′
	49+00	=	16°11′ + $D/2$ =	18°18′
	50+00	=	18°18′ + $D/2$ =	20°26′
PT	50+65.55	=	20°26′ + (0.6555)(2.125) =	21°49.5′

Field notes for this curve are shown in Table 18.1.

17. TRANSIT AT POINT ON CURVE

On long highway curves, it is often impossible to locate the entire curve from one point. Obstructions along the arc or along the line of sight from the transit to a station may prevent location of the curve from one point. In these situations, part of the curve can be located from the PC and then the transit can be moved to a point on the curve that has been located from the PC. The balance of the curve can then be located. On extremely long curves, part of the curve can be located from the PC and part from the PT. Any error will be located where the two parts meet.

In all cases, when it is necessary to move up on a curve, the deflection angles are computed as if the curve were to be located from the PC. When it is decided to move the transit, tack points are set at the station that is to be occupied and at the station that is be used as a backsight, unless the PC can be used as a backsight. In either case, to orient the transit at the new station, the deflection angle of the station that is to be used for the backsight is set on the horizontal circle with the upper clamp before setting on the backsight with the lower clamp. If the PC is to be used for the backsight, the vernier would be set on 0°00′. After orientation, deflection angles as originally computed are turned for the balance of the curve.

Table 18.1 *Field Notes for Alignment of Reisel-Mart Highway*

point	station	deflection angle	calculated bearing	curve data
PT	50+65.55	21°49.5′	N 71°54′ E	
	50+00.00	20°26′		
	49+00.00	18°18′		
	48+00.00	16°11′		
	47+00.00	14°03′		
				$\Delta = 43°39'$R
	46+00.00	11°56′		$D = 4°15'$
PI	45+78.39			$T = 539.90$ ft
	45+00.00	9°48′		$L = 1027.06$ ft
				$R = 1348.14$ ft
	44+00.00	7°41′		
	43+00.00	5°33′		
	42+00.00	3°26′		
	41+00.00	1°18′		
PC	40+38.49	0°00′	N 28°15′ E	

This procedure can be used in locating *culverts* on curves. The culvert station is occupied, the line of sight is made tangent at this point, and 90° is turned off this tangent. This puts the centerline of the culvert on a radial line.

Example 18.3

Sta 41+00, 42+00, and 43+00 for the curve tabulated in Table 18.1 have been set from the PC, but because of an obstruction in the line of sight, sta 44+00 cannot be located. The transit is to be moved to sta 43+00 for continuation of the curve location; the PC is to be used for the backsight. Locate sta 44+00.

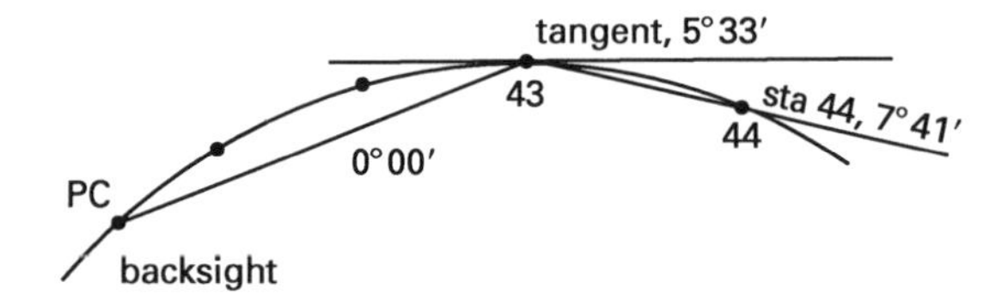

Solution

step 1: Set the transit on sta 43+00.

step 2: Set 0°00′ on the *A* vernier.

step 3: Backsight on the PC with the telescope inverted.

step 4: Plunge the telescope and continue the location using the deflection angles as originally computed. (Note: When the vernier reads 5°33′, the line of sight is tangent to the curve at the point occupied.)

Example 18.4

Sta 41 through 46 of the curve tabulated in Table 18.1 have been set from the PC. The balance of the curve is to be located from sta 46+00, from which the PC is not visible. Sta 43+00 is to be used as a backsight. Outline the procedure.

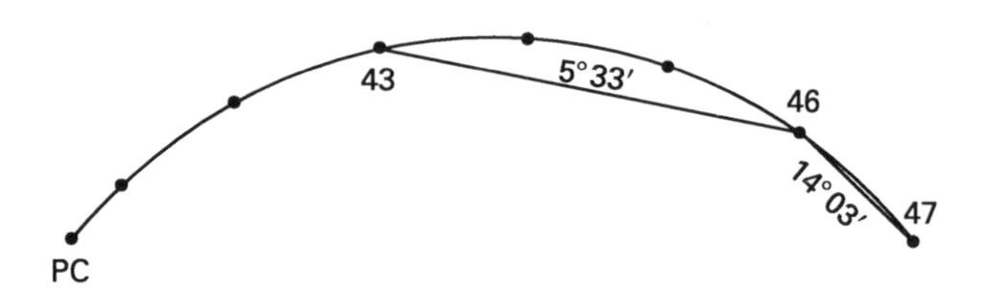

Solution

step 1: Set the transit on sta 46+00.

step 2: Set 5°33′ on the *A* vernier that is the deflection angle for sta 43+00.

step 3: Backsight on sta 43 | 00 with the telescope inverted.

step 4: Plunge the telescope and continue the location using deflection angles as originally computed. (The vernier is set on the deflection angle of the station sighted.)

18. COMPUTING TRANSIT STATIONS FOR HIGHWAY LOCATION

As has been mentioned, initial locations of highways usually consist of a series of tangents. Curves are inserted later to connect two intersecting tangents. The original tangents are chained from PI to PI, and deflection angles are measured. When curves are inserted, the original stations at PIs must be corrected. Stations of PCs and PTs must be computed by considering distances along the curves and not along the tangents. Obviously, the total length of the project will be less than the lengths of the tangents.

Example 18.5

A preliminary highway location has been made. PI stations, deflection angles, and the end of the line are as shown in the following table. (The beginning station is 0+00.) Degree of curve for each curve has been selected. Stations for each PC and PT and the station for the end of the line are to be computed.

PI number	original station	Δ	D
1	12+24.31	21°28′R	1°30′
2	31+12.48	40°56′L	2°30′
3	51+90.13	22°12′R	2°00′
end	65+56.03	end of line	

Solution

A sketch is drawn showing curves connecting tangents. Distances from PI to PI, tangent distances and lengths of curves are then computed and tabulated. Using the tabulated distances and the sketch, stations are computed in sequence.

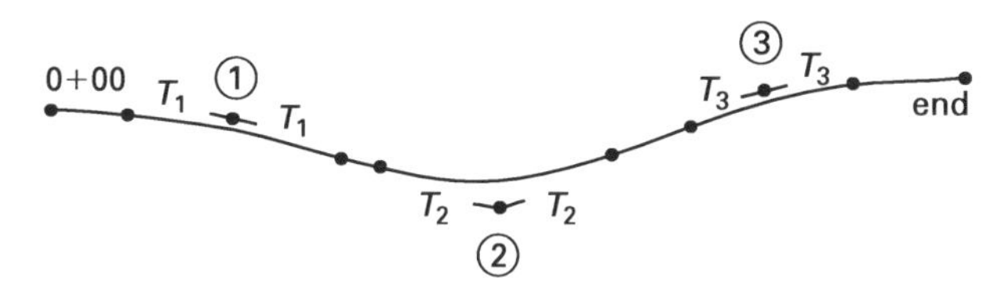

$$\text{distance } \text{PI}_2 - \text{PI}_1 = 3112.48 - 1224.31 = 1888.17$$

$$\text{distance } \text{PI}_3 - \text{PI}_2 = 5190.13 - 3112.48 = 2077.65$$

$$\text{distance end} - \text{PI}_3 = 6556.03 - 5190.13 = 1365.90$$

$$\text{PC}_1 = \text{PI}_1 - T_1 = 12{+}24.31 - 724.05 = 5{+}00.26$$

$$\begin{aligned}\text{PT}_1 &= \text{PC}_1 + L_1 = 5{+}00.26 + 1431.11 \\ &= 19{+}31.37\end{aligned}$$

$$\begin{aligned}\text{PC}_2 &= \text{PT}_1 + \text{PI}_2 - \text{PI}_1 - T_1 - T_2 \\ &= 19{+}31.37 + 1888.17 - 724.05 - 855.36 \\ &= 22{+}40.13\end{aligned}$$

$$\begin{aligned}\text{PT}_2 &= \text{PC}_2 + L_2 = 22{+}40.13 + 1637.32 \\ &= 38{+}77.46\end{aligned}$$

$$\begin{aligned}\text{PC}_3 &= \text{PT}_2 + \text{PI}_3 - \text{PI}_2 - T_2 - T_3 \\ &= 38{+}77.46 + 2077.65 - 855.36 - 562.05 \\ &= 45{+}37.70\end{aligned}$$

$$\begin{aligned}\text{PT}_3 &= \text{PC}_3 + L_3 = 45{+}37.70 + 1110.00 \\ &= 56{+}47.70\end{aligned}$$

$$\begin{aligned}\text{end} &= \text{PT}_3 + \text{end} - \text{PI}_3 - T_3 \\ &= 56{+}47.70 + 1365.90 - 562.05 = 64{+}51.55\end{aligned}$$

The stations can be computed by successive additions.

$$\begin{aligned}&1224.31 - 724.05 + 1431.11 + 3112.48 - 1224.31 \\ &\quad - 724.05 - 855.36 + 1637.33 + 5190.13 - 3112.48 \\ &\quad - 855.36 - 562.05 + 1110.00 + 6556.03 - 5190.13 \\ &\quad - 562.05 = 6451.11 \text{ (end)}\end{aligned}$$

The data can now be summarized.

PI	original station	T	L	PC station	PT station
1	12+24.31	724.05	1431.11	5+00.26	19+31.37
	1888.17				
2	31+12.48	855.36	1637.33	22+40.13	38+77.46
	2077.65				
3	51+90.13	562.05	1110.00	45+37.70	56+47.70
	1365.90				
4	65+56.03				end = 64+51.55

19. LOCATING CURVE WHEN PI IS INACCESSIBLE

It is often necessary to find the deflection angle Δ and locate the PC and PT of a curve when the PI is inaccessible as shown in Fig. 18.7.

Figure 18.7 *Inaccessible PI*

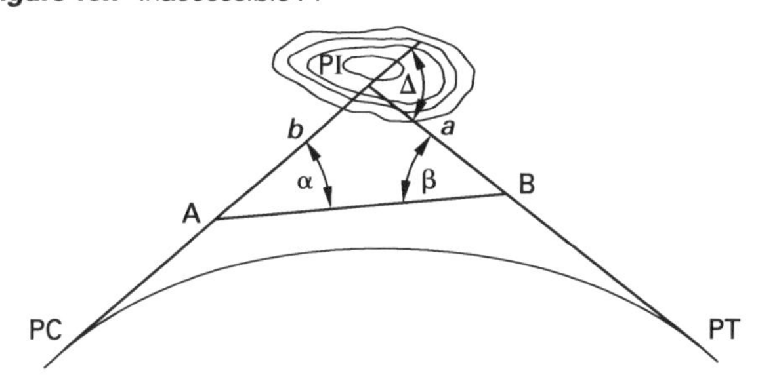

To solve the problem, a *point on tangent* (POT) is established on the back tangent at any point A that is visible from a point B on the forward tangent. The station number of point A is established by chaining, and the angles α and β (the sum of which equals Δ) and the distance AB are measured.

Distances a and b in triangle PI-A-B are computed from the law of sines. The station number of the PI is found by adding distance b to the station number of point A. The tangent distance T is computed, and the PC and PT are then located by measuring from points A and B.

Example 18.6

Two tangents of a highway location intersect in a lake, which makes the PI inaccessible. A POT has been established at sta 23+45.67 and designated as point A. Locate a 4° curve on the ground.

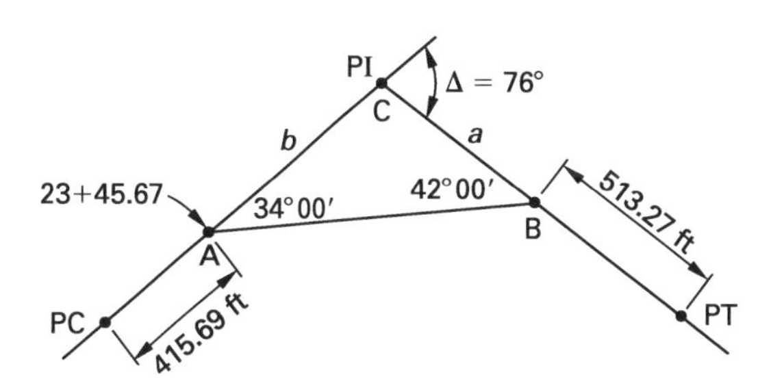

Plane Survey Calculations

Solution

step 1: Point B is located on the forward tangent so that it is visible from point A on the back tangent.

step 2: Angle A is measured and found to be 34°00′.

step 3: Angle B is measured and found to be 42°00′.

step 4: Distance AB is measured and found to be 1020.00 ft.

step 5: The deflection angle Δ is the sum of angles A and B, which equals 76°00′.

step 6: From the law of sines, in triangle PI-A-B,

$$a = \frac{(1020.00 \text{ ft})(\sin 34°)}{\sin 104°} = 587.84 \text{ ft}$$

$$b = \frac{(1020.00 \text{ ft})(\sin 42°)}{\sin 104°} = 703.41 \text{ ft}$$

step 7: Using $\Delta = 76°00'$ and $D = 4°$, $T = 1119.10$ ft.

step 8: Adding distance b to the station of point A, the PI is at sta 30+49.08. The PC is at sta 19+29.98.

step 9: The PC is located by measuring 415.69 ft from point A.

step 10: The PT is located by measuring 531.26 ft from point B.

20. SHIFTING FORWARD TANGENT

Route locations often require changes in curve locations. One such case involves shifting the forward tangent to a new location parallel to the original tangent and keeping the back tangent in its original location. This produces a change in both the PC and PT stations.

Example 18.7

The forward tangent of the highway curve shown is to be shifted outward so that it will be parallel to and 100 ft from the original tangent. Curve data for the original curve are shown in the figure. The degree of curve is to remain unchanged. Compute the PC and PT stations for the new curve.

Solution

$$\text{PI}_1 - \text{PI}_2 = \frac{100 \text{ ft}}{\sin 60°} = 115.47 \text{ ft}$$

$$\text{PI}_2 = 28{+}97.00 + 115.47 = 30{+}12.47$$

$$\text{PC}_2 = 30{+}12.47 - 551.33 = 24{+}61.14$$

$$\text{PT}_2 = 24{+}61.14 + 1000.00 = 34{+}61.14$$

Illustration for Example 18.7

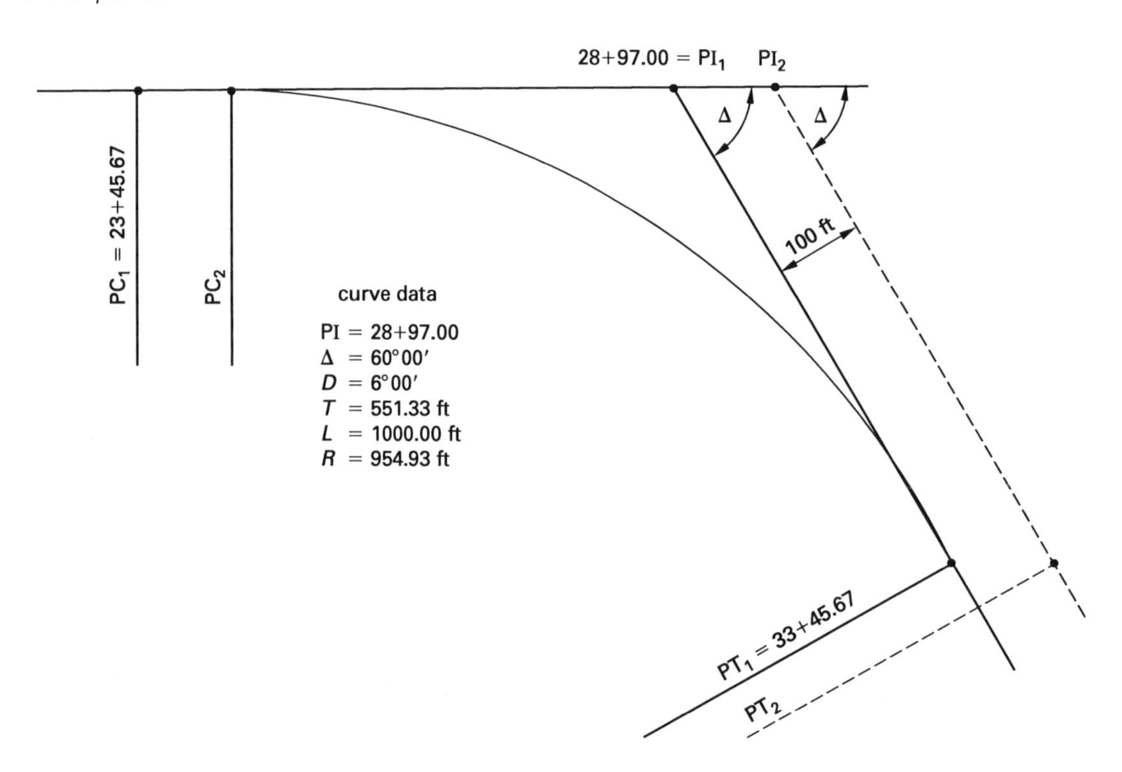

21. EASEMENT CURVES

Where simple curves are used, the tangent changes to a curved line at the PC. This means that a vehicle on a tangent arriving at the PC changes direction to a curved path instantaneously. At high speed this is impossible. What actually happens in an automobile is that the driver adjusts the steering wheel to make a gradual transition from a straight line to a curved line.

For railroad cars traveling at high speed, the problem is more acute than for automobiles. The rigidity and length of a railroad car cause a sharp thrust on the rails by the wheel flanges. To alleviate this situation, curves that provide a gradual transition from tangent to circular curve (and back again) are inserted between the tangent and the circular curve. These transition curves are known as *easement curves*. These curves also provide a place to increase superelevation from zero to the maximum required for the circular curve. The spiral is a curve that fulfills the requirements for the transition.

22. SPIRALS

The simple *spiral* has certain features that make it useful as an easement curve.

- The degree of curve of the spiral increases from zero at the beginning to the degree of curve of the circular arc where they meet. Likewise, the radius decreases from infinity at the beginning to the radius of the circular arc.
- Because the degree of curve changes uniformly, the central angle of the spiral equals the length of the spiral in stations times the average degree of curve.

$$\theta_s = \left(\frac{L_s}{100}\right)\left(\frac{D}{2}\right) = \frac{L_s D_c}{200} \qquad 18.11$$

- Spiral central angles are directly proportional to the squares of the lengths from the beginning of the spiral, as are the deflection angles.

23. LENGTH OF SPIRAL

The spiral provides a transition for superelevation. The length required to attain maximum superelevation is a function of the speed of the vehicle. Experiments have provided a formula for the length of the spiral (where L_s= length of spiral (in feet), v = design speed (in miles per hour), and R = radius of circular arc (in feet)).

$$L_s = \frac{1.6\text{v}^3}{R_c} \qquad 18.12$$

24. COMPUTATIONS AND PROCEDURE FOR STAKING

Spirals can be computed by using spiral tables. Symbols used on spirals should be memorized to facilitate the use of such tables in other texts. These symbols are illustrated in Fig. 18.8.

Figure 18.8 shows that the simple curve has been shifted inward in order to insert the easement curve, while the radius of the simple curve has been maintained.

Example 18.8

A circular curve with spiral transitions is to be computed and staked. Curve data are as follows.

$$\text{PI} = \text{sta } 72{+}58.00$$
$$\Delta = 42°00'$$
$$D_c = 5°$$
$$\text{v} = 60 \text{ mph}$$

Solution

(This solution requires the use of spiral curve tables.)

$$R_c = 1145.92 \text{ ft}$$

$$L_s = \frac{1.6\text{v}^3}{R_c} = \frac{(1.6)\left(60\ \frac{\text{mi}}{\text{hr}}\right)^3}{1145.92 \text{ ft}} = 301 \text{ ft (300 ft)}$$

$$\theta_s = \frac{L_s D_c}{200} = \frac{(300 \text{ ft})(5°)}{200} = 7°30'$$

$$\Delta_c = \Delta - 2\theta_s = 42° - (2)(7°30') = 27°00'$$

$$L_c = (100 \text{ ft})\left(\frac{\Delta_c}{D_c}\right) = (100 \text{ ft})\left(\frac{27°}{5°}\right) = 540 \text{ ft}$$

$$p = 3.27 \text{ ft};\ k = 149.91 \text{ ft}$$

$$T_s = (R_c + p)\tan\frac{\Delta}{2} + k = (1145.92 \text{ ft} + 3.27 \text{ ft})(0.38386) + 149.91 \text{ ft} = 591.04 \text{ ft}$$

The transit stations are as follows.

PI	=	72+58.00
T_s	=	−5+91.04
TS	=	66+66.96
L_s	=	+3+00.00
SC	=	69+66.96
L_c	=	+5+40.00
SC	=	75+06.96
L_s	=	+3+00.00
ST	=	78+06.06
LT	=	200.18 ft
ST	=	100.16 ft
LC	=	299.77 ft

Plane Survey Calculations

Figure 18.8 *Spiral Curve Nomenclature*

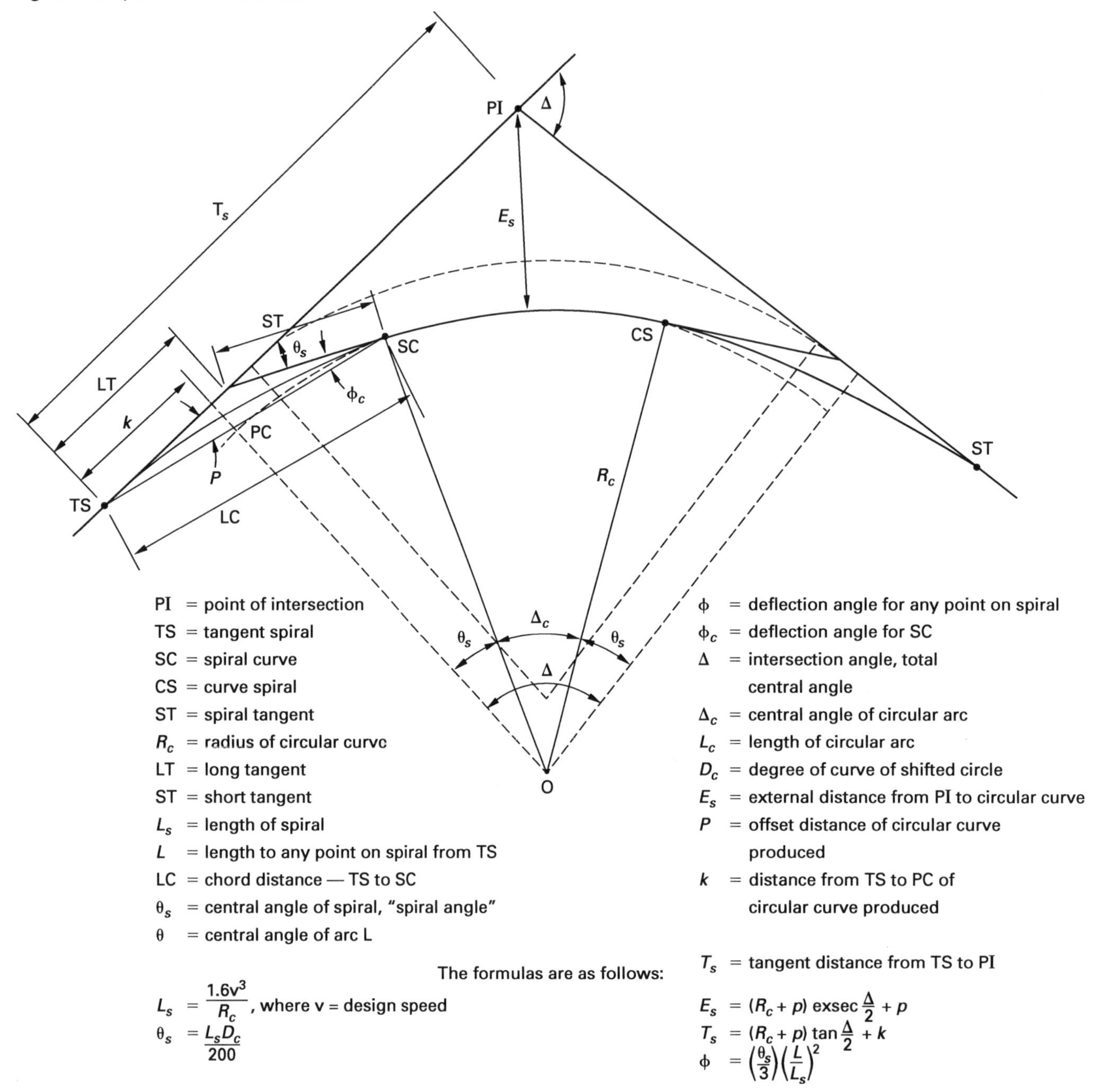

For the deflection angles for the spiral,

$$\phi = \left(\frac{\theta_s}{3}\right)\left(\frac{L}{L_s}\right)^2 = \left(\frac{1^\circ}{3}\right)\left(\frac{7.5L^2}{(300 \text{ ft})^2}\right) = 0.0000278L^2$$

point	station	L	L^2	ϕ		ϕ
TS	66+66.96	0.00	0	0°	=	0°
	67+00	33.04	1092	0.0303°	=	0°02′
	67+50	83.04	6896	0.1915°	=	0°11.5′
	68+00	133.04	17,700	0.4916°	=	0°29.5′
	68+50	183.04	33,503	0.9306°	=	0°56′
	69+00	233.04	54,307	1.5082°	=	1°30.5′
	69+50	283.04	80,111	2.2253°	=	2°13.5′
SC	69+66.96	300.00	90,000	2.5000°	=	2°30′

The field procedure is:

step 1: Locate the TS by measuring from the PI.

step 2: Locate the intersection of the LT and ST (200.18 ft from TS).

step 3: Set up at the intersection of LT and ST; sight on PI. Locate the SC by turning 7°30′ (θ_s) and measuring 100.16 (ST).

step 4: Locate the CS as in step 2 and step 3.

step 5: Set up on TS and sight on PI. Locate points on the spiral using deflection angles.

step 6: Locate points on the circular curve: (a) Set up on the SC. (b) Sight on the intersection of LT and ST with the telescope inverted and 0°00′ on the *A* vernier. (c) Set points on circular curve in normal manner.

step 7: Set up on the ST and locate spiral as in step 5.

25. STREET CURVES

Because *street curves* are usually short in radius, field procedures in staking them may differ from those used for highway curves. All formulas used for highway curves are valid, but the choice of formulas may vary. Street curves are computed by using the arc definition, just like highway curves, but stakes are usually set at 25 ft or 50 ft stations. The difference in the arc length and the chord length is much more pronounced on curves of short radius. Highway curves are usually designated by their degree of curve. Street curves are usually designated by a round-number radius, and the degree of curve concept is not used.

From Eq. 18.4, the long chord for a highway curve is equal to $2R\sin(\Delta/2)$, Δ being the central angle. Accordingly, $C = 2R\sin(D/2)$ for an arc of 100 ft. For street curves, the formula (where δ is the central angle for any arc) is expressed as

$$C = 2R\sin\frac{\delta}{2} \qquad 18.13$$

Some street curves are divided into 3, 4, or more equal arcs, but usually they are staked on the half- or quarter-stations. Chord lengths from quarter-station to quarter-station (or half-station to half-station) are usually measured, as is done on highway curves. For short curves, chords from the PC to each station can also be measured.

26. STREET CURVE COMPUTATIONS

The degree of curve concept is not used. Therefore, the deflection angle will be expressed in terms of the central angle, δ. This is illustrated in Ex. 18.9.

Example 18.9

Computations are to be made for a street curve with a deflection angle Δ of 50°00′ and a centerline radius of 120 ft. The PI is at sta 8+72.43.

Solution

$$\begin{aligned} T &= R\tan\frac{\Delta}{2} = (120\text{ ft})\left(\tan\frac{50^\circ}{2}\right) \\ &= 55.96\text{ ft} \end{aligned}$$

$$\begin{aligned} L &= \frac{2\pi R\Delta}{360^\circ} = \frac{(2)\pi(120\text{ ft})(50^\circ)}{360^\circ} \\ &= 104.72\text{ ft} \end{aligned}$$

$$\begin{aligned} \text{PI} &= 8{+}72.43 \\ T &= \phantom{+8{+}}-55.96 \\ \hline \text{PC} &= 8{+}16.47 \\ L &= +1{+}04.72 \\ \hline \text{PT} &= 9{+}21.19 \end{aligned}$$

The deflection angles are

$$\text{PC}8{+}16.47 = 0^\circ 00'$$

$$\begin{aligned} \delta_{8+25} &= \left(\frac{25.00 - 16.47}{104.72}\right)\left(\frac{50^\circ}{2}\right) \\ &= \left(\frac{8.53}{104.72}\right)(25^\circ) \\ &= 2.0364^\circ \\ &= 2^\circ 02' \end{aligned}$$

$$\begin{aligned} \delta_{8+50} &= \left(\frac{8.53 + 25.00}{104.72}\right)\left(\frac{50^\circ}{2}\right) \\ &= \left(\frac{33.53}{104.72}\right)(25^\circ) \\ &= 8.0047^\circ \\ &= 8^\circ 00' \end{aligned}$$

$$\begin{aligned} \delta_{8+75} &= \left(\frac{8.53 + 50.00}{104.72}\right)\left(\frac{50^\circ}{2}\right) \\ &= \left(\frac{58.53}{104.72}\right)(25^\circ) \\ &= 13.9730^\circ \\ &= 13^\circ 58' \end{aligned}$$

Illustration for Example 18.9

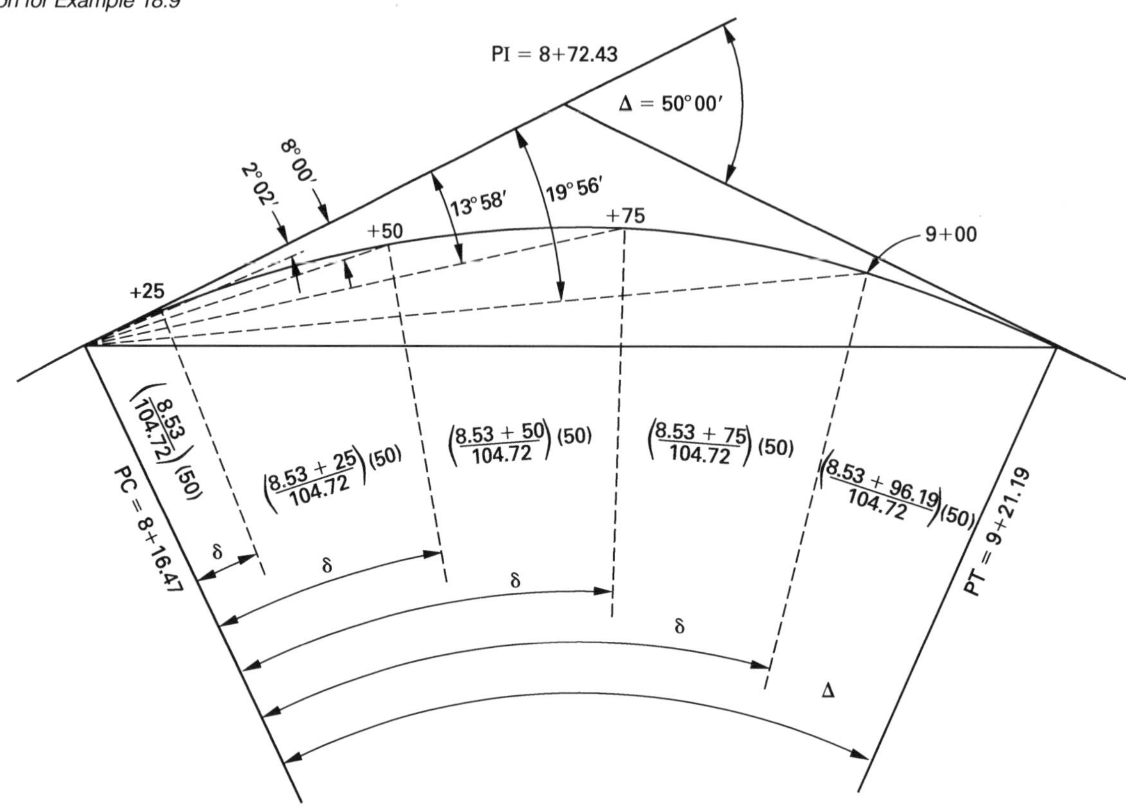

$$\delta_{9+00} = \left(\frac{8.53 + 75.00}{104.72}\right)\left(\frac{50°}{2}\right)$$
$$= \left(\frac{83.53}{104.72}\right)(25°)$$
$$= 19.9413°$$
$$= 19°56'$$

$$\text{PT } 9{+}21.19 = \left(\frac{8.53{+}96.19}{104.72}\right)\left(\frac{50°}{2}\right)$$
$$= \left(\frac{104.72}{104.72}\right)(25°)$$
$$= 25.0000°$$
$$= 25°00'$$

The chord lengths $\left(C = 2R\sin(\delta/2)\right)$ are

8+16.47 to 8+25.00:

$$C = (240 \text{ ft})(\sin 2.0364°) = 8.53 \text{ ft}$$

8+25.00 to 8+50.00:

$$C = (240 \text{ ft})\sin(8.0047° - 2.0364°) = 24.95 \text{ ft}$$

9+00.00 to 9+21.19:

$$C = (240 \text{ ft})\sin(25.0000° - 19.9413°) = 21.16 \text{ ft}$$

The field notes are as follows.

point	station	$\delta/2$	deflection angle	C	curve data
PT	9+21.19		25°00′	21.16	
	9+00	19.9412°	19°56′	24.95	$\Delta = 50°00'$
	8+75	13.9730°	13°58′	24.95	$R = 120$ ft
	8+50	8.0047°	8°00′	24.95	$T = 55.96$ ft
	8+25	2.0364°	2°02′	8.53	$L = 104.72$ ft
PC	8+16.7	0.0000°	0°00′		

The length of the curve is only 104.72 ft. Therefore, the curve should be staked by measuring all chords from the PC. These chord lengths are computed as follows.

PC to 8+25.00: $C = (240 \text{ ft})(\sin 2.0364°) = 8.53$ ft
PC to 8+50.00: $C = (240 \text{ ft})(\sin 8.0047°) = 33.42$ ft
PC to 8+75.00: $C = (240 \text{ ft})(\sin 13.9730°) = 57.95$ ft
PC to 9+00.00: $C = (240 \text{ ft})(\sin 19.9412°) = 81.85$ ft
PC to 9+21.19: $C = (240 \text{ ft})(\sin 25.0000°) = 101.43$ ft

27. PARALLEL CIRCULAR ARCS

The design radius for a curve is usually to the centerline. However, it is often necessary to locate a parallel curve such as a right-of-way line, the edge of pavement, or a curb line. Because the central angle is the same for *parallel arcs*, deflection angles are the same.

The chord lengths are a function of the radius of the arc, and they will be different for arcs of different radius. Length of the curve is also a function of the radius. The PC stations for all parallel arcs will be the same. Arc lengths for inside and outside curves will be different, but the PT stations will be the same.

Example 18.10

Construction stakes are to be set for the curve in the preceding example. The street width is 34 ft. Stakes are to be set on 3 ft offsets, inside and outside of the edge of pavement.

Solution

Deflection angles are given in the field notes. Chord lengths are computed from the formula $C = 2R\sin(\delta/2)$ and are also shown in the field notes.

point	station	deflection angle	$R = 100$ ft C inside	$R = 140$ ft C outside	curve data
PT	9+21.19	25°00′			
			17.64	24.69	$\Delta = 50°00'$
	9+00	19°56′			
			20.80	29.11	$R = 120$ ft
	8+75	13°58′			
			20.80	29.11	$T = 55.96$ ft
	8+50	8°00′			
			20.80	29.11	$L = 104.72$ ft
	8+25	2°02′			
			7.11	9.95	
PC	8+16.47	0°00′			

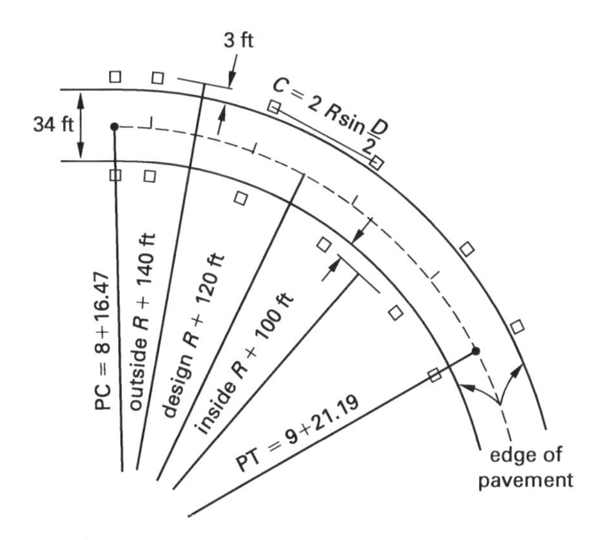

28. CURB RETURNS AT STREET INTERSECTIONS

Curb returns are the arcs made by the curbs at street intersections. The radius of the arc is selected by the designer with consideration given to the speed and volume of traffic. A radius of 30 ft to the back of the curb is common. Streets that intersect at right angles have curb returns of one quarter circle.

Figure 18.9 *Curb Returns*

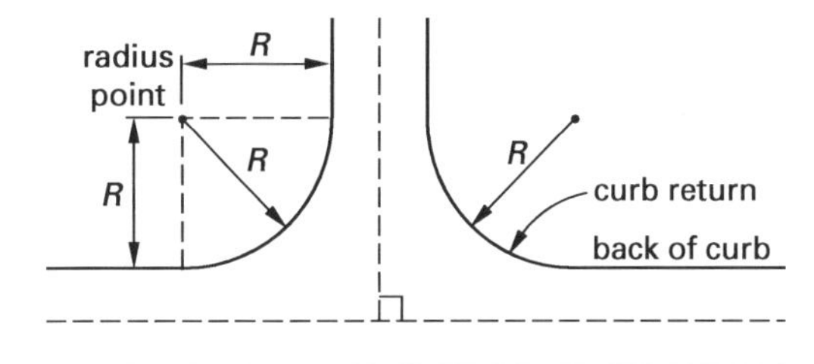

The arcs can be swung from a *radius point*, the center of the circle. The radius point can be located by finding the intersection of two lines, each of which is parallel to one of the curb lines and at a distance equal to the radius. One stake at the radius is sufficient for the curb return.

Curb returns for streets that do not intersect at right angles are shown in Ex. 18.11. Computations for the location of PCs, PTs, and radius points are shown in Ex. 18.11.

Example 18.11

Elm Street and 23rd Street intersect as shown. The pavement width for each street is 36 ft. The radius to the edge of the pavement is 30 ft. Compute the location of the radius points.

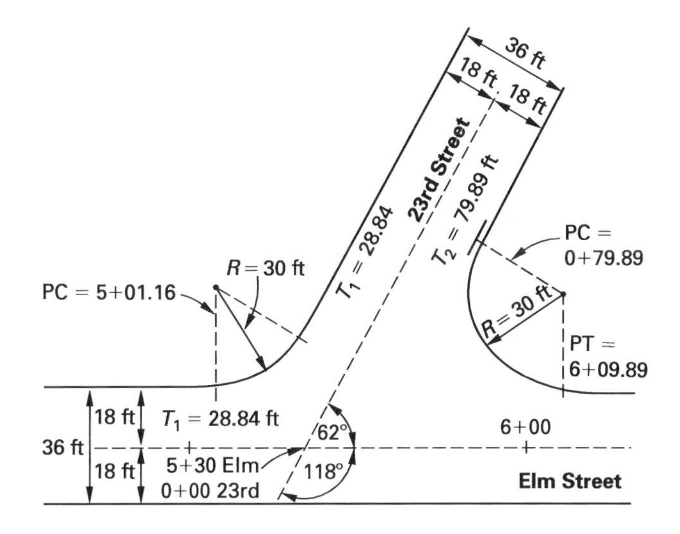

Solution

The PC stations along Elm Street are

$$\Delta_1 = 62°00'$$

$$T_1 = R\tan\frac{\Delta}{2} = (48)\left(\tan\frac{62°}{2}\right) = 28.84 \text{ ft}$$

PI	=	5+30.00
T_1	=	− 28.84
PC	=	5+01.16

$$\Delta_2 = 118°00'$$

$$T_2 = R\tan\frac{\Delta}{2} = (48)\left(\tan\frac{118°}{2}\right)$$

$$= 79.89 \text{ ft}$$

PI	=	5+30.00
T_2	=	+ 79.89
PC	=	6+09.89

The PT stations along 23rd Street are

PI	=	0+00.00
T_1	=	+ 28.84
PT	=	0+28.84
PI	=	0+00.00
T_2	=	+ 79.89
PT	=	0+79.89

29. COMPOUND CURVES

A *compound curve* consists of two or more simple curves with different radii joined together at a common tangent point. Their centers are on the same side of the curve.

Compound curves are not generally used for highways except in mountainous country, because an abrupt change in degree of curve causes a serious hazard even at moderate speeds. They are sometimes used for curvilinear streets in residential subdivisions, however.

In Fig. 18.10, the subscript 1 is used for the curve of longer radius, and the subscript 2 is used for the curve of shorter radius. The point of common tangency is called the *point of common curvature*, PCC. The short tangents for the two curves are designated as t_1 and t_2.

Figure 18.10 *Compound Curve*

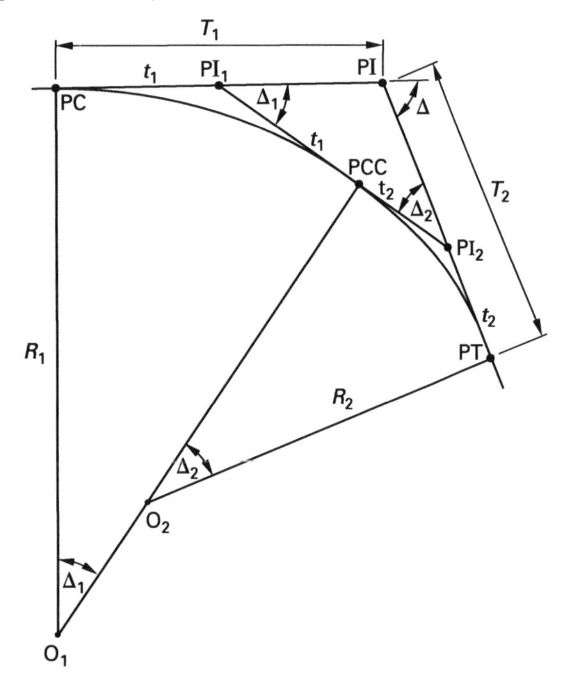

There are seven major parameters of a compound curve: Δ, Δ_1, Δ_2, R_1, R_2, T_1, and T_2. Four of these must be known before computations can be made. Usually, Δ is measured. R_1, R_2, and either Δ_1 or Δ_2 are given. If Δ, Δ_1, R_1, and R_2 are given, Eqs. 18.14 through 18.17 can be used.

$$\Delta = \Delta_1 + \Delta_2 \quad \textit{18.14}$$

$$\Delta_2 = \Delta - \Delta_1 \quad \textit{18.15}$$

$$t_1 = R_1 \tan\frac{\Delta_1}{2} \quad \textit{18.16}$$

$$t_2 = R_2 \tan\frac{\Delta_2}{2} \quad \textit{18.17}$$

The triangle PI-PI$_1$-PI$_2$ can be solved by the law of sines. The sine of the angle at the PI equals the sine of Δ because they are related angles.

$$\text{PI} - \text{PI}_1 = \frac{\sin\Delta_2(t_1 + t_2)}{\sin\Delta} \quad \textit{18.18}$$

$$\text{PI} - \text{PI}_2 = \frac{\sin\Delta_1(t_1 + t_2)}{\sin\Delta} \quad \textit{18.19}$$

$$T_1 = t_1 + \text{PI} - \text{PI}_1 \quad \textit{18.20}$$

$$T_2 = t_2 + \text{PI} - \text{PI}_2 \quad \textit{18.21}$$

To stake out the curve, the PC and PT are located as is done for a simple curve. The PCC is located by establishing the common tangent from either point PI$_1$ or PI$_2$. Deflection angles for the two curves are computed separately. The first curve is staked from the PC, and the second curve is staked from the PCC using the common tangent for orientation with the vernier set on 0°00′.

Example 18.12

PC, PCC, and PT stations, deflection angles, and chord lengths are to be computed from the following information.

$$\text{PI} = \text{sta } 15{+}56.32$$

$$\Delta = 68°00'$$

$$\Delta_1 = 35°00'$$

$$R_1 = 600 \text{ ft}$$

$$R_2 = 400 \text{ ft}$$

Solution

$$\Delta_2 = \Delta - \Delta_1 = 68°00' - 35°00'$$

$$= 33°00'$$

$$t_1 = R_1 \tan\frac{\Delta_1}{2} = (600 \text{ ft})(\tan 17°30')$$

$$= 189.18 \text{ ft}$$

$$t_2 = R_2 \tan\frac{\Delta_2}{2} = (400 \text{ ft})(\tan 16°30')$$

$$= 118.49 \text{ ft}$$

$$\text{PI} - \text{PI}_1 = \frac{\sin \Delta_2(t_1 + t_2)}{\sin \Delta} = \frac{(\sin 33^\circ)\begin{pmatrix} 189.18 \text{ ft} \\ + 118.49 \text{ ft} \end{pmatrix}}{\sin 68^\circ}$$
$$= 180.73 \text{ ft}$$

$$\text{PI} - \text{PI}_2 = \frac{\sin \Delta_1(t_1 + t_2)}{\sin \Delta} = \frac{(\sin 35^\circ)\begin{pmatrix} 189.18 \text{ ft} \\ + 118.49 \text{ ft} \end{pmatrix}}{\sin 68^\circ}$$
$$= 190.33 \text{ ft}$$

$$T_1 = t_1 + \text{PI} - \text{PI}_1 = 189.18 \text{ ft} + 180.73 \text{ ft}$$
$$= 369.91 \text{ ft}$$

$$T_2 = t_2 + \text{PI} - \text{PI}_2 = 118.49 \text{ ft} + 190.33 \text{ ft}$$
$$= 308.82 \text{ ft}$$

$$L_1 = \frac{2\pi R_1 \Delta_1}{360^\circ} = \frac{(2)\pi(600 \text{ ft})(35^\circ)}{360^\circ}$$
$$= 366.52 \text{ ft}$$

$$L_2 = \frac{2\pi R_2 \Delta_2}{360^\circ} = \frac{(2)\pi(400 \text{ ft})(33^\circ)}{360^\circ}$$
$$= 230.38 \text{ ft}$$

PI	=	15+56.32
T_1	=	−3+69.91
PC	=	11+86.41
L_1	=	+3+66.52
PCC	=	15+52.93
L_2	=	+2+30.38
PT	=	17+83.31

The deflection angles are as follows.

point	station	deflection angles
PC	11+86.41	
	12+00	$\left(\frac{13.59}{366.52}\right)\left(\frac{35}{2}\right) = 0.6489^\circ = 0^\circ 39'$
	13+00	$\left(\frac{113.59}{366.59}\right)\left(\frac{35}{2}\right) = 5.4235^\circ = 5^\circ 25'$
	14+00	$\left(\frac{213.59}{366.52}\right)\left(\frac{35}{2}\right) = 10.1981^\circ = 10^\circ 12'$
	15+00	$\left(\frac{313.59}{366.52}\right)\left(\frac{35}{2}\right) - 14.9728^\circ = 14^\circ 59'$
PCC	15+52.93	$\left(\frac{366.52}{366.52}\right)\left(\frac{35}{2}\right) = 17.5000^\circ = 17^\circ 30'$
	16+00	$\left(\frac{47.07}{230.38}\right)\left(\frac{33}{2}\right) = 3.3712^\circ = 3^\circ 22'$
	17+00	$\left(\frac{147.07}{230.38}\right)\left(\frac{33}{2}\right) = 10.5332^\circ = 10^\circ 32'$
PT	17+83.31	$\left(\frac{230.38}{230.38}\right)\left(\frac{33}{2}\right) = 16.5000^\circ = 16^\circ 30'$

The chord lengths are

$$C = (1200 \text{ ft})(\sin 0.6489^\circ) = 13.59 \text{ ft}$$
$$C = (1200 \text{ ft})(\sin 4.7746^\circ) = 99.88 \text{ ft}$$
$$C = (1200 \text{ ft})(\sin 2.5272^\circ) = 52.91 \text{ ft}$$
$$C = (800 \text{ ft})(\sin 3.3712^\circ) = 47.04 \text{ ft}$$
$$C = (800 \text{ ft})(\sin 7.1620^\circ) = 99.74 \text{ ft}$$
$$C = (800 \text{ ft})(\sin 5.9668^\circ) = 83.16 \text{ ft}$$

The field notes showing the results of these computations are shown as follows.

Field Notes for Elm Street

point	station	deflection angle	chord	calculated bearing	curve data
PT	17+83.31	16°30′			
			83.16′		
	17+00.00	10°32′			
			99.74′		
	16+00.00	3°22.3′			$\Delta = 68°00'$
PI	15+56.32		47.04′		$R_1 = 600$ ft
PCC	15+52.93	17°30′			$\Delta_1 = 35°00'$
			52.91′		$R_2 = 400$ ft
	15+00.00	14°58.7′			$\Delta_2 = 33°00'$
			99.88′		$T_1 = 369.91$ ft
	14+00.00	10°12.1′			$T_2 = 308.81$ ft
			99.88′		$L_1 = 366.52$ ft
	13+00.00	5°25.5′			$L_2 = 230.38$ ft
			99.88′		
	12+00.00	0°38.9′			
			13.59′		
PC	11+86.41	0°00′			

PRACTICE PROBLEMS

1. Provide the missing word or words in each sentence.

(a) Highway curves are most often _____ arcs known as simple curves.

(b) An inscribed angle is an angle that has its vertex on a _____ and that has _____ for its sides.

(c) An inscribed angle is measured by _____ its intercepted arc.

(d) An angle formed by a tangent and a chord is measured by _____ its intercepted arc.

(e) The radius of a circle is _____ to a tangent at the point of tangency.

(f) A perpendicular bisector of a chord passes through the _____ of the circle.

(g) By the arc definition, degree of curve, D, is the central angle that subtends a 100 ft _____ .

Plane Survey Calculations

(h) By the chord definition, degree of curve, D, is the central angle that subtends a 100 ft ____ .

(i) By the arc definition, the radius R of a 1° curve is ____ ft.

(j) The deflection angle for a full station for a 1° curve is ____ .

2. Place all symbols pertinent to a circular curve on the following figure.

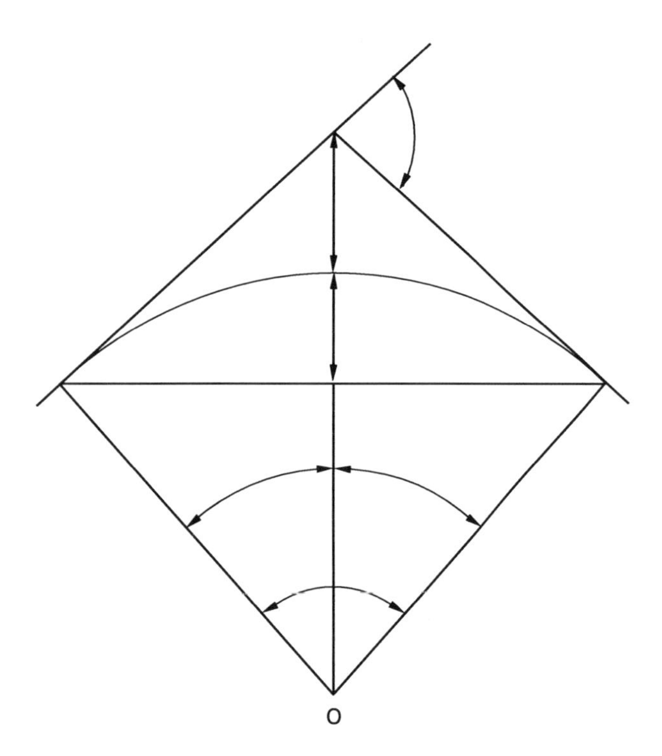

3. Using the information given, compute the PC and PT stations and the deflection angles for each full station of the simple highway curve. Round off T to the nearest foot.

$$\text{PI} = \text{sta } 25{+}01$$
$$\Delta = 10°$$
$$D = 1°$$

4. Using the information given, compute the PC and PT stations and the deflection angles for each full station of the simple highway curve. Express length to two decimal places.

$$\text{PI} = \text{sta } 45{+}11.75$$
$$\Delta = 30°$$
$$D = 3°$$

5. Using the following information, prepare field notes to be used in staking the centerline of a simple horizontal curve for a highway.

$$\text{PI} = 29 + 62.78$$
$$\Delta = 40°21'\text{L}$$
$$D = 5°15'$$
$$\text{back tangent bearing} = \text{N}\,56°12'\,\text{W}$$

6. A preliminary highway location has been made by locating tangents. Deflection angles have been measured at each PI, and the station number of each PI has been established by measuring along the tangents. Circular curves have not been located, but the degree of curve has been established for each curve. The beginning point is at sta 0+00. Using these data, make necessary computations to establish stations for PCs and PTs of the curves and for the end of the line.

PI no.	original station	Δ	D
1	10+35.27	13°34′R	1°30′
2	36+15.44	15°18′L	2°30′
3	52+98.40	18°05′R	3°00′
end	61+32.77	end of line	

7. In locating a highway, the PI of two tangents falls in a lake and is inaccessible. Point A on the back tangent has been established at sta 26+52.61. Point B has been established on the forward tangent and is visible from point A. Angle A has been measured and found to be 23°13′; angle B has been measured and found to be 19°55′. The length of AB has been found to be 434.87 ft. Find the deflection angle Δ and the PC and PT stations for a 3° curve.

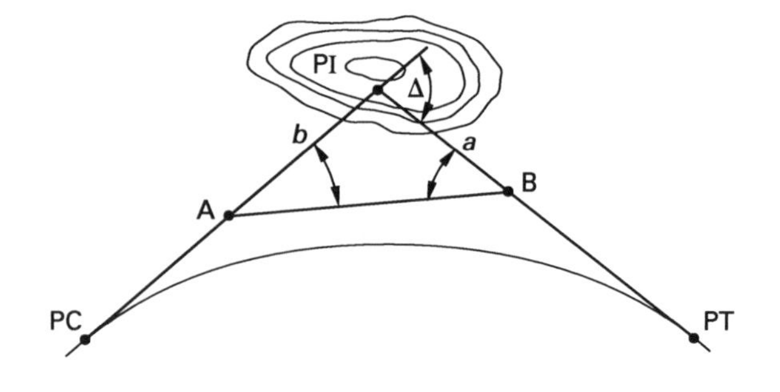

8. The forward tangent of the highway curve shown is to be shifted outward so that it will be parallel to and 50 ft from the original tangent. Data for the original curve are shown in the figure. The degree of curve is to be unchanged. Find the PC and PT stations for the new curve.

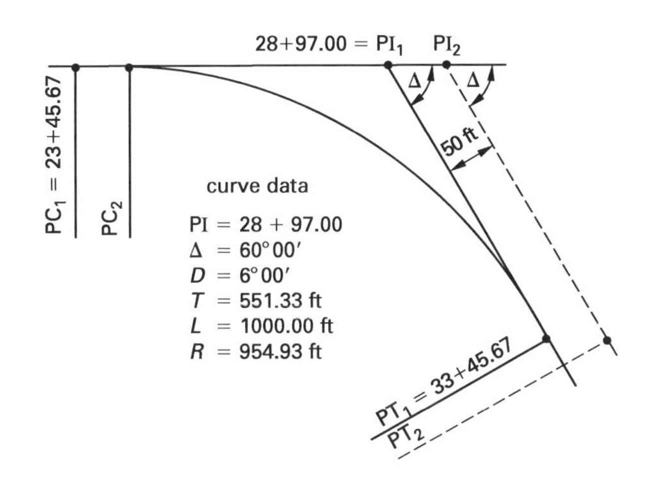

9. Find L for Δ and D indicated.

(a) $\Delta = 32°18'$

$D = 2°30'$

(b) $\Delta = 41°27'$

$D = 3°15'$

10. Find T for Δ and D indicated.

(a) $\Delta = 41°51'$

$D = 1°45'$

(b) $\Delta = 39°14'$

$D = 2°15'$

11. Find E for Δ and D indicated.

(a) $\Delta = 31°30'$

$D = 1°30'$

(b) $\Delta = 42°21'$

$D = 2°45'$

12. Find D for the nearest full degree for Δ indicated and T approximately as indicated.

(a) $\Delta = 32°56'$

$T = 600$ ft

(b) $\Delta = 40°10'$

$T = 1000$ ft

13. Use trigonometric equations to solve the following problems.

(a) Find R for $D = 2°$.

(b) Find D for $R = 1909.86$ ft.

(c) Find T for $\Delta = 34°44'$ and $R = 800$ ft.

(d) Find E for $\Delta = 37°20'$ and $R = 650$ ft.

(e) Find M for $\Delta = 42°51'$ and $R = 800$ ft.

(f) Find LC for $\Delta = 32°55'$ and $R = 850$ ft.

(g) Find the chord length for $D = 8°$, $R = 716.20$ ft, and arc $= 50$ ft.

14. Compute deflection angles and chord lengths for quarter stations (25 ft) for a street curve. Chords are to be measured from quarter-station to quarter-station.

PI = sta 8+78.22

$\Delta = 28°$

$R = 250$ ft

15. Prepare field notes to be used in staking the centerline of a horizontal street curve on the quarter-stations.

PI = sta 8+47.52

$\Delta = 36°00'$Right

$R = 400$ ft

16. Compute PC, PCC, and PT stations and deflection angles for full stations for the compound curve with the information given.

PI = sta 14+78.32

$\Delta = 68°00'$

$\Delta_1 = 36°00'$

$R_1 = 400$ ft

$R_2 = 300$ ft

17. Prepare field notes to be used in staking the centerline of the compound curve on full stations.

PI = sta 12+65.35

$\Delta = 70°00'$

$\Delta_1 = 36°00'$

$R_1 = 900$ ft

$R_2 = 600$ ft

18. Compute the area of the traverse to the nearest tenth of an acre.

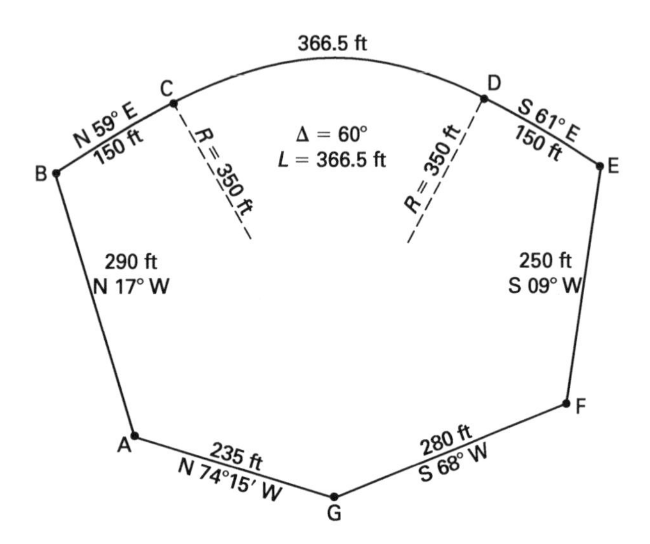

SOLUTIONS

1. (a) circular (b) circle chords
(c) one-half (d) one-half
(e) perpendicular (f) center
(g) arc (h) chord
(i) 5729.58 (j) 0°30′

2.

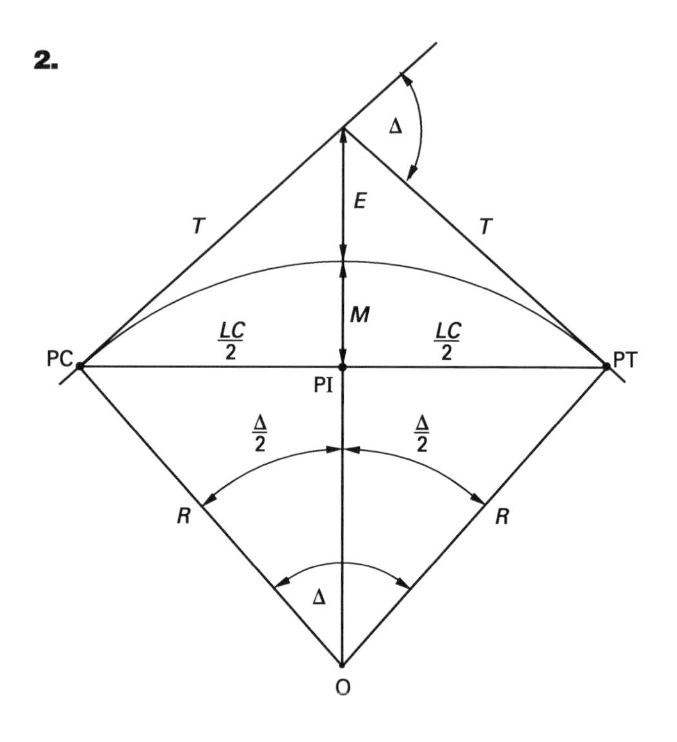

symbol	name	definition
PI	point of intersection	where two tangents intersect
Δ	deflection angle	central angle at PI; angle at center
D	degree of curve	central angle that subtends 100 ft arc
L	length of curve	distance from PC to PT along the arc
T	tangent distance	distance from PI to PI; from PI to PT
R	radius	
LC	length of long chord	
C	length of chord	chord length for angle *D*
PC	point of curvature	beginning of curve
PT	point of tangency	end of curve
M	middle ordinate	length from middle of LC to middle of curve
E	external distance	distance from PI to middle of curve

3.

$$L = (100\ \text{ft})\left(\frac{\Delta}{D}\right) = (100\ \text{ft})\left(\frac{10^\circ}{1^\circ}\right) = 1000\ \text{ft}$$

$$\boxed{T = 501\ \text{ft}}$$

$$R = 5729.58\ \text{ft}$$

The PC and PT stations are

$$\text{PI} = 25{+}01$$
$$T = 5{+}01$$
$$\boxed{\text{PC} = 20{+}00}$$
$$L = 10{+}00$$
$$\boxed{\text{PT} = 30{+}00}$$

The deflection angles are as follows.

point	station	deflection angle
PC	20+00	0°00′
	21+00	0°30′
	22+00	1°00′
	23+00	1°30′
	24+00	2°00′
	25+00	2°30′
	26+00	3°00′
	27+00	3°30′
	28+00	4°00′
	29+00	4°30′
PT	30+00	5°00′

4.

$$L = (100\ \text{ft})\left(\frac{\Delta}{D}\right) = (100\ \text{ft})\left(\frac{30^\circ}{3^\circ}\right) = 1000\ \text{ft}$$

$$T = 511.75\ \text{ft}$$

$$R = 1909.86\ \text{ft}$$

The PC and PT stations are

$$\text{PI} = 45{+}11.75$$
$$T = 5{+}11.75$$
$$\boxed{\text{PC} = 40{+}00}$$
$$L = 10{+}00$$
$$\boxed{\text{PT} = 50{+}00}$$

The deflection angles are as follows.

point	station	deflection angle
PC	40+00	0°00′
	41+00	1°30′
	42+00	3°00′
	43+00	4°30′
	44+00	6°00′
	45+00	7°30′
	46+00	9°00′
	47+00	10°30′
	48+00	12°00′
	49+00	13°30′
PT	50+00	15°00′

5.

point	station	deflection angle	chord	calculated bearing	curve data
PT	33+30.35	20°10.5′		S 83°27′ W	
	33+00	19°22.5′			
	32+00	16°45′			
					$\Delta = 40°21'$L
	31+00	14°07.5′			$D = 5°15'$
					$T = 401.00$ ft
	30+00	11°30′			$L = 768.57$ ft
	29+00	8°52.5′			$R = 1091.35$ ft
	28+00	6°15′			
	27+00	3°37.5′			
	26+00	1°00′			
PC	25+61.78	0°00′		N 56°12′ W	

6.

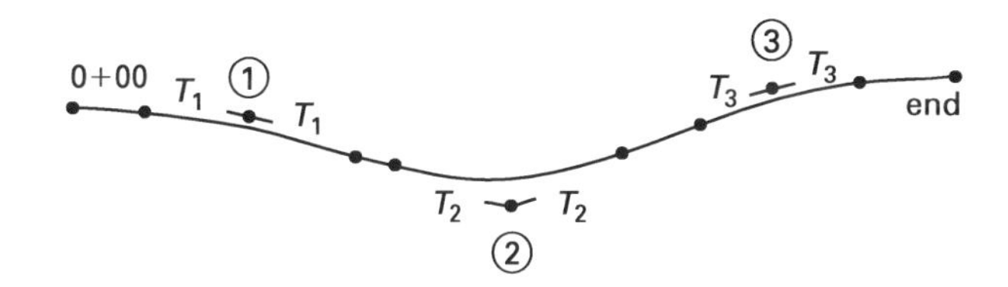

PI no.	original station	T	L	PC station	PT station	end
1	10+35.27	454.35	904.44	5+80.92	14+85.36	
2	36+15.44	307.83	612.00	33+03.35	39+15.35	
3	52+98.40	303.92	602.78	49+86.56	55+89.34	
end						61+19.79

7.

$$\Delta = 23^\circ 13' + 19^\circ 55' = \boxed{43^\circ 08'}$$

$$L = (100\ \text{ft})\left(\frac{\Delta}{D}\right) = (100\ \text{ft})\left(\frac{43^\circ 08'}{3^\circ}\right) = 1437.78\ \text{ft}$$

$$T = R\tan\frac{\Delta}{2}$$

$$= \left(\frac{5729.58}{3}\right)\left(\tan\frac{43°08'}{2}\right) = 754.88 \text{ ft}$$

$$a = \frac{(434.87 \text{ ft})(\sin 23°13')}{\sin(180° - 43°08')} = 250.74 \text{ ft}$$

$$b = \frac{(434.87 \text{ ft})(\sin 19°55')}{\sin(180° - 43°08')} = 216.67 \text{ ft}$$

$$\text{PC} = 26{+}52.61 + 216.67 - 754.88 = \boxed{21{+}14.40}$$

$$\text{PT} = 21{+}14.40 + 1437.78 = \boxed{35{+}52.18}$$

8. $\text{PI}_1 - \text{PI}_2 = \dfrac{50 \text{ ft}}{\sin 60°} = 57.74 \text{ ft}$

$$\text{PC} = 23{+}45.67 + 57.74 = \boxed{24{+}03.41}$$

$$\text{PT} = 24{+}03.41 + 1000.00 = \boxed{34{+}03.41}$$

9. (a) $L = (100 \text{ ft})\left(\dfrac{\Delta}{D}\right)$

$$= (100 \text{ ft})\left(\frac{32°18'}{2°30'}\right) = \boxed{1292.00 \text{ ft}}$$

(b) $L = (100 \text{ ft})\left(\dfrac{\Delta}{D}\right)$

$$= (100 \text{ ft})\left(\frac{41°27'}{3°15'}\right) = \boxed{1275.38 \text{ ft}}$$

10. (a) $T = R\tan\dfrac{\Delta}{2}$

$$= \left(\frac{5729.58}{1°45'}\right)\left(\tan\frac{41°51'}{2}\right)$$

$$= \boxed{1251.87 \text{ ft}}$$

(b) $T = R\tan\dfrac{\Delta}{2}$

$$= \left(\frac{5729.58}{2°15'}\right)\left(\tan\frac{39°14'}{2}\right)$$

$$= \boxed{907.60 \text{ ft}}$$

11. (a) $E = \dfrac{223.51 \text{ ft}}{1°30'} = \boxed{149.01 \text{ ft}}$

(b) $E = \dfrac{414.86 \text{ ft}}{2°45'} = \boxed{150.86 \text{ ft}}$

12. (a) $D = \left(\dfrac{5729.58}{600 \text{ ft}}\right)\left(\tan\dfrac{32°56'}{2}\right) = \boxed{3°}$

(b) $D = \left(\dfrac{5729.58}{1000 \text{ ft}}\right)\left(\tan\dfrac{40°10'}{2}\right) = \boxed{2°}$

13. (a) $R = \dfrac{5729.58}{2} = \boxed{2864.79 \text{ ft}}$

(b) $D = \dfrac{5729.58}{1909.86 \text{ ft}} = \boxed{3°}$

(c) $T = R\tan\dfrac{\Delta}{2} = (800 \text{ ft})\left(\tan\dfrac{34°44'}{2}\right)$

$$= \boxed{250.19 \text{ ft}}$$

(d) $E = (650 \text{ ft})\left(\dfrac{1}{\cos\dfrac{37°20'}{2}}\right) = \boxed{36.09 \text{ ft}}$

(e) $M = R\left(1 - \cos\dfrac{\Delta}{2}\right)$

$$= (800 \text{ ft})\left(1 - \cos\frac{42°51'}{2}\right)$$

$$= \boxed{55.28 \text{ ft}}$$

(f) $\text{LC} = (2)(850 \text{ ft})\left(\sin\dfrac{32°55'}{2}\right)$

$$= \boxed{481.64 \text{ ft}}$$

(g) $\text{chord} = (2)(716.20 \text{ ft})\left(\sin\dfrac{4°}{2}\right)$

$$= \boxed{49.99 \text{ ft}}$$

14. $T = R\tan\dfrac{\Delta}{2} = (250 \text{ ft})\left(\tan\dfrac{28°}{2}\right)$

$$= \boxed{62.33 \text{ ft}}$$

$$L = \frac{2\pi R\Delta}{360°}$$

$$= \frac{(2)\pi(250 \text{ ft})(28°)}{360°}$$

$$= \boxed{122.17 \text{ ft}}$$

The PC and PT stations are

$$\text{PI} = 8{+}78.22$$

$$T = 62.33 \text{ ft}$$

$$\text{PC} = 8{+}15.89$$

$$L = 1{+}22.17$$

$$\text{PT} = 9{+}38.06$$

The deflection angles are as follows.

point	station			deflection angle
PC	8+15.89		0°00′	
	8+25	$\left(\frac{9.11}{122.17}\right)\left(\frac{28}{2}\right)$	1.0440°	1°03′
	8+50	$\left(\frac{34.11}{122.17}\right)\left(\frac{28}{2}\right)$	3.9088°	3°55′
	8+75	$\left(\frac{59.11}{122.17}\right)\left(\frac{28}{2}\right)$	6.7737°	6°46′
	9+00	$\left(\frac{84.11}{122.17}\right)\left(\frac{28}{2}\right)$	9.6385°	9°38′
	9+25	$\left(\frac{109.11}{122.17}\right)\left(\frac{28}{2}\right)$	12.5034°	12°30′
	9+38.06	$\left(\frac{122.17}{122.17}\right)\left(\frac{28}{2}\right)$	14.0000°	14°00′

15.

Elm Street

point	station	deflection angle	chord	calculated bearing	curve data
PT	9+68.88	18°00′	18.88		
	9+50	16°39′			
			25.00		
	9+25	14°51′			
			25.00		
	9+00	13°04′			
			25.00		$\Delta = 36°00'$
	8+75	11°17′			$T = 129.97$ ft
			25.00		$L = 251.33$ ft
	8+50	9°29′			$R = 400$ ft
	8+25	7°42′			
			25.00		
	8+00	5°54′			
			25.00		
	7+75	4°07′			
			25.00		
	7+50	2°19′			
			25.00		
	7+25	0°32′			
PC	7+17.55	0°00′	7.45		

16.

$$\Delta_2 = 68° - 36° = 32°00'$$

$$t_1 = (400 \text{ ft})(\tan 18°) = 129.97 \text{ ft}$$

$$t_2 = (300 \text{ ft})(\tan 16°) = 86.02 \text{ ft}$$

$$\text{PI} - \text{PI}_1 = \frac{(215.99 \text{ ft})(\sin 32°)}{\sin 68°} = 123.45 \text{ ft}$$

$$\text{PI} - \text{PI}_2 = \frac{(215.99 \text{ ft})(\sin 36°)}{\sin 68°} = 136.93 \text{ ft}$$

$$T_1 = 129.97 + 123.45 = 253.42 \text{ ft}$$

$$T_2 = 86.02 + 136.93 = 222.95 \text{ ft}$$

$$L_1 = \left(\frac{36}{360}\right) 2\pi(400) = 251.33 \text{ ft}$$

$$L_2 = \left(\frac{32}{360}\right) 2\pi(300) = 167.55 \text{ ft}$$

$$\text{PI} = 14{+}78.32$$

$$T_1 = 2{+}53.42$$

$$\boxed{\text{PC} = 12{+}24.90}$$

$$L_1 = 2{+}51.33$$

$$\boxed{\text{PCC} = 14{+}76.23}$$

$$L_2 = 1{+}67.55$$

$$\boxed{\text{PT} = 16{+}43.78}$$

point	station			deflection angle
PC	12+24.90			0°00′
	13+00	$\left(\frac{75.10}{251.33}\right)\left(\frac{36°}{2}\right)$	= 5.3786°	5°23′
	14+00	$\left(\frac{175.10}{251.33}\right)\left(\frac{36°}{2}\right)$	= 12.5405°	12°32′
PCC	14+76.23	$\left(\frac{251.33}{251.33}\right)\left(\frac{36°}{2}\right)$	= 18.0000°	18°00′
	15+00	$\left(\frac{23.77}{167.55}\right)\left(\frac{32°}{2}\right)$	= 2.2699°	2°16′
	16+00	$\left(\frac{123.77}{167.55}\right)\left(\frac{32°}{2}\right)$	= 11.8193°	11°49′
PT	16+43.78	$\left(\frac{167.55}{167.55}\right)\left(\frac{32°}{2}\right)$	= 16.0000°	16°00′

Plane Survey Calculations

17.

point	station	Elm Street deflection angle	curve data
PT	16+11.28	17°00′	
	16+00	16°28′	
	15+00	11°41′	
	14+00	6°55′	
			$\Delta = 70°$
	13+00	2°08′	$R_1 = 900$ ft
			$R_2 = 600$ ft
PCC	12+55.23	18°00′	$\Delta_1 = 36°$
			$\Delta_2 = 34°$
	12+00	16°15′	$T_1 = 575.61$ ft
			$T_2 = 481.10$ ft
	11+00	13°04′	$L_1 = 565.49$ ft
			$L_2 = 356.05$ ft
	10+00	9°53′	
	9+00	6°42′	
	8+00	3°31′	
	7+00	0°20′	
PC	6+89.74		

18.

line	bearing	distance	latitude	departure	DMD	area (ft)^2
AB	N 17° W	290	+277.3	−84.8	84.8	+23,515
BC	N 59° E	150	+77.3	+128.6	128.6	+9941
CD	N 89° E	350	+6.1	+349.9	607.1	+3703
DE	S 61° E	150	−72.7	+131.2	1088.2	−79,112
EF	S 09° W	250	−246.9	−39.1	1180.3	−291,416
FG	S 68° W	280	−104.9	−259.6	881.6	−92,480
GA	N 74°15′ W	235	+63.8	−226.2	395.8	+ 25,252
			0.0	0.0		200,298
						2) 400,597

$$\text{area of segment: } \tfrac{1}{2}R^2(\theta - \sin\theta) = \left(\frac{1}{2}\right)(350 \text{ ft})^2\left(\frac{\pi}{3} - 0.8660\right) = \underline{11{,}098}$$

$$211{,}396 \text{ ft}^2$$

$$\text{total area} = \frac{211{,}396 \text{ ft}^2}{43{,}560 \ \frac{\text{ft}^2}{\text{ac}}} = \boxed{4.9 \text{ ac}}$$

19 Vertical Alignment

Nomenclature

e	ordinate at PI in feet
g	gradient
g_1	gradient of first connecting tangent
g_2	gradient of second connecting tangent
L	length of curve in stations
PC	beginning of curve
PI	point of intersection of two tangents
PT	end of curve
VC	vertical curve
x_1	horizontal distance in stations from PC
x_2	horizontal distance in stations from PT
y_1	ordinate of any station less than PI station
y_2	ordinate of any station greater than PI station

1. GRADE (STEEPNESS)

The *grade*, or *steepness*, of a road is the ratio of elevation to horizontal distance. If a highway rises 6 ft for every 100 ft of horizontal distance, the grade of the highway is 6 ft/100 ft = 0.06 ft/ft.

Figure 19.1 *Calculation of Grade*

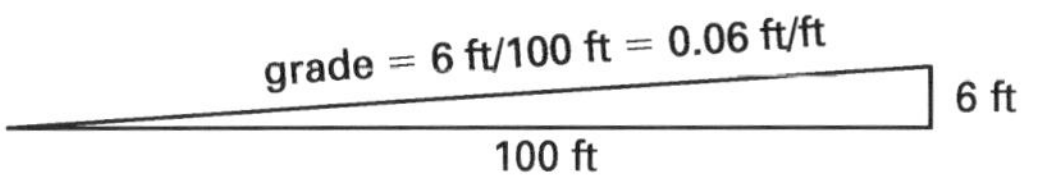

grade = 6 ft/100 ft = 0.06 ft/ft

6 ft

100 ft

2. SLOPE OF A LINE

The grade of a line representing the profile of a highway is also known as the *slope of the line.*

$$\text{slope} = \frac{\text{rise}}{\text{run}} \qquad 19.1$$

Figure 19.2 *Slope of a Line*

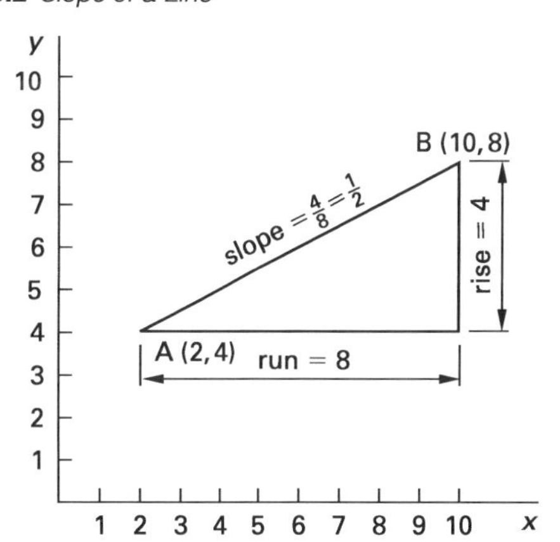

In Fig. 19.2, points A (2, 4) and B (10, 8) are connected by the straight line AB.

$$\begin{aligned}\text{slope of AB} &= \frac{\text{ordinate of B} - \text{ordinate of A}}{\text{abscissa of B} - \text{abscissa of A}} \\ &= \frac{8-4}{10-2} = \frac{4}{8} = \frac{1}{2}\end{aligned} \qquad 19.2$$

The symbol m is used to denote the slope of the line between points 1 and 2.

$$m = \frac{y_2 - y_1}{x_2 - x_1} \qquad 19.3$$

Figure 19.3 *Line Between Two Points*

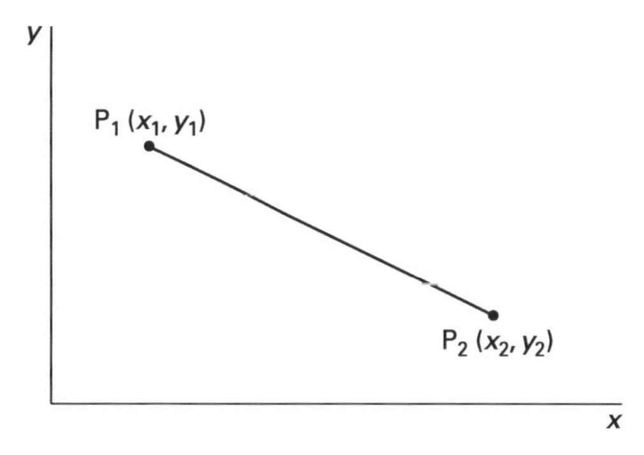

In Fig. 19.4, the slope of line AB is

$$m = \frac{y_2 - y_1}{x_2 - x_1} = \frac{8-2}{6-3} = 2$$

Line BC has a slope of

$$m = \frac{y_2 - y_1}{x_2 - x_1} = \frac{2-8}{9-6} = -2$$

Figure 19.4 Positive and Negative Slopes

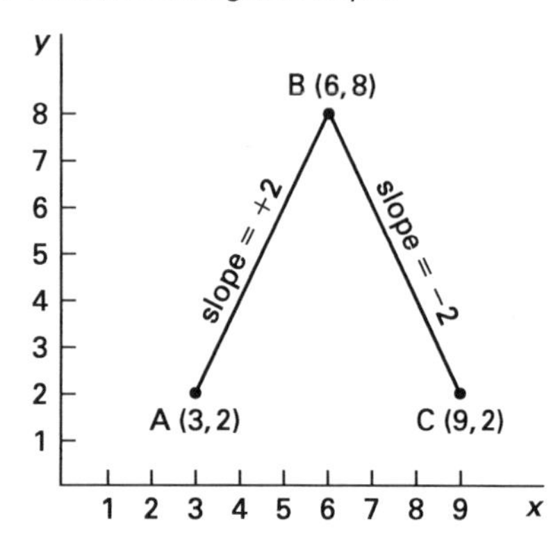

A line rising from left to right has a positive slope, and a line falling from left to right has a negative slope.

A horizontal line has a slope of zero, as shown in Fig. 19.5.

Figure 19.5 Slope of Horizontal Line

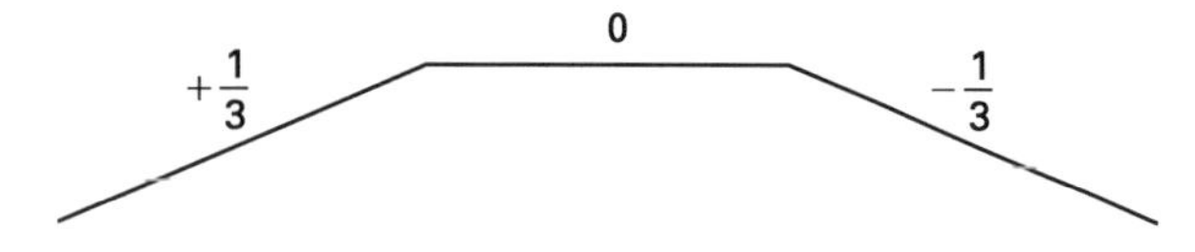

3. GRADE (GRADIENT)

In highway construction, the slope of the line that is the profile of the centerline is known as the *grade* or *gradient*, g.

The grade of a highway is computed in the same way as the slope of a line is computed. Horizontal distances are usually expressed in stations; vertical distances are expressed in feet.

Example 19.1

Determine the gradient, g, of a highway that has a centerline elevation of 444.50 ft at sta 20+75.00 and a centerline elevation of 472.20 ft at sta 32+25.00.

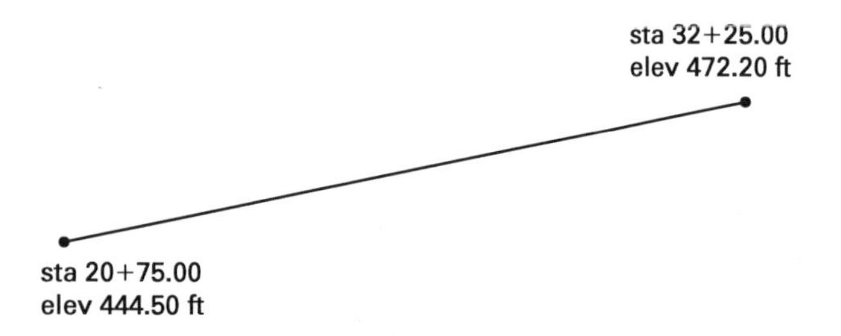

Solution

$$g = \frac{472.20 \text{ ft} - 444.50 \text{ ft}}{3225.00 \text{ ft} - 2075.00 \text{ ft}} = \frac{27.70 \text{ ft}}{1150.00 \text{ ft}}$$
$$= +0.02409 \text{ ft/ft}$$

If the numerator is expressed in feet and the denominator is expressed in stations, the decimal point in the gradient will move two places to the right. The gradient can then be expressed as a percent. In this form, gradient expresses change in elevation per station. For Ex. 19.1,

$$g = \frac{27.70 \text{ ft}}{11.50 \text{ sta}} = 2.409 \text{ ft/sta} \quad (+2.409\%)$$

4. POINTS OF INTERSECTION

Vertical alignment for a highway is located similarly to horizontal alignment. Straight lines are located from point to point, and vertical curves are inserted. The points of intersecting gradients are known as *points of intersection*. These lines, after vertical curves have been inserted, are the centerline profile of the highway. Usually the profile is the finish elevation (pavement) profile, but it may be the subgrade (earthwork) profile.

5. TANGENT ELEVATIONS

After points of intersection have been located and connected by tangents (straight lines), elevations of each station on the tangent need to be determined before finding elevations on the vertical curve.

The gradient is the change in elevation per station. If the station number and elevation of each PI is known, the elevation at each station can be calculated. The gradient should be computed to three decimal places in percent. Finish elevations should be computed to two decimal places in feet (hundredths of a foot).

Example 19.2

From the information shown, compute the gradient of each tangent and the elevation at each full station on the tangents.

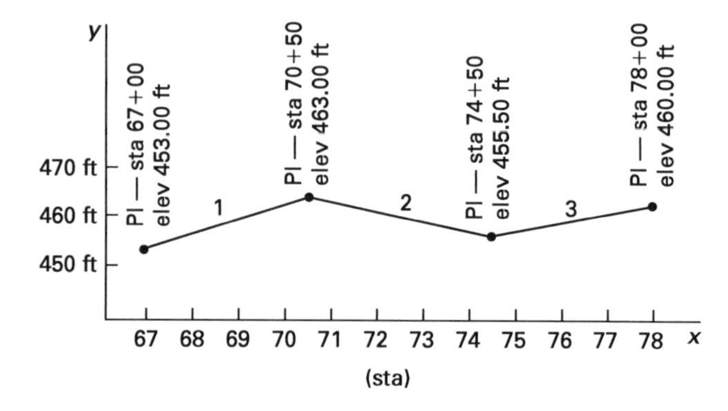

Solution

$$g_1 = \frac{463.00\text{ ft} - 453.00\text{ ft}}{70.50\text{ sta} - 67.00\text{ sta}}$$
$$= 2.857\text{ ft/sta} \quad (+2.857\%)$$
$$g_2 = \frac{455.50\text{ ft} - 463.00\text{ ft}}{74.50\text{ sta} - 70.50\text{ sta}}$$
$$= -0.01875\text{ ft/sta} \quad (-1.875\%)$$
$$g_3 = \frac{460.00\text{ ft} - 455.50\text{ ft}}{78.00\text{ sta} - 74.50\text{ sta}}$$
$$= 0.01286\text{ ft/sta} \quad (+1.286\%)$$

Elevation Computations

point	station	computations	elevation (ft)
PI	67+00		= 453.00
	68+00	453.00 + (1)(2.857)	= 455.86
	69+00	453.00 + (2)(2.857)	= 458.71
	70+00	453.00 + (3)(2.857)	= 461.57
PI	70+50	453.00 + (3.5)(2.857)	= 463.00
	71+00	463.00 − (0.5)(1.875)	= 462.06
	72+00	463.00 − (1.5)(1.875)	= 460.19
	73+00	463.00 − (2.5)(1.875)	= 458.31
	74+00	463.00 − (3.5)(1.875)	= 456.44
PI	74+50	463.00 − (4.0)(1.875)	= 455.50
	75+00	455.50 + (0.5)(1.286)	= 456.14
	76+00	455.50 + (1.5)(1.286)	= 457.43
	77+00	455.50 + (2.5)(1.286)	= 458.72
PI	78+00	455.50 + (3.5)(1.286)	= 460.00

6. VERTICAL CURVES

Just as horizontal curves connect two tangents in horizontal alignment, vertical curves connect two tangents in vertical alignment. However, although the horizontal curve is usually an arc of a circle, the vertical curve is usually a parabola.

The simplest equation of a parabola is $y = ax^2$, where y is vertical distance and x is horizontal distance. This means that the vertical distances y from tangent to curve vary as the square of the horizontal distances x, measured from either the PC or the PT.

Figure 19.6 *Vertical Curve*

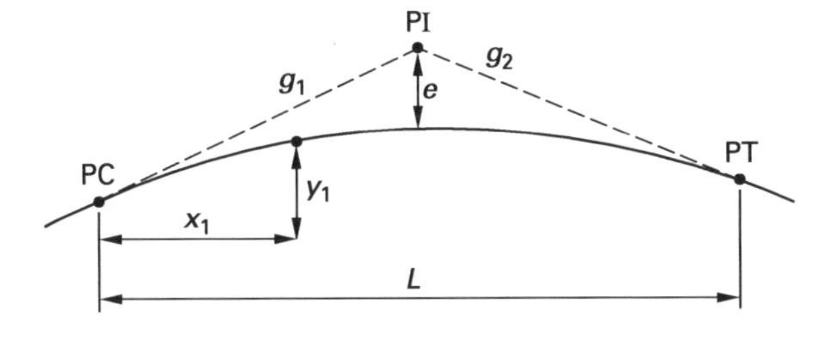

Equations for computing y-distances can be derived as follows.[1]

$$e = \frac{(g_1 - g_2)L}{8} \quad \text{19.4}$$

$$a = \frac{g_1 - g_2}{2L} \quad \text{19.5}$$

$$y = ax^2 \quad \text{19.6}$$

7. COMPUTATIONS FOR FINISH ELEVATIONS

Elevations along the tangent are first computed, and y-distances are then computed for stations on the vertical curve and added to or subtracted from the tangent elevations to determine the finish elevation. The y-distance is added on sag curves (Fig. 19.7) and subtracted on crest curves (Fig. 19.6).

Figure 19.7 *Sag Curve*

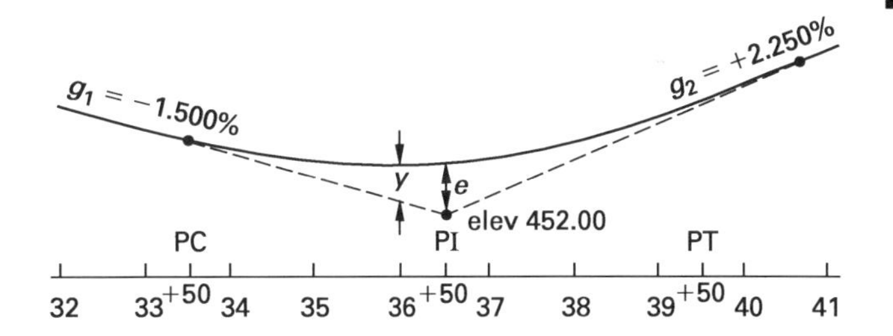

Example 19.3

A −1.500% grade meets a +2.250% grade at sta 36+50, elev 452.00 ft. A vertical curve of length 600 ft (six stations) will be used. Compute the finish elevations for each full station and the PC and PT from sta 32+00 to sta 41+00. Refer to Fig. 19.7.

Solution

From Eqs. 19.4 and 19.5,

$$a = \frac{g_1 - g_2}{2L}$$
$$= \frac{-1.500\% - 2.250\%}{(2)(6\text{ sta})}$$
$$-0.312\%/\text{sta}$$
$$e = \frac{(g_1 - g_2)L}{8}$$
$$= \frac{(-1.500\% - 2.250\%)(6\text{ sta})}{8}$$
$$= 2.81\text{ ft}$$

[1]The symbol C is often used in place of a in Eq. 19.6. E may also be used instead of e.

Tangent Elevation Computations

point	station	computations	elevation
	32+00	452.00 + (4.5)(1.500)	= 458.75
	33+00	452.00 + (3.5)(1.500)	= 457.25
PC	33+50	452.00 + (3.0)(1.500)	= 456.50
	34+00	452.00 + (2.5)(1.500)	= 455.75
	35+00	452.00 + (1.5)(1.500)	= 454.25
	36+00	452.00 + (0.5)(1.500)	= 452.75
PI	36+50		= 452.00
	38+00	452.00 + (1.5)(2.250)	= 455.37
	39+00	452.00 + (2.5)(2.250)	= 457.62
PT	39+50	452.00 + (3.0)(2.250)	= 458.75
	40+00	452.00 + (3.5)(2.250)	= 459.88
	41+00	452.00 + (4.5)(2.250)	= 462.13

Computations for $y = 0.312x^2$

station	computations	y (ft)
33+50		= 0.00
34+00	= $(0.312)(0.5)^2$	= 0.08
35+00	= $(0.312)(1.5)^2$	= 0.70
36+00	= $(0.312)(2.5)^2$	= 1.95
36+50		= 2.81
37+00	= same as 36+00	= 1.95
38+00	= same as 35+00	= 0.70
39+00	= same as 34+00	= 0.08
39+50		= 0.00

The curve is symmetrical about the PI, so the distance x to sta 37+00 (from the PT) is the same as the distance x to sta 36+00 (from the PC). Therefore, the y-distances are the same. Likewise, the y-distances for sta 38+00 and 35+00 are the same, as well as for sta 39+00 and 34+00.

Finish Elevations

point	station	tangent elevation (ft)	y (ft)	finish elevation (ft)
	32+00	458.75		458.75
	33+00	457.25		457.25
PC	33+50	456.50	0.00	456.50
	34+00	455.75	+0.08	455.83
	35+00	454.25	+0.70	454.95
	36+00	452.75	+1.95	454.70
PI	36+50	452.00	+2.81	454.81
	37+00	453.12	+1.95	455.07
	38+00	455.37	+0.70	456.07
	39+00	457.62	+0.08	457.70
PT	39+50	458.75	0.00	458.75
	40+00	459.88		459.88
	41+00	462.13		462.13

8. PLAN-PROFILE SHEETS

On construction plans, a profile of the natural ground along the centerline of a highway project and the profile of the finish grade along the centerline are shown on *plan-profile sheets*. A plan view of the centerline with surrounding topography is shown on the top half of the sheet. The profiles are shown on the bottom half, together with PIs, PCs, PTs, gradients, and finish elevations. A plan-profile sheet is shown in Fig. 19.8.

Plan-profile sheets are also used in construction plans for streets, sanitary sewers, and storm sewers.

Example 19.4

Using the following information, compute finish elevations for each full station. Plot the finish elevation profile and show pertinent information needed for construction.

- gradient, sta 32+00 to 34+00 = −2.000%
- gradient, sta 39+50 to 43+00 = −2.125%
- PI, sta 34+00, elev 470.00, 300 ft (vertical curve)
- PI, sta 39+50, elev 482.00, 500 ft (vertical curve)

Solution

$$g = \frac{482.00 \text{ ft} - 470.00 \text{ ft}}{39.5 \text{ sta} - 34.00 \text{ sta}}$$

$$= 0.02182 \text{ ft/sta} \quad (+2.182\%)$$

$$a_1 = \frac{g_1 - g_2}{2L} = \frac{-2.000\% - 2.182\%}{(2)(3.00 \text{ sta})}$$

$$= 0.697\%/\text{sta}$$

$$a_2 = \frac{g_1 - g_2}{2L} = \frac{2.182\% - (-2.125\%)}{(2)(5.00 \text{ sta})}$$

$$= 0.431\%/\text{sta}$$

$$e_1 = \frac{(g_1 - g_2)L}{8} = \frac{(-2.000\% - 2.182\%)(3.00 \text{ sta})}{8}$$

$$= 1.57 \text{ ft}$$

$$e_2 = \frac{(g_1 - g_2)L}{8} = \frac{\big(2.182\% - (-2.125\%)\big) \times (5.00 \text{ sta})}{8}$$

$$= 2.69 \text{ ft}$$

Prior to performing finish elevation computations, the approximate finish elevation profile should be plotted on the plan-profile sheet. Vertical curves are symmetrical. To locate the PC and PT, measure one-half the length of the vertical curve in each direction from the PI. The midpoint of the vertical curve can be found by drawing a straight line from the PC to the PT and measuring one-half the distance from the PI to this line.

In determining whether y-distances should be added to or subtracted from tangent elevations, look at the plotted profile to see whether the curve is higher or lower than the tangent at a particular station.

Figure 19.8 would normally show the profile of the natural ground along the centerline, and the elevation at each station would be shown just under the finish elevation at the bottom of the sheet.

Figure 19.8 *Plan-Profile Sheet*

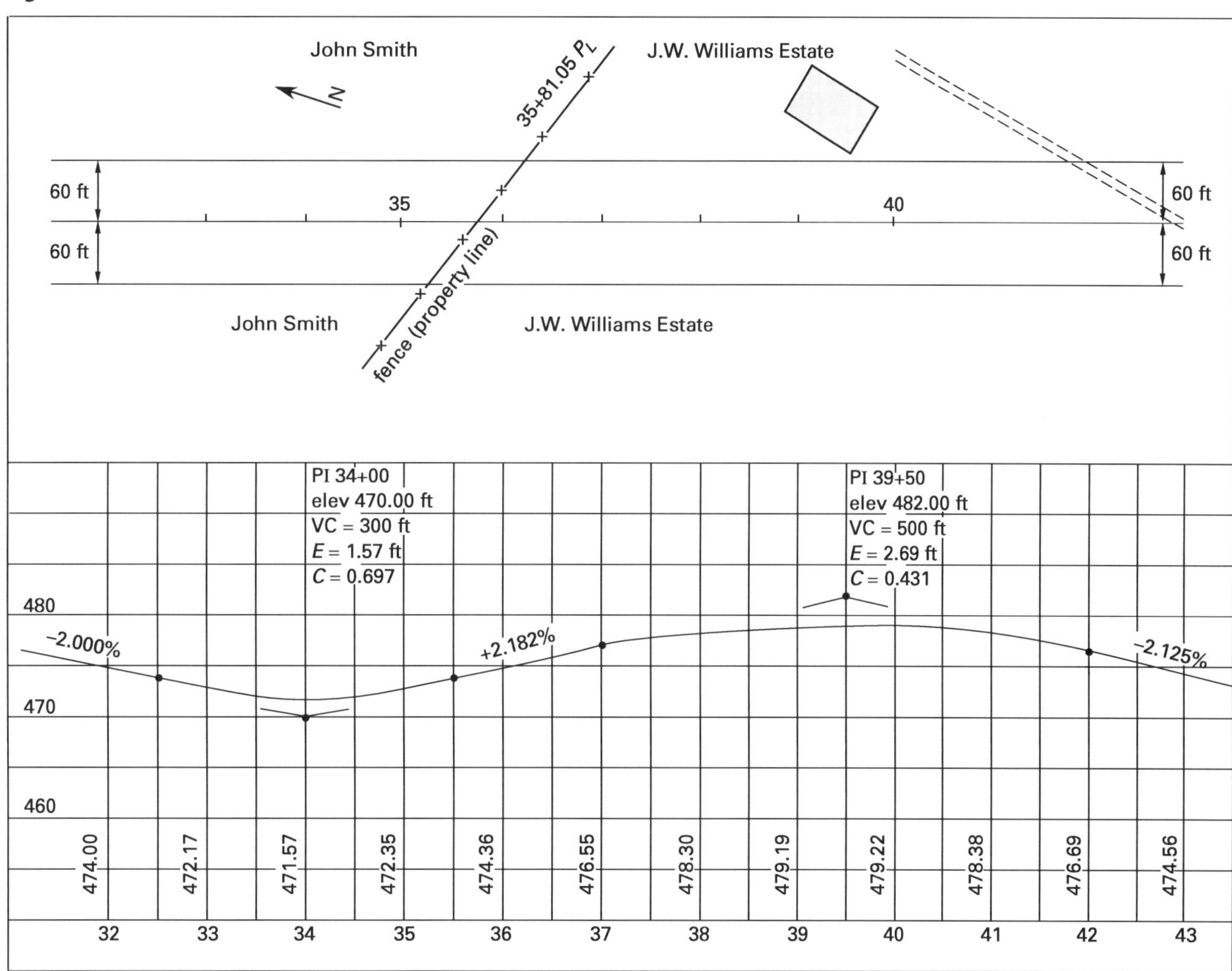

Computations for Finish Elevations

point	station	x (sta)	x^2 (sta^2)	tangent elevation (ft)	y (ft)	finish elevation (ft)
	32+00			474.00		474.00
PC	32+50					473.00
	33+00	0.5	0.25	472.00	+0.17	472.17
PI	34+00	1.5	2.25	470.00	+1.57	471.57
	35+00	0.5	0.25	472.18	+0.17	472.35
PT	35+50					
	36+00			474.36		474.36
PC	37+00			476.55		476.55
	38+00	1.0	1.00	478.73	−0.43	478.30
	39+00	2.0	4.00	480.91	−1.72	479.19
PI	39+50	2.5	6.25	482.00	−2.69	479.31
	40+00	2.0	4.00	480.94	−1.72	479.22
	41+00	1.0	1.00	478.81	−0.43	478.38
PC	42+00			476.69		476.69
	43+00			474.56		474.56

9. TURNING POINT ON SYMMETRICAL VERTICAL CURVE

The highest point of a crest curve (or the lowest point on a sag curve) is not usually vertically below (or above) the PI. This point is called the *turning point*. The distance x from the PC to the turning point can be found from Eq. 19.7.

$$x = \frac{g_1 L}{g_1 - g_2} \tag{19.7}$$

Example 19.5

A +1.500% grade meets a −2.500% grade at sta 12+50. Determine the distance from the PC to the turning point if a 600 ft vertical curve is used.

Solution

$$x = \frac{g_1 L}{g_1 - g_2} = \frac{(1.500\%)(6\text{ sta})}{+1.500\% - (-2.500\%)} = 2.25\text{ sta}\quad(225\text{ ft})$$

PRACTICE PROBLEMS

1. Determine the gradients between the points on the highway profiles in percent to three decimal places.

Example:

PI = 5+50

elev = 452.00 ft

PI = 8+50

elev = 455.00 ft

PI = 11+00

elev = 453.00 ft

Solution:

$$g_1 = \frac{455.00\text{ ft} - 452.00\text{ ft}}{8.50\text{ sta} - 5.50\text{ sta}} = \boxed{0.01\text{ ft/sta}\quad(+1.000\%)}$$

$$g_2 = \frac{455.00\text{ ft} - 453.00\text{ ft}}{11.00\text{ sta} - 8.50\text{ sta}} = \boxed{-0.008\text{ ft/sta}\quad(-0.800\%)}$$

(a) PI = 20+70
elev = 504.00 ft
PI = 23+50
elev = 498.00 ft
PI = 26+60
elev = 503.00 ft

(b) PI = 40+00
elev = 461.00 ft
PI = 46+00
elev = 459.00 ft
PI = 52+00
elev = 465.00 ft

(c) PI = 55+00
elev = 474.00 ft
PI = 59+00
elev = 469.00 ft
PI = 67+00
elev = 477.50 ft
PI = 67+00
elev = 477.50 ft
elev = 453.00 ft
PI = 70+50
elev = 463.00 ft
PI = 74 + 50
elev = 455.50 ft
PI = 79+00
elev = 461.70 ft

(e) PI = 29+25
elev = 445.00 ft
PI = 32+50
elev = 432.00 ft
PI = 37+75
elev = 432.00 ft
PI = 41+00
elev = 437.70 ft

2. Compute the gradient for each tangent of the highway profile and elevation of each full station on the tangents.

point	station	tangent elevation (ft)
PI	25+00	466.00
PI	31+00	458.00
PI	35+50	472.00
PI	39+00	465.00
PI	43+00	472.00

3. Compute a, e, and y for each station on the vertical curve.

(a) A +2.234% grade meets a −1.875% grade at sta 28+50, elev 436.00 ft. The vertical curve is 600 ft.

(b) A −3.467% grade meets a +2.250% grade at sta 45+00, elev 515.00 ft. The vertical curve is 800 ft.

4. From the information given, compute the finish elevation for each full station.

PI = 20+00
elev = 455.00 ft
no vertical curve

PI = 23+50
elev = 448.50 ft
400 ft vertical curve

PI = 33+00
elev = 469.00 ft
1000 ft vertical curve

PI = 40+00
elev = 455.50 ft
no vertical curve

5. Determine the distance x from the PC of the symmetrical curve to the high point of the crest curve or to the low point of the sag curve.

(a) A +4.000% grade meets a −3.000% grade at sta 35+00. The vertical curve is 600 ft in length.

(b) A +3.125% grade meets a −2.250% grade at sta 12+00. The vertical curve is 800 ft in length.

(c) A −2.750% grade meets a +3.500% grade at sta 22+00. The vertical curve is 500 ft in length.

(d) A −1.275% grade meets a +3.250% grade at sta 15+00. The vertical curve is 600 ft in length.

6. From the information given, complete the profile half of the plan-profile sheet.

station	finish elevation (ft)
32+00	476.00
33+00	474.00
34+00	472.18
35+00	471.65
36+00	472.58
37+00	474.80
38+00	476.82
39+00	478.09
40+00	478.61
41+00	478.37
42+00	477.37
43+00	475.63

Illustration for Problem 6

gradient sta 32+00 to sta 35+00: −2.000% $e_1 = 1.65$ ft $a_1 = 0.733\%/\text{sta}$
gradient sta 35+00 to sta 40+00: +2.400% $e_2 = 3.39$ ft $a_2 = 0.377\%/\text{sta}$
gradient sta 40+00 to sta 43+00: −2.125% $\text{VC}_1 = 300$ ft $\text{VC}_2 = 600$ ft
PI = 35+00, elev 470.00 ft
PI = 40+00, elev 482.00 ft

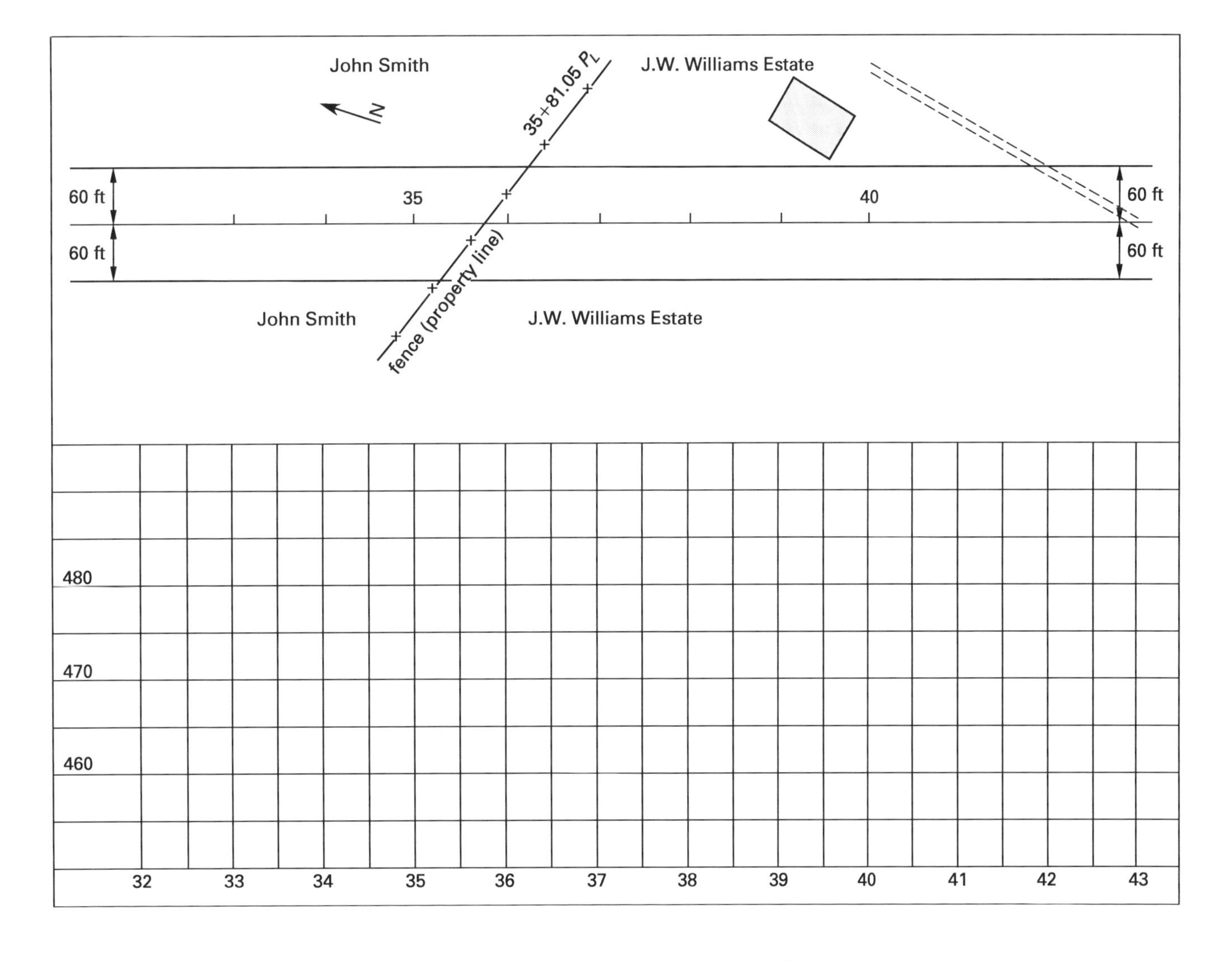

Plane Survey Calculations

SOLUTIONS

1. (a)
$$g_1 = \frac{504.00\text{ ft} - 498.00\text{ ft}}{23.50\text{ sta} - 20.70\text{ sta}} = \boxed{-0.02143\text{ ft/sta}\quad(-2.143\%)}$$

$$g_2 = \frac{503.00\text{ ft} - 498.00\text{ ft}}{26.60\text{ sta} - 23.50\text{ sta}} = \boxed{0.01613\text{ ft/sta}\quad(+1.613\%)}$$

(b)
$$g_1 = \frac{461.00\text{ ft} - 459.00\text{ ft}}{46.00\text{ sta} - 40.00\text{ sta}} = \boxed{-0.00333\text{ ft/sta}\quad(-0.333\%)}$$

$$g_2 = \frac{465.00\text{ ft} - 459.00\text{ ft}}{52.00\text{ sta} - 46.00\text{ sta}} = \boxed{0.01000\text{ ft/sta}\quad(+1.000\%)}$$

(c)
$$g_1 = \frac{474.00\text{ ft} - 469.00\text{ ft}}{59.00\text{ sta} - 55.00\text{ sta}} = \boxed{-0.01250\text{ ft/sta}\quad(-1.250\%)}$$

$$g_2 = \frac{477.50\text{ ft} - 469.00\text{ ft}}{64.00\text{ sta} - 59.00\text{ sta}} = \boxed{0.01700\text{ ft/sta}\quad(+1.700\%)}$$

$$g_3 = \frac{477.50\text{ ft} - 477.50\text{ ft}}{67.00\text{ sta} - 64.00\text{ sta}} = \boxed{0\text{ ft/sta}\quad(0\%)}$$

(d)
$$g_1 = \frac{463.00\text{ ft} - 453.00\text{ ft}}{70.50\text{ sta} - 67.00\text{ sta}} = \boxed{0.02857\text{ ft/sta}\quad(+2.857\%)}$$

$$g_2 = \frac{463.00\text{ ft} - 455.50\text{ ft}}{74.50\text{ sta} - 70.50\text{ sta}} = \boxed{-0.01875\text{ ft/sta}\quad(-1.875\%)}$$

$$g_3 = \frac{461.70\text{ ft} - 455.50\text{ ft}}{79.00\text{ sta} - 74.50\text{ sta}} = \boxed{0.01378\text{ ft/sta}\quad(+1.378\%)}$$

(e)
$$g_1 = \frac{445.00\text{ ft} - 432.00\text{ ft}}{32.50\text{ sta} - 29.25\text{ sta}} = \boxed{-0.04000\text{ ft/sta}\quad(-4.000\%)}$$

$$g_2 = \frac{432.00\text{ ft} - 432.00\text{ ft}}{37.75\text{ sta} - 32.50\text{ sta}} = \boxed{0\text{ ft/sta}\quad(0\%)}$$

$$g_3 = \frac{437.70\text{ ft} - 432.00\text{ ft}}{41.00\text{ sta} - 37.75\text{ sta}} = \boxed{0.01754\text{ ft/sta}\quad(+1.754\%)}$$

2.

point	station	tangent elevation (ft)
PI	25+00	466.00
	26+00	464.67
	27+00	463.33
$g_1 = -1.333\%$	28+00	462.00
	29+00	460.67
	30+00	459.33
PI	31+00	458.00
	32+00	461.11
	33+00	464.22
$g_2 = +3.111\%$	34+00	467.33
	35+00	470.44
PI	35+50	472.00
	36+00	471.00
$g_3 = -2.000\%$	37+00	469.00
	38+00	467.00
PI	39+00	465.00
	40+00	466.75
$g_4 = +1.750\%$	41+00	468.50
	42+00	470.25
PI	43+00	472.00

3. (a)
$$a = \frac{g_1 - g_2}{2L} = \frac{2.234\% - (-1.875\%)}{(2)(6\text{ sta})} = \boxed{0.342\%/\text{sta}}$$

$$e = \frac{(g_1 - g_2)L}{8} = \frac{\big(2.234\% - (-1.875\%)\big)(6\text{ sta})}{8} = \boxed{3.08\text{ ft}}$$

$$y\text{ @ sta }26{+}00 = \left(0.342\ \frac{\%}{\text{sta}}\right)(0.5\text{ sta})^2 = \boxed{0.09\text{ ft}}$$

$$y\text{ @ sta }27{+}00 = \left(0.342\ \frac{\%}{\text{sta}}\right)(1.5\text{ sta})^2 = \boxed{0.77\text{ ft}}$$

$$y \text{ @ sta } 28{+}00 = \left(0.342\ \frac{\%}{\text{sta}}\right)(2.5\ \text{sta})^2 = \boxed{2.14\ \text{ft}}$$

$$y \text{ @ sta } 29{+}00 = \left(0.342\ \frac{\%}{\text{sta}}\right)(2.5\ \text{sta})^2 = \boxed{2.14\ \text{ft}}$$

$$y \text{ @ sta } 30{+}00 = \left(0.342\ \frac{\%}{\text{sta}}\right)(1.5\ \text{sta})^2 = \boxed{0.77\ \text{ft}}$$

$$y \text{ @ sta } 31{+}00 = \left(0.342\ \frac{\%}{\text{sta}}\right)(0.5\ \text{sta})^2 = \boxed{0.09\ \text{ft}}$$

(b)
$$a = \frac{g_1 - g_2}{2L} = \frac{-3.467\% - 2.250\%}{(2)(8\ \text{sta})} = \boxed{0.357\%/\text{sta}}$$

$$e = \frac{(g_1 - g_2)L}{8} = \frac{(3.467\% - 2.250\%)(8\ \text{sta})}{8} = \boxed{5.72\ \text{ft}}$$

$$y \text{ @ sta } 42{+}00 = \left(0.357\ \frac{\%}{\text{sta}}\right)(1\ \text{sta})^2 = \boxed{0.36\ \text{ft}}$$

$$y \text{ @ sta } 43{+}00 = \left(0.357\ \frac{\%}{\text{sta}}\right)(2\ \text{sta})^2 = \boxed{1.43\ \text{ft}}$$

$$y \text{ @ sta } 44{+}00 = \left(0.357\ \frac{\%}{\text{sta}}\right)(3\ \text{sta})^2 = \boxed{3.21\ \text{ft}}$$

$$y \text{ @ sta } 45{+}00 = \left(0.357\ \frac{\%}{\text{sta}}\right)(4\ \text{sta})^2 = \boxed{5.72\ \text{ft}}$$

$$y \text{ @ sta } 46{+}00 = \left(0.357\ \frac{\%}{\text{sta}}\right)(3\ \text{sta})^2 = \boxed{3.21\ \text{ft}}$$

$$y \text{ @ sta } 47{+}00 = \left(0.357\ \frac{\%}{\text{sta}}\right)(2\ \text{sta})^2 = \boxed{1.43\ \text{ft}}$$

$$y \text{ @ sta } 48{+}00 = \left(0.357\ \frac{\%}{\text{sta}}\right)(1\ \text{sta})^2 = \boxed{0.36\ \text{ft}}$$

4.

point	station	tangent elevation (ft)	y (ft)	finish elevation (ft)
PI	20+00	455.00		455.00
	21+00	453.14		453.14
PC	21+50	452.21		452.21
	22+00	451.29	0.13	451.42
	23+00	449.43	1.13	450.56
PI	23+50	448.50	2.01	450.51
	24+00	449.58	1.13	450.71
	25+00	451.74	0.13	451.87
PT	25+50	452.82		452.82
	26+00	453.90		453.90
	27+00	456.05		456.05
PC	28+00	458.21		458.21
	29+00	460.37	0.20	460.17
	30+00	462.53	0.82	461.71
	31+00	464.69	1.84	462.85
	32+00	466.84	3.26	463.58
PI	33+00	469.00	5.11	463.89
	34+00	467.07	3.26	463.81
	35+00	465.14	1.84	463.30
	36+00	463.21	0.82	462.39
	37+00	461.28	0.20	461.08
PT	38+00	459.36		459.36
	39+00	457.43		457.43
PI	40+00	455.50		455.50

5. (a)
$$x = \frac{g_1 L}{g_1 - g_2} = \frac{(4.0\%)(6\ \text{sta})}{4\% + 3\%} = 3.43\ \text{sta}\ (343\ \text{ft})$$

(b)
$$x = \frac{g_1 L}{g_1 - g_2} = \frac{(3.125\%)(8\ \text{sta})}{3.125\% + 2.250\%} = 4.65\ \text{sta}\ (465\ \text{ft})$$

(c)
$$x = \frac{g_1 L}{g_1 - g_2} = \frac{(-2.750\%)(5\ \text{sta})}{-2.750\% - 1.500\%} = 2.20\ \text{sta}\ (220\ \text{ft})$$

(d)
$$x = \frac{g_1 L}{g_1 - g_2} = \frac{(1.275\%)(6\ \text{sta})}{-1.275\% - 3.250\%} = 1.69\ \text{sta}\ (169\ \text{ft})$$

6.

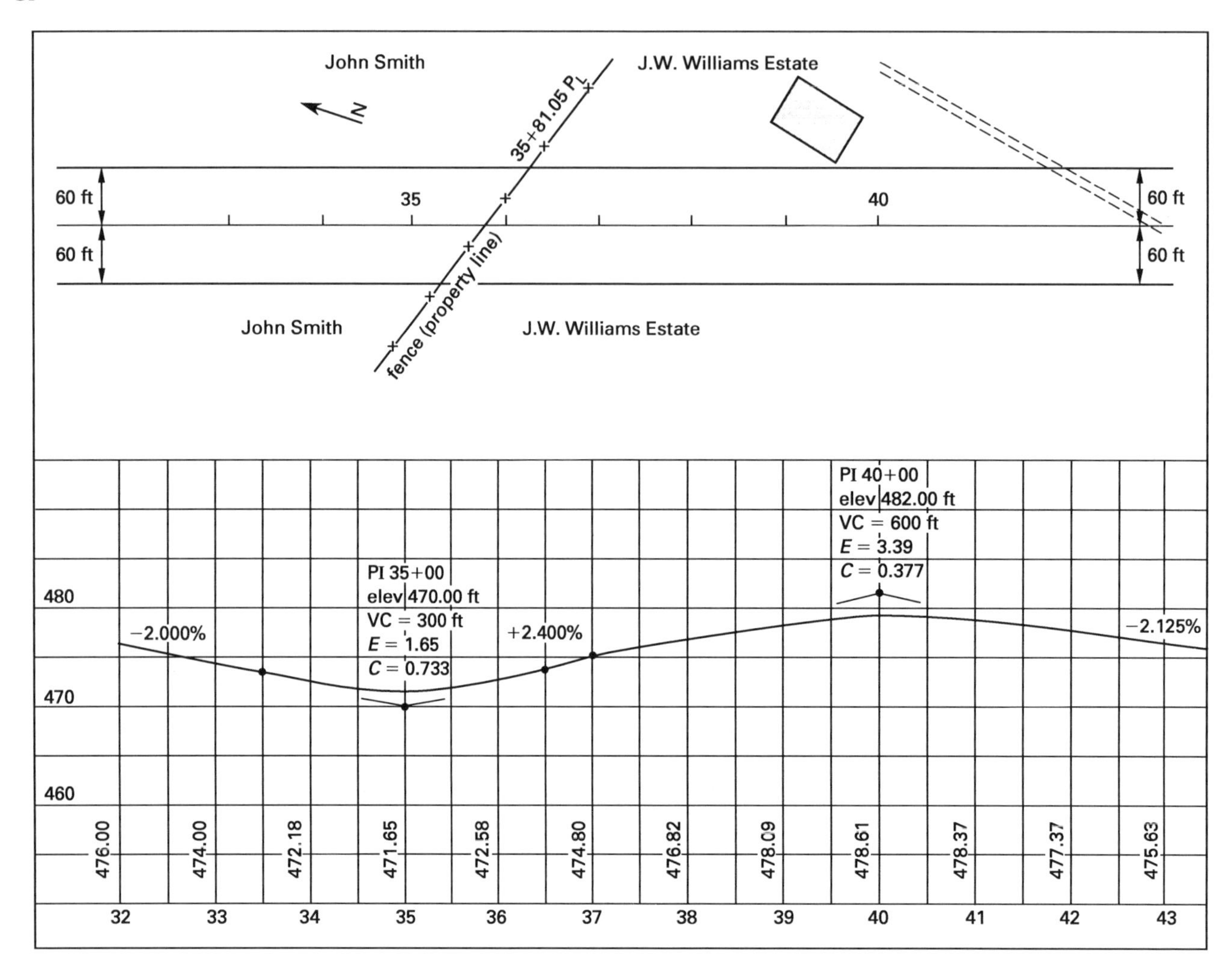
John Smith
J.W. Williams Estate
N
35+81.05 PL
60 ft
35
40
60 ft
60 ft
60 ft
fence (property line)
John Smith
J.W. Williams Estate
PI 40+00
elev 482.00 ft
VC = 600 ft
E = 3.39
C = 0.377
PI 35+00
elev 470.00 ft
VC = 300 ft
E = 1.65
C = 0.733
480
470
460
−2.000%
+2.400%
−2.125%
476.00
474.00
472.18
471.65
472.58
474.80
476.82
478.09
478.61
478.37
477.37
475.63
32
33
34
35
36
37
38
39
40
41
42
43

Topic IV: Geodesy and Survey Astronomy

20 Astronomic Observations

1. ANCIENT ASTRONOMERS

People of ancient times were aware of their dependence on the cycles of earth and sky. The sun's life-giving warmth and radiance, the mysterious waxing and waning of the moon, and the regular movements of the stars and planets inspired reverence and gave rise to the belief that the heavenly bodies were gods who controlled the universe.

Temple priests were assigned the task of observing the movements of these deities. They were also responsible for noting other celestial events and interpreting them. Colorful stories about these gods and their interactions with humankind evolved.

A great deal of astronomical knowledge was accumulated and passed on orally through these myths. Only the Babylonians and Egyptians recorded their findings for future civilizations.

The Greeks, from 700 B.C. to A.D. 200 first gave serious thought to the size and shape of the earth, and they were the first to conclude that the earth is a sphere. Aristotle noticed that the positions of the stars change, and that during an eclipse the shadow of the earth on the moon is curved. Eratosthenes calculated the circumference of the earth and found it to be 25,000 miles. Ptolemy, the great cartographer, devised a system for locating a point on the earth using a system similar to present-day latitude and longitude.

After Ptolemy, civilization slipped into the Dark Ages. For more than a thousand years the theory that the earth is a sphere was forgotten.

In the 15th century, the beginning of the European Renaissance, there was renewed interest in astronomy. Copernicus concluded that the earth revolved around the sun, directly contradicting the beliefs of the Catholic Church. In the 17th century, Kepler discovered that the orbits of the planets around the sun are elliptical, with the sun at one focus of the ellipse. He also discovered that a line joining a planet and the sun sweeps out equal areas of space in equal time, proving that a planet moves fastest when nearest the sun and slowest when farthest away.

Galileo, who was the first to appreciate the significance of the telescope, accepted Copernicus' theories and was tried by the Inquisition for his heretical ideas.

2. THE EARTH

The earth rotates from west to east on its polar axis and revolves about the sun in an elliptical orbit with the sun at one focus of the ellipse. It completes one revolution in a period of 365.2564 days. The inclination of the earth ($23^1/_2°$ with the perpendicular to the orbital plane), combined with its revolution around the sun, causes the lengths of day and night to change and also causes the seasons (see Fig. 20.1).

Figure 20.1 *Revolution of the Earth about the Sun*

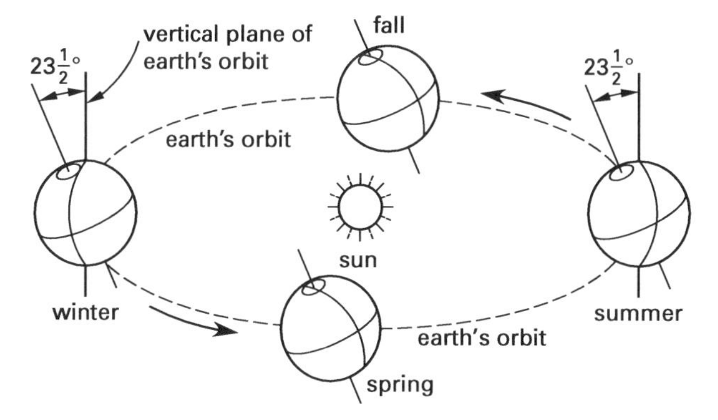

On March 21 and September 23, the light from the sun reaches from one pole to the other. On these dates, shown as spring and fall in Fig. 20.1, the sun is directly overhead at the equator. On June 21, shown as summer in Fig. 20.1, the sun is directly overhead at points $23^1/_2°$ above (north of) the equator. On December 22, shown as winter, the sun is overhead at points $23^1/_2°$ below (south of) the equator.

A line around the earth, parallel to and $23^1/_2°$ north of the equator, where the sun is directly overhead at its northernmost position, is known as the *Tropic of Cancer.* A line around the earth, parallel to and 23½° south of the equator, where the sun is directly overhead at its southernmost position, is known as the *Tropic of Capricorn.*

March 21 and September 23, when the sun crosses the equator and day and night are everywhere of equal length, are known as *equinoxes* (equal nights). The vernal equinox (March 21) marks the beginning of spring, and the autumnal equinox (September 23) marks the beginning of fall. June 21, the longest day of the year in the northern hemisphere, is known as the *summer solstice.* December 22, the shortest day of the year in the northern hemisphere, is known as the *winter solstice.*

In the southern hemisphere, the seasons are opposite those in the northern hemisphere.

3. GEODETIC NORTH OR GEODETIC AZIMUTH

The direction of a line is determined by the horizontal angle between the line and a reference line, usually true (geodetic) north. Determining the true azimuth of a line involves observations on a celestial body such as the sun or another star. The star usually used in the northern hemisphere is Polaris, the North Star. It is selected because it is very near true north from any point north of the equator. In making observations on celestial bodies, a surveyor is not interested in the distance to these bodies from the earth, but merely in their angular positions from the surveyor's observation point.

In addition to determining true azimuth, a surveyor can determine the latitude and longitude of a position by making observations on celestial bodies.

Observations on celestial bodies include making angular measurements, both horizontal and vertical. Determining true azimuth requires using an ephemeris, which is discussed in following sections.

4. PRACTICAL ASTRONOMY

Since ancient times, it has been convenient in practical astronomy to regard all celestial bodies as being fixed onto a sphere of infinite radius whose center is the earth's center. (However, stars are not the same distance from the earth, and they are much farther away than they appear to be.) This sphere of infinite radius is called the *celestial sphere* (see Fig. 20.2). To the observer, this sphere appears to be rotating about an axis, but it is not; the rotation of the earth causes the stars to be in rotation. In practical astronomy, it is assumed that the earth is stationary and that the celestial sphere revolves about the earth from west to east.

Figure 20.2 *Celestial Sphere*

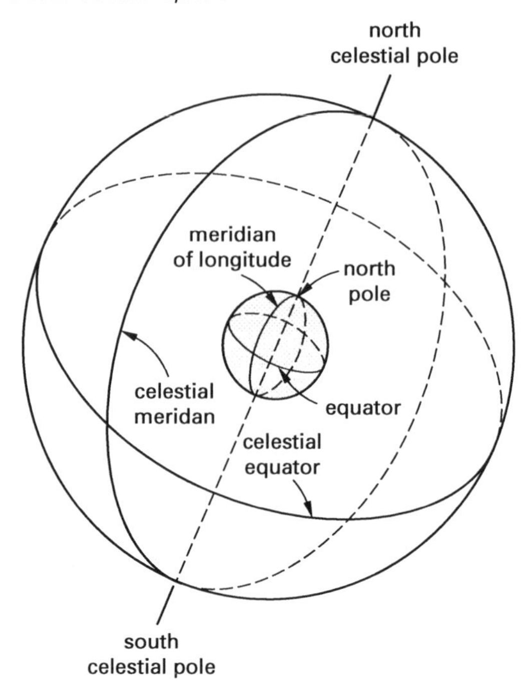

The celestial sphere is assumed to rotate about the celestial axis, which is a prolongation of the earth's polar axis. The north and south poles become the *north celestial pole* and the *south celestial pole.*

The plane of the earth's equator extended to the celestial sphere becomes the *celestial equator.*

Comparable with the parallels of latitude of the earth are the *parallels of declination* of the celestial sphere. They measure the angular distance from the celestial equator to the north and south celestial poles (see Fig. 20.5).

Comparable with meridians of longitude of the earth are the *hour circles* of the celestial sphere, all of which converge at the celestial poles. They are also known as *celestial meridians,* as shown in Figs. 20.2 and 20.3.

The observer's position on the earth is located by latitude and longitude. If the observer's plumb line is extended upward to the celestial sphere, the point of intersection with the celestial sphere is the observer's *zenith.* If the plumb line is extended downward to the celestial sphere, the intersection is the observer's *nadir.*

The relative positions of an observer on the earth and on the celestial sphere are shown in Fig. 20.3. It can be seen that the angle at the center of the earth that measures latitude is the same for the earth and for the celestial sphere. Likewise, the angle that measures longitude is the same for the earth and for the celestial sphere.

In Fig. 20.3, the observer's position on earth is latitude 35° N and longitude 98° W. The arc distance (in degrees) between the zenith of the observer and the celestial equator is also 35° N. The arc distance (in degrees) along the celestial equator between the planes of the Greenwich meridian and the meridian of the observer, extended to the celestial sphere, is 98° W. This arc distance is also the angle at the celestial north pole between the planes of the two meridians.

Figure 20.3 *Relation Between Earth and Celestial Sphere*

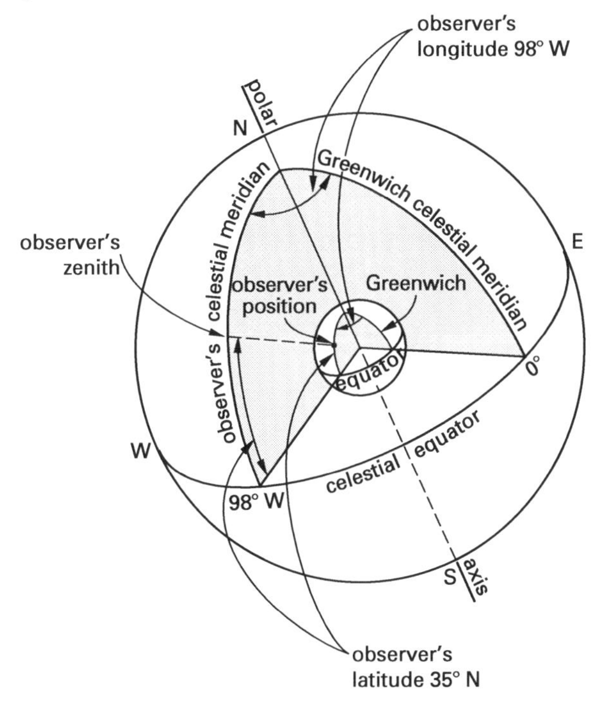

By using latitude and longitude and projecting it to the celestial sphere, the position of the observer's instrument is fixed at a point on the celestial sphere.

To identify the location of a celestial body on the celestial sphere, a coordinate system similar to latitude and longitude was devised (see Fig. 20.4). The celestial coordinates are called *declination* and *right ascension.* Using declination and right ascension as coordinates, the location of any celestial body on the celestial sphere can be determined with respect to the observer's zenith on the celestial sphere and to the north celestial pole.

Declination is the star's angular distance north or south of its celestial equator measured along the hour circle of the star (see Fig. 20.4). North declination is plus (+); south declination is minus (−). Declination corresponds to latitude on the earth.

Right ascension is the arc distance measured eastward along the celestial equator from the vernal equinox to the hour circle of the star. It may be measured in degrees, minutes, and seconds of arc or in hours, minutes, and seconds of time (see Fig. 20.4). Right ascension corresponds with longitude on the earth.

As with latitude and longitude, a celestial coordinate system requires points of origin, such as the equator for latitude and the Greenwich meridian for longitude. The celestial coordinate system uses the *celestial equator*, the *ecliptic*, and the *vernal equinox* in locating points of origin.

Figure 20.4 *Ecliptic and Vernal Equinox*

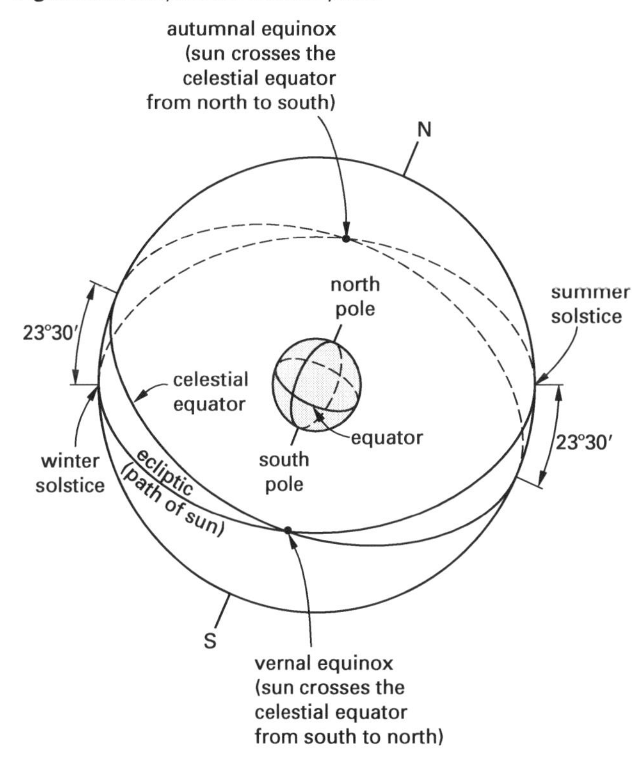

Because of the tilt of the earth as it follows its orbit around the sun, the sun traces a path called the *ecliptic* on the celestial sphere. The path of the sun moves from the southern hemisphere of the celestial sphere to the northern hemisphere and back (see Fig. 20.5).

The point where the sun crosses the celestial equator on its movement each year from south to north along the ecliptic is known as the *vernal equinox* (see Fig. 20.4). Astronomers designated the vernal equinox as the point of reference for right ascension.

Figure 20.5 *Celestial Coordinates*

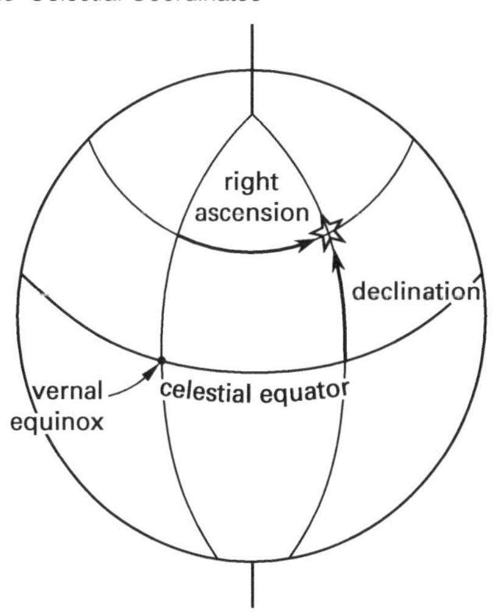

The vernal equinox is a point on the celestial sphere of infinite distance from the earth. Its location in time, relative to the Greenwich meridian, is known.

The point where the sky and earth meet, as seen by an observer on earth, is known as the *horizon*. The horizon for any place on the earth's surface is the great circle formed on the celestial sphere by the extension of the plane of the observer's horizon. In practical astronomy, the horizon is the plane tangent to the earth at the observer's position, perpendicular to the plumb line and extended to the celestial sphere. It is used as a reference for determining the altitude of a celestial body.

The *altitude* (h) of a celestial body is the angular distance measured from the horizon to a celestial body. It is the vertical angle measured by the observer from the horizon to the body (see Fig. 20.9).

Figure 20.7 *Three Sides of the PZS Triangle*

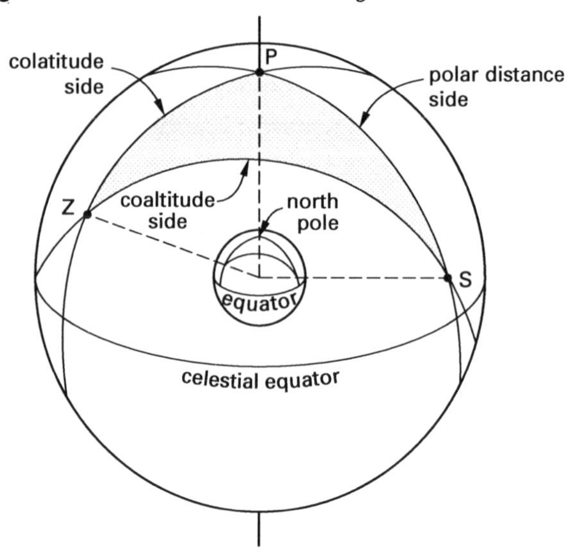

5. THE ASTRONOMICAL TRIANGLE

To determine azimuth or latitude and longitude, the surveyor needs to be able to solve a spherical triangle on the celestial sphere known as the *astronomical triangle*, or the celestial triangle PZS.

The vertices of the PZS triangle are the north celestial pole (P), the observer's zenith (Z), and the position of the star or the sun (S), as shown in Fig. 20.6.

The sides of the triangle are arcs of great circles on the celestial sphere that pass through any two of the vertices, measured in degrees or hours. The angular value of each side is determined by the angle that the side subtends on the earth (see Fig. 20.7).

Figure 20.6 *PZS Triangle*

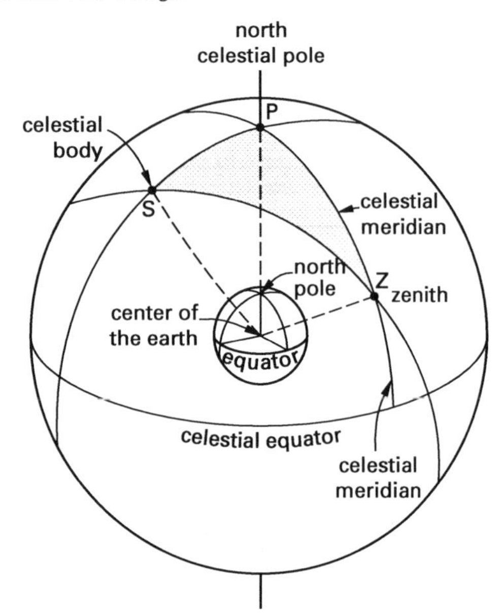

The three sides of the PZS triangle are known as (1) the polar distance, (2) the coaltitude, (3) the colatitude.

(1) The *polar distance* is the side PS. It is determined from the declination of the star or the sun. (Recall that declination is defined as the angular distance from a celestial body to the celestial equator.) When the celestial body lies north of the celestial equator, the declination has a positive sign; when it lies south of the celestial equator, it has a negative sign. The polar distance is determined algebraically by subtracting the declination of the celestial body from 90°. In Fig. 20.8, the polar distance is $90° - (-20) = 110°$. Observations on stars south of the equator are seldom made from the northern hemisphere.

Figure 20.8 *Polar Distance Side of the PZS Triangle*

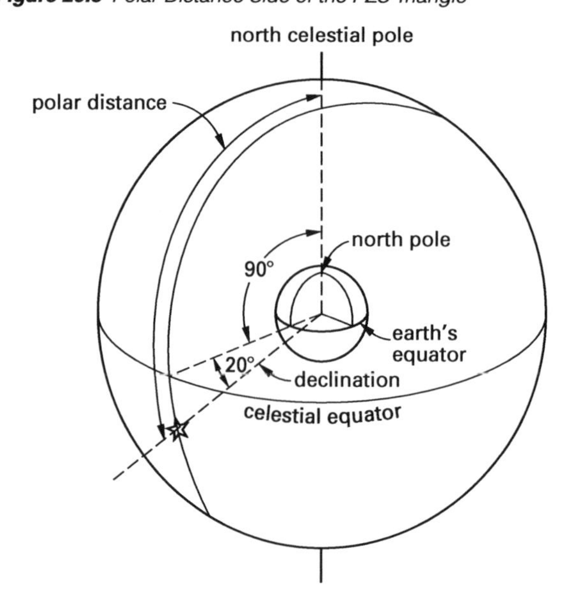

(2) The *coaltitude* is the side SZ, the arc distance from the celestial body to the observer's zenith. It is determined by subtracting the observed altitude of the celestial body (corrected for refraction and parallax) from 90°. Altitude is the vertical angle measured from the observer's horizon to the celestial body (see Fig. 20.9).

(3) The *colatitude* is the side PZ. It is the distance from the celestial north pole to the zenith. It is determined by subtracting the latitude of the observer from 90° (see Fig. 20.10).

The three interior angles of the PZS triangle are known as (1) the parallactic angle, (2) the azimuth or zenith angle, and (3) the angle t (see Fig. 20.11).

(1) The *parallactic angle* (angle S) is formed by the polar distance side and the coaltitude side.

(2) The *azimuth angle* (angle Z) is formed by the coaltitude side and the colatitude side. It is used to find the azimuth from the observer to the celestial body. When the celestial body is in the east, the azimuth angle is equal to the true azimuth. When the celestial body is in the west, true azimuth equals 360° minus the azimuth angle.

(3) The angle at the pole P formed by the colatitude side and polar distance side is known as the angle t, or the angle P.

Figure 20.9 *Coaltitude Side of the PZS Triangle*

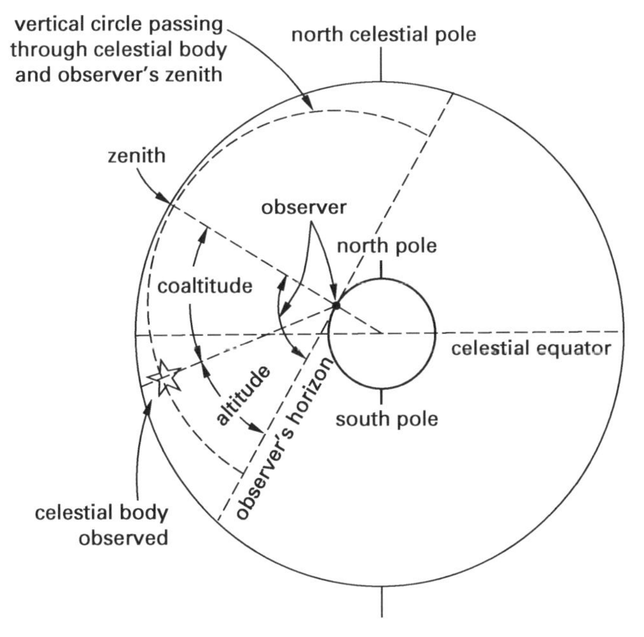

Figure 20.10 *Colatitude Side of the PZS Triangle*

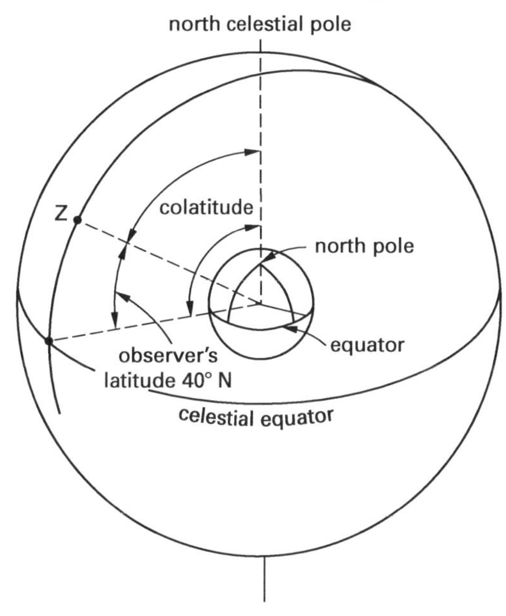

Figure 20.11 *Interior Angles of the PZS Triangle*

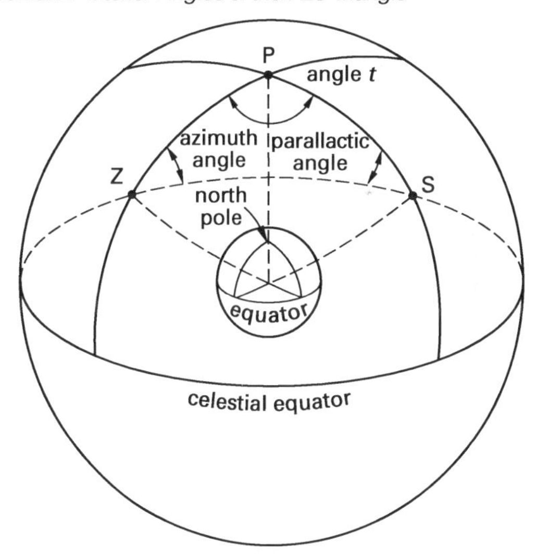

If any three elements of the PZS triangle are known, the other elements can be found by spherical trigonometry. However, each astronomical triangle is changing constantly because of the apparent rotation of the celestial sphere. The observer, then, must know the position of the PZS triangle at the time of the observation. Information concerning the position of the celestial bodies can be found in an ephemeris. (An *ephemeris* is similar to an almanac. It contains tables showing the positions of celestial bodies on certain dates in sequence. Typically, ephemerides are published by manufacturers of surveying instruments.)

6. TIME

Because all celestial bodies are in constant apparent motion with respect to the observer, it is extremely important to know the precise time of an observation on a celestial body.

In practical astronomy, there are two categories of time: *sun time* and *sidereal time*. Both categories of time are based on the rotation of the earth with respect to a standard reference line.

Because the earth revolves around the sun in the plane of its orbit once each year, the reference line to the sun is changing constantly and the length of one solar day is not the true time of one rotation of the earth.

In practical astronomy, the true time of one rotation of the earth, which is known as the *sidereal day*, is based on one rotation with respect to the vernal equinox.

Explaining the difference between the solar day and the sidereal day requires temporarily abandoning the theory that the earth is stationary and that the celestial sphere is revolving and returning to the true condition that the earth revolves around the sun.

Figure 20.12 Difference Between Solar Day and Sidereal Day

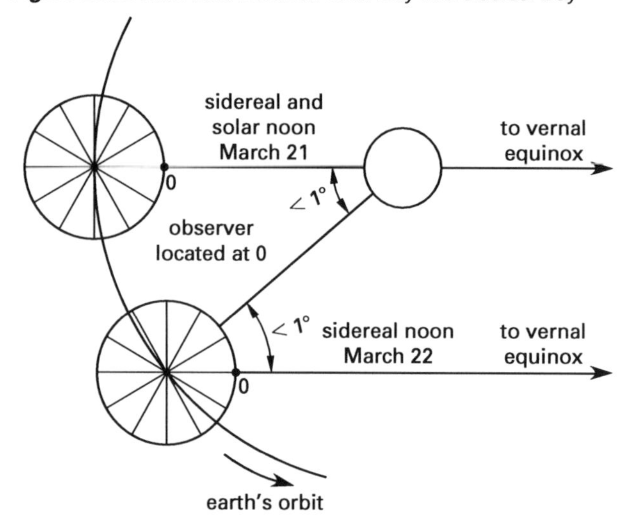

The earth completes one revolution around the sun in 365.2564 days, although one year is 365 calendar days. In Fig. 20.12, an observer at zero on the earth on March 21 (the vernal equinox) would find the sun directly overhead at noon. From March 21 until the summer solstice, the observer would find that the sun is not directly overhead at noon, but advances about 1° north per day. This motion of the sun makes the intervals between the sun's transits (stated above) of the observer's meridian greater by about three minutes, fifty-six seconds than the interval between transits of the vernal equinox of the observer's meridian. Therefore, the solar day is about three minutes, fifty-six seconds longer than the sidereal day.

Because one apparent rotation of the celestial sphere is completed in one sidereal day, a star rises at nearly the same sidereal time throughout the year. On solar time, it rises about four minutes earlier from night to night, or two hours earlier from month to month. Thus, observed at the same hour night by night, the stars seem to move slowly westward across the sky as the year lengthens.

An apparent solar day is the interval between two successive transits of the sun over the same meridian. Because of the earth's tilt and the variation in the earth's velocity about the sun, the interval between two transits of the sun over the same meridian varies from day to day. This makes it impossible to use the variable *day* as a basis for accurate time. Therefore, a fictitious, or mean, sun was devised that is imagined to move at a uniform rate in its apparent path around the earth. It makes one apparent revolution around the earth in one year, the same as the actual sun. The average apparent solar day was used as a basis for the mean day. The time indicated by the position of the actual sun is called *apparent solar time.*

The difference between mean solar time and apparent solar time is called the *equation of time* (EOT). It varies from minus 14 min to plus 6 min (see Fig. 20.13). The value of EOT for any day can be found in an ephemeris.

Figure 20.13 Equation of Time

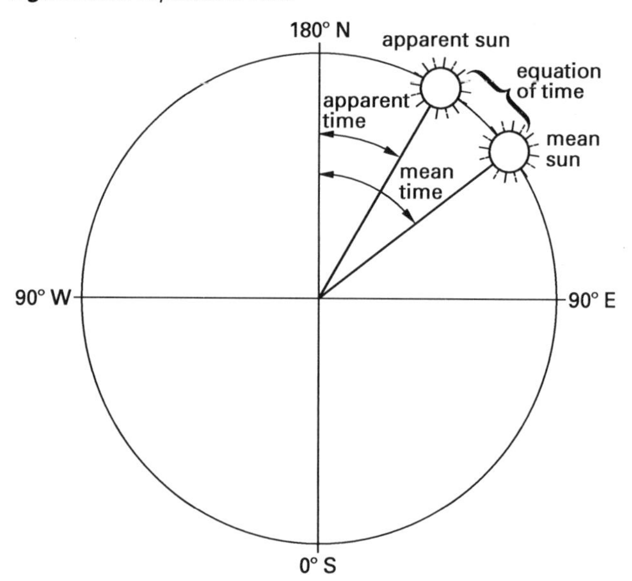

In mean solar time, the length of the year is divided into 365.2422 mean solar days. Because the mean sun appears to revolve around the earth every twenty-four hours of the mean time, the apparent rate of movement of the mean sun is 15° of arc, or of longitude, per hour ($360° \div 24$ hr $= 15°/\text{hr}$).

In the system of latitude and longitude on the earth, the zero reference for latitude is the equator; the zero reference for longitude is the meridian that passes through

Greenwich, England (the prime meridian, or 0° longitude). Using the Greenwich meridian as a basis for reference, time at a point 15° west of the Greenwich meridian is one hour earlier than the time at the Greenwich meridian because the sun passes the Greenwich meridian one hour before it crosses the meridian lying 15° to the west. The opposite is true along the meridian lying 15° to the east, where time is one hour later, because the sun crosses this meridian one hour before it arrives at the Greenwich meridian. Therefore, the difference in local time between the two places equals their difference in longitude (see Fig. 20.14).

Figure 20.14 *Apparent Motion of the Sun*

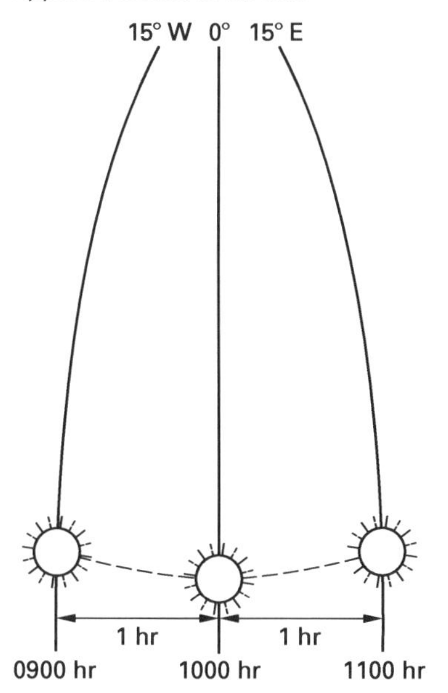

Because the mean solar day has been divided into twenty-four equal units of time (hours), there are twenty-four time zones, each 15° wide, around the earth. Using the Greenwich meridian as the central meridian of a time zone and as the zero reference for the computation of time zones, each 15° zone extends $7^1/_2°$ east and west of the zone's central meridian (see Fig. 20.15).

Figure 20.15 *Time Zone Boundaries*

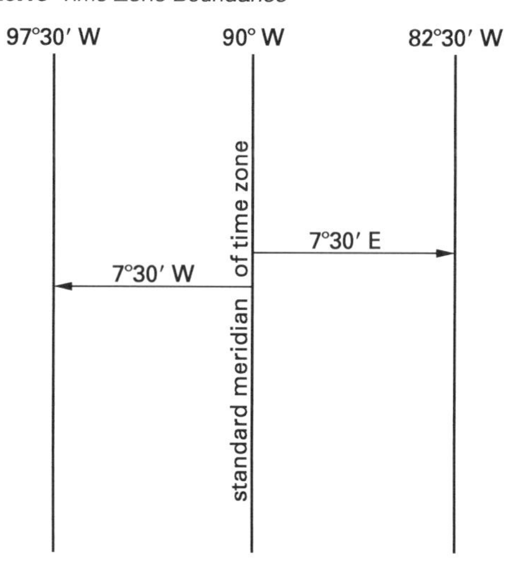

The central meridian of each time zone, east or west of Greenwich, is a multiple of 15°. For example, the time zone of the 90° meridian extends from 82°30′ to 97°30′. Each 15° meridian or multiple thereof east or west of the Greenwich meridian is called a *standard time meridian.* Four of these meridians (75°, 90°, 105°, 120°) cross the continental United States (see Fig. 20.16).

Figure 20.16 *United States Standard Time Zones*

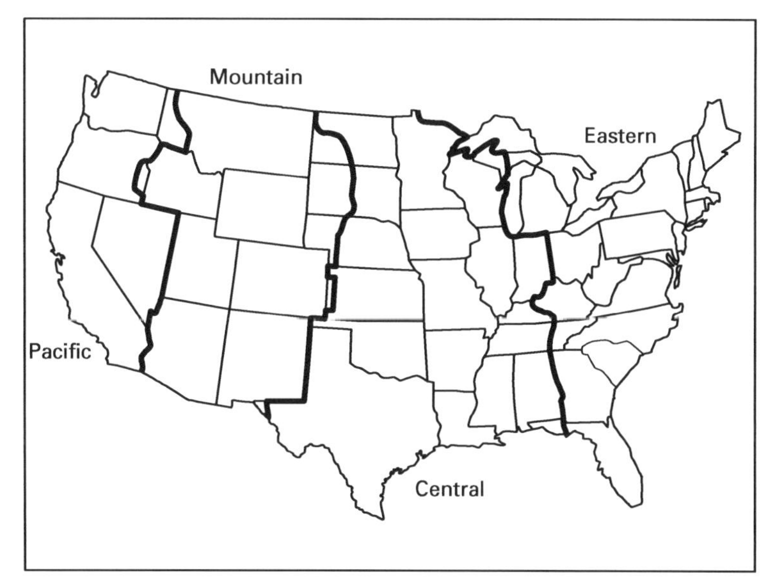

Geodesy/Survey Astronomy

Figure 20.17 World Time Zone Map

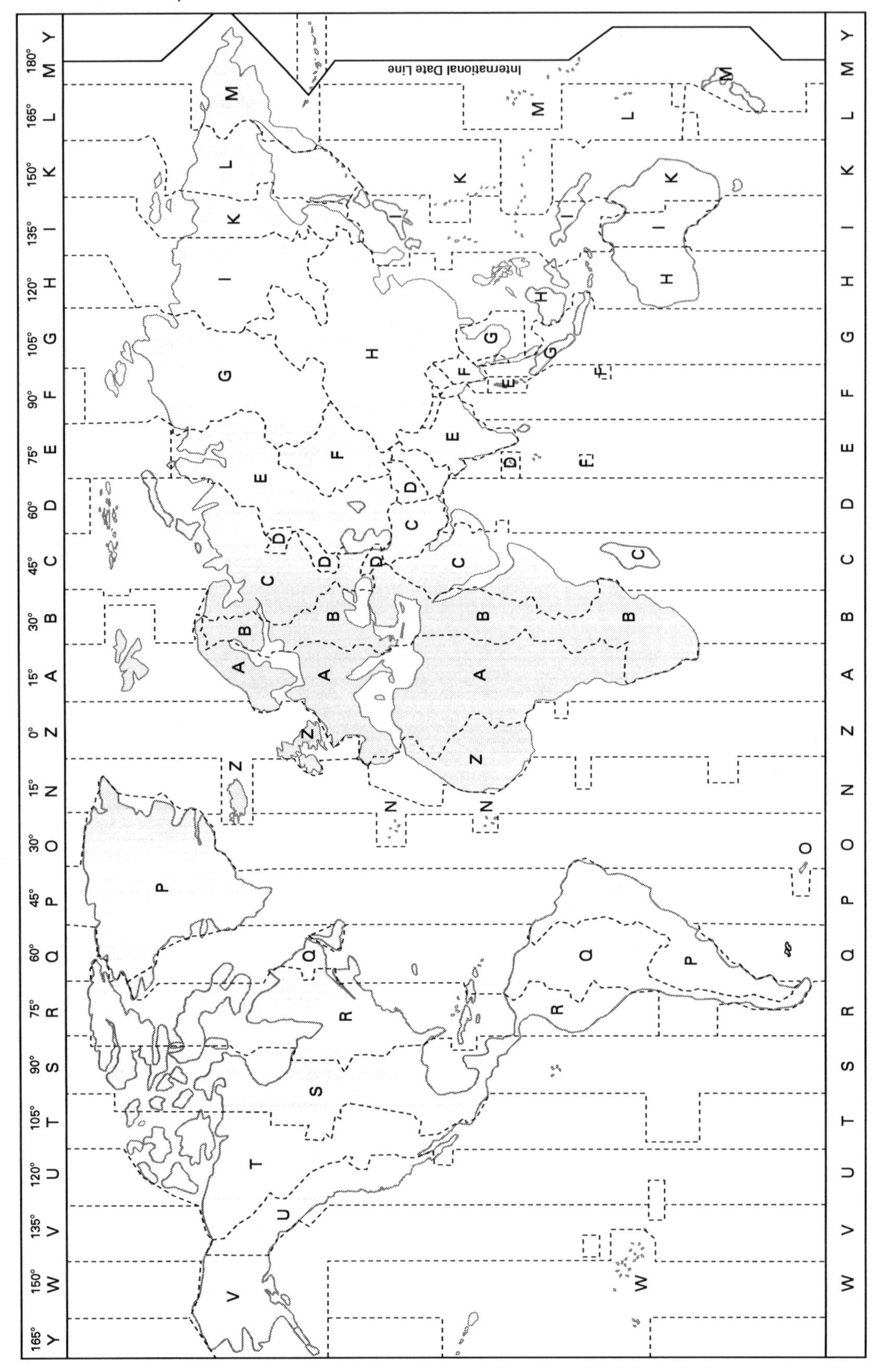

Standard time zones in the United States are named Pacific, Mountain, Central, and Eastern. Standard time zone boundaries often run along state boundaries so that time is the same over a single state. In Fig. 20.17, time zones of the world are designated by letters of the alphabet.

Standard time in any zone is referred to a *local mean time* (LMT). It is clock time where the observer's position is located. It does not take into account daylight savings time.

Rather than establishing a reference meridian for measuring time in each time zone, it was decided to establish a single reference line for all parts on the earth. Standard time zone Z (see Fig. 20.17), which uses the Greenwich meridian as its basic time meridian, was chosen for computing data pertaining to mean solar time. Greenwich standard time is also *Greenwich mean time* (GMT). It is referred to as *Universal Time* in an ephemeris. *Central Standard Time* (CST) is standard time zone S.

Time can be expressed as the reading of the standard twenty-four hour clock at the Greenwich Observatory at the moment an observation is made on a celestial body; therefore, it is the same time throughout the world. Because the observer's watch is usually set to the local standard time (local mean time, or LMT), a conversion must be made from LMT to GMT. Data in an ephemeris are based on the Greenwich meridian and zero-hour Greenwich mean time.

To convert LMT to GMT when the observer is located in west longitude, divide the value of the central meridian of the time zone in degrees of longitude by 15°. This equals the time zone correction in hours. The difference in time between the standard time zone of the observer's position and GMT must be added to the LMT to arrive at the Greenwich mean time of observation (see Fig. 20.18). If the result is greater than twenty-four hours, the amount over twenty-four hours is dropped and one day is added to obtain the Greenwich time and date. If the observer is located in east longitude, the difference is subtracted.

When observers sight the sun, it is obvious that they observe the apparent sun and not the mean sun on which their time is based. Therefore, they must convert mean time to apparent time, which is done by converting GMT at the point of observation to *Greenwich apparent time* (GAT). The observers first convert LMT to GMT. This is done by adding the time zone correction (see Fig. 20.18) to LMT. In the Central time zone, six hours would be added. To obtain GAT, the equation of time is added to GMT for observations in west longitude and subtracted from GMT for east longitude. The equation is found in an ephemeris, using the date and time of observation. In summary, for west longitude,

local mean time

+ time zone correction

= GMT

+ equation of time 0 hr (from ephemeris)

+ daily change (from ephemeris)

= GAT

Figure 20.18 *Time Zone Correction*

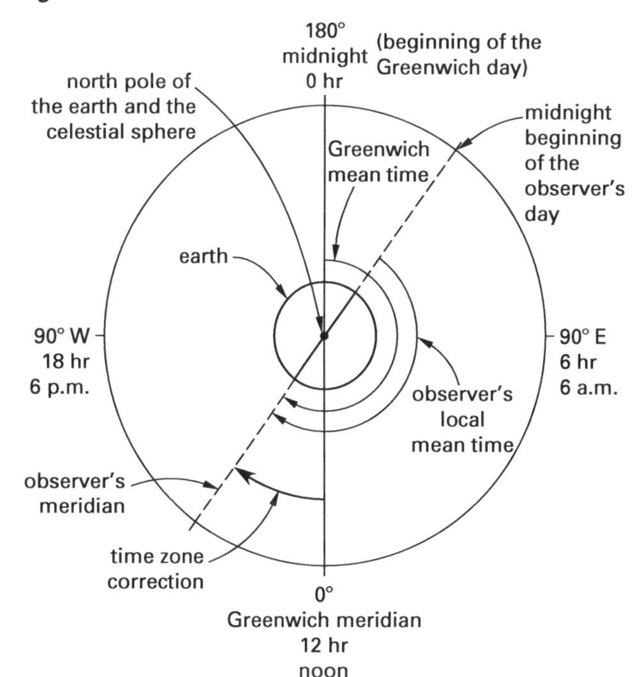

An *hour angle* is any great circle on the celestial sphere that passes through the celestial poles. It corresponds to a meridian on the earth.

The *observer's meridian* is the great circle on the celestial sphere that passes through the celestial poles and the observer's zenith.

The *hour angle* of a celestial body is the angle at the celestial poles between the plane of the meridian of the observer and the plane of the hour angle of the celestial body. In Fig. 20.19, it is shown as the angle P; it is also known as the angle t.

The *Greenwich hour angle* (GHA) of a celestial body is the time that has elapsed since the body crossed the Greenwich meridian (projected on the celestial sphere).

The *local hour angle* (LHA) of a celestial body is the angle measured along the plane of the celestial equator from the meridian of the observer (zenith) to the meridian of the celestial body (projected on the celestial sphere).

For west longitude, where λ is the longitude of the observer's position,

$$\text{LHA} = \text{GHA} - \text{W}\lambda \qquad 20.1$$

Geodesy/Survey Astronomy

Example 20.1

The observer's longitude is 98°30′00″ W, and the local mean time of observation is 09 hr 00 min 00 sec CST, 24 April 68. Determine the angle t for the solution of the PZS triangle.

Solution

	09 hr	00 min	00 sec	
+	06	00	00	time zone correction (90° W ÷ 15° = +6)
	15	00	00	GMT of observation
+		01	44	equation of time for 0 hr GMT*
+		00	07	equation of time for partial day*
	15	01	51	GAT of observation
−	12	00	00	GHA measured from noon
	03	01	51	(GHA) = 45°27′45″
	45°	27′	45″	
	360°			(if necessary)
	405°	27′	45″	
−	98°	30′	00″	longitude of observer
	306°	57′	45″	LHA of the sun
	53°	02′	15″	angle t (360° − 306°57′45″)

*from ephemeris

The following factors can be used in converting hours to degrees and degrees to hours.

24 hr	= 360°	360°	= 24 hr
1 hr	= 15°	1°	= 4 min
1 min	= 15′	1′	= 4 sec
1 sec	= 15″	1″	= 0.067 sec

Figure 20.19 *Time, Hour-Angle Relationship, and West Longitude*

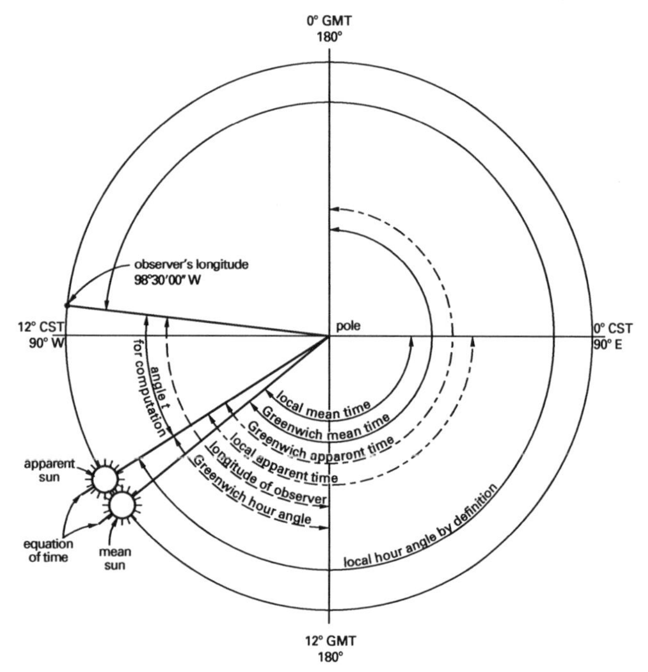

As mentioned previously, Greenwich mean time, which is also Greenwich standard time, is referred to in an ephemeris as *Universal Time*, or UTC (Coordinated Universal Time). This time is broadcast by the radio station WWV of the National Bureau of Standards and can be received on receivers that are pretuned to WWV.

The time signals can be used to determine a more precise time called UT1. UT1 is obtained by adding a small correction called DUT (the difference between UTC and UT1; UT1 = UTC + DUT).

The DUT correction can be determined by listening carefully to the WWV time signal. Following a minute tone, there will be a number of double ticks. Each double tick represents a correction of 0.1 sec and is positive for the first 7 sec. Beginning with the ninth second, each double tick is a negative correction. For example, a voice on the radio will announce "15 hr 36 min." Just after this will be a minute tone followed by the double ticks. This occurs for each minute.

The DUT correction changes 0.1 sec periodically, but not uniformly; it does not change rapidly. It may remain constant for a week or more in some instances.

Ephemerides are based on UT1 time for observations on the sun. Data needed for azimuth calculations are tabulated in an ephemeris for each day of the year for 0 hr Universal Time, so that it is possible for the day of the month at Greenwich to be one day later than the date of the observation. At 6:00 p.m. CST it is midnight at Greenwich, so that for observations on Polaris after 6:00 p.m. CST, one day would be added to the local date to find the Greenwich date to enter the tables in an ephemeris. For most sun shots, Greenwich date and local date will be the same.

The *sidereal day* is defined by the time interval between successive passages of the vernal equinox over the upper meridian of a given location. The *sidereal year* is the interval of time required for the earth to orbit the sun and return to the same position in relation to the stars. Because the sidereal day is 3 min 56 sec shorter than the solar day, this differential in time results in the sidereal year being one day longer than the solar year, or a total of 366.2422 sidereal days. And, because the vernal equinox is used as a reference point to mark the sidereal day, the sidereal time for any point at any instant is the number of hours, minutes, and seconds that have elapsed since the vernal equinox passed the meridian of the point.

The general steps for converting local mean time of observation to local hour angle (then to the interior angle t at the pole) from sidereal time are as follows.

(1) Greenwich mean time of observation is determined the same way as solar time is determined.

(2) Sidereal time for 0 hr GMT plus the correction for GMT (from an ephemeris) determines the Greenwich sidereal time of observation.

(3) Greenwich sidereal time of observation minus the right ascension of the star (from an ephemeris) equals the Greenwich hour angle.

(4) Greenwich hour angle plus the observer's longitude if in east longitude (or minus the observer's longitude if in west longitude) is the local hour angle of the star.

(5) Angle t, the interior angle of the PZS triangle at the pole, equals the local hour angle when the star is the east. The specific steps performed in determining the local hour angle = angle t are as follows.

corrected watch time
+ time zone correction
= GMT
+ sidereal time for 0 hr (from ephemeris)
+ correction for partial day (from ephemeris)
= Greenwich sidereal time
– right ascension from star (from ephemeris)
= GHA
– W longitude (+for E longitude)
= LHA = angle t

In general, it can be stated that observations on the sun involve apparent solar time, while observations on the stars are based on sidereal time. The computations using either apparent solar time or sidereal time are similar in that they do nothing more than fix the location of both the celestial body and the observer in relation to the Greenwich meridian. Once a precise relationship has been established, it is a simple matter to complete the determination of azimuth to the celestial body.

7. METHODS AND TECHNIQUES OF DETERMINING AZIMUTH

There are two methods of determining azimuth by astronomical observations: the *altitude method* and the hour *angle method*. Both methods require a horizontal angle from an azimuth mark on the ground to the observed body (sun or star) in order to establish azimuth on the ground. The basic difference between the two methods is that the altitude method requires an accurate vertical angle measurement but does not require precise time; the hour angle method requires precise time but does not require a vertical angle measurement.

In both methods, it is extremely important that the instrument be leveled carefully. The vertical axis must be truly vertical; if it is not, the error caused by the inclination of the horizontal plane will not be eliminated by a reversal of the telescope between sights. Centering the plate level bubbles on the instrument makes the vertical axis of the instrument truly vertical, provided that the plate levels are in perfect adjustment.

Figure 20.20 *Measuring Horizontal and Vertical Angles*

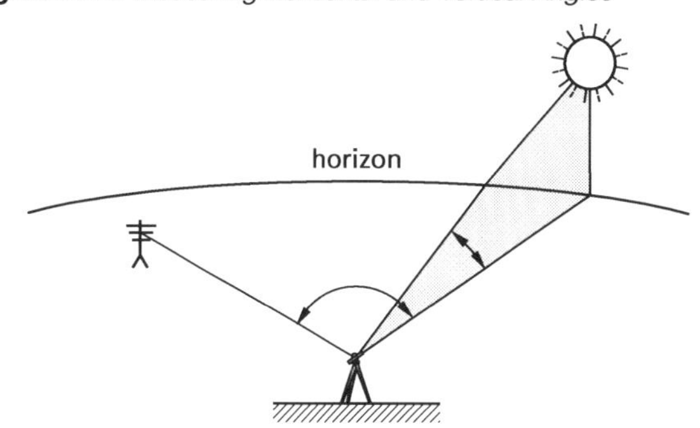

In the altitude method of determining azimuth, the PZS triangle is solved by using the three sides of the triangle. In addition to the horizontal angle from a ground point to the celestial body, three elements are necessary and must be determined: (1) the latitude for determining the colatitude side (colatitude equals 90° minus latitude), (2) the declination of the celestial body (angular distance from the celestial equator to the celestial body) for determining the polar distance side of the PZS triangle (polar distance equals 90° minus declination), and (3) the observed altitude (vertical angle) to the celestial body for determining the coaltitude side of the triangle (coaltitude equals 90° minus corrected altitude).

In the hour angle method of determining azimuth, the azimuth angle is determined from two sides and the included angle of the PZS triangle. The sides are the polar distance and the colatitude, as explained in the altitude method. The angle at the north celestial pole, the angle t, is determined as explained in Sec. 6.

The principle advantage of the altitude method is that precise time is not required. Until recent years, timepieces that make UT1 time possible in the field were not available. Because of this, the altitude method has been widely used.

Figure 20.21 *Altitude Method*

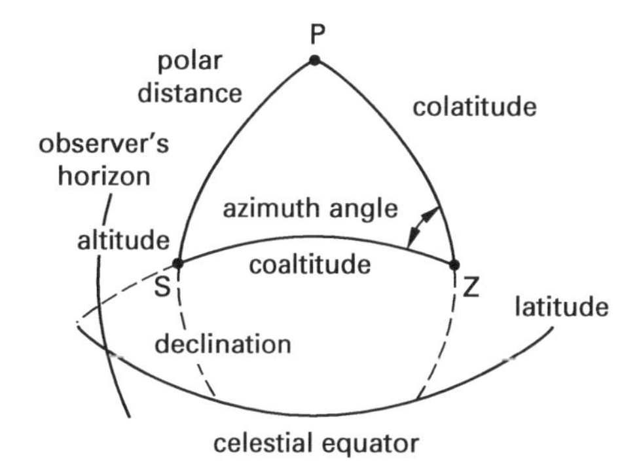

A disadvantage of the altitude method is that a vertical angle is required for observation on both the sun and the stars, which makes it necessary to set both the horizontal and vertical crosshairs tangent to the sun

Geodesy/Survey Astronomy

simultaneously. Also, measuring a vertical angle introduces the necessity of making corrections for parallax and refraction.

Advantages of the hour angle method counter the disadvantages of the altitude method. Bringing the vertical crosshair tangent to the sun without concern for the horizontal crosshair is much less difficult than simultaneous tangency. Also, eliminating corrections for parallax and refraction (for sun sights) contributes to more accurate results.

Disadvantages of the hour angle method are the cost of timepieces and the additional training required to use them.

Both methods can be used for observation on either the sun or the stars. Both methods require the determination of the latitude and longitude of the point of observation. All in all, however, the hour angle method seems to be the preferred method.

Figure 20.22 *Hour Angle Method*

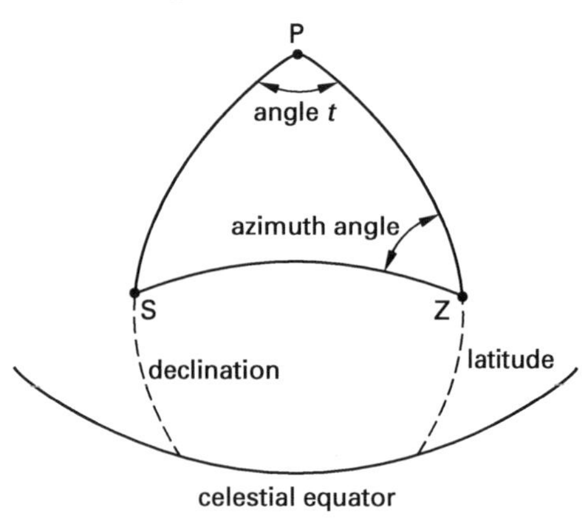

In determining the azimuth of a line, the general equation is

$$\text{az line} = \text{az sun or star} + 360° - \text{angle right} \quad 20.2$$

Example 20.2

Find the azimuth of a line when the azimuth to the sun is 78°31′24.6″ and the angle right is 346°20′18.1″.

Solution

$$\text{az line} = 78°31'24.6'' + 360° - 346°20'18.1''$$
$$= 92°11'06.5''$$

8. MAPS AND MAP READING

Topographic maps of various scales are available from the United States Geological Survey (USGS).

USGS quadrangle series maps cover areas bounded by parallels of latitude and meridians of longitude. Standard edition maps are produced at 1:24,000 scale in either 7.5′ by 7.5′ or 7.5′ by 15′ format. The 7.5′ quadrangle map is satisfactory for determining latitude and longitude by scaling in determining azimuth.

The scale of the map (see Fig. 20.23), 1:24,000, is very convenient because 24,000 in = 2000 ft exactly.

The east side of the Elm Mott map is bounded by a line representing the meridian of 97°00′ west longitude; the west side is bounded by a line representing the meridian of 97°07′30″ west longitude. The south side is bounded by a line representing the parallel of 31°37′30″ north latitude; the north side is bounded by a line representing the parallel of 31°45′ north latitude. Thus, the quadrangle formed is 7.5′ on each side. The east and west lines of the map are marked by ticks at 2.5′ intervals (31°40′ and 31°42′30″) north latitude, and the north and south lines are marked by ticks at 2.5′ intervals (97°02′30″ and 97°05′) west longitude. Connecting corresponding ticks on the east and west lines and connecting corresponding ticks on the north and south lines with lines divides the quadrant into nine 2.5′ by 2.5′ subquadrants.

If the latitude and longitude of Monument BM no. 498 are needed, first select the subquadrant that contains the monument and determine the lines of latitude and longitude that bound the subquadrant.

The borders of the Elm Mott map are not shown to scale in Fig. 20.24. The subquadrant that contains the monument is shown in Fig. 20.25.

Figure 20.23 *Scale of the Elm Mott Map*

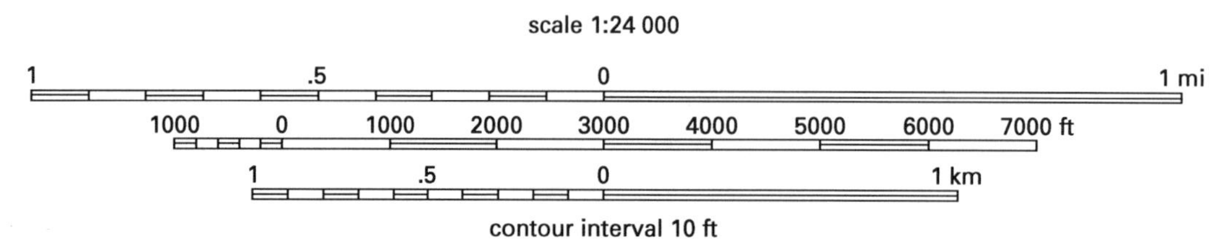

Figure 20.24 *Outline of the Elm Mott Map*

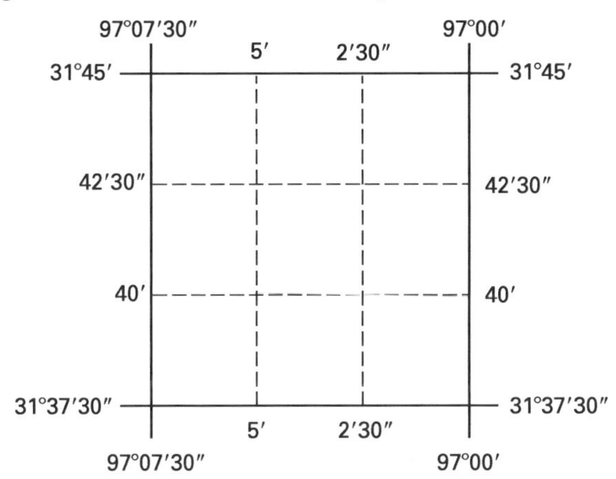

After the parallels and meridians for the subquadrant have been drawn (see Fig. 20.25), the geographic interval (angular distance between two adjacent lines) must be determined. Examination of the tick marks gives the interval. On the 7.5′ quad map, the interval is $2'30'' = 150''$. Any scale with 150 divisions may be used to find latitude and longitude. The twenty scale of an engineer's scale fits this requirement.

To find the longitude of the Monument BM no. 498, place the 0 mark of the twenty scale on meridian 97°00′ (east side because longitude increases from east to west) and the 150 mark of the scale on the meridian 97°02′30″ (west side) with the scale just above the monument. Then slide the scale downward, keeping the 0 and 150 marks on the lines, until the edge of the scale just touches the point of the monument. The reading of the scale will give the number of seconds west longitude from the meridian 97°00′. The scale reading is 66, so the longitude of the monument is 97°01′06″ west. Following the same procedure for latitude, the scale reading is 76, so the latitude of the monument is 31°43′46″ north.

Figure 20.25 *Scaling Longitude*

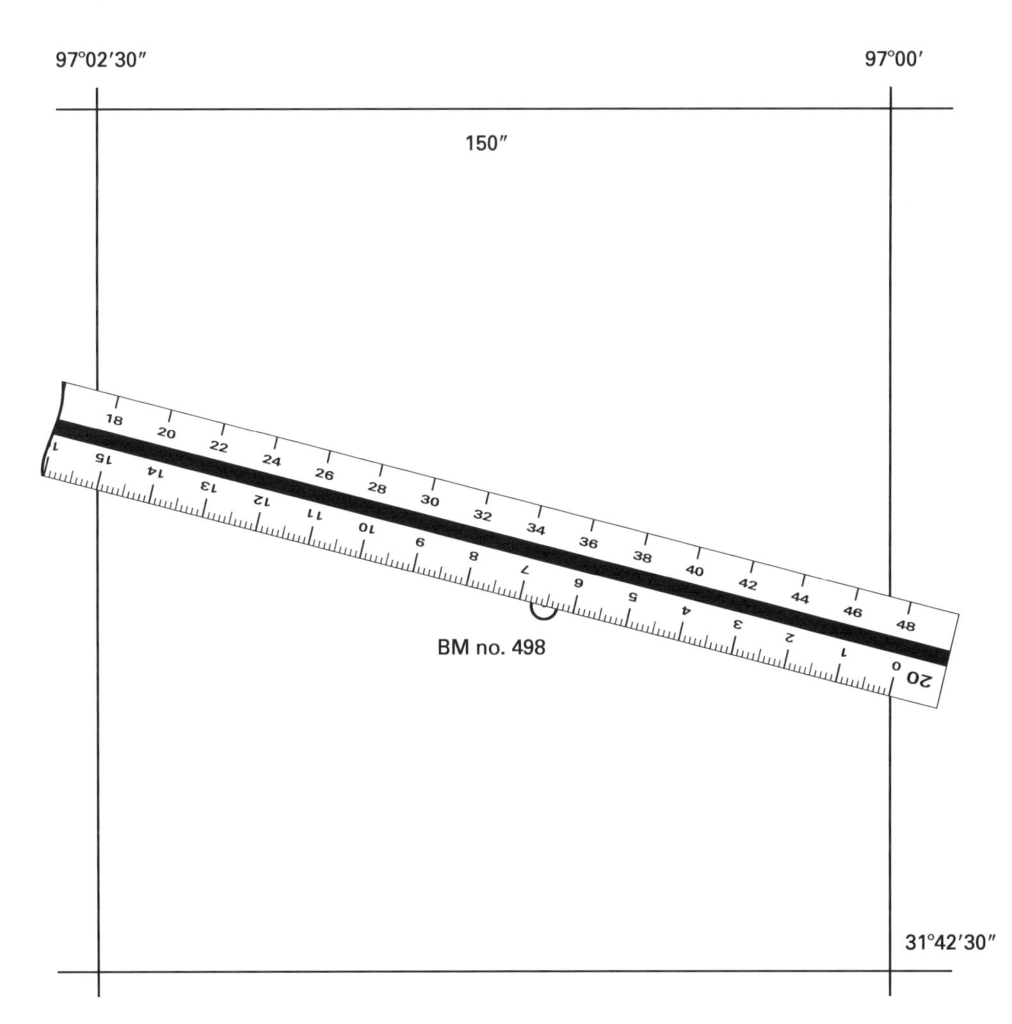

9. LOCATING POLARIS

Ancient astronomers identified some groups of stars with mythological characters, animals, and everyday objects, and named the groups accordingly. These groups of stars are called *constellations*. Two of the most well-known constellations are Ursa Major and Ursa Minor. Ursa Major translates to Great Bear, and contains the prominent configuration of stars known as the *Big Dipper*. Ursa Minor is the Little Bear and contains the Little Dipper.

As mentioned in Sec. 6, because of the difference in solar time and sidereal time, a star apparently rises about four minutes earlier from night to night, or two hours earlier from month to month. Thus, at the same hour, night by night, a constellation seems to move slowly westward across the sky.

Because of the motion of the earth, the celestial sphere appears to rotate once a day, causing a constant movement of the constellations across the sky.

Stars that never set are said to be *circumpolar*. A star with declination greater than 90° minus an observer's latitude is circumpolar at that latitude. The stars in Ursa Major can be seen as far south as 30° south latitude, while the constellation Orion straddles the celestial equator and can be seen from anywhere in the world.

Polaris appears to move in a small, counterclockwise, circular orbit around the celestial north pole. Because Polaris stays so close to the north celestial pole, it is visible throughout most of the northern hemisphere. When the Polaris hour angle is 0 hr or 12 hr, the star is said to be in its *upper* or *lower culmination*. When the Polaris local hour angle is 6 hr or 18 hr, the star is said to be in its *western* or *eastern elongation*. When Polaris is near western or eastern elongation, it appears to move slowly and vertically.

Polaris is the brightest star in Ursa Minor, which is near Ursa Major and the constellation Cassiopeia. It is the end star of the three stars making up the handle of the Little Dipper. Polaris can also be identified with respect to the Big Dipper and Cassiopeia. The two stars forming the side of the bowl farthest from the handle of the Big Dipper are called the *pointer stars*. A line through the pointer stars toward Cassiopeia nearly passes through the celestial north pole. The distance from the nearest pointer star to Polaris is about five times the distance between the two pointer stars. Polaris and Cassiopeia are on the same side of the north celestial pole.

The vertical angle to Polaris is nearly the same as the latitude of the observer, depending on the position of Polaris in its orbit. In searching for Polaris, this angle can be set on the observing instrument. The telescope should first be focused on any bright star.

When the telescope is directed at Polaris, the observer will see two other stars nearby that are not visible to the naked eye. However, Polaris will be the only star visible when the crosshairs are lighted.

The orbit of Polaris is shown in Fig. 20.26, and the constellations that serve to identify Polaris are shown in Fig. 20.27.

Figure 20.26 *Orbit of Polaris*

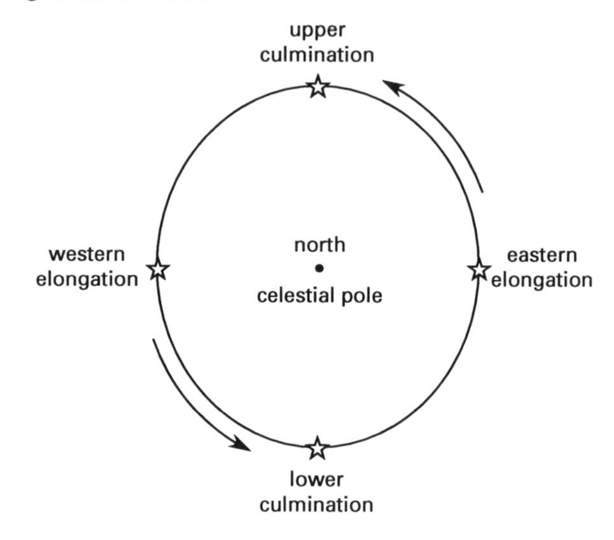

Figure 20.27 *Identification of Polaris*

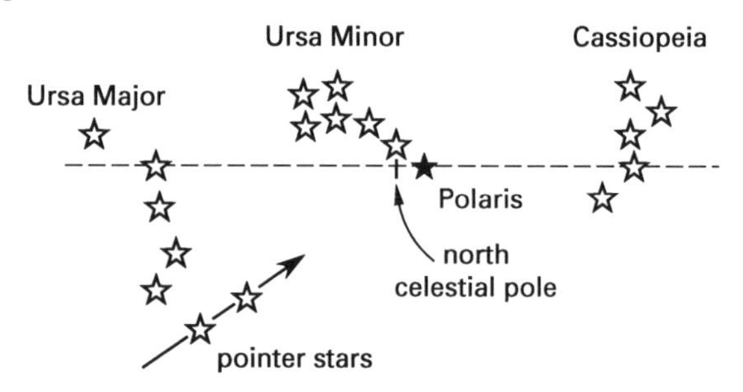

10. A SIMPLER METHOD OF DETERMINING AZIMUTH

In determining azimuth from astronomical observations, the work involved has been made easier by present-day ephemerides. Determining the Greenwich hour angle as in Ex. 20.1 has been simplified.

Ephemerides tabulate the GHA and the declination of the sun and Polaris at 0 hr Universal Time in degrees, minutes, and seconds for each day of the year, thus eliminating some of the computations in Ex. 20.1. Interpolation for an exact time of day is, of course, still necessary. The LHA is determined by Eq. 20.1 ($\text{LHA} = \text{GHA} - \text{W}\lambda$).

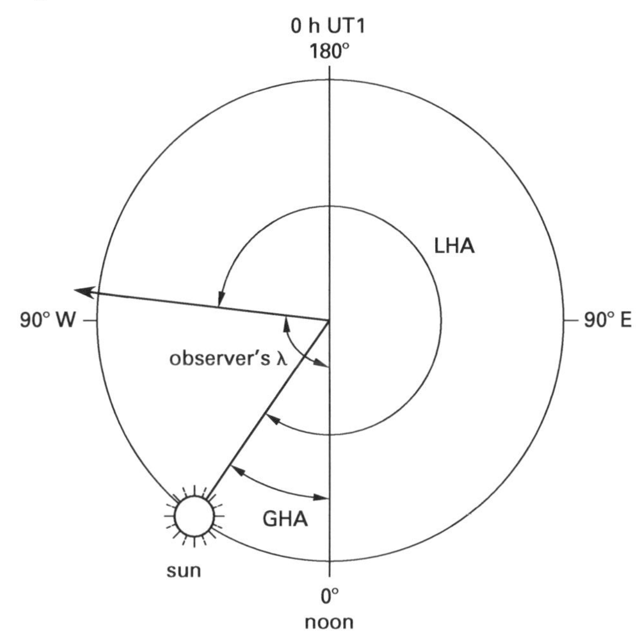

Figure 20.28 *Observations on Sun*

While single observations for azimuth can be made, it is better to increase accuracy by making at least six pointings on the sun or star and averaging the best of the computed azimuths.

There are two basic procedures in making a set of observations. The multiple foresight (MFS) procedure consists of a backsight on the mark, three foresights on the sun/star, three telescope direct, three foresights on the sun/star with the telescope reversed, then a final backsight on the mark with the telescope reversed. This method assumes that a sight on a stationary mark is more accurate than on a rapidly moving celestial body. The single foresight (SFS) procedure is a special case of the MFS procedure in that only one foresight is taken per backsight. An observation normally consists of three sets (i.e., six pointings on both the mark and the sun/star). This method is frequently used for Polaris and other large declination stars where the star motion is slow.

Once an observation has been completed, an azimuth is computed for each pointing on the celestial body. These azimuths are compared with each other with acceptable values being averaged.

Example 20.3

A solar observation is taken on June 6, 2005, by observing the left edge of the sun. All angles are turned to the right (clockwise).

Given data:

latitude = 36°04′00″ N
longitude = 94°10′08″ W
UTC when stopwatch was started = 13 hr 34 min 02 sec, DUT = −0.5 sec
stopwatch time of pointing = 0 hr 15 min 42.0 sec
backsight on mark = 180°00′05″
foresight on sun = 171°56′10″

GHA, declination, and semidiameter are taken from the SOKKIA ephemeris table shown in Table 20.1.

The following symbols are defined as:

az = azimuth of sun
az′ = azimuth of line
decl and δ = declination
ϕ = latitude
h = computed slope angle to sun
dH = semidiameter correction
DUT = UTC − UT1 (correction)

Solution

Correction to stopwatch equals UT1 when stopwatch was started.

$$\begin{aligned} \text{UTC} &= 13 \text{ hr } 34 \text{ min } 02.0 \text{ sec} \\ \text{DUT} &= -0.5 \text{ sec} \\ \hline \text{UT1} &= 13 \text{ hr } 34 \text{ min } 01.5 \text{ sec (at stopwatch} = 0{:}00{:}00) \end{aligned}$$

From Table 20.1,

$$\begin{aligned}
\text{GHA 0 hr} &= 180°21'25.6'' \\
\text{GHA 24 hr} &= 180°18'39.3'' \\
\text{decl 0 hr} &= 22°38'27.2'' \\
\text{decl 24 hr} &= 22°44'29.8'' \\
\text{semidiameter} &= 0°15'47.2'' \\
\text{UT1} &= 15 \text{ min } 42.0 \text{ sec} + 13 \text{ hr } 34 \text{ min } 01.5 \text{ sec} \\
&= 13 \text{ hr } 49 \text{ min } 43.5 \text{ sec} \\
\text{GHA} &= \text{GHA 0 hr} + (\text{GHA 24 hr} \\
&\quad - \text{GHA 0 hr} + 360°)\left(\frac{\text{UT1}}{24 \text{ hr}}\right) \\
&= 387°45'42.2'' \\
&= 27°45'42.2'' \\
\text{LHA} &= \text{GHA} - \text{W}\lambda \\
&= -66°24'25.8'' \\
&= 293°35'34.2'' \\
\text{decl} &= \text{decl 0 hr} + (\text{decl 24 hr} - \text{decl 0 hr}) \\
&\quad \times \left(\frac{\text{UT1}}{24 \text{ hr}}\right) + (0.000\,039\,5)(\text{decl 0 hr}) \\
&\quad \times \sin 7.5(\text{UT1}) \\
&= 22°41'56.1'' + 0°00'03.1'' \\
&= 22°41'59.2'' \\
\text{az} &= \tan^{-1} \frac{-\sin \text{LHA}}{\cos\phi \tan\delta - \sin\phi \cos \text{LHA}} \\
&= 83°37'04.8''
\end{aligned}$$

LHA	correction: if az is positive	correction: if az is negative
0° to 180°	180°	360°
180° to 360°	0°	180°

Table 20.1 *Sample Page from Ephemeris*

JUNE 2005

Greenwich Hour Angle for the Sun and Polaris for 0 Hr Universal Time

day	GHA (sun)			declination			eq. of time appt-mean		semi-diam.		GHA (Polaris)			declination			Greenwich transit		
	deg	min	sec	deg	min	sec	M	S	min	sec	deg	min	sec	deg	min	sec	H	M	S
1 W	180	33	59.9	22	02	19.5	02	15.99	15	47.9	210	36	54.6	89	17	07.75	9	55	54.
2 Th	180	31	40.5	22	10	19.8	02	06.70	15	47.7	211	35	47.6	89	17	07.48	9	52	00.
3 F	180	29	15.0	22	17	56.9	01	57.00	15	47.6	212	34	38.2	89	17	07.22	9	48	05.
4 Sa	180	26	43.8	22	25	10.6	01	46.92	15	47.4	213	33	26.7	89	17	06.96	9	44	10.
5 Su	180	24	07.3	22	32	00.7	01	36.48	15	47.3	214	32	13.6	89	17	06.71	9	40	16.
6 M	180	21	25.6	22	38	27.2	01	25.71	15	47.2	215	30	59.5	89	17	06.49	9	36	21.
7 Tu	180	18	39.3	22	44	29.8	01	14.62	15	47.0	216	29	45.3	89	17	06.29	9	32	27.
8 W	180	15	48.6	22	50	08.5	01	03.24	15	46.9	217	28	31.6	89	17	06.11	9	28	33.
9 Th	180	12	53.9	22	55	23.1	00	51.60	15	46.8	218	27	18.7	89	17	05.95	9	24	38.
10 F	180	09	55.6	23	00	13.4	00	39.71	15	46.7	219	26	07.0	89	17	05.80	9	20	43.
11 Sa	180	06	54.1	23	04	39.4	00	27.61	15	46.6	220	24	56.4	89	17	05.65	9	16	49.
12 Su	180	03	49.6	23	08	41.0	00	15.31	15	46.5	221	23	46.7	89	17	05.50	9	12	54.
13 F	180	00	42.7	23	12	18.1	00	02.84	15	46.4	222	22	37.5	89	17	05.34	9	08	59.
14 Tu	179	57	33.5	23	15	30.5	−00	09.76	15	46.3	223	21	28.4	89	17	05.17	9	05	05.
15 W	179	54	22.6	23	18	18.3	−00	22.49	15	46.2	224	20	18.6	89	17	04.98	9	01	10.
16 Th	179	51	10.2	23	20	41.4	−00	35.32	15	46.2	225	19	07.7	89	17	04.79	8	57	15.
17 F	179	47	56.8	23	22	39.8	−00	48.22	15	46.1	226	17	55.0	89	17	04.59	8	53	21.
18 Sa	179	44	42.5	23	24	13.4	−01	01.17	15	46.0	227	16	40.2	89	17	04.39	8	49	26.
19 Su	179	41	27.9	23	25	22.1	−01	14.14	15	45.9	228	15	23.1	89	17	04.20	8	45	32.
20 M	179	38	13.1	23	26	06.1	−01	27.13	15	45.9	229	14	04.3	89	17	04.03	8	41	38.
21 Tu	179	34	58.5	23	26	25.4	−01	40.10	15	45.8	230	12	44.5	89	17	03.90	8	37	44.
22 W	179	31	44.4	23	26	19.9	−01	53.04	15	45.8	231	11	25.0	89	17	03.80	8	33	50.
23 Th	179	28	31.0	23	25	49.7	−01	05.94	15	45.7	232	10	07.1	89	17	03.73	8	29	56.
24 F	179	25	18.6	23	24	54.8	−01	18.76	15	45.7	233	08	51.6	89	17	03.68	8	26	01.
25 Sa	179	22	07.5	23	23	35.2	−01	31.50	15	45.6	234	07	38.3	89	17	03.62	8	22	07.
26 Su	179	18	58.0	23	21	51.0	−02	44.13	15	45.6	235	06	26.5	89	17	03.55	8	18	12.
27 M	179	15	50.4	23	19	42.2	−02	56.64	15	45.5	236	05	15.0	89	17	03.45	8	14	18.
28 Tu	179	12	44.9	23	17	08.7	−03	09.01	15	45.5	237	04	02.5	89	17	03.32	8	10	23.
29 W	179	09	41.9	23	14	10.7	−03	21.21	15	45.4	238	02	48.2	89	17	03.18	8	06	29.
30 Th	179	06	41.7	23	10	48.3	−03	33.22	15	45.4	239	01	31.8	89	17	03.04	8	02	35.

Since LHA is between 180° and 360°, and az is positive, the normalized correction equals 0°.

$$\text{az to sun} = 83°37'04.8''$$

$$\begin{aligned}\text{angle rt from mark} &= 171°56'10'' - 180°00'05'' \\ &= -8°03'55'' \\ &= 351°56'05''\end{aligned}$$

$$\begin{aligned}h &= \sin^{-1}(\sin\phi\sin\delta + \cos\phi\cos\delta\cos\text{LHA}) \\ &= 31°42'44.0''\end{aligned}$$

$$\begin{aligned}\text{dH} &= \frac{\text{semidiameter}}{\cos h} \\ &= 0°18'33.4''\end{aligned}$$

left edge pointed D&R (down and right); therefore, correction dH is positive.

$$\begin{aligned}\text{angle rt from mark} &= 351°56'05'' + 0°18'33.4'' \\ &= 352°14'38.4''\end{aligned}$$

$$\begin{aligned}\text{az to mark} &= \text{az to sun} + 360° - \text{angle rt from mark} \\ &= 83°37'04.8'' + 360° - 352°14'38.4'' \\ &= 91°22'26.4''\end{aligned}$$

This same calculation procedure is used for each pointing on the sun.

Example 20.3 works equally well for Polaris and other stars. In the case of stars, the correction for semidiameter would be zero.

Computer software is now available that computes azimuths from observational data either using ephemeris tables or by generating the ephemeris internally. The latter permits azimuth calculations from the early 1900s to well into the twenty-first century.

Example 20.4

Given the following information, determine the azimuth of Polaris.

local date: June 7, 2005
UTC time: 01:56:31
latitude: N 31°38′20″
longitude: W 97°04′46″

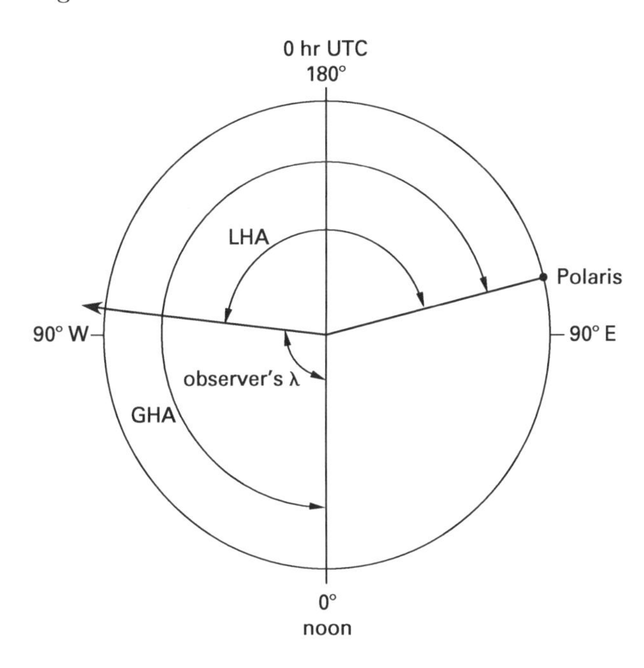

Observation on Polaris

Solution

Greenwich date is one day later than local date.

GHA = 246°26′39.9″ (from ephemeris corrected for observed time as in Ex. 20.3)

LHA = GHA − Wλ
= 246°26′39.9″ − 97°04′46.0″
= 149°21′53.9″

decl = 89°17′06.4″ (from ephemeris corrected for observed time as in Ex. 20.3)

az to Polaris = 0°27′47.0″(west of north)

11. GEODETIC AZIMUTH

The azimuth obtained from a celestial observation is known as *astronomical azimuth.* This can be converted to geodetic azimuth using Eq. 20.3.

$$\text{geodetic azimuth} = \begin{array}{l}\text{astronomic azimuth} \\ + \text{ Laplace correction}\end{array} \qquad 20.3$$

Except for mountainous areas, the Laplace correction is relatively small. Consequently, geodetic and astronomic azimuth are frequently considered to be the same. If Laplace corrections are necessary, they can be obtained from software furnished by the National Geodetic Survey (NGS).[1]

[1]Formerly a component of the U.S. Coast and Geodetic Survey.

Geodesy/Survey Astronomy

21 Global Positioning System

Nomenclature

d	distance	mi	m
h	height	ft	m
N	geoid separation	ft	m
t	time	sec	s
v	speed of light	mi/sec	m/s

1. SATELLITE GEODESY

Surveys measuring for the spatial relationships between various landmarks have long been performed. For surveys over large areas such as continents, this has traditionally been accomplished through the process of triangulation. In the United States, a high-accuracy triangulation network, with stations located on mountain peaks and on tall steel observing towers, has served as the geodetic foundation for our mapping systems for many years. The network was created by the U.S. Coast and Geodetic Survey. That agency, founded in 1807 under the administration of Thomas Jefferson, originally was charged with mapping our coasts. After it was realized that comprehensive mapping was impossible without a "skeleton" or control network, the agency's mission was expanded to include a *geodetic control network* for the nation.

Similar networks have existed for a long time within many other developed countries and continents, allowing accurate mapping within each network. Since this type of survey requires line of sight between stations, the networks typically have been limited in coverage. Furthermore, the relationship between networks was only roughly known prior to the latter part of the 1900s. Absolute positioning using conventional astronomic observations was adequate for coarse navigation, but it did not allow precise positioning. Therefore, by the middle of the 20th century, the world had a number of precise local networks with only coarsely known interrelationships.

With the advent of the space age, a new tool for mappers emerged: *satellite geodesy*. Even though the early satellites were relatively unsophisticated, they offered high, widely visible targets. Using these targets, a three-dimensional triangulation process was developed by the Coast and Geodetic Survey. Under that scheme, a passive, sun-illuminated satellite could be photographed simultaneously from three widely spaced locations, with relative positions for two of the locations being known. Using stars in the background of all three photographs, the precise orientation of each camera could be found, and subsequently the relative location of the third point could be calculated. This process required that the shutters on all three cameras "chop" the image of the satellite trail at times synchronized to within a few millionths of a second.

Figure 21.1 *Satellite Triangulation Concept*

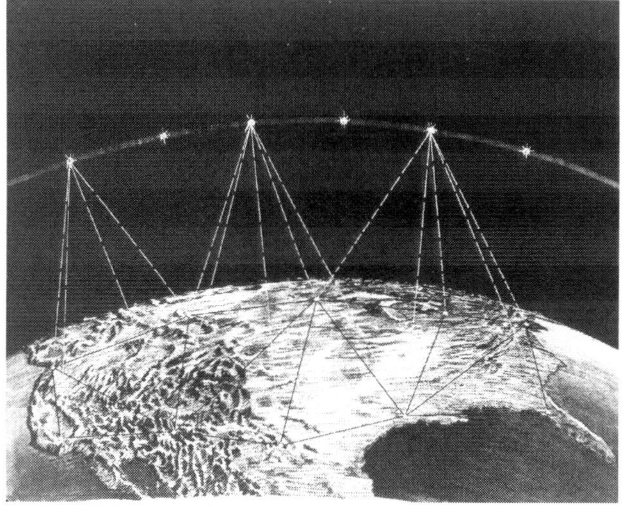

Source: U.S. National Geodetic Survey

Figure 21.2 *Typical Satellite Triangulation Station*

Source: U.S. National Geodetic Survey

Geodesy/Survey Astronomy

Beginning in 1965, with funding and support from the U.S. Department of Defense, this approach was used in an international program to create a world-wide network connecting all of the continents and many of the major islands of the world. Survey parties journeyed to some of the earth's most remote outposts, carrying with them elaborate cameras on theodolite bases and precise timing synchronization systems, to photograph an orbiting balloon against a starry night sky. Occupation of each site normally lasted six months or more, since repeated photographic observations of the satellite had to be made simultaneously with those at other stations. Those observations had to be made when the satellite was illuminated by the sun and when the skies were clear at each of the three stations. The program required tremendous effort to obtain a geographic position with an accuracy of a few meters, but it provided the first world-wide geodetic control network.

Yet even before that program was complete, an improved process of satellite geodesy was developed. With the introduction of more sophisticated satellites, the Department of Defense built the Navy Navigational Satellite System, also called TRANSIT. The primary purpose of the system was navigation. TRANSIT consisted of seven satellites in polar orbits at an altitude of about 660 mi. The satellites would pass overhead about once every 90 min, allowing at those times the determination of a position somewhere below, using the radio signals generated by the satellites as opposed to the photographic observations of satellite triangulation. TRANSIT used the Doppler shift of the satellites' signal broadcast to determine the position of the observation point. Tests indicated the system was capable of accuracies of about one meter with occupations of several days and post processing, and the equipment was only the size of a suitcase. It offered a significant improvement over the satellite triangulation process in length of occupation, size of equipment, practicality, economy, and accuracy.

2. THE GLOBAL POSITIONING SYSTEM

In the early 1980s, satellite geodesy made an even greater step forward with the *Global Positioning System* (GPS). The system was developed by the Department of Defense to allow continuous determination of geographic position anywhere in the world. The system consists of twenty-four operating satellites and several backup satellites, together with five tracking stations located around the world. The satellites orbit the earth at an approximate altitude of 12,000 mi. They are spaced to allow visibility of several satellites from any point on the earth at all times to allow continuous use. Each satellite takes about twelve hours to complete an orbit. The satellites transmit radio signals that are used for positioning determination.

SI units are frequently used for GPS, satellite geodesy, and EDM measurements (see Ch. 12). Reference data often use these units.

GPS was originally conceived as a means of precise navigation, and it has made a huge impact in that area. In addition to its application for military navigation and weaponry, the system has been made available to the public and become the standard mode of navigation for almost all aircraft and ocean vessels. It has become an essential component in fleet management for land transportation and emergency response activities. GPS has also had a significant impact on other activities of our society. With the extreme portability of modern GPS receivers and the system's continuous availability, GPS has assumed an important role in recreational activities such as hunting, fishing, hiking, and biking. It has given the public a tool for better understanding geographical positioning.

In addition to the navigational uses of GPS, the system has revolutionized the field of surveying. It has allowed for the first time the measurement of centimeter and even sub-centimeter level geographic positions without line of sight measurements to established control points. It provides a practical means of determining the geographical positions of almost all boundaries. A surveyor's role has changed from determining positions relative to local landmarks to determining geodetic positions relative to a global network. In the words of one GPS equipment manufacturer, it allows you to "locate anything, any time, anywhere."

3. GPS CONCEPTS

The basic concept of GPS is to simultaneously determine the distances, or ranges, between a point on earth and a number of satellites with known positions. Thus, the system uses the principle of resection. The distances are determined by measuring the time between signal generation at the satellites and reception at the receiver. The knowledge of this time and the velocity of the signal (considered to be the same as the speed of light, which has traditionally been 186,000 mi/sec or more precisely defined as 299 792 459 m/s).

$$d = \mathrm{v}t \qquad 21.1$$

The time delays are measured by observation of the phase shift in similar signal pulses generated simultaneously at the satellite and the receiver. Ranges from three satellites will provide a single solution for a three-dimensional point in space. Since there may be a mistake in such a solution due to any timing errors on the part of the GPS receiver, a range from a fourth satellite is necessary to determine a corrected position. Four satellites must therefore be visible to the receiver to determine a valid position. Real-time kinematic GPS surveys require at least five satellites.

The GPS satellites transmit signals on two frequencies. The primary frequency (L1) is at 1575.42 MHz, and the secondary frequency (L2) is at 1227.60 MHz. The signals include two types of code: the C/A-Code (coarse/acquisition) that is available to all users, and the P-Code (precise) that is encrypted and available only to the military. In addition to the code used by GPS receivers for measuring time delay, the satellites also broadcast a navigational message. That message includes a correction value to adjust the satellite's atomic clock to universal time coordinated (UTC), an ephemeris with information on the satellite's position in its orbit, an abbreviated ephemeris for all twenty-four satellites that allows the receiver to locate and track any visible satellites, information about atmospheric effects on the satellite signal, and information on the satellite's condition.

Example 21.1

For a signal transit time of 100 ms, determine the approximate distance from the GPS satellite to the receiver.

Solution

Using Eq. 21.1,

$$\begin{aligned} d &= \mathrm{v}t \\ &= \left(186{,}000\ \frac{\text{mi}}{\text{sec}}\right)(0.100\ \text{sec}) \\ &= 18{,}600\ \text{mi} \end{aligned}$$

4. GPS RECEIVERS

Survey-grade GPS receivers are available in a wide range of capabilities. The intended application should be carefully considered in the selection of a receiver.

One of the most significant factors that determines the accuracy of a receiver is whether it allows the collection of carrier phase measurements or only code phase measurements. GPS satellites transmit both code phase and carrier phase signals. The use of the code phase signal alone limits accuracy to the meter or sub-meter range. The addition of carrier phase measurements allows centimeter-level accuracy.

Another key performance feature for GPS receivers is dual-frequency reception. As the radio waves generated by the GPS satellites travel through the earth's atmosphere, they are deflected and slowed by *atmospheric effects*, resulting in receiver errors as large as several meters. Since the deflection is frequency dependent, the reception of both GPS frequencies allows the receiver to calculate and remove such errors in the solution for the position. Dual-frequency receivers require shorter occupation times and provide greater accuracy than single-frequency receivers. However, some single-frequency receivers have the capability to use the atmospheric information contained in the navigation message to remove some of the errors due to atmospheric deflection.

Another desirable feature is the receiver's ability to reduce or eliminate interfering signals. Even though the frequencies used by GPS are dedicated to that use, electrical interference still occurs. Better receivers can collect good observational data despite such interference.

There are a number of other features to consider, depending on the application of the receiver. These include the ability to track a large number of satellites simultaneously, the ability to receive differential corrections, the ability to set elevation masks to exclude observations from satellites low on the horizon where atmospheric effects are greatest, and the ability to output data in a format suitable for software programs or navigation.

5. GPS SURVEYING PROCESSES

Autonomous GPS is the use of a stand-alone receiver without differential corrections or post processing. Such an application is capable of accuracies of a few meters. While this application may provide sufficient accuracies for navigation, it is usually considered inadequate for survey applications.

Differential GPS requires the use of a second set of GPS observations, taken simultaneously with the first at a known position that is used as a base station. Individual corrections are determined from signals from all satellites tracked at the base station and applied to observations at the subordinate point or points. For differential GPS, it is necessary to take observations at precisely the same time at the base and subordinate stations and from the same satellites. It is important that the receivers at the subordinate points be set at the data collection rate, or at an exact multiple of that rate, at the base station to ensure this. In addition, it is important that elevation masks at the subordinate points are set at a higher elevation than the elevation used by the base to ensure that all satellites tracked at the subordinate station are also tracked at the base. Typically, the masks are set at 10° above the horizon at the base station and at 15° at the subordinate points.

For surveying applications, differential corrections are typically applied by post processing, in which the base and unknown station points are processed simultaneously by one of various software programs designed for this purpose. For carrier phase applications, observations are typically made in a network of points planned around several existing high-order control points (see Fig. 21.3). Such a network is typically designed with baseline lengths consistent with project specifications for the order of work being performed. The results of the observations are a series of baselines, or three-dimensional vectors, between the various points in the network. The spatial relationship among the points in the network are typically determined by a least-squares network adjustment, holding the control points fixed.

In many states, *high-accuracy reference networks* (HARN) have been established especially for control of GPS projects. These networks are frequently accomplished through cooperative programs between the National Geodetic Survey and the state. With a HARN available, the usual approach of a differential GPS project is to place one receiver on a control point as a base and then place other receivers at nearby points that are constrained to the base.

Figure 21.3 *Typical GPS Observation Network*

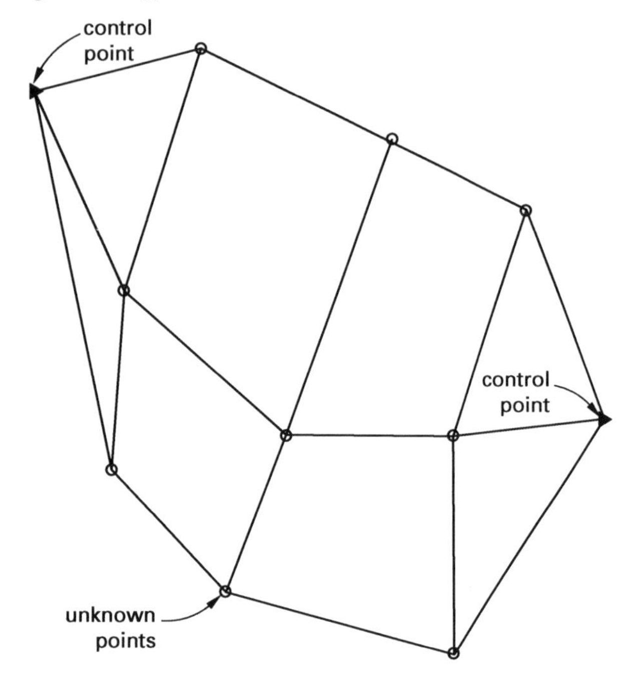

Another method of differential GPS uses data from *continuously operating reference stations* (CORS), operated as a multi-agency effort to provide accurate GPS observation data at no charge. The data are available on the internet. Often, those stations can be used as base stations, thus offering a more efficient survey operation.

Another differential GPS survey process involves the *Online Positioning User Service* (OPUS) of the National Geodetic Survey. Under that program, standalone GPS observations with durations of at least 2 hr may be submitted to the National Geodetic Survey via email. The observations are automatically processed by that agency against CORS with the corrected observations immediately returned by email. With longer observational periods of 4–5 hr, the process reportedly provides centimeter-range accuracy for horizontal coordinates and accuracy within a few centimeters for vertical coordinates.

Real-time differential GPS is a variation of differential GPS that eliminates all or most post processing. The process involves equipping the base station with a radio transmitter to broadcast the corrections and equipping the subordinate point with a radio receiver to accept such corrections. Many GPS receivers are equipped with modems and receivers for such an application, and data transmission systems for this purpose are readily available. In addition, a number of governmental and private operations broadcast GPS corrections for real-time differential applications. It is important that surveyors realize there is considerable variation in the accuracy of the various operations. For example, there is a satellite-based corrections broadcast system called Wide-Area Augmentation System (WAAS) developed by the Federal Aviation Administration and its parent the U.S. Department of Transportation to provide a horizontal position within 3 m for aviation navigation and landing control. A beacon-based U.S. Coast Guard system provides corrections in coastal areas to a design accuracy of 1–2 m. Some private satellite-based correction signals are designed to provide a sub-meter accuracy, and some CORS transmit real-time corrections that allow a centimeter-level accuracy.

An adaptation of real-time differential GPS called *real-time kinematic* (RTK) makes use of carrier-phase observation at a base station and rover stations connected with a communication link. Such systems allow precise position determination within a few seconds and are widely used for topographic surveys, construction staking, and boundary surveying in locations where a clear horizon is available and where the rover stations are sufficiently close to the base station.

Example 21.2

Using a local, publicly operated base station that collects GPS data 24 hr a day at 10 sec increments, what would be an appropriate data collection rate at which to set a receiver at an unknown station?

Solution

The collection rate could not be set at a more frequent rate, such as 1 sec or 5 sec, but it could be set at 10 sec, or a multiple of 10 sec such as 20 sec, to ensure the availability of a correction factor for each observation.

6. ELEVATION DETERMINATION WITH GPS

The concepts of geoid and ellipsoid heights are essential to understanding the application of GPS to elevations. GPS positions are determined in reference to the ellipsoid, which is a mathematically smooth surface created by rotating an ellipsoid about its smaller axis. In contrast, survey and map data are typically expressed in terms of orthometric heights. Orthometric heights refer to the geoid, which is a surface of consistent gravitational values closely coincident with the elevation of mean sea level. The ellipsoid is a predictable, mathematically perfect shape, while the geoid is an irregular surface that depends on local observations for its definition and has considerable local variation. It is necessary to know the *geoid separation* (the distance between the ellipsoid and the geoid) to convert a GPS elevation to an orthometric elevation at any given point. Those separations vary globally by well over 328 ft (100 m). Several models of the geoid are available to allow such conversion.

With a knowledge of the geoid separation, an orthometric height may be calculated from a GPS-derived ellipsoid height using the following equation. N is the geoid separation, and h is height.

$$h_{\text{ortho}} = h_{\text{ellipsoid}} - N \qquad 21.2$$

Figure 21.4 *Ellipsoid, Geoid, and Orthometric Heights*

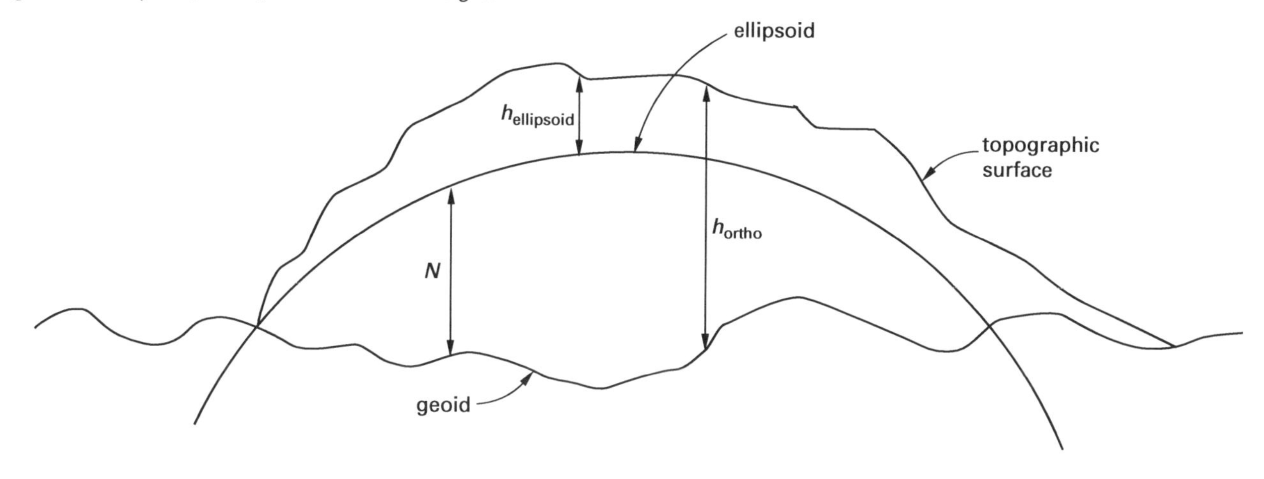

Example 21.3

With an ellipsoid height for a point derived by GPS of 407.5 ft and a geoid separation of 66.6 ft, what is the orthometric elevation for the point?

Solution

Using Eq. 21.2,

$$\begin{aligned} h_{\text{ortho}} &= h_{\text{ellipsoid}} - N \\ &= 407.5 \text{ ft} - 66.6 \text{ ft} \\ &= 340.9 \text{ ft} \end{aligned}$$

7. SOURCES OF ERROR

Surveyors using GPS should have an understanding of possible errors in the system. While GPS can be an extremely useful tool, the surveyor needs to take precuations against error and make necessary corrections to ensure the results meet the required accuracy standards.

Vague *satellite geometry* is a frequent source of errors. The greatest accuracy results when the GPS satellites are widely separated in the sky to provide well defined geometry, also known as strength of figure. If the satellites are tightly grouped, ambiguity in the intersections of the resulting distances may contribute to a weak figure. Most survey-grade receivers provide a measurement of the strength of figure called *position dilution of precision* (PDOP). Ideally, observations should be made when the PDOP is low and the number of available satellites is high. Most receivers allow an operator to set a maximum acceptable value for PDOP and will not accept observations if the PDOP exceeds that value. As a general guideline, observations should not be made with a PDOP greater than four. Prediction programs are available that allow an operator to select observation times when optimum conditions exist.

Multipath error is caused by deflection of the satellite signal by nearby objects. This can be especially problematic in urban areas with tall structures and in mountainous areas. The result of such deflections is that the measured distance between the satellite and receiver is longer than the actual distance. This type of error may be avoided by selecting locations that are clear of tall objects and terrain. An antenna equipped with a ground plane may reduce this effect, as will extended observation times that allow sampling of satellite signals from a wide range of directions.

Figure 21.5 *Multipath Error*

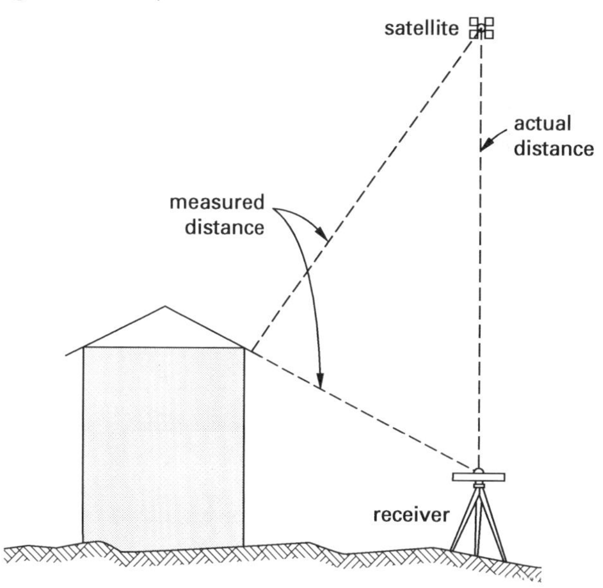

As previously mentioned in this chapter, *atmospheric effects* are another source of errors in GPS observations. These errors may be corrected using differential GPS to make sure the paths of the signals to the unknown points are consistent with those to the base station. The use of dual-frequency receivers is another means of reducing atmospheric errors.

Ephemeris errors and *satellite clock errors* can cause false readings of GPS positions. The concept of GPS is based on knowing the precise position of the satellite when it transmits a signal. While the satellites travel in predictable orbits, small discrepancies in the ephemeris are expected and may result in errors in the derived position. Similarly, while each GPS satellite has a precise atomic clock, a bias between the clock's time and UTC may result in error. The satellites are monitored from five stations around the world. Signals received at those stations are used to create precise orbital models and clock corrections, and that data is transmitted in the navigation messages. The best observational technique to reduce these types of errors is redundancy. Observations should be made using a number of different satellites and the longest duration possible.

All of these errors may be reduced through the use of conservative standards for GPS observations. There is no substitute for redundant observations under a variety of conditions, appropriate limits for baseline length, proximity of base stations, and observational locations with clear horizons. Government agencies have promulgated specifications that define safe parameters and assist surveyors in establishing accurate GPS positions.

PRACTICE PROBLEMS

1. Multipath errors are most likely to occur in a receiver in which of the following locations?

(A) open field of crops
(B) dense forest
(C) urban area with numerous tall buildings
(D) right of way of an interstate highway

2. Elevations determined by GPS are referenced to which of the following?

(A) mean sea level
(B) geoid
(C) NAVD88
(D) ellipsoid

3. Assuming the speed of light to be 186,000 mi/sec and a signal transit time of 92 ms, what is the approximate distance from the GPS satellite to the receiver?

(A) 15,000 mi
(B) 17,000 mi
(C) 19,000 mi
(D) 21,000 mi

4. Using GPS, a point's ellipsoidal elevation is determined to be 592.8 ft. From the GEOID03 model, the geoid separation for that point is −139.4 ft. What is the orthometric height above the geoid at that point?

(A) 453 ft
(B) 515 ft
(C) 627 ft
(D) 732 ft

5. Which of the following conditions would be expected with an increasing value for PDOP?

(A) more multipath errors
(B) a decrease in positional accuracy
(C) a greater number of visible satellites
(D) an increase in positional accuracy

6. The most suitable elevation masks for differential GPS surveys are

(A) 5° at the base and subordinate station
(B) 10° at the base and subordinate station
(C) 10° at the base and 15° at the subordinate station
(D) 15° at the base and 10° at the subordinate station

7. Compared with single-frequency receivers, dual-frequency receivers generally are associated with which of the following?

(A) lower PDOP values
(B) ability to track a greater number of satellites
(C) fewer errors due to atmospheric effect
(D) fewer ephemeris errors

8. With a base station data collection interval of 2 sec, which of the following would be an acceptable collection rate for a GPS receiver at a subordinate station?

(A) 1 sec
(B) 3 sec
(C) 4 sec
(D) 5 sec

9. In an area where the geoid is considerably above the earth's surface, which of the following is probably true?

(A) The area has a significant deflection of the vertical.
(B) The area is mountainous or on a high plateau.
(C) The area is below sea level.
(D) The ellipsoid is above the geoid.

SOLUTIONS

1. Of the listed options, multipath errors are most likely to occur in an urban area with numerous tall buildings. The signal can be deflected off the structures, resulting in the measured transit time being significantly longer than the time for a straight path from satellite to receiver.

The answer is (C).

2. Elevations determined by GPS are referenced to the ellipsoid.

The answer is (D).

3. Using Eq. 21.1,

$$\begin{aligned} d &= \text{v}t \\ &= \left(186{,}000\ \frac{\text{mi}}{\text{sec}}\right)(0.092\ \text{sec}) \\ &= 17{,}112\ \text{mi} \quad (17{,}000\ \text{mi}) \end{aligned}$$

The answer is (B).

4. Using Eq. 21.2,

$$\begin{aligned} h_{\text{ortho}} &= h_{\text{ellipsoid}} - N \\ &= 592.8\ \text{ft} - (-139.4\ \text{ft}) \\ &= 732.2\ \text{ft} \quad (732\ \text{ft}) \end{aligned}$$

The answer is (D).

5. With increasing PDOP, a decrease in positional accuracy would be expected.

The answer is (B).

6. With differential GPS, elevation masks of 10° at the base and 15° at the subordinate station is the best configuration to ensure all satellites tracked at the subordinate station are also tracked at the base.

The answer is (C).

7. Dual-frequency receivers generally have fewer errors due to atmospheric effects than single-frequency receivers.

The answer is (C).

8. With a base station collecting GPS observations at 2 sec intervals, the subordinate station should collect at the same rate, or at an integral multiple of the same rate, to ensure that observations at the subordinate station are made at the exact time that an observation is made at the base.

The answer is (C).

9. The geoid is the equipotential surface that best fits mean sea level. Therefore, where the earth's surface is considerably lower than the geoid, the area is probably below sea level.

The answer is (C).

22 Map Projections and Plane Coordinate Systems

Nomenclature

h ellipsoid height
H orthometric height
N geoidal separation

1. INTRODUCTION TO MAP PROJECTIONS

Most land surveying processes are ultimately used to create maps of the features surveyed. Therefore, understanding how to create precise maps based on measurements from the earth's curved face is essential. As early as the sixth century B.C., Greek scholars were proposing methods to project the earth's curvature onto a plane surface.

Many scholars have contributed to the current understanding of map projection, but possibly the greatest contribution was made by Claudius Ptolemaus (AD 100–170), also known as Ptolemy. As the last great geographer of the Roman Empire, he is often called the father of modern surveying and geography. Ptolemy is believed to have invented the terms *latitude* and *longitude* and used them to record location. In his most famous geographical work, *Guide to Geography*, he devised how to project the earth's curved face onto a plane surface.

2. THE FIGURE OF THE EARTH

The earth is often considered spherical. However, though it approximates a sphere, it flattens near the polar regions, bulges nearer the equator, and has an irregular surface even over the oceans. One approach to describing this irregular surface is to use an equipotential surface (i.e., a surface closely approximating the mean sea level where the gravity potential is equal). This surface, which can also be considered as the level of the undisturbed sea, is called the *geoid*. The uneven distribution of the earth's mass makes the geoid irregular. It has no complete mathematical expression, although models for selected areas of the geoid have been constructed.

Since it is impractical to describe the geoid mathematically, a more simple figure is used for coordinate calculation purposes. That figure is an *ellipsoid of revolution*, one that would result if an ellipse is spun about its minor axis as shown in Fig. 22.1. Ellipsoids are typically described by two factors: the flattening factor and the semi-major axis. The *flattening factor*, f, is the ratio of the difference between the length of the semi-major axis, a, and the length of the semi-minor axis, b, to the semi-major axis. It can be calculated using Eq. 22.1.

$$f = \frac{a-b}{a} \qquad 22.1$$

Figure 22.1 *Ellipse*

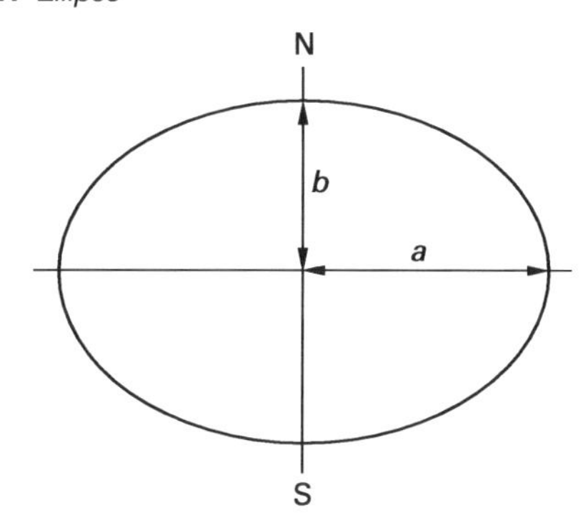

Prior to satellites, the *Clarke 1866 Ellipsoid* was used as a basis for geodetic coordinates. It worked well for North America, Central America, and Greenland, and was used as the basis for datums (i.e., reference frames used to make position measurements) such as the North American Datum of 1927 (NAD 27). In South America, the *South America 1969 Ellipsoid* served the same purpose.

With the advent of satellites, it was necessary to create an ellipsoid to fit the entire earth. As a result, an earth-centered ellipsoid identified as the *Geodctic Rcference System of 1980* (GRS 80) was developed. GRS 80 has a semi-major axis length of 6,378,137 m and a flattening factor, f, of 1/298.257222101. The United States Department of Defense uses the *World Geodetic System of 1984* (WGS 84), which has the same semi-major axis length, but a slightly different flattening factor (1/298.257223563). For most practical purposes, the differences between GRS 80 and WGS 84 are minimal and the two ellipsoids can be considered equivalent.

Most elevations used for topographic purposes are *orthometric heights*, H, or heights above the mean sea level. They are referenced to the irregular surface of the geoid. Since the height above the geoid closely approximates the height above mean sea level, it is extensively used in surveying and mapping, and is widely displayed on most charts and maps. Mean sea level varies with location; therefore, reference planes are different in each country—a factor that must be considered in international projects, such as with bridges and tunnels between countries.

Elevations produced by the *global positioning system* (GPS) are referenced to the ellipsoid since it is a predictable, mathematically perfect shape. As a result, there may be significant differences between elevations referenced to the ellipsoid (ellipsoid height, h) as determined by GPS, and elevations referenced to the geoid (orthometric height, H). It is necessary to know the *geoidal separation*, N, or the distance between the ellipsoid and the geoid, to convert a GPS elevation to an orthometric elevation at any given point. Those differences vary globally by well over a hundred meters. Models of the geoid surface relative to the ellipsoid may be derived from gravity observations or by observation of satellite orbits. World-wide models, such as the *Earth Geopotential Model 1996* (EGM 96), provide global coverage, although they are limited to accuracies of about one meter. More precise models have been developed in countries with dense gravity observation networks to provide a difference between the geoid and ellipsoid within a few centimeters. The current objective of the National Oceanic and Atmospheric Administration's (NOAA) U.S. National Geodetic Survey (NGS) is to develop a model for North America with a precision of one centimeter.

Knowing the geoidal separation, an orthometric height can be calculated from a GPS derived ellipsoid height using Eq. 22.2.

$$H = h - N \qquad 22.2$$

Figure 22.2 *Typical Relationship of the Ellipsoid and Geoid in the United States*

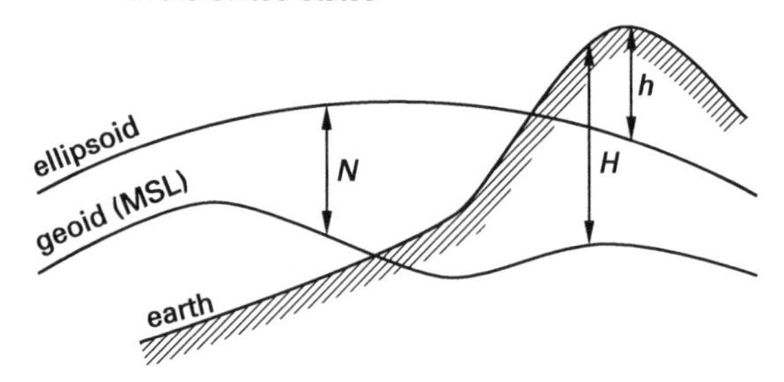

3. GEOGRAPHIC COORDINATE SYSTEMS

Most models of surveyed features use either a graphic or digital format. Survey measurements are typically related to a spatial coordinate system, even if the system is relative and not related to a geodetic datum. However, with modern survey practice, surveys using relative coordinate systems are becoming increasingly rare as most current survey data must be geo-referenced. Technological developments such as GPS, geographic information systems (GIS), and the availability of remote sensing data taken from satellites, make it increasingly important to not only produce survey data related to a geographic coordinate system, but also to be able to convert data among these various systems.

The most familiar geospatial coordinate system is the *plane rectangular Cartesian coordinate system*, which involves linear measurements in two directions from a pair of perpendicular fixed axes. Most plane surveying calculations are based on this system and on the assumption that the earth is a plane. While a plane rectangular system works well with relatively small areas, when large geographical areas are involved, or when a local survey must be related to other areas, allowances must be made for the earth's true shape.

Latitude and Longitude

A *spherical* or *spheroidal grid system* is used when it is necessary to describe a location's features using a global perspective, or when it is necessary to allow for the earth's curvature over a large area. This type of system involves a parallel system of lines of latitude measuring the distance north or south of the equator and a second system of longitude lines measuring the distance east or west of a designated north-south reference line as shown in Fig. 22.3.

Latitude is the angle measured from the center of the earth (or a spheroid representing the earth) northerly or southerly from the equator. Latitude varies from zero at the equator to 90° at the poles and must be designated as either north or south. Latitude grid lines are called *parallels*. Each degree of latitude represents approximately 69 statute miles on the surface of the earth.

Longitude is the angle measured from the center of the earth (or a spheroid representing the earth) easterly or westerly from a north-south line of reference, known as the prime meridian. Longitude lines are called *meridians*. The *prime meridian* is the meridian passing through the Royal Observatory in Greenwich, England, and is used by international agreement. Along the equator, each degree of longitude represents approximately 69 statute miles on the surface of the earth. However, that distance becomes increasingly smaller towards either pole due to convergence of the meridians. The length of a degree or distance along a meridian at any given latitude can be approximated using Eq. 22.3.

$$\text{distance} = (69 \text{ mi})\cos(\text{latitude}) \qquad 22.3$$

Although the ellipsoid defines the shape of the earth when used for a geodetic coordinate system, a horizontal datum is needed to identify the ellipsoid's origin and orientation of the coordinate system. While all regions of the earth may use a common geocentric ellipsoid (e.g., GRS 80) typically each country or region utilizes a different datum—one that best fits between the geoid and the ellipsoid in that area. A point of origin is typically chosen where the geoid-ellipsoid separation is zero; therefore, local datums are usually better than datums covering a larger region.

Figure 22.3 *Lines of Latitude and Longitude*

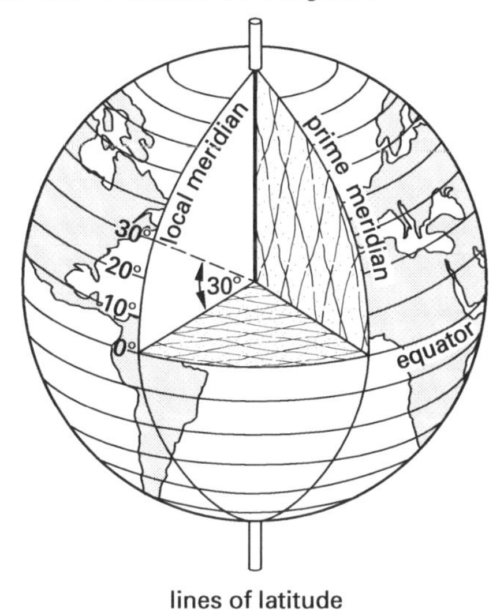

lines of latitude

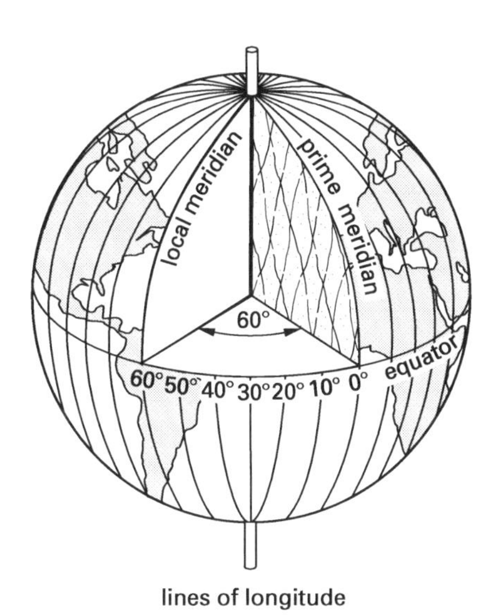

lines of longitude

Based on the GRS 80 ellipsoid, the NGS developed the *North American Datum of 1983* (NAD 83). Rather than fitting it to one point, NAD 83 was fit to the ellipsoid using observations, then taken at over 266,000 control points. The points, located in the United States, Canada, Mexico, Central America, Greenland, Hawaii, and the Caribbean Islands were fit to the ellipsoid using simultaneous least-squares adjustment creating the *National Spatial Reference System* (NSRS). Coordinates for the current adjustment of the NSRS are designated as NAD 83 (NSRS 2007). Nations in other regions of the world have developed similarly constructed horizontal datums.

When using spherical coordinates, care must be taken to ensure that all coordinates are based on the same datum. Conversions between datums can be made by using a variety of commercially available software.

National or regional vertical datums have also been developed for the standardization of heights. Vertical datums are typically referenced to the geoid, and established by reference to one or more tidal gauge stations. Because of the geographic variation in mean sea level, vertical datums vary from country to country. With the advent of GPS, ellipsoidal heights referenced to the ellipsoid are often used in lieu of heights referenced to a local datum. For many applications, ellipsoidal heights are preferable, especially when precise elevations are required, or when used for global or international applications. However, ellipsoidal heights are typically not suitable for topographic surveys since they are not based on the gravity-based geoid. Ellipsoidal coordinates can sometimes show a higher elevation where water flows downhill.

4. PROJECTIONS TO PLANE SURFACES

A systematic means is needed to project known ellipsoidal coordinates onto a two-dimensional surface, such as a map or a computer screen. (See Fig. 22.4.) Such a projection is desirable for many surveying and mapping calculations since calculations with plane coordinates are less complicated, and the results more easily used in computer-aided drafting (CAD) and GIS applications. Map projections used for this purpose typically mathematically define the relationship between coordinates on the ellipsoid and a plane.

With map projections, a plane is the ultimate surface where measurements on the surface of the earth are reduced. Although some types of projections involve projecting directly to a plane, most involve projecting to an intermediate mathematical surface, such as a cylinder or cone. That intermediate surface is then cut open and laid flat onto a plane and a coordinate grid overlaid on the plane.

Two approaches are used with the projection to the intermediate surface. The first approach is a *tangent projection* where the intermediate surface is tangent to the spheroid. A single line of contact exists around the spheroid that becomes the standard meridian or parallel for the projection. The second approach is a *secant projection* where the intermediate surface slices through a small portion of the spheroid. Two lines of contact exist with the surface of the spheroid, creating two standard parallels or meridians.

The projection process involves calculating X (east) and Y (north) Cartesian coordinates for each pair of latitude and longitude coordinates, although there are several alternative approaches used for this process. Worldwide, the two most widely used projections are the *Lambert conformal conic projection* and the *transverse Mercator projection.* The U.S. State Plane Coordinate System (See Sec. 22.5) uses either Lambert or Mercator projections, depending on the configuration of the area.

The worldwide Universal Transverse Mercator System (UTM) uses a Mercator projection.

Figure 22.4 *Basic Projection Surfaces*

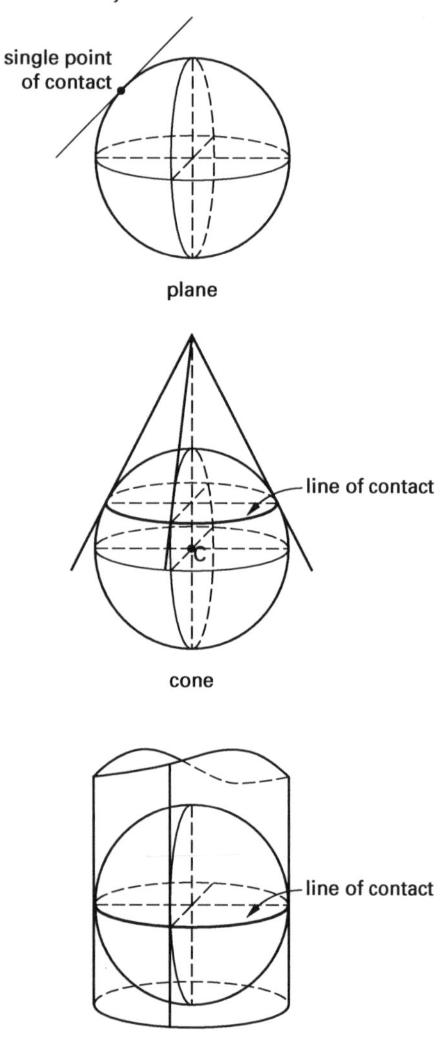

Lambert Conformal Conic Projection

The Lambert conformal conic projection, as shown in Fig. 22.5, was developed by Johann Heinrich Lambert in 1772. It has two standard parallels, and meridians are converging straight lines that meet at a point outside of the map limits, while parallels are arcs of concentric circles. The meridians and parallels meet at right angles. Scale is true only along the two standard parallels and is compressed between those lines and expanded outside of them. As a result, this projection is best for regions of greater east-west extent.

Figure 22.5 *Lambert Conformal Conic Projection*

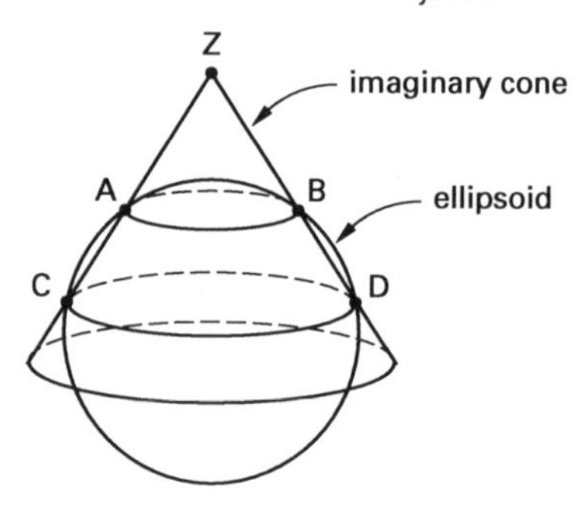

Transverse Mercator Projection

The basic Mercator projection was created by Gerardus Mercator in 1569. Then in 1772, the same year he developed the conformal conic project, Lambert developed the transverse Mercator projection as shown in Fig. 22.6. The transverse Mercator projection uses a cylinder with an east-west orientation. As used in the State Plane Coordinate System, the cylinder cuts the sphere along two meridians, parallel to the central meridian. The scale is true along the two standard meridians and is compressed betweens those lines and expanded outside of them. As a result, this projection is best for regions of greater north-south extent.

Figure 22.6 *Transverse Mercator Projection*

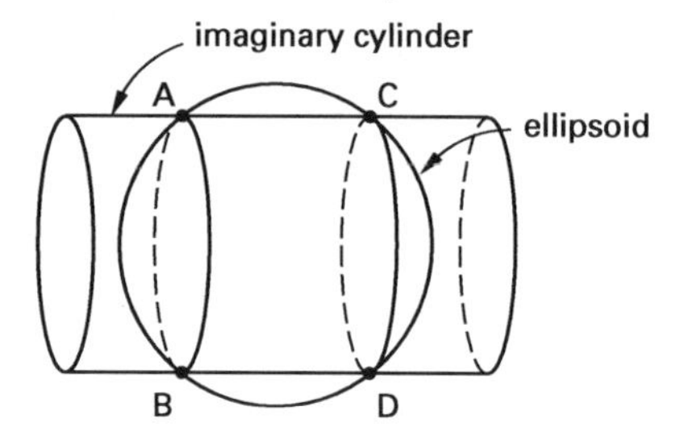

The projection process is often described as being similar to taking the peel from an orange and spreading it on a flat surface. Just as it is impossible to flatten the orange peel without tearing it, it is impossible to project a section of a spheroid surface without some distortion. Nevertheless, the distortion can be minimized by the design of the projection system and the restriction of the geographic area covered.

5. STATE PLANE COORDINATE SYSTEM

The *State Plane Coordinate System* (SPCS) is widely used in the United States, as well as in Puerto Rico and the Virgin Islands. It comprises a series of multiple zones that minimize distortion because of its projection process. Each state is given one or more zones, depending on its size and configuration. NGS provides programs on its website (www.ngs.noaa.gov) to obtain the correct zone for a given location, as well programs that perform conversions to and from the coordinate systems. Software can be downloaded, or the website's "geodetic tool kit" can be used interactively online.

With the SPCS, the Lambert conformal conic projection is used in states or portions of states that are longer in the east-west dimension. This is because it provides the best fit for projecting a rectangular zone greatest in east-west extent from a spherical surface to a plane. The transverse Mercator projection is used for areas that are longer in the north-south dimension since it provides the best results in such areas. One zone, 0101 in Alaska, uses an oblique Mercator projection where the cylinder is rotated to a predetermined azimuth for better alignment with the shape of the coverage area. Table 22.1 provides a list of the state plane coordinate zones. Information on the origin and standard lines for each zone, such as illustrated in Table 22.2, is available from the NGS.

The current SPCS is based on the NAD 83 datum and the GRS 80 ellipsoid. Coordinate units are in meters. This system supersedes the 1927 State Plane Coordinate System, which was based on the NAD 27 datum and the Clarke 1866 spheroid. The 1927 system's coordinates were in U.S. survey foot units, and calculated by multiplying the distance in meters by 3937/1200. There is no direct mathematical relationship between NAD 83 and NAD 27. Conversions between the two systems must be accomplished by comparison of a data base of known points. Public domain software known as CORPSCON, available from the U.S. Army Corps of Engineers, can be used for such conversions.

Mapping Angles and Scale Factors

The mapping angle and the scale factor are two parameters in addition to coordinates that are essential for correctly using the state plane coordinate system. In any plane projection, the grid north and geodetic north will not coincide except along the central meridian. This is because the meridians converge towards the poles, while north-south grid lines are parallel to the central meridian. The *mapping angle* (also known as the *convergence angle, grid declination,* or *variation*) is the angular difference between grid north and geodetic north. (See Fig. 22.7.) Mapping angles are either positive or negative depending on whether the location is east or west of the central meridian, regardless of the projection used.

The grid *scale factor* is a measure of the lineal distortion associated with the projection of ellipsoidal distances onto a plane surface. The scale factor is equal to 1.0 along the standard lines of the projection, but is either less or greater than 1.0 in other areas. For Lambert projections, the standard lines are the standard parallels that represent the lines of contact between the conic surface and the surface of the ellipsoid. The scale is true along the two standard parallels and is compressed between those lines and expanded outside of them. Therefore, the scale factor varies with latitude with Lambert projections. For transverse Mercator projections, the standard lines are the standard meridians that represent the lines of contact between the cylinder and the surface of the ellipsoid. The scale is true along the two standard meridians and is compressed between those lines and expanded outside of them. Therefore, the scale factor varies with longitude with transverse Mercator projections. Figure 22.8 shows scale factors in Lambert and transverse Mercator projections.

Figure 22.7 *Angle (Convergence) in a State Plane Coordinate System*

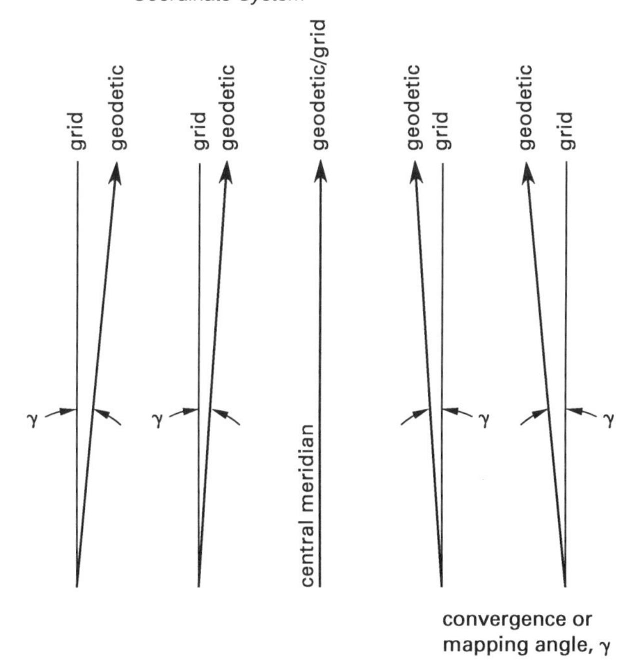

Reduction of Distances from Ground to Grid

Since measurements on the surface of the earth are arc distances and can be taken at elevations other than on the reference spheroid, reductions must be made to use them in plane coordinate systems to ensure a precise survey. This reduction involves correction for elevation above the ellipsoid, as well as using the scale factor to reduce the distance from an ellipsoid arc distance to a plane distance on the grid. The relationship between the grid, geodetic, and ground distances is shown in Fig. 22.9. Since elevations are typically expressed as orthometric elevations (in reference to the geoid), it is necessary to use the location's *geoidal separation value*, N, for the reduction. The geoidal separation value is readily obtained from benchmark data sheets, or from NGS's online geoid calculators. The formula for reducing observed distances for elevation above the ellipsoid, can be expressed as

$$S = D\left(\frac{R}{R+N+H}\right) \qquad 22.4$$

In Eq. 22.4, S is the distance on the ellipsoid, D is the observed distance, R is the mean radius of ellipsoid (a value of 20,906,000 ft is sufficiently accurate), N is the geoidal separation, and H is the elevation above the sea level (or geoid).

Table 22.1 State Zones for Plane Coordinates (NAD 83)

state	zone	projection	central meridian deg	central meridian min of arc
Alabama	E	TM	85	50
	W	TM	87	30
Alaska	1	TM	Oblique	
	2	TM	142	00
	3	TM	146	00
	4	TM	150	00
	5	TM	154	00
	6	TM	158	00
	7	TM	162	00
	8	TM	166	00
	9	TM	170	00
	10	L	176	00
Arizona	E	TM	110	10
	C	TM	111	55
	W	TM	113	45
Arkansas	N	L	92	00
	S	L	92	00
California	1	L	122	00
	2	L	122	00
	3	L	120	30
	4	L	119	00
	5	L	118	00
	6	L	116	15
Colorado	N	L	105	30
	C	L	105	30
	S	L	105	30
Connecticut		L	72	45
Delaware		TM	75	25
Florida	N	L	84	30
	E	TM	81	00
	W	TM	82	00
Georgia	E	TM	82	10
	W	TM	84	10
Hawaii	1	TM	155	30
	2	TM	156	40
	3	TM	158	00
	4	TM	159	30
	5	TM	160	10
Idaho	E	TM	112	10
	C	TM	114	00
	W	TM	115	45
Illinois	E	TM	88	20
	W	TM	90	10
Indiana	E	TM	85	40
	W	TM	87	05
Iowa	N	L	93	30
	S	L	93	30
Kansas	N	L	98	00
	S	L	98	30
Kentucky	N	L	84	15
	S	L	85	45
Louisiana	N	L	92	30
	S	L	91	20
	OF	L	91	20
Maine	E	TM	68	30
	W	TM	70	10
Maryland		L	77	00
Massachusetts	M	L	71	30
	IS	L	70	30
Michigan	N	L	87	00
	C	L	84	22
	S	L	84	22
Minnesota	N	L	93	06
	C	L	94	15
	S	L	94	00
Mississippi	E	TM	88	50
	W	TM	90	20
Missouri	E	TM	90	30
	C	TM	92	30
	W	TM	94	30
Montana		L	109	30
Nebraska		L	100	00
Nevada	E	TM	115	35
	C	TM	116	40
	W	TM	118	35
New Hampshire		TM	71	40
New Jersey		TM	74	30
New Mexico	E	TM	104	20
	C	TM	106	15
	W	TM	107	50
New York	E	TM	74	30
	C	TM	76	35
	W	TM	78	35
	LI	L	74	00
North Carolina		L	79	00
North Dakota	N	L	100	30
	S	L	100	30
Ohio	N	L	82	30
	S	L	82	30
Oklahoma	N	L	98	00
	S	L	98	00
Oregon	N	L	120	30
	S	L	120	30
Pennsylvania	N	L	77	45
	S	L	77	45
Rhode Island		TM	71	30
South Carolina		L	81	00
South Dakota	N	L	100	00
	S	L	100	20
Tennessee		L	86	00
Texas	N	L	101	30
	NC	L	98	30
	C	L	100	20
	SC	L	99	00
	S	L	98	30
Utah	N	L	111	30
	C	L	111	30
	S	L	111	30
Vermont		TM	72	30
Virginia	N	L	78	30
	S	L	78	30
Washington	N	L	120	50
	S	L	120	30
West Virginia	N	L	79	30
	S	L	81	00
Wisconsin	N	L	90	00
	C	L	90	00
	S	L	90	00
Wyoming	E	TM	105	10
	EC	TM	107	20
	WC	TM	108	45
	W	TM	110	05
Puerto Rico		L	66	26

Abbreviations

TM = transverse Mercator
L = Lambert
N = north
NC = north central
E = east
EC = east central
C = central
M = mainland
OF = offshore
S = south
SC = south central
W = west
WC = west central
LI = Long Island
IS = island

Source: U.S. National Geodetic Survey

Table 22.2 *Typical Lambert Projection Tables*

Lambert Conformal Conic Projection Tables
Arkansas north

latitude deg	min of arc	R (m)	difference	k
34	20	9062395.198	30.81870	1.00017199
34	21	9060546.076	30.81859	1.00016576
34	22	9058696.961	30.81848	1.00015961
34	23	9056847.852	30.81838	1.00015354
34	24	9054998.749	30.81828	1.00014755
34	25	9053149.652	30.81819	1.00014165
34	26	9051300.561	30.81809	1.00013583
34	27	9049451.475	30.81800	1.00013010
34	28	9047602.395	30.81791	1.00012445
34	29	9045753.320	30.81783	1.00011888
34	30	9043904.250	30.81774	1.00011339
34	31	9042055.186	30.81766	1.00010799
34	32	9040206.126	30.81758	1.00010267
34	33	9038357.071	30.81751	1.00009743
34	34	9036508.020	30.81744	1.00009228
34	35	9034658.974	30.81737	1.00008721
34	36	9032809.932	30.81730	1.00008222
34	37	9030960.894	30.81723	1.00007732
34	38	9029111.861	30.81717	1.00007250
34	39	9027262.830	30.81711	1.00006776
34	40	9025413.804	30.81705	1.00006311
34	41	9023564.781	30.81700	1.00005854
34	42	9021715.761	30.81694	1.00005405
34	43	9019866.744	30.81689	1.00004965
34	44	9018017.731	30.81685	1.00004533
34	45	9016168.720	30.81680	1.00004109
34	46	9014319.712	30.81676	1.00003694
34	47	9012470.707	30.81672	1.00003287
34	48	9010621.703	30.81668	1.00002889
34	49	9008772.702	30.81665	1.00002498
34	50	9006923.704	30.81662	1.00002116
34	51	9005074.707	30.81659	1.00001743
34	52	9003225.711	30.81656	1.00001377
34	53	9001376.718	30.81654	1.00001021
34	54	8999527.726	30.81651	1.00000672
34	55	8997678.735	30.81650	1.00000332
34	56	8995829.745	30.81648	1.00000000
34	57	8993980.756	30.81647	0.99999677
34	58	8992131.768	30.81645	0.99999361
34	59	8990282.781	30.81645	0.99999055

latitude deg	min of arc	R (m)	difference	k
35	0	8988433.794	30.81644	0.99998756
35	1	8986584.808	30.81644	0.99998466
35	2	8984735.821	30.81644	0.99998185
35	3	8982886.835	30.81644	0.99997911
35	4	8981037.849	30.81644	0.99997646
35	5	8979188.862	30.81645	0.99997390
35	6	8977339.875	30.81646	0.99997142
35	7	8975490.888	30.81647	0.99996902
35	8	8973641.900	30.81649	0.99996670
35	9	8971792.910	30.81650	0.99996447
35	10	8969943.920	30.81652	0.99996233
35	11	8968094.929	30.81655	0.99996026
35	12	8966245.936	30.81657	0.99995828
35	13	8964396.941	30.81660	0.99995639
35	14	8962547.945	30.81663	0.99995458
35	15	8960698.948	30.81666	0.99995285
35	16	8958849.948	30.81670	0.99995121
35	17	8957000.946	30.81674	0.99994965
35	18	8955151.942	30.81678	0.99994817
35	19	8953302.935	30.81682	0.99994678
35	20	8951453.925	30.81687	0.99994547
35	21	8949604.913	30.81692	0.99994425
35	22	8947755.898	30.81697	0.99994311
35	23	8945906.880	30.81702	0.99994205
35	24	8944057.859	30.81708	0.99994108
35	25	8942208.834	30.81714	0.99994019
35	26	8940359.806	30.81720	0.99993939
35	27	8938510.774	30.81726	0.99993867
35	28	8936661.738	30.81733	0.99993803
35	29	8934812.698	30.81740	0.99993748
35	30	8932963.654	30.81747	0.99993701
35	31	8931114.606	30.81755	0.99993663
35	32	8929265.553	30.81763	0.99993633
35	33	8927416.495	30.81771	0.99993611
35	34	8925567.433	30.81779	0.99993598
35	35	8923718.366	30.81787	0.99993594
35	36	8921869.294	30.81796	0.99993597
35	37	8920020.216	30.81805	0.99993609
35	38	8918171.133	30.81814	0.99993630
35	39	8916322.044	30.81824	0.99993659

zone 0301 (Arkansas north)	
constants	
ϕ_o	34°20′
λ_{CM}	92°00′
N_o	0 m
E_o	400,000 m
R_b	9,062,395.1981
$\sin B_o$	0.581899128039

Source: U.S. National Geodetic Survey

Figure 22.8 Scale Factors in Lambert and Transverse Mercator Projections

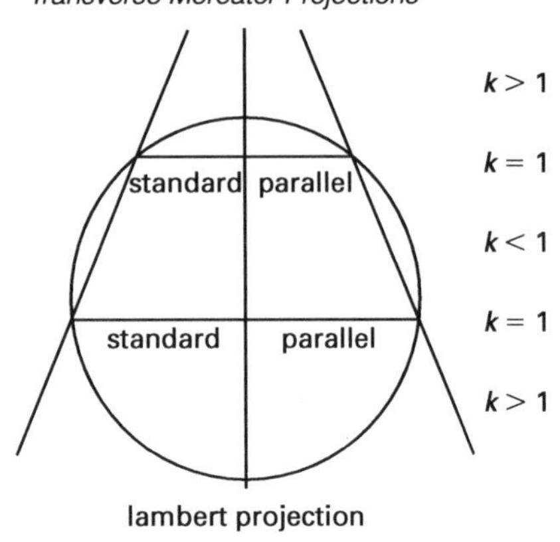

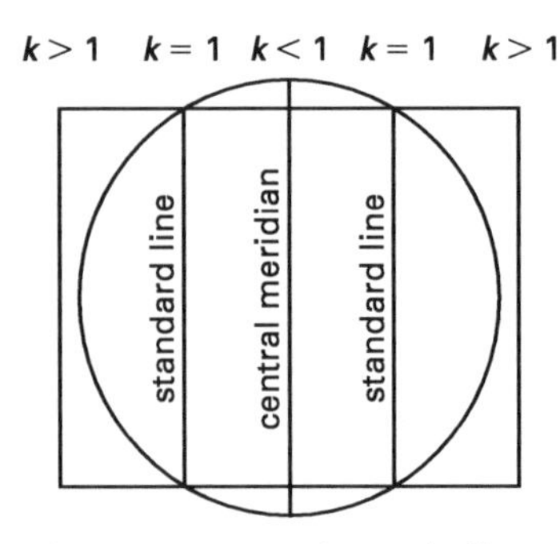

Figure 22.9 Relationship Between Grid, Geodetic (Ellipsoidal), and Ground Distances

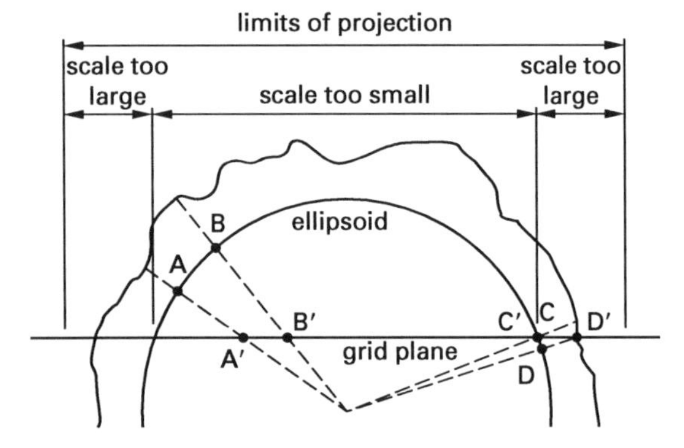

The distance A′ to B′ is smaller than geodetic distance A to B.
The distance C′ to D′ is larger than geodetic distance C to D.

In addition to the elevation correction, observed distances should be further reduced to grid distances using a location's *average scale factor*. The scale factor can be obtained from NGS, or from a projection table such as Tables 22.2, 22.3, or 22.4. The grid distance can be calculated using Eq. 22.5.

$$\text{grid distance} = (\text{scale factor})(\text{ellipsoidal distance}) \qquad 22.5$$

The scale factor is often combined with the elevation factor to form a single multiplier for reducing observed distances to plane coordinate grid distances. The two factors can be combined by taking the product of the two factors. That product, when multiplied by the observed horizontal distance, will provide the grid distance. The area for a parcel of land at ground elevation can be calculated by dividing the plane coordinate area by the square of the combined factor used to reduce measured distances.

Table 22.3 Scale Factor—East and Central Zones—Missouri

E' (m)	k	E' (m)	k	E' (m)	k
1000	0.9999333	41 000	0.9999540	81 000	1.0000141
2000	0.9999334	42 000	0.9999550	82 000	1.0000161
3000	0.9999334	43 000	0.9999561	83 000	1.0000181
4000	0.9999335	44 000	0.9999572	84 000	1.0000202
5000	0.9999336	45 000	0.9999583	85 000	1.0000223
6000	0.9999338	46 000	0.9999594	86 000	1.0000244
7000	0.9999339	47 000	0.9999605	87 000	1.0000265
8000	0.9999341	48 000	0.9999617	88 000	1.0000287
9000	0.9999343	49 000	0.9999629	89 000	1.0000309
10 000	0.9999346	50 000	0.9999641	90 000	1.0000331
11 000	0.9999348	51 000	0.9999654	91 000	1.0000353
12 000	0.9999351	52 000	0.9999666	92 000	1.0000375
13 000	0.9999354	53 000	0.9999679	93 000	1.0000398
14 000	0.9999357	54 000	0.9999692	94 000	1.0000421
15 000	0.9999361	55 000	0.9999706	95 000	1.0000444
16 000	0.9999365	56 000	0.9999719	96 000	1.0000468
17 000	0.9999369	57 000	0.9999733	97 000	1.0000492
18 000	0.9999373	58 000	0.9999747	98 000	1.0000516
19 000	0.9999378	59 000	0.9999762	99 000	1.0000540
20 000	0.9999383	60 000	0.9999777	100 000	1.0000564
21 000	0.9999388	61 000	0.9999791	101 000	1.0000589
22 000	0.9999393	62 000	0.9999807	102 000	1.0000614
23 000	0.9999398	63 000	0.9999822	103 000	1.0000639
24 000	0.9999404	64 000	0.9999838	104 000	1.0000665
25 000	0.9999410	65 000	0.9999853	105 000	1.0000691
26 000	0.9999417	66 000	0.9999870	106 000	1.0000717
27 000	0.9999423	67 000	0.9999886	107 000	1.0000743
28 000	0.9999430	68 000	0.9999903	108 000	1.0000769
29 000	0.9999437	69 000	0.9999919	109 000	1.0000796
30 000	0.9999444	70 000	0.9999937	110 000	1.0000823
31 000	0.9999452	71 000	0.9999954	111 000	1.0000850
32 000	0.9999459	72 000	0.9999972	112 000	1.0000878
33 000	0.9999467	73 000	0.9999989	113 000	1.0000905
34 000	0.9999476	74 000	1.0000007	114 000	1.0000933
35 000	0.9999484	75 000	1.0000026	115 000	1.0000962
36 000	0.9999493	76 000	1.0000044	116 000	1.0000990
37 000	0.9999502	77 000	1.0000063	117 000	1.0001019
38 000	0.9999511	78 000	1.0000082	118 000	1.0001048
39 000	0.9999521	79 000	1.0000102	119 000	1.0001077
40 000	0.9999530	80 000	1.0000121	120 000	1.0001106

Source: U.S. National Geodetic Survey

Table 22.4 Scale Factor—West Zone—Missouri

E' (m)	k	E' (m)	k	E' (m)	k
1000	0.9999412	41 000	0.9999619	81 000	1.0000219
2000	0.9999412	42 000	0.9999629	82 000	1.0000240
3000	0.9999413	43 000	0.9999639	83 000	1.0000260
4000	0.9999414	44 000	0.9999650	84 000	1.0000280
5000	0.9999415	45 000	0.9999661	85 000	1.0000301
6000	0.9999416	46 000	0.9999672	86 000	1.0000322
7000	0.9999418	47 000	0.9999684	87 000	1.0000344
8000	0.9999420	48 000	0.9999695	88 000	1.0000365
9000	0.9999422	49 000	0.9999707	89 000	1.0000387
10 000	0.9999424	50 000	0.9999720	90 000	1.0000409
11 000	0.9999427	51 000	0.9999732	91 000	1.0000431
12 000	0.9999429	52 000	0.9999745	92 000	1.0000454
13 000	0.9999433	53 000	0.9999758	93 000	1.0000476
14 000	0.9999436	54 000	0.9999771	94 000	1.0000500
15 000	0.9999439	55 000	0.9999784	95 000	1.0000523
16 000	0.9999443	56 000	0.9999798	96 000	1.0000546
17 000	0.9999447	57 000	0.9999812	97 000	1.0000570
18 000	0.9999452	58 000	0.9999826	98 000	1.0000594
19 000	0.9999456	59 000	0.9999840	99 000	1.0000618
20 000	0.9999461	60 000	0.9999855	100 000	1.0000643
21 000	0.9999466	61 000	0.9999870	101 000	1.0000668
22 000	0.9999471	62 000	0.9999885	102 000	1.0000693
23 000	0.9999477	63 000	0.9999900	103 000	1.0000718
24 000	0.9999483	64 000	0.9999916	104 000	1.0000743
25 000	0.9999489	65 000	0.9999932	105 000	1.0000769
26 000	0.9999495	66 000	0.9999948	106 000	1.0000795
27 000	0.9999502	67 000	0.9999964	107 000	1.0000821
28 000	0.9999508	68 000	0.9999981	108 000	1.0000848
29 000	0.9999515	69 000	0.9999998	109 000	1.0000874
30 000	0.9999523	70 000	1.0000015	110 000	1.0000901
31 000	0.9999530	71 000	1.0000032	111 000	1.0000929
32 000	0.9999538	72 000	1.0000050	112 000	1.0000956
33 000	0.9999546	73 000	1.0000068	113 000	1.0000984
34 000	0.9999554	74 000	1.0000086	114 000	1.0001012
35 000	0.9999563	75 000	1.0000102	115 000	1.0001040
36 000	0.9999571	76 000	1.0000123	116 000	1.0001068
37 000	0.9999580	77 000	1.0000142	117 000	1.0001097
38 000	0.9999590	78 000	1.0000161	118 000	1.0001126
39 000	0.9999599	79 000	1.0000180	119 000	1.0001155
40 000	0.9999609	80 000	1.0000200	120 000	1.0001184

Source: U.S. National Geodetic Survey

Example 22.1

An observed horizontal distance of 2640.00 ft is measured in the Arkansas North State Plane Coordinate zone at an average elevation of 1400 ft. The approximate position for the measurement is 35°30′ latitude and 92°10′ longitude. The geoidal separation is −90.47 ft. Using Table 22.2, determine the grid distance.

Solution

From Table 22.2 for 35°30′ latitude, the scale factor is 0.99993701.

Using Eqs. 22.4 and 22.5,

$$\begin{aligned}\text{ellipsoidal distance} &= D\left(\frac{R}{R+N+H}\right)\\ &= 2640.00\ \text{ft}\left(\frac{20{,}906{,}000\ \text{ft}}{20{,}906{,}000\ \text{ft}+(-90.47\ \text{ft})+1400\ \text{ft}}\right)\\ &= 2639.83\ \text{ft}\end{aligned}$$

$$\begin{aligned}\text{grid distance} &= (\text{scale factor})(\text{ellipsoidal distance})\\ &= (0.99993701)(2639.83\ \text{ft})\\ &= 2639.66\ \text{ft}\end{aligned}$$

Conversion from Geodetic to Grid Azimuth

When calculating in state plane coordinates using geodetic azimuths, the geodetic azimuth must be converted to grid azimuth using Eq. 22.6.

$$\begin{aligned}\text{grid azimuth} &= \text{geodetic azimuth} - \text{mapping angle}\\ &\quad + \text{arc-to-chord correction}\end{aligned} \qquad 22.6$$

For Lambert projections, the mapping angle, γ, can be calculated using data from published tables for each zone, as well as Eq. 22.7.

$$\gamma = (\lambda_o - \lambda)\sin B_o \qquad 22.7$$

For Eq. 22.7, λ is the longitude of point where mapping angle is needed, λ_o is the longitude of grid origin, and $\sin B_o$ is the sine of latitude of central parallel.

Calculation of mapping angles for transverse Mercator projections is more complex and is beyond the scope of this chapter. Such calculations may be accomplished using various computer software packages, including those from NGS.

In Eq. 22.6, the *arc-to-chord correction*, also known as the *second term correction*, represents the difference between the projected geodetic and grid azimuths caused by the curvature of the meridians. This term is small and may be neglected for most surveys. For large area surveys, however, this term may be significant. Calculation of the term requires use of the coordinates for both ends of the line. The value of the term may be estimated by

$$\begin{array}{l}\text{arc-to-chord}\\ \quad\text{correction (seconds)}\end{array} = 2.36(\Delta N)(\Delta E)\times 10^{-10} \qquad 22.8$$

For Lambert projections, ΔE is the difference (in meters) in easting coordinates of the end points of the line, and ΔN is the distance (in meters) between the midpoint of the line and the central parallel of the projection. That is, $(2.36\times 10^{-10})\Delta N\Delta E$.

Geodesy/Survey Astronomy

For transverse Mercator projections, ΔN is the difference (in meters) in northing coordinate between the end points of the line, and ΔE is the distance (in meters) between the midpoint of the line and the central meridian of the projection.

Note that the arc-to-chord correction varies with the length and orientation of the line, and with the distance from the line to the central line for the projection.

Example 22.2

Use Table 22.2 to determine the grid azimuth for a line in the Arkansas North State Plane Coordinate zone that has a geodetic azimuth of 241°12′37″. The approximate position for the line is 35°30′ latitude and 92°10′ longitude. Ignore the arc-to-chord correction.

Solution

From Table 22.2, the longitude of the central meridian, λ_o, for this zone is 92°00′ and $\sin B_o = 0.581899128039$.

For 92°10′ longitude, the mapping angle is

$$\begin{aligned}\gamma &= (\lambda_o - \lambda)\sin B_o \\ &= (92°00' - 92°10')(0.581899128039) \\ &= -5.81899128039' \\ &= -5'49.1''\end{aligned}$$

Using Eq. 22.6,

$$\begin{aligned}\text{grid azimuth} &= \text{geodetic azimuth} - \text{mapping angle} \\ &= 241°12'37'' - (-5'49.1'') \\ &= 241°18'26.1''\end{aligned}$$

Converting Between Geodetic and Grid Coordinates

Conversions between geodetic and grid coordinates can be calculated for Lambert projections using Eqs. 22.9 and 22.10 and data from published tables for each zone.

For geodetic to state plane coordinates,

$$E = R\sin\gamma + E_o \qquad 22.9$$

$$N = R_b - R\cos\gamma + N_o \qquad 22.10$$

For state plane to geodetic coordinates,

$$R = \frac{R_b - (N - N_o)}{\cos\gamma} \qquad 22.11$$

$$\lambda = \lambda_o - \frac{\gamma}{\sin B_o} \qquad 22.12$$

Conversion of coordinates for transverse Mercator projections is more complex and is beyond the scope of this chapter. Such calculations may be accomplished using various computer software packages, including those from NGS.

Example 22.3

Use Table 22.2 to convert the given latitude and longitude to plane coordinates.

latitude = 34°44′47.7648″
longitude = 92°17′20.7595″
zone 0301 Arkansas north

Solution

From Table 22.2,

$$\begin{aligned}N_o &= 0.000 \text{ m} \\ E_o &= 400{,}000.000 \text{ m} \\ \lambda_o &= 92°00'00'' \\ R_b &= 9{,}062{,}395.1981 \\ \sin B_o &= 0.581899128039\end{aligned}$$

From Table 22.2 with latitude 34°44′47.7648″, the mapping radius, R, is

$$\begin{aligned}R &= 9{,}018{,}017.731 \text{ m} \\ \text{difference} &= -30.81685 \text{ m} \\ R &= 9{,}018{,}017.73 \text{ m} + (47.7648'')(-30.81685 \text{ m}) \\ &= 9{,}016{,}545.770 \text{ m}\end{aligned}$$

From Eq. 22.7, the mapping angle, γ, is

$$\begin{aligned}\gamma &= (\lambda_o - \lambda)\sin B_o \\ &= (92°00' - 92°17'20.7595'')(0.581899128039) \\ &= (-17'20.7595'')(0.581899128039) \\ &= -10.09361743' = -10'05.6''\end{aligned}$$

From Eq. 22.9,

$$\begin{aligned}E &= R\sin\gamma + E_o \\ &= (9{,}016{,}545.770 \text{ m})\sin(-10'05.6'') + 400{,}000.000 \text{ m} \\ &= 373{,}526.300 \text{ m}\end{aligned}$$

From Eq. 22.10,

$$\begin{aligned}N &= R_c - R\cos\gamma + N_o \\ &= 9{,}062{,}395.1981 \text{ m} \\ &\quad - (9{,}016{,}545.770 \text{ m})\cos(-10'05.6'') \\ &\quad + 0.000 \\ &= 45{,}888.293 \text{ m}\end{aligned}$$

6. THE UTM COORDINATE SYSTEM

The *Universal Transverse Mercator* (UTM) System was designed as a worldwide plane coordinate system. The UTM system uses the GRS 80 ellipsoid in North America, but uses various ellipsoids elsewhere. The system has 60 zones. Each zone's longitude has a width of six degrees, starting with zone 1 for the zone from 180° west to 174° west, and increasing eastward to zone 60 for the zone from 174° east to 180° east. Thus, the boundaries of each zone are on those meridians that are integral multiples of six degrees. Table 22.5 gives a list of the UTM zones.

The grid origin for each UTM zone is located at the intersection of the equator and the central meridian for that zone. The scale factor at the central meridian for each zone is 0.9996 and increases with longitude to either side of the central meridian. As with SPCS, the mapping angle is zero at the central meridian, negative to the west of the central meridian, and positive to the east of that meridian. The UTM projection is not used for latitudes above 84° north or below 80° south. The *Universal Polar Stereographic* projection is typically used for polar regions.

The UTM scale factor corrections (between ground and grid distances), as well as the mapping angle corrections (between geodetic and grid azimuths), are handled in the same manner as with SPCS corrections. Software to calculate corrections, as well as to convert between geodetic and UTM coordinates is available on the NGS website and from various commercial sources.

Table 22.5 *Universal Transverse Mercator*

UTM zone numbers and corresponding longitude of central meridians

western hemisphere				eastern hemisphere			
zone	central meridian	zone	central meridian	zone	central meridian	zone	central meridian
1	177	16	87	31	−03	46	− 93
2	171	17	81	32	−09	47	− 99
3	165	18	75	33	−15	48	−105
4	159	19	69	34	−21	49	−111
5	153	20	63	35	−27	50	−117
6	147	21	57	36	−33	51	−123
7	141	22	51	37	−39	52	−129
8	135	23	45	38	−45	53	−135
9	129	24	39	39	−51	54	−141
10	123	25	33	40	−57	55	−147
11	117	26	27	41	−63	56	−153
12	111	27	21	42	−69	57	−159
13	105	28	15	43	−75	58	−165
14	99	29	09	44	−81	59	−171
15	93	30	03	45	−87	60	−177

Note: Each central meridian is in the center of a 6° wide zone.
Scale factor is 0.9996 at each central meridian.

PRACTICE PROBLEMS

1. Parallels, or lines of latitude, measure the arc distance (in degrees)

(A) east and west of the Prime Meridian
(B) westward from the Prime Meridian
(C) northerly and southerly from the equator to the poles
(D) northerly and southerly from the poles to the equator

2. Meridians, or lines of longitude, measure the arc distance (in degrees)

(A) east and west of the Prime Meridian
(B) westward from the Prime Meridian
(C) northerly and southerly from the equator to the poles
(D) northerly and southerly from the poles to the equator

3. A Mercator projection uses which of the following?

(A) plane
(B) cone
(C) cylinder
(D) ellipse

4. A Lambert projection uses which of the following?

(A) plane
(B) cone
(C) cylinder
(D) ellipse

5. A Lambert conic secant projection uses which of the following?

(A) parallel lines of longitudes
(B) two cones
(C) parallel lines of latitude
(D) one cone that slices the earth at two parallels

6. The mapping angle in a Lambert projection varies with which of the following?

(A) longitude
(B) latitude
(C) latitude and longitude
(D) none of the above

7. The mapping angle in a transverse Mercator projection varies with which of the following?

(A) longitude
(B) latitude
(C) latitude and longitude
(D) none of the above

8. The scale factor in a Lambert projection varies with which of the following?

(A) longitude
(B) latitude
(C) latitude and longitude
(D) none of the above

9. The scale factor in a transverse Mercator projection varies with which of the following?

(A) longitude
(B) latitude
(C) latitude and longitude
(D) none of the above

10. An observed horizontal distance of 2640.00 ft is measured at an average orthometric elevation of 1400 ft. The geoidal separation is −131.23 ft and the scale factor is 0.9994. What grid distance should be used for calculating state plane coordinates with this measurement? Use 20,906,000 ft for the radius of the earth.

(A) 2638.26 ft
(B) 2639.84 ft
(C) 2640.22 ft
(D) 2641.58 ft

11. What grid azimuth is needed to calculate state plane coordinates for a line that has a geodetic azimuth of 118°22′15″? The mapping angle is +4′ 32.1″. Ignore the arc-to-chord correction.

(A) 118°16′41.1″
(B) 118°17′42.9″
(C) 118°22′45.1″
(D) 118°26′47.1″

12. What are the state plane coordinates that correspond to the following latitude and longitude located in the Arkansas north zone?

latitude = 35°15′29.5″
longitude = 92°12′20.2″

(A) N = 380,224.567 m, E = 100,678.213 m
(B) N = 100,253.116 m, E = 380,943.546 m
(C) N = 381,287.347 m, E = 100,806.696 m
(D) N = 102,624.87 m, E = 381,291.14 m

13. What is the expected ground distance between one point with state plane coordinates of N = 202,400 m and E = 305,600 m, and a second point with coordinates of N = 203,300 m and E = 306,500 m? The scale factor is 0.99992, the orthometric elevation is 900 m and the geoidal separation is −36.2 m. Use a value of 6372 km for the radius of the earth.

(A) 1257.243 m
(B) 1273.063 m
(C) 1562.378 m
(D) 1772.937 m

SOLUTIONS

1. Parallels, or lines of latitude, measure the arc distance (in degrees) as northerly and southerly from the equator to the poles.

The answer is (C).

2. Meridians, or lines of longitude, measure the arc distance (in degrees) as east and west of the Prime Meridian.

The answer is (A).

3. A Mercator projection uses a cylinder as a projection surface.

The answer is (C).

4. A Lambert projection uses a cone as a projection surface.

The answer is (B).

5. A Lambert conic secant projection uses one cone that slices the earth at two parallels.

The answer is (D).

6. The mapping angle in a Lambert projection varies with longitude.

The answer is (A).

7. The mapping angle in a transverse Mercator projection varies with longitude.

The answer is (A).

8. The scale factor in a Lambert projection varies with latitude.

The answer is (B).

9. The scale factor in a transverse Mercator projection varies with longitude.

The answer is (A).

10. From the problem statement,

observed horizontal distance, $D = 2640.00$ ft
orthometric elevation, $H = 1400$ ft
geoidal separation, $N = -131.23$ ft
scale factor, $k = 0.9994$
radius of earth, $R = 20{,}906{,}000$

Using Eq. 22.4,

$$\begin{aligned}\text{ellipsoidal distance} &= D\left(\frac{R}{R+N+H}\right)\\ &= (2640.00\text{ ft})\left(\frac{20{,}906{,}000\text{ ft}}{20{,}906{,}000\text{ ft} + (-131.23\text{ ft}) + 1400\text{ ft}}\right)\\ &= 2639.84\text{ ft}\end{aligned}$$

$$\begin{aligned}\text{grid distance} &= (\text{scale factor})(\text{ellipsoidal distance})\\ &= (0.9994)(2639.84\text{ ft})\\ &= 2638.26\text{ ft}\end{aligned}$$

The answer is (A).

11. From the problem statement,

geodetic azimuth $= 118°22'15''$
mapping angle $= +\ 4'32.1''$

Using Eq. 22.6,

$$\begin{aligned}\text{grid azimuth} &= \text{geodetic azimuth} - \text{mapping angle}\\ &= 118°22'15'' - 4'32.1''\\ &= 118°17'42.9''\end{aligned}$$

The answer is (B).

12. From the problem statement,

latitude $= 35°15'29.5''$
longitude $= 92°12'20.2''$

From Table 22.2 with a latitude of $35°15'29.5''$,

$N_o = 0.000$ m
$E_o = 400{,}000.000$ m
$\lambda_o = 92°00'00''$
$R_b = 9{,}062{,}395.1981$
$\sin B_o = 0.581899128039$
$R = 8{,}960{,}698.948$
difference $= 30.81666$

The mapping radius, R, is

$$\begin{aligned}R &= 8{,}960{,}698.948 - (29.5'')(30.81666)\\ &= 8{,}959{,}789.856\text{ m}\end{aligned}$$

From Eq. 22.7, the mapping angle, γ, is

$$\begin{aligned}\gamma &= (\lambda_o - \lambda)\sin B_o = (92°00'00'' - 92°12'20.2'')\\ &\quad\times (0.581899128039)\\ &= (-12'20.2'')(0.581899128039)\\ &= -7.178695576' \quad (-7'10'')\end{aligned}$$

From Eq. 22.9,

$$\begin{aligned}E &= R\sin\gamma + E_o\\ &= (8{,}959{,}789.856\text{ m})\sin(-7'10.7'') + 400{,}000.000\text{ m}\\ &= 381{,}291.14\text{ m}\end{aligned}$$

From Eq. 22.10,

$$\begin{aligned}N &= R_b - R\cos\gamma + N_o\\ &= 9{,}062{,}395.1981\text{ m}\\ &\quad - (8{,}959{,}789.856\text{ m})\cos(-7'10.7'') + 0.000\\ &= 102{,}624.87\text{ m}\end{aligned}$$

The answer is (D).

13. From the problem statement,

point 1: $N_1 = 202{,}400$ m, $E_1 = 305{,}600$ m
point 2: $N_2 = 203{,}300$ m, $E_2 = 306{,}500$ m
scale factor: $k = 0.99992$
orthometric elevation: $H = 900$ m
geoidal separation: $N = -36.2$ m
radius of earth $= 6372$ km

The grid distance (using Eq. 15.11) is

$$\begin{aligned}\text{grid distance} &= \sqrt{(N_2 - N_1)^2 + (E_2 - E_1)^2}\\ &= \sqrt{(203{,}300\text{ m} - 202{,}400\text{ m})^2 + (306{,}500\text{ m} - 305{,}600\text{ m})^2}\\ &= 1272.792\text{ m}\end{aligned}$$

Rearranging from Eq. 22.5,

$$\begin{aligned}\text{ellipsoidal distance} &= \frac{\text{grid distance}}{\text{scale factor}} = \frac{1272.792\text{ m}}{0.99992}\\ &= 1272.894\text{ m}\end{aligned}$$

Using Eq. 22.4,

$$\begin{aligned}\text{ground distance} &= (\text{ellipsoid distance})\left(\frac{R+N+H}{R}\right)\\ &= (1272.89\text{ m})\left(\frac{(6372\text{ km})\left(1000\ \frac{\text{m}}{\text{km}}\right) + -36.2\text{ m} + 900\text{ m}}{(6372\text{ km})\left(1000\ \frac{\text{m}}{\text{km}}\right)}\right)\\ &= 1273.063\text{ m}\end{aligned}$$

The answer is (B).

Topic V: Cadastral and Boundary Law

Chapter

23 History and Origins of Title

1. EARLY HISTORY OF PROPERTY LAW

The earliest record of property ownership goes back to the Babylonians in 2500 B.C. The Bible also furnishes many references to property ownership. Numerous references are found in the book of Genesis, including a passage relating to the purchase of land by Abraham on which to bury his wife, Sarah.

In the book of Jeremiah is "Thou shall not remove thy neighbor's landmark which they of old times have set in thine inheritance which thou shall inherit in the land that the Lord thy God giveth thee to possess it." Also in the book of Jeremiah is "Cursed be to he that removeth his neighbor's landmark and all the people shall say Amen."

Historical records show that in about 1400 B.C. the king of Egypt divided the land into squares of equal size and gave each Egyptian one square. The king in turn levied taxes on each person. This land was in the fertile Nile valley where the river overflowed and destroyed parts of these plots. The owner of the destroyed land was required to report his loss to the king and request a reduction of taxes. The need to determine the actual losses of property resulted in the beginning of surveying in this part of the world. These early surveyors are referred to as *rope stretchers* in the Bible. Drawings on the walls of tombs show these rope stretchers accompanied by officials who recorded the measurements.

The Greeks and Romans also recognized individual property ownership and the Romans used taxes on land to support the cost of government.

At the beginning of the Christian era in Europe, the ownership of land was usually determined by conquest. After conquest, the ruler took over all the land and earlier land titles were extinguished. Often, sovereign rights were vested in the ruler by the Pope, but in non-Christian lands, the practice of the ruler having rights to all the land was much the same. The land belonged to the Sovereign, often referred to as the *Crown*.

2. FEUDAL SYSTEM

In the 11th century Great Britain was conquered by William the Conquerer (William I). He claimed all the land and ruled over all the people. Later, he introduced the feudal system, which was the social and political system in both Great Britain and Europe during the 11th, 12th, and 13th centuries. The feudal system was the basis of real property law in medieval times. Many of its principles found their way into American law.

In England, the system was an arrangement between the king, noblemen, and vassals. The arrangement included an intricate set of rules for the tenure and transfer of real property. The noblemen ruled over the tenants (*vassals*) of the land. These landlords protected their vassals but expected them to pay rent on the land they used and to pay allegiance to their lord. Often, they were required to fight for their lord. The land was in possession of the lord and could not be sold. It passed to the eldest son by inheritance. Vassals had no chance to own property.

3. COMMON LANDS

Almost all the peasant class at the time of William the Conqueror was engaged in farming. Tenant farmers acquired strips of land from the lord for row crops and for producing hay for the cattle. While the crops were growing, these farmers needed pasture for their cattle, and they acquired from their lord *right in common* on land used for permanent pasture. These lands eventually became known as *commons*, and the word was brought to this country by early settlers. In the United States, "commons" means a park.

4. DOMESDAY BOOK

In 1086, William the Conqueror made a survey for tax purposes that included every farm, every farm owner, and all common land. This was known as the *Domesday Book*. Besides being useful for collecting taxes, it allowed him to demand allegiance from everyone in his kingdom. The Domesday Book was a complete record for doomsday, the modern spelling.

A modern edition of the Domesday Book, which lists these common lands, has recently been published. Over the centuries, they have become private property subject to certain rights by claimants to rights in common. Parliament, with the new book, is trying to pin down just what may be done to this land (which comprises 4% of England).

5. TREND TO PRIVATE OWNERSHIP

As people turned from agriculture to crafts and trade in villages and cities, they curbed the power of kings and began to think of ownership of land without fealty to the lord, without obligations of service, and with the right to dispose of the land as they saw fit. This is known as *fee ownership.*

Early grants of the kings were not recorded. The grantee received a packet of papers as evidence of his ownership. These documents became so voluminous that loss of them became common. Parliament began to take steps to correct the confusion caused by lost or stolen documents. Laws were established that abolished the practice of passing title from father to eldest son, and steps were taken to better describe the property transferred.

6. MAGNA CARTA

In 1215, powerful English noblemen forced the king of England to sign the *Magna Carta*, a document that forced the king to share authority with the nobles. By 1700, Parliament had gained supremacy over the king.

7. STATUTE OF FRAUDS

In 1677, the English Parliament passed the *Statute of Frauds*, which, among other things, prohibited any transfer of land or any transfer of interest in land by oral agreement. All conveyances were required to be in writing, a requirement that is very much a part of our present-day law.

8. PROPERTY LAW IN THE UNITED STATES

Ideas of property ownership in existence in England at the time were brought to this country by English settlers. Lands in America were granted to settlers by kings and queens of England. The idea of ownership by the sovereign still prevailed, and some tribute to the crown was still required.

The settlers themselves still had the idea of ownership by the conqueror as they displaced the Native American Indians from their home and lands with little compensation to them.

Because land was cheap in the early days of our country, exact descriptions and exact locations were not necessary. Some grants to the original colonists extended from the Atlantic to the Pacific.

9. *STARE DECISIS* (PRECEDENT)

The doctrine of *stare decisis* established the principle that when a court hands down a decision regarding certain facts, it will adhere to that decision in deciding all future cases where the facts are substantially the same.

10. TYPES OF PROPERTY

Property is divided into two classes: real property and personal property. *Real property* is immovable and can be recovered. It has been defined as "the interest that a man has in lands, tenements, or hereditaments." These terms include land, buildings, trees, and the right to use them. Anything that grows on the land or any structure that is fixed to the land is real property.

Real property law is, for the most part, state law rather than federal law. It, therefore, varies among the states.

Personal property is movable and often cannot be recovered. Action to recover such things as money and valuable goods is often taken against the person who removed them illegally.

11. DEFINITION

Title is the right to own real property and the evidence of that right. Right to ownership is not enough, however. There must also be possession of property. Title, then, is the outward evidence of the right to ownership.

12. CLEAR TITLE, GOOD TITLE, MERCHANTABLE TITLE

The terms *clear title, good title,* and *merchantable title* are essentially synonymous. Clear title means the property is free from encumbrances. Good title is a title free from litigation.

13. RECORD TITLE

A title entered on the public records is referred to as a *record title.*

14. COLOR OF TITLE

Any written instrument, such as a forgery, that appears to convey title but in fact does not, establishes *color of title*. A consecutive chain of transfers of title down to the person in possession in which one or more of the written instruments is not registered may also establish color of title.

15. CLOUD ON TITLE

A claim on land that would, if valid, impair the title to the land creates a *cloud on title*. The claim may be any encumbrance such as a lien, judgment, tax-levy, mortgage, or conveyance.

16. CHAIN OF TITLE

The change in ownership of a piece of property in sequence is known as the *chain of title*. Any defective conveyance of title in the chain adversely affects the title from that point on.

17. ABSTRACT OF TITLE

Before buying real property, a buyer should institute a search of title—a review of all documents affecting the ownership of the property to determine if the person selling the property has a good and clear title. A compilation of abstracts of deeds, deeds of trust, or any other estate or interest, together with all liens or liabilities that affect the title to the property, may be obtained from an abstract or title company. This condensed history of the title to the land in chronological order is known as an *abstract of title* or simply an *abstract*.

18. ATTORNEYS' OPINION

After attorneys secure an abstract of title, they examine the various transfers of title and write an opinion for their client as to whether they think the grantor has a good and clear title to the property. The attorneys do not guarantee the title; they merely state their opinion from the facts shown in the abstract. They cannot guarantee that there has not been fraud or forgery.

Attorneys may point out errors that were made in the execution of conveyances but that, in their opinion, will not affect the title. For instance, a deed was dated October 5, 1938, and the acknowledgment was dated October 4, 1938. The attorneys' opinion was that the instrument had been recorded for more than ten years and the acknowledgment was cured by limitation. They further stated that, if necessary, an affidavit could be secured from the notary public involved to the effect that a stenographic error had been made in dating the acknowledgment.

19. AFFIDAVIT

An *affidavit* is a statement made under oath in the presence of a notary public or other authorized person. In the case of the misdated acknowledgment mentioned in Sec. 18, the notary public made a sworn statement, in the presence of another notary public, to the effect that the acknowledgment was actually made after the grantor had signed the instrument.

20. TITLE INSURANCE POLICY

In recent years, the practice of preparing an abstract of title and the practice of submitting an opinion on the title has been replaced by the issuance of a *title insurance policy*, often referred to as a *title policy*. Title insurance policies assure purchasers of real property that they have good title to the land they have purchased. These policies are issued by title abstract companies operating under the insurance laws of the state.

The amount the assureds are guaranteed is usually limited to the amount they are paying for the property. If the property is enhanced in later years, the assureds would receive no extra compensation for the increased value of the property if their title were defeated.

The policy is usually issued subject to certain exceptions: taxes, easements, encumbrances, oil royalties, and so on.

21. HOMESTEAD RIGHTS

Many of the colonists in Stephen F. Austin's colony had left the United States because of a financial crisis of the time. Austin knew they needed time to establish themselves in Texas, and he appealed to the legislature of Coahuila-Texas for legislation to protect them from property seizure for old debts in the United States. A bill was passed in 1829 that exempted, without limitation, lands acquired by virtue of a colonization law from seizure for debts incurred before the acquisition of the land. The law had some precedence in the laws of Spain in the 15th century.

In 1839, the Third Congress of the Republic of Texas passed a law that read as follows:

> Be it enacted... that from and after the passage of this act, there shall be reserved to every citizen or head of a family in this Republic, free and independent of the power of a writ of fire facies, or other execution issuing from any court of competent jurisdiction whatever, fifty acres of land or one town lot, including his or her homestead, and improvements not exceeding five hundred dollars in value, all household and kitchen furniture (provided it does not exceed in value two hundred dollars), all implements of the husbandry (provided they shall not exceed fifty dollars in value), all tools, apparatus and books belonging to the trade or profession of any citizen, five milch cows, one yoke of work oxen or one horse, twenty hogs, and one year's provisions...

This was the first law of its kind. It has since been adopted by most of the states, with realistic revisions in the protected quantities.

PRACTICE PROBLEMS

1. The Statute of Frauds provides that

(A) ownership of land by private individuals is fraudulent

(B) a father may not pass title to his eldest son

(C) transfer of title to land must be in writing

(D) all of the above

2. Property is divided into which of the following classes?

(A) farmland and ranchland

(B) urban land and rural land

(C) real and personal

(D) none of the above

3. A title abstract is

(A) a conveyance that transfers title

(B) a description of a piece of property

(C) a history of transfers of title in chronological order

(D) an insurance policy

4. Color of title is

(A) any written instrument that appears to convey title but in fact does not

(B) a claim on land that, if valid, would impair the title to the land but that can be proven invalid

(C) an estate in real property

(D) all of the above

5. Title, in the legal sense, is

(A) the right to own real property

(B) the evidence of the right to own real property

(C) the legal instrument constituting the evidence to the right of ownership

(D) all of the above

SOLUTIONS

1. The Statute of Frauds provides that transfer of title to land must be in writing.

The answer is (C).

2. Property is divided into real and personal classes.

The answer is (C).

3. A title abstract is a history of transfers of title in chronological order.

The answer is (C).

4. Color of title is any written instrument that appears to convey a title, but in fact, does not.

The answer is (A).

5. Title, in the legal sense, is the right to own property, the evidence of the right to own real property, and the legal instrument constituting the evidence to the right of ownership.

The answer is (D).

24 Transfer of Ownership

1. CONVEYANCE

A *conveyance* is a written instrument that transfers ownership of property. It includes any instrument that affects the ownership of property. The term not only refers to a written document, but also means a method of transfer of property.

Changes in title law in early U.S. history were accompanied by changes in methods of conveying land. Deeds had to be in writing, but they were shortened and simplified. No longer was it necessary that deeds be written by lawyers learned in English law.

2. ESTATE

An *estate* in real property is an interest in real property. It can be complete and inclusive without limit or duration; it can be partial and of limited duration; it can be for the life of one person or for the life of several; it can include surface and all minerals below, or surface and no minerals below, or minerals but not the surface. It can be acquired in many ways: by purchase, by inheritance, by power of the state, or by gift.

Several people may hold an interest in the same property. Consider a person who buys a house with a mortgage and then leases the house to someone else. The lessee, the owner, the mortgagee, and various taxing agencies have an interest in the house.

3. FEE

The word *fee* comes from the feudal era and refers to an estate in land. The true meaning of the word is the same as that of "feud" or "fief." Under the feudal system, a freehold estate in lands came from a superior lord as a reward for services and on the condition that services would be rendered in the future. A fee and a freehold estate are the same.

4. FEE TAIL

Under the feudal system, a fee or freehold estate was passed on to the eldest son on the death of the fee holder. An estate in which there is a fixed line of heirs to inherit the estate is known as a *fee tail*.

5. ESTATE IN FEE SIMPLE ABSOLUTE

In U.S. law, an *estate in fee simple absolute* (also called an *estate in fee simple*) is the highest type of interest. It is an estate limited absolutely to a person and the heirs, and assigns forever without limitation. In other words, a person who owns a parcel of land "in fee" can hold it, sell it, or divide it without limitations.

6. DEED

The most important document in the transfer of ownership of real property is the *deed*, which is evidence in writing of the transfer of an estate. A deed is a formal document. It needs not only to be in writing but also to be written by a person versed in the law. Deeds are of two principle types: warranty deeds and quitclaim deeds.

Warranty Deeds

In a *warranty deed*, the grantors proclaim that they are the lawful owners of the real estate and bind themselves, bind their heirs, and assigns to warrant and forever defend the property unto the grantees and their heirs, and assigns against every person who lawfully claims it or any part of it. The warranty deed is the instrument used to convey an estate in fee simple absolute (in fee).

Quitclaim Deeds

The *quitclaim deed* passes on to the grantees whatever interest the grantors have. If the grantors have a complete title, they pass on a complete title. If their title is incomplete, they pass on whatever interest they have.

7. ESSENTIALS OF A DEED

Because a deed is evidence of the transfer of an estate, the evidence must be clear and concise. The wording of the deed must clearly state the intent of the parties involved in the transfer. It is not sufficient that the grantors and grantees understand the terms of the transfer. In order to protect the rights of the real property owners and to establish an orderly method of transfer of real property, state legislatures and courts have adopted requirements for conveyance of such interest.

- A deed must be in writing. As previously mentioned, this requirement originated in the *statute of frauds* and now is found in the statutes of all states.
- A deed must be in legal terminology.
- Parties to a deed must be competent. A person of unsound mind or a minor cannot execute a deed.
- There must be a grantor and a grantee, and they must be clearly identified.
- There must be a valid consideration, although the total amount of the consideration need not be shown. Deeds containing the phrase "ten dollars and other consideration" provide evidence that the grantor received remuneration for the property.
- A deed must contain a description of the property being conveyed and clearly show the interest conveyed.
- A deed must be signed. In the case of joint ownership by husband and wife, both must sign.
- A deed must be acknowledged. The signer or signers of the deed must sign in the presence of a registered notary public who must know the identity of the signer or signers. The notary must sign the acknowledgment and affix a seal to it.
- A deed must be delivered. Centuries ago, land was conveyed by a ceremony known as *livery of seisin*. Parties to the transfer of ownership met on the property to be conveyed and performed such acts as handing over twigs and soil, driving stakes in the ground and shouting. The ceremony was practiced in England as late as 1845. Today, delivery of the deed is considered to be the delivery of the property.

8. RECORDING DEEDS

It is important that deeds be recorded in order to constitute notice to the public. Unrecorded deeds may be valid, but to avoid future controversy, deeds should be recorded as soon after execution as possible. It is not necessary for the grantor to actually carry the deed to the grantee.

9. PATENT

A *patent* is a conveyance or deed from the sovereign for the sovereign's interest in a tract of land. Most, but not all, land in the United States was patented by the United States. The original thirteen colonies received grants from the king of England. Owners of land in Texas have received patents (grants) from the king of Spain, the Republic of Mexico, the Republic of Texas, and the State of Texas. The lands of Texas have never come under the ownership of the United States, and no patents have been conveyed from that source.

10. WILL

A *will* is a declaration of a person's wishes for the distribution of his or her property after death. These wishes are carried out by a probate court. A *devise* transfers real property, whereas a *bequest* transfers personal property. The devisee is the person receiving the real property. The *probate court* will distribute the property according to the wishes of the *testator* (the deceased) if a will exists. If no will exists, the court will distribute the property in accordance with the law of descent and distribution. Widows, widowers, and children come first in this succession.

A will may devise certain property to certain individuals, or it may devise an entire estate to several heirs. In the latter case, the heirs will own the undivided property jointly.

Before property can be transferred under the terms of the will, the heirs must submit the will to a probate court or a county court that has probate jurisdiction. If the will designates an *executor*, the court will recognize him or her. If no will exists, the court will appoint an administrator. Heirs of an estate must then file an inventory of the property of the estate. Public notice must be given to creditors of the estate and these claims must be paid, if valid. State and federal taxes must also be paid before final settlement of the estate.

In Louisiana, the *forced heirship law*, based on the Napoleonic Civil Code, decrees that children are entitled to a testator's property regardless of the terms of the will. Under this law, children are entitled to one-fourth of the estate. If there are no children, parents are entitled to this one-fourth.

11. HOLOGRAPHIC WILL

A *holographic will* is a will in the handwriting of the deceased.

12. EASEMENT

An *easement* is the right that the public or an individual has in the lands of another. An easement does not give the grantee a right to the land—only a right to use the land for a specified purpose. The owner of the land may also use it for any purpose that does not interfere with the specified use by the grantee.

Utilities wishing to install power lines, underground pipe, canals, drainage ditches, and so on, sometimes do not require fee title to land but need only the use of the land to install and maintain the facility. The owner of the land retains title to it, subject to the terms of the easement.

13. LEASE

A *lease* must be for a certain term, and there must be a consideration. It is a contract for exclusive possession of lands or tenements though use may be restricted by reservations. The person who conveys is known as the *lessor* and the person to whom the property is conveyed is known as the *lessee*. Both parties must be named in the lease.

In many cases a tenant holds real estate without a lease, paying rent each week, month, or year. This is known as *tenancy without lease*.

In general, whatever buildings or improvements stand upon the land and whatever grows upon the land belongs to the landlord. Under a lease, the tenant is entitled to the crops of annual planting.

14. SHARECROPPER'S LEASE

A lease of farmland wherein the landlord and tenant each receive a predetermined share of the total income from crops on the land is known as a *sharecropper's lease*. The share to each is usually determined by custom in certain areas, but it can be set at any figure by agreement between the two.

15. OIL LEASE

An oil company or private individual may enter into an agreement with a land owner to remove oil, gas, or other minerals from the land and to share the profits from the sale of the minerals with the landowner. This is known as a *mineral lease* or *oil lease*. The shares are usually set by the company that removes the minerals, and this share has been accepted by custom.

An oil lease is for a definite number of years (often five), stipulating rental on a per-acre, per-year basis. In addition to the yearly rental, the agreement usually includes a bonus paid by the oil company at the beginning of the lease period. This also is usually on a per-acre basis. If drilling has not commenced by the end of a specified period, the lease expires.

16. MORTGAGE

A *mortgage* is a conditional conveyance of an estate as a pledge for the security of a debt. People borrowing money to purchase property guarantee that they will repay the lender by making a conditional conveyance to the lender. If they repay the loan as specified, the mortgage becomes null and void. If they do not pay the loan as specified (a *default*), they must deliver the property to the lender.

17. DEED OF TRUST

A *deed of trust* is a mortgage that gives the creditor the right to sell property, in case of default, through a third person known as the *trustee*. Early American law regarding mortgages included a complicated system of equitable foreclosure to give the debtor protection. It included the debtor's "equity," which gave him the right to redeem his land after it had been foreclosed on. This "equity" created difficulties for the lender, and in time, laws in many states were modified so that if the debtor agrees in advance, the creditor can sell the property through a third person, known as the *trustee*, without going through court, in case of default.

18. CONTRACT OF SALE

Often the sale of a large estate involves many complexities that are time consuming for the parties involved and their attorneys. A thorough examination of the complexities of the transaction in advance can save time that otherwise might be spent in court settling a dispute.

In order that buyers may express their intent to buy and sellers may express their intent to sell, the two parties may enter into a *contract of sale*, which describes the property involved and the terms of the sale and specifies a date, not later than which the transfer of property must be completed. This contract usually stipulates that the sellers will furnish a good and merchantable title to the buyers by a warranty deed, and that if the sellers cannot furnish such a deed, the contract is null and void. The contract also often provides for an *escrow fund*, which the buyers will forfeit if they do not carry out the terms of the contract.

In some instances, real property is sold by contract of sale and all payments are made before the deed is executed. Throughout the period of the payments, the title remains in the name of the sellers, and their names appear on the tax roll as owners of the property. Contracts of sale are frequently not recorded.

19. UNWRITTEN TRANSFERS OF LAND OWNERSHIP

The statute of frauds requires that all conveyances of real property be in writing. But, if that statute would deprive the rightful owners of their property, then the law may be set aside and an unwritten transfer of real property may take place.

This transfer may take place by expressed or implied agreement such as by the principle of recognition and acquiescence over a long period of time, by dedication, by adverse (hostile) relationships, or by acts of nature.

A legal unwritten transfer of title supersedes written title and will extinguish written title. Evidence to prove the location of a written title will not overturn a legal unwritten title.

20. RECOGNITION AND ACQUIESCENCE

Acquiescence in a boundary line is evidence from which it may be inferred that the parties by agreement established a line as the true line. From such acquiescence, a jury or court may find that the line used is the true line. Acquiescence in a line other than the true line will not support a finding of an agreement establishing the line as the boundary when there is no evidence of agreement other than acquiescence and where it is shown that the use of the line resulted, not from agreement, but only from a mistaken belief of the parties that it was the true line.

21. DEDICATION

Dedication is the giving of land or rights in land to the public. It must be given voluntarily, either expressed or implied. It may be written or unwritten, but there must be acceptance of the dedication. A consideration is not necessary.

Common law dedication may be expressed, as when the intention to dedicate is expressed by a written document or by an act that makes the intent obvious. It may be implied, as when some act or acts of the donor make it reasonable to infer that he or she intended to dedicate.

Dedication made in accordance with the provisions of a statute is called *statutory dedication* and usually requires that the donor sign and acknowledge the dedication.

Developers of a subdivision may subdivide a tract of land, lay out streets, lay sewer lines and water lines, and pave streets. They may then turn over the use of these facilities to the public. The facilities must be accepted for use by the public by the state, city, town, or other governing body. Other examples of land dedicated to the public include parks, cemeteries, and schools.

22. ADVERSE POSSESSION AND TITLE BY LIMITATION

Transfer of property may occur without the agreement of the owner by the method known as *adverse possession*. Adverse possession is the acquisition of title to property belonging to another by performing certain acts. The rights to acquire property in this manner are often referred to as *squatter's rights*.

Requirements for transfer of title by adverse possession vary among the states but are essentially the same.

- Possession of the land by the person claiming it from another must be such that the owner will be aware of the possession if he or she visits the property.
- Possession must be open and notorious. Possession so open, visible, and notorious that it will raise the presumption of an adverse claim is the equivalent of actual knowledge. The land must be occupied in a straightforward, not clandestine, manner.
- Possession must be continuous. Statutes do vary among the states as to the period necessary to establish title from adverse possession, but the land in question must be held continuously for the period required by statute.
- Possession is required to be exclusive. This means that the person making the claim cannot share the possession with the owner or others. He or she must have complete control of the property.
- The possession must be hostile. The claimant must possess the land as if he or she were the owner in defiance of the owner.

23. ADVERSE POSSESSION USED TO CLEAR TITLE

In modern times, people seldom squat on land with the intention to acquire title by adverse possession (although there have been many instances in the past when this has occurred). The importance of adverse possession today is in its use to clear up defects in title or to settle boundary disputes between adjacent land owners.

Honest differences may occur between adjacent owners as to where the boundary between them actually is. Monuments and landmarks may be obliterated; changes in the location of fences, ditches, and roadways may have occurred. Adverse possession provides a means of clarifying an obliterated boundary line.

24. RIGHT OF THE STATE AGAINST ADVERSE POSSESSION

Title to state or public land generally cannot be acquired by adverse possession.

25. TRESPASS TO TRY TITLE ACTION

The action usually taken by the record owner of land against a person in adverse possession of land is known as *trespass to try title*. The record owner brings suit against the person in possession for recovery of the land and for damages for any trespass committed. If the court rules in favor of the plaintiff, the person in possession is evicted, but if the court rules in favor of the defendant, the defendant acquires a good title to the land.

26. PRESCRIPTION

The method of obtaining easement rights from long usage is known as *prescription*. A person may travel across a tract or parcel of land for a period of time required by the statute of limitations and acquire a right to continue the act of using the land. The act of using the land must have been open, continuous, and exclusive for the period of time required.

A highway right of way can be acquired by the state if it has been used by the public for a long period of time. As with individual acquisition, the use must be open and continuous for the required period of time.

27. RIGHT OF EMINENT DOMAIN

When the owner of land refuses to sell and the improvement is of public character, the law allows that land shall be taken under what is called the right of *eminent domain*. Eminent domain gives the state, or others delegated, the right and power to condemn private property for public use.

The constitution of the United States and state laws limit eminent domain. Owners are guaranteed adequate compensation for their property and they may not be deprived of their property without *due process of law*.

The power to exercise eminent domain must be authorized by the state legislature by statute, and the legislature may delegate this power to such agencies as it deems proper. Counties, incorporated cities and towns, water districts, and school districts have been delegated the power of eminent domain. The power is also given to private corporations that are engaged in public service.

Owners of condemned property must be fully compensated for the property. When only a part of their property is taken, they are entitled to compensation for *consequential damage*. (A highway that cuts off access to a watering tank for cattle might create consequential damage.)

The owner of the land is entitled to know the precise boundaries of the land to be condemned. It is the obligation of the agency executing the acquisition to furnish an adequate description of the boundaries.

Before the right of eminent domain can be exercised, it is essential that no purchase agreement be reached between the parties. It is necessary that the state or city or other governing body make the owners an offer that they refuse, and that the owners shall name their price, which the state or city refuses, or else that the owners refuse to name the price. There must be a definite failure to agree. After disagreement, the state or other body must initiate *condemnation proceedings*.

28. ENCROACHMENT

An *encroachment* is a gradual, stealthy, illegal, acquisition of property. By moving a fence a small amount over a period of years, an adjoining owner may acquire from the lawful owner a strip of land.

29. ACTION TO QUIET TITLE

Where the boundary between adjacent landowners is not clear or where there is a dispute over the location of the boundary line, one of the parties can sue the other to determine the location of the line. Either party may employ a surveyor as an expert witness. The judgment in the lawsuit becomes a public record and will be reflected in abstracts of title.

30. COVENANT

An agreement on the part of the grantee to perform certain acts or to abstain from performing certain acts regarding the use of property that has been conveyed to him or her is known as a *covenant*. Developers of residential property, in order to assure buyers that their neighborhood will be pleasing to the eye and pleasant to live in, require the buyer to accept certain restrictions as to the use of property he or she buys. These restrictions include such things as type of building construction, minimum distance between house and property line, minimum number of square feet in floor plan, use of the property, and kinds of animals allowed on premises. These covenants are sometimes called *deed restrictions*.

31. LIEN

A *lien* is a claim or charge on property for payment of a debt or obligation. It is not the right of possession and enjoyment of property, but it is the right to have the property sold to satisfy a debt. Mortgages and deeds of trust constitute liens.

32. TAX LIEN

States, counties, and other governing bodies impose taxes on real property, which gives them a first lien on the property. Failure to pay taxes gives the governing body the right to have the property sold to satisfy the tax debt. Failure to pay income taxes gives the federal government the same right. Before purchasing the real estate, the buyer may obtain a *tax certificate* in which the tax collector certifies that there are no unpaid taxes on the property up to a certain date.

33. PROMISSORY NOTE

A *promissory note* is the written promise of the borrower to pay the lender a sum of money with interest. The principal sum, the interest rate, and a schedule of dates of payment are included on the face of the note. Also included in some notes is a listing of the *security* for the note—the property that is to be mortgaged to guarantee payment of the note. The same information as to principal, interest, and schedule of payments shown on the note is, where applicable, shown in the deed of trust, and reference is made in the deed of trust to the promissory note between the two parties involved.

A deed of trust does not necessarily accompany every promissory note, and a promissory note need not list any security. Lenders may, if they wish, lend simply on a person's *personal note*, which is a promise to pay. But this does not prevent lenders from taking legal action to collect the amount of the note or obtaining property of equal value in the event the borrower does not pay the note.

PRACTICE PROBLEMS

1. A conveyance is

(A) a written instrument that transfers property
(B) an estate in real property
(C) a method of transferring property from the deceased to the heirs
(D) a release of lien

2. An estate can be acquired by

(A) purchase
(B) inheritance or gift
(C) power of the state
(D) all of the above

3. An estate in fee simple absolute

(A) implies a complete title
(B) is the highest type of interest
(C) gives the owner the right to dispose of the property in any way the owner sees fit
(D) all of the above

4. A deed

(A) is the most important document in the transfer of real property
(B) is evidence in writing of the transfer of property
(C) must be in writing
(D) is all of the above

5. Conveyance of fee title to land is usually by

(A) deed of trust
(B) quitclaim deed
(C) warranty deed
(D) contract of sale

6. Delivery of the property sold is considered to be

(A) delivery of the deed
(B) a statement by the seller that the seller has vacated the property
(C) a notice to the buyer from the county clerk that the transaction is complete
(D) the ceremony of livery of seisin

7. The grantee is

(A) the person to whom a grant is made
(B) the person who makes the grant
(C) a notary public
(D) the mortgagor

8. A patent is

(A) a conveyance of the sovereign's interest in a tract of land
(B) the state's warranty that title to a tract of land is good and clear
(C) a Spanish or Mexican grant
(D) the title to vacant land

9. An easement is

(A) title to surface rights only
(B) a lease or "estate for years"
(C) a fee simple title
(D) the right that the public or an individual has in the lands of another

10. A mortgage is a

(A) promissory note held by a lending agency
(B) conditional conveyance of an estate as a pledge of security of a debt
(C) request for payment of a note
(D) transfer of title

11. A deed of trust is a

(A) conditional conveyance of an estate as a pledge of security of a debt
(B) deed for excess acreage
(C) conveyance held in the hands of another for safekeeping
(D) conveyance made to a charitable organization

12. Dedication is

(A) a means of transferring title to property
(B) the acquisition of property without the owner's consent
(C) a ceremony marking the completion of streets
(D) a city ordinance relating to street specifications

13. Which of the following is NOT a requirement for obtaining possession by adverse possession?

(A) possession must be open and notorious
(B) possession must be hostile
(C) possession must be intermittent
(D) possession must be actual

14. Prescription is

(A) a method of acquiring right to use of property
(B) a tax lien
(C) an encumbrance
(D) a rendition of property

15. The power to exercise eminent domain must be authorized by the

(A) governor of the state
(B) U.S. Congress
(C) state legislature
(D) state department of highways

16. The Constitution of the United States provides that, before the power of eminent domain can be exercised,

(A) the owner must be guaranteed adequate compensation for the property
(B) the improvement must be for a public use
(C) the owner may not be deprived of property without due process of law
(D) all of the above

17. An encroachment is

(A) a cloud on title to property
(B) a second lien
(C) a written instrument that conveys title
(D) a gradual, stealthy, illegal acquisition of property

18. A quitclaim deed

(A) is a deed for excess acreage
(B) warrants a good and clear title
(C) passes any title the grantor may leave
(D) passes title to minerals only

19. Which of the following is not an essential of a deed?

(A) A deed must be signed.
(B) A deed must be acknowledged.
(C) A deed must contain a description of the property.
(D) A deed must be recovered.

20. A holographic will

(A) devises real property only
(B) is written entirely in the testator's own hand
(C) leaves all property to the eldest son
(D) cannot be overturned

21. Unwritten transfer of title to property implies

(A) transfer by a holographic will
(B) transfer of property from husband to wife by oral agreement
(C) acquisition of title by adverse possession
(D) none of the above

SOLUTIONS

1. A conveyance is a written statement that transfers property.

The answer is (A).

2. An estate can be acquired by purchase, inheritance or gift, or by the power of the state.

The answer is (D).

3. An estate in fee simple absolute implies a complete title, is the highest type of interest, and gives the owner the right to dispose of the property in any way the owner sees fit.

The answer is (D).

4. A deed is the most important document in the transfer of real property, is evidence of the transfer of property, and must be in writing.

The answer is (D).

5. Conveyance of fee title to land is usually by warranty deed.

The answer is (C).

6. Delivery of the property sold is considered to be delivery of the deed.

The answer is (A).

7. The grantee is the person to whom a grant is made.

The answer is (A).

8. A patent is a conveyance of the sovereign's interest in a tract of land.

The answer is (A).

9. An easement is the right that the public or an individual has in the lands of another.

The answer is (D).

10. A mortgage is a conditional conveyance of an estate as a pledge of security of a debt.

The answer is (B).

11. A deed of trust is a conditional conveyance of an estate as a pledge of security of a debt.

The answer is (A).

12. Dedication is a means of transferring title to property.

The answer is (A).

13. To obtain possession by adverse possession, the possession must be open and notorious, hostile, and actual. It is not required that the possession be intermittent.

The answer is (C).

14. Prescription is a method of acquiring the right to use a property.

The answer is (A).

15. The power to exercise eminent domain must be authorized by the state legislature.

The answer is (C).

16. The Constitution of the United States provides that before the power of eminent domain can be exercised, the owner may not be deprived of property without due process of law.

The answer is (C).

17. An encroachment is a gradual, stealthy, illegal acquisition of property.

The answer is (D).

18. A quitclaim deed passes any title the grantor may leave.

The answer is (C).

19. It is not essential that a deed be recovered.

The answer is (D).

20. A holographic will is written entirely in the testator's own hand.

The answer is (B).

21. Unwritten transfer of title to property implies acquisition of title by adverse possession.

The answer is (C).

25 Water Boundaries

1. LEGAL BACKGROUND

In the United States, the public of each state holds title to the submerged lands under navigable waters within their respective state boundaries. That ownership is based on the *public trust doctrine*. The earliest mention of this doctrine occurs in the *Roman Civil Code* of Emperor Justinian I around AD 500. It stated the sea and rivers were considered to be *res communes*, or "commonly owned by all."

Since the Roman Empire, various societies developed differing approaches to determine the boundaries between waters in the public trust and the bordering uplands that are subject to private ownership. These approaches differ depending on whether the boundary is in tidally affected waters or non-tidal waters, or whether the water is navigable or non-navigable. Therefore, it is important for land surveyors to have a thorough knowledge of the prevailing state legal history and doctrine when surveying water boundaries.

2. TIDAL WATERS UNDER ANGLO-AMERICAN COMMON LAW

Because the British Isles have a relatively large range of daily tides, English common law used the daily tides to define coastal boundaries. It held that the limits of the Crown's claim to the submerged lands of the kingdom extended to the average reach of the daily tides. One early authority referred to "ordinary tides," and an early case instructed that the boundary be found by use of "the average of medium tides." Legal systems derived from English common law also use the daily tide to define coastal boundaries.

Most U.S. states follow English common law, but tidal boundaries were not clarified until 1935 with the Supreme Court's decision in *Borax Consolidated Ltd. v. City of Los Angeles*. This case highlighted the English use of the average reach of daily tides, but called for scientific techniques to precisely define that boundary. It recognized that because the astronomical cycle affecting tides has a period of 18.6 years, a statistically valid mean high tide should involve averaging all the high tides over such a period. Therefore, the court concluded that high tides should be averaged over such a period to find the *mean high water line* used to fix the boundary of tidal waters. That line, where the elevation of mean high water meets the rising shoreline, is considered the boundary between the publicly owned submerged lands and the uplands subject to private ownership.

Case law in most of the U.S. coastal states has followed the Borax decision for defining tidal boundaries. Sixteen states, including Alabama, Alaska, California, Connecticut, Florida, Georgia, Maryland, Mississippi, New Jersey, New York, North Carolina, Oregon, Rhode Island, South Carolina, Texas, and Washington, use the mean high water line as the boundary. However, several states, especially those in New England, recognize the *mean low water line* as the boundary and some use the civil law definitions.

The use of mean high water represents an attempt to use the *average upper reach* of the daily tide as the boundary. The tidal range at any given location varies considerably from day to day due to the relationships of various astronomical cycles affecting the tides. Some of these cycles, such as the daily rotation of the moon and sun about the earth, have periods of one day or less. Other cycles affecting the tide, such as those related to the proximity to the earth and the declination of the moon and sun, have periods as long as one month or year. The longest commonly recognized cycle—that dealing with the moon's nodal cycle—has a period of 18.6 years. Although somewhat complex, the use of such an average value results in a boundary that has the advantages of objectivity and mathematical certainty, which are critical for a modern boundary.

The classic definition of the mean high water is the average of all of the high tides over a period of 19 years.[1] This period is referred to as a *tidal epoch*. Published U.S. tidal data is based on a standard tidal epoch to allow for secular differences between such periods.

In the United States, the National Oceanic and Atmospheric Administration (NOAA) maintains a database for a network of tidal stations that provide the elevation

[1] 19 yrs is used by rounding up 18.6 years to include an integral number of the annual periods associated with the sun's declination.

of local mean high water. Since a tidal datum, such as mean high water, may vary considerably with location, it is critical that it only be used in the immediate vicinity of where it was established. Frequently, it is necessary to conduct a new tidal study in an area where the local elevation of mean high water is needed for a water boundary. When this is the case, a simultaneous comparison procedure may be used to adjust short-period observations to the equivalent of a 19-year mean value by simultaneously comparing it to observations of a known tide. Chapter 38 provides details for calculating the local elevation of mean high water by that process.

Once the local elevation of mean high water is determined, the line where that elevation meets the rising shoreline can be mapped. The most common method to achieve this mapping assumes that the mean high water line is a topographic contour in the area around the datum. This contour line is then located by leveling and mapped by horizontal surveying procedures.

Another procedure allows the water itself to define the line. For example, assume that the correct reading on a staff for mean high water has been previously determined. The staff is observed for an incoming tide that is predicted to reach or exceed that level. When the water level reaches the predetermined staff reading for mean high water, personnel in the area around the staff place a series of stakes at frequent intervals along the incoming edge of the water. These stakes defining the tidal datum line then may be mapped by any of various horizontal surveying procedures. The obvious advantage of this method is that the surveyor actually sees the line on the ground and can identify all inflection points. Tide-coordinated, black and white infrared aerial photography or LiDAR may also be used for mapping the mean high water line when that line is not obscured by dense vegetation.

3. TIDAL WATERS UNDER ROMAN AND SPANISH CIVIL LAW

In areas where land title has roots in grants made under Roman or Spanish civil law (as opposed to English common law) different rules generally prevail. Some areas of the United States recognize civil law, including Louisiana and parts of Texas where the root of title stems from Spanish or Mexican grants.

Roman civil law does not define the coastal boundary in terms of the daily tide, but rather in terms of seasonal water levels, possibly because it was developed near the Mediterranean Sea where there is minimal daily tidal range. The *Roman Civil Code* of Emperor Justinian I states, *The sea-shore, that is, the shore as far as the waves go at furthest, was considered to belong to all men.... The sea shore extends as far as the greatest winter floods runs up.*

Similar to Roman civil law, thirteenth century Spanish law contained in *Las Siete Partidas* (the seven parts), defines the boundary as the highest reach of the water throughout the year, but regardless of season. It states, *... e todo aquel lugar es llamado ribera de la mar quanto se cubre el agua de ella, quanto mas crece en todo el año, quier en tiempo del invierno o verano...* (Partida 3, Title 28, Law 4).

The correct translation of this code has been greatly debated. Some interpretations say it calls for an average annual high tide, while other interpretations suggest it calls for the highest tide in the nodal cycle. A generalized translation is "...and all places are called the shore which are covered by the highest reach of the water whether it be in winter or summer."

However, regardless of the interpretation, the civil law expressed in both the *Roman Civil Code* and *Las Siete Partidas* calls for a line that can be higher than that defined under the English common law.

4. NON-TIDAL WATERS

The public trust doctrine also extends to navigable non-tidal waters in most of the United States, although there are a few coastal states whose laws consider only tidal waters to be public trust waters. With the lack of a predictable rising and falling water levels found in tidal waters, different definitions apply for boundaries in non-tidal waters. To distinguish from the mathematically derived boundary of tidal waters (i.e., the mean high water), the boundary of non-tidal waters is generally called the *ordinary high water line* or the *ordinary high water mark.*

The approach typically suggested by U.S. courts for determining the ordinary high water mark is the *physical fact test,* whose leading definition in federal case law stems from the 1851 case of *Howard v. Ingersoll.* This case gave the following instructions to determine the boundary of such waters using the physical fact test.

> This line is to be found by examining the bed and banks and ascertaining where the presence and action of waters are so common and usual and so long continued in all ordinary years, as to mark upon the soil of the bed a character distinct from that of the banks, in respect to vegetation, as well as in respect to the nature of the soil itself.

Thus, to survey a non-tidal water boundary, a surveyor must examine the banks to find where the preponderance of evidence indicates a long-continued presence of water has created a distinction between upland and submerged lands. Land surveyors look for a variety of indicators in the soil or vegetation to help find this distinction. Examples of *vegetative indicators* are typically the lower limit of terrestrial plant life, or the complete lack of vegetative growth, which is often found in a narrow zone slightly below the ordinary high water line (wave action in this zone tends to prevent the growth of either terrestrial or aquatic species, and there is often a non-vegetated sand bottom at such places). *Soil indicators* can include changes in soil composition or in the organic content of the soil.

Other types of evidence for locating the ordinary high water line are various geomorphological features such as natural levees and escarpments. *Natural levees* are low ridges that parallel a river course. They owe their greater height near the stream channel to the cumulative effects of sudden losses in transporting power when a river overspreads its banks. Therefore, the ordinary high water level is usually on the steep or river side and below the crest of such features. *Escarpments* are eroded features that are often found on the river side of levees.

In Texas, the use of natural levees has been refined to very specific techniques for determining boundaries of streams called the *gradient boundary*. The gradient boundary calls for locating the lowest "accretion bank" (levee) in the area and then determining a point that is halfway between the top and bottom of that bank. The difference of elevation between that point and the current surface of the water is then used over the length of the stream to map the boundary. Water level records are another potentially valuable class of evidence to determine non-tidal water boundaries. Resolution of the best method for interpretation of such data, however, is still an unsettled question.

When weighing the evidence to determine the correct boundary regardless of its nature, the surveyor must be careful to ensure that the line chosen is the limit of the submerged land, and does not include bordering swamp and overflowed lands. Such wetlands, while they may be subject to regulation, are considered to be in private ownership and should not be included with the public trust submerged lands.

5. NON-NAVIGABLE WATERS

Generally, non-navigable waters are not considered publicly owned under the public trust doctrine. Therefore, ownership of uplands bordering non-navigable streams or lakes typically includes the adjoining submerged lands.

Streams

An elementary case of a water boundary along non-navigable waters involves a stream as the boundary between two parcels of land. The center of the stream is the boundary, but there are two complicating questions: What constitutes the "center" of the stream, and how should the lateral boundaries between the upland boundaries and the center of the stream be run? Some courts hold that the "center of the stream" should be the *thread of the stream*, or the line lying equidistant between the banks. Other courts hold that the correct line is the *thalweg*, or deepest part of the channel. It would seem that the latter approach is more equitable in that it allows access to the water even if the stream drops to an extremely low level.

The simplest method to determine how the lateral boundary should be extended to the center of the stream is to use a straight line extension of the upland boundary. However, if the stream is especially sinuous, the straight line approach could result in inequitable boundaries. Therefore, some authorities suggest that partition lines should be drawn perpendicular to the *line of navigation* (i.e., running along the deepest channel) when this is the case. Others suggest running lines perpendicular to the center of stream. A third approach apportions the length of a median line according to the length of a meander line from the adjacent riparian lot's shoreline.

Lakes and Ponds

More complicated boundary problems occur when a non-sovereign lake or pond forms a boundary line between two or more parcels due to a non-linear shape and lack of a main channel or current. For lakes and ponds that are substantially round, the general rule holds that a center point at the geographical center should be used. Partition lines are then run from the ends of the upland boundaries to this center point forming "pie slices." Arguably, a center point in the deepest portion in the lake or pond (as opposed to the geographical center) should be used if the water's floor has significant variations in depth. (This ensures equal access to the water, even if the water drops to a low level.) For long or irregular bodies of water, the methods described for streams should be applied to the main portion of the lake or pond, and methods described for round lakes applied to the ends.

Generally, changes in non-navigable waters result in corresponding boundary shifts as they do with navigable waters. For example, if a body of water shifts position by the process of erosion on one side and accretion on the other, then the water's center—and with it the boundary—shifts with the water. However, most states hold that riparian owners may not benefit from artificial erosion and accretion caused by the riparian owner. As with navigable water boundaries, sudden changes due to avulsion generally do not result in a boundary shift.

PRACTICE PROBLEMS

1. Title to the beds of navigable streams within most states is held by the

(A) riparian owner
(B) federal government
(C) state department of environmental protection
(D) public of that state

2. In most states, the boundary between navigable tidal waters and bordering uplands is the line of the

(A) mean sea level
(B) highest tide in the winter
(C) mean high water
(D) mean low water

3. Submerged lands under navigable tidally affected waters are owned by the

(A) riparian owner
(B) federal government
(C) state department of environmental protection
(D) public of that state

4. The boundary between navigable non-tidal waters and bordering uplands in most states is the

(A) mean high water line
(B) ordinary high water mark
(C) mean low water line
(D) ordinary low water mark

5. Submerged lands under non-navigable, non-tidal waters are generally owned by the

(A) riparian owner
(B) federal government
(C) state department of environmental protection
(D) public of that state

SOLUTIONS

1. A state's public hold title to the beds of navigable streams.

The answer is (D).

2. The line of the mean high water is the boundary between navigable tidal waters and bordering uplands in most states.

The answer is (C).

3. A state's public own submerged lands under navigable, tidally affected waters.

The answer is (D).

4. The ordinary high water mark is the boundary between navigable non-tidal waters and bordering uplands in most states.

The answer is (B).

5. The riparian owner generally owns the submerged lands under non-navigable, non-tidal waters.

The answer is (A).

26 Riparian and Littoral Rights

1. RIPARIAN AND LITTORAL RIGHTS

Riparian and littoral rights apply solely to land abutting a water body and do not apply to other adjacent properties. *Riparian rights* refer to rivers and streams, while *littoral rights* refer to oceans, seas, or lakes. Riparian and littoral rights may include a wide range of privileges (subject to state law) associated with the use and enjoyment of waterfront property, including reasonable use of water for domestic, irrigation, and livestock use; construction of docks and mills; scenic views across the water; and ownership of gradually deposited new land. Because riparian and littoral rights are similar in nature, the terms are often used interchangeably.

2. CHANGES IN SHORELINES AND WATER BODIES

Shorelines are dynamic in nature. Over time, their location changes with the rise and fall of the water, and with the water's interaction with the fast land. These changes are generally classified as follows.

Erosion

Erosion is the gradual wearing away of the soil by the action of water, wind, or other elements.

Alluvium

Material resulting from erosion is transported by the water and often deposited along the shoreline in other locations. These deposits are known as *alluvium.*

Accretion

When land is formed slowly and imperceptibly by alluvium, the process is known as *accretion.* The buildup of alluvium can be on the banks of rivers and streams, or on the shores of lakes and tidal waters.

Reliction

The gradual withdrawal of water that leaves land uncovered is known as *reliction.*

Water Level Rise

The gradual rise of the water level, such as occurring with a long-term rise in sea level, can cover upland, resulting in a landward movement of the shoreline. This is known as *water level rise*, and is the opposite of reliction.

Avulsion

The sudden and perceptible change in a shoreline, or the change in the course of a river or stream, is known as *avulsion.*

3. IMPACT OF SHORELINE CHANGES ON BOUNDARIES

Water boundaries do not remain fixed; rather, they change as shorelines change. Generally, riparian or littoral owners gain title to new land created by reliction or accretion, and lose title to land that is submerged by a water level rise or lost by erosion.

Changes that occur suddenly, such as during storms, are considered avulsion and do not change the boundary. One possible exception occurs when avulsion takes place along a navigable stream that results in a change in the channel. In this situation, ownership of the upland between the new and old channels does not change, but it has been held that the state gains title to the bed of both channels.

4. DIVISION LINES FOR RIPARIAN AND LITTORAL RIGHTS

Accretion or reliction often results in newly formed upland that adjoins multiple ownerships. When this occurs, the newly formed upland must be divided among those owners. Furthermore, division lines must be determined for waters adjacent to shorelines in order to plan for the construction of docks and ensure navigational access and view. The method for determining division lines varies significantly with the type of water body and with the prevailing state law. Nevertheless, the accepted procedures follow some general principles.

One well-accepted principle holds that lines for dividing accreted areas and riparian-rights areas over water, generally do not involve extending upland boundaries without a change of direction. Rather, other approaches are used to extend the upland lateral boundary over the accreted land or the water for equitable division of the new upland or water areas. Exceptions to this principle include some cases in states covered by the U.S. Public Land Survey where the land boundary is divided by public land survey sections or aliquot parts of sections.

Another well-accepted principle regards the points of departure for the dividing lines. For dividing riparian rights areas over submerged lands, the departure points should intersect at the upland lateral boundary and the shoreline, typically considered to be the mean high water lines in tidal waters or the ordinary high water line in non-tidal waters. For division of accreted uplands, the departure points should intersect at the lateral upland boundary and the shoreline at the time of subdivision.

For rivers and streams, the most equitable division of riparian rights for areas over water is typically achieved by projecting the dividing lines in a direction perpendicular to the "thread" or center of the stream. In wide rivers, bays, or the ocean, most case law suggests using lines run perpendicular to the shoreline or similar baseline, such as a bulkhead line. When the shoreline curves or is irregular, such as in coves, the procedure generally holds that each riparian tract will receive a proportional share based on the ratio of the frontage of each lot to the total frontage, applied to some outer line (e.g., a channel line).

When riparian rights on lakes are involved, a different approach is required. For round lakes, most case law suggests dividing lines be drawn to a center point to create pie-slice shaped areas. For long lakes, a centerline with dividing lines running perpendicular to it must be created. At the ends of the centerline, dividing lines are drawn to a center point, as in the round lake solution.

For any type of water body, when newly formed uplands are proportioned, most case law suggests that the land be proportioned based on the frontage length prior to the accretion. Generally, the appropriate procedure is to proportion the accreted land based on the ratio of the original frontage of each lot to the total original frontage applied to the new shoreline.

PRACTICE PROBLEMS

1. Land uncovered by reliction is owned by the

(A) public of the state
(B) federal government
(C) riparian or littoral owner
(D) U.S. Department of the Interior

2. A riparian or littoral owner loses title to land by which of the following?

(A) accretion
(B) avulsion
(C) erosion
(D) reliction

3. A riparian or littoral owner gains title to land by which of the following?

(A) accretion
(B) avulsion
(C) erosion
(D) water level rise

4. A riparian owner's rights associated with the river or stream is based on

(A) acquisition of a permit from the state
(B) acquisition of a permit from the federal government
(C) the fact that land abutting the stream is owned
(D) English common law

5. When avulsion occurs on a navigable stream, title to the bed of the new channel belongs to the

(A) public of the state
(B) federal government
(C) riparian owner
(D) Department of the Interior

6. When a navigable stream cuts a new channel while still continuing to flow in the old channel, the island formed belongs to the

(A) public of the state
(B) federal government
(C) riparian owner
(D) Department of the Interior

7. Riparian rights include the use of water for

(A) domestic use
(B) livestock
(C) irrigation
(D) all of the above

8. In the illustration shown, line 1 represents the boundary between the privately owned upland and the publicly owned bed of a navigable stream in 1940. Line 2 represents the boundary at the present time.

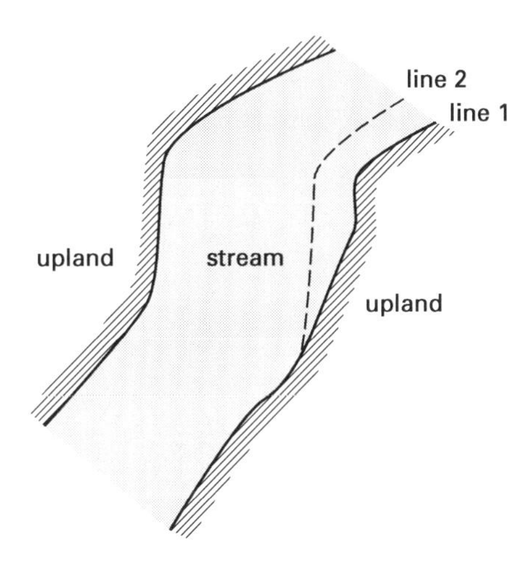

The area between the two lines belongs to which of the following?

(A) the public of the state
(B) the upland owner(s)
(C) it depends whether the original upland belonged to one or more owners
(D) none of the above

9. When dividing accreted land, which of the following principles generally prevails?

(A) The dividing lines should be extensions of the upland lateral boundaries without change of direction.
(B) The dividing lines should be normal to the shoreline.
(C) The accreted land's proportioning should be based on frontage length prior to the accretion.
(D) none of the above

SOLUTIONS

1. Land uncovered by reliction is owned by the riparian or littoral owner.

The answer is (C).

2. A riparian or littoral owner loses title to land by erosion.

The answer is (C).

3. A riparian or littoral owner gains title to land by accretion.

The answer is (A).

4. A riparian owner's rights associated with the river or stream is based on the fact that the land abutting the stream is owned.

The answer is (C).

5. When avulsion occurs on a navigable stream, title to the bed of the new channel belongs to the public of the state.

The answer is (A).

6. When a navigable stream cuts a new channel while still continuing to flow in the old channel, the island formed belongs to the riparian owner.

The answer is (C).

7. Riparian rights include a wide range of privileges, including domestic use, livestock, and irrigation.

The answer is (D).

8. The area between lines 1 and 2 represents accreted lands. Therefore, the riparian owner(s) would gain ownership to the newly formed lands.

The answer is (B).

9. When dividing accreted land, the accreted land's proportioning should be based on the frontage length prior to the accretion.

The answer is (C).

27 Public Land Survey System

1. GENERAL

In 1785, Congress enacted a law that provided for the subdivision of the public lands into townships 6 mi^2 with townships subdivided into 36 sections, most of which are 1 mi on a side. Sections were subdivided into half-sections, quarter-sections, and quarter-quarter sections (the quarter-quarter section being 40 ac in area). Thirty states of the United States were subdivided into tracts by this system, known as the Public Land Survey System or the U.S. System of Rectangular Surveys.

The other states did not pass title to vacant lands to the United States. These states are Texas, West Virginia, Kentucky, Tennessee, the Colonial States, and the other New England and Atlantic Coast states except Florida.

2. QUADRANGLES

A *quadrangle* is approximately 24 mi^2 and consists of 16 townships. Quadrangles were laid out from an initial point through which was established a *principle meridian* and a baseline extending east and west that is a true parallel of latitude. All north-south township lines are true meridians, and all east-west township lines are circular curves that are parallels of latitude. Because of the convergence of meridians, quadrangle corners do not coincide except along the principal meridian.

3. SUBDIVISION OF TOWNSHIPS

Under the 1785 law, townships were divided into sections numbered from 1 to 36 beginning in the northeast corner and ending in the southeast corner as shown in Fig. 27.2. As many sections as possible with 1 mi on a side (640 ac) were laid out in the township. But, due to convergence of the east and west boundaries of a township, it was impossible for all 36 sections to be 1 mi on each side. East and west section boundaries were laid out parallel and not as true meridians. They were laid out parallel to the east boundary of the township. This made it impossible for all sections to be 1 mi on a side and at the same time coincide with the township lines.

To produce as many sections as possible 1 mi on a side, the sections along the north township line and the west township line were of varying dimensions to compensate for errors and the convergence of the west township line. The errors were actually thrown into the north one-half of the sections along the north township line and into the west one-half of the sections along the west township line. Thus, sections 1–6, 7, 18, 19, 30, and 31 were not regular sections. When a section was limited by a lake, river, or old survey, part of it was eliminated, but the existing section was numbered as if the whole section were laid out.

Figure 27.1 *Quadrangle Divided into Townships*

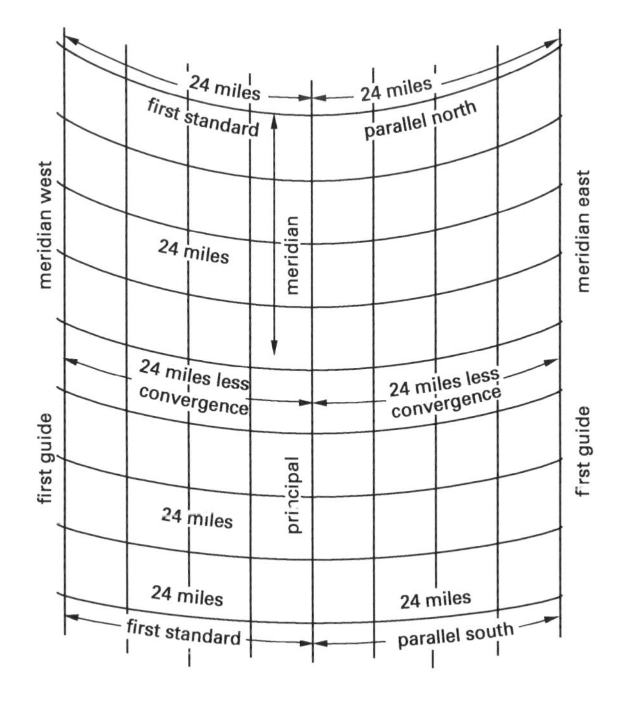

Cadastral and Boundary Law

4. SUBDIVISION OF SECTIONS

Sections may be divided into half-sections, quarter-sections, half-quarter sections, or quarter-quarter sections.

Figure 27.2 *Township Subdivided into Sections*

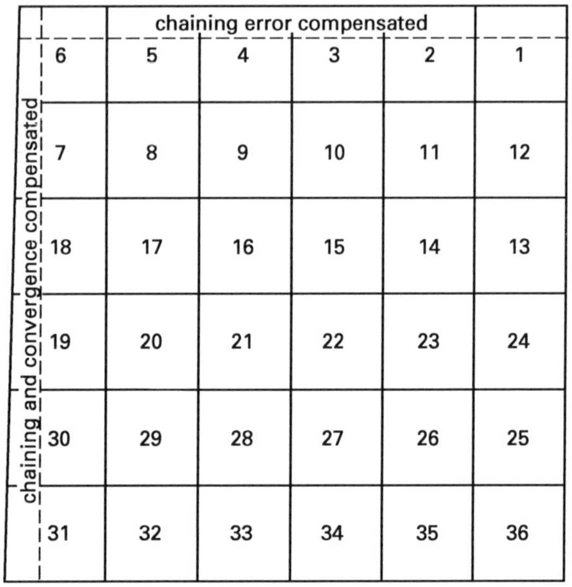

chaining error compensated

chaining and convergence compensated

6	5	4	3	2	1
7	8	9	10	11	12
18	17	16	15	14	13
19	20	21	22	23	24
30	29	28	27	26	25
31	32	33	34	35	36

28 Restoration of Lost and Obliterated Public Land Survey Corners

1. JURISDICTION

The U.S. Bureau of Land Management, under the supervision of the Secretary of the Interior, has complete jurisdiction over the survey and resurvey of the public lands of the United States.

After title to a piece of land is granted by the United States, jurisdiction over the property passes to the state. The federal government retains its authority only with respect to the public lands in federal ownership. Where the lands are in private ownership, it is a function of the county or local surveyor to restore lost corners and to subdivide the sections. Disputes concerning these questions must come before the local courts unless settled by joint survey or agreement. It should be understood, however, that no adjoining owner can make a valid encroachment upon the public lands.

2. RESURVEYS

Public and privately owned lands may both be resurveyed by the Bureau of Land Management in certain cases, under the authority of an act of Congress approved March 3, 1909, and amended June 25, 1910:

> That the Secretary of the Interior may, in his discretion, cause to be made, as he may deem wise under the rectangular system now provided by law, such resurveys or retracements of the surveys of public lands as, after full investigation, he may deem essential to properly mark the boundaries of the public lands remaining undisposed of.

The 1909 act is generally invoked where the lands are largely in federal ownership and where there may be extensive obliteration or other equally unsatisfactory conditions.

Another act of Congress approved September 21, 1918, provides authority for the resurvey by the government of townships previously ineligible for resurvey by reason of the disposals' being in excess of 50% of the total area.

The 1918 act may be invoked where the major portion of the township is in private ownership, where it is shown that the need for retracement and remonumentation is extensive, and especially if the work proposed is beyond the scope of ordinary local practice. The act requires that the proportionate costs be carried by the landowners.

3. PROTECTION OF BONA FIDE RIGHTS

Under the Bureau of Land Management laws discussed in Sec. 2, and in principle as well, it is required that no resurvey or retracement shall be so executed as to impair the bona fide rights or claims of any claimant, entryman, or owner of land so affected.

Likewise in general practice, local surveyors should be careful not to exercise unwarranted jurisdiction, nor to apply an arbitrary rule. They should note the distinction between the rules for original surveys and those that relate to retracements. The disregard for these principles, or for acquired property rights, may lead to unfortunate results.

In unusual cases where the evidence of the survey cannot be identified with ample certainty to enable the application of the regular practices, the surveyors may submit their questions to the proper state office of the Bureau of Land Management, or to the Director of the Bureau of Land Management.

4. ORIGINAL SURVEY RECORDS

The township plat furnishes the basic data relating to the survey and the description of all areas in the particular township. All title records within the area of the former public domain are based upon a government grant or patent, with description referred to an official plat. The lands are identified on the ground through the retracement, restoration, and maintenance of the official lines and corners.

The plats are developed from the field notes. Both are permanently filed for reference purposes and are accessible to the public for examination or making of copies.

Many supplemental plats have been prepared by protraction to show new or revised lottings within one or more sections. These supersede the lottings shown on the original township plat. There are also many plats of the survey of islands or other fragmentary areas of public land that were surveyed after the original survey of the township.

These plats should be referred to as governing the position and a description of the subdivision should be shown on them.

5. RESURVEY RECORDS

The plats and field notes of resurveys that become a part of the official record fall into two principal classes according to the type of resurvey.

A *dependent resurvey* is a restoration of the original survey according to the record of that survey, based upon the identified corners of the original survey and other acceptable points of control, and the restoration of lost corners in accordance with proportional measurement as described herein. Normally, the subdivisions shown on the plat of the original survey are retained on the plat of the dependent resurvey, although new designations and areas for subdivisions still in public ownership at the time of the resurvey may be shown to reflect true areas.

An *independent resurvey* is designed to supersede the original survey and creates new subdivisions and lottings of the vacant public lands. Provision is made for the segregation of individual tracts of privately owned lands, entries, or claims that may be based upon the original plat, when necessary for their protection, or for their conformation, if feasible, to the regular subdivisions of the survey.

6. RECORDS TRANSFERRED TO STATES

In those states where the public land surveys are considered as having been completed, the field notes, plats, maps, and other papers relating to those surveys have been transferred to an appropriate state office for safekeeping as public records. No provision has been made for the transfer of the survey records to the State of Oklahoma, but in the other states the records are filed in offices where they may be examined and copies made or requested.

7. GENERAL PRACTICES

The rules for the restoration of lost corners have remained substantially the same since 1883, when they were first published. These rules are in harmony with the leading judicial opinions and the most approved surveying practice. They are applicable to the public land rectangular surveys and to the retracement of those surveys (as distinguished from the running of property lines that may have legal authority only under state law, court decree, or agreement).

In the New England and Atlantic coast states except Florida, and in Pennsylvania, West Virginia, Kentucky, Tennessee, and Texas, jurisdiction over the vacant lands remained in the states. The public land surveys were not extended in these states, and it follows that the practices that are outlined herein are not applicable there, except as they reflect sound surveying methods.

The practices outlined herein are in accord with the related provisions of the BLM *Manual of Surveying* (1973). They have been segregated for convenience to separate them from the instructions pertaining only to the making of original surveys.

For clarity, the practices, as such, are set in boldface type. The remainder of the text is explanatory and advisory only, the purpose being to exemplify the best general practice.

In some states, the substance of the practices for restoration of lost or obliterated corners and subdivision of sections as outlined herein has been enacted into law. It is incumbent on the surveyor engaged in practice of land surveying to become familiar with the provisions of the laws of the state, both legislative and judicial, that affect his or her work.

8. GENERAL RULES

The general rules followed by the Bureau of Land Management, which affect all public lands, are summarized in the following paragraphs.

- **The boundaries of the public lands, when approved and accepted, are unchangeable.**
- **The original township, section, and quarter-section corners must stand as the true corners that they were intended to represent, whether in the place shown by the field notes or not.**
- **Quarter-quarter section corners not established in the original survey shall be placed on the line connecting the section and quarter-section corners, and midway between them, except on the last half-mile of section lines closing on the north and west boundaries of the township, or on the lines between fractional or irregular sections.**
- **The center lines of a section are to be straight, running from the quarter-section corner on one boundary to the corresponding corner on the opposite boundary.**
- **In a fractional section where no opposite corresponding quarter-section corner has been or can be established, the center line must be run from the proper quarter-section corner as nearly in a cardinal direction to the meander line, reservation, or other boundary of such fractional section, as due parallelism with the section boundaries will permit.**

Corners established in the public land surveys remain fixed in position and are unchangeable, and lost or obliterated corners of those surveys must be restored to their original locations from the best available evidence of the official survey in which such corners were established.

9. RESTORATION OF LOST OR OBLITERATED CORNERS

The restoration of lost corners should not be undertaken until after all control has been developed. Such control includes both original and acceptable collateral evidence. However, the methods of proportionate measurement will be of material aid in the recovery of evidence.

An existent corner is one whose position can be identified by verifying the evidence of the monument, or its accessories, by reference to the description that is contained in the field notes, or where the point can be located by an acceptable supplemental survey record, some physical evidence, or testimony.

Even though its physical evidence may have entirely disappeared, a corner will not be regarded as lost if its position can be recovered through the testimony of one or more witnesses who have a dependable knowledge of the original location.

An obliterated corner is one at whose point there are no remaining traces of the monument, or its accessories, but whose location has been perpetuated, or the point that may be recovered beyond reasonable doubt, by the acts and testimony of the interested landowners, competent surveyors, or other qualified local authorities, or witnesses, or by acceptable record evidence.

A position based upon collateral evidence should be duly supported, generally through proper relation to known corners, and agreement with the field notes regarding distances to natural objects, stream crossings, line trees, and off-line tree blazes, and so on, or unquestionable testimony.

A lost corner is a point of survey whose position cannot be determined, beyond reasonable doubt, either from traces of the original marks or from acceptable evidence or testimony that bears upon the original position, and whose location can be restored only by reference to one or more interdependent corners.

If there is some acceptable evidence of the original location of the corner, that position will be employed.

The decision that a corner is lost should not be made until every means has been exercised that might aid in identifying its true original position. The retracements, which are usually begun at known corners and run according to the record of the original survey, will indicate the probable position for the corner, and show what discrepancies may be expected. Any supplemental survey record or testimony should then be considered in light of the facts thus developed. A line will not be regarded as doubtful if the retracement affords recovery of acceptable evidence.

In cases where the probable position for a corner cannot be made to harmonize with some of the calls of the field notes, due to errors in description or to discrepancies in measurement developed in the retracement, it must be ascertained which of the calls for distances along the line are entitled to the greater weight. Aside from the technique of recovering traces of the original marks, the main problem is one that treats with the discrepancies in alignment and measurement.

Existing original corners cannot be disturbed; consequently, discrepancies between the new and the record measurements will not in any manner affect the measurements beyond the identified corners, but the difference will be distributed proportionally within the several intervals along the line between the corners.

10. PROPORTIONATE MEASUREMENT

The ordinary field problem consists of distributing the excess or deficiency in measurement between existent corners in such a manner that the amount given to each interval shall bear the same proportion to the whole difference as the record length of the interval bears to the whole record distance. After having applied the proportionate difference to the record length of each interval, the sum of the several parts will equal the new measurements of the whole distance.

A proportionate measurement is one that gives concordant relation between all parts of the line—that is, the new values given to the several parts, as determined by the remeasurement, shall bear the same relation to the record lengths as the new measurement of the whole line bears to that record. Lengths of proportioned lines are comparable only when reduced to their cardinal equivalents.

Discrepancies in measurement between those recorded in the original survey and those developed in the retracements should be carefully verified with the object by placing each such difference properly where it belongs. This is quite important at times, because, if disregarded, the result may be the fixing of a corner position where it is obviously improper. Accordingly, wherever possible, the manifest errors in the original measurements should be segregated from the general average difference and placed where the blunder was made. The accumulated surplus or deficiency that then remains is the quantity that is to be uniformly distributed by the methods of proportionate measurement.

11. SINGLE PROPORTION

The term "single proportionate measurement" is applied to a new measurement made on a line to determine one or more positions on that line.

In *single proportionate measurement*, the position of two identified corners controls the direction of the line between those corners, and intermediate positions on that line are determined by proportionate measurement between those controlling corners. The method is sometimes referred to as a *two-way proportion*. Examples are: a quarter-section corner on the line between two section corners; all corners on standard parallels; and all intermediate positions on any township boundary line.

12. DOUBLE PROPORTION

The term "double proportionate measurement" is applied to new measurement made between four known corners, two each on intersecting meridional and latitudinal lines, for the purpose of relating the intersection of both.

By *double proportionate measurement*, the lost corner is reestablished on the basis of measurement only, disregarding the record directions. An exception will be found in those cases where there is some acceptable survey record, some physical evidence, or testimony that may be brought into the control. The method may be referred to as a *four-way proportion*. Examples are a corner common to four townships, or one common to four sections within a township.

The double proportionate measurement is the best example of the principle that existent or known corners to the north and to the south should control any intermediate latitudinal position, and that corners east and west should control the position in longitude.

As between single or double proportionate measurement, the principle of precedence of one line over another of less original importance is recognized in order to harmonize the restoring process with the method followed in the original survey, thus limiting control.

13. STANDARD PARALLELS AND TOWNSHIP BOUNDARIES

Standard parallels will be given precedence over other township exteriors, and ordinarily the latter will be given precedence over subdivisional lines; section corners will be relocated before the position of lost quarter-section corners can be determined.

To restore a lost corner of four townships, a retracement will first be made between the nearest known corners on the meridional line, north and south of the missing corners, and upon that line a temporary stake will be placed at the proper proportionate distance; this will determine the latitude of the lost corner.

Next, the nearest corners on the latitudinal line will be connected, and a second point will be marked for the proportionate measurement east and west; this point will determine the position of the lost corner in departure (or longitude).

Then, through the first temporary stake run a line east or west, through the second temporary stake a line north or south, as relative situations may determine; the intersection of these two lines will fix the position for the restored corner.

Figure 28.1 *Standard Parallels and Township Boundaries*

In Fig. 28.2, points A, B, C, and D represent four original corners. Point E represents the proportional measurement between A and B; and similarly, F represents the proportional measurement between C and D. Point X satisfies the first control for latitude and the second control for departure.

A lost corner of four townships should not be restored, nor the township boundaries reestablished, without first considering the full field note record of the four intersecting lines and the plats of the township involved. In most cases there is a fractional distance in the half-mile to the east of the township, corner, and frequently in the half-mile to the south. The lines to the north and to the west are usually regular—that is, quarter-section and section corners at normal intervals of 40.00 and 80.00 chains—but there may be closing-section corners on any or all of the boundaries. So, it is important to verify all of the distances by reference to the field notes.

14. INTERIOR CORNERS

A lost interior corner of four sections will be restored by double proportionate measurement.

When a number of interior corners of four sections, and the intermediate quarter-section corners, are missing on all sides of the one sought to be reestablished, the entire distance between the nearest identified corners both north and south, east and west, must be measured. Lost section corners on the township exteriors, if required for control, should be relocated.

15. RECORD MEASUREMENT

Where the line has not been established in one direction from the missing township or section corner, the record distance will be used to the nearest identified corner in the opposite direction.

Figure 28.2 *Lost Township Corner*

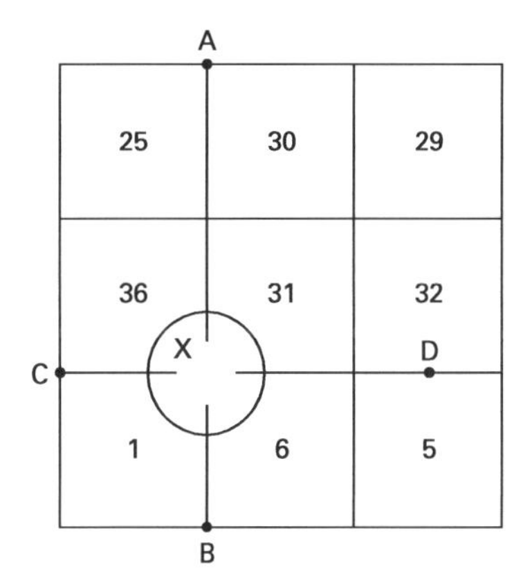

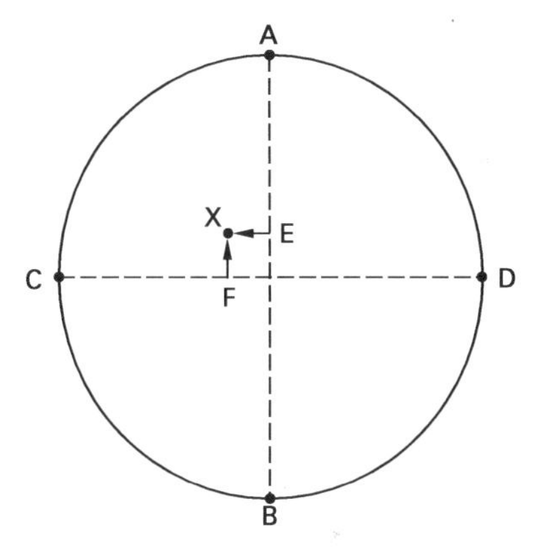

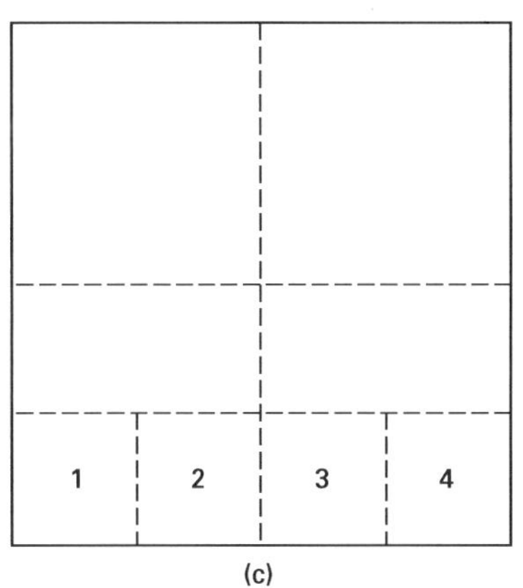

Thus, in Fig. 28.2, if the latitudinal line in the direction of point D has not been established, the position of point F in departure would have been determined by reference to the record distance from point C. Point X would then be fixed by cardinal offsets from points E and F as already explained.

Where the intersecting lines have been established in only two of the directions, the record distances to the nearest identified corners on

these two lines will control the position of the temporary points; then from the latter the cardinal offsets will be made to fix the desired point of intersection.

16. TWO SETS OF CORNERS

In many surveys the field notes and plats indicate two sets of corners along township boundaries and section lines where parts of the township were subdivided on different dates. In such cases, there are usually corners of two sections at regular intervals, and closing section corners established later upon the same line. The quarter-section corners on such lines usually are controlling for one side only.

In the more recent surveys, where the record calls for two sets of corners, those that are the corners of the two sections first established and the quarter-section corners relating to the same sections will be employed for the retracement, and they will govern both the alignment and the proportional measurements along that line. The closing section corners set at the intersections will be employed in the usual way—that is, to govern the direction of the closing lines.

17. RESTORATION BY SINGLE PROPORTION

The method of single proportionate measurement is generally applicable to the restoration of lost corners on standard parallels and other lines established with reference to definite alignment in one direction only. Intermediate corners on township exteriors and other controlling boundary lines are to be included in this class.

To restore a lost corner by single proportionate measurement, a retracement will be made connecting the nearest identified regular corners on the line in question. A temporary stake will be set on the trial line at the original distance. The total distance and the falling at the objective corner will be measured.

On meridional township lines, an adjustment will be made at each temporary stake for the proportional distance along the line. The temporary stake will be set over to the east or to the west for falling, counting its proportional part from the point of beginning.

On east-and-west township lines and on standard parallels, the proper adjustment should be made at each temporary stake for the proportional distance along the line for the falling. (The temporary stake will either be advanced or set back for the proportional part of the distance between the record distance and the new measurement. It will then be set over for the curvature of the line, and eventually corrected for the proportional part of the true falling.)

The adjusted position is thus placed on the true line that connects the nearest identified corners, and at the same proportional interval from either as existed in the original survey. Any number of intermediate lost corners may be located on the same plan by setting a temporary stake for each when making the retracement.

Lost standard corners will be restored to their original positions on a baseline, standard parallel or correction line, by single proportionate measurement on the true line connecting the nearest identified standard corners on opposite sides of the missing corner or corners, as the case may be.

The term *standard corners* will be understood to mean all corners that were established on the standard parallel during the original survey of that line, including, but not limited to, standard township, section, quarter-section, meander, and closing corners. Closing corners, or other corners purported to be established on a standard parallel after the original survey of that line, will not control the initial restoration of lost standard corners.

Corners on baselines are to be regarded as the same as those on standard parallel. In the older practice, the term *correction line* was used for what later has been called the *standard parallel.* The corners first set in the running of a correction line will be treated as original standard corners. Those that were set afterwards at the intersection of a meridional line will be regarded as closing corners.

All lost section and quarter-section corners on the township boundary lines will be restored by single proportionate measurement between the nearest identified corners on opposite sides of the missing corner, north and south on a meridional line, or east and west on a latitudinal line, after the township corners have been identified or relocated.

An exception to this rule will be found in the case of any exterior the record of which shows deflections in alignment between the township corners.

A second exception to this rule is found in those occasional cases were they may be persuasive proof of a deflection in the alignment of the township boundary, even though the record shows the line to be straight. For example, measurements east and west across a range line, or north and south across a latitudinal township line, counting from a straight-line exterior adjustment, may show distances to the nearest identified subdivisional corners to be materially long in one direction and correspondingly short in the opposite direction as compared to the record measurements. This condition, when supported by corroborative collateral evidence as might generally be expected, would warrant an exception to the straight-line or two-way adjustment under the rules for the acceptance of evidence—that is, the evidence outweighs the record. The rules for a four-way or double proportionate measurement would then apply, provided there is conclusive proof.

All lost quarter-section corners on the section boundaries within the township will be restored by single proportionate measurement between the adjoining section corners, after the section corners have been identified or relocated.

This practice is applicable in the majority of the cases. However, in those instances where other corners such as meander corners, sixteenth-section corners, and so on, were originally established between the quarter-section and the section corners, such minor corners, when identified, will exercise control in the restoration of lost quarter-section corners.

Lost meander corners, originally established on a line projected across the meanderable body of water and marked upon both sides, will be relocated by single proportionate measurement, after the section or quarter-section corners upon the opposite sides of the missing meander corner have been duly identified or relocated.

Under ordinary conditions, the actual shore line of a body of water is considered the boundary of lands included in an entry and patent, rather than the meander line returned in the field notes. It follows that the restoration of a lost meander corner would be required only infrequently. Under favorable conditions, a lost meander corner may be restored by treating the shoreline as an identified natural feature controlling the measurement to the point for the corner. This is particularly applicable where it is evident that there has been no change in the shore line.

A lost closing corner will be reestablished on the true line that was closed upon, and at the proper proportional interval between the nearest regular corners to the right and left.

To reestablish a lost closing corner on a standard parallel or other controlling boundary, the line that was closed upon will be retraced, beginning at the corner from which the connecting measurement was originally made. A temporary stake will be set at the record connecting distance, and the total distance and falling will be noted at the next regular corner on that line on the opposite side of the missing closing corner. The temporary stake will then be adjusted as in single proportionate measure.

A closing corner not actually located on the line that was closed upon will determine the direction of the closing line, but not its legal terminus. The correct position is at the true point of intersection of the two lines.

18. IRREGULAR EXTERIORS

Some township boundaries, not established as straight lines, are termed *irregular* exteriors. Parts were surveyed from opposite directions, and the intermediate portion was completed later by random and true line, leaving a fractional distance. Such irregularity follows some material departure from the basic rules for the establishment of original surveys. A modified form of single proportionate measurement is used in restoring lost corners on such boundaries. This is also applicable to a section line or township line that has been shown to be irregular by a previous retracement.

To restore one or more lost corners or angle points on such irregular exteriors, a retracement between the nearest known corners is made on the record courses and distances to ascertain the direction and length of the closing distance. A temporary stake is set for each missing corner or angle point. Thc closing distance is then reduced to its equivalent latitude and departure.

On a meridional line, the latitude of the closing distance is distributed among the courses in proportion to the latitude of each course. The departure of the closing distance is distributed among the courses in proportion to the length of each course. That is, after the excess or deficiency of latitude is distributed, each temporary stake is moved east or west an amount proportional to the total distance from the starting point.

On a latitudinal line, the temporary stakes should be placed to suit the usual adjustments for curvature. The departure of the closing distance is distributed among the courses in proportion to the departure of each course. Then, each temporary stake is moved north or south an amount proportional to the total distance from the starting point.

Angle points and intermediate corners are treated alike.

19. ONE-POINT CONTROL

Where a line has been terminated with measurement in one direction only, a lost corner will be restored by record bearing and distance, counting from the nearest regular corner, the latter having been duly identified or restored.

Examples will be found where lines have been discontinued at the intersection with large meandrous bodies of water, or at the border of what was classified as impassable ground.

20. INDEX ERRORS FOR ALIGNMENT AND MEASUREMENT

Where the original surveys were faithfully executed, it is to be anticipated that retracement of many miles of the lines in a given township will develop a definite and consistent difference in measurement and in a bearing between original corners. Under such conditions, it is proper that allowance be made for the average differences in the restoration of a lost corner where control is lacking in one direction. The adjustment will be taken care of automatically where there is a suitable basis for proportional measurement.

21. SUBDIVISION OF SECTIONS

The ordinary unit of administration of the public lands under the rectangular system of surveys is the quarter-quarter-section of 40 ac. Usually the sections are not subdivided on the ground in the original survey. The boundaries of the legal subdivisions generally are shown by protraction on the plats.

On the plat of the original survey of a normal township, it is to be expected that the subdivision of sections will be indicated (by protraction) according to standard procedures.

The sections bordering the north and west boundaries of the township, except section 6, are subdivided into two regular quarter-sections, two regular half-quarter sections, and four fractional quarter-quarter units, which are usually designated as lots. In section 6, the subdivision will show one regular quarter-section, two regular half-quarter sections, one regular quarter-quarter section, and seven fractional quarter-quarter units. This is the result of the plan of subdivision, whereby the excess or deficiency in measurement is placed against the north and west boundaries of the township.

The plan of subdivisions and controlling measurements employed is illustrated in Fig. 28.4.

In a normal section that is subdivided by protraction into quarter-sections, it is not considered necessary to indicate the boundaries of the quarter-quarter sections on the plat. Such subdivisions are aliquot parts of the quarter-sections, based on midpoint protraction.

Sections that are invaded by meanderable bodies of water or by private claims that do not conform to the regular legal subdivisions are subdivided by protraction into regular and fractional parts as nearly as practicable in conformity with the uniform plan already outlined. The meander lines, and the boundary lines of the private claims, are platted according to the field note record. The subdivision-of-section lines are terminated at the meander line or claim boundary, as the case may be, but their positions are controlled precisely as though the section had been completed regularly. For the purpose of protracting the subdivisional lines in a section whose boundary lines are partly within the limits of a meanderable body of water or private claim, the fractional section boundaries are completed in theory.

The protracted position of the subdivision-of-section line is controlled by the theoretical points so determined (see Fig. 28.3).

Figure 28.3 *Examples of Subdivision by Protraction*

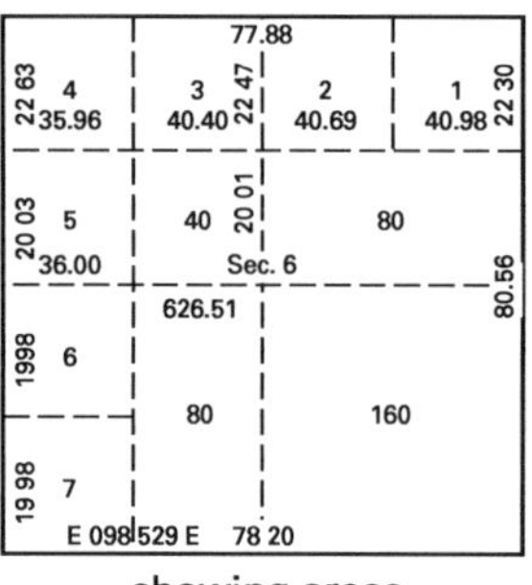

showing areas

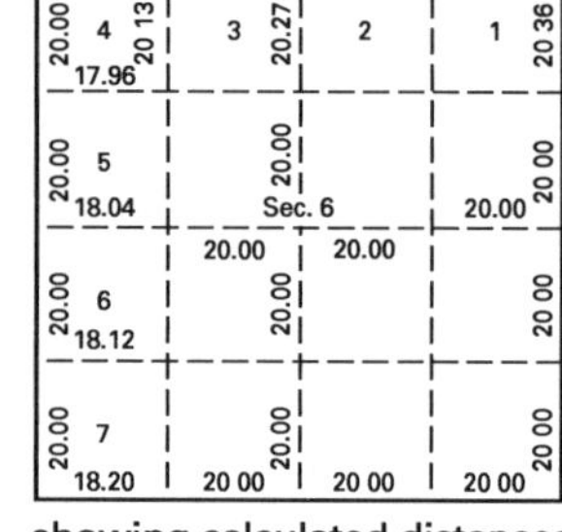

showing calculated distances

Figure 28.4 Examples of Subdivision of Fractional Sections

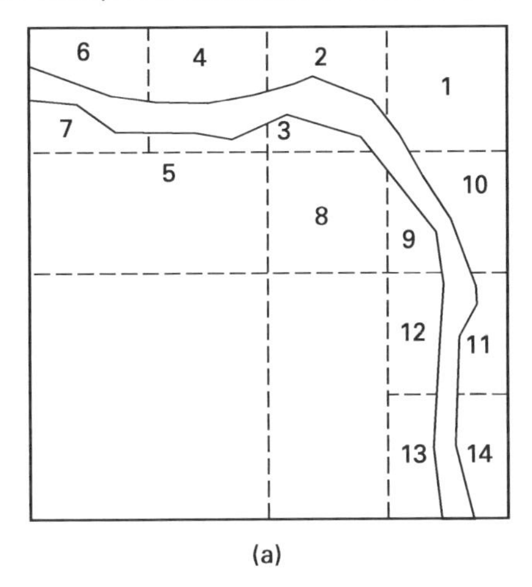

(a)

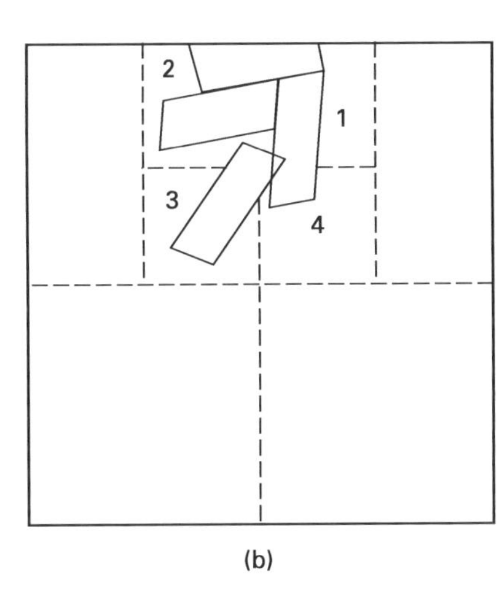

(b)

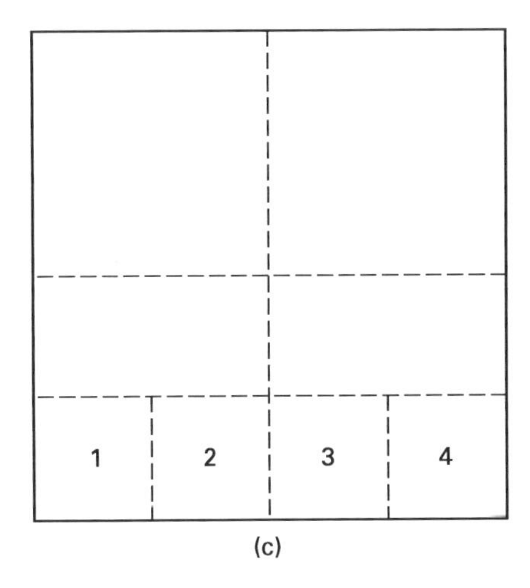

(c)

22. ORDER OF PROCEDURE IN SURVEY

The order of procedure is first to identify or reestablish the corners on the section boundaries, including determination of the points for the necessary one-sixteenth section corners. Next, fix the boundaries of the quarter-section. Finally, form the quarter sections or small tracts by equitable and proportionate division.

23. SUBDIVISION OF SECTIONS INTO QUARTER-SECTIONS

To subdivide a section into quarter-section, run straight lines from the established quarter-section corners to the opposite quarter-section corners. The point of intersection of the lines thus run will be the corner common to the several quarter-sections, or the legal center of the section.

Upon the lines closing on the north and west boundaries of a regular township, the quarter-section corners were established originally at 40 chains to the north or west of the last interior section corners. The excess or deficiency in measurement was thrown into the half-mile next to the township or range line, as the case may be. If such quarter-section corners are lost, they should be reestablished by proportionate measurement based upon the original record.

Where there are double sets of section corners on township and range lines, the quarter-section corners for the sections south of the township line and east of the range line usually were not established in the original surveys. In subdividing such sections, new quarter-section corners are required. They should be placed to suit the calculations of the areas that adjoin the township boundary, as indicated upon the official plat, adopting proportional measurements where the new measurements of the north or west boundaries of the section differ from the record distances.

24. SUBDIVISION OF FRACTIONAL SECTIONS

The law provides that where opposite corresponding quarter-section corners have not been or cannot be fixed, the subdivision-of-section lines shall be ascertained by running from the established corners north, south, east, or west, as the case may be, to the water course, reservation line, or other boundary of such fractional section, as represented upon the official plat.

This law presumes that the section lines are either due north and south, or east and west lines, but usually this is not the case. Hence, to carry out the spirit of the law, it will be necessary in running the center lines through fractional sections to adopt mean courses, where the section lines are not on due cardinal, or to run parallel to the east, south, west, or north boundary of the section, as conditions may require, where there is no opposite section line.

25. SUBDIVISION OF QUARTER-SECTIONS

Preliminary to the subdivision of quarter-sections, the quarter-quarter, or sixteenth-section, corners will be established at points midway between the section and quarter-section corners, and the center of the section, except on the last half mile of the lines closing on township boundaries, where they should be placed at 20 chains, proportionate measurement, counting from the regular quarter-section corner.

The quarter-quarter, or sixteenth-section, corners having been established as directed previously, the center lines of the quarter-section will be run straight between opposite corresponding quarter-quarter, or intersection of the lines thus run will determine the legal center of a quarter-section (see Fig. 28.5).

Figure 28.5 *Example of Subdivision by Survey*

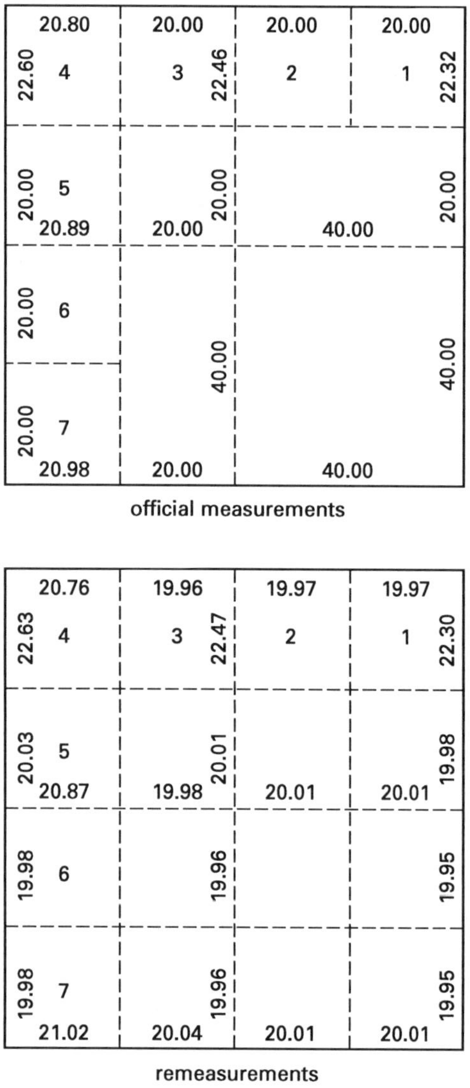

official measurements

remeasurements

26. SUBDIVISION OF FRACTIONAL QUARTER-SECTIONS

The subdivisional lines of fractional quarter-sections will be run from properly established quarter-quarter, or sixteenth-section corners, with courses governed by the conditions represented upon the official plat, to the lake, watercourse, reservation, or other irregular boundary that renders such sections fractional.

What has been written on the subject of subdivision of sections relates to the procedure contemplated by law, and refers to the methods to be followed in the initial subdivision of the areas, prior to development and improvement. Care should be exercised to avoid disturbing satisfactory improvements such as roads, fences, or other features making subdivision-of-section lines and that may define the extent of property rights.

27. RETRACEMENTS

When it is necessary to retrace the lines of the rectangular public-land surveys, the first step is to assemble copies of field notes and plats and determine the names of the owners who will be concerned in the retracement and survey. A thorough search and inquiry with regard to the record of any additional surveys that have been made since the approval of the original survey should be made. The county surveyor, county clerk, register of deeds, practicing engineers and surveyors, landowners, and others who may furnish useful information should be consulted as to such features.

The matter of boundary disputes should be carefully reviewed, particularly as to whether claimants have based their locations upon evidence of the original survey and a proper application of surveying rules. If there has been a boundary suit, the record testimony and the court's opinion and decree should be carefully examined insofar as these may have a bearing upon the problem at hand.

The law requires that the position of original corners must not be changed. There is a penalty for defacing corner marks, or for changing or removing a corner. The corner monuments afford the principle means for identification of the survey, and accordingly, the courts attach the greatest weight to the evidence of their location. Discrepancies that may be developed in the directions and lengths of lines, as compared with the original record, do not warrant any alteration of a corner position.

Obviously, on account of roadways or other improvements, it is frequently necessary to reconstruct a monument in some manner in order to preserve its position. Alterations of this type are not regarded as changes in willful violation of the law, but rather as being in complete accordance with the legal intent to safeguard the evidence.

Therefore, whatever the purpose of the retracement may be—if it calls for the recovery of the true lines of the original survey, or for the running of the subdivisional lines of a section—the practices outlined require some or all of certain definite steps.

step 1: Secure a copy of the original plat and field notes.

step 2: Secure all available data regarding subsequent surveys.

step 3: Secure the names and contact the owners of the property adjacent to the lines that are involved in the retracement.

step 4: Find the corners that may be required:

- first, by the remaining physical evidence;
- second, by collateral evidence, supplemental survey records, or testimony, if the original monument is regarded as obliterated, but not lost, or;
- third, by application of the rules for proportionate measurement, if lost.

step 5: Reconstruct the monuments as required, including the placing of reference markers where improvements of any kind might interfere, or if the site is such as to suggest the need for supplemental monumentation.

step 6: Note the procedure for the subdivision of sections where these lines are to be run.

step 7: Prepare and file a suitable record of what was found, the supplemental data that were employed; a description of the methods used, the direction and length of lines, the new markers, and any other facts regarded as important.

A knowledge of the practices and instructions in effect at the time of the original survey will be helpful. These should indicate what was required and how it was intended that the original survey should be made.

The data used in connection with the retracements should not be limited to the section or sections under immediate consideration. They should also embrace the areas adjacent to those sections. The plats should be studied carefully. Fractional parts of sections should be located on the ground as indicated on the plats.

28. DOUBLE SETS OF CORNERS

The methods of determining true meridian and true latitudinal curve were developed many years after the inception of the rectangular survey system. Without these refinements, accumulated discrepancies were bound to develop.

As a result, in order to maintain rectangularity in some of the older surveys, two sets of corners were established on the township boundaries. The section and quarter-section corners established in the survey of the boundary itself are the corners to be adopted in retracement and for control of proportionate measurements. These corners control the subdivisions on only one side of the township boundary. The second set of corners on these boundaries are the closing section corners for the subdivisional surveys on the opposite side of the boundary. The descriptions of these closing corners, and the connecting distances to the regular township boundary corners, will be found in the field notes of the subdivisional survey in which they were established. These closing section corners should be considered and evaluated as evidence in the solution of the whole problem.

Where the section corners of the township boundaries are of minimum control, the quarter-section corners have the same status for the same side of the boundary. In the older surveys, quarter-section corners usually were not established for the opposite side of the boundary. Subsequent to 1919, it was the practice to establish the second set of quarter-section corners. These are at midpoint for distances between the closing section corners, except where the plan of subdivision dictates otherwise, in which event the quarter-section corner is placed at 40.00 chains from the controlling closing corner.

These conditions merit careful study of the plats to the end that the subdivisions shown on the plats be given proper protection. The plats will indicate whether these quarter-section corners should be at midpoint between the closing corners, or if they should be located with regard to a fractional distance. The surveyor should make sure that the position is determined for all corners necessary for control in his or her work.

There is nothing especially different or complicated in the matter of one or two sets of corners on the township boundary lines. It is merely a matter of assembling complete data and of making a proper interpretation of the status of each monument.

The same principles should be applied in the consideration of the data of the subdivisional surveys, where for any of the several causes, there may be two sets of interior corners.

29. THE NEEDLE COMPASS AND SOLAR COMPASS

Very simple needle-compass equipment in the hands of persons skilled as surveyors, coupled with natural woodcraft and faithfulness in doing their work, satisfied the requirements of the early public-land surveys.

A large proportion of surveys made prior to 1890 are of the needle-compass type. It should be noted that retracements may be made—that is, the evidence of the marks can be developed—by needle-compass methods if properly employed. Some surveyors maintain that you can "follow the steps" of the original surveyor more closely by use of the needle compass than by more precise methods.

In addition to the uncertainties of local attraction and temporary magnetic disturbances, the use of the needle compass is exceedingly unreliable in the vicinity of power lines, pipelines, steel rails, steel-framed structures of all kinds, wire fences, and so on. Its use is now much more restricted because of these improvements. The needle compass is rapidly becoming obsolete because it fails to satisfy the present need for more exact retracements.

30. EXCESSIVE DISTORTION

The needle-compass surveys, before being discontinued, had extended into the region of magnetic ore deposits of the Lake Superior watershed in northern Michigan, Wisconsin, and Minnesota. Here many townships were surveyed, and the lands patented, in which the section boundaries are now found to be grossly distorted. There is no way in which to correct these lines, nor to make an estimate (except by retracement), of the extent of the irregularities, which involve excessive discrepancies both in the directions and lengths of lines.

Considerable experience is required in retracing and successfully developing the evidence of the lines and corners in these areas of excessive distortion. However, the procedure for restoration of lost corners and for the subdivision of sections is the same as in areas of more regularity.

Another feature to be considered in connection with the retracements is that the record may show that one surveyor ran the south boundary, a second the east boundary, and others the remaining exteriors and subdivisional lines. All of these lines may be reported on cardinal, but may not be exactly comparable—that is, the east boundary may not be truly normal to the south boundary, and so on. It was customary to retrace one or more miles of the east boundary of the township to determine the "variation" of the needle. This value was then adopted in the subdivision of the township. It follows that the meridional section lines should be found to be reasonably parallel with the east boundary. Under that plan of operation, it should be anticipated that the latitudinal section lines should be found to be reasonably parallel with the south boundary. However, discrepancies in measurement on the meridional lines frequently affect such parallelism. For these reasons, the index corrections for bearings may not be the same for the east and west as for the north and south lines. The two classes should be considered separately in this respect.

Before 1900, most lines were measured with the Gunter's link chain. The present surveyor must realize the difficulties of keeping that chain at standard length and the inaccuracy of measuring steep slops by this method. It is to be expected that the retracements will show various degrees of accuracy in the recorded measurements, which were intended to reflect true horizontal distances.

31. INDEX ERRORS

Where the original surveys were faithfully made, there will generally be considerable uniformity in the directions and lengths of the lines. Frequently, this uniformity is so definite as to indicate *index errors* which, if applied to the record bearings and distances, will place the trial lines in close proximity to the true positions and aid materially in the search for evidence. With experience, the present surveyor will become familiar with the work of the original surveyor and know about what to expect in the way of such differences.

32. COLLATERAL EVIDENCE

The identified corners of the original survey constitute the main control for the surveys to follow. After those corners have been located, and before resorting to proportionate measurement for restoration of lost corners, the other calls of the field notes should be considered. The recorded distances to stream crossings or to other natural objects that can be identified often lead to the position for a missing corner. At this stage, the question of acceptance of later survey marks and records, the location of roads and property fences, and the reliability of testimony are to be considered.

A line tree, connection to some natural object, or an improvement recorded in the original field notes that can be identified may fix a point of the original survey. The calls of the field notes for the various items of topography may assist materially in the recovery of the lines. The mean position of a blazed line, when identified as the original line, will identify a meridional line for departure or a latitudinal line for latitude. These are matters that require the exercise of considerable judgment.

33. ORIGINAL MARKS

Original line-tree marks, off-line tree blazes, and scribe marks on bearing trees and tree corner-monuments whose age exceeds 100 years are found occasionally. Such marks of later surveys are recovered in much greater number. Different surveyors used distinctive marks. Some surveyors used hacks instead of blazes, and some used hacks over and under the blazes; some employed distinctive forms of letters and figures. All of these will be recognized while retracing the lines of the same survey and will serve to verify the identification of the work of a particular surveyor.

The field notes give the species and the diameter of the bearing trees and line trees. Some of the smooth-barked trees were marked on the surface, but most of the marks were made on a flat smoothed surface of the live wood tissue. The marks remain as long as the tree is sound.

The blaze and marks will be covered by a gradual overgrowth, showing a scar for many years. The overgrowth will have a lamination similar to the annual rings of the tree, which may be counted in order to verify the date of marking, and to distinguish the original marks from later marks and blazes. On the more recent surveys, it is to be expected that the complete quota of marks should be found, clear cut and plainly legible. This cannot be expected in the older surveys, however.

It is advisable not to cut into a marked tree except as necessary to secure proof. The evidence is frequently so abundant, especially in the later surveys, that the proof is conclusive without inflicting additional injury that would hasten the destruction of the tree.

Finding original scribe marks, line-tree hacks, and off-line tree blazes furnishes the most convincing identification that can be desired.

It is not intended to disturb satisfactory local conditions with respect to roads and fences. Surveyors have no authority to change a property right that has been acquired legally. On the other hand, they should not accept the location of roads and fences as evidence of the original survey without something to support these locations. This supporting evidence may be found in some intervening survey record, or the testimony of individuals who may be acquainted with the facts.

34. RULES ESTABLISHED BY STATE LAW OR DECISIONS

Other factors that require careful consideration are the rules of the state law and the state court decisions, as distinguished from the methods followed by the Bureau of Land Management. Under state law, property boundaries may be fixed by agreement between owners, acquiescence, or adverse possession. Such boundaries may be defined by roads, fences, or survey marks, disregarding exact conformation with the original section lies. The rights of adjoining owners may be limited to such boundaries.

In many cases, due care has been exercised to place the property fences on the lines of legal subdivision. It has been the general practice in the prairie states to locate the public roads on the section lines. These are matters of particular interest to the adjoining owners. It is reasonable to presume that care and good faith were exercised in placing such improvements with regard to the evidence of the original survey in existence at the time. Obviously, the burden of proof to the contrary must be borne by the party claiming differently. In many cases, at the time of construction of a road, the positions for the corners were preserved by subsurface deposits of marked stones or other durable material. These are to be considered as exceptionally important evidence of the position of the corner, when duly recovered and verified.

The replacement of those corners that are regarded as obliterated, but not lost, should be based on such collateral evidence as has been found acceptable. All lost corners can be restored only by reference to one or more interdependent corner.

35. ADEQUATE MONUMENTATION ESSENTIAL

Surveyors will appreciate the great extent to which their successful retracements have depended upon an available record of the previous surveys, and upon the markers that were established by those who preceded them. The same appreciation will apply in subsequent retracements. It is essential to the protection of the integrity and accuracy of the work, the reputation of the surveyors, and the security of the interested property owners, that durable new corner markers be constructed in all places where required, and that a record of the survey as executed be filed.

The preferred markers are stone; concrete block; glazed sewer-tile filled with concrete, cast-iron, or galvanized-iron pipe; and similar durable materials. Many engineers and surveyors, counties, and landowners employ specially designed markers with distinctive lettering, including various cast-iron plates or bronze tablets.

The Bureau of Land Management has adopted a standard monument for use on the public-land surveys. This is made of wrought-iron pipe, zinc-coated, $2^1/_2$ in diameter and 28 in long, with one end split and spread to form flanges or foot plates. A brass cap is securely attached to the top on which appropriate markings for the particular corner are inscribed by use of steel dies.

Frequently, on account of roadway or other improvements, it is advisable to set a subsurface marker and, in addition, to place a reference monument where it may be found readily, selecting a site that is not likely to be disturbed.

36. MEANDER LINES AND RIPARIAN RIGHTS

The traverse run by a survey along the bank of a stream or lake is termed a *meander line*. The meander line is not generally a boundary in the usual sense, as ordinarily the bank itself marks the limits of the survey. All navigable bodies of water are meandered in the public land surveying practice, as well as many other important streams and lakes that have not been regarded as navigable in the broader sense.

All navigable rivers within the territory occupied by the public lands remain, and are deemed to be, public highways. Unless otherwise reserved for federal purposes, the beds of these waters were vested in the states at time of statehood. Under federal law, in all cases where the opposite banks of any stream not navigable belong to different persons, the stream and the bed thereof become common to both.

Grants by the United States of its public lands, including lands bounded by streams or other waters, are construed as to their effect according to federal law. This includes lands added to the grants by accretion.

The government conveyance of title to a fractional subdivision fronting upon a nonnavigable stream, unless specific reservations are indicated in the patent from the federal government, carries ownership to the middle of the stream.

Where surveys purport to meander a body of water where no such body exists, or where the meanders may be considered grossly erroneous, the United States may have a continuing public land interest in the lands within the segregated areas.

Where partition lines are to be run *across accretions*, the ordinary federal rule is to apportion the new frontage along the water boundary in the same ratio as the frontage along the line of the record meander courses. There are many variations to this rule where local conditions prevail and the added lands are not of great width or extent. The application of any rule, when surveying private lands, should be brought into harmony with the state law.

Where there is occasion to define the partition lines *within the beds* of nonnavigable streams, the usual rule is to begin at the property line at its intersection with the bank. From that point, run a line normal to the medial line of the stream that is located midway between the banks. Where the normals to the medial lines are deflecting rapidly, owing to abrupt changes in the course of the stream, suitable locations are selected above and below the doubtful positions, where acceptable normals may be placed. The several intervals along the medial line are then apportioned in the same ratio as the frontage along the bank.

The partition of the bed of nonnavigable lakes, whether water-covered or relicted, presents a more difficult problem because of the wide range of shapes of lake beds. In the simplest case of a circular bed, the partition lines can be run to the centroid, thus creating pie-shaped tracts fronting the individual holdings at the edge of the lake bed. Where odd-shaped beds are concerned, ingenuity will be required to divide the lake bed in such a manner that each shore proprietor will receive an equitable share of land in front of his or her holding. Any consideration of riparian rights inuring to private lands should be brought into line with appropriate state laws or decisions.

PRACTICE PROBLEMS

1. State whether single proportionate measure (S), single proportionate measure on a latitudinal (curved east-west) line (SC), or double proportionate measure (D) is used in restoring the following corners (assume none of the township boundaries are standard parallels).

(a) northeast corner of section 10

(b) northeast corner of section 12

(c) south quarter-section corner of section 12

(d) southeast corner of section 36

(e) southwest corner of section 36

Problems 2–9 are based on illustrations (a) through (d) shown on the following page.

2. Compute the field measurement for the east side of the southeast quarter of the fractional northeast quarter of section 3 shown in illustration (a).

3. Compute the field measurement for the west side of the southwest quarter of section 3 in illustration (a).

4. Compute the field measurement for the north side of the fractional northeast quarter of the northeast quarter of section 3 in illustration (a).

5. Compute the field coordinates to restore the lost south quarter-section corner of section 18 shown in illustration (b).

6. Compute the field coordinates to restore the lost northwest corner of section 16 shown in illustration (c).

7. Compute the coordinates to restore the lost north quarter-section corner of section 16 shown in illustration (c).

8. Compute the coordinates for the legal center of section 10 in illustration (d).

9. From the coordinate values for the center of section 10 in illustration (d), compute the coordinates for the center of the northeast quarter of the section.

(Note: Distances shown in parentheses are from the GLO plat; "o" indicates proven GLO corner.)

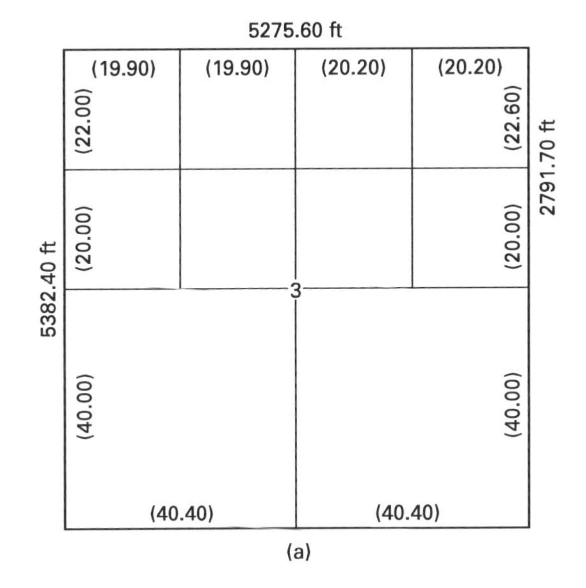

(a)

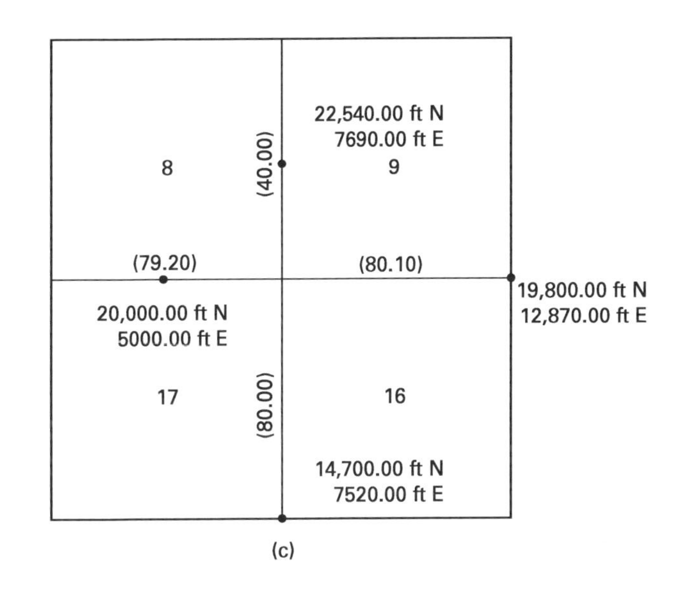

(c)

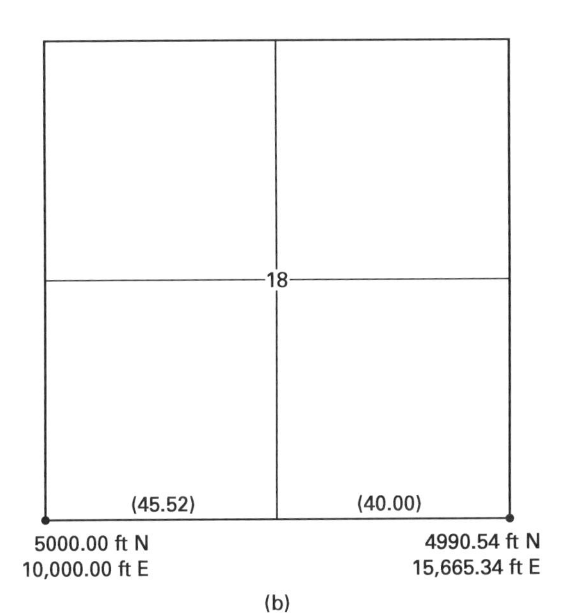

(b)

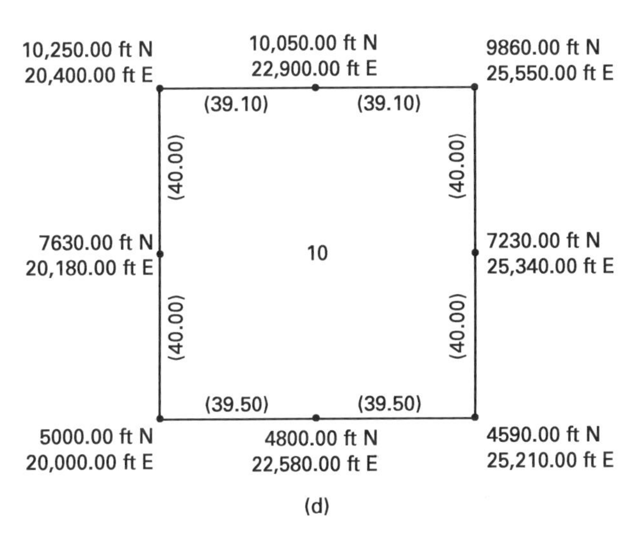

(d)

Cadastral and Boundary Law

SOLUTIONS

1. (a) D

(b) S

(c) S

(d) S

(e) SC

2.

$$\frac{\text{field}}{2791.70\ \text{ft}} = \frac{20.00\ \text{ch}}{20.00\ \text{ch} + 22.60\ \text{ch}}$$

$$\text{field} = \left(\frac{20.00\ \text{ch}}{42.60\ \text{ch}}\right)(2791.70\ \text{ft}) = \boxed{1310.66\ \text{ft}}$$

3.

$$\frac{\text{field}}{5382.40\ \text{ft}} = \frac{40.00\ \text{ch}}{40.00\ \text{ch} + 20.00\ \text{ch} + 22.00\ \text{ch}}$$

$$\text{field} = \left(\frac{40.00\ \text{ch}}{82.00\ \text{ch}}\right)(5382.40\ \text{ft}) = \boxed{2625.56\ \text{ft}}$$

4.

$$\frac{\text{field}}{5275.60\ \text{ft}} = \frac{20.20\ \text{ch}}{19.90\ \text{ch} + 19.90\ \text{ch} + 20.20\ \text{ch} + 20.20\ \text{ch}}$$

$$\text{field} = \left(\frac{20.20\ \text{ch}}{80.20\ \text{ch}}\right)(5275.60\ \text{ft}) = \boxed{1328.77\ \text{ft}}$$

5. Field coordinates can be proportioned similar to field distances.

Proportioning left to right:

$$\frac{\Delta\text{N}}{4990.54\ \text{ft} - 5000.00\ \text{ft}} = \frac{45.52\ \text{ch}}{45.52\ \text{ch} + 40.00\ \text{ch}}$$

$$\Delta\text{N} = \left(\frac{45.52\ \text{ch}}{85.52\ \text{ch}}\right)(-9.46\ \text{ft}) = -5.04\ \text{ft}$$

$$\frac{\Delta\text{E}}{15{,}665.34\ \text{ft} - 10{,}000\ \text{ft}} = \frac{45.52\ \text{ch}}{45.52\ \text{ch} + 40.00\ \text{ch}}$$

$$\Delta\text{E} = \left(\frac{45.52\ \text{ch}}{85.52\ \text{ch}}\right)(5665.34\ \text{ft}) = 3015.51\ \text{ft}$$

$$\text{N} = 5000.00\ \text{ft} - 5.04\ \text{ft} = \boxed{4994.96\ \text{ft}}$$

$$\text{E} = 10{,}000.00\ \text{ft} + 3015.51\ \text{ft} = \boxed{13{,}015.51\ \text{ft}}$$

6. Using double proportionate measure, the northing of the restored corner will be the northing obtained by single proportionate measure in the north-south direction. Likewise, the easting will be the easting obtained by single proportionate measure in the east-west direction.

$$\frac{\Delta\text{N}}{22{,}540.00\ \text{ft} - 14{,}700.00\ \text{ft}} = \frac{80.00\ \text{ch}}{80.00\ \text{ch} + 40.00\ \text{ch}}$$

$$\Delta\text{N} = \left(\frac{80.00\ \text{ch}}{120.00\ \text{ch}}\right)(7840.00\ \text{ft}) = 5226.67\ \text{ft}$$

$$\frac{\Delta\text{E}}{12{,}870.00\ \text{ft} - 5000.00\ \text{ft}} = \frac{\frac{79.20\ \text{ch}}{2}}{\frac{79.20\ \text{ch}}{2} + 80.10\ \text{ch}}$$

$$\Delta\text{E} = \left(\frac{39.60\ \text{ch}}{119.70\ \text{ch}}\right)(7870.00\ \text{ft}) = 2603.61\ \text{ft}$$

$$\text{N} = 14{,}700.00\ \text{ft} + 5226.67\ \text{ft} = \boxed{19{,}926.67\ \text{ft}}$$

$$\text{E} = 5000.00\ \text{ft} + 2603.61\ \text{ft} = \boxed{7603.61\ \text{ft}}$$

7. A quarter corner is to be restored by single proportionate measure between adjacent section corners after they have been found or *restored.*

Using the restored section corner coordinates obtained in Prob. 6, and with the quarter corner being at a midpoint on a regular (normal) section,

$$N = \frac{19{,}926.67 \text{ ft} + 19{,}800.00 \text{ ft}}{2} = \boxed{19{,}863.33 \text{ ft}}$$

$$E = \frac{7603.61 \text{ ft} + 12{,}870.00 \text{ ft}}{2} = \boxed{10{,}236.80 \text{ ft}}$$

8. The center of the section is located at the intersection of the lines connecting the opposite quarter corners (bearing-bearing intersection).

In the following calculations, the subscripts N, E, W, and S represent the four half-section corners, and the subscript C represents the middle (center) of the section.

The inverse from the west quarter corner to the east quarter corner yields an azimuth of

$$\begin{aligned} az_{W\text{-}C} = az_{W\text{-}E} &= \arctan\frac{\Delta E}{\Delta N} \\ &= \arctan\left(\frac{25{,}340 \text{ ft} - 20{,}180 \text{ ft}}{7230 \text{ ft} - 7630 \text{ ft}}\right) \\ &= 94°25'58'' \end{aligned}$$

The inverse from the south quarter corner to the north quarter corner yields an azimuth of

$$\begin{aligned} az_{S\text{-}C} = az_{S\text{-}N} &= \arctan\frac{\Delta E}{\Delta N} \\ &= \arctan\left(\frac{22{,}900 \text{ ft} - 22{,}580 \text{ ft}}{10{,}050 \text{ ft} - 4800 \text{ ft}}\right) \\ &= 3°29'17'' \end{aligned}$$

The inverse from the west quarter corner to the south quarter corner yields an azimuth of

$$\begin{aligned} az_{W\text{-}S} &= \arctan\frac{\Delta E}{\Delta N} = \arctan\left(\frac{22{,}580 \text{ ft} - 20{,}180 \text{ ft}}{4800 \text{ ft} - 7630 \text{ ft}}\right) \\ &= 139°42'01'' \end{aligned}$$

Calculate angles in the triangle west-center-south.

$$\begin{aligned} \angle_{C\text{-}W\text{-}S} &= az_{W\text{-}S} - az_{W\text{-}C} = 139°42'01'' - 94°25'58'' \\ &= 45°16'03'' \end{aligned}$$

$$\begin{aligned} \angle_{W\text{-}S\text{-}C} &= az_{S\text{-}C} - az_{S\text{-}W} \\ &= (360° + 3°29'17'') - (180° + 139°42'01'') \\ &= 43°47'16'' \end{aligned}$$

$$\begin{aligned} \angle_{W\text{-}C\text{-}S} &= 180° - (\angle_{C\text{-}W\text{-}S} + \angle_{W\text{-}S\text{-}C}) \\ &= 180° - (45°16'03'' + 43°47'16'') \\ &= 89°03'19'' \end{aligned}$$

By the law of sines, the distance from the west corner to the center is

$$\begin{aligned} \text{distance}_{W\text{-}C} &= \frac{\text{distance}_{W\text{-}S} \sin \angle_{W\text{-}S\text{-}C}}{\sin \angle_{W\text{-}C\text{-}S}} \\ &= \frac{(3710.65 \text{ ft}) \sin 43°47'16''}{\sin 89°03'19''} \\ &= 2568.08 \text{ ft} \end{aligned}$$

The coordinates for the legal center of section are

$$\begin{aligned} N_C &= N_W + \text{distance}_{W\text{-}C} \cos az_{W\text{-}C} \\ &= 7630 \text{ ft} + (2568.08 \text{ ft}) \cos 94°25'58'' \\ &= \boxed{7431.51 \text{ ft}} \end{aligned}$$

$$\begin{aligned} E_C &= E_W + \text{distance}_{W\text{-}C} \sin az_{W\text{-}C} \\ &= 20{,}180 \text{ ft} + (2568.08 \text{ ft}) \sin 94°25'58'' \\ &= \boxed{22{,}740.40 \text{ ft}} \end{aligned}$$

9. The center of a quarter section is located at the intersection of lines connecting opposite quarter-quarter corners. For a regular (normal) section, quarter-quarter corners are to be located at midpoints between adjacent quarter corners.

For the northeast quarter section:

The north quarter-quarter corner is

$$N = \frac{10{,}050.00 \text{ ft} + 9860.00 \text{ ft}}{2} = 9955.00 \text{ ft}$$

$$E = \frac{22{,}900.00 \text{ ft} + 25{,}550.00 \text{ ft}}{2} = 24{,}225.00 \text{ ft}$$

The east quarter-quarter corner is

$$N = \frac{9860.00 \text{ ft} + 7230.00 \text{ ft}}{2} = 8545.00 \text{ ft}$$

$$E = \frac{25{,}550.00 \text{ ft} + 25{,}340.00 \text{ ft}}{2} = 25{,}445.00 \text{ ft}$$

Using the values determined in Sol. 8, the south quarter-quarter corner is

$$N = \frac{7431.51 \text{ ft} + 7230.00 \text{ ft}}{2} = 7330.755 \text{ ft}$$

$$E = \frac{22{,}740.40 \text{ ft} + 25{,}340.00 \text{ ft}}{2} = 24{,}040.2 \text{ ft}$$

Using the values determined in Sol. 8, the west quarter-quarter corner is

$$N = \frac{7431.51 \text{ ft} + 10{,}050.00 \text{ ft}}{2} = 8740.785 \text{ ft}$$

$$E = \frac{22{,}740.40 \text{ ft} + 22{,}900.00 \text{ ft}}{2} = 22{,}820.2 \text{ ft}$$

The coordinates of the intersection of lines connecting opposite midpoints on a four-sided figure will be at the average of the coordinates of these opposite midpoints.

Therefore, for east-west,

$$N = \frac{8545.00 \text{ ft} + 8740.785 \text{ ft}}{2} = 8642.89 \text{ ft}$$

$$E = \frac{25{,}445.00 \text{ ft} + 22{,}820.2 \text{ ft}}{2} = 24{,}132.6 \text{ ft}$$

Check:

For north-south,

$$N = \frac{9955.00 \text{ ft} + 7330.755 \text{ ft}}{2} = 8642.88 \text{ ft}$$

$$E = \frac{24{,}225.00 \text{ ft} + 24{,}040.2 \text{ ft}}{2} = 24{,}132.6 \text{ ft}$$

29 Land Descriptions

1. LAND DESCRIPTIONS

Throughout history, mankind has created representative systems, such as written music notation and currency, that have allowed great advances in civilization. One such system, the representation of land with a title document, provides security of land ownership and allows land to be considered as a commodity that can be bought and sold. In addition, a title allows the use of land as collateral for credit, which is one of the most important sources of funds for new businesses in the United States. The representation of land by a written description is often considered to be the key to the remarkable economies of the Western world. Therefore, one of the surveyor's most important responsibilities is that of preparing land descriptions.

Because of the important role that land descriptions play in the security of land title as well as in the use of land as capital, it is incumbent upon the surveyor to use utmost care in the crafting of these descriptions. Descriptions must allow the precise and unambiguous location of the boundaries of the tracts of land for generations after they are written.

2. TYPES OF LAND DESCRIPTIONS

Public Land Survey System Descriptions

Public Land Survey System descriptions are based on the *Public Land Survey System* (also known as the *U.S. System of Rectangular Surveys*), which today is administered by the U.S. Bureau of Land Management. These descriptions have traditionally been some of the most commonly used descriptions in the United States. Thirty states use the rectangular system (see Ch. 27). Almost all of the privately owned lands in those states were originally conveyed by reference to that system.

The original public lands act (Act of May 18, 1796, 1 Stat. 464) that created the Public Land Survey System specified that public lands were to be sold by the section and descriptions were to reference land surveys. Each section was a square mile, with a standard size of 640 ac. Since each section had its own identity, conveyance of any of these lands involved a relatively simple description providing the pertinent section, township, and range. In addition, the description contained references to the base meridian from which the section had been surveyed and to the official plat that depicted the section. The following is an example of a Public Land Survey System description.

> Section 6, township 1 north, range 2 east, Tallahassee Baseline as depicted on the official General Land Office plat approved June 1, 1842.

Due to concern that the minimum size of a 640 ac section for land purchases restricted land ownership to the wealthy, successive legislation reduced the minimum size available for purchase to standardized fractions of a section, called aliquot parts. In 1800, the half section, with a standard size of 320 ac, became the minimum size. Four years later, the sale of land by the quarter section, with a standard size of 160 ac, was authorized. In 1820, sale of land by the half quarter section, with a standard size of 80 ac, was approved. Then in 1832, the minimum area for public land sale became the quarter of a quarter section, with a standard size of 40 ac. That tract became the standard unit of settlement for the United States.

The public land surveys only traversed the perimeter of sections. The boundaries of aliquot parts of sections were not surveyed. Nevertheless, the description of such tracts is a simple process. Figure 29.1 illustrates the convention used. Note that when attempting to visualize the location of aliquot parts of sections, it is best to start at the end of the description and work backward.

Figure 29.1 *Aliquot Parts of Public Land Sections*

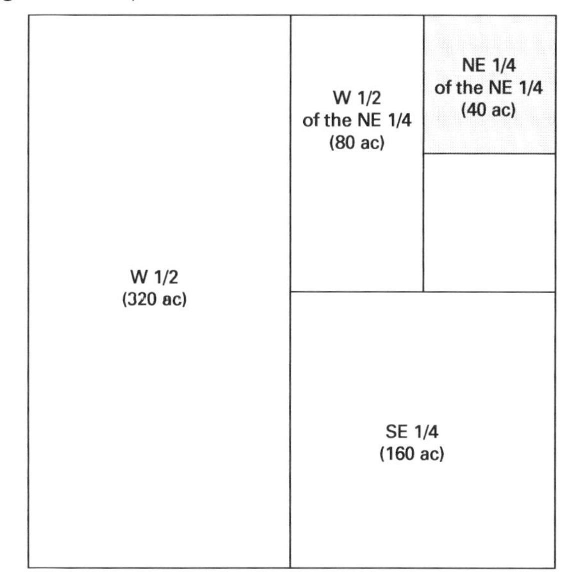

Cadastral and Boundary Law

In the platting process for public lands surveys, it was the practice to divide fractional sections along navigable water bodies into small lots. Typically, the boundaries of such lots were drawn along the lines of the fractional parts of the section. This was a cartographic process, and the boundaries of such government lots were not surveyed except for those boundaries coincident with section lines and those coincident with meander lines along the water body. As with descriptions for Public Land Survey System sections and aliquot parts of sections, government lots may be described simply by reference to the lot number, section, township, range, base meridian, and pertinent official plat. The following is an example of a government lot description.

> Government lot 3, section 12, township 1 north, range 2 east, Tallahassee Meridian as depicted on the official General Land Office plat approved June 1, 1842.

Descriptions Referencing Other Subdivisions

A second type of land description references other subdivisions of land. These descriptions are similar to those involving sections and aliquot parts of sections of the Public Land Survey System. Both types of descriptions involve a reference to a subdivision of land where a plat of survey has been filed in the public records. However, this second type of description may reference subdivisions performed by government agencies outside of the Public Land Survey System, or it may reference subdivisions created by private landowners and recorded in the public records. Such recorded subdivision plats are typically filed in each county courthouse. Most states have statutes that provide standards for the survey and platting of such subdivisions. In addition, many counties have subdivision ordinances with their own requirements.

As a result, modern day subdivision plats are relatively standardized. Since such subdivision plats are available in the public records, a description on such a plat requires only that the lot or lots be identified along with the subdivision name and the location of the plat record, as shown in the following example.

> Lots 100 and 101 of Dreblow & Company's Silver Lake Subdivision as recorded at Plat Book A, p. 3 of the public records of Jefferson County, Florida.

Metes and Bounds Descriptions

In the 20 states not subdivided under the Public Land Survey System, most of the surveys for the original private conveyances were made by the metes and bounds method. *Metes and bounds descriptions* are also used to describe irregular areas that are severed from lands originally subdivided under the Public Land Survey System.

A bounds description defines the boundaries of a tract of land by identifying adjoiners or monuments, but does not typically provide a direction, as shown in the following example.

> All of that land lying north of State Road 99; bounded on the north by land of Albert Bowie, bounded on the east by lands of Betty Anderson; and bounded on the west by Trout Creek.

A metes description identifies a beginning point and then describes each course in sequence around the perimeter of the tract until the point of beginning is reached again to complete the description of the perimeter. Such a description includes a distance and direction for each course, as shown in the following example.

> A tract of land in section 12, township 3 north, range 6 east in Jefferson County, Florida, more particularly described as follows: For a point of beginning, commence at an old axle marking the southeast corner of said section 12; then go N00°01′E for 200 ft; then go N89°59′W for 400 ft; then go S00°01′W for 200 ft; then go S89°59′E for 400 ft to the point of beginning.

As may be seen from the examples, metes and bounds descriptions are somewhat more complex than descriptions that merely reference a plat of survey in the public records, such as those descriptions from the Public Land Survey System or those that describe a lot in a private subdivision. Since the survey that forms the foundation of a metes or bounds description is not typically part of official records, the description must, in effect, communicate the results of the survey.

Other Types of Descriptions

There are a number of other types of descriptions used for conveying land. An example of these is a *strip description*. This type is frequently used to describe a road easement or right of way. Other common descriptions are those where a well described tract of land is divided based on a man-made or natural *monument* such as a road or stream, based on a certain distance or width, based on a certain area, or based on a certain fraction of the total area of the tract. Another type of description that is increasingly used today relies on *geographic coordinates*. Many other descriptions are a combination of these types. Descriptions may also include *qualifying clauses* or *augmentation clauses* that take away or add something to a tract, respectively. Examples of these other types of descriptions follow.

- strip description: a right of way for ingress and egress purposes across a strip of land lying 30 ft on each side of the described center line
- division by monument: all of section 12 lying northerly of U.S. Highway 90
- division by distance: the westerly 50 ft of lot 2
- division by area: the southern 10 ac of government lot 2...
- division by fraction: the western one-half of lot 4

- geographic coordinates: For a point of beginning, commence at an old concrete monument marking the south quarter corner of said section 24 and having a north coordinate of 1,972,048.50 and an east coordinate of 563,589.10; then go N00°01′E for 200 ft to a point having a north coordinate of 1,972,248.62 and an east coordinate of 563,589.16; then go N89°59′W for 400 ft to a point having a north coordinate of 1,972,248.50 and an east coordinate of 563,189.16; then go S00°01′W for 200 ft to a point having a north coordinate of 1,972,048.50 and an east coordinate of 563,189.10; then go S89°59′E for 400 ft to the point of beginning. Coordinates are in feet and are based on the Florida State Plane Coordinate System, North Zone, NAD88.

3. ESSENTIALS OF LAND DESCRIPTION WRITING

The following are some essential considerations in preparing land descriptions.

- format: A land description should include both a *caption* that provides the general location, city, county, and state and a *body* that provides a detailed description of the area. In addition, the land description may include qualifying clauses that take away from the area outlined in the body or augmenting clauses that add something to the area outlined in the body.

- monuments: Any monuments called, especially monuments called as the reference points for metes and/or bounds descriptions, should be permanent in character, visible, and stable.

- directions: The basis for any directions called should be stated. For example, bearings should be identified as based on astronomic coordinates, state plane coordinate grids, magnetic north, or a specifically defined line.

- coordinates: If coordinates of corners are called, the datum for the coordinates should be stated.

- curves: For any curves called, at least two elements of the curve should be stated. The most frequently used elements are the radius, arc length, central angle, and tangent length. In addition, the relationship of the curve to the previous line, the direction of the curve (e.g., concave to the south) and the direction of travel along the curve (e.g., easterly) should be stated, as described in the following example.

  Then go 1000 ft along a tangent curve to the left, said curve being concave to the east and having an interior angle of 45° and a radius of 2000 ft.

4. INTERPRETING LAND DESCRIPTIONS

Most surveyors have encountered problems in boundary determination due to a land description with ambiguous or conflicting calls within it, or due to conflicts between that description and another. Fortunately, case law over the years has resulted in a body of guidelines for the best action in such situations. The most salient of these guidelines follow.

Priority of Calls

In the absence of a clear intention to the contrary, the order of priority for conflicting calls is as follows.

1. calls for natural objects such as rivers, creeks, and mountains
2. calls for artificial objects such as concrete monuments and adjoining lands
3. calls for distance
4. calls for course
5. calls for area

Senior and Junior Rights

Many original metes and bounds surveys resulted in overlaps or gaps between surveys. It became the rule of law in the early history of the United States that where an overlap occurred, the holder of the first grant or patent, the senior awardee, retained ownership of the overlap area, and the holder of the latter grant or patent, the junior awardee, lost the area. These rights are known as senior rights and junior rights. A common expression summarizes the priority of these rights: "The first deed is the best deed."

Where two parties have title to the same land, the party holding the senior conveyance has the right of possession. As an example, Smith owns 200 ac of land, or thinks that he owns that amount. He sells one half of his land to Jones by a metes and bounds description that calls for 100 ac. Later, Smith moves to sell the other half of his land to Brown with a description that calls for one-half of the original tract. At that time, a survey reveals he had only 195 ac originally. He cannot sell more than 95 ac to Brown. Jones has the senior deed and is entitled to the full 100 ac that Smith sold to him. Smith cannot recover any of Jones's land. This principle in law assures the first buyer that his land can not be taken from him as it was conveyed. It is the basis of senior and junior rights.

Distances

Unless otherwise stated, distances are horizontal and measured in a straight line along the shortest distance.

Excess and Deficiency in Subdivision

Distances and directions between found original measurements take priority over plat distances and directions. Excess or deficiency of such lines between original monuments should be distributed proportionally throughout the line within each block.

Calls that Use the Word "To"

When calls use the word "to"—such as "to a concrete monument" or "to a river" or "to an adjoining property boundary"—it should be interpreted that the actual direction and distance to the object called takes priority over any direction and distance cited in the description. When the call is to an adjoiner, this means that the adjoining boundary must be located before the description can be correctly interpreted.

Indeterminate Fractional or Area Calls

Indeterminate fractional calls or area calls are those where the direction of the dividing line is not given or implied. Principles have been developed for such situations.

- A called area of land on the side of a tract, such as "the northern 80 ac," should be interpreted as including the called area in the form of a parallelogram with the dividing line parallel to the side of the tract on which the area is located.
- A called area of land in a specified corner of a tract should be interpreted as being a corner quadrangle with equal sides.
- If the easterly and westerly sides of a lot are nearly parallel and either the easterly or westerly half is described, the dividing line should be considered to be on a mean bearing between the two sides.
- If the easterly and westerly sides of a lot are not parallel and either the easterly or westerly half is described, the dividing line should be considered to be north-south.
- If the eastern boundary of a lot is nearly north-south, and if the easterly half of the lot is conveyed and a description calls the lot except for the eastern half, the dividing line should be parallel to the eastern side.

PRACTICE PROBLEMS

1. What is the area of the southeast quarter of the northeast quarter of the southwest quarter of a typical public land survey section?

(A) 5 ac
(B) 10 ac
(C) 20 ac
(D) 40 ac

2. A tract of land is described as the southeast quarter of the northeast quarter of the southwest quarter of a typical public land survey section. In which direction would this tract of land lie in relation to the southwest quarter of the northwest quarter of the southwest quarter of the same section?

(A) north
(B) east
(C) south
(D) west

3. Which is the controlling call in the following description?

N54°E for a distance of 298 ft to the shore of Wolf Creek

(A) N54°E
(B) 298 ft
(C) the thread of Wolf Creek
(D) the shore of Wolf Creek

4. The recorded plat of the Silver Lake Subdivision shows lot 2 to be a 100 ft by 200 ft rectangular lot. The owner conveyed the northern one-half of the lot and later conveyed the southern 100 ft of the lot. A survey conducted for the second conveyance revealed that the western and eastern lines of the lot are 198.5 ft long. Where should the northeast corner of the later conveyance be placed?

(A) 89.25 ft from the southeast corner of lot 2
(B) 98.50 ft from the southeast corner of lot 2
(C) 99.25 ft from the northeast corner of lot 2
(D) 100.00 ft from the southeast corner of lot 2

5. In the event of conflict among calls in a description, which of the following calls has the lowest priority?

(A) call for an artificial monument
(B) call for distance
(C) call for a natural monument
(D) call for a course

6. In the event of conflict among calls in a description, which of the following calls has the second highest priority?

(A) call for an artificial monument
(B) call for distance
(C) call for a natural monument
(D) call for a course

7. In the following description, what would be considered an example of a bounds?

A parcel of land situated in Jefferson County, Florida being part of Section 12, Township 3 North, Range 5 East and being more particularly described as follows: Begin at an old axle marking the southwest corner of said Section 12; then go N80°42′E for 542.0 ft to the western right of way of County Road 246; then go N°2°35′W for 120.0

ft along said right of way; then go N50°15′W for 502.5 ft to an old iron pipe; then go S15°09′W for 547.8 ft to the point of beginning.

(A) N80°42′E
(B) 542.0 ft
(C) westerly right of way of County Road 246
(D) There are no bounds in this description.

8. Referring to the illustrated recorded plat for Wild Turkey Subdivision, a survey has found no interior lot corners fronting on Wing Road, but finds the original front block corners 500.80 ft apart. At what distance from the block corners should the adjacent lot corners be set?

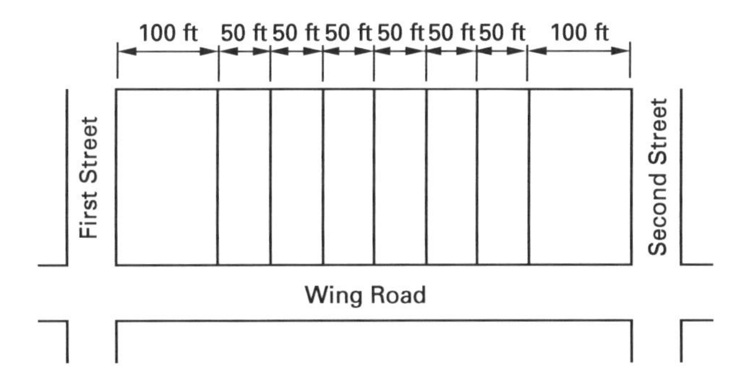

(A) 100.00 ft
(B) 100.10 ft
(C) 100.16 ft
(D) 100.80 ft

9. The answer for Prob. 8 is correct because the

(A) excess should be placed in the corner remnant lots
(B) excess is distributed equally in each lot
(C) excess is distributed by proportional measurement
(D) plat dimensions control remonumentation

SOLUTIONS

1. The total section area is 640 ac.

$$\text{quarter section} = \frac{640 \text{ ac}}{4} = 160 \text{ ac}$$

$$\begin{array}{l}\text{quarter of a} \\ \text{quarter section}\end{array} = \frac{160 \text{ ac}}{4} = 40 \text{ ac}$$

$$\begin{array}{l}\text{quarter of a} \\ \text{quarter of a} \\ \text{quarter section}\end{array} = \frac{40 \text{ ac}}{4} = 10 \text{ ac}$$

The answer is (B).

2. For the following illustration, the described tract (southeast quarter of the northeast quarter of the southwest quarter) lies easterly of the southwest quarter of the northwest quarter of the southwest quarter.

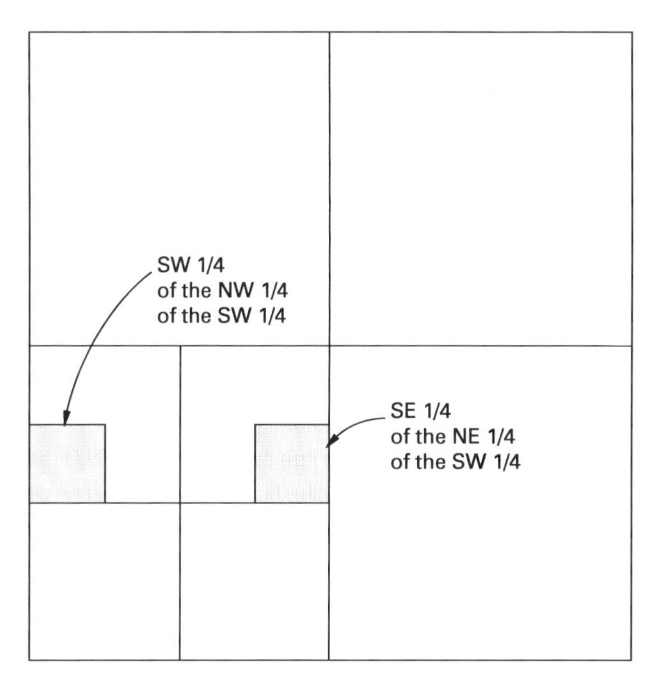

The answer is (B).

3. The controlling call is "to the shore of Wolf Creek." The creek is a natural monument that has precedence over the calls for distance and bearing.

The answer is (D).

4. The senior deed called for one-half of the lot. The east and west lines would be divided in half to set the monument.

$$\frac{198.5 \text{ ft}}{2} = 99.25 \text{ ft}$$

The answer is (C).

Cadastral and Boundary Law

5. In general, calls for course should be given the lowest priority of the options given.

The answer is (D).

6. Calls for artificial monuments should be given the second highest priority of the options given.

The answer is (A).

7. In the description provided, the westerly right of way of County Road 246 is an example of a bounds, which is a monument or adjoiner called in the deed.

The answer is (C).

8. Excess and deficiency should be proportioned within the block. The total frontage excess within the block is

$$\begin{aligned}\text{measured distance}\\ -\text{ recorded distance} &= 500.80 \text{ ft} - 500.00 \text{ ft}\\ &= 0.80 \text{ ft}\end{aligned}$$

For a 100 ft lot, the excess is

$$\left(\frac{100 \text{ ft}}{500 \text{ ft}}\right)(0.80 \text{ ft}) = 0.16 \text{ ft}$$

The required lot corner should be set 100.16 ft from the block corner.

The answer is (C).

9. Excess between found original monuments in a subdivision is distributed by proportional measurement.

The answer is (C).

30 Colonial History and the U.S. Legal System

1. ENGLISH COMMON LAW

English common law consists of those ideas of right and wrong determined by court decisions over many centuries. Such ideas have been accepted by generations trying to establish rules to meet social and economic needs. Sir William Blackstone, eighteenth century author of *Commentaries on the Law of England*, called it "unwritten law" in the sense that it was not enacted by a legislative body. It was legal custom expressed by the decisions of judges.

English common law was evolutionary. It changed slowly. Judges made decisions based on former decisions, but they also modified their decisions to reflect changing times.

Some of the important parts of our present law that came from English common law are the grand jury, trial by jury, freedom of press, habeas corpus, and oral testimony.

2. STATUTE LAW

Laws enacted by legislative bodies are known as *statute laws*. In contrast with common law, it is written law. Laws of France, Germany, Spain, and other countries in the continent of Europe are largely statute laws. After the French Revolution, France adopted the Code Napoleon to clarify its laws.

3. COLONIAL LAW

Colonial law was not evolutionary because there was nothing on which to base precedence. Therefore, settlers of the original thirteen colonies adopted English common law. But they did not adopt it in its entirety. It did not fit their new social economic environment entirely. Many of their laws were statutory (written).

Furthermore, the same parts of English common law were not adopted in each of the colonies. The parts of the law that seemed to fit the needs of the particular situation in each colony were the parts that were adopted.

4. SPAIN AND FRANCE IN THE NEW WORLD

Spain acquired title to land in the New World by grant from Pope Alexander VI in 1493. The grant conveyed all lands not held by a Christian prince on Christmas day of 1492 from a meridian known as the *Line of Demarcation*, 100 leagues west of the Azores and Cape Verde Islands. Later, a treaty with Portugal moved the Line of Demarcation 270 leagues west, with Portugal to have rights to the east and Spain to the west. Possession came from conquest, and by 1600, the territory extended from New Mexico and Florida on the North to Chile and Argentina on the south. The first seat of government was at Santa Domingo.

In 1511, Don Diego Velasquez led an expedition for the conquest of Cuba that was accomplished without serious opposition. Velasquez was appointed Governor of the island. During his rule, he promoted settlement of the land by Spaniards.

Hernandez de Cordova set out in 1517 from Cuba on an expedition to the Bahama Islands to obtain Indian slaves, but a storm drove him off his course. Three weeks later, he landed on the coast of Yucatan. Cordova returned to Cuba with tales of a more advanced civilization than previously found in the New World. And of great interest to Velasquez, he brought tales of gold and fine cotton garments. In 1518 Velasquez sent his nephew, Juan de Grijalva, on an expedition to explore the coast of Mexico. Grijalva also landed on Yucatan and was impressed by the advanced civilization. He returned to Cuba with many gold ornaments he had received in trade.

After approval from Spain, Velasquez decided on the conquest and colonization of the new land. He chose Hernando Cortes as the commander of his expedition.

After many skirmishes and battles, Cortes reached the capital of the Aztecs, Tenochtitlan, now Mexico City, on November 8, 1519. He quickly conquered the forces of Montezuma, the Aztec emperor, and placed him under house arrest where he was treated very cordially by Cortes and allowed to retain his many luxuries.

After the conquest of the new land, Spanish statute law was introduced. Legislation for this new land was codified in *Recopilacion de las Leyes de Indias* in 1680. The Crown of Spain had complete authority that was administered through the Minister of the Indies, and all the land belonged to the king of Spain.

Spain ruled Mexico for 300 years. However, in 1821, the people of Mexico revolted and declared their independence from Spain. Augustin de Iturbide, leader of the revolt, was crowned as Augustine I, Emperor of Mexico in 1822. In 1824, he was deposed by Lopez de Santa Anna who established a constitutional government.

France established Quebec in 1608. From there, settlement moved south along the Mississippi to its mouth. René Robert Cavalier, Sieur de la Salle, claimed all the Mississippi Valley for France and named it Louisiana in honor of Louis XIV.

Napoleon, in 1803, sold all of Louisiana to the United States, "with the same extent that it now has in the hands of Spain, and that it had when France possessed it." President Thomas Jefferson claimed Texas as part of the purchase but Spain protested vigorously. The boundary between Texas and the United States was established by treaty between the king of Spain and the United States in 1819.

An act barring emigration from the United States and other restrictive land laws caused unrest in Texas, which resulted in Texas winning its independence from Mexico. In 1836, the Republic of Texas was established.

5. ORGANIZATION

Each state in the United States is a sovereign state. Except for those powers delegated to the federal government, each state writes its own laws, including those pertaining to land titles. Each state has a constitution, a legislative body, and a court system.

Many state constitutions are patterned after the U.S. Constitution and contain a bill of rights. Many establish English common law as the basis of law.

In most states, the legislative body is known as the *legislature*, but in some states is known as the *general assembly* or the *general court*. The legislature in most states consists of two bodies: the senate and the house of representatives.

6. CRIMINAL AND CIVIL COURTS

Except for the lowest courts, most courts handle criminal cases but not civil cases or civil cases but not criminal cases. Cases that pertain to land titles are tried in civil courts. In general, civil courts are either trial courts (courts of the first instance) or courts of civil appeal (appellate courts).

7. TRIAL COURTS

Trial courts determine any questions of fact in dispute and then apply rules of law. Evidence is presented by both the plaintiff and the defendant.

If the parties do not ask for a jury trial, the court (meaning the judge) hears the proceedings and hands down conclusions and delivers the judgment (the official and authentic decision of the court of justice). If either party demands a jury, the judge charges the jury on the law of the case, and the jury decides the case.

8. COURTS OF CIVIL APPEAL

Courts of civil appeal (appellate courts) do not hear evidence from witnesses; they only decide questions of law from the record of the trial court. A suit in an appellate court is not a continuation of a suit to which it relates; it is a new suit to set aside judgment in a lower court. It is brought for supposed error in law apparent in the record.

The party who appeals a court decision is known as the appellant, whether he or she was the plaintiff or defendant in the lower court. The party who defends against the appeal is known as the *respondent.*

9. PETITION FOR WRIT OF ERROR

The appellant usually makes his or her application to an appeals court by a petition for writ of error, in which he or she contends that certain errors were made in the lower court.

The appellate court may take one of four actions on an application for writ of error:

(1) It may grant the request, which means it will hear the case.

(2) It may refuse the request, which means that the court will not hear the case. This indicates that the appellate court believes that the question of law was correctly decided in the lower court.

(3) It may refuse the request *no reversible error* (n.r.e.). This indicates that the court is not satisfied that the opinion of the lower court, in all respects, has correctly declared law, but it is of the opinion that the application presents no error that may require a reversal.

(4) It may dismiss the application because the parties have settled out of court or because the appellate court lacks jurisdiction.

10. OPINION OF THE COURT

The *opinion of the court* is an explanation of the court's decision. When the judges in a case reach a decision, one of them writes the opinion. A concurring opinion may be written by a judge in the majority who agrees with the decision but disagrees with the reasoning of the opinion. If the opinion is not unanimous, the judge who writes the opinion is selected from the judges in the majority. A dissenting opinion may be written by a judge in the minority. Another dissenting opinion may be written by a judge in the minority who agrees with the dissenting opinion but disagrees with the reasoning in that opinion. A *per curiam* opinion is an opinion of the whole court as opposed to an opinion written by a specific judge. The words "decision" and "judgment" are synonymous.

11. ELEMENTS OF A COURT DECISION

Reports of cases (i.e., *citations*) follow, in general, the same outline:

(1) the title (sometimes called the *style*) of the case

(2) the case (or docket) number

(3) the name of the court that hears the case

(4) the date of the decision

(5) a summary of points of law involved

(6) a synopsis of lower court findings

(7) the names of counsel for both parties

(8) the name of the judge writing the opinion

(9) the opinion of the court

(10) the decision: affirmed or reversed

PRACTICE PROBLEMS

1. English common law

(A) was referred to as unwritten law

(B) was evolutionary

(C) evolved from court decisions over hundreds of years

(D) all of the above

2. Colonial law was

(A) evolutionary

(B) based on English common law to fit the colonies' new social economic environment

(C) adopted in full by each colony

(D) statutory

3. Trial by jury, the grand jury, and freedom of the press had their beginning in

(A) the Magna Carta

(B) Roman law

(C) English common law

(D) none of the above

SOLUTIONS

1. English common law was referred to as unwritten law, was evolutionary, and evolved from court decisions over hundreds of years.

The answer is (D).

2. Colonial law was statutory (written). It was not evolutionary because there was nothing on which to base precedence. While colonial law was based on English common law, the English common law was not adopted in its entirety because it did not fit the colonies' new economoic environment. Furthermore, each colony adopted only the parts of the colonial law that fit its particular needs.

The answer is (D).

3. Trial by jury, the grand jury, and freedom of the press had their beginnings in English common law.

The answer is (C).

Topic VI: Land Planning and Development

Chapter

31 Subdivisions

1. DEFINITION

The act of subdivision is the division of any tract or parcel of land into two or more parts for the purpose of sale or building development.

2. REGULATION

Resurveys are made to determine what has taken place in the past. On the other hand, surveys made for the purpose of planning subdivisions are creative in nature, and the care and imagination used in planning affect the entire community for many years to come. There are many instances in the United States in the past where developers have created subdivisions that are a credit to their foresight and integrity without any regulatory laws. But, increasing population and decreasing availability of land for development have made it necessary for the states and the federal government to adopt laws regulating land divisions.

3. SUBDIVISION LAW AND PLATTING LAW

Subdivision law includes regulations for land use, types of streets and their dimensions, arrangement of lots and their sizes, land drainage, sewage disposal, protection of nature, and many other details. *Platting law* includes regulations for recording the subdivision plat, monumenting the parcels, establishing the accuracy of the survey, and means of identifying the parcels and their dimensions.

4. PURPOSE OF SUBDIVISION LAW

Creation of a subdivision involves more than furnishing a location for a home for a new member of a community and a profit for the developer. It requires planning for traffic, transportation, the location of schools, churches, and shopping centers, and the health and happiness of the citizens of the community. Poorly planned subdivisions have caused cities to spend excessive amounts for street widening and resurfacing, reconstruction of sewer lines, establishment of additional drainage facilities and increase in size of water mains. Poor planning in the past has caused an acute awareness among state, county, and municipal officials of the need for the regulation of subdivision development.

5. THE CITY AS THE REGULATORY AUTHORITY

The authority to enforce subdivision regulations lies for the most part with cities. Counties have legal authority to regulate subdivisions, but most of the regulation is enforced by cities. Cities have the authority to enforce regulations by means of an ordinance adopted by the city council.

6. CERTAINTY OF LAND LOCATION

Monuments set by the original surveyor and called for by the conveyance have no error of position. An interior monument in a subdivision, after it is used, is correct, but boundary monuments of a subdivision marking the line between an adjacent owner, if not original monuments, can be located in error. Establishing a subdivision does not take away the rights of the adjoiner. It is the surveyor's role to eliminate future boundary difficulties. An important part in eliminating future disputes is the establishment of a precise control traverse. It should be established before any detailed planning is made.

7. BOUNDARY SURVEY

The first step in subdividing is the establishment of the control traverse, which is the base for determining the boundaries of the tract. Investigation as to any conflict with senior title holders should be made. After certainty of location is established, the corners should be monumented.

8. TOPOGRAPHIC MAP

The next step is to establish a system of bench marks for vertical control and to prepare a topographic map. The map is essential in planning the subdivision, especially in regard to drainage and sanitary sewer plans.

9. THE PLANNING COMMISSION

Most cities deal with developers of subdivisions through a planning commission. The next step for the developer is to contact the planning commission of the city (or any other designated approving agency) for consultation. Many planning commissions require, at the outset, a subdivider's data sheet that indicates the general features of the developer's plan and a location map that locates the proposed subdivision in relation to zoning regulation and to existing community facilities.

10. GENERAL DEVELOPMENT PLAN

After preliminary discussion, the developer should submit a penciled sketch showing contours, street locations, and lots. If the planning commission feels that the development plans fit into the city's overall plan and into the surrounding neighborhood, it will ask for a preliminary plat.

11. PRELIMINARY PLAT

The *preliminary plat* is actually the detailed plan for the subdivision and includes names of the subdivider, engineer, or surveyor; a legal description of the tract; location and dimensions of all streets, lots, drainage structures, parks, and public areas; easement, lot, and block numbers; contour lines; scale of map; north arrow; and date of preparation. After approval of the preliminary plat, the developer can proceed with stake-out and construction operations.

12. MONUMENTS

When the control traverse has been established and preliminary approval of street right-of-way widths and locations has been obtained from municipal authorities, monuments should be set on property corners and street right-of-way lines based on the control survey. Control monuments that are to be used to relocate lost corners should be permanent and indestructible.

13. FINAL PLAT

The *final plat* conforms to the preliminary plat except for any changes imposed by the planning commission. It must be prepared and filed for record in accordance with platting laws of the state. It establishes a legal description of the streets, residential lots, and other sites in the subdivision.

32 Residential Planning

1. STORM DRAINAGE

The first step in planning subdivision drainage is a careful study of the contour map. Storm sewers and sanitary sewers are designed for gravity flow. Streets act as drainage collectors for storm runoff. Therefore, a general plan for storm sewers and sanitary sewers should be formulated before the streets are finally located.

Lots should drain to the street or to some open drainage system in the area. Lots should never receive drainage from the street. The Federal Housing Administration has set up requirements for lot drainage for various types of topography where this agency guarantees loans. These requirements call for the lot to slope away from the house for some distance in all directions. The ideal situation is for the lot to slope from the house to the street in the front of the lot and from the house to some drainage collector at the rear of the lot if one exists. Concrete alleyways with inverted crowns sometimes serve this purpose. Topography sometimes makes it necessary for the lot to slope from front to rear or from rear to front, but the house is always located so that the slope is away in all directions.

Storm water is carried along the gutter, which necessitates a curb. To avoid water collecting in pools, streets should have a minimum grade of 0.3% and preferably 0.5%. Water should not run across streets in valleys. Inlets should be planned for both sides of the streets at intersections.

The design of storm sewers depends on the amount of surface runoff to be carried in the storm sewers. The amount of rainfall that is absorbed by the surface soil depends on the perviousness of the soil, the intensity of the rainfall, the duration of the storm, and the slope of the surface. Water that is not absorbed is known as *surface runoff.* Prior to subdivision, farm or grazing land may absorb a large percentage of a given rainstorm. Streets, driveways, sidewalks, and rooftops, however, will cause most of the rain water to run off rather than be absorbed by the soil. The developer must be conscious of this fact in order to prevent damage to the development and to property downstream.

Factors included in the design of storm drains include area of drainage area in acres, shape of drainage area, slope of land in drainage area, use of land in drainage area (present and future), maximum intensity of rainfall, and frequency of maximum intensity.

2. SANITARY SEWERAGE

Septic tanks are undesirable in subdivisions and should not be used. The most desirable solution to sewage disposal is a collecting system connected to a municipal sewage disposal system. The second most desirable solution is a private collecting system and disposal plant.

Sanitary sewers flow by gravity, and their location depends on the topography of the land. They are often located under the streets, but they also may be located at the back of lots or in alleyways. The slope of the sewer should be such that the velocity will be between 2 ft/sec and 10 ft/sec. Manholes must be located at change of grade, junctions, and other points for inspection. Sewers must be laid in a straight line between manholes.

3. STREETS

Of major importance in the subdivision is the street system. Streets not only furnish an avenue for vehicle passage, they also furnish access to property for pedestrians, right-of-way for utility lines, channels for drainage, and access to fire plugs, garbage cans, and so on. The right street in the right place contributes to pleasant living. Insufficient street width creates traffic hazards, while excess width adds to the cost of construction and uses land that could be used for lots.

Local residential streets (also called *minor streets*) are designed to furnish access to private property. They are not designed to carry through traffic and should be designed to discourage it. Curved streets, loops, and cul-de-sacs all discourage speed. Long, curvilinear streets with block lengths up to 1800 ft have been found to be satisfactory. Where off-street parking is adequate, a roadway width of 26 ft (27 ft back-to-back of curbs) and a right-of-way of 50 ft are adequate for single-family residential neighborhoods. For multifamily neighborhoods, street widths should be 31 ft minimum (32 ft back-to-back of curbs) with 60 ft right-of-way. A street width of 26 ft will not provide parking on both sides of the street with two lanes for traffic, but many homeowners find weaving in and out between parked cars slows traffic.

Collector streets carry traffic between local streets and arterial streets. They also furnish access to private property along the street. A width of 36 ft (37 ft back-to-back of curbs) will furnish two parking lanes and two traffic lanes. Right-of-way width should be 60 ft.

Arterial streets (also called *major streets*) move heavy traffic at relatively high speeds. Intersections are usually controlled by traffic lights. Many arterial streets provide six moving lanes with no parking. A 100 ft right-of-way will provide two 33 ft (back-to-back of curbs) lanes and a 14 ft median.

Cul-de-sacs are dead-end streets with turnarounds at the end. They are popular for single-family residences because of the privacy and freedom from noise they provide. They should not be more than 1000 ft in length because of the long turnaround. The turnaround circle should be not less than 40 ft in radius.

Loop streets have the same advantages as cul-de-sacs, plus the advantage of better circulation for fire trucks, delivery trucks, and police cars. Loop streets and cul-de-sacs can be used in odd corners of subdivisions.

4. BLOCKS

Until recent years, most subdivisions were laid out on a gridiron pattern. Modern subdivisions use curved streets with blocks 1400 ft to 1800 ft in length. These have proven to be more economical, using less street area by eliminating many cross-streets, and safer because of fewer intersections and slower speed on curved streets. They also provide more lots per undeveloped acre.

5. LOTS

Lot size should depend on the type of development, the topography of the land, and the expected cost per housing unit. Lot dimensions vary in different parts of the United States, but a minimum width of 60 ft and a minimum depth of 100 ft (or about 6000 ft^2) is considered desirable. Large, ranch-style houses with multicar garages opening on the front require more than 75 ft in width.

6. COVENANTS

To protect the interests of future property owners, developers often include certain restrictive *covenants* as a part of the deed to the property. These include such things as type of construction, minimum size and cost, setback distance, restrictions against advertising signs, raising of animals, parking of mobile homes, conducting certain types of commercial enterprises, and any other restrictions that will insure that the neighborhood is used for the purpose for which it was designed.

7. SETBACK LINES

To prevent locating buildings in such a way as to mar the general view of the neighborhood, restrictive covenants often designate the minimum distance from the front property line to the front of the building. To prevent monotony, staggered setback lines are sometimes used.

8. DENSITY ZONING

Many cities that have formerly controlled crowding in new neighborhoods by controlling the dimensions of lots have now adopted *density control*, which controls the maximum number of dwellings per acre. This has allowed more imagination in planning and better use of natural features.

9. CLUSTER PLANNING

A recent development in subdivision planning is the *cluster pattern*, where residences are clustered together in small, private sections of the subdivision with common open space. Cul-de-sacs and loop streets are adaptable to the cluster plan.

Topic VII: Mapping

Chapter

33 Topographic Surveying and Mapping

1. CARTOGRAPHY

Cartography is the profession of making maps. Topographic maps provide a plan view of a portion of the earth's surface showing natural and constructed features such as rivers, lakes, roads, buildings, and canals. The shape, or relief, of the area is shown by contour lines, hachures, or shading.

2. USES OF TOPOGRAPHIC MAPS

The planning of most construction begins with the *topographic map*, sometimes referred to as the *contour map*. A study of a topographic map should precede the planning of highways, canals, subdivisions, shopping centers, airports, golf courses, and other improvements.

3. TOPOGRAPHIC SURVEYS

Topographic surveys are made to determine the relative positions of points and objects so that the map maker can accurately represent their positions on the map.

4. TYPES OF MAPS

There are two basic types of maps: the strip map and the area map. The *strip map* is used in the development of highways, railroads, pipelines, powerlines, canals, and other projects that are narrow in width and long in length. The *area map* is used in the development of subdivisions, shopping centers, airports, and other localized projects.

5. CONTROL FOR TOPOGRAPHIC SURVEYS

Of great importance in topographic surveys is horizontal and vertical control. *Control* is the means of transferring the relative positions of points and objects on the surface of the earth to the surface of the map.

6. HORIZONTAL CONTROL

Relative position in the horizontal plane is maintained by *horizontal control.* Horizontal control consists of a series of points accurately fixed in position by distance and direction in the horizontal plane. For most topographic surveying, traverses furnish satisfactory control. For strip maps, the open traverse is used. For area maps, the closed traverse is used. The open traverse can be tied to fixed points at each end. The closed traverse can be closed to form a net that is accurate to the degree required.

For large areas, such as states, triangulation or trilateration furnish the most economical control.

7. VERTICAL CONTROL

Relative position in the vertical plane can be maintained by *vertical control*, a series of bench marks in the map area. These bench marks are referred to a known *datum*, usually mean sea level.

8. HORIZONTAL TIES

After the traverse is closed to the required specifications, objects that are to be included on the map are *tied* to the traverse. These *horizontal ties* are sometimes called the *detailing*. For large surveys, ties are made by photogrammetry. For smaller surveys, ties are made on the ground.

At least two measurements are required to tie one point to the traverse.

9. METHODS OF LOCATING POINTS IN THE FIELD

The two measurements required to tie one point to the traverse may consist of two horizontal distances, an angle and a horizontal distance, or two angles. There are several methods used to locate a point in the field. Only the four most common will be discussed.

10. RIGHT-ANGLE OFFSET METHOD OF TIES

The *right-angle offset method* of ties is the most common method used in route surveying for preparing strip maps. The ties are made after the centerline (or traverse line) has been established. Usually, stakes are driven at each station on the centerline, a 100 ft steel tape is stretched between successive stations with the 100 ft mark on the tape forward, and points on either side of the tape are tied to the traverse before the tape is moved forward to the next two stakes. To tie in the corners of the house shown in Fig. 33.1, surveyors move along the tape to a point on the line where they estimate a perpendicular line from the traverse line would strike a corner of the building. They observe the plus at this point by glancing at the tape on the ground and then measure the distance from the traverse line to the corner with another (usually cloth) tape, and record both measurements in the field book. With the 100 ft mark of the steel tape forward, pluses are read directly. This procedure is repeated for the next corner. All sides of the house, including the side between the tied corners, are then measured with the cloth tape. A sketch of the house showing the dimensions of all sides of the house is placed in the field book.

Figure 33.1 *Right-Angle Offsets*

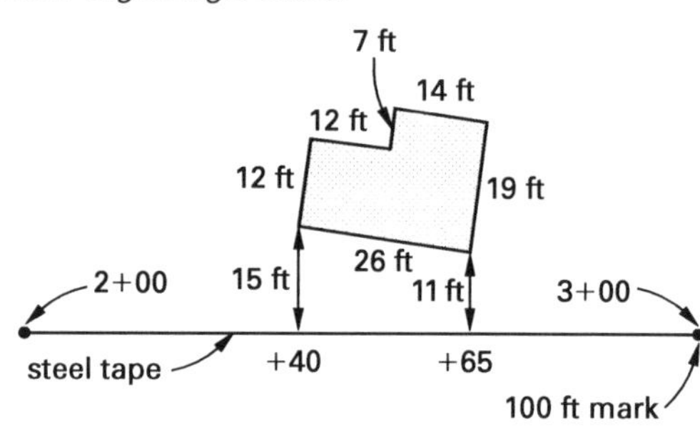

Unless the scale of the map is very large, measurements for ties are recorded to the nearest foot. It is usually impossible to scale a distance on the map for a tenth of a foot.

A right-angle mirror prism is convenient in establishing right angles. As a less accurate method, the surveyor may stand on the transit line facing the point to be tied with arms outstretched on each side pointing along the traverse line, and then bring both arms to the front of the body. If they do not point to the object to be tied, the surveyor should move along the traverse line until they do.

11. ANGLE AND DISTANCE METHOD OF TIES

The *angle and distance method* of ties, also called the *azimuth-stadia method*, is the most common of those used in preparation of area maps. The azimuth-stadia method allows direction and distance measurements to be made almost simultaneously. If the object is a house or building, two corners must be tied to the traverse, and all sides must be measured and recorded in the field book. The transit does not have to be confined to the traverse stations. Intermediate stations can be set from the traverse stations and ties made to the intermediate station. This method, using stadia for horizontal distance, is usually the most efficient.

Figure 33.2 Angle/Distance Ties and Two Distance Ties

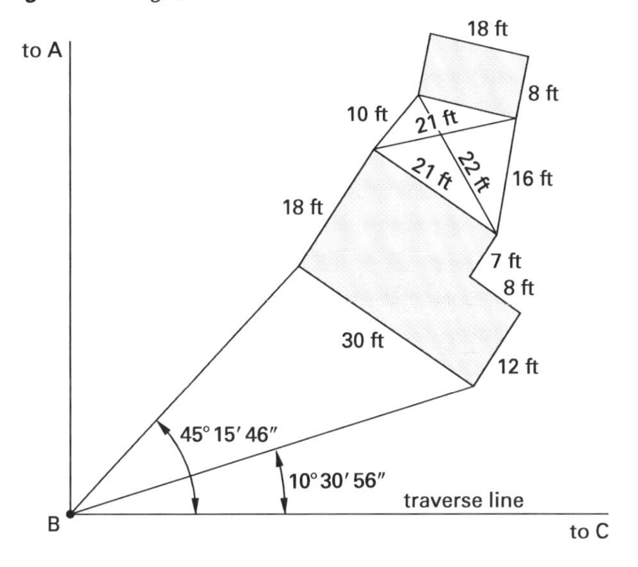

12. TWO-DISTANCES METHOD OF TIES

The *two-distances method* of ties can be used in conjunction with the angle and distance method. Where barns or out-buildings lie behind a house and are obscured from view, the house can be tied to the traverse by the angle and distance method. The out-building can be tied to the house by the two distances method. Two horizontal distances are required to locate one corner of the out-building. Two corners must be tied to the house as shown in Fig. 33.2. All sides of the out-building are measured and recorded in the field book.

13. TWO-ANGLE METHOD OF TIES

The *two-angle method* of ties is used in special cases where the object to be tied is inaccessible because it lies across a river, lake, or busy highway. The object is tied to the traverse by turning an angle to the object from two different points on the traverse.

14. STRENGTH OF TIES

When making horizontal ties, certain practices can be followed to reduce the error in locating points on the map.

- When the two distances method is used, the two distances should be as nearly at right angles as possible.
- When the two angles method is used, the two lines of sight to the object should be as nearly at right angles as possible.

15. VERTICAL TIES

Vertical ties can be made simultaneously with horizontal ties by stadia or by leveling. Leveling is used for strip maps with *cross-sectioning* as the usual method. For area maps, the *grid system* is common. This consists of laying out the area into a grid with 50 ft or 100 ft intervals and determining elevations at the grid intersections.

16. SUMMARY OF HORIZONTAL AND VERTICAL TIES

No one method of making ties excludes the possible use of others. The azimuth-stadia method is very efficient, does not require a large party, and is accurate enough for most work. Combinations of methods may be used. Where ditches or streams run through the map area, a combination of azimuth-stadia and cross-sectioning may by economical. The size of the area, slope of the terrain, and amount and size of vegetation also influence the selection of a method.

17. NOTEKEEPING

Examples of field notes for a right-angle offset survey are shown in Fig. 33.3. Most measurements are shown on the right. Transit stations and full stations are shown on the left. It is not necessary that the right half be drawn to scale, but it is often very convenient to do so. Each line space on the left represents 20 ft and the smallest line space on the right represents 10 ft. This scale makes for rapid plotting, but different topographic details require different scales. It is accepted practice to vary the scale from page to page if the amount of necessary detailing varies.

18. STADIA METHOD

The Greek word *stadia* denotes a unit of measure for horizontal distance. In surveying, the term is used to denote a system for measuring horizontal distances based on the optics of the transit telescope, theodolite, or level. This system eliminates the need for horizontal taping, and while not as accurate, it is satisfactory for making the horizontal measurements for topographic maps. The system is also known as *tacheometry*.

When employing the stadia method to obtain horizontal distances, the horizontal circle of the transit or theodolite indicates direction, and the vertical circle and a level rod determine elevation. Horizontal and vertical ties to the traverse can therefore be made simultaneously.

Figure 33.3 Typical Field Notes for Right-Angle Offset Survey

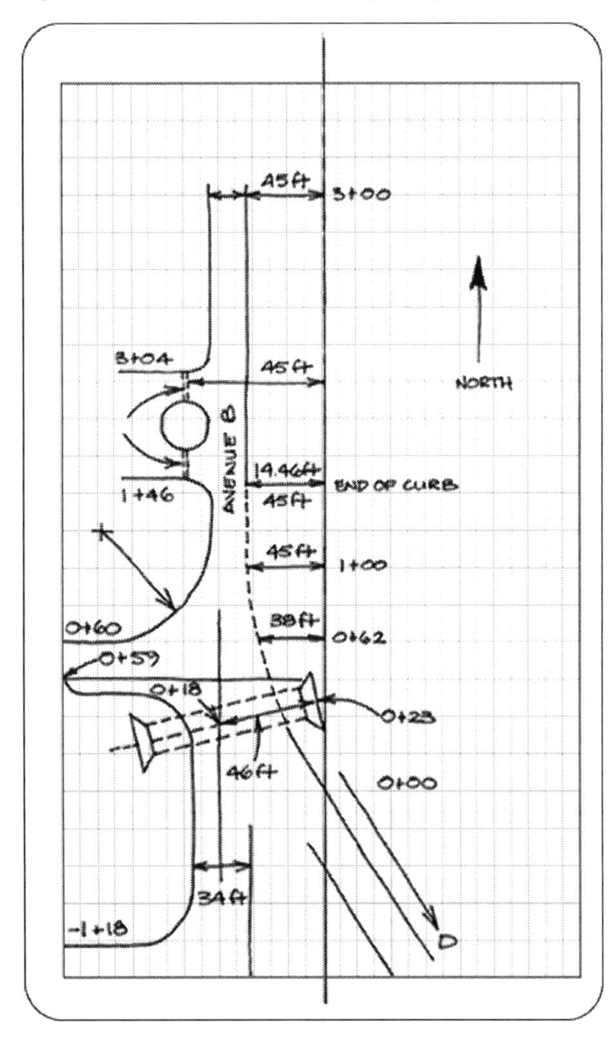

19. STADIA

Stadia sighting depends on two horizontal crosshairs, known as *stadia hairs*, within the telescope. These hairs are parallel to the horizontal crosshair and are equally spaced above and below it. The stadia hairs are shortened so that they will not be confused with the middle horizontal crosshair, although this may not be true in older transits (see Fig. 33.4).

The instrument operator sees the stadia hairs imposed on the stadia rod, as shown in Fig. 33.4. The distance on the rod is known as the *intercept.* If the rod is vertical and the telescope is horizontal, the distance from the center of the instrument to the rod is 100 times the intercept. Actually, the diverging lines of sight to the stadia hairs are from the vertex, which may be 1 in or 2 in from the center of the instrument. This discrepancy is ignored in topographic surveying. In older transits, a constant of 1 ft is added to the stadia distance because the vertex is about 1 ft forward of the center of the instrument.

The stadia principle is illustrated in Fig. 33.5. V_1, V_2, and V_3 represent rod intercepts at varying distances from the vertex of the transit. H_1, H_2, and H_3 represent corresponding horizontal distances. As the lines of sight from the vertex and the intercepts form similiar triangles, it can be seen that $H_1/V_1 = H_2/V_2 = H_3/V_3$. The stadia hairs are so constructed that when the stadia rod is 100 ft from the vertex, the intercept is 1.00 ft. Thus, $H_1 = 100V_1$ and $H_2 = 100V_2$. This means that the horizontal distance from the vertex to the rod is 100 times the intercept.

Figure 33.4 Stadia Hairs and Use

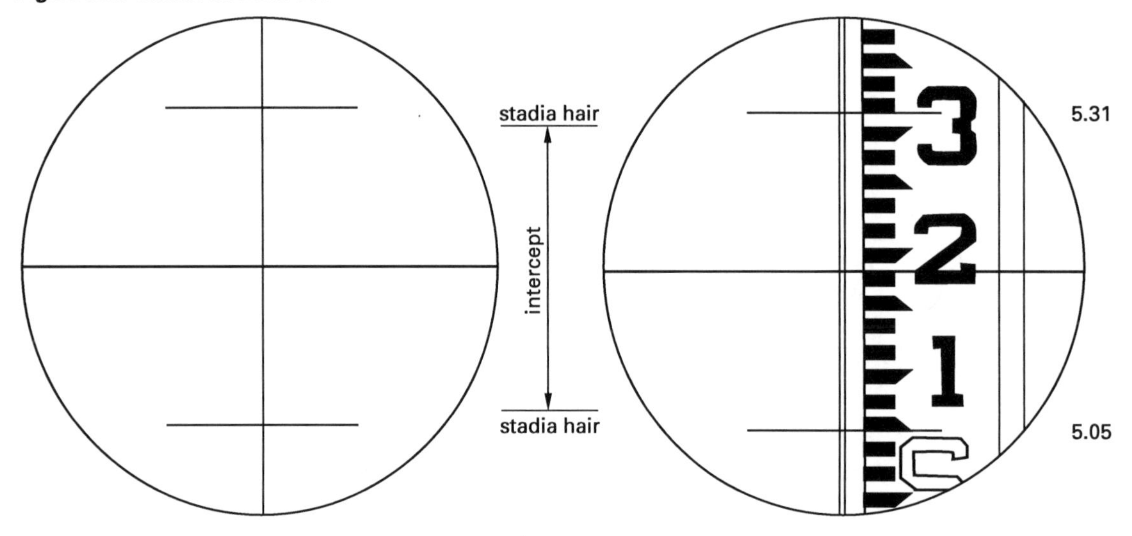

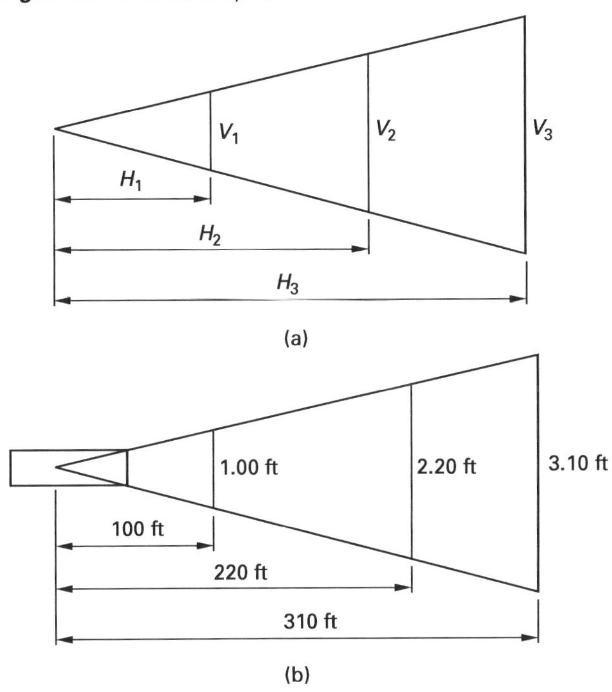

Figure 33.5 Stadia Principles

20. READING THE INTERCEPT

The intercept in Fig. 33.6(a) is 5.31 ft − 5.05 ft = 0.26 ft. The intercept in Fig. 33.6(b) is also 0.26 ft. In Fig. 33.6(b), the lower stadia hair has been placed on the nearest full foot by lowering the line of sight with the vertical tangent screw. The intercept can be determined easily by subtracting the full foot value from the reading at the upper stadia hair. This slight adjustment does not cause an appreciable error in the horizontal distance.

21. HORIZONTAL DISTANCE FROM INCLINED SIGHTS

In terrain steep enough to prevent all readings being made with the telescope horizontal, the horizontal distance is computed by trigonometry, using the rod intercept and the vertical angle of the line of sight.

Figure 33.7 indicates that the rod intercept is greater on inclined sights than it would be if the line of sight were perpendicular to the rod. As it would be impractical to hold the rod perpendicular to the line of sight, the rod is held plumb and the normal intercept is computed by trigonometry.

In Fig. 33.7, the horizontal crosshair is placed on a rod reading equivalent to the *height of the instrument, h.i.* (h.i. is the distance from the hub to the center of the instrument. It is not to be confused with HI, which is elevation above a datum.) This makes the vertical angle α at the instrument equal to the vertical angle α at the hub. In reading the intercept, the bottom stadia hair is set on a full foot mark such that the middle hair is approximately on the h.i. After the intercept is read, the middle hair is set on the h.i. with the tangent screw and the angle α is read on the vertical circle.

If S' is the intercept actually read, S is the intercept normal to the line of sight, D is the slope distance, and H is horizontal distance, then

$$S = S' \cos\alpha \qquad 33.1$$

$$D = 100S = 100S' \cos\alpha \qquad 33.2$$

$$H = D\cos\alpha = 100S' \cos^2\alpha \qquad 33.3$$

Figure 33.6 Reading Intercepts

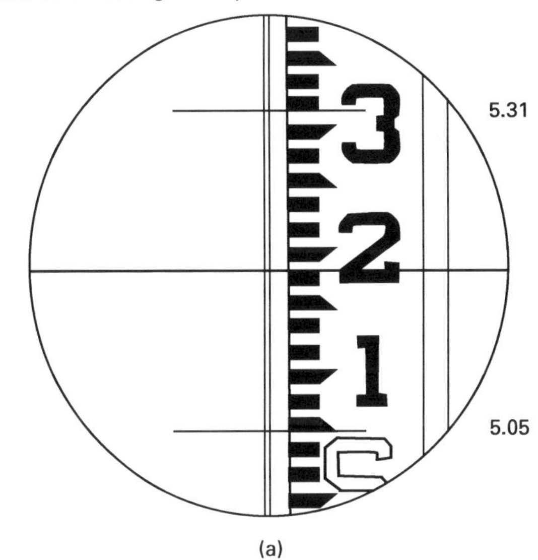

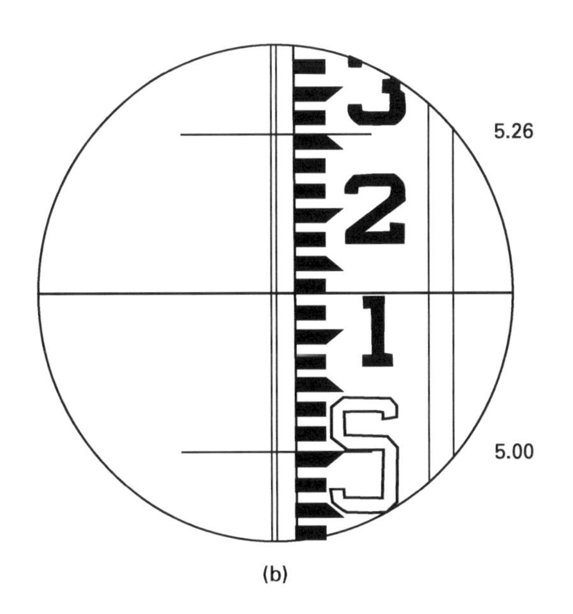

Mapping

Figure 33.7 Using Stadia Rod to Measure Horizontal Distance

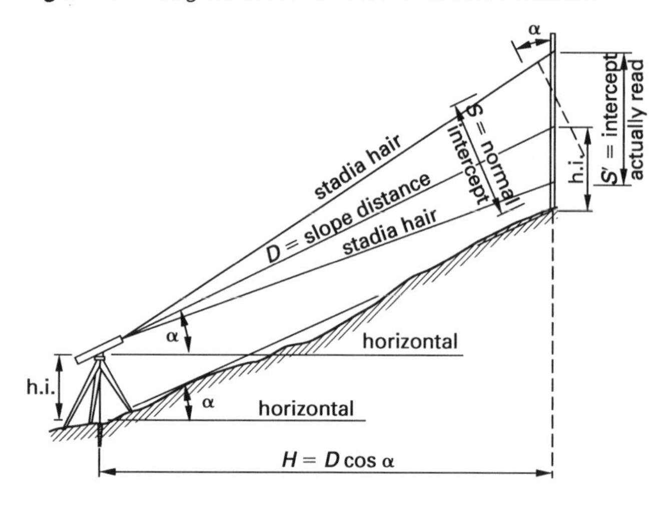

22. VERTICAL DISTANCE TO DETERMINE ELEVATION

The difference in elevation of the hub and any point can also be determined by trigonometry, using the angle α and the rod intercept. It can be seen in Fig. 33.8 that if the middle crosshair is on the h.i., the vertical distance V from the horizontal through the center of the instrument to this h.i. reading on the rod is the same as the vertical distance from the hub to the point on the ground where the rod rests.

$$V = D\sin\alpha = 100S'\cos\alpha\sin\alpha \qquad 33.4$$

Figure 33.8 Using Stadia Rod to Measure Vertical Distance

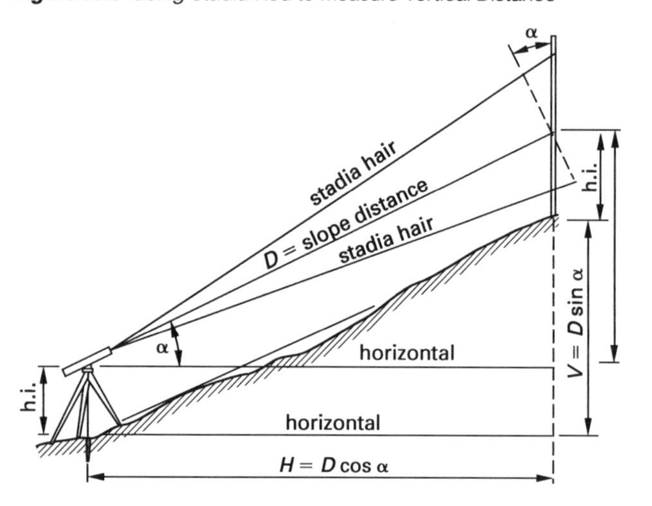

23. USE OF STADIA REDUCTION TABLES

To avoid calculating $\cos^2\alpha$ and $\cos\alpha\sin\alpha$, the horizontal distance H and the vertical distance V can be found in tables similar to Table 33.1.

Table 33.1 Typical Stadia Reductions

	4°		5°	
minutes	horizontal distance	vertical distance	horizontal distance	vertical distance
0	99.51	6.96	99.24	8.68
2	99.51	7.02	99.23	8.74
4	99.50	7.07	99.22	8.80
6	99.49	7.13	99.21	8.85
8	99.48	7.19	99.20	8.91
10	99.47	7.25	99.19	8.97
12	99.46	7.30	99.18	9.03
14	99.46	7.36	99.17	9.08

Table 33.1 lists horizontal and vertical distances for a rod intercept of 1.00 ft for 4° and 5° vertical angles. Vertical angles to the nearest full degree are shown at the top of the columns, and minutes are shown in the left column. Interpolation is required for angles of an odd number of minutes. To find horizontal and vertical distance for a particular vertical angle and intercept not equal to 1.00 ft, the intercept is multiplied by the distance found in the table.

Example 33.1

Find the horizontal distance H and the vertical distance V for a rod intercept of 3.68 ft and vertical angle α of $4°12'$.

Solution

$$H = (3.68 \text{ ft})(99.46) = 366 \text{ ft}$$
$$V = (3.68 \text{ ft})(7.30) = 26.9 \text{ ft}$$

It can be seen from the table that for vertical angles up to 4°, the slope distance is very nearly the horizontal distance. For topographic detailing, no correction is needed. Of course, the vertical distance must still be computed.

24. AZIMUTH

Azimuth is the most efficient method of determining direction where a number of shots are taken from one station. By measuring all angles from the same reference line, plotting points on the map is simplified. A full circle protractor is oriented to north on each transit station. Azimuth readings from each transit station are plotted with one setting of the full circle protractor.

After the control traverse has been closed, the direction of all legs of the traverse should be recorded in azimuth in the field book. These directions are used in orienting the horizontal circle of the transit at each transit station.

To orient the horizontal circle, first set the vernier on the azimuth of the line along which the transit is to be sighted.

Set the vertical crosshair on that line by using the lower clamp, and then release the upper clamp. The circle will be oriented, but, as a check, the alidade is turned until the vernier reads 0° and the direction of the telescope is observed to see that it is pointing to the north. In Fig. 33.9, the transit is set up on point B. The azimuth of AB is now 150°00′. The horizontal circle is to be oriented for backsighting on point A.

To orient the circle, the vernier must be set on the azimuth of the line from the transit station B to the backsight A. The azimuth of BA is the back-azimuth of AB, so the vernier must be set on 330°00′ before the vertical crosshair is set on point A with the lower clamp.

Figure 33.9 *Azimuth Measurments*

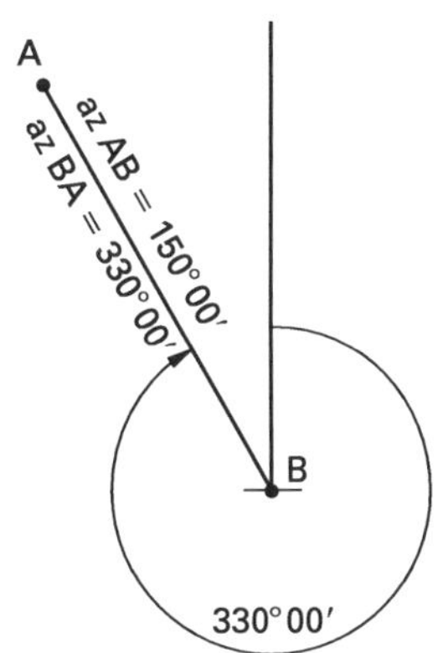

In locating points to be tied to the control net by azimuth, it is not necessary to read the vernier to determine the azimuth. In plotting with a protractor, it is impossible to plot to one-minute accuracy. Plotting to the nearest quarter of a degree is all that is practical. The horizontal circle can be read to 10′ without the use of the vernier. In establishing intermediate transit points not previously located, more care should be exercised in reading the circle.

After moving the transit to a new station, the circle will be oriented by setting on the back-azimuth of the line just established. The intercept from the new station to the previous station should be read as a check against the previous intercept. It is also good practice to take a shot on a known point as a check.

Where a number of shots are taken from one station, it is good practice to observe the backsight again before moving to a new station to see that the transit has not been disturbed.

25. ALGEBRAIC SIGN OF VERTICAL ANGLE

An *angle of elevation* is given a plus sign, and an *angle of depression* is given a minus sign. For small vertical angles, care should be taken that the wrong sign is not recorded. As a check, if the telescope bubble is forward, the sign is plus.

26. ELEVATION

As an example of determining vertical distance from the transit station to a point, assume in Fig. 33.8 that the elevation of the hub is 465.8 ft, the h.i. is 5.2 ft, the rod intercept is 4.22 ft, and the vertical angle is $+4°12'$ with the middle crosshair on the h.i. Using the tables,

$$V = (7.30)(4.22\text{ ft}) = 30.8\text{ ft}$$

$$\begin{aligned}\text{elevation} &= 465.8\text{ ft} + 5.2\text{ ft} + 30.8\text{ ft} - 5.2\text{ ft}\\ &= 465.8\text{ ft} + 30.8\text{ ft}\\ &= 496.6\text{ ft}\end{aligned}$$

It can be seen that when the middle horizontal crosshair is on the h.i. when the vertical angle is read, the two h.i.'s will cancel. V can be added (or subtracted when α is minus) to the elevation of the transit station to obtain the elevation of a point. However, the reading of the h.i. on the rod may be obscured. In this case, the middle hair can be placed on the next higher (or lower) full foot mark and the vertical angle read. The h.i. and V will be added to the hub elevation and the rod reading will be subtracted.

Where possible, the rod readings should be taken with the telescope horizontal to eliminate work in reducing field notes. In this case, the h.i. is added to the elevation of the station and the rod reading is subtracted from this elevation just as in leveling. This eliminates the need to read the vertical angle.

27. FINDING THE h.i.

The h.i. can be found by placing the rod near the transit and moving a target along the rod until the horizontal line of the target lines up with the horizontal axis of the telescope.

28. SELECTING POINTS TO BE USED IN LOCATING CONTOURS

Contours are located on the map by assuming that there is a uniform slope between any two points that have been recorded in the notes. Elevations of points are written on the map and contour lines are interpolated between any two points. To ensure that the slope between any two points is uniform, shots must be taken in the field at certain key points.

29. KEY POINTS FOR CONTOURS

Key points are any points that will show breaks in the slope of the ground, just as in cross-sectioning. The most important of these are

- summits or peaks
- stream beds or valleys
- saddles (between two summits)

- depressions
- ridge lines
- ditch bottoms and tops of cuts
- tops of embankments and toes of slope

30. SPECIAL SHOTS

It is sometimes impossible for the rodperson to place the rod exactly on the point to be located, such as a corner of a house with a wide roof overhang. In this case, the rod is held near the house in such a position that the rod is the same distance from the transit as the house corner, the intercept is read at this point, and the house corner is used to determine horizontal direction.

For long shots where the intercept does not fall entirely on the rod or portion of the rod observed, the intercept between the upper crosshair and the middle crosshair can be observed and doubled for the intercept. On long shots, a stadia rod with bold markings and different colors is more suitable than a level rod.

31. EFFICIENCY OF THE SURVEY PARTY

The *efficiency* of the field party depends on the number of points located during a period of time. Time can be saved by following the order of taking readings on a point as follows.

step 1: Set the vertical crosshair on the rod.

step 2: Set the lower stadia crosshair on a full foot mark such that the middle crosshair is approximately on the h.i.

step 3: Read the upper crosshair, subtract the reading of the lower crosshair, and record the intercept.

step 4: With the tangent screw, place the middle crosshair on the h.i.

step 5: Wave the rodperson to the next point.

step 6: Read and record the horizontal angle.

step 7: Read and record the vertical angle.

It is important that the rodperson be waved on (step 5) before the two angle readings are made and recorded so that he can be moving to a new location while the instrument person is reading the angles. Many shots in a day's work will be lost by forgetting to do this.

32. COMPUTATIONS FROM FIELD NOTES

Field notes for an azimuth-stadia survey are shown in Table 33.2. In the notes, the transit is at station B at the beginning of the day. A backsight is made on station A (which was set at the close of the preceding day). Referring to the field notes of the preceding day, the azimuth of AB was 316°22′. Therefore, when the sight on station A is made, the vernier is set on 136°22′, which is the azimuth of BA and the back-azimuth of AB.

The intercept at station A is read as a check against what was read at B from A on the previous day. The vertical angle, −1°18′, is also read as a check.

The rod reading for station 1 was made with the telescope level. The elevation of station 1 is

$$467.2 \text{ ft} + 5.0 \text{ ft} - 8.0 \text{ ft} = 464.2 \text{ ft}$$

The rod reading on station 4 was made with a vertical angle of +4°34′. This is large enough to require a correction for horizontal distance. The correction factor for 4°34′, found in stadia reduction tables, is 99.37. The horizontal distance B to 4 is

$$H = (99.37)(3.40 \text{ ft}) = 338 \text{ ft}$$

The vertical angle was read when the horizontal crosshair was set on the h.i. If it had not been, the notes would indicate this fact. The factor for the vertical distance for an angle of 4°34′ is found to be 7.94. The vertical distance is

$$V = (7.94)(3.40 \text{ ft}) = 27.0 \text{ ft}$$

The elevation of station 4 is

$$467.2 \text{ ft} + 27.0 \text{ ft} = 494.2 \text{ ft}$$

Table 33.2 *Field Notes for an Azimuth-Stadia Survey*

sta	rod int	az	∠ of rod	H-dist.	V-dist.	elev.
		⊼ @ B elev. 467.2 h.i. 5.0				
A	6.82	136°22′	−1°18′			
1	1.15	88°10′	8.0			
2	1.84	96°30′	7.8			
3	2.28	124°45′	+3°21′			
4	3.40	206°20′	+4°34′			
5	3.25	318°00′	+4°20′			
6	4.36	345°30′	+2°47′			
C	6.25	318°52′	+1°12′			
		⊼ @ C elev. 480.3 h.i. 5.2				
B	6.24	138°52′	−1°13′			
1	1.78	96°10′	+2°41′			
2	2.49	128°45′	9.2			
3	3.12	186°00′	−4°56′			
4	3.40	232°40′	9.5			
D	7.18	285°30′	−1°16′			
		⊼ @ D elev. 464.4 h.i. 5.1				
C	7.18	105°30′	+1°16′			
1	2.42	145°15′	7.7			
2	3.18	180°20′	10.6			

The horizontal crosshair could not be placed on 5.0 (the h.i.) at station 4 because a limb of a tree obstructed the view. The horizontal crosshair was placed on 6.0, and the vertical angle was $+4°45'$. Then,

$$V = (8.25)(3.40 \text{ ft}) = 28.0 \text{ ft}$$

$$\begin{aligned} \text{elevation} &= 467.2 \text{ ft} + 5.0 \text{ ft} + 28.0 \text{ ft} - 6.0 \text{ ft} \\ &= 494.2 \text{ ft} \end{aligned}$$

33. CONTOURS AND CONTOUR LINES

A *contour* is an imaginary line on the surface of the earth that connects points of equal elevation. A *contour line* is a line on a map that represents a contour on the ground.

34. CONTOUR INTERVAL

The *contour interval* of a map is the vertical distance between contour lines. The contour interval is selected by the mapmaker. In flat country, it may be 1 ft and in mountainous country it may be 100 ft, depending on the scale of the map and the character of the terrain. The contour interval can be too small, making the map a maze of lines that are not legible; the contour interval can be too large, not showing the true relief. The more accurate the contours, the more costly the map. The intended use of the map is a basic consideration in the selection of the contour interval.

Fig. 33.10 shows that the vertical distance between contour lines is constant, but the horizontal distance varies with the steepness of the ground.

Figure 33.10 *Contour Line Intervals*

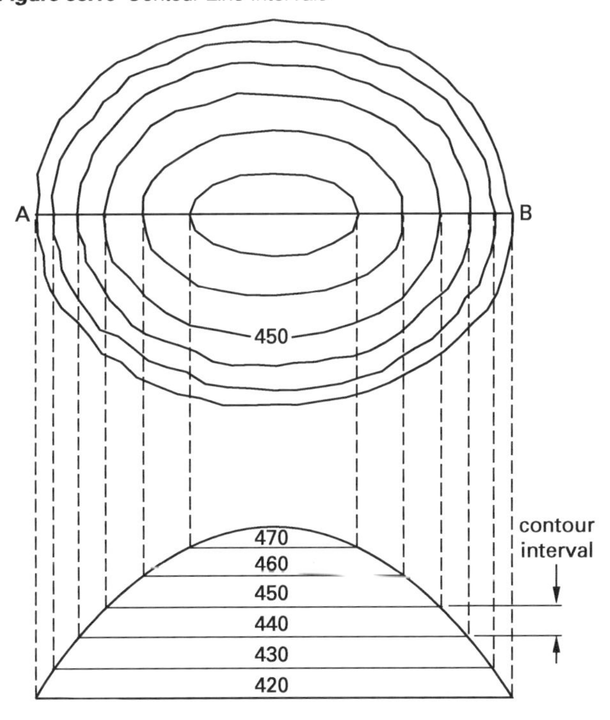

35. INDEX CONTOURS

To facilitate reading a topographic map, every fifth contour line may be darker. The elevation of that contour line is written in a break in the line as shown in Fig. 33.11. There are five spaces between any two heavier lines, called *index contours*, so that the contour interval can be computed by dividing the difference in elevation between two index contours by five. In Fig. 33.11, there are five spaces between the 700 ft contour and the 750 ft contour. The contour interval is

$$\frac{750 \text{ ft} - 700 \text{ ft}}{5} = 10 \text{ ft}$$

Figure 33.11 *Typical Contour Map*

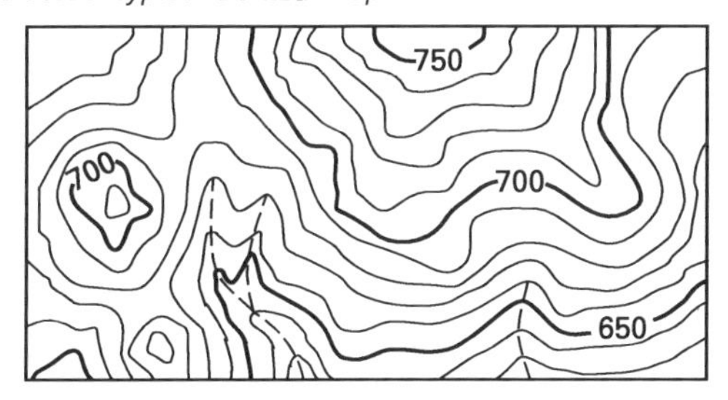

36. CLOSED CONTOUR LINES

Contour lines that are closed represent either a hill or depression. Whether the closed contour lines represent a hill or depression can be determined by reading the elevation of the index contours. For a hill, the elevations increase as the contour lines become shorter. Depressions are often indicated by short hachures on the down slope side of the contour line.

37. SADDLE

The name *saddle* is given to the shape of contours that define two summits in the same vicinity. A profile view of the two summits looks somewhat like the profile view of a horse saddle, as shown in Fig. 33.12.

Figure 33.12 *Saddle*

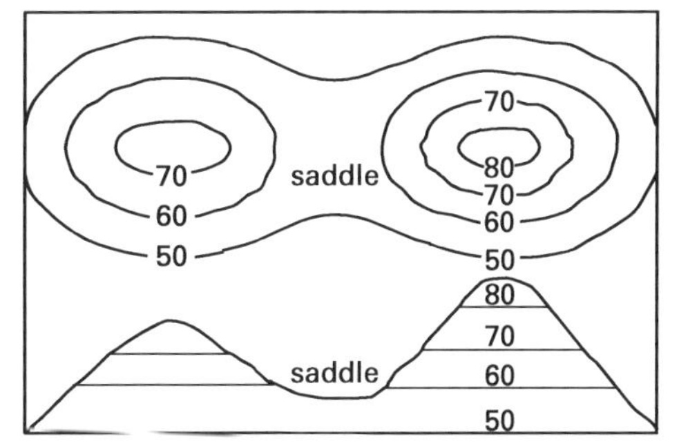

38. CHARACTERISTICS OF CONTOURS

Certain fundamental characteristics of contours should be kept in mind when plotting contour lines and reading a contour map. The characteristics are illustrated in Fig. 33.13.

Mapping

Figure 33.13 Typical Contour Map Features

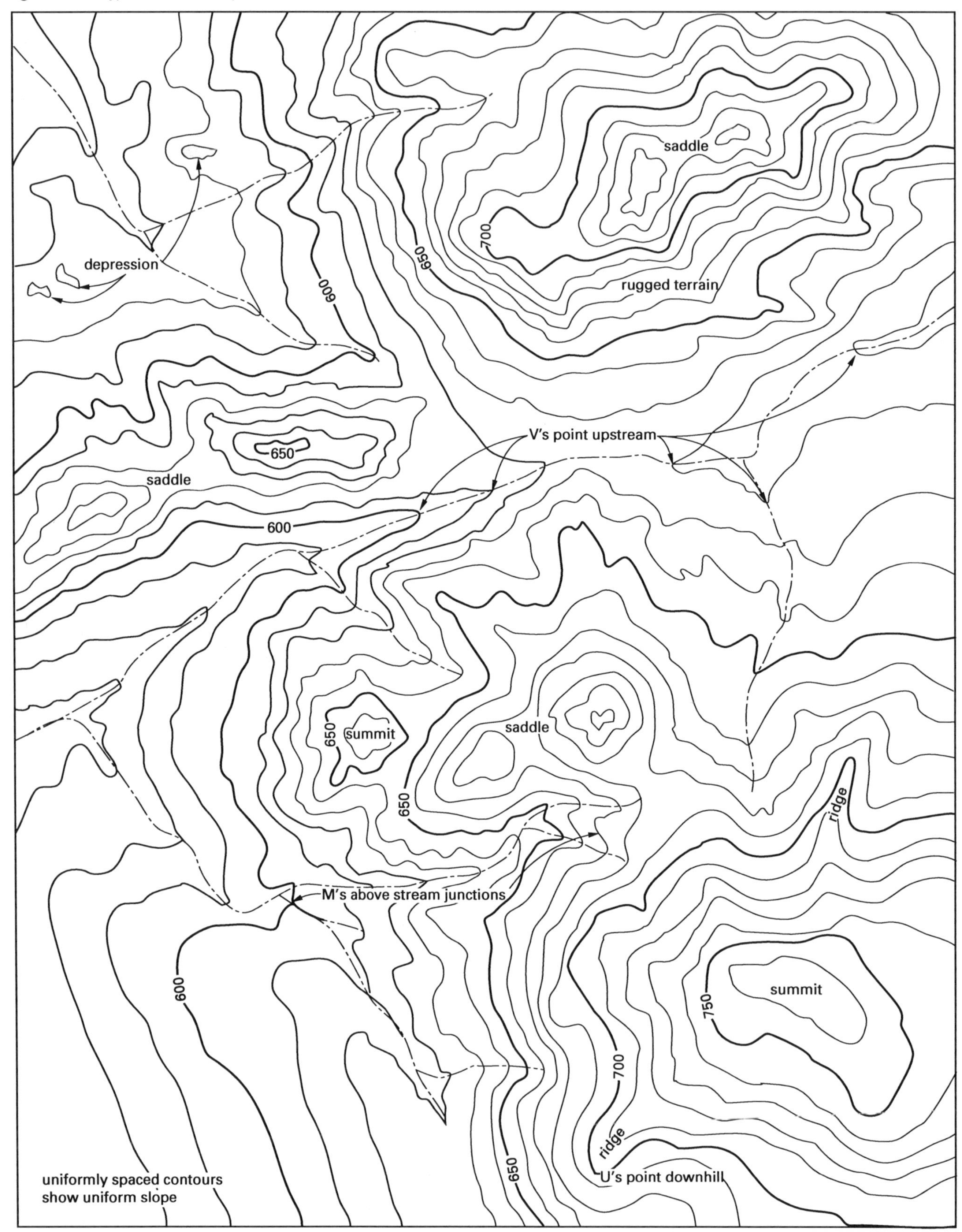

- Each contour line must close upon itself either within or outside the borders of the map. Since all land areas on the surface of the earth rise above the sea, it can be seen that each contour will, no matter how long, finally close. This means that a contour line cannot end abruptly on a map.
- Contour lines cannot cross or meet, except in unusual cases of waterfalls or cliff overhangs. (If it were possible for two contour lines to cross, the intersection would represent two elevations for the same point.)
- A series of closed contour lines represents either a hill or depression. The elevations of the index contour lines will indicate which series represents a hill and which represents a depression.
- Contour lines crossing a stream form Vs that point upstream.
- Contour lines crossing a ridge form Us that point down the ridge.
- Contour lines tend to parallel streams. Rivers usually have a flatter gradient than do intermittent streams. Therefore, contour lines along rivers will be more nearly parallel than contours along intermittent streams, and they will run parallel for a longer distance.
- Contour lines form Ms just above stream junctions.
- Contour lines are uniformly spaced on uniformly sloping ground.
- Irregularly spaced contour lines represent rough, rugged ground.
- The horizontal distance between contour lines indicates the slope of the ground. Closely spaced contour lines represent steeper ground than widely spaced contour lines.
- Contours are perpendicular to the direction of maximum slope. The direction of rainfall run-off in a map area can be determined from this characteristic.

39. METHODS OF LOCATING CONTOURS

Several methods of locating contours are used in topographic surveying and mapping. They are the grid method, controlling points method, cross-section method, and tracing contours method. All methods depend on the assumption that there is a uniform slope between any two ground points located in the field.

40. GRID METHOD

The *grid method* is very effective in locating contours in a relatively small area of fairly uniform slope. The area is divided into squares or rectangles of 25 ft to 100 ft (depending on the scale of the map and the contour interval desired). Stakes are set at each intersection and at any points of slope change, such as at ridge lines or valleys.

The location of the point where the contour line crosses each side of each square is determined by interpolation (either by estimation or mathematical proportion).

In Fig. 33.14(a), contours are to be plotted on a 2 ft interval. Starting at A-1, Fig. 33.14(b), it can be seen that the 440 ft contour will cross between A-1 and B-1. The vertical distance between A-1 and B-1 is 4 ft, so the 440 ft contour will cross halfway between A-1 and B-1. The 442 ft contour will cross at B-1, so a mark is placed at each of these points.

The 444 ft contour will cross between B-1 and C-1. The vertical distance between B-1 and C-1 is 444.3 ft − 442.0 ft = 2.3 ft. The vertical distance between B-1 and the 444 ft contour is 2.0 ft. Therefore, the horizontal distance will be 2.0/2.3 of the way, or about 0.9 of the way. A mark is made at this point.

The 440 ft, the 442 ft, and the 444 ft contours will cross between A-2 and B-2. The crossing points are found in a similar manner and marked.

After the crossing points are located by interpolation, the crossing points for each contour are connected as shown in Fig. 33.14(c).

After all crossing points are connected, the contour lines are smoothed. Small irregularities are taken out so that the contour lines are more like the contours on the ground.

In following a particular contour line using the grid method, an inspection must be made of each grid line between each intersection to see if the contour line can cross. If it cannot, another line must be inspected. For example, in Fig. 33.14(c), the 444 ft contour line can cross between B-1 and C-1, between B-1 and B-2, between A-2 and B-2, or between B-2 and B-3, but not between B-2 and C-2. Each contour line must close or reach the border of the map at two points.

After contour lines are smoothed, index contour lines must be made heavier than the other lines, and the elevation of the index contour must be written in a break in the line.

41. CONTROLLING POINTS METHOD

The *controlling points method* is suitable for maps of large area and small scale. The selection of ground points is very important. The accuracy of the contours depends on the knowledge and experience of the survey party. Shots should be taken at stream junctions, at intermediate points in stream beds between junctions, and along ridge lines. Field notes should indicate these points so that ridges and streams can be plotted before interpolations are made. Interpolations are made in much the same way as they are made using the grid method.

Mapping

Figure 33.14 *Grid Method of Contour Location*

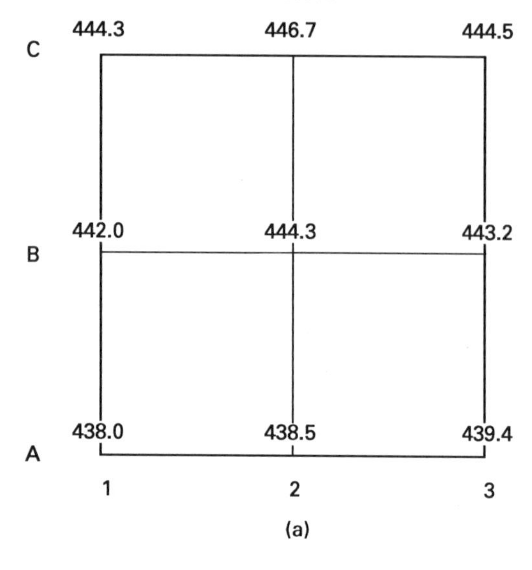

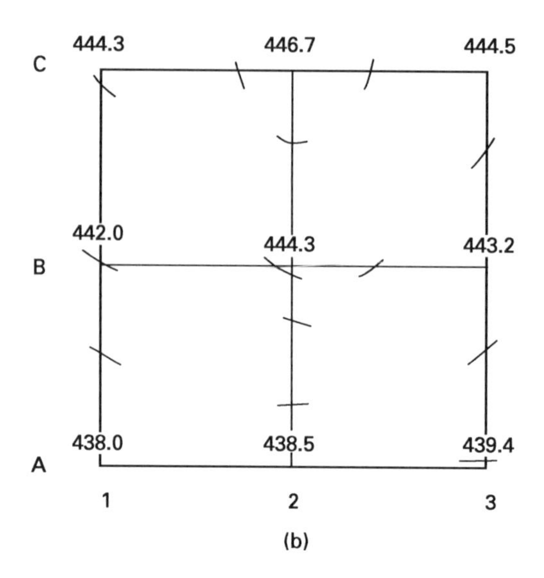

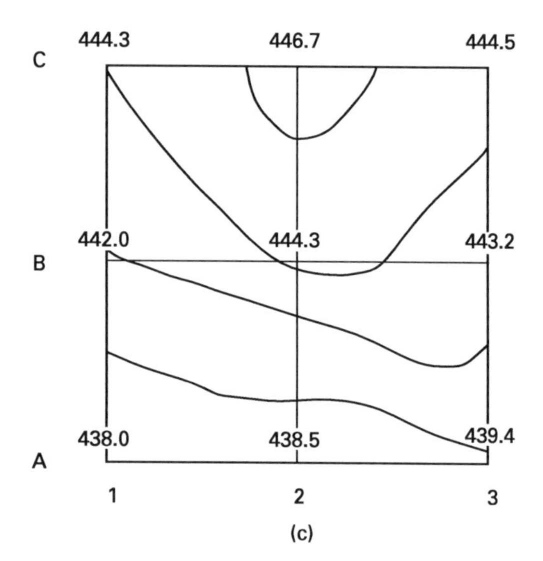

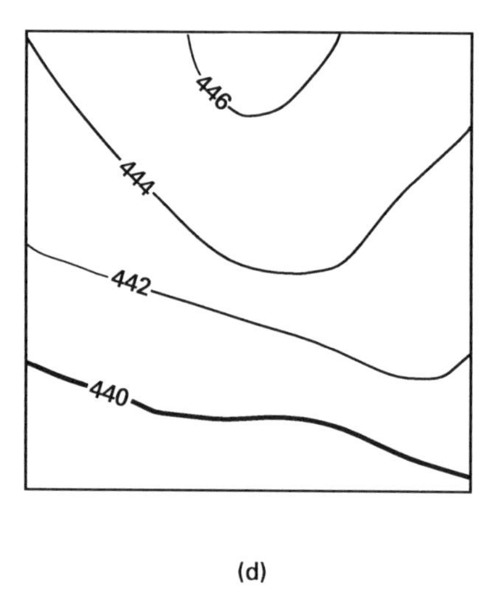

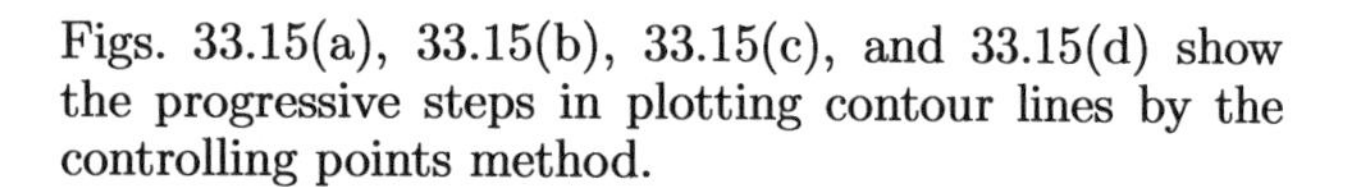

Figs. 33.15(a), 33.15(b), 33.15(c), and 33.15(d) show the progressive steps in plotting contour lines by the controlling points method.

42. CROSS-SECTION METHOD

The *cross-section method* is satisfactory for the preparation of strip maps. It can be accomplished by using level and tape or azimuth-stadia. Cross sections are taken at right angles to a centerline or baseline. Elevations of each cross-section shot are written on the strip map, and interpolation of contour lines is performed as in the grid and controlling points methods.

43. TRACING CONTOURS METHOD

The *tracing contours method* is used when the exact location of a particular contour line is needed. It is effectively performed by use of the plane table, but can be done by the azimuth-stadia survey method.

44. MAPPING

The first step in preparing a map is the selection of a scale and a contour interval. This selection is influenced by the size of the sheet to be used, the purpose of the map, and the required accuracy.

The traverse can be plotted by the coordinate method, the tangent method, or the protractor method.

45. COORDINATE METHOD

The *coordinate method* is the most accurate for plotting a traverse. Any error in plotting one point does not affect the location of the other points. Each point is plotted independently of the others.

Coordinates of each traverse station are computed prior to plotting. Each point is plotted on the grid system using the coordinates.

Figure 33.15 Controlling Points for Contour Location

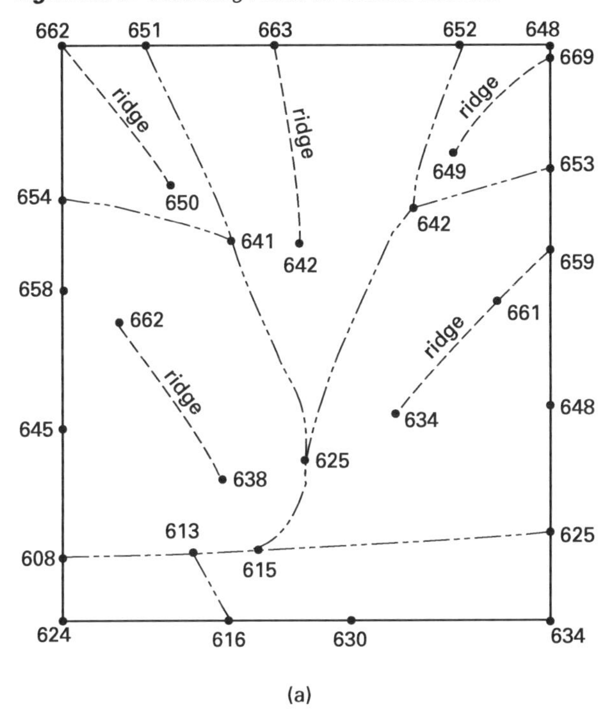

(a)

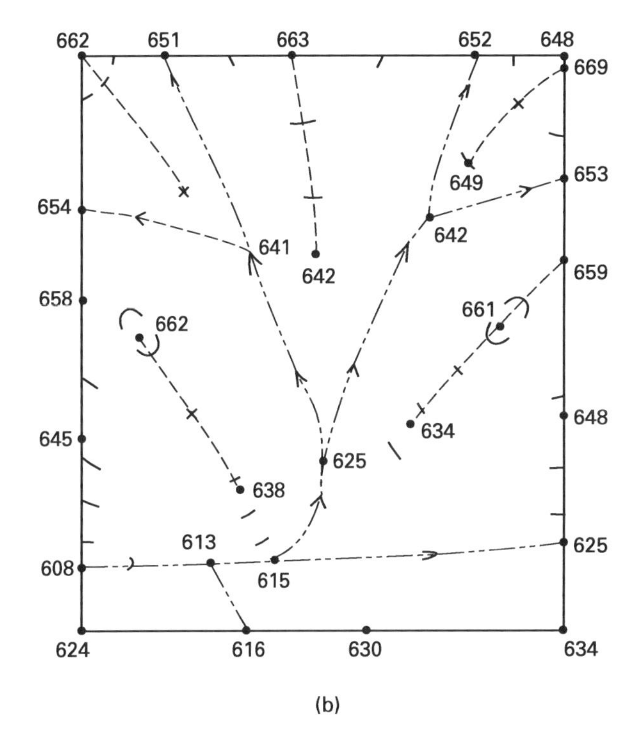

(b)

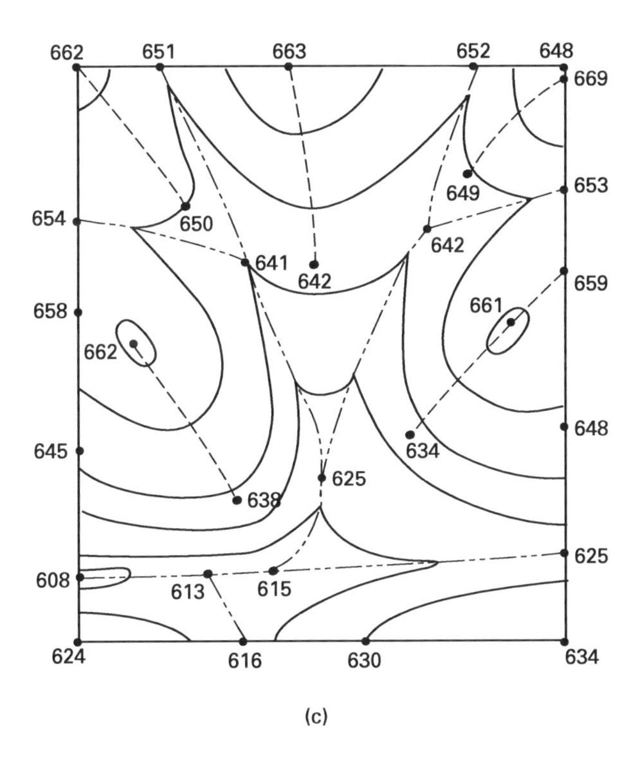

(c)

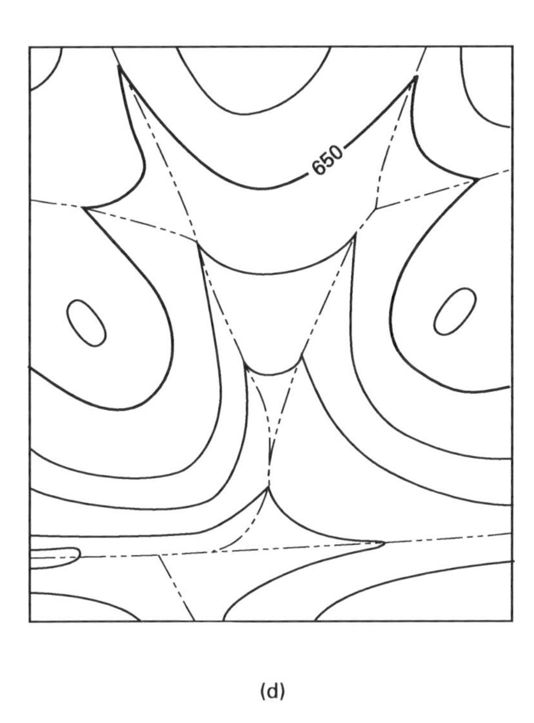

(d)

Before a traverse is plotted, the sheet is laid out with perpendicular grid lines. The distance between grid lines can be 50 ft, 100 ft, 500 ft, 1000 ft, or any other multiple that suits the scale of the map.

As an example, Fig. 33.16 shows a grid system laid out to scale of 1 in = 500 ft with grid lines 1000 ft apart.

The point P has the coordinates $x = 1660, y = 4705$. In plotting the point, the 50 scale is laid on the paper horizontally so that the 10 mark on the scale lines up with the vertical grid line marked 1000. A pencil dot is then made at 1660 on the scale. With a straight edge, a temporary vertical line is drawn through the pencil dot. The scale is then laid vertically along this vertical line with the 40 mark on the scale lined up with the horizontal line on the paper marked 4000. A pencil dot is carefully made on the vertical line at 4705 on the scale (as close as can be read).

Figure 33.16 Coordinates of a Traverse Point

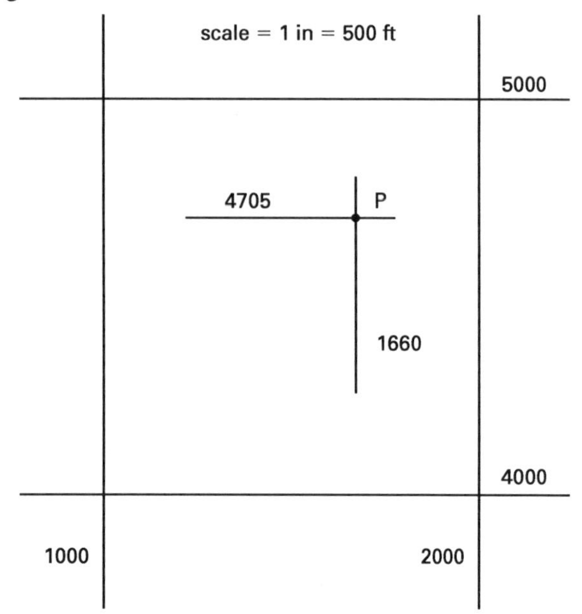

Figure 33.17 Typical Map Symbols

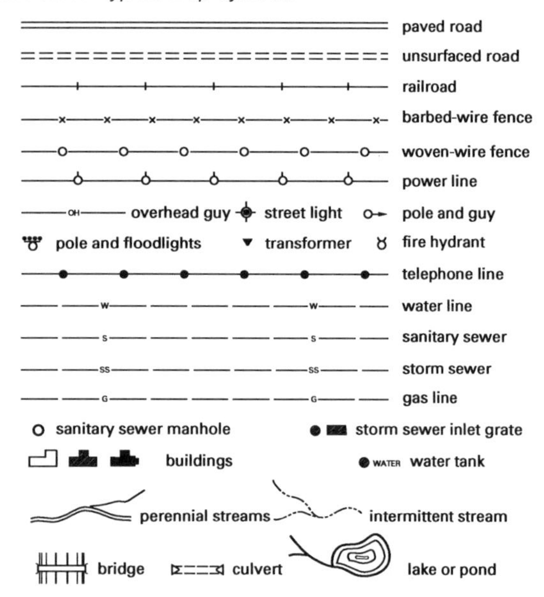

After all points of the traverse are plotted, they are connected with lines. Distances between points are scaled and checked against distances that were recorded in the field.

Detail points can be plotted with a protractor using grid lines to orient the protractor.

46. TANGENT METHOD

The *tangent method* is very convenient and accurate where deflection angles have been turned for a route survey. At any traverse station, the back line is produced past the points. A convenient distance (such as 10 in) is laid off from the traverse point along this prolongation. A perpendicular line is determined at this point, and the tangent distance for the deflection angle is marked on this line. A line from the traverse point defines the next leg of the traverse. Any error made in this plotting will be carried on to the next plotting.

47. PROTRACTOR METHOD

The *protractor method* is the fastest but least accurate method of plotting a traverse. Any error in plotting an angle or a distance will be carried on throughout the traverse. The protractor is commonly used for detailing and, for this, it is sufficiently accurate.

PRACTICE PROBLEMS

1. Complete the topography field notes using the right-angle offset method for ties to the road, stream, and buildings. Consider the enclosed area to be the right half of the page in the field notes. The vertical line is the baseline of the survey. The scale is 1/2 in = 100 ft.

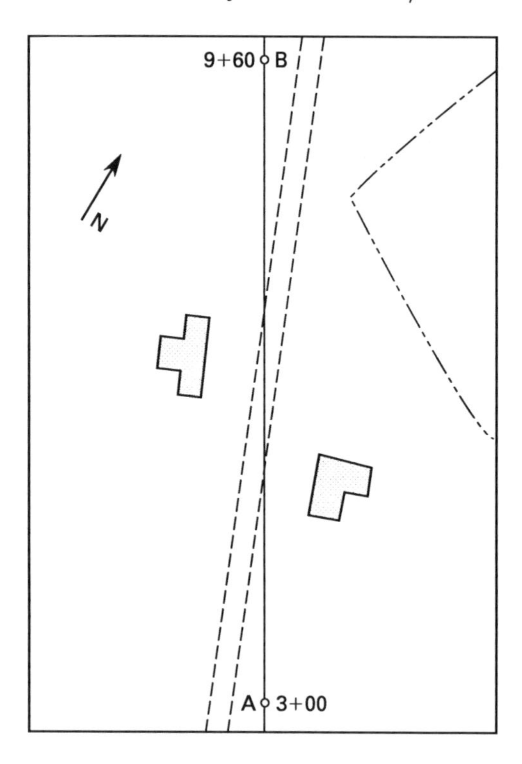

2. Complete the topography field notes using the angle and distance method and the two distances method. Use the two distances method to tie the small building to the adjacent building; use the angle and distance method for other ties. Consider the transit to be set up at station A on the baseline with the foresight on station B on the baseline. The scale is 1/2 in = 100 ft.

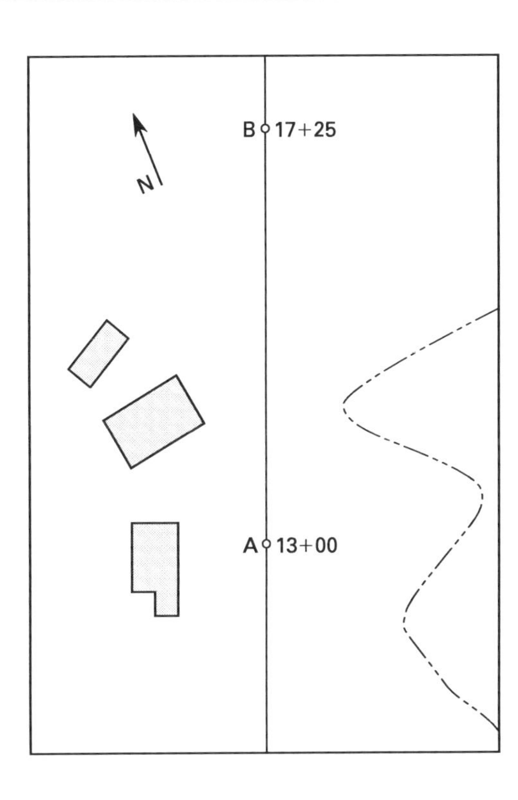

3. Tie the streets to the baseline by the right angle offset method. The scale is 1/2 in = 100 ft.

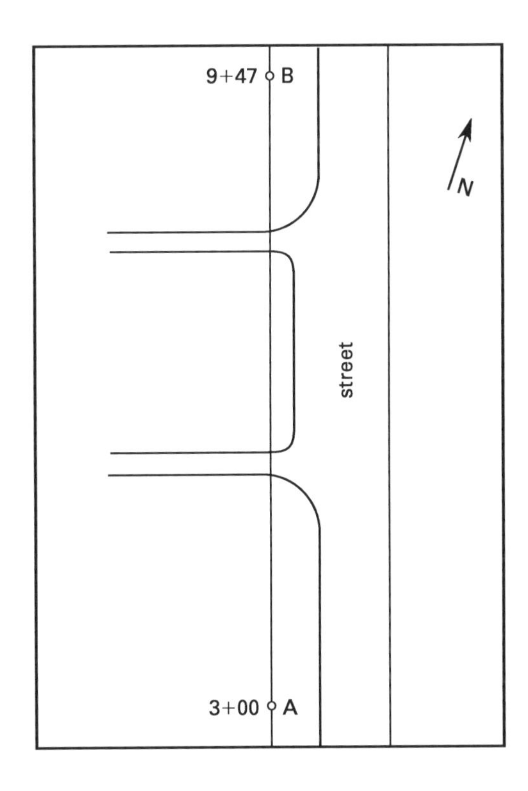

4. Find the horizontal distance H and the vertical distance V between two points when the stadia rod intercept and vertical angle α are as shown.

	rod intercept	α
(a)	3.48	4°00′
(b)	4.68	5°09′
(c)	5.75	5°12′
(d)	3.91	4°11′

5. Indicate the angle to be set on the vernier to orient the transit for an azimuth-stadia survey for each station of the traverse ABCDEA. AB: S 53°18′ E; BC: N 42°00′ E; CD: N 72°47′ W; DE: N 36°27′ W; and EA: S 30°39′ W.

6. Considering the transit to be set up at station A, elev 463.2 ft, with h.i. = 5.1 ft, compute the horizontal distance from A to points 1, 2, 3, and 4 and the elevation of points 1, 2, 3, and 4. (Consider middle crosshair on h.i. when vertical angle is read.)

point	intercept	V angle
1	3.85	+4°05′
2	2.98	+5°08′
3	2.56	−5°09′
4	5.44	−4°03′

7. Complete the following azimuth-stadia survey field notes.

azimuth-stadia survey

sta	int	az	V ang or rod	Hdist	elev
	TRAN @ B ELEV 467.2 h.i. 5.0				
A	6.74	148°04′	−4°14′		
1	0.91	90°45′	8.1		
2	1.66	120°20′	−5°12′		
3	2.10	135°15′	−5°07′		
4	3.25	143°00′	+4°11′		
C	4.00	60°10′	+5°08′		
	TRAN @ C ELEV 502.8 h.i. 5.2				
B	4.01	240°10′	−5°08′		
5	3.15	286°00′	−4°00′		
6	2.21	36°20′	9.2		

8. Plot 1 ft contours. Indicate index contours.

451.7 452.5 453.1 453.7 453.7 453.0 451.9 451.2
452.8 453.8 454.5 455.0 454.8 453.9 452.6 451.6
453.9 455.0 456.0 456.6 455.4 454.0 452.5 450.9
454.4 455.9 457.5 456.2 455.0 453.7 452.2 451.9
454.1 455.7 456.1 455.0 454.2 453.2 453.1 452.9
453.3 454.2 454.1 453.9 453.9 454.1 454.5 453.9
452.4 453.1 452.9 453.7 454.6 455.9 456.1 454.8
451.4 451.6 452.8 454.1 455.2 456.1 455.3 454.8
450.2 451.7 453.1 453.8 454.5 454.8 454.3 453.7

9. Plot 1 ft contours. Indicate index contours.

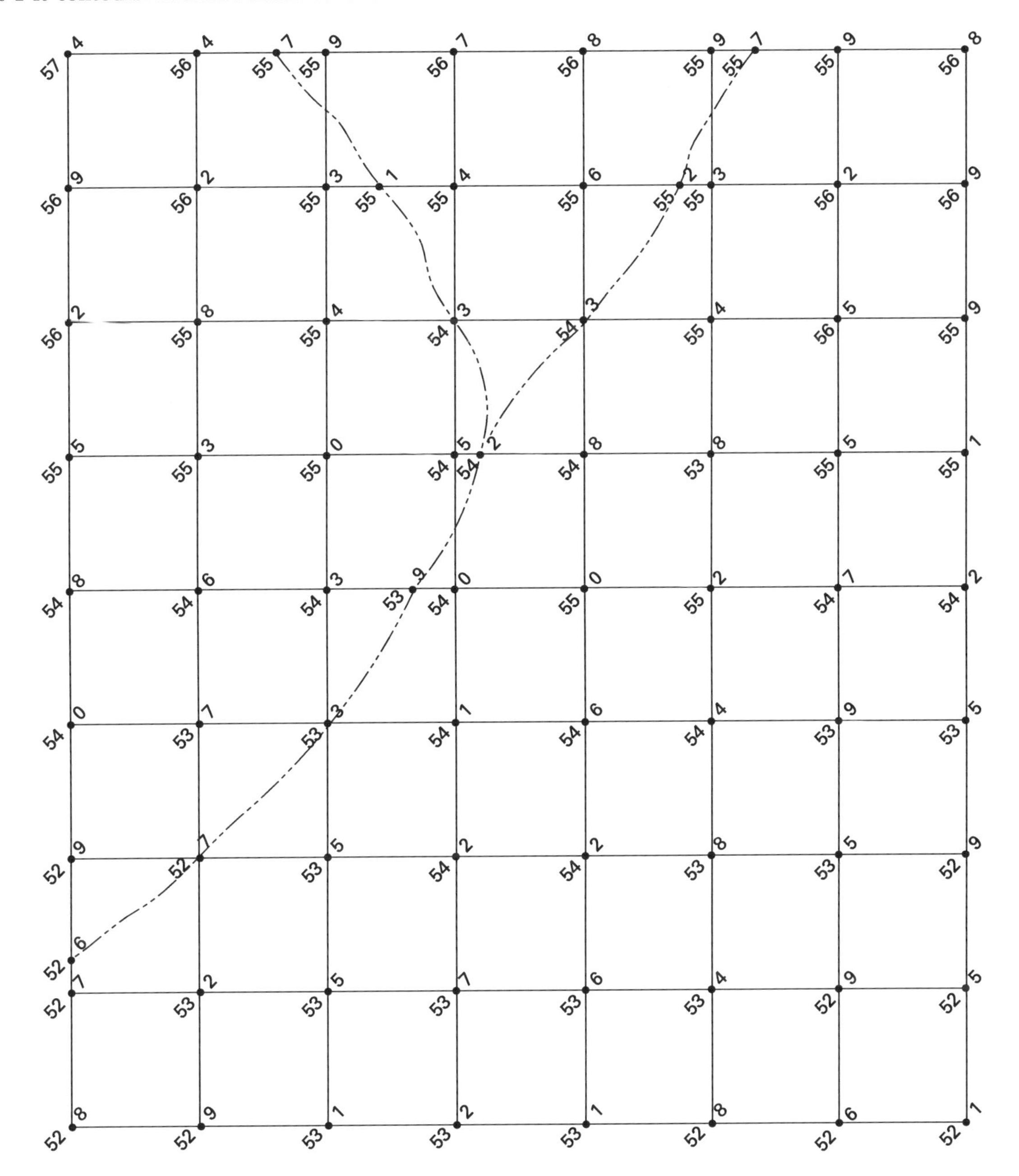

Mapping

10. Plot 5 ft contours.

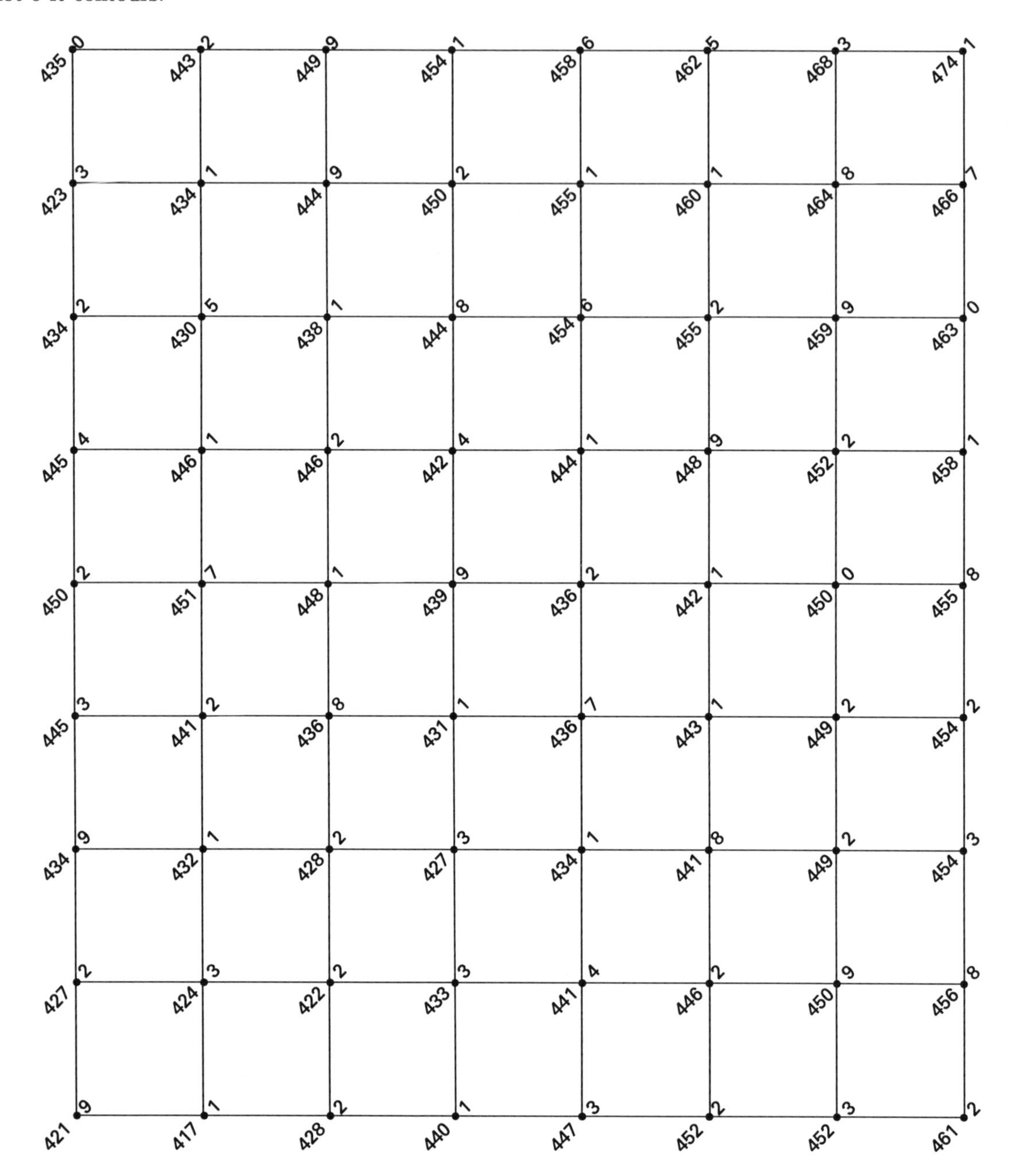

11. Plot 5 ft contours.

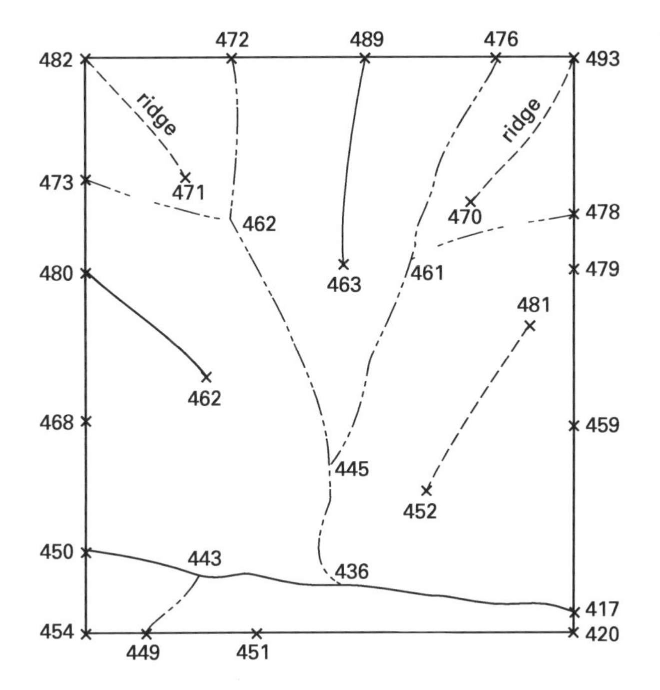

SOLUTIONS

1. Use an engineer's scale to determine distances. (See Fig. 33.17 in the text for map symbols.)

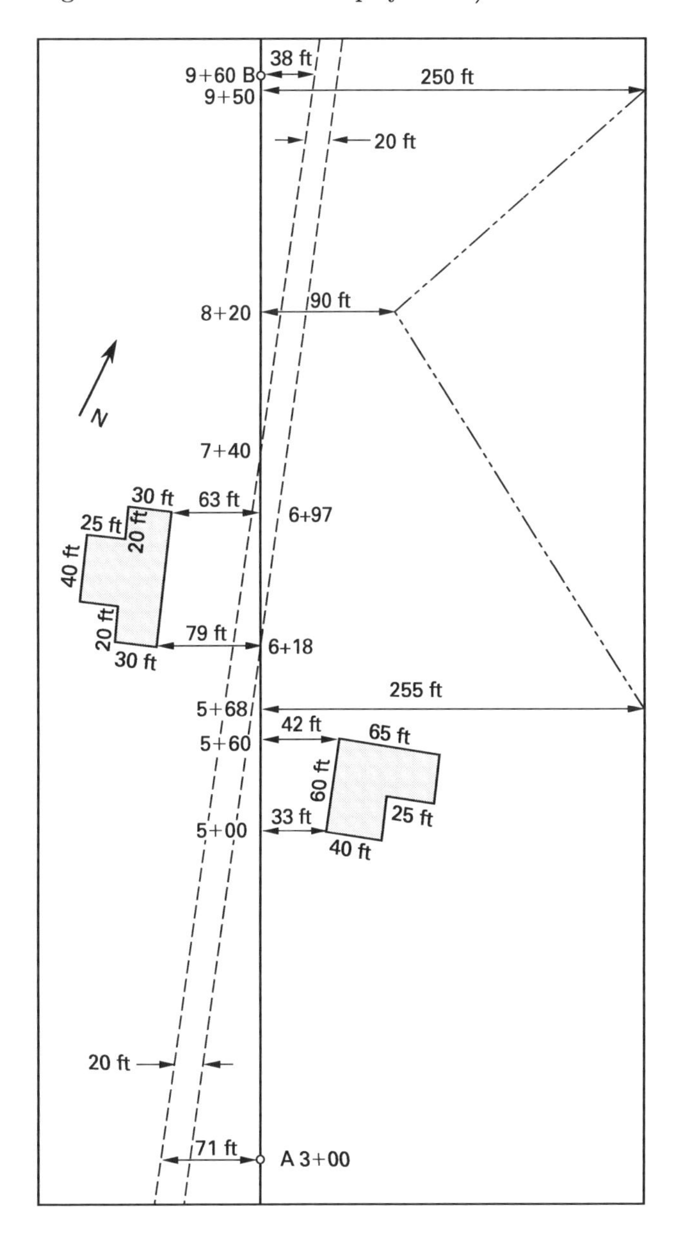

not to scale

2. Measure angles with a protractor; measure distances with an engineer's scale.

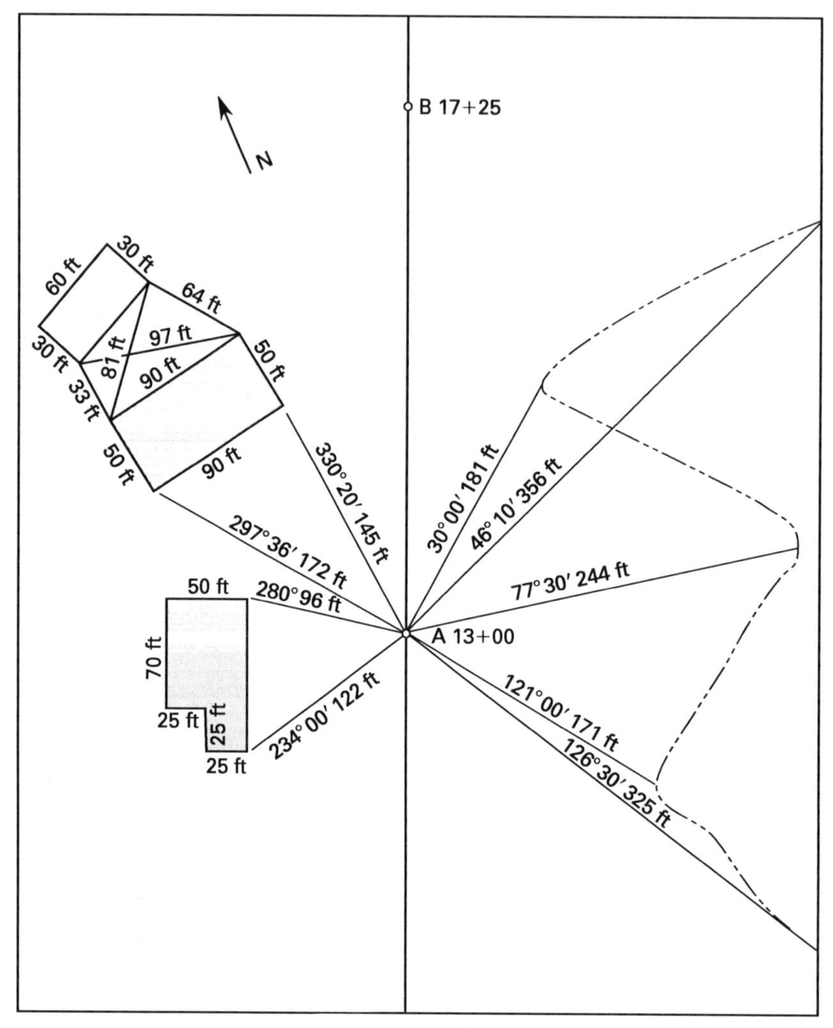

not to scale

3. Use an engineer's scale for distances.

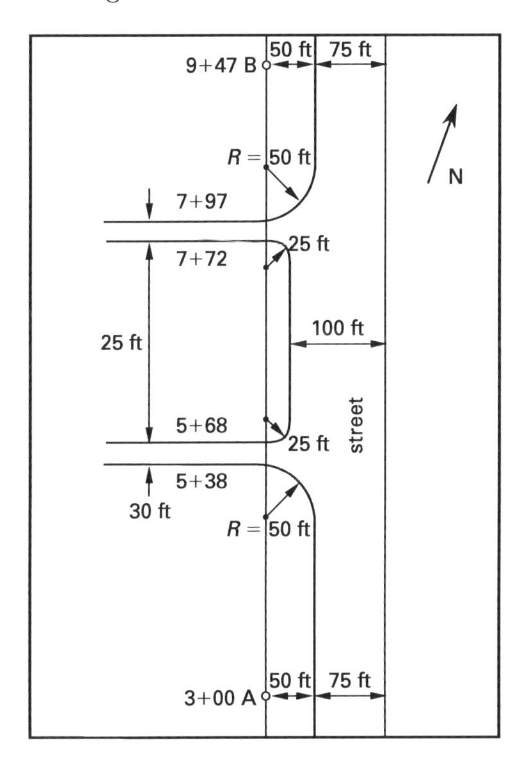

4. (a) $H = (3.48 \text{ ft})(99.51) = \boxed{346 \text{ ft}}$

$V = (3.48 \text{ ft})(6.96) = \boxed{24.2 \text{ ft}}$

(b) $H = (4.68 \text{ ft})(99.20) = \boxed{464 \text{ ft}}$

$V = (4.68 \text{ ft})(8.94) = \boxed{41.8 \text{ ft}}$

(c) $H = (5.75 \text{ ft})(99.08) = \boxed{570 \text{ ft}}$

$V = (5.75 \text{ ft})(9.03) = \boxed{51.9 \text{ ft}}$

(d) $H = (3.91 \text{ ft})(99.46) = \boxed{389 \text{ ft}}$

$V = (3.91 \text{ ft})(7.28) = \boxed{28.5 \text{ ft}}$

5.

transit at	sight on	vernier on
A	B	126°42′
B	A	306°42′
C	D	287°13′
D	E	323°33′
E	D	143°33′

6.

pt	int	*V* ang	*H* dist	elev
1	3.85	+4°05′	383	490.5
2	2.98	+5°08′	296	489.8
3	2.56	−5°09′	254	440.3
4	5.44	−4°03′	541	424.9

7.

azimuth-stadia survey

sta	int	az	*V* ang or rod	*H* dist	elev
	TRAN @ B ELEV 467.2 h.i. 5.0				
A	6.74	148°04′	−4°14′	670	417.6
1	0.91	90°45′	8.1	91	464.1
2	1.66	120°20′	−5°12′	164	452.2
3	2.10	135°15′	−5°07′	208	448.6
4	3.25	143°00′	+4°11′	323	490.8
C	4.00	60°10′	+5°08′	397	502.8
	TRAN @ C ELEV 502.8 h.i. 5.2				
B	4.01	240°10′	−5°08′	398	467.2
5	3.15	286°00′	−4°00′	313	480.9
6	2.21	36°20′	9.2	221	498.8

8.

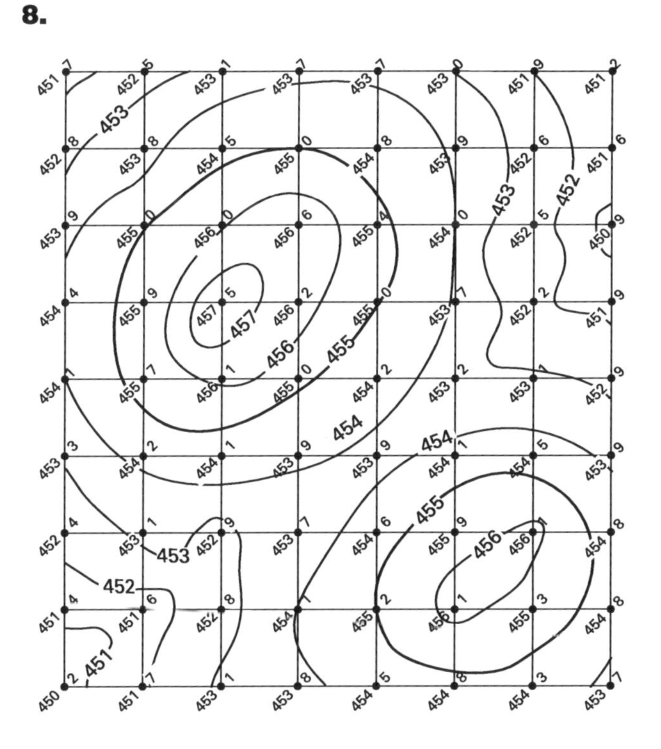

9.

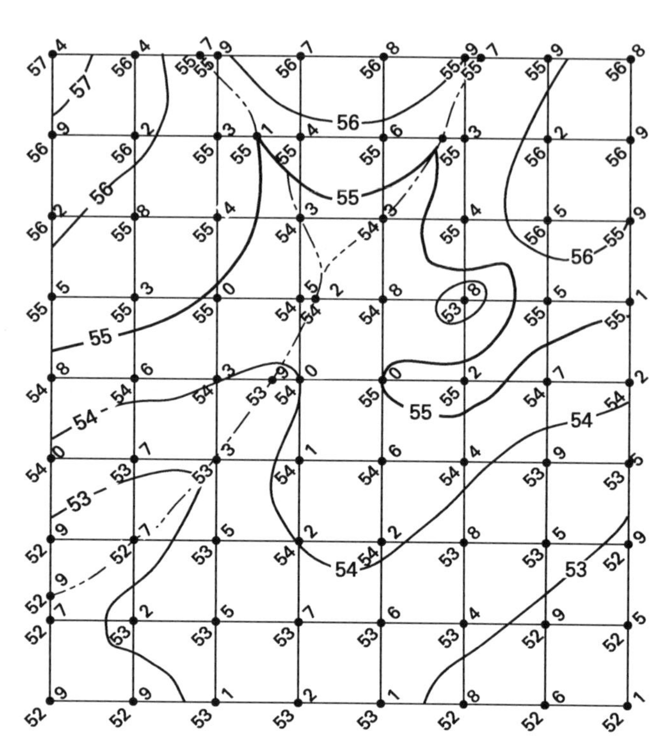
57
56
55
54
53
56
55
56
55
54
55
54
53
54
53

11.

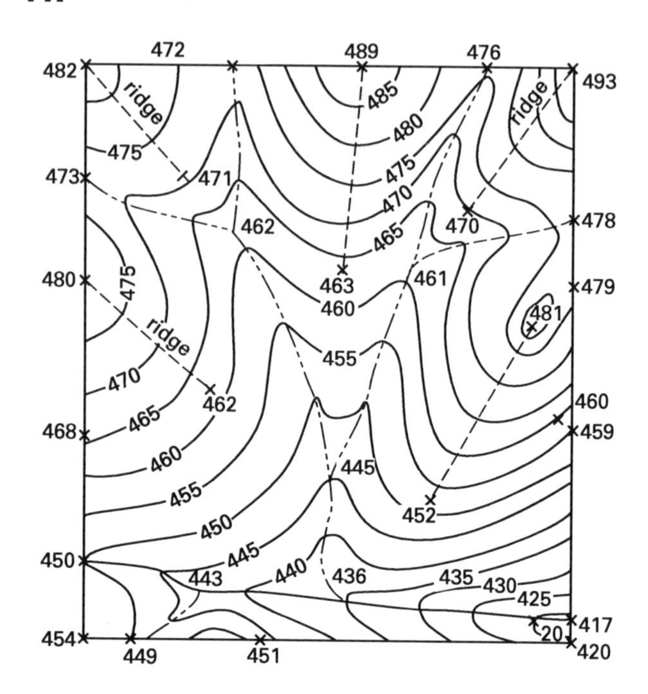
472
489
476
482
493
ridge
485
480
475
473
471
462
470
465
475
470
478
480
463
461
479
481
460
455
ridge
470
462
465
460
468
459
460
445
455
452
450
450
445
440
443
436
435
430
425
417
420
454
449
451

10.

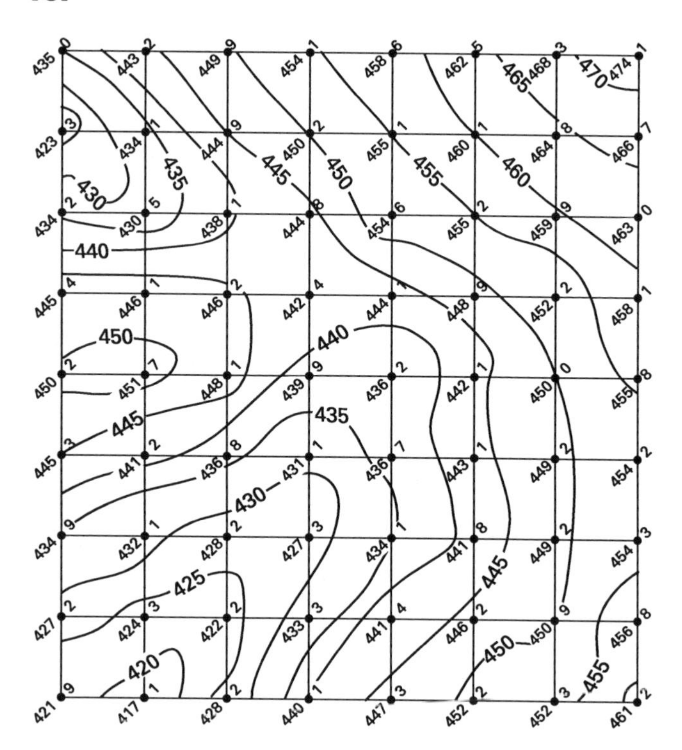
465
470
430
435
445
450
455
460
440
450
440
445
435
430
425
420
445
450
455

34 Geographic Information Systems

1. INTRODUCTION

In the revolution that computers have brought to surveying, geographic information systems (GIS) are one of the most rapidly evolving and most significant developments. The capacity of GIS to collect, store, analyze, and display a great diversity of information about the earth's surface and its occupants allows the systems to be sophisticated models of the real world. With such models, the interrelationships of various layers of information can be examined, and trends observed and quantified. GIS has the potential to provide significant increases in our understanding of human beings and the environment.

Although the roots of GIS lie in the 1960s, it was not until 1988 that the first true GIS was developed by the government of Canada to support the mapping and assessment of its land resources. Today, GIS technology has developed into a large industry and impacts nearly all aspects of our lives. Among its many applications, GIS has become an essential component of property tax assessment. It is widely used in agriculture to allow precise application of fertilizers at the right location and at the right time for maximum efficiency and maximum yields. It is used by almost all utilities to manage the distribution of their products, and it is used by most transportation industries for route planning and fleet management.

GIS is especially important to the surveying profession, since surveyors work almost exclusively with geographic data. Many surveyors today are actively involved in the production of these systems. Even those surveyors who are not should realize that almost all of the survey data they produce will end up in some type of GIS. In the near future, it is likely that all survey data will be presented in a GIS environment. For these reasons, it is important that surveyors have a thorough understanding of this technology.

2. WHAT IS GIS?

Many definitions of GIS have been suggested over the years, although none seems totally adequate to describe the multiple facets of GIS. A definition used by the U.S. Geological Survey states that GIS is "a computer system capable of capturing, storing, analyzing, and displaying geographically referenced information." Another leading definition is "a database in which every object has a precise geographical location, together with software to perform functions of input, management, analysis, and output." Despite the lack of consensus on an all-encompassing definition, there is agreement on several key requirements of a GIS.

1. It must have the capability to integrate data from different sources and at different scales.
2. It must allow the graphic display of multiple layers of geographic information.
3. It must link to databases of various attribute information tied to geographic positions.
4. It must be able to analyze the spatial relationship between various features and respond to queries regarding those relationships.

3. TYPES OF GIS DATA MODELS

The heart of a GIS is the data model that represents objects and processes. Most systems use either a *raster* or *vector* data model, or combinations and variations of those models. A raster data model consists of arrays of grid cells, or pixels, usually referenced by row and column numbers, along with an identifier representing the attribute being mapped. With such a model, points are represented by a single cell; lines are represented by a string of neighboring cells, and areas are represented by collections of neighboring cells. A digitized aerial photograph is an example of a raster data model.

A vector data model assumes a continuous coordinate space, not quantified as with a raster model, and therefore allows positions, lines and area boundaries to be more precisely defined. With such a model, points on a plane are defined by x- and y-coordinate pairs, as opposed to grid cells in a raster model. Line segments on a plane are defined by a pair of x- and y-coordinates for the beginning point and end point, allowing a precise calculation of their length and bearing. Areas are similarly defined by a series of x- and y-coordinate pairs, one pair for each inflection point on the perimeter. A computer-aided design and drafting (CADD) file is an example of a vector data model.

Both types of data models are valid methods for representing spatial data, and there are advantages and disadvantages associated with each. Traditionally, the raster model has been considered more useful for spatial analysis due to the ease of operations such as polygon intersection of features. Disadvantages of the raster model include the relatively large computer memory requirements for obtaining an acceptable level of spatial resolution. The vector model poses technical difficulties in operations such as polygon intersections, but it does allow greater precision. In addition, some types of spatial analysis processes, such as transport network analysis, are only possible with vector models. With the recent development of more efficient computer hardware, improved computational methods, and improved routines for data transformation between model types, distinctions between the two types have become less significant, and some systems utilize both types of data models.

There are other, less commonly used data models, and one of special note is the *triangulated irregular network* (TIN) data model. It is frequently used for three-dimensional data. A TIN is composed of a series of cells as with the raster data model, but unlike raster data, cells in a TIN are irregular in shape. A TIN model can be best visualized as triangles connecting a series of sample points with measured elevations that are irregularly spaced across a given area. The points typically are measured at inflection points in the topography. The structure of a TIN makes it easy to calculate topographic slope and aspect. This strength means TIN models are widely used in applications such as volume calculation for earthwork and drainage studies.

4. DATA CAPTURE FOR GIS

A diversity of geographic data from different sources can be integrated into a GIS. As a result, there is a variety of methods used to capture data for these systems.

One frequently used method of data capture for raster models is *remote sensing*. This process measures properties of objects without direct physical contact. Remote sensing uses passive sensors that rely on reflected solar radiation or emitted terrestrial radiation and active sensors, such as radar, that generate their own sources of electromagnetic radiation. Remote sensing is generally associated with measurements from earth-orbiting satellites, but also includes aircraft-based surveys. Though often considered separately from remote sensing, aerial photography is an example of remote sensing using a passive sensor. *Light detection and ranging technology* (LiDAR) is an example of an active sensor used in remote sensing. All remote sensing processes measure a horizontal position together with one or more attributes associated with that position. The attributes derived by remote sensing may be a spectral signature or, in the case of a system using lidar, may be an elevation.

Land surveying is an example of data capture for a vector data model. Data capture occurs in surveys using the Global Positioning System (GPS), conventional leveling, and total station surveys. GPS is used most frequently for data capture for GIS, since this technology can directly produce geographic coordinates.

In addition to data capture for GIS from primary sources such as remote sensing and land surveying, geographic data may be captured from secondary sources. An example of secondary data capture for raster models are scans of existing maps or photographs. Examples of secondary data capture for vector models are tablet digitizing, stereo photogrammetry, and coordinate geometry (COGO) data entry.

5. SPATIAL ANALYSIS

Most GIS software packages provide capability for various spatial analysis processes. In many ways, that capability represents the greatest value of a GIS. It adds value to the raw geographic data by revealing patterns and anomalies not otherwise visible.

The most basic type of spatial analysis is the *query*, in which the GIS finds answers to simple questions. Typical questions might be "How many people live within 5 mi of a potential hazard?" or "What areas in a county have elevations greater than 200 ft?" Other types of simple spatial analysis include *measurements* of properties such as length, area, shape, and distance or direction between objects; and *transformations* that use simple geometric rules to generate new data sets based on existing data.

An important type of transformation is *spatial interpolation*, such as that used to develop contour lines from a series of irregularly spaced ground elevation points. One process used for spatial interpolation is *inverse distance weighing* (IDW). This process is used to estimate elevations at regularly spaced points on a fine grid covering an area and then systematically thread elevation contours through the grid points. The estimated elevation at each grid point is determined by finding the sum of the elevations at all random data points and multiplying by a parameter based on the inverse of the distance from the grid point to the random point.

Another process that is used for spatial interpolation is *Kriging*. This is more theoretically sound than IDW, but considerably more complex. Kriging interpolates the attribute value at unobserved points across a grid by modeling the attenuating effect of distance and fitting the points to smooth, continuous curves or surfaces described by mathematical functions.

Another example of spatial interpolation related to topographic elevations is the calculation of slope and aspect. This process can be used to predict runoff volume and direction. The process calculates the elevation difference between each cell and the eight surrounding cells. From this information, both the slope and the direction that surface water would flow may be determined.

6. ERRORS IN GIS

Two types of error occur in GIS: mistakes in positional accuracy and mistakes in attribute accuracy. *Positional accuracy* is the accuracy of absolute or relative mapping coordinates for a particular feature. *Attribute accuracy* is the accuracy of the qualitative or quantitative values attached to the feature. Note that elevation, while it could also be thought of as positional data, is normally considered to be an attribute of a horizontal position, and therefore an error in elevation values would be classified in the attribute accuracy category.

The most common sources of inaccurate information in GIS fall into four general categories: errors in original field measurements, data input errors, processing errors, and other errors.

Original field measurement errors: These errors, which may affect both positional and attribute accuracy, are typically caused by weak measurement techniques and judgment or by faulty equipment.

Data input errors: These errors in the digital representation of graphic data are often encountered in the digitizing or geocoding processes, conversions between raster and vector data types, and feature coding and topological matching.

Processing errors: These mistakes are often rounding errors, errors caused by limitation of the computer hardware or software, errors in interpolation due to assumption of linearity for non-linear applications, and errors due to overlay and boundary intersections.

Other errors: Mistakes in this category may be due to changes in the data since they were measured, use of data at scales significantly different from the compilation scale, and insufficient density of data.

7. TYPICAL APPLICATIONS

A study of a simple GIS model will make it easier to understand the processes and capabilities of the technology. This example will demonstrate the basic concepts and some potential applications using data familiar to the surveyor.

The basic information for the GIS was established by a county property appraisal office in Jefferson County, Fla.[1] The base map layer, or shape file, illustrated in Fig. 34.1, is a land parcel map of the county. The base map was created in two steps: a survey location of many of the controlling public land survey section corners, followed by the use of COGO and CADD routines to create a map from recorded subdivision plats and deeds of record. Aerial photography for the county is available as another layer of information, shown in Fig. 34.2.

Figure 34.1 *GIS Base Map*

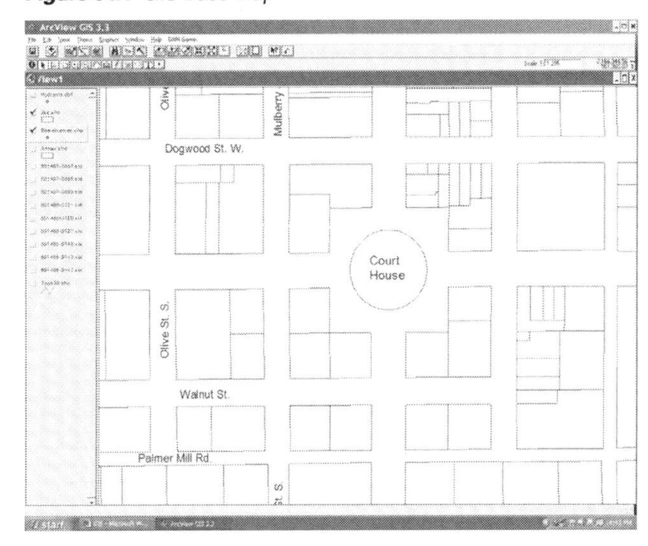

Source: Jefferson County (Florida) Property Appraiser

Figure 34.2 *GIS Aerial Photo*

Source: Florida Department of Transportation

Each parcel on the base map is linked by parcel number to a database of information. Information for each parcel includes the name and mailing address of each owner of record, area of the parcel, appraised value, tax district, the location of the deed in the public records, and a number of additional items associated with the appraisal process. A small portion of the database is shown in Table 34.1.

The basic information in this GIS allows a complete overview of land ownership and land value in the county. Such a GIS is obviously of value to land surveyors, in addition to tax officials and real estate professionals. All parcels can be readily viewed on the parcel map as well as in the aerial photo, and information for each parcel is available with a click of a mouse, as shown in Fig. 34.3.

[1]The illustrated base map and database was developed by the Jefferson County (Florida) Property Appraiser's Office under the direction of County Property Appraiser David Ward.

Table 34.1 Portion of GIS Database

area (ft^2)	parcel identification	just value	assessed value
20,828,481.31320	01-3N-3E-0000-0010-0000	$1,581,000	$167,183
19,104,179.05649	01-3N-3E-0000-0010-0000	$1,581,000	$167,183
329,532.29695	01-3N-3E-0000-0020-0000	$38,961	$38,961
898,197.80266	01-3N-3E-0000-0030-0000	$18,000	$1,404
3,700,148.56154	01-3N-4E-0000-0060-0000	$175,862	$39,111
29,880,471.89741	01-3N-4E-0000-0010-0000	$1,532,444	$278,222
367,263.93574	01-3N-4E-0000-0040-0000	$10,600	$10,600
502,491.70116	01-3N-4E-0000-0190-0000	$22,922	$22,922
204,572.19697	01-3N-4E-0000-0030-0000	$27,220	$2922

Source: Jefferson County (Florida) Property Appraiser

Figure 34.3 Individual Parcel Information

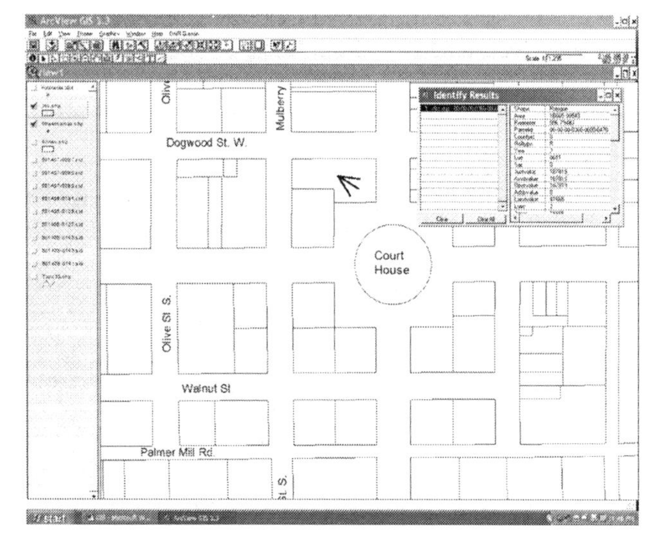

Source: Jefferson County (Florida) Property Appraiser

In addition, the GIS allows queries to be made regarding information in the system. In response to a query, the relevant parcel is highlighted on the map as well as in the database. Some examples of possible queries follow.

- Where is a tract of land with a given parcel number?
- What parcels are owned by a given individual?
- Which parcels are appraised at a value greater than a given figure?

The system can serve many other functions in a typical community. The system can serve as a base map for land regulation and zoning. With the addition of a street address field to the parcel database, the system can serve as a base map for emergency response. An emergency dispatcher or an emergency responder can immediately locate an emergency with a query for a street address. On the photo layer, the dispatcher can view structures, terrain, and access points for the property and determine at a glance if there are nearby locations for landing emergency helicopters. The addition to the GIS database of a file of phone numbers tied to street addresses allows the system to be used with 911 emergency response systems that capture the phone numbers of emergency calls.

With the addition of a simple database of surveyed coordinates for fire hydrants, such as those shown in Table 34.2, the location of the closest fire hydrant can be viewed as well as information regarding that hydrant. This would provide guidance for dispatching fire emergency units. A typical view of how such features may be graphically displayed is provided in Fig. 34.4, with the hydrant locations represented by targets and identification codes. Note that, as depicted in Table 34.2, a database can be as basic as a file of x- and y-coordinates along with a feature identifier. Such files may be easily prepared in commonly used spreadsheet programs. Conversely, a database may be much more complex and include many additional attributes associated with the position.

Table 34.2 Fire Hydrant Coordinates

northing	easting	hydrant no.
562,516	2,166,635	A1
563,077	2,166,516	A10
563,068	2,165,398	A11
563,504	2,166,524	A13
563,761	2,166,207	A14
563,751	2,165,500	A15
564,265	2,165,692	A17

Figure 34.4 Map of Fire Hydrants

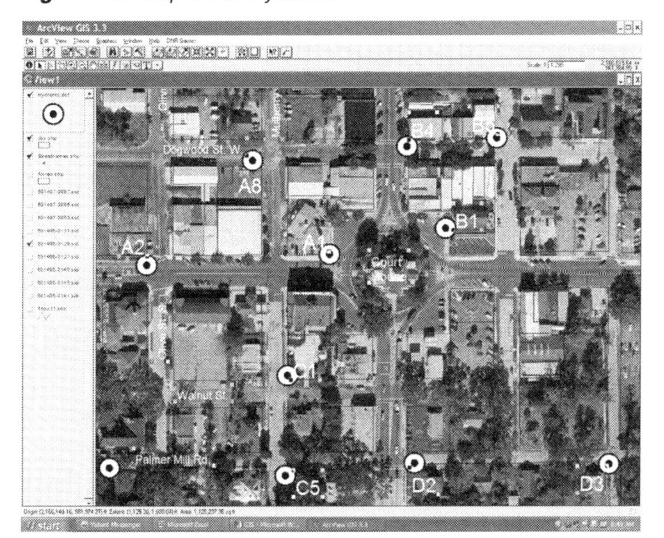

Source: Florida Department of Transportation

Law enforcement agencies can also use the system as a graphic database of crime. This can be accomplished by adding a database containing geographic coordinates for crime scenes measured with an inexpensive hand-held GPS receiver. Such a system provides a view of geographic trends in crime, as well as a simple means of retrieving information regarding certain crimes.

Local road departments can use the GIS for creating a graphic inventory of signs or drainage structures. Those features could be located using a hand-held GPS receiver. The addition of public domain shape files and associated databases of soil types, topographic elevations (see Fig. 34.5), and land use could be added to the system to make it invaluable to agricultural, engineering, construction, and land development interests. Examples of typical queries that could be addressed with these additions follow.

- Where are tracts of land of a given size with soil suitable for a certain crop?
- Where are wetland soils located?
- Where are tracts of land with high elevations suitable for cell tower location?
- Where are vacant tracts of a given size with soil suitable for a certain type of construction?

These example queries represent relatively simple applications. With the addition of census tract data, environmental data, transportation statistics, and similar information, the potential applications for such a community-based GIS are almost endless. GIS offers a means of using survey technology to create models of the earth that can be used in almost all aspects of society.

Figure 34.5 Topographic Map

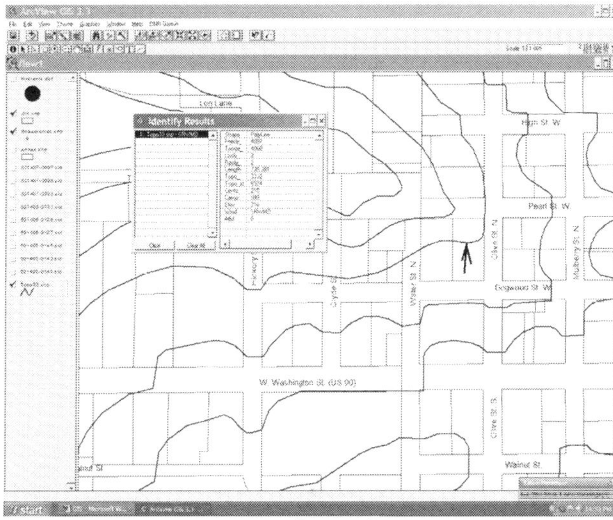

Map source: Jefferson County (Florida) Property Appraiser
Contour source: U.S. Geological Survey

PRACTICE PROBLEMS

1. All of the assessor plats for a municipality are scanned into a file for a GIS. What would be the probable structure for the resulting file?

(A) vector data structure
(B) metadata structure
(C) raster data structure
(D) none of the above

2. A CADD file is an example of which of the following?

(A) vector data structure
(B) metadata structure
(C) raster data structure
(D) none of the above

3. Inverse distance weighing and Kriging are examples of which of the following?

(A) geographic data models
(B) database linkage methods
(C) processes for visualization of geographic data
(D) spatial interpolation

4. A TIN is often used for which of the following?

(A) photographic data
(B) census data
(C) topographic data
(D) land use data

5. An error in elevation is an example of which of the following?

(A) positional error
(B) attribute error
(C) qualitative value error
(D) none of the above

SOLUTIONS

1. Scanning typically creates a grid or raster structure file.

The answer is (C).

2. A CADD file is an example of a vector data structure.

The answer is (A).

3. Inverse distance weighing and Kriging are examples of spatial interpolation.

The answer is (D).

4. A TIN, or triangulated irregular network, is often used for topographic data.

The answer is (C).

5. An error in elevation is an example of an attribute error. While elevation is also representative of position, it is more typically considered an attribute of a horizontal position.

The answer is (B).

35 Aerial Mapping

Nomenclature[1]

d	photo displacement	in	mm
D	ground displacement or length of ground shadow	ft	m
f	focal length	in	mm
h	object height	ft	m
H	mean flying height AGL (above ground level)	ft	m
R	radial distance from isocenter	ft	m
S	photo scale	ft/ft	m/m
t	tilt angle	deg	deg
y	separation of image point and isocenter	in	mm

Subscripts

t	tilt
p	principal

1. INTRODUCTION TO AERIAL MAPPING

Aerial mapping is the process of creating maps from measurements made with photography or other types of remote sensing from an airborne platform. While it is usually considered a recent surveying and mapping innovation, aerial mapping has been used for almost two centuries. It is widely used for topographic surveys since it is generally more economical than ground surveys except for relatively small projects.

The first aerial mapping photographs were taken from hot-air balloons in the late nineteenth century. Kites were also used as platforms for aerial photography during this period. For example, after San Francisco's destruction in the 1906 earthquake, the city was mapped with a camera suspended from seven kites.

[1]Generally, uppercase letters are used to represent locations and distance on the earth, while lowercase letters are used to represent points and distances on the photograph or within the camera.

Fortunately today, more controllable aerial platforms are available, including both aircraft and satellites. In recent years, *unmanned aerial vehicles* (UAVs) have been increasingly used for all modes of airborne remote sensing, although regulation of their use is still evolving. The versatility and economy of UAVs make them an ideal aerial platform for modern miniaturized equipment.

Data acquisition has also evolved, advancing from relatively crude cameras to photographic systems with advanced optics as well as digital cameras, airborne LiDAR (see Sec. 9), and other remote sensing systems. Although some aerial mapping still uses traditional photographic film technology for data acquisition, the use of digital cameras and airborne LiDAR sensors is becoming more and more prevalent. The processing of aerial mapping data today, whether acquired on film or as digital data, is generally performed using digital methods. Typically, when film is used, it is scanned to allow digital processing. This trend to digital processing avoids the costly optical-mechanical components previously associated with aerial mapping and allows maximum use of modern computers. In addition, having both the original data and resulting map data in digital format allows more efficient use of GIS and other computer applications.

2. PHOTOGRAPHIC OPTICS

Photographic optics are based on the *refraction of light*, an optical effect used since the Middle Ages to project images. When light rays pass from one medium into another, they change speed and bend, or refract. Light rays will bend toward the normal if the speed of light is less in the second medium (such as when passing from air into a camera's glass lens). This refraction produces a reduced-scale copy of the image as it passes through the lens, which can be projected onto a plane at a focal length dependant on the lens' characteristics.

3. AERIAL CAMERAS

Aerial cameras are normally mounted on a stabilized platform using gimbals, or pivoted supports, to maintain a vertical orientation. The stabilized platform is attached to the aircraft by isolators. Most modern aerial film-type cameras are composed of three main parts: the lens cone, the camera body, and the film magazine. (See Fig. 35.1.)

Figure 35.1 *Typical Configuration of a Film-Type Aerial Camera*

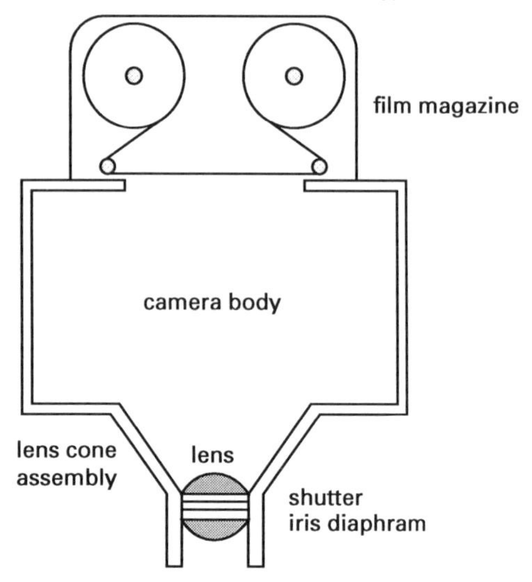

The lens cone assembly (*lens cone*) is the camera's most critical component. Cameras used specifically for aerial mapping have lens cones with alignment tabs embedded around the perimeter. These cause fiducial marks to be imprinted onto each image, which are used to determine the principal (or center) point of the photograph. The lens cone is also important because it restricts the light striking the focal plane so that it passes through the camera lens. The lens itself typically consists of two parts, one on either side of the shutter. The *iris diaphragm* is used to regulate the amount of light admitted to the lens.

Most aerial cameras also have a "leaf-type" shutter located between the two lenses that opens during each exposure. The focal length of the camera, as established by the lens, varies with different camera models. A 6 in (155 mm) focal length lens is the most common, although lenses with 3.5 in (88 mm), 8.25 in (210 mm), and 12 in (305 mm) focal lengths are commonly available.

The *camera body* contains the drive motor and mechanisms that operate the shutter, advance the film, and perform other camera operations. The camera body also typically contains a recording chamber that prints pertinent information on each exposure such as the photo number, date, time, altitude, level bubble, and so on.

The *film magazine* holds the unexposed and exposed film. Most aerial cameras use a detachable magazine that can be changed or removed as needed. Typically, the magazine contains a vacuum system that flattens the film against the focal plane. Many cameras are also equipped with a *forward motion compensation* system that, during the exposure, moves the film backward at the same rate as the aircraft's forward motion to prevent blurring caused by the aircrafts movements.

All aerial mapping camera systems need to be periodically calibrated in order to precisely determine the processing parameters of the camera's photographs. In the United States, the U.S. Geological Survey provides camera calibration services.

Film-Type Cameras

Most film-type aerial photography is taken on film with a 9 in by 9 in (23 cm by 23 cm) square format. 5 in by 5 in (13 cm by 13 cm) is also used. The light-sensitive emulsion is coated on a flexible transparent polyester film base. Then the emulsion is coated with a thin protective layer of clear gelatin to shield the emulsion from scratches.

There are four general types of film emulsions used in film-type aerial photography. These include panchromatic black and white, natural color, black and white infrared, and false color infrared. *Panchromatic black and white film* detects a wide spectrum of light and is widely used for aerial mapping and interpretation. The emulsion on *natural color film* is composed of three primary color layers to cover the full color spectrum. *Black and white infrared film* is sensitive to longer light wavelengths and is capable of penetrating haze. It is used to delineate water bodies, as well in for military and intelligence photographs because it can detect camouflage. With *false color infrared film*, different colors are arbitrarily used to represent different portions of the light spectrum. It is often used to detect agricultural crop diseases and to monitor pollution.

Digital Aerial Cameras

Digital aerial cameras replace the focal plane's film with an array of light-sensitive detectors that record images in digital form. Data can be directly transferred to, and used in, digital processing systems. Some digital cameras have the ability to capture black and white, natural color, and infrared data simultaneously to significantly increase efficiency. Aerial digital imaging cameras generally use one of two technologies to capture data: frame imaging or "push-broom sensing." *Frame imaging* uses a shutter device similar to that used with film-type cameras that allows light to strike sensors at a predetermined interval and duration to achieve the desired overlap. *Push-broom sensing* uses three sets of sensors orientated forward, down, and back. These are generally fixed to provide a 67% overlap so that one third of the image is viewed simultaneously by all three sensors. Ground sample dimensions are defined by the flying velocity and altitude, and instead of using shutters, a continuous "carpet," or rows, of pixels are captured by each set of sensors. The rows are then combined to form a long image of the aircraft's entire flight path.

Image Centers

Three "center" points are used when correcting aerial photographs for tilt and displacement. These three points coincide in untilted photographs, but they diverge in tilted photographs. The *principal point* is the geometric center of the photograph. It is located at the intersection of lines connecting the fiducial side (or, in some cases, corner) marks. The *nadir point* (also known as the *vertical point* and *plumb point*), N, is directly below the camera. There are no indications on the photograph where the nadir point is located, although it can be located optically by extending lines from perfectly vertical objects in the photograph toward a common intersection point. The nadir is always located on the downward side of a tilted photograph. Relief displacement is radial from the nadir point and is proportional to the distance from it. The *isocenter*, I, is located on a line connecting the nadir and principal points, midway between them.

4. PLANNING FOR DATA ACQUISITION

To obtain good data, it is important to consider exactly when and where aerial photography will take place. When mapping bare-earth elevations in temperate zones with deciduous trees, aerial photography should be scheduled during the winter months when there is better visibility because trees have fewer leaves, and there are smoother flying conditions and clearer air because cooler, dryer air prevails. In tropical zones, aerial photography is typically scheduled during the dry season, generally January through March in the northern hemisphere, so that photographs are not obstructed by rain.

In addition to the time of year, time of day must also be considered for aerial photography. A relatively high sun angle minimizes large shadows, though small shadows help delineate detail and generally increase the photograph's quality. A 45° sun angle is generally considered most desirable. Solar angular altitude diagrams and several computer programs are available to assist in selecting the most appropriate schedule for mapping a given location. Time of day is not a concern with active remote sensors such as LiDAR since sunlight and shadows are not factors with those systems.

It is also necessary to plan the route of the mapping aircraft. Aerial photography is generally taken along a series of preplanned flight lines where successive photographs are exposed such that a predetermined amount of overlap occurs. Generally, 60% forward overlap and 20–40% (30% nominal average) side lap is used.

5. SCALE

The *scale* of a photograph is typically expressed as the ratio of the dimension of the image of an object on the photograph to the dimension of that object on the ground. As an example, the scale of 1:24,000 indicates that an object measuring one inch in length on a photograph would measure 24,000 inches or 2000 feet on the ground. If the focal length, f, of the camera and the flying height, H, is known, the scale of a photograph may be calculated as the ratio of those parameters.

$$S = \frac{f}{H} \qquad 35.1$$

When the desired scale of the photography and the focal length of the camera are known, the required *flying height* can be calculated by rearranging Eq. 35.1.

$$H = \frac{f}{S} \qquad 35.2$$

Example 35.1

What is the required flying height necessary to produce a 1:24,000 negative scale if a land surveyor is using a camera with a 6 in focal length?

Solution

Using Eq. 35.2,

$$H = \frac{f}{S} = \frac{(6\text{ in})\left(\frac{1\text{ ft}}{12\text{ in}}\right)}{\frac{1}{24{,}000}} = 12{,}000\text{ ft}$$

6. TILT AND RELIEF DISPLACEMENT

When photogrammetric measurements are made from an aerial photograph, not all images will have the correct spatial relationship to each other. Tilt in the photograph, as well as variation in terrain elevation, will cause image displacement as well as scale variation. Figure 35.2 shows *tilt displacement*, d_t, of a point A, which causes points downslope of the isocenter to be displaced radially away from the isocenter, while points upslope would be displaced towards the isocenter. The tilt displacement for a specific point is calculated as

$$d_t = \frac{y^2}{\left(\frac{f}{\sin t}\right) - y} \qquad 35.3$$

t is the tilt angle, f is camera lens focal length, and y is the distance between the image point and the isocenter. The *principal line* is the line of intersection between the plane in which the tilt angle is measured and the plane of the tilted photograph.

Figure 35.2 *Image Displacement due to Tilt in Aerial Photograph*

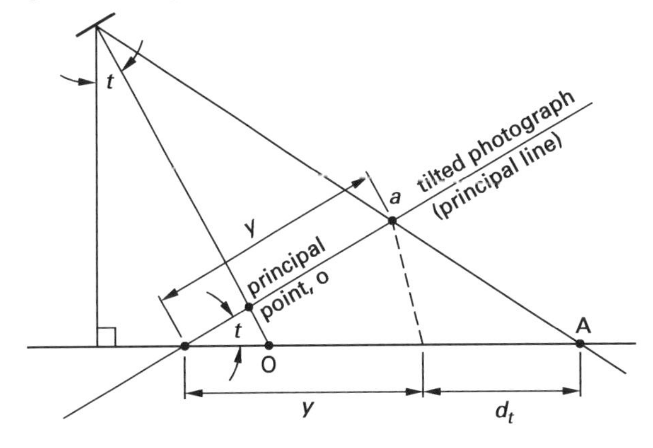

Mapping

In Fig. 35.3, the principal point, O, is the only point in a truly vertical photograph where there is no *relief displacement*. Relief displacement is especially visible when tall buildings or towers are visible in aerial photographs. When viewing those images, the radial displacement from the photo center of the tops may be readily seen. Relief displacement, d, is calculated as

$$d = \frac{Df}{H} \quad \text{35.4}$$

Similarly, Eq. 35.4 can be used to calculate the actual height, D, of a vertical object from its measured (on the photograph) distance (displacement) from its top and bottom, d. f is the focal length of the camera, H is the elevation of the camera, and D is the length of the ground image's shadow. D can be calculated by

$$D = \frac{hR}{H - h} \quad \text{35.5}$$

H is the camera elevation above mean terrain, h is the object's ground elevation above mean terrain, and R is the radial distance on the ground from the principal point to the object.

Figure 35.3 *Image Displacement due to Relief*

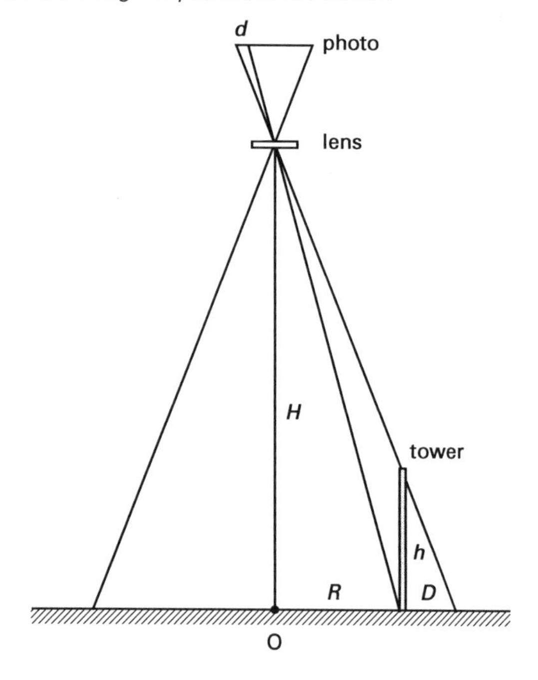

Example 35.2

A tower's height is 300 ft above mean terrain. It has a 1000 ft radial distance from the center of a vertical aerial photograph. The photograph was taken 2000 ft above mean terrain using a camera with a 6 in focal length. What is the relief displacement for the top of the tower?

Solution

Use Eq. 35.5 to calculate the length of the tower's shadow on the ground.

$$\begin{aligned} D &= \frac{hR}{H - h} \\ &= \frac{(300\ \text{ft})(1000\ \text{ft})}{2000\ \text{ft} - 300\ \text{ft}} \\ &= 176.47\ \text{ft} \end{aligned}$$

Use Eq. 35.4 to find the relief displacement.

$$\begin{aligned} d &= \frac{Df}{H} \\ &= \frac{(176.47\ \text{ft})(6\ \text{in})\left(\frac{1\ \text{ft}}{12\ \text{in}}\right)}{2000\ \text{ft}} \\ &= 0.0441\ \text{ft} \end{aligned}$$

7. MAP COMPILATION

Photogrammetric mapping is traditionally performed using a *stereoplotter*, which is an optical-mechanical instrument that reconstructs the depicted terrain's spatial geometry in overlapping aerial photographs. It allows the measurement of a terrain's planimetric and topographic attributes by correcting for tilt and relief displacement.

Stereoplotters use the concept of *stereoscopy*, which is the ability to perceive a photographic image in three dimensions. The process involves using overlapping photographs that depict an area as viewed from two different locations in the air. As each eye sees a different photograph (binocular viewing), the two images merge into a central image that has depth. This is because the closer a point is to the eye, the larger the convergence angle is between the eyes and the point. Thus, when looking at the same feature on two photographs taken from two aerial locations, the highest point would have a larger convergence angle and appear taller.

In the past, the most commonly used steroplotter was the *direct-projection plotter*. This plotter uses diapositive photographs (i.e., prints on glass or mylar) the same size as the original photographs to project enlarged images of the two overlapping photographs onto a tracing table. The two projected images are distinguished from each other by projecting one image through a red filter and the other image through a blue-green filter. The operator then views the images using a pair of glasses with separate red and blue-green lenses, thus allowing each image to be viewed by one eye. The direct-projection plotter has been superseded by the analytical plotter and softcopy processing and is rarely used today. The *analytical stereoplotter* uses a computer to mathematically align images so they line up correctly. It also stores data on a hard drive and can redraw the data to any scale.

Stereoplotters are typically equipped with a "floating-mark" that can be lowered or raised within the three-dimensional view of the model to provide an elevation readout. As part of the orientation process, the floating-mark and its readout are referenced to ground control points in order to map the elevation contours or obtain spot elevations.

The stereoplotter is rapidly being superseded by computerized photogrammetric processes. These processes are often called *softcopy photogrammetry* due to the physical (hard copy) form of the photograph being replaced with a digital version. Soft copy processes use specialized software to digitally manipulate the spatial geometry of the view depicted in the overlapping photographs. Since all manipulation is performed digitally, the elaborate optical-mechanical components of the stereoplotter are eliminated. Thus, soft copy processes have far less equipment requirements. Further, such processes offer the advantages of speed and accuracy with the resulting data more easily integrated into other mapping processes. Softcopy systems also allow the aero-triangulation of data in addition to compilation and editing. In addition, these programs allow the operator to superimpose the resulting topographic map over aerial photography and to view the map in three dimensions.

As an alternative to photogrammetric map compilation, stereo-pairs of aerial photographs may be used to create a corrected form of photography called an *orthophoto*. An orthophoto shows photographic detail without errors caused by tilt, relief displacement, or scale variation. It can be used to measure true distances since they show an image's correct planimetric position and have uniform scale, and are often used with geographic information systems (GIS). They also display all visible features, whereas conventional line maps only show the digitized features. Orthophotographs are generally less expensive than line maps, which involve extensive manual digitizing.

8. STEREO MODEL ORIENTATION

Before the mapped terrain's elements (e.g., roads, sidewalks, buildings, vegetation) are digitized and compiled, a pair of overlapping photographs must be correctly orientated relative to each other, to the ground coordinate system, and to the camera coordinate system. There are three orientations—inner, relative, and absolute. After these orientations are completed, map elements can be digitized and transformed in real-time to the ground coordinate system.

Inner Orientation

The camera's coordinate system has its origin at the photograph center, and with the x- and y-axes parallel to the sides of the photo. The camera also has fiducial marks located at the four corners of the exposure area and at the middle of each of the four sides to total eight points. These fiducial marks must be digitized, and a linear transformation computed to transform the digitized map elements into the camera's coordinate system.

Relative Orientation

The aerial photograph is similar to a bundle of lines passing through the lens's projection center and through each point appearing on the photo. Six points are selected that appear on both overlapping photos to generate a six-point bundle for each photo. One photo is moved and rotated in 3D until all pairs of corresponding lines intersect. When all lines intersect, a perfect stereo model is formed and can be viewed through the stereoplotter microscope.

Absolute Orientation

The final (absolute) orientation is the translation rotation and scaling of the stereo model into the ground coordinate system. Surveyed ground points must be identified in the model and have x-, y-, and z-coordinates in the mapping ground coordinate system. These points are usually marked on the ground by a white cross that is clearly visible in the photographic model. The absolute orientation involves a seven-parameter transformation containing three rotations, three translations, and a scale factor.

9. LiDAR MAPPING

Airborne *LiDAR* (Light Detection and Ranging) is a relatively new aerial mapping technology. A form of remote sensing, the LiDAR system is mounted on an aircraft and, during the flight, transmits high frequency laser pulses toward the earth. The LiDAR sensor records the time difference between the pulse's initial transmission and the reflected laser pulse's return to the aircraft. The laser pulse is projected by a mirror that rapidly rotates from side-to-side along the flight line.

Airborne LiDAR systems include a geodetic grade airborne GPS receiver that measures the aircraft's position every second and an *inertial measurement unit* (IMU) that determines the aircraft's orientation (e.g., pitch, yaw, and roll) approximately 200 times a second. The integration of the data from these components allows the calculation of precise horizontal and vertical coordinates for each point. Airborne LiDAR typically provides vertical accuracies better than 15 cm RMSE[1] and horizontal accuracies better than 50 cm RSME.

While the LiDAR laser beam is only approximately one micron wide at its source, it can expand wider than two feet during its transmission to the ground. Therefore,

[1]RMSE, or *root-mean-square error*, is a commonly used measure of precision and is calculated as the square root of the mean of the errors.

Mapping

the beam can hit several surfaces, such as tree branches, during the descent, and as a result, several different reflections may be recorded from a single pulse.

Some modern LiDAR systems are capable of operating at up to 100 kHz (100,000 pulses per second), resulting in millions of points being defined by three-dimensional geographical coordinates. The number of points is increased because there are often multiple returns for a particular pulse. (See Fig. 35.4.) Thus, a large "point cloud" of data is produced by LiDAR systems, which can be problematic because it is not always obvious if a particular point was reflected off the ground or off some other above-ground feature. Therefore, creating layers by correctly classifying the points is a significant challenge in using LiDAR data. Various spatial analysis processes have been developed to help classify data points into layers, such as bare earth, vegetation, and structures, based on the trends of the points. Nevertheless, a certain amount of human editing and ground truthing remains necessary to map a bare-earth surface beneath vegetation. When mapping ground beneath vegetation, there may always be some uncertainty. For example, where there are certain sharp ground features hidden by tree cover, these may be misinterpreted as above ground features.

Figure 35.4 *Multiple Returns from a Single LiDAR Pulse*

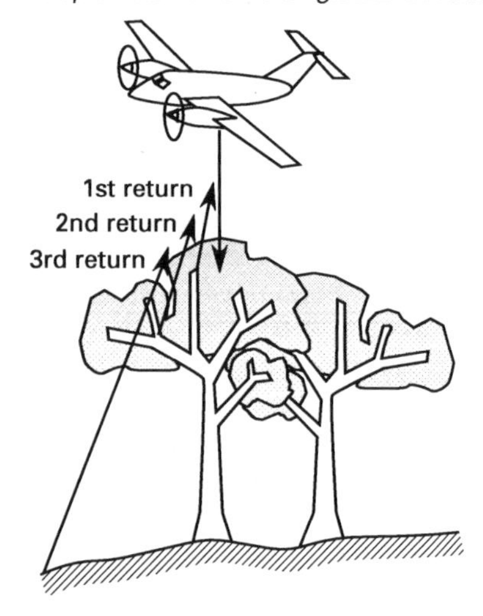

Unlike aerial photography, which is passive, LiDAR involves an "active" sensor that can collect data during the day, night, or even under cloud cover. LiDAR missions are often flown at night for better weather and air conditions, and lighter air traffic. However, while LiDAR data acquisition is less weather dependent than aerial photography, weather conditions must still be considered.

The most common use for LiDAR data is *bare-earth digital terrain models* (DTM), although it is also used to map various planimetric features. (See Fig. 35.5.)

Because LiDAR pulses detect above-ground features, prominent vegetation may also be mapped for quantitative analyses (e.g., tree counts, vegetation heights), or for volume analyses (e.g., to monitor agricultural crop yields). LiDAR systems also record an *intensity factor* (the amplitude of the returned wave) as an attribute of each return. The intensity factor can be used to delineate water bodies, as well as various cultural features, and to help distinguish between various classes of data points.

Hydrographic mapping, the mapping of underwater terrain, is a more recent and still-evolving application of airborne LiDAR. Hydrographic mapping is more expensive than topographic mapping and is limited to shallower, less turbid waters. Nevertheless, the use of LiDAR for this purpose can reduce operational costs and increase productivity considerably in comparison to sounding with hydrographic vessels. Results suggest that a precision level of about 0.2 ft may be achieved. Figure 35.6 shows a typical cross section of a spring basin that has been mapped through the use of airborne LiDAR.

Mapping submerged terrain requires a LiDAR system designed for *bathymetry*, the measurement of underwater depth. Most commercial LiDAR systems designed for land applications use a single laser with a wavelength between 800 nm and 1550 nm, and the pulses from such a system tend to be reflected from the water surface.[2] In contrast, a LiDAR system designed for bathymetry uses two lasers, one in the infrared spectrum with a typical wavelength of 1064 nm and another in the green spectrum with a typical wavelength of 532 nm. The infrared laser pulse is reflected by the water's surface, while the green laser pulse penetrates the water and is reflected by the submerged land surface. In this way the water's surface and depth can be measured at the same time.

Not only are two lasers needed instead of one, but the green laser pulse is weakened by absorption and scattering as it passes through the water, so a bathymetric LiDAR system needs considerably more energy than a system designed for land applications alone. However, the infrared LiDAR also works in the topographic mode, so both the submerged land and the bordering upland topography may be surveyed using the same system.

One important consideration with bathymetric LiDAR systems is *refraction*. Light travels more slowly in water than it does in air, and this causes light waves to change direction as they enter water (see Fig. 35.7). In calculating a bathymetric measurement, then, it is necessary to account for the angle of refraction by applying a correction to the azimuth from the aircraft to the submerged ground point.[3] In addition, a correction to the range is needed to account for the slower speed of light along the portion of its path that is in water.

[2]Nanometers, abbreviated nm, are billionths of a meter.

[3]The angle of refraction can be calculated with *Snell's law*, which states that the ratio of the sines of the angles of incidence and refraction (θ_1 and θ_2 in Fig. 35.7, respectively) is equivalent to the ratio of the velocities of the light in air and water (v_1 and v_2, respectively). This can be stated algebraically as $\sin\theta_1/\sin\theta_2 = v_1/v_2$.

Figure 35.5 Cross Section of LiDAR Data Showing Power Lines and Vegetation Layers

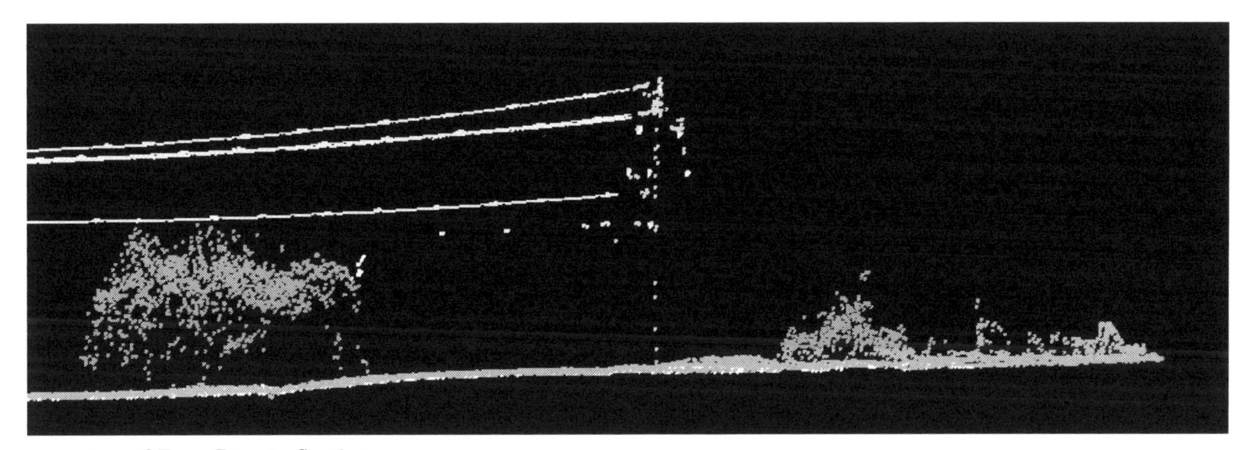

Image courtesy of Terra Remote Sensing.

Figure 35.6 Typical Cross Section of a Spring Basin Using Airborne LiDAR

Image courtesy of Suwannee River Water Management District.

Figure 35.7 Refraction

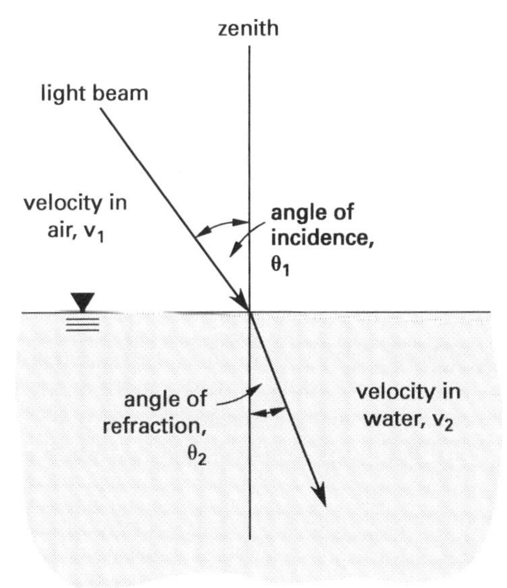

The primary challenge in using airborne LiDAR is in classifying the return pulses. In turbid water, some last-return pulses will be reflected from particles suspended in the water column. To avoid mistaking these for pulses reflected from the bathymetric ground, various attributes of the return pulses, such as intensity, must be considered. Making frequent ground truth comparisons will also help with classification.

10. CONTROL FOR AERIAL MAPPING

Traditional photogrammetric processes that use stereoplotters require *ground control points* to scale and orient each model. Control points are usually marked by aerial targets made of a material whose color contrasts with the background terrain. The optimum size for aerial targets depends on the photograph's scale. Three dimension coordinates are needed for the targets and may be established by conventional or GPS surveying.

In lieu of targets, photo-identification points that can be clearly and precisely identified on the photograph and ground may be used. For example, sharp cultural features such as the intersection of sidewalks and paved roads are ideal photo-identification points.

When mapping large areas involving multiple stereo models, it is not always necessary to establish several targets or photo-identification points on each model. Rather, fewer control points can be established and those points densified by aerotriangulation. *Aerotriangulation* is a process that establishes the geometric relationships between overlapping photographs to determine artificial horizontal and vertical control points. This substantially reduces the requirement for ground surveys and minimizes the cost of photogrammetric mapping. Sophisticated computer software is used to analytically transfer scale and orientation between successive stereo models, and to compute the precise orientation and position of each aerial photo.

Airborne GPS also significantly reduces the need for ground control points since the aircraft's precise position can be determined. However, for mapping accuracy, differential GPS corrections from a nearby GPS ground station must be applied. Therefore, a GPS base station with a known geodetic control point must be operated in the vicinity of the aerial survey. LiDAR surveys do not typically require additional ground control points to supplement the differentially corrected airborne GPS positions. However, for aerial photography, it is generally considered necessary for some ground control points, although the requirements are significantly reduced with the use of airborne GPS.

For any type of aerial mapping, a certain amount of ground verification of the mapping product is important for quality control. For traditional mapping products (hard copy maps), the *National Mapping Accuracy Standards* require that the maps be tested by independent ground surveys to evaluate the accuracy of both the horizontal positions of mapped features and the elevation contours. Those standards require that for maps with scales greater than 1:20,000, no more than 10% of the points tested can have a horizontal error greater than 1/30 inch (0.8 mm). For maps with scales less than 1:20,000, no more than 10% of the points tested can have a horizontal error greater than 1/50 inch (0.5 mm). Regarding vertical accuracy, those standards require that no more than 10% of the points tested can have a vertical error greater than one-half the contour interval.

For modern mapping products, which are primarily digital (see Sec. 35.11) and therefore are essentially files with x-, y-, and z-coordinates for each point, different standards for evaluating the mapping accuracy are necessary since these products have no scale. For these products, a statistical measure of accuracy, such as the *root-mean-square error* (RMSE) is typically used. As an example, the U.S. Federal Emergency Management Agency (FEMA) specifications for LiDAR flood plain mapping require that a minimum of 20 test points for each major vegetation category (e.g., bare earth and low grass, high grass and crops, brush lands and low trees, fully covered by trees) be checked by ground surveying methods to verify the accuracy of the LiDAR elevation measurements. These specifications require that the RMSE for the total survey not exceed 15 cm.

Therefore, regardless of aerial mapping process used, some ground surveying is usually necessary. Control is required for the actual production of the mapping products and, in addition, independent survey points are necessary for quality control of the process.

11. AERIAL MAPPING PRODUCTS

A number of different types of mapping products can be created using aerial mapping. Traditionally, the products have been in two categories: prints of photographic images and hard-copy maps. The maps were generally classified into two categories: planimetric or topographic. *Planimetric maps* depict the horizontal position of ground features such as buildings, roads, water courses, and so on. *Topographic maps* have more expansive content and generally depict the ground relief using contours, as well as various planimetric features.

The products of most aerial mapping projects are digital. The modern alternative to prints of photographic images is the *orthophotograph*, which is a spatially corrected photograph typically provided in digital form. Hard-copy maps have generally been replaced by digital elevation models (DEM) or digital terrain models (DTM). These products consist of a file of x-, y-, and z-coordinates, with an identifying attribute for each point mapped.

There are significant advantages to the change from hard-copy to digital mapping products. In addition to being generally more economical, digital versions may be plotted at any desired scale. In addition, other geo-referenced information, such as boundary surveys, zoning maps, photographic images, and so on, may be easily merged with the mapped data.

PRACTICE PROBLEMS

1. Which type of film emulsion would most effectively detect agricultural crop diseases?

(A) panchromatic black and white
(B) natural color
(C) black and white infrared
(D) false color infrared

2. Which type of film emulsion would most effectively delineate water bodies?

(A) panchromatic black and white
(B) natural color
(C) black and white infrared
(D) false color infrared

3. An aerial photograph is taken from an altitude of 3000 ft using a camera with a focal length of 6 in. The photograph's scale is most nearly

(A) 1 in = 400 ft
(B) 1 in = 500 ft
(C) 1 in = 1200 ft
(D) 1 in = 4800 ft

4. What is most nearly the minimum negative scale needed to cover a standard public land survey section, with one photograph, using a camera with a 6 in focal length and 9 in by 9 in format film?

(A) 1:8000
(B) 1:10,000
(C) 1:12,000
(D) 1:14,000

5. Most nearly, what height above the mean terrain should a photograph be taken to achieve a negative scale of 1:18,000 using a camera with a focal length of 6 in?

(A) 4000 ft
(B) 6000 ft
(C) 9000 ft
(D) 12,000 ft

6. A 300 ft tall tower is located at the southwestern corner of the southeast quarter of the northeast quarter of a standard public land survey section. Assuming that a vertical aerial photograph, taken with a 6 in focal length camera at an altitude of 9000 ft, is centered on the section, what is most nearly the relief displacement for the top of the tower?

(A) 0.03 in
(B) 0.1 in
(C) 0.3 in
(D) 0.7 in

7. If the relief displacement for the top of a tower located 1000 ft from the center of a vertical photograph was 0.1 in on the photograph and the flying height with a six inch focal length camera was 6000 ft, what would be the approximate height of the tower?

(A) 345 ft
(B) 445 ft
(C) 545 ft
(D) 645 ft

8. If a vertical photograph is taken at a negative scale of 1:15,000, the expected measurement of the distance on the photograph between two adjacent public land survey section corners would most nearly be

(A) 0.42 in
(B) 2.2 in
(C) 3.3 in
(D) 4.2 in

SOLUTIONS

1. False color infrared film emulsion would most effectively detect agricultural crop diseases.

The answer is (D).

2. Black and white infrared film emulsion would most effectively delineate water bodies.

The answer is (C).

3. Using a focal length of 6 in (0.5 ft) and a flying height of 3000 ft with Eq. 35.1, the scale is

$$S = \frac{f}{H} = \frac{0.5\text{ ft}}{3000\text{ ft}} = \frac{1}{6000}$$

This may be restated as

$$1\text{ in} = 6000\text{ in} = \frac{6000\text{ in}}{12\ \frac{\text{in}}{\text{ft}}} = 500\text{ ft}$$

The answer is (B).

4. Scale of a photograph may be expressed as the ratio of the dimension of an image on the photograph to the dimension of that object on the ground. Since a standard public land survey section is one mile (5280 ft) on each side, the scale may be expressed as

$$9\text{ in} = (5280\text{ ft})\left(12\ \frac{\text{in}}{\text{ft}}\right) = 63{,}360\text{ in}$$

This may be restated as 9 in = 63,360 in, or

$$\begin{aligned}1\text{ in} &= \frac{63{,}360\text{ in}}{9\text{ in}}\\ &= 7040\text{ in}\quad (1{:}8000)\end{aligned}$$

The answer is (A).

5. Using the scale of 1:18,000 and a 6 in (0.5 ft) focal length with Eq. 35.2,

$$H = \frac{f}{S} = \frac{0.5\text{ ft}}{\frac{1\text{ in}}{18{,}000\text{ in}}} = 9000\text{ ft}$$

The answer is (C).

6. Assuming the section is of standard size, the distance from the center of the section to the tower is 1320 ft. The length of the tower's "shadow," D, on the ground may be calculated using Eq. 35.5.

$$D = \frac{hR}{H-h} = \frac{(300\text{ ft})(1320\text{ ft})}{9000\text{ ft} - 300\text{ ft}} = 45.52\text{ ft}$$

Mapping

Then, using Eq. 35.4, the relief displacement on the photograph may be calculated.

$$d=\frac{Df}{H}=\frac{(45.52\text{ ft})(0.5\text{ ft})}{9000\text{ ft}}=0.0025\text{ ft}\quad(0.03\text{ in})$$

The answer is (A).

7. Rearranging Eq. 35.4,

$$D=\frac{dH}{f}=\frac{\left(\dfrac{0.1\text{ in}}{12\ \dfrac{\text{in}}{\text{ft}}}\right)(6000\text{ ft})}{0.5\text{ ft}}=100\text{ ft}$$

Rearranging Eq. 35.5,

$$\begin{aligned}h&=\frac{DH}{R+D}=\frac{(100\text{ ft})(6000\text{ ft})}{1000\text{ ft}+100\text{ ft}}\\&=545.5\text{ ft}\quad(545\text{ ft})\end{aligned}$$

The answer is (C).

8. The width of a standard public land section is 5280 ft. The photo scale of 1:15,000 may be restated as

$$1\text{ in}=15{,}000\text{ in or }\frac{15{,}000\text{ in}}{12\ \dfrac{\text{in}}{\text{ft}}}=1250\text{ ft}$$

Therefore, the photo distance is

$$\frac{5280\text{ ft}}{1250\ \dfrac{\text{ft}}{\text{in}}}=4.22\text{ in}\quad(4.2\text{ in})$$

The answer is (D).

36 Laser Scanning

1. LASER SCANNING

Laser scanning, also known as *high-definition surveying* and *three-dimensional imaging*, is the process of using a laser to measure existing conditions in the built or natural environment. It enables high-definition mapping of topography, structures, mechanical systems, overhead power lines, and other surveyed objects requiring great detail. Lasers can measure thousands of points per second with a high level of detail and precision, significantly reducing the time required to produce three-dimensional models of surveyed objects or topography. Laser scanning also requires minimal personnel for data acquisition, allowing roadways, cliffsides, bridges, buildings, and similar hazardous areas to be mapped without endangering personnel.

The purpose of laser scanning is to create a dense cloud of points on the surface of the surveyed object to allow extrapolation of its shape and dimensions. A *point cloud* is a three-dimensional model composed of scanned points, with each point defined by x-, y-, and z-coordinates. Although the points comprising the model may have been scanned from various locations on the ground, the model can be viewed from any vantage point. Figure 36.1, an example of a point cloud defining a bridge, shows the level of detail of a virtual model created by a laser scanning point cloud.

Laser scanning measurements can also include *return pulse intensity*, or the amplitude of the return signal, which provides information about the nature of the object being scanned. Many laser scanners also incorporate digital photography or video to capture imagery for correlation with the positional information.

In essence, a laser scanner is a programmable, automated, total station equipped with a reflectorless electronic distance meter (EDM). (See Fig. 36.2.) It uses an active sensing system with its own energy source.

Figure 36.1 *Typical Point Cloud*

Image courtesy of Nobles Consulting Group.

Figure 36.2 *Typical Laser Scanner*

Because it does not rely on reflected or naturally emitted radiation, it can be operated in daylight. Unlike photography, it can measure distances to places in shadows and even in underground locations such as mines and tunnels.

The maximum range of a laser scanner varies from under 30 ft to over 0.5 mi. Measurement errors vary from thousandths of an inch to a couple of inches, with longer-range instruments generally having greater errors. With their many advantages over traditional total stations, laser scanners have become standard surveying tools. The availability of laser scanners has led to changed design processes to take advantage of this form of modeling.

2. PRINCIPLES OF OPERATION

As with any total station, a laser scanner is used to measure horizontal and vertical angles to specific points, and it includes a laser range finder to measure distances to those points. The difference between a traditional total station and a laser scanner is that the laser scanner is automated. With a traditional total station, the surveyor aims the instrument at an object and then takes the angular and distance measurements to that object, repeating the process for each point needing location. With a laser scanner, angles and distances are measured at thousands of points per second without the need for an operator to point the device.

Laser scanners rotate around the horizon and measure angles and distances to points on a customizable grid. For example, a laser scanner could be set up to measure points on a 0.2 ft by 0.2 ft grid at a distance of 150 ft. Points at a distance closer than 150 ft would be closer together, and points at a distance greater than 150 ft would be farther apart.

Most laser scanners will scan a complete 360° around the horizon and 90° or more vertically. As a laser scanner rotates, a scanning mirror deflects the line of sight up and down, taking measurements at a rate as high as 100,000 points per second. Distance, horizontal angles, and vertical angles are recorded for each measurement, similar to a total station. These measurements, together with knowledge of the position and directional orientation of the instrument, allow calculation of x-, y-, and z-coordinates for each point measured. Although the designs vary by manufacturer, the orientation of the measurement axes in some laser scanners is based on a compensator, similar to the compensators used on levels and conventional total stations.

Two types of technology are used to measure range in laser scanners: time of flight and phase shift measurement systems.

Time of flight measurement systems use sub-nanosecond timing and pulse detection circuitry to measure the time it takes a light pulse to travel from the laser scanner to the surface of an object and back. The precision of range measurements that use time of flight technology is about 0.4 in or better, depending on the distance to the target. Time of flight measurement systems are the systems most commonly used for measuring the range of laser scanners, since they allow measurement of distances at ranges of several hundreds of meters. However, these systems cannot capture high-density scans as quickly as phase shift systems can.

Phase shift measurement systems[1] have a measurement speed up to 100 times faster than that of time of flight scanners. These measurement systems utilize a light source that is continually on—rather than pulsed—and modulated to a known frequency and wavelength. The range of the laser scanner is determined by counting the number of wavelengths in the double path. The integral number of wavelengths is determined by redundant measurements using multiple frequencies. Fractional wavelengths are measured by comparing the phase of the transmitted signal to that of the reflected signal. Knowledge of the integral number of wavelengths, the fractional portion of a wavelength from the phase comparison, and the wavelength of the modulated signal allows calculation of the range. The precision of the measured distances using this technology is within a tenth of an inch, although the range is restricted to a maximum of about 230 ft to 330 ft.

Positional and directional orientation for laser scanners are typically determined by special targets (such as the ones shown in Fig. 36.3) set on control points that can be recognized by the laser scanning system. In the setup procedure, a laser scanning system is typically programmed to use higher resolution scans for the area around the target and other critical points, which allows the center of the target to be precisely located.

Figure 36.3 *Typical Laser Scanning Targets*

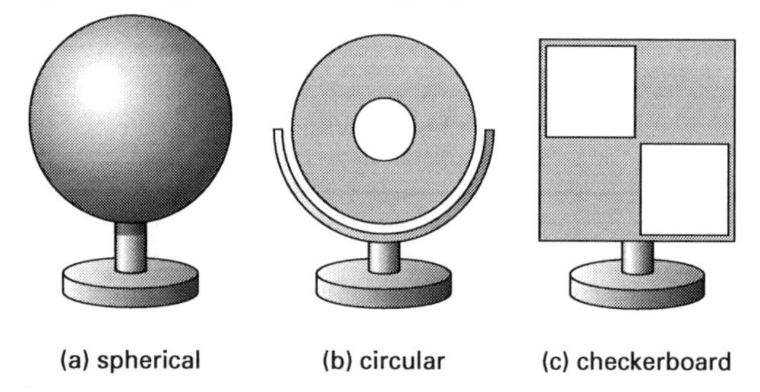

3. DATA ACQUISITION

Control Points

Although approaches will vary with each project, a critical part of the fieldwork for a typical laser scanning project is establishing control points. For each scan, position and orientation could be controlled by setting up the laser scanner on a control point with known coordinates and setting up a backsight target on a second control point with known coordinates. However, while this is the most straightforward method, it is not the best. It is preferable to determine the position of the laser scanner using *resection*, which involves sighting three or more targets on points with known coordinates established by traditional traverse or GPS observations. The resection approach is best because it eliminates the need to measure the height of instrument for the laser scanner.

There are three main approaches to controlling subsequent scans: cloud-to-cloud point matching, three-point resection from additional independent control points,

[1]Phase shift measurement systems are also used in traditional EDM instruments.

and traversing.[2] *Cloud-to-cloud point matching* involves using at least three points common to both scans. Such points do not need to have coordinates established by an independent process; they can be targets set on random points. The coordinates of those points from the first scan allow the point cloud for the second scan to be translated to the datum of the first. *Traversing* requires laser scanners to be equipped with a compensator, and then the laser scanner can be used for traditional traverse, with control being carried from one laser scanning setup to the next. (See Chap. 15 for information on traverse calculations.)

A minimum of two control points with coordinates from an independent source are needed for a laser scanning project, and three or more are desirable. Targets should be set on such control points for positive identification in the point cloud. For larger and linear projects, several control points should be established at each end of the project. In addition to these minimum requirements, additional targeted control points on random points are recommended. This allows for redundancy in the event that there is a problem with the data for one or more of the targets.

Laser Scanner Setup

Setup locations should be carefully selected to minimize shadows, to ensure a clear view and optimum range of desired features and targeted points, and to ensure an adequate angle of incidence. An adequate angle of incidence is an important factor in obtaining quality data. Since laser scanners are typically set up at eye height, the angle of incidence a few hundred feet away is rather small. Typically, users report satisfactory precision using tripod-mounted laser scanners scanning at a distance of 300 ft, which provides an angle of incidence of about one degree.

Programming Scan Parameters

A laser scanner is usually connected to a laptop computer for operation; the computer controls the laser scanning operation and allows the operator to view the resulting data. Once a laser scanner is set up, the scan parameters are programmed into it. These include the limits of the desired field of view and the scan density. Limits can be set visually, using photography taken at the setup. Typically, areas around targets and other key features should be scanned at greater density, especially with time of flight scanners.

Laser Scanning and Data Review

Once the laser scanner is programmed and turned on, it automatically scans the defined areas. As the areas are scanned, the results are available for immediate viewing as a point cloud, with each point in its correct relative position. Digital photographs taken at the setup can be merged with the measured positions to aid in field examination of the data.

[2]For subsequent scans, control points with coordinates from an independent source may not be necessary, depending upon the approach used for extending control.

4. DATA PROCESSING

Laser scanning collects a large amount of data in a relatively short time. However, while the data collection time is significantly reduced, the vast amount of data that is collected shifts the burden of work from the field to the office. Software can facilitate the move from a raw point cloud to a fully parametric model, but such software lags behind the data collection process.

For most situations, a single scan will not produce a complete model of the subject. Multiple scans from several different directions are usually required to obtain information about all sides of the subject, which must be merged using a common reference system to create a complete model. This process is called *registration of the scans*. As long as the scans have been referenced to a control point network with all points on a common datum, this is a straightforward process.

In the registration process, the imaging data cannot be scaled. As a result, depending upon the elevation and the grid scale factor for the plane projection being used, there may be slight disagreement between control plane coordinate values and measured distances. Allowance for such disagreement should be made.

Once the scans are registered, the point clouds produced by three-dimensional laser scanners can be used directly for measurement and visualization. Software is available that allows point clouds to be viewed from multiple vantage points. The surveyor can virtually walk through the surveyed site and extract coordinates for desired features. Additionally, measurements can be taken for the creation of CADD drawings. For example, points on features, such as the curb, gutter, and pavement edge of roadways, may be selected visually to create CADD drawings of those features.

Laser scanner data should not be used for precise visual selection of the edges of features. Laser scanning samples points on a preset grid, so the exact location of the edge may be missed. A modeling process is typically used to fit the three-dimensional data to surfaces for the CADD drawings. For example, with a scan of a building, points covering each side of the building can be fit to planes whose intersection would accurately depict the limits of the building for CADD drawings. Similarly, the center of a round utility pole can be determined by fitting the points on the surface of the pole to a cylinder. A specific group of points belonging to a defined surface is selected. Then, cloud fitting algorithms are used to create the best-fit geometric shape in CADD format. Most scan processing software includes modeling tools to fit points to a wide variety of surfaces, such as planes, spheres, cones, and cylinders.

5. QUALITY CONTROL

Evaluation of point cloud data relies on identifiable points for which three-dimensional coordinates from a point cloud can be compared to independently measured coordinates. This requires the use of either formal targets, for which point cloud coordinates and independently measured coordinates can be determined, or other identifiable points, for which point cloud coordinates can be determined by modeling and for which independently measured coordinates can be obtained. When points other than formal targets are used, the selected points should be readily identifiable points on surfaces that can be modeled using the point cloud data. Examples of points meeting these criteria include the center of a circular object, the center of a spherical object, the intersection of three planes (e.g., the intersection of a corner of a building with a concrete slab), or the intersection of the center line of a cylinder with a plane, such as the intersection of a cylindrical pole with a grade.

Most specifications recommend that a minimum of 20 points be checked to measure the quality of a point cloud. These can be a combination of formal targets and other identifiable points. The test points should be carefully selected to obtain a sample representative of the study area. Ideally, the points would be randomly spaced over the study area and over different planes. National spatial data specifications use the *root-mean-square error* (RMSE) to estimate both horizontal and vertical accuracy.

The RMSE is the square root of the average of the set of squared differences between the scanned data coordinates and the coordinates measured at the same points by an independent check of higher accuracy.

$$\text{RMSE} = \sqrt{\frac{\sum_{i=1}^{n}(x_i - x_i')^2}{n}} \qquad 36.1$$

In Eq. 36.1, x_i is the scanned measurement, x_i' is the check measurement, and n is the number of measurements.

Federal Emergency Management Agency (FEMA) specifications for LiDAR use a value of 1.96 times the RMSE as the required accuracy for a project at a 95% confidence level. With a sampling of the elevation of 20 points, no more than one point should have an error greater than 1.96 times the RMSE to meet this standard.

The North Carolina Department of Transportation's standards for laser scanning surveying for design surveys provide an example of how quality control procedures are applied in practice. These standards require each scan setup to have a minimum of four control targets with a control point located on both sides of the project at distances no greater than 500 ft. The length of usable data in each scan is restricted to 600 ft. For quality control, each scan must have at least three visible validation points, such as painted dots on the pavement. Both the control points and the validation points must be locatable within a tolerance of plus or minus 0.03 ft by traditional methods. Using the scanned data, a digital terrain model (DTM) is created using points on a grid spaced 5 ft apart, as well as points delineating breaklines on the back, face, and flow line of the curb, the edge of the pavement, and the paint striping. After the data is processed using the targets as a control, the scanned elevations of the validation points must be compared with the independently measured elevations. The RMSE of all such validation points must be 0.05 ft or less, with no individual validation point having an error in excess of 0.10 ft. Surveys meeting these standards should result in at least 95 percent of the measured elevations being within 0.10 ft (1.96 times 0.05 ft) of the true elevations.

6. MOBILE LASER SCANNING

Laser scanners are sometimes used from moving platforms (see Fig. 36.4), especially for linear projects such as roadway and coastal mapping projects. Such applications add complexity to processing in that the sensor is at a different location for almost every point that is measured, as opposed to being in one location during the entire scan. To compensate, this process requires the addition of a high-accuracy GPS device to determine the position at the time of each pulse, as well as an inertial measurement unit (IMU) for measuring acceleration and rotational changes in the three axes as the platform vehicle moves. The process and equipment for mobile laser scanning are the same as for airborne LiDAR, as discussed in the chapter on aerial mapping. The equipment is considerably more expensive, but linear projects can be mapped much more quickly.

Figure 36.4 *Typical Mobile Scanning Setup*

PRACTICE PROBLEMS

1. A laser scanner is set up on a tripod on level ground. The instrument height is 4 ft 6 in. The angle of incidence (the angle at which the scanner is pointed below the horizontal) for a measurement to a point on the ground at a 150 ft distance is most nearly

(A) 0°45′
(B) 1°15′
(C) 1°43′
(D) 2°05′

2. Which of the following processes for controlling laser scanning is typically considered to be the best?

(A) setups of laser scanner on known control points
(B) resection from three or more control points
(C) photo identification of control points
(D) none of the above

3. On the site of a topographic survey performed with a laser scanner, the elevations of 20 points were independently measured using RTK GPS for quality control purposes. The comparative measurements are provided in the following table.

pt. no.	scanned elev. (ft)	check elev. (ft)	pt. no.	scanned elev. (ft)	check elev. (ft)
1	150.2	150.1	11	116.7	116.5
2	98.6	98.8	12	114.2	114.4
3	125.3	125.3	13	89.9	89.9
4	87.7	87.6	14	78.2	78.2
5	102.4	102.5	15	101.7	101.6
6	115.6	115.4	16	121.2	121.3
7	88.9	88.9	17	122.3	122.2
8	92.4	92.5	18	111.7	111.9
9	103.5	103.5	19	103.4	103.4
10	151.4	151.3	20	98.6	98.7

What is the approximate root-mean-square error for the comparison?

(A) 0.12 ft
(B) 0.24 ft
(C) 0.29 ft
(D) 0.34 ft

4. In regard to the data for Prob. 3, to meet FEMA standards, the error for 95% of the measurements must be less than

(A) 0.12 ft
(B) 0.24 ft
(C) 0.29 ft
(D) 0.34 ft

SOLUTIONS

1. The angle of incidence can be calculated as the arc tangent of the height of instrument divided by the distance.

$$\alpha = \arctan \frac{H_I}{\text{dist}} = \arctan \frac{4.5 \text{ ft}}{150 \text{ ft}} = 1°43'$$

The answer is (C).

2. Resection from three or more coordinated control points is usually considered the best method for controlling laser scanning, as it eliminates the need for measuring the height of instrument of the laser scanner.

The answer is (B).

3.

pt. no.	scanned elev. (ft)	check elev. (ft)	$x_i - x_i'$ (ft)	$(x_i - x_i')^2$ (ft^2)
1	150.2	150.1	0.1	0.01
2	98.6	98.8	−0.2	0.04
3	125.3	125.3	0	0
4	87.7	87.6	0.1	0.01
5	102.4	102.5	−0.1	0.01
6	115.6	115.4	0.2	0.04
7	88.9	88.9	0	0
8	92.4	92.5	−0.1	0.01
9	103.5	103.5	0	0
10	151.4	151.3	0.1	0.01
11	116.7	116.5	0.2	0.04
12	114.2	114.4	−0.2	0.04
13	89.9	89.9	0	0
14	78.2	78.2	0	0
15	101.7	101.6	0.1	0.01
16	121.2	121.3	−0.1	0.01
17	122.3	122.2	0.1	0.01
18	111.7	111.9	−0.2	0.04
19	103.4	103.4	0	0
20	98.6	98.7	−0.1	0.01
			sum	0.29

Using Eq. 36.1,

$$\text{RMSE} = \sqrt{\frac{\sum_{i=1}^{n} (x_i - x_i')^2}{n}} = \sqrt{\frac{0.29 \text{ ft}^2}{20}} = 0.12 \text{ ft}$$

The answer is (A).

4. The FEMA specifications require that the elevation of 95% of sampled points be within 1.96 times the RMSE for the sample data set. That value is

$$1.96(\text{RSME}) = (1.96)(0.12 \text{ ft}) = 0.24 \text{ ft}$$

The answer is (B).

Mapping

Topic VIII: Specialty Surveying Areas

Chapter

37 Construction Staking

1. DEFINITION

Construction surveying involves locating and marking locations of structures that are to be built. It is often referred to as *giving line and grade*. A transit or theodolite is used in establishing line (horizontal alignment), and a level is used in establishing grade (elevation).[1]

2. CONVERSION BETWEEN INCHES AND DECIMALS OF A FOOT

Construction stakes are usually set to the nearest hundredth of a foot for concrete, asphalt, pipelines, and so on. For earthwork, stakes are set to the nearest tenth of a foot. Constructors use foot and inch rules. So, in deference to them, surveyors set the stakes for their convenience.

In converting measurements in feet, inches, and fractions to feet, it may be easier to convert the inches and fractions of an inch separately, and then to add the parts.

[1]The word *grade* is not used consistently, sometimes meaning slope and sometimes meaning elevation above a datum. In this text, *gradient* will be used for rate of slope, and *finish elevation* will be used for the elevation above a datum to which a part of the structure is to be built.

Example 37.1

Convert the measurements to feet and decimals of a foot.

(a) 1 ft, 4 in

(b) 11 ft, $9^{1}/_{8}$ in

(c) 7 ft, $5^{3}/_{4}$ in

(d) 2 ft, $8^{7}/_{8}$ in

(e) 5 ft, $11^{1}/_{2}$ in

Solution

(a) 1.33 ft

(b) 11.76 ft

(c) 7.48 ft

(d) 2.74 ft

(e) 5.96 ft

Example 37.2

Convert the following measurements to feet and inches.

(a) 3.79 ft

(b) 6.34 ft

(c) 5.65 ft

(d) 3.72 ft

Solution

(a) 3 ft, $9^{1}/_{2}$ in

(b) 6 ft, $4^{1}/_{8}$ in

(c) 5 ft, $7^{3}/_{4}$ in

(d) 3 ft, $8^{5}/_{8}$ in

3. STAKING OFFSET LINES FOR CIRCULAR CURVES

Stakes set for the construction of pavement or curbs must be set on an offset line so that they will not be destroyed by construction equipment. However, they must be close enough for short measurements to the actual line. The offset line may be 3 ft or 5 ft, or any convenient distance from the edge of pavement or back of curb. Stakes are set at 25 ft or 50 ft intervals, and tacks are set in the stakes to designate the offset line.

In setting stakes on a parallel circular arc, the central angle is the same for parallel arcs. The radius to the centerline of the street or road is usually the design radius. The PC of the design curve and the PC of a parallel offset curve, whether right or left, will fall on the same radial line. Likewise, the PTs of the parallel arcs will fall on the same radial line. In computing the stations for PCs and PTs, the design curve data (design radius) should be used. Then, the PC and PT stations for an offset line will be the same as for the design curve (centerline of road or street), even though the lengths of offset curves will not be the same as the length of the centerline curve.

The design curve data will be used to compute deflection angles. These angles will be the same for offset lines, since the central angle between any two radii is the same for the parallel arcs.

Because chord lengths are a function of the radius of an arc $\left(C = 2R\sin(\Delta/2)\right)$, the chord length between two stations on the design curve and two corresponding stations on the offset line will not be the same.

By using design curve data in computing PC and PT stations, deflection angles for curves can be recorded in the field book. Such angles will be the same whether a right offset line, a left offset line, or both right and left offset lines are used.

In performing field work, centerline PIs, PCs, and PTs are located on the ground before construction. Offset PCs and PTs are located at right angles to the centerline PCs and PTs. Offset PCs and PTs should be carefully referenced.

Example 37.3

Stakes are to be set on 4 ft offsets for each edge of pavement, which is 36 ft wide. The curve has a deflection angle Δ of 60° to the right, and a centerline radius of 300 ft, PI is at station 12+44.32, and stakes are to be set for each full-station, half-station, and at the PC and PT.

Compute PC and PT stations, deflection angles, and chord lengths. Set up field notes for the curve.

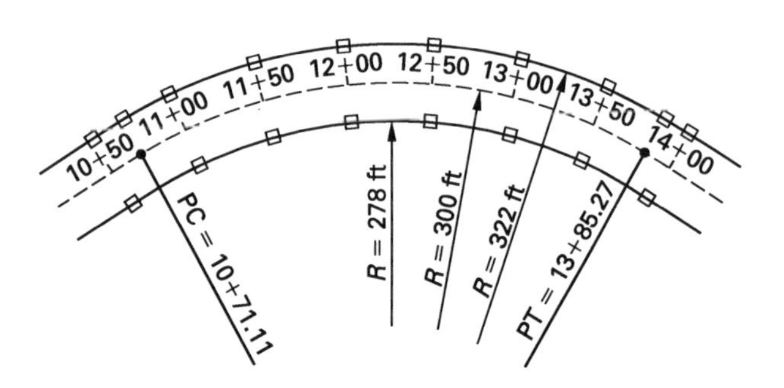

Solution

$$\Delta = 60°00'$$

$$T = R\tan\frac{\Delta}{2} = (300\text{ ft})(\tan 30°) = 173.21\text{ ft}$$

$$L = \left(\frac{60°}{360°}\right)2\pi R = \left(\frac{60°}{360°}\right)2\pi(300\text{ ft}) = 314.16\text{ ft}$$

$$\begin{aligned} \text{PI} &= 12{+}44.32 \\ T &= \underline{-1{+}73.21} \\ \text{PC} &= 10{+}71.11 \\ L &= \underline{3{+}14.16} \\ \text{PT} &= 13{+}85.27 \end{aligned}$$

Deflection Angle Computations

sta 10+71.11: 0°00′

$$\text{sta } 11{+}00\text{: } \left(\frac{11{+}00 - 10{+}71.11}{314.16\text{ ft}}\right)\left(\frac{60°}{2}\right) = \left(\frac{28.89\text{ ft}}{314.16\text{ ft}}\right)(30°) = 2.7588° \quad (2°45')$$

$$\text{sta } 11{+}50\text{: } \left(\frac{28.89 + 50\text{ ft}}{314.16\text{ ft}}\right)\left(\frac{60°}{2}\right) = \left(\frac{78.89\text{ ft}}{314.16\text{ ft}}\right)(30°) = 7.5334° \quad (7°32')$$

$$\text{sta } 12{+}00\text{: } \left(\frac{128.89\text{ ft}}{314.16\text{ ft}}\right)(30°) = 12.3081° \quad (12°18')$$

$$\text{sta } 12{+}50\text{: } \left(\frac{178.89\text{ ft}}{314.16\text{ ft}}\right)(30°) = 17.0827° \quad (17°05')$$

$$\text{sta } 13{+}00\text{: } \left(\frac{228.89\text{ ft}}{314.16\text{ ft}}\right)(30°) = 21.8573° \quad (21°51')$$

$$\text{sta } 13{+}50\text{: } \left(\frac{278.89\text{ ft}}{314.16\text{ ft}}\right)(30°) = 26.6320° \quad (26°38')$$

$$\text{sta } 13{+}85.27\text{: } \left(\frac{314.16\text{ ft}}{314.16\text{ ft}}\right)(30°) = 30.0000° \quad (30°00')$$

Outside Chord Lengths

station to station	computation	length
10+71.11 to 11+00:	(2)(322 ft)(sin 2.7588°)	= 31.00 ft
11+00 to 11+50:	(2)(322 ft)(sin 4.7746°)	= 53.60 ft
13+50 to 13+85.27:	(2)(322 ft)(sin 3.3680°)	= 37.83 ft

Inside Chord Lengths

station to station	computation	length
10+71.11 to 11+00:	(2)(278 ft)(sin 2.7588°)	= 26.76 ft
11+00 to 11+50:	(2)(278 ft)(sin 4.7746°)	= 46.28 ft
13+50 to 13+85.27:	(2)(278 ft)(sin 3.3680°)	= 32.66 ft

Field Notes

point	station	angle	C-out (ft)	C-in (ft)	curve data
PT	13+85.27	30°00′			
			37.83	32.66	
	+50.00	26°38′			
			53.60	46.28	
	13+00.00	21°51′			$\Delta = 60°00'$
			53.60	46.28	
	+50.00	17°05′			$R = 300$ ft
			53.60	46.28	
	12+00.00	12°18′			$T = 173.21$ ft
			53.60	46.28	
	+50.00	7°32′			$L = 314.16$ ft
			53.60	46.28	
	11+00.00	2°45′			
			31.00	26.76	
PC	10+71.11	0°00′			

4. CURB RETURNS AT STREET INTERSECTIONS

Curb returns are the arcs made by the curbs at street intersections. The radius of the arc is selected by the designer with consideration given to the speed and volume of traffic. A radius of 30 ft to the back of curb is common. Streets that intersect at a right angle have curb returns of one-quarter circle. The arcs can be swung from a radius point (center of circle). The radius point can be located by finding the intersection of two lines, each of which is parallel to one of the centerlines of the streets and at a distance from the centerline equal to half the street width plus the radius, as shown in Fig. 37.1. One stake at the radius point is sufficient for a curb return.

Figure 37.1 *Curb Returns*

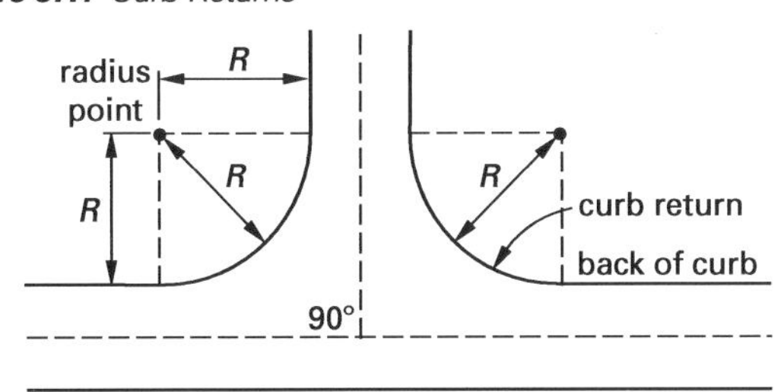

5. STAKING OFFSET LINES AT STREET INTERSECTIONS

In setting stakes for curb and gutter for street construction on an offset line, stations for the PC and PT of a curb return at a street intersection are computed along the centerline of the street. In Ex. 37.4, stakes are to be set on a 5 ft offset line from the back of the left curb along Elm Street. In setting stakes at the intersection of 24th Street, the PCs and PTs of the curb return are computed from the centerline stations. However, it must be remembered that in computing the long chord on the offset line, the radius R is not the design radius, but is the design radius minus the offset distance. Stakes are set at the PC, the PT, and the radius point of each curb return arc.

At street intersections not at 90°, there will be two deflection angles, one being the supplement of the other.

Example 37.4

Elm Street and 24th Street intersect as shown. Both streets are 26 ft wide, back to back of curb.

Compute the PC and PT stations, deflection angles from PC to PT, and long chord measured from the offset line for each return.

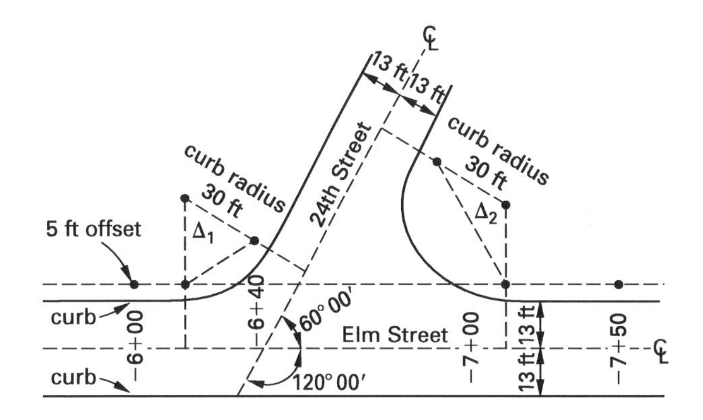

Solution

PC Stations Along Elm Street

$$\Delta_1 = 60°00'$$

$$T_1 = R_1 \tan \frac{\Delta_1}{2}$$

$$= (43 \text{ ft}) \tan\left(\frac{60°}{2}\right)$$

$$= 24.83 \text{ ft}$$

$$\text{PI} = 6{+}40.00$$

$$T_1 = \underline{-24.83}$$

$$\text{PC} = 6{+}15.17$$

$$\Delta_2 = 120°00'$$

$$T_2 = R_2 \tan \frac{\Delta_2}{2}$$

$$= (43 \text{ ft}) \tan\left(\frac{120°}{2}\right)$$

$$= 74.48 \text{ ft}$$

$$\text{PI} = 6{+}40.00$$

$$T_2 = \underline{+74.48}$$

$$\text{PC} = 7{+}14.48$$

Specialty Surveying Areas

PT Stations Along 24th Street

$$\begin{aligned} \text{PI} &= 0{+}00.00 \\ T_1 &= \underline{\;+24.83\;} \\ \text{PT} &= 0{+}24.83 \\ \text{PI} &= 0{+}00.00 \\ T_2 &= \underline{\;+74.48\;} \\ \text{PT} &= 0{+}74.48 \end{aligned}$$

Deflection Angles and Long Chords

$$\begin{aligned} \text{LC} &= 2R\sin\frac{\Delta_1}{2} \\ &= (2)(25\text{ ft})\sin\left(\frac{60°}{2}\right) \\ &= 25.00\text{ ft} \\ \text{deflection angle} &= \frac{60°}{2} = 30° \\ \text{LC} &= 2R\sin\frac{\Delta_2}{2} \\ &= (2)(25\text{ ft})\sin\left(\frac{120°}{2}\right) \\ &= 43.30\text{ ft} \\ \text{deflection angle} &= \frac{120°}{2} = 60° \end{aligned}$$

6. ESTABLISHING FINISH ELEVATIONS OR "GRADE"

Establishing the elevation above a datum to which a structure, or part of a structure, is to be built is usually accomplished by the following steps.

step 1: Set the top of a grade stake to the exact elevation (nearest one hundredth). Mark the top of the stake with blue keel.

step 2: Set the top of a grade stake at an exact distance above or below finish elevation. Mark the top of it with blue keel, and mark this exact distance above or below (called *cut* or *fill*) on another stake known as a *guard stake*, usually driven at an angle, beside the grade stake.

step 3: Use the line stake as a grade stake driven to a random elevation. Compute the difference in elevation between that elevation and the finish elevation, and mark this difference as cut or fill on a guard stake.

Marks on a wall, such as the wall of forms for a concrete structure, may be used instead of the tops of stakes. This is illustrated in Ex. 37.5.

7. GRADE ROD

A *grade rod* is the rod reading determined by finding the difference in elevation between the height of instrument (height of the level) and the finish elevation. In Fig. 37.2, the finish elevation is 441.23 ft, and the HI of the level is 445.55 ft. The grade rod is the difference in these two numbers, 4.32. A stake is driven so that when the level rod is placed on the top of it, the rod reading is 4.32.

Figure 37.2 *Use of Grade Rod*

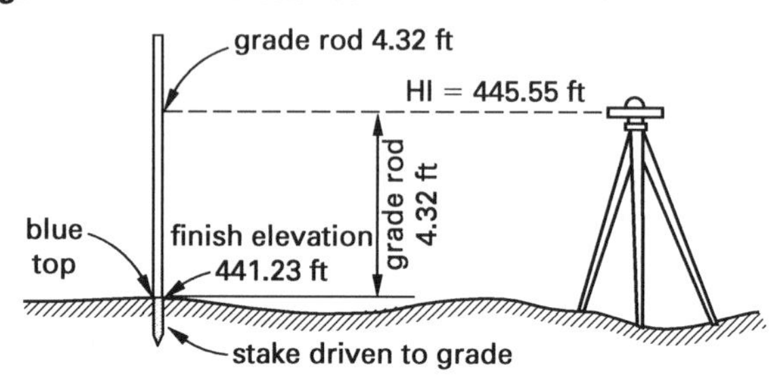

The procedure to set the stake is to place the rod on the ground where the stake is to be driven and determine the distance (in tenths) that the top of the stake should be above the ground. Then drive the stake until the top of the stake is at the finish elevation, stopping to check the rod reading so that the top of the stake will not be too low. When the grade rod reading is reached, the top of the stake is marked with blue keel, and thus the name *blue top* is given to this type of stake. A guard stake is driven beside the blue top in a slanting position. The guard stake is marked "G" to indicate the stake is driven to grade (finish elevation).

If the finish elevation is just below ground level, the blue top can be left above ground. The cut from the top of the stake to the finish elevation should be marked on a guard stake.

Where line stakes are also used as grade stakes driven to random elevations, a grade rod is not used. The elevation of the top of the stake is determined by leveling. The difference in elevation between the top of the stake and the finish elevation is determined and marked on the guard stake.

Example 37.5

The finish elevation is to be marked on the inside wall of the form for the concrete cap of a bridge. The finish elevation of the cap is 466.97 ft, and the HI is 468.72 ft.

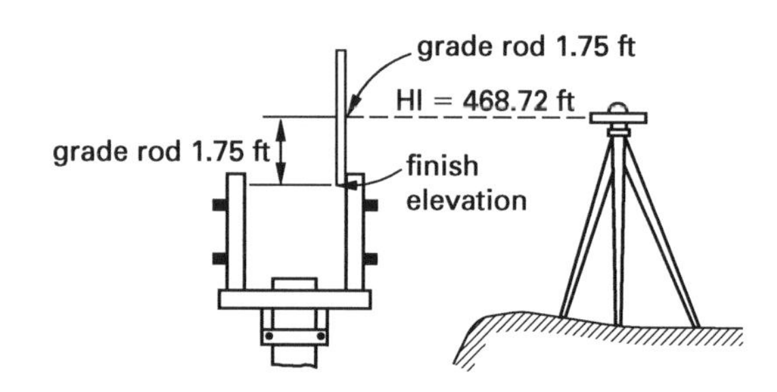

Solution

$$\text{HI} = 468.72 \text{ ft}$$
$$\text{finish elevation} = \underline{466.97 \text{ ft}}$$
$$\text{grade rod} = 1.75 \text{ ft}$$

The rod is held against the side of the form and raised or lowered until the rod reading is 1.75 ft. A nail is driven at the bottom of the rod, and the rod is placed on the nail so that the rod reading can be checked to see that the nail is correctly placed. Another finish elevation nail is driven at the other end of the form and a string line is drawn between the two nails, then is chalked and snapped to mark the grade line on the form. A chamfer strip is nailed on the form along this line, and the strip is used to finish the concrete to grade (finish elevation).

8. SETTING STAKES FOR CURB AND GUTTER

Separate stakes are often set for line and grade. In Fig. 37.3, a line stake is driven so that a tack is exactly 3 ft from the back of a curb. These line stakes are set on any convenient offset to avoid disturbance by construction equipment. A separate grade stake is driven so that the top of the stake is either at finish elevation or at an elevation that makes it an exact distance above or below finish elevation.

A guard stake is driven near the grade stake and marked to show this exact distance as cut or fill, and the top of the grade stake is marked with blue keel. Grade stakes can be driven so that the cut or fill is in multiples of a half foot. If the grade stake is driven to finish elevation, the guard stake is marked "G" for grade. The cut or fill can be determined by considering the finish elevation and a ground rod reading at each station.

Figure 37.3 *Line and Grade Stakes for Curb and Gutter*

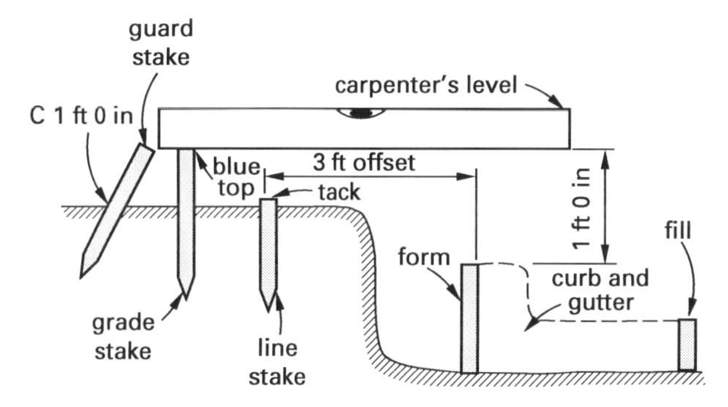

The top of the grade stake should be above ground. Then, the builder can lay a carpenter's level on top of the stake and measure from the established level line to establish the top of curb forms. In Ex. 37.6, the guard stake is marked for a cut as "C 1 ft, 0 in" so that the builder will measure 1 ft, 0 in down from the level line to the top of the forms. Horizontal alignment will be maintained by measuring 3 ft from each tack point to the back of curb line.

Table 37.1 *Field Notes for Curb and Gutter Grades (all measurements are in feet)*

station	+	HI	−	rod	elevation	finish elevation	grade rod	ground	mark stake
BM no. 1	3.42	455.78			452.36	r.r. spike in 12″ oak-100′ lt. sta 0+00			
0+00				0.18	455.60	454.10	1.68	0.4	C 1 ft, 6 in
0+50				1.20	454.58	454.58	1.20	1.6	grade
1+00				2.72	453.06	455.06	0.72	3.2	F 2 ft, 0 in
1+50				2.75	453.03	455.53	0.25	3.3	F 2 ft, 6 in
2+00				2.27	453.51	456.01	−0.23	2.5	F 2 ft, 6 in
2+50				1.79	453.99	456.49	−0.71	1.8	F 2 ft, 6 in
3+00				0.31	455.47	455.97	−1.19	0.6	F 1 ft, 6 in
T.P.	8.21	460.24	3.75		452.03				
3+50				3.30	456.94	457.44	2.80	3.6	F 0 ft, 6 in
4+00				1.41	458.83	457.83	2.41	1.7	C 1 ft, 0 in
4+50				0.70	459.54	458.04	2.20	1.0	C 1 ft, 6 in
5+00				1.16	459.08	458.08	2.16	1.5	C 1 ft, 0 in
5+50				0.31	459.93	457.93	2.31	0.7	C 2 ft, 0 in
6+00				1.04	459.20	457.70	2.54	1.4	C 1 ft, 6 in
BM no. 2			2.06		458.18				
	11.63		5.81						
	11.63		458.18						
	5.81		452.36						
	5.82		5.82						

Example 37.6

Grade stakes for curb and gutter have been driven to grade or to a multiple of 6 in above or below grade. Part of the level notes recorded in setting the stakes is shown. Also shown are finish elevations that have been taken from construction plans. Computations for grade rod and for the cut or fill marks on guard stakes are to be made and recorded. Rod readings on grade stakes as driven are to be recorded in the column marked "rod." (Note: A more detailed explanation of this procedure can be found in Sec. 14.)

Curb and Gutter Grades
(all measurements are in feet)

station	+	–	elevation	finish elevation	ground rod
BM no. 1	3.42		452.36		
0+00				454.10	0.4
+50				454.58	1.6
1+00				455.06	3.2
+50				455.53	3.3
2+00				456.01	2.5
+50				456.49	1.8
3+00				456.97	0.6

Curb and Gutter Grades (continued)
(all measurements are in feet)

station	+	–	elevation	finish elevation	ground rod
TP	8.21	3.75			
+50				457.44	3.6
4+00				457.83	1.7
+50				458.04	1.0
5+00				458.08	1.5
+50				457.93	0.7
6+00				457.70	1.4
BM no. 2		2.06			

Solution

Finished field notes are shown in Table 37.1. A graphical solution is shown in Ex. 37.6, the finished grade line. Solutions for sta 0+00, 0+50, 1+00, and 2+50 are shown in Ex. 37.6, the sample staking.

Finished Grade Line for Ex. 37.6

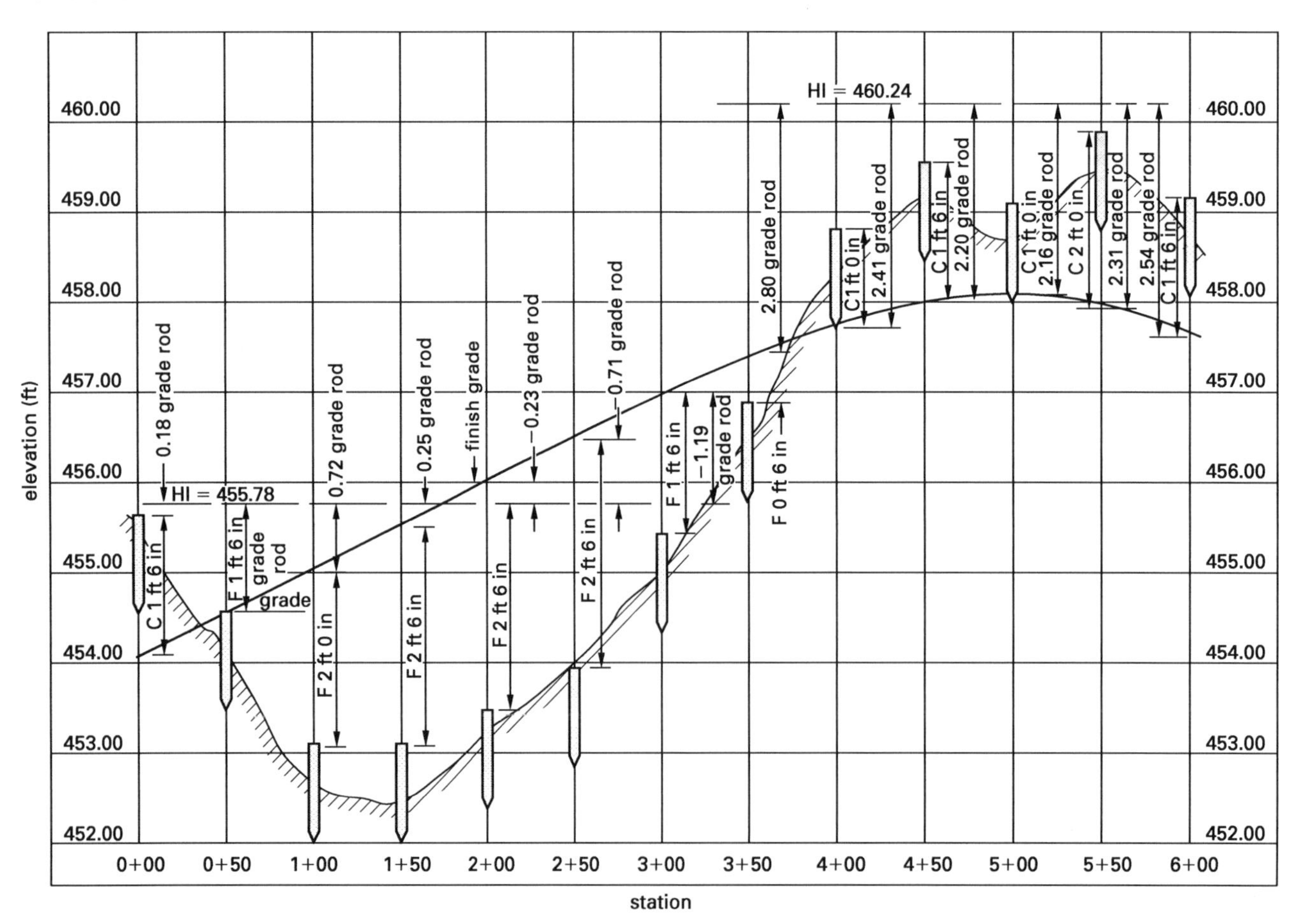

Sample Staking for Ex. 37.6

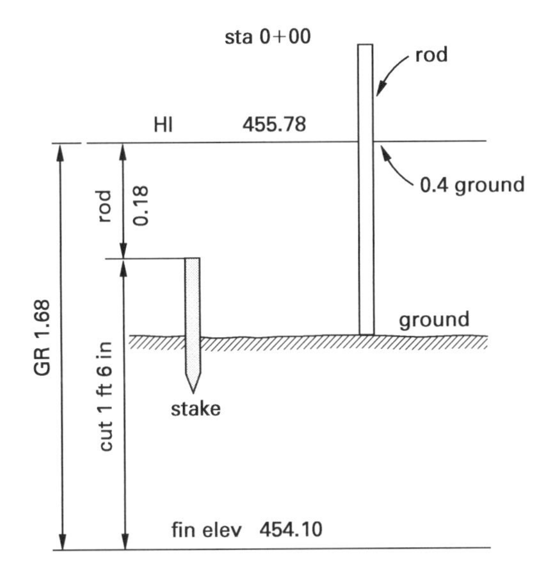

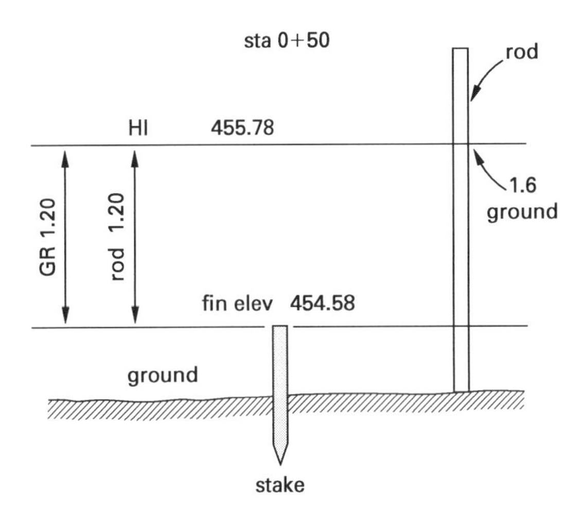

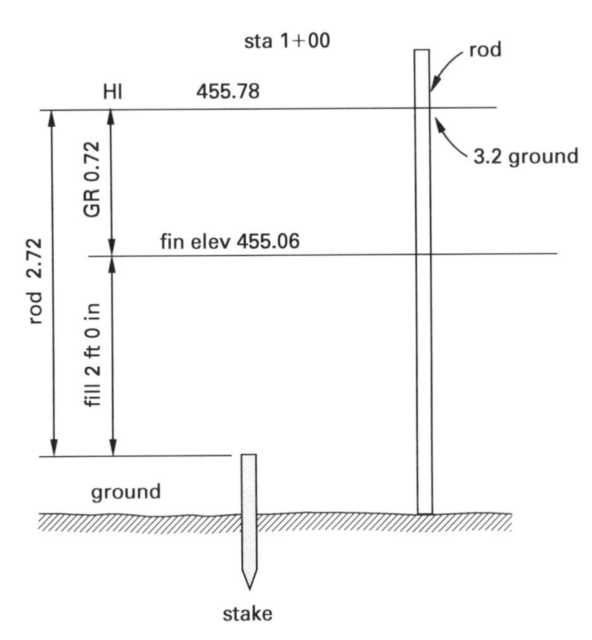

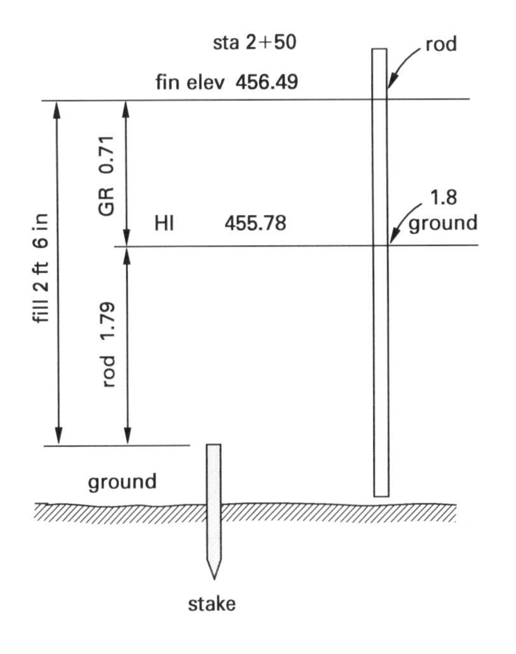

9. STAKING CONCRETE BOX CULVERTS ON HIGHWAYS

Tack points for concrete box culverts can be set on offsets from the outside corners of the culvert headwalls. For normal culverts (centerline of culvert at right angle to centerline of roadway), the distance from the centerline of the roadway to the outside of the headwall is equal to one-half the clear roadway width plus the width of the headwall. Tacks should also be set on the centerline of the roadway, offset from the outside of the culvert wall. The offset distance from the outside walls depends on the depth of cut. Stakes for wingwalls and aprons are not necessary, although stakes to establish the centerline of the culvert can be set if desired. Cuts to the flowline of the culvert can be marked on guard stakes at the tack points.

In staking skewed culverts, the distance from the centerline of the roadway to the outside of headwall (along the centerline of culvert) is equal to one half the clear roadway plus the headwall thickness divided by the cosine of the skew angle. The *skew angle* is the angle between the normal and the centerline of culvert.

Figure 37.4 Staking Box Culverts (Plan View)

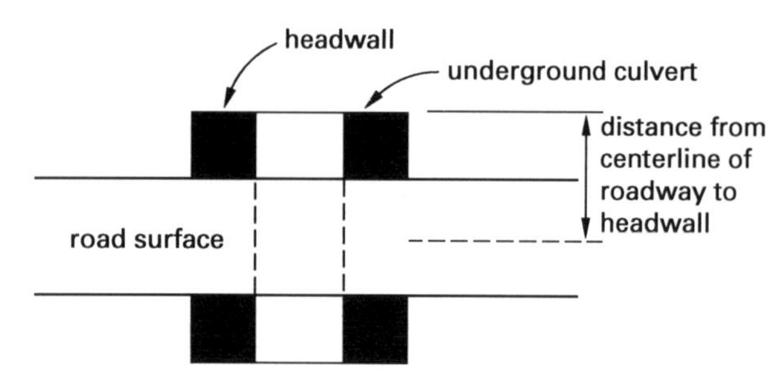

(a) normal culvert (headwall perpendicular to culvert)

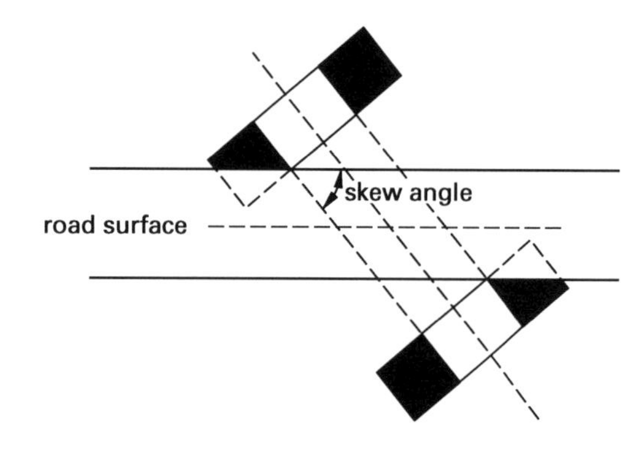

(b) skewed culvert (headwall perpendicular to culvert)

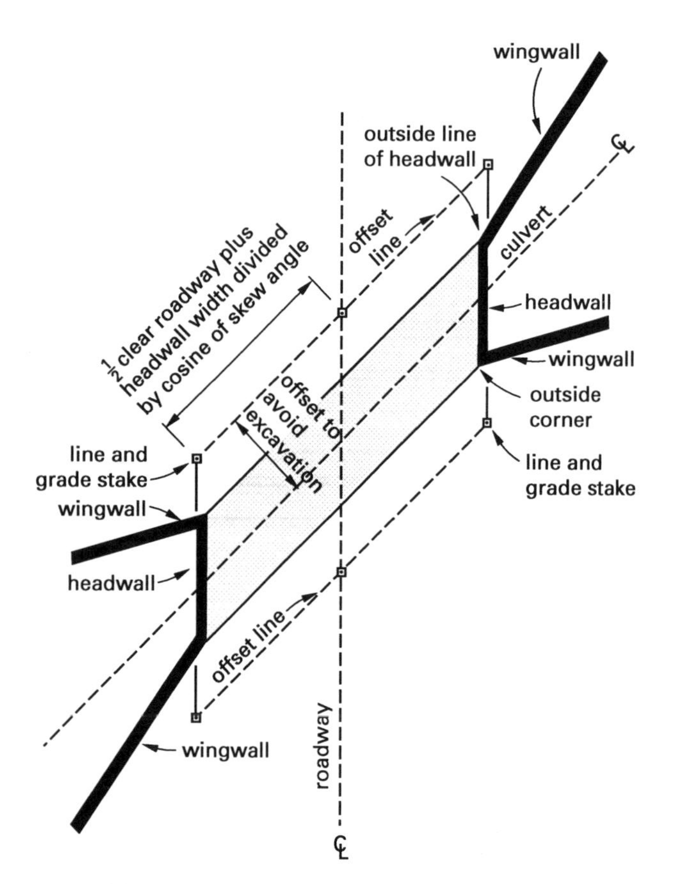

(c) skewed culvert (headwalls parallel to roadway)

10. SETTING SLOPE STAKES

Before earthwork construction is started, the extremities of a cut or fill must be located at numerous places for the benefit of machine operators engaged in the earthwork.

With the centerline as a reference, the edge (*toe*) of a fill must be established on the natural ground. This point is known as the *toe of slope*. Likewise, the top edge of a cut must be established on the natural ground.

Figure 37.5 Slope Staking

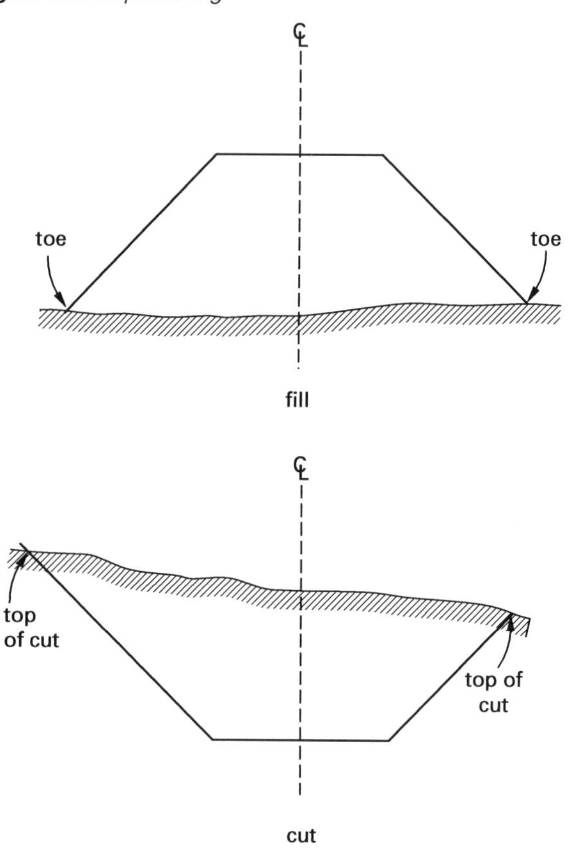

Because the natural ground may slope from left to right or from right to left, the distance from the centerline to the left toe of slope of a fill is usually different from the distance from the centerline to the right toe of slope at any particular station. The same is true of the top of a cut. This fact, plus the fact that the height of fill or depth of cut varies along the centerline, makes toe and top lines irregular when seen in plan view, as is shown in Fig. 37.6.

The toe of a fill or the top of a cut is found by a measure-and-try method. The horizontal distance from centerline to toe or top is determined by horizontal tape measurements combined with vertical distance measurements derived by use of level and rod.

Dimensions of the top of a fill or bottom of a cut, and the slope of the sides of the fill or cut must be known. These are used in the measure and try method. They are shown on the "Typical Sections Sheet" of construction plans, as in Fig. 37.7.

The side slopes of a fill and the back slopes of a cut are expressed as a ratio of horizontal to vertical distance. Thus, a 4:1 slope means a rise or fall of 1 ft for each 4 ft of horizontal distance. Slopes of 1:1, 2:1, and 3:1 are illustrated in Fig. 37.8.

With the centerline finish elevation, width of top of fill or bottom of cut, and side slopes all known, the intersection of the side slopes and the natural ground is located at each station or intermediate point.

Figure 37.6 *Fill Views*

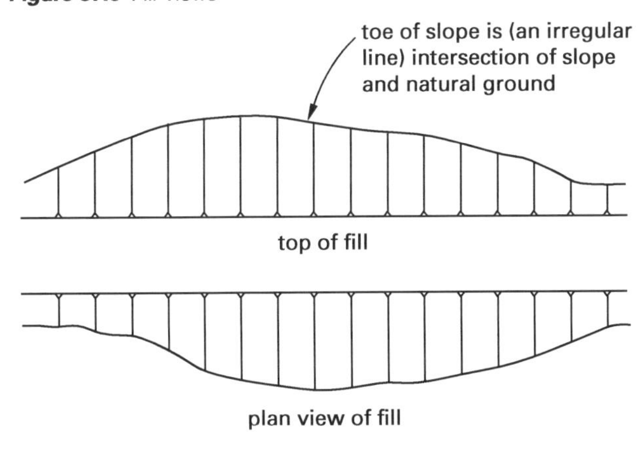

Figure 37.7 *Fill and Cut Dimensioning*

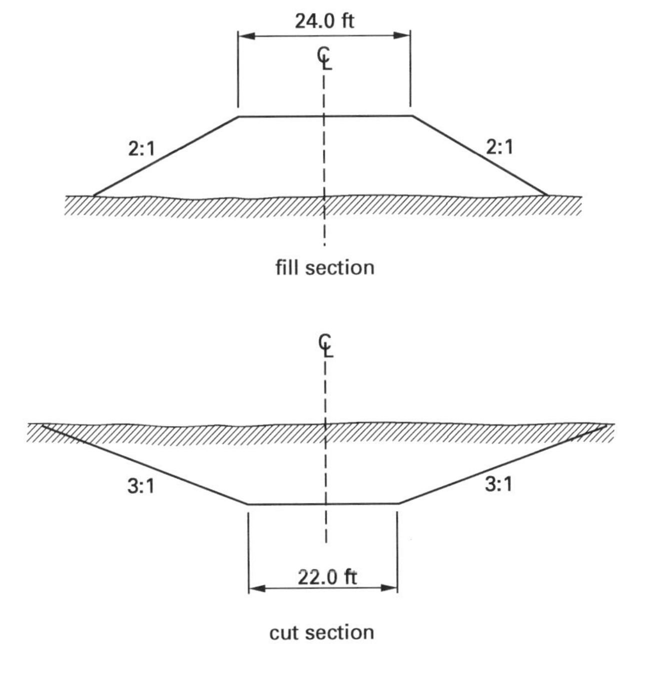

When the intersection is found, it is marked by a *slope stake*. The stake is driven so that it slopes away from the fill or cut and is marked with its horizontal distance (left or right) from the centerline and the vertical distance from the ground at the stake to the finish elevation. A stake marked "C 3.2-48.2" means that the stake is 48.2 ft from the centerline, and the ground at the stake is 3.2 ft above the finish elevation. The station number is shown on the side of the stake facing the ground.

Figure 37.8 *Calculation of Slopes*

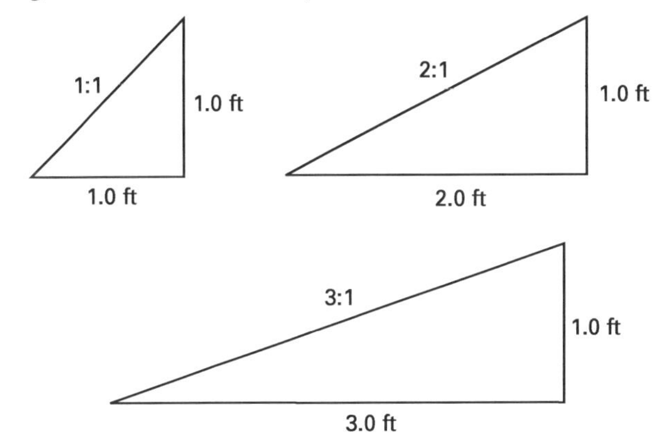

Figure 37.9 *Stake Orientations*

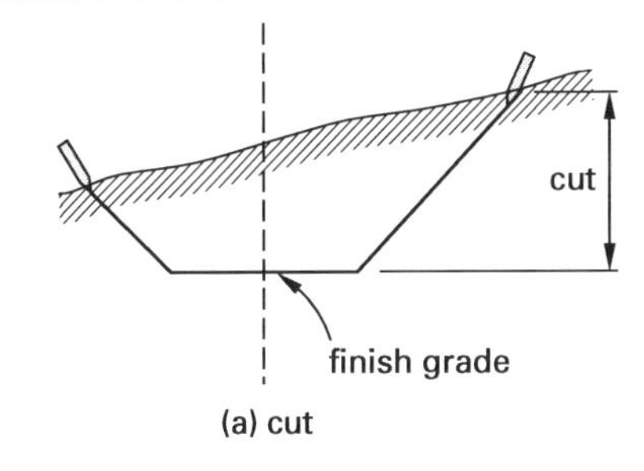

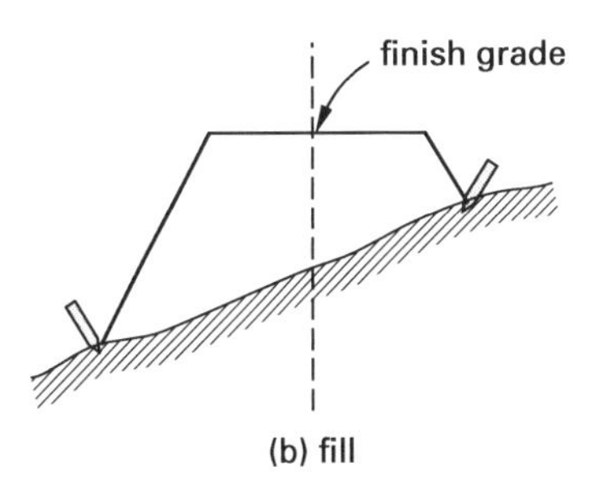

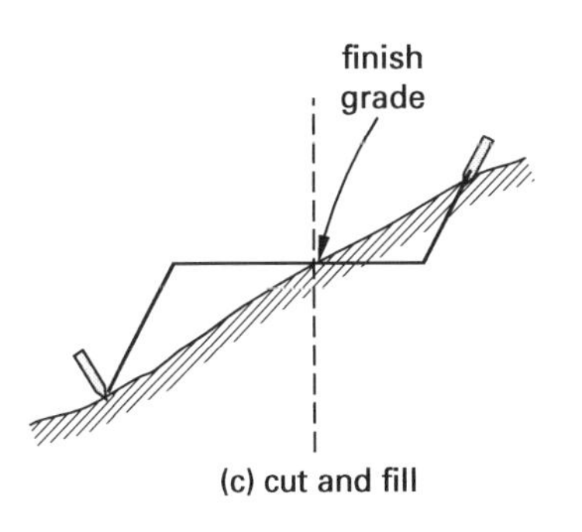

11. GRADE ROD

In setting slope stakes, as in setting finish elevation for pavement, sewer lines, and so on, the *grade rod* is used to determine the difference in elevation between the HI and the finish elevation. To determine the cut at a particular point, the rod is read on the ground, and the ground rod is subtracted from the grade rod at that point. To determine the fill at a particular point, the grade rod is subtracted from the ground rod if the HI is above the finish elevation. The grade rod is added to the ground rod if the HI is below the finish elevation. (See Fig. 37.10, Ex. 37.9, and Ex. 37.10.)

Figure 37.10 *Use of Grade Rod to Determine Cut and Fill*

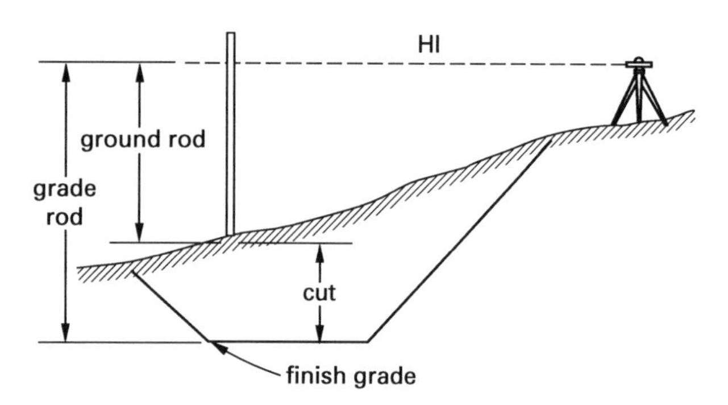

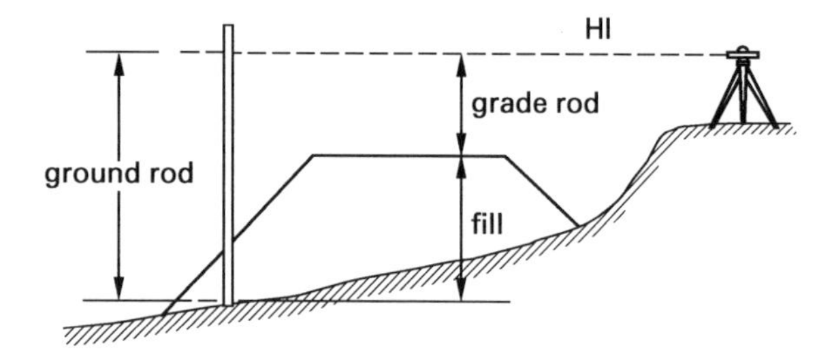

12. SETTING SLOPE STAKES AT CUT SECTIONS

An explanation of setting slope stakes without the benefit of a demonstration in the field is difficult. In Ex. 37.7, a scale drawing is used at a cut section at which the HI and finish elevation are known and plotted on the drawing. The width of the ditch bottom and the side slopes (also referred to as *back slopes*) are also known. In this example, the level and rod are replaced by the plotted HI and an engineer's scale. The scale is used to measure vertical distance from HI to ground, just as the level and rod are used.

Example 37.7

The illustration on the following page shows the ground cross section at a station at which slope stakes are to be set for a ditch to be excavated. HI has been established and finish elevation, width of ditch bottom, and side slopes have been obtained from construction plans. Centerline of ditch has also been established at this station. Two unsuccessful attempts have been made to locate the stakes, as shown. Known information is tabulated.

finish elevation of ditch bottom = 470.45 ft
bottom width = 12 ft
side slopes = 2:1
HI = 479.24 ft

Solution

step 1: Compute the grade rod (GR).

$$\begin{aligned}\text{GR} &= \text{HI} - \text{finish elevation}\\ &= 479.24 \text{ ft} - 470.46 \text{ ft} = 8.78 \text{ ft}\end{aligned}$$

step 2: Read the rod on the ground (use scale) at the centerline. A rod reading on the ground is known as a *ground rod*. This centerline ground rod helps find the cut (vertical distance from ground to finish elevation) at the centerline and the horizontal distance from the centerline to the slope stake (on each side) if the ground were level. This distance will be used as a guide to find the actual distance to the slope stake where the ground is not level. The ground rod is 4.1 ft.

$$\begin{aligned}\text{cut at centerline} &= \text{GR} - \text{ground rod}\\ &= 8.78 \text{ ft} - 4.1 \text{ ft}\\ &= 4.7 \text{ ft}\end{aligned}$$

step 3: Find the cut and horizontal distance from centerline to slope stake on the left side.

(a) The horizontal distance from centerline to left stake is equal to one half the width of the ditch bottom plus the horizontal distance from the left edge of the ditch bottom to the stake.

(b) The slope is 2:1. Therefore, the side slope will rise (from ditch bottom) 1 ft vertically for each 2 ft horizontally. For level ground, the vertical rise is the cut at the centerline, which has been found to be 4.7 ft. Therefore, the horizontal distance for level ground is

$$(2)(4.7 \text{ ft}) = 9.4 \text{ ft}$$

The distance from the centerline is

$$6 \text{ ft} + (2)(4.7 \text{ ft}) = 15.4 \text{ ft}$$

(c) The ground is not level. The slope is down from left to right, and the left slope stake will be at a greater distance from the centerline than the right slope stake.

(d) Use the horizontal distance computed for level ground (15.4 ft) as a guide. Make a first try beyond it because of the slope of the ground.

(e) Try a distance of 19.0 ft (chosen arbitrarily) from the centerline and read the rod on the ground at this point. (Use scale for rod.) The ground rod is 2.7 ft. Then,

$$\text{cut} = 8.78 \text{ ft} - 2.7 \text{ ft} = 6.1 \text{ ft}$$

$$\text{distance from centerline} = 6 \text{ ft} + (2)(6.1 \text{ ft}) = 18.2 \text{ ft}$$

This is not the correct location because the measured distance (19.0 ft) does not agree with the computed distance (18.2 ft).

(f) For the next try, move toward the centerline because 19.0 ft was too far. Try 17.0 ft where the ground rod is 3.2 ft. Then,

$$\text{cut} = 8.78 \text{ ft} - 3.2 \text{ ft} = 5.6 \text{ ft}$$

$$\text{distance from centerline} = 6 \text{ ft} + (2)(5.6 \text{ ft}) = 17.2 \text{ ft}$$

(g) Try 17.2 ft, where the ground rod is 3.2 ft again.

$$\text{distance from centerline} = 6 \text{ ft} + (2)(5.6 \text{ ft}) = 17.2 \text{ ft}$$

The correct location for the slope stake has been found, so mark it "C 5.6 @ 17.2" on one side and the station number on the other. Drive the stake with the station number down and sloping away from the cut.

Cut Cross Section for Ex. 37.7

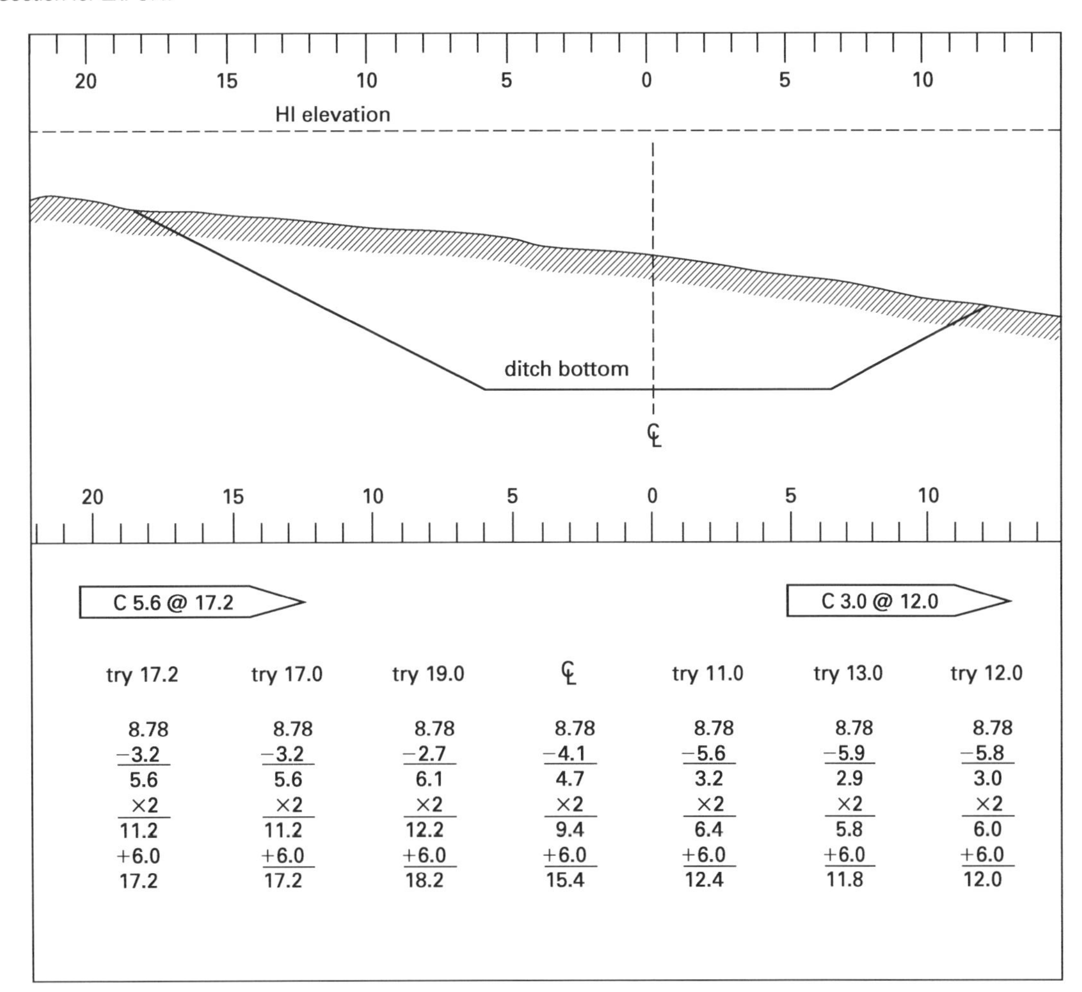

try 17.2	try 17.0	try 19.0	℄	try 11.0	try 13.0	try 12.0
8.78	8.78	8.78	8.78	8.78	8.78	8.78
−3.2	−3.2	−2.7	−4.1	−5.6	−5.9	−5.8
5.6	5.6	6.1	4.7	3.2	2.9	3.0
×2	×2	×2	×2	×2	×2	×2
11.2	11.2	12.2	9.4	6.4	5.8	6.0
+6.0	+6.0	+6.0	+6.0	+6.0	+6.0	+6.0
17.2	17.2	18.2	15.4	12.4	11.8	12.0

step 4: Find the cut and horizontal distance on the right side.

The slope stake on the right is set in the same manner. In arbitrarily selecting a horizontal distance for the first try, select a distance less than the 15.4 computed for level ground because the slope is down from left to right.

The correct cut and distance for the right slope stake is shown on the stake marking in Ex. 37.7.

Example 37.8

Slope stakes are to be set at sta 3+00. The bottom of the cut is to be at elev 462.00 ft and is 10 ft wide. The side slopes are 2:1 (All measurements are in feet.).

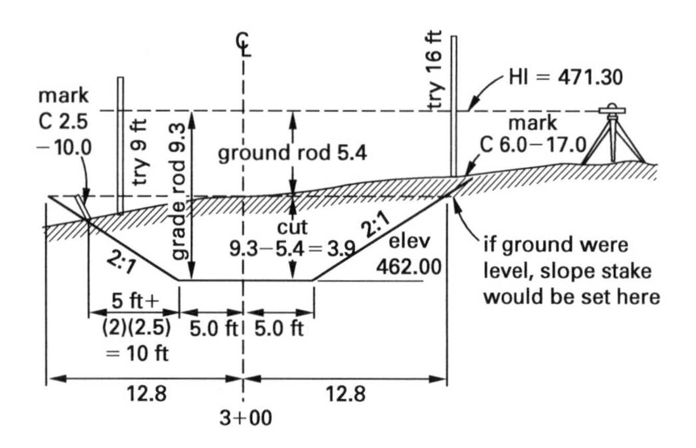

The computed distance is

$$5 \text{ ft} + (2)(2.5 \text{ ft}) = 10.0 \text{ ft}$$

Solution

step 1: Establish the level near sta 3+00 and determine HI (471.30 ft in this example).

step 2: Compute the grade rod by subtracting the elevation at the bottom of the cut from HI.

$$471.30 \text{ ft} - 462.00 \text{ ft} = 9.30 \text{ ft}$$

step 3: Determine the ground rod by placing the rod on the ground at centerline. Read 5.4.

step 4: Compute the cut at the centerline by subtracting the ground rod from the grade rod.

$$9.3 \text{ ft} - 5.4 \text{ ft} = 3.9 \text{ ft}$$

step 5: Compute the distance to the left slope stake from the centerline as if the ground were level at this station.

$$5 \text{ ft} + (2)(3.9 \text{ ft}) = 12.8 \text{ ft}$$

step 6: Note that the ground on the left slopes down and the side of the cut slopes up, indicating that the distance to the stake will be less than that for level ground.

step 7: Try a distance less than 12.8 ft, say 9.0 ft, and read rod at this distance. The rod reading is 6.6 ft.

$$\begin{aligned} \text{grade rod} - \text{ground rod} &= 9.3 \text{ ft} - 6.6 \text{ ft} \\ &= 2.7 \text{ ft} \end{aligned}$$

The distance computed from this rod reading is

$$5 \text{ ft} + (2)(2.7 \text{ ft}) = 10.4 \text{ ft}$$

Move toward 10.4 ft; try 10.0 ft. (Move less because slopes are opposite.)

step 8: The ground rod at 10.0 ft is 6.8 ft.

$$9.3 \text{ ft} - 6.8 \text{ ft} = 2.5 \text{ ft}$$

The computed distance is

$$5 \text{ ft} + (2)(2.5 \text{ ft}) = 10.0 \text{ ft}$$

The computed distance agrees with the measured distance.

step 9: Set the stake at 10.0 ft left of centerline and mark "C 2.5 @ 10.0" on top face of stake and "3+00" on bottom.

step 10: Move to the right side. Try a distance greater than that for level ground because the ground and sides both slope up.

step 11: Try 16.0; the ground rod is 3.4 ft.

$$9.3 \text{ ft} - 3.4 \text{ ft} = 5.9 \text{ ft}$$
$$5 \text{ ft} + (2)(5.9 \text{ ft}) = 16.8 \text{ ft}$$

step 12: Try 17.0 ft. Move beyond 16.8 ft because the slopes are in the same direction. The ground rod is 3.3 ft.

$$9.3 \text{ ft} - 3.3 \text{ ft} = 6.0 \text{ ft}$$
$$5 \text{ ft} + (2)(6.0 \text{ ft}) = 17.0 \text{ ft}$$

step 13: Set the stake at 17.0 ft. Mark "C 6.0 @ 17.0."

13. SETTING SLOPE STAKES AT FILL SECTIONS

In setting slope stakes for fills, two situations may arise: (a) the HI may be below the finish elevation as shown in Exs. 37.9 and 37.10, or (b) the HI may be above the finish elevation as shown in Fig. 37.10. If the HI is below the finish elevation, the fill is the sum of the grade rod and the ground rod. If the HI is above the finish elevation, the fill is the difference between the ground rod and the grade rod.

Example 37.9

The illustration on the following page shows the ground cross section at a station at which slope stakes are to be set for a fill. The HI has been established and the finish elevation, width of top of fill, and side slopes have been obtained from the construction plans. The centerline of fill has also been established. Known information is tabulated as follows.

finish elevation of top of fill = 452.36 ft
top of fill width = 4 ft
side slopes = 2:1
HI = 450.54 ft

Solution

The solution is shown on the following page. The correct cut and distance can be found on the marked stake.

Example 37.10

In the following illustration, slope stakes are to be set at sta 10+00. The top of fill is to be at elevation 468.00 ft and is 10 ft wide. Side slopes are $1^1/_2$:1.

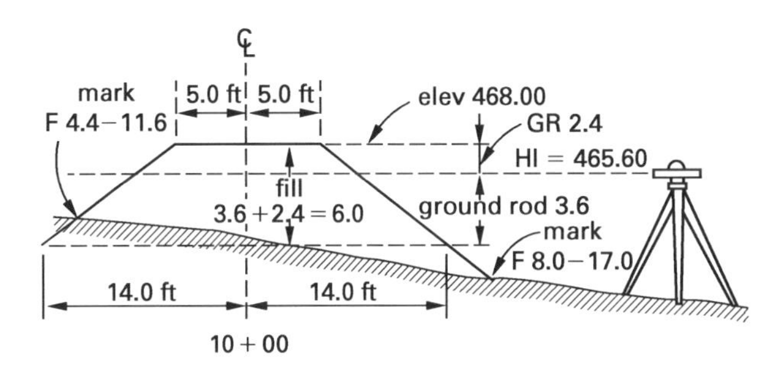

Fill Section for Example 37.9

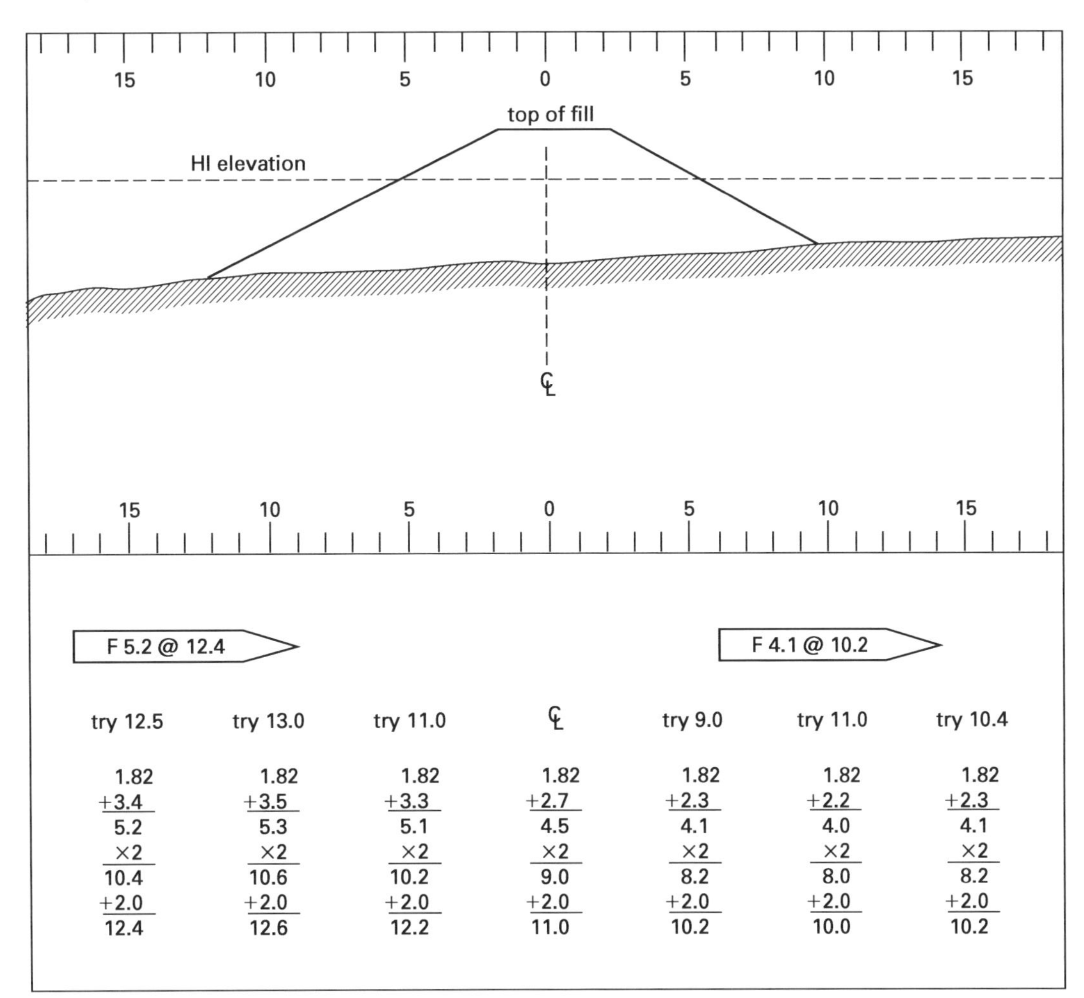

Specialty Surveying Areas

Solution

step 1: Establish the level near sta 10+00 and determine the HI (465.60 ft in this example).

step 2: Compute grade rod by subtracting the HI from the elevation of the top of the fill.

$$468.00 \text{ ft} - 465.60 \text{ ft} = 2.4 \text{ ft}$$

step 3: Determine ground rod by placing the rod on the ground at the centerline. Read 3.6 ft.

step 4: Compute the fill at the centerline by adding the grade rod and the ground rod.

$$3.6 \text{ ft} + 2.4 \text{ ft} = 6.0 \text{ ft}$$

step 5: Compute the distance to the left slope stake from the centerline as if the ground were level at this station.

$$5 \text{ ft} + (1.5)(6.0 \text{ ft}) = 14.0 \text{ ft}$$

step 6: Note that slopes are opposite, indicating that the distance will be less than that for level ground.

step 7: Try 11.0 ft. The rod reads 2.2 ft.

$$2.2 \text{ ft} + 2.4 \text{ ft} = 4.6 \text{ ft}$$

$$5 \text{ ft} + (1.5)(4.6 \text{ ft}) = 11.9 \text{ ft}$$

Move toward 11.9 ft, but less because slopes are opposite.

step 8: Try 11.5. The ground rod is 2.0 ft.

$$2.0 \text{ ft} + 2.4 \text{ ft} = 4.4 \text{ ft}$$

$$5 \text{ ft} + (1.5)(4.4 \text{ ft}) = 11.6 \text{ ft}$$

This is close enough.

step 9: Set the stake at 11.6 ft left of centerline and mark "F 4.4 @ 11.6." Mark "10+00" on the bottom face of the stake.

step 10: Move to the right side. Try a distance greater than that for level ground because ground and slope are in the same direction.

step 11: Try 15.0 ft. The ground rod is 5.3 ft.

$$5.3 \text{ ft} + 2.4 \text{ ft} = 7.7 \text{ ft}$$

$$5 \text{ ft} + (1.5)(7.7 \text{ ft}) = 16.6 \text{ ft}$$

Move toward 16.6 ft and beyond because the slopes are both down.

step 12: Try 17.0 ft. The ground rod is 5.6 ft.

$$5.6 \text{ ft} + 2.4 \text{ ft} = 8.0 \text{ ft}$$

$$5 \text{ ft} + (1.5)(8.0 \text{ ft}) = 17.0 \text{ ft}$$

step 13: Set the stake at 17.0 ft and mark "F 8.0 @ 17.0."

14. SETTING STAKES FOR UNDERGROUND PIPE

Stakes for line and grade for underground pipe, like stakes for roads and streets, are set on an offset line. One hub stake with tack can be used at each station for both line and grade, or separate stakes can be set for line and grade. If only one stake is to be used, the elevation of the top of that stake is determined. Then, the cut from the top of the stake to the flowline (*invert*) is computed and marked on a guard stake. This method is faster.

It is often desirable to set a grade stake close to the tacked line stake. This may be set so that the cut from the top of stake to the flowline is at some multiple of a half-foot.

In setting a cut stake for underground pipe, the surveyor first sets up the level and determines its HI. Using the flowline of the pipe at a particular station, the grade rod at that station is computed and recorded in the field book. A rod reading on the ground is taken at the point where the stake is to be driven. This is the ground rod. Using the grade rod and the ground rod, the rod reading on top of the stake that will give a half-foot cut from the top of the stake to the flowline is computed. A stake is driven to the rod reading that gives this cut. The stake is blued, and the cut is marked on the guard stake.

Example 37.11

A grade stake is to be set to show the cut to the flowline of a pipe. The HI is 472.36 ft, the flowline is 462.91 ft, and the ground rod at the point of stake is 5.1 ft. Determine the grade rod and rod reading that will give a half-foot cut to the flowline.

Solution

$$\begin{aligned} \text{HI} &= 472.36 \text{ ft} \\ \text{flowline} &= \underline{462.91 \text{ ft}} \\ \text{GR} &= \quad 9.45 \text{ ft} \end{aligned}$$

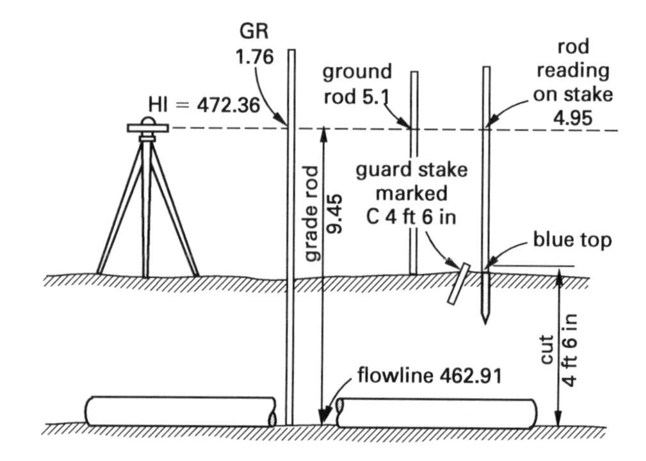

The rod reading on the grade stake to give a half-foot cut from the top of the stake to the flowline could be 8.95 ft, 8.45 ft, ..., 5.45 ft, 4.95 ft, and so on, and the corresponding cuts would be 0 ft 6 in, 1 ft 0 in, 4 ft 0 in, 4 ft 6 in, and so on. The ground rod is 5.1 ft; therefore, the cut is approximately 9.5 ft−5.1 ft = 4.4 ft. Therefore, the rod reading for this cut will be either 5.45 ft (which will give a cut of 4 ft 0 in) or 4.95 ft (which will give a cut of 4 ft 6 in).

A rod reading of 5.45 ft cannot be used because the top of the stake would be 0.3 ft below the surface of the ground. A rod reading of 4.95 ft would place the top of the stake about 0.1 ft above ground, which is satisfactory. The stake is driven so that the rod reading is 4.95 ft. The top is marked with blue keel, and the guard stake is marked "C 4 ft, 6 in" since 9.45 ft − 4.95 ft = 4.50 ft (4 ft 6 in). It can be seen that the rod reading on the stake must be less than the ground rod in order that the top of the stake be above ground.

15. FLOWLINE AND INVERT

The bottom inside of a drainage pipe is known as the *flowline*.[2] It is also referred to as the *invert*. Invert is more commonly used to describe the bottom of the flow channel within a manhole.

Vertical control is of prime importance in laying pipe for gravity flow, especially sanitary sewer pipe. In order to facilitate vertical alignment, excavation of the trench often extends a few inches below the bottom of the pipe so that a bedding material, such as sand, can be placed in the trench for the pipe to lie on. Because of various methods of using bedding material in laying pipe, stakes are always set for the flowline, or invert, of the pipes. Excavation depth allows for the amount of bedding specified.

16. MANHOLES

Sanitary sewers are not laid along horizontal or vertical curves. Horizontal and vertical alignments are straight lines. Where a change in horizontal alignment or a change in slope is necessary, a *manhole* is required at the point of change. Therefore, a vertical drop within the manhole is needed. In staking, two cuts are often recorded on guard stakes: one for the incoming sewer and one for the outgoing sewer.

Gravity lines, such as sanitary sewers, flow only partially full. The slope of the sewer determines the flow velocity, and the velocity and size of the pipe determine the quantity of flow. Manholes are used to provide a point of change in conditions. Sewers must be deep enough below the surface of the ground to prevent freezing of their contents and damage to the pipe by construction equipment.

[2] *Flowlines* are the lines used as finish elevation for pipes.

PRACTICE PROBLEMS

1. Stakes are to be set on 4 ft offsets for each edge of pavement (which is 28 ft wide), for a curve that has a deflection angle of 55°00′ and a centerline radius of 250 ft. The PI is at station 8+56.45. Stakes are to be set on full-stations and half-stations, and at the PC and PT. (a) Calculate T and L. (b) Determine the deflection angles used to stake the curve. (c) Calculate the outside and inside chord lengths.

2. Prepare a set of field notes to be used in staking a street curve on the quarter-stations from 3 ft offset lines on both sides of the street.

$$\text{PI} = 4+55.00$$
$$\Delta = 60°00' \quad \text{(angle to the left)}$$
$$R = 100 \text{ ft} \quad \text{(centerline)}$$
$$\text{pavement width} = 28 \text{ ft}$$

3. The intersection of Ash Lane and 32nd Street is to be staked for paving from an offset line 4 ft left of the left edge of pavement. The pavement width is 28 ft, and the radius to the edge of the pavement is 30 ft. From this information and information shown on the following sketch, compute PC and PT stations and deflection angles along with chord lengths from PC to PT. The scale is 1/2 in = 30 ft.

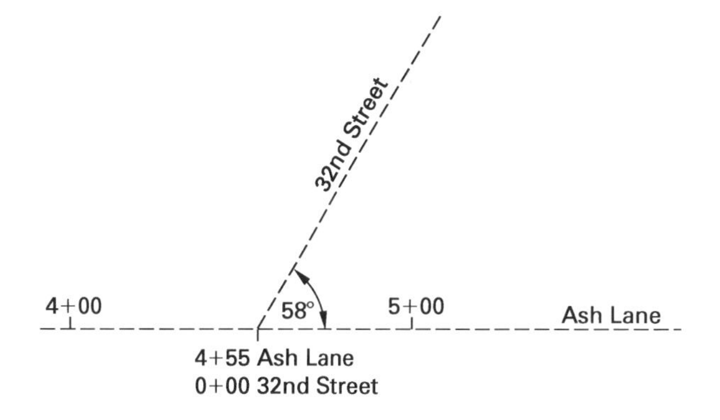

Specialty Surveying Areas

SOLUTIONS

1. (a)

$$T = (250 \text{ ft}) \tan\left(\frac{55^\circ}{2}\right) = \boxed{130.14}$$

$$L = \frac{(55^\circ)2\pi(250 \text{ ft})}{360^\circ} = \boxed{239.98 \text{ ft}}$$

$$\text{PI} = 8{+}56.45$$
$$T = 1{+}30.14$$
$$\text{PC} = 7{+}26.31$$
$$L = 2{+}39.98$$
$$\text{PT} = 9{+}66.29$$

(b)

point	station		deflection angle
PC	7+26.31		0° (0°00′)
	7+50	$\left(\frac{23.69 \text{ ft}}{239.98 \text{ ft}}\right)(27^\circ 30')$	2.7147° (2°43′)
	8+00	$\left(\frac{73.69 \text{ ft}}{239.98 \text{ ft}}\right)(27^\circ 30')$	8.4443° (8°27′)
	8+50	$\left(\frac{123.69 \text{ ft}}{239.98 \text{ ft}}\right)(27^\circ 30')$	14.1740° (14°10′)
	9+00	$\left(\frac{173.69 \text{ ft}}{239.98 \text{ ft}}\right)(27^\circ 30')$	19.9036° (19°54′)
	9+50	$\left(\frac{223.69 \text{ ft}}{239.98 \text{ ft}}\right)(27^\circ 30')$	25.6333° (25°38′)
PT	9+66.29	$\left(\frac{239.98 \text{ ft}}{239.98 \text{ ft}}\right)(27^\circ 30')$	27.5000° (27°30′)

(c) The outside chord lengths are

$$7{+}26.31 \text{ to } 7{+}50\text{: } (2)(268 \text{ ft})(\sin 2.7147^\circ) = \boxed{25.39 \text{ ft}}$$
$$7{+}50 \text{ to } 8{+}00\text{: } (2)(268 \text{ ft})(\sin 5.7296^\circ) = \boxed{53.51 \text{ ft}}$$
$$9{+}50 \text{ to } 9{+}66.29\text{: } (2)(268 \text{ ft})(\sin 1.8667^\circ) = \boxed{17.46 \text{ ft}}$$

The inside chord lengths are

$$7{+}26.31 \text{ to } 7{+}50\text{: } (2)(232 \text{ ft})(\sin 2.7147^\circ) = \boxed{21.98 \text{ ft}}$$
$$7{+}50 \text{ to } 8{+}00\text{: } (2)(232 \text{ ft})(\sin 5.7296^\circ) = \boxed{46.32 \text{ ft}}$$
$$9{+}50 \text{ to } 9{+}66.29\text{: } (2)(232 \text{ ft})(\sin 1.8667^\circ) = \boxed{15.11 \text{ ft}}$$

2.

Elm Street

point	station	deflection angle	C-in (ft)	C-out (ft)	curve data
PT	5+01.98	30°00′			
			1.65	2.33	
	5+00	29°26′			$\Delta = 60^\circ$ to the left
			20.70	29.17	$R = 100$ ft
	4+75	22°16′			$T = 57.72$ ft
			20.70	29.17	$L = 104.72$ ft
	4+50	15°07′			
			20.70	29.17	
	4+25	7°57′			
			20.70	29.17	
	4+00	0°47′			
			2.27	3.19	
PC	3+97.26	0°00′			

3.

$$\Delta_1 = 58^\circ 00' \text{left}$$
$$T_1 = (44 \text{ ft}) \tan\left(\frac{58^\circ}{2}\right) = 24.39 \text{ ft}$$
$$\text{PI} = 4{+}55.00$$
$$T_1 = 24.39 \text{ ft}$$
$$\text{PC} = \boxed{4{+}30.61}$$
$$\text{PI} = 0{+}00.00$$
$$T_1 = 24.39 \text{ ft}$$
$$\text{PT} = \boxed{0{+}24.39}$$
$$\text{LC}_1 = (2)(26 \text{ ft}) \sin\left(\frac{58^\circ}{2}\right) = \boxed{25.21 \text{ ft}}$$
$$\text{deflection angle} = \boxed{29^\circ 00'}$$
$$\Delta_2 = 122^\circ 00'$$
$$T_2 = (44 \text{ ft}) \tan\left(\frac{122^\circ}{2}\right) = 79.38 \text{ ft}$$
$$\text{PI} = 4{+}55.00$$
$$T_2 = 79.38 \text{ ft}$$
$$\text{PC} = \boxed{5{+}34.38}$$
$$\text{PI} = 0{+}00.00$$
$$T_2 = 79.38 \text{ ft}$$
$$\text{PT} = \boxed{0{+}79.38}$$
$$\text{LC}_2 = (2)(26 \text{ ft}) \sin\left(\frac{122^\circ}{2}\right) = \boxed{45.48 \text{ ft}}$$
$$\text{deflection angle} = \boxed{61^\circ 00'}$$

38 Earthwork

1. DEFINITION

Earthwork is the excavation, hauling, and placing of soil, rock, gravel, or other material found below the surface of the earth. This definition also includes the measurement of such material in the field, the computation in the office of the volume of such material, and the determination of the most economical method of performing such work.

2. UNIT OF MEASURE

The *cubic yard* (i.e., the "yard") is the unit of measure for earthwork. However, the volume and density of earth changes under natural conditions and during the operations of excavation, hauling, and placing.

3. SWELL AND SHRINKAGE

A cubic yard of earth measured in its natural position will be more than a cubic yard after it is excavated. If the earth is compacted after it is placed, the volume may be less than a cubic yard.

The volume of the earth in its natural state is known as *bank-measure*. The volume in the vehicle is known as *loose-measure*. The volume after compaction is known as *compacted-measure*.

The change in volume of earth from its natural to loose state is known as *swell*. Swell is expressed as a percent of the natural volume.

The change in volume of earth from its natural state to its compacted state is known as *shrinkage*. Shrinkage also is expressed as a percent decrease from the natural state.

As an example, 1 yd^3 in the ground may become 1.2 yd^3 loose-measure and 0.85 yd^3 after compaction. The swell would be 20%, and the shrinkage would be 15%. Swell and shrinkage vary with soil types.

4. CLASSIFICATION OF MATERIALS

Excavated material is usually classified as *common excavation* or *rock excavation*. Common excavation is soil.

In highway construction, common road excavation is soil found in the roadway. *Common borrow* is soil found outside the roadway and brought in to the roadway. Borrow is necessary where there is not enough material in the roadway excavation to provide for the embankment.

5. CUT AND FILL

Earthwork that is excavated, or is to be excavated, is known as *cut*. Excavation that is placed in embankment is known as *fill*.

Payment for earthwork is normally either for cut and not for fill, or for fill and not for cut. In highway work, payment is usually for cut; in dam work, payment is usually for fill. To pay for both would require measuring two different volumes and paying for moving the same earth twice.

6. FIELD MEASUREMENT

Cut and fill volumes can be computed from slope-stake notes, from plan cross sections, or by photogrammetric methods.

7. CROSS SECTIONS

Cross sections are profiles of the earth taken at right angles to the centerline of an engineering project (such as a highway, canal, dam, or railroad). A cross section for a highway is shown in Fig. 38.1.

Figure 38.1 *Typical Highway Cross Section*

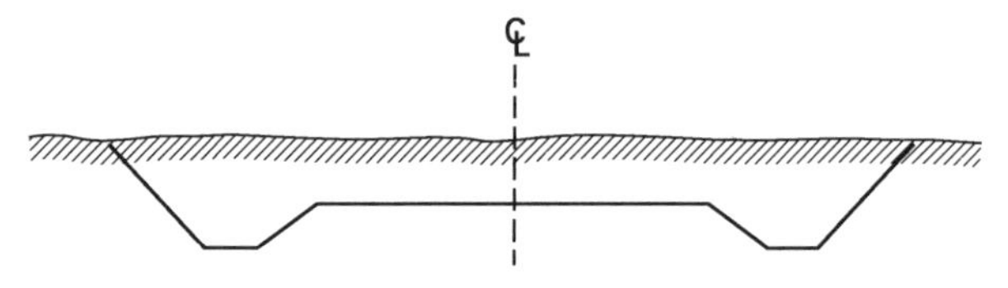

8. ORIGINAL AND FINAL CROSS SECTIONS

To obtain volume measurement, cross sections are taken before construction begins and after it is completed. By plotting the cross section at a particular station both before and after construction, a sectional view of the change in the profile of the earth along a certain line is obtained. The change along this line appears on the plan as an area. By using these areas at various intervals along the centerline, and by using distance between the areas, volume can be computed.

9. ESTIMATING EARTHWORK

Earthwork quantities for a highway, canal, or other project can be estimated by superimposing a template on the original plotted cross section, which is drawn to represent the final cross section. The template is obtained from the *typical section sheet* of the construction plans.

10. TYPICAL SECTIONS

Typical sections show the cross section view of the project as it will look on completion, including all dimensions. Highway projects usually show several typical sections including cut sections, fill sections, and sections showing both cut and fill. Interstate highway plans also show access-road sections and sections at ramps.

11. DISTANCE BETWEEN CROSS SECTIONS

Cross sections are usually taken at each full station and at breaks in the ground along the centerline. In taking cross sections, it must be assumed that the change in the earth's surface from one cross section to the next is uniform, and that a section halfway between the cross sections is an average of the two. If the ground breaks appreciably between any two full-stations, one or more cross sections between full-stations must be taken. This is referred to as *taking sections at pluses.* Figure 38.3 shows the stations at which cross sections should be taken.

Figure 38.3 *Cross-Section Distances*

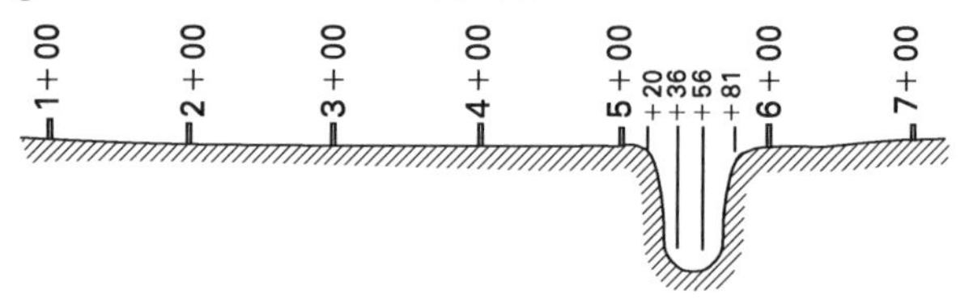

In rock excavation, or any other expensive operation, cross sections should be taken at intervals of 50 ft or less. Cross sections should always be taken at the PC and PT of a curve. Plans should also show a section on each end of a project (where no construction is to take place) so that changes caused by construction will not be abrupt.

Where a cut section of a highway is to change to a fill section, several additional cross sections are needed. Such sections are shown in Fig. 38.4.

Figure 38.2 *Typical Completed Section*

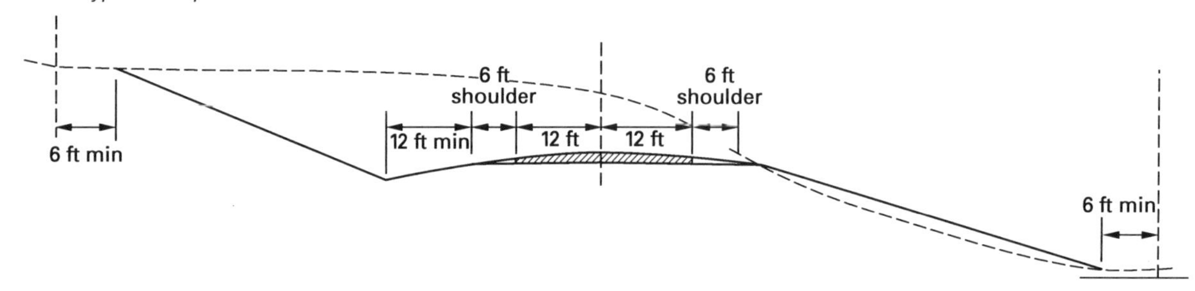

Figure 38.4 Cut Changing to Fill

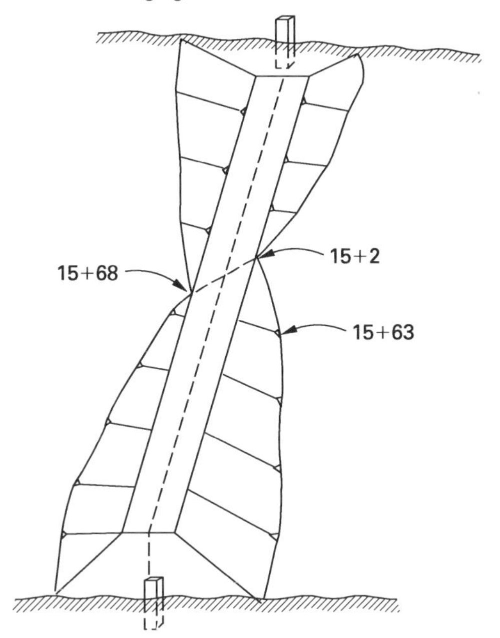

12. GRADE POINT

The point where a fill section meets the natural ground (where a cut section begins) is known as a *grade point.*

13. METHODS FOR COMPUTING VOLUME

The most common method for computing volume of earthwork is by the *average end area method.* (The *Prismoidal formula* furnishes more accurate results but is more complex.) The average end area method is accurate enough for most work.

14. AVERAGE END AREA METHOD

The average end area method is based on the assumption that the volume of earthwork between two vertical cross sections A_1 and A_2 is equal to the average of the two end areas multiplied by the horizontal distance L between them. Area is expressed in square feet, and distance is expressed in feet. So, the volume in cubic yards is

$$V = \frac{L(A_1 + A_2)}{(2)\left(27\ \frac{\text{ft}^3}{\text{yd}^3}\right)} \qquad 38.1$$

15. FIELD NOTES

Figure 38.5 shows a sample sheet from cross-section field notes. The left half of the page is the same as for any set of level notes. The right half shows rod readings

Figure 38.5 Typical Field Notes for Cross-Section Work

station	+	HI	−	rod elevation					℄				
BM no. 12	3.30	468.21		464.91	r.r. spike in 12 in elm 125 rt sta 16+75								
11+00					$\frac{4.4}{50}$	$\frac{7.1}{15}$	$\frac{4.9}{12}$	$\frac{7.3}{5}$	7.9	$\frac{9.1}{20}$	$\frac{12.0}{25}$	$\frac{9.7}{30}$	$\frac{11.2}{50}$
TP$_4$	5.27	462.97	10.51	457.70									
12+00					$\frac{2.0}{50}$	$\frac{4.0}{20}$	$\frac{1.3}{15}$	$\frac{4.5}{10}$	5.0	$\frac{6.0}{15}$	$\frac{10.1}{20}$	$\frac{5.7}{27}$	$\frac{8.0}{50}$
13+00					$\frac{4.8}{50}$	$\frac{6.0}{25}$	$\frac{4.2}{20}$	$\frac{7.7}{15}$	9.9	$\frac{9.8}{8}$	$\frac{12.6}{15}$	$\frac{11.0}{24}$	$\frac{13.0}{50}$
TP$_5$	1.76	458.22	6.51	456.46									
13+50					$\frac{2.3}{50}$	$\frac{4.1}{30}$	$\frac{1.2}{25}$	$\frac{5.0}{20}$	6.7	$\frac{7.1}{3}$	$\frac{12.2}{10}$	$\frac{8.3}{19}$	$\frac{11.1}{50}$
14+00					$\frac{5.2}{50}$	$\frac{6.0}{42}$	$\frac{10.2}{35}$	$\frac{7.9}{20}$	10.1	$\frac{11.0}{20}$	$\frac{8.1}{50}$		
15+00					$\frac{5.0}{50}$	$\frac{5.8}{48}$	$\frac{9.6}{40}$	$\frac{7.3}{20}$	9.2	$\frac{10.1}{22}$	$\frac{7.5}{50}$		
BM no. 13			5.15	453.07									

over horizontal distance measured from the centerline for each point on the ground that requires a reading. These readings should always include shots on the centerline, at each break in the ground, and at the right-of-way on each side.

16. PLOTTING CROSS SECTIONS

When drawn by hand, cross sections are plotted on specially printed cross-section paper. A scale of 1 in = 5 ft is usually used for both the horizontal and vertical. For wide sections, a scale of 1 in = 10 ft or 1 in = 20 ft can be used. The vertical scale can also be exaggerated if necessary.

A vertical line in the center of the sheet is drawn to represent the centerline of the project. Shots taken in the field are plotted to the proper elevation and distance from the centerline.

Each cross section is plotted as a separate section. Sufficient space is allowed between cross sections so that they do not overlap. The station number for each cross section is recorded just under the centerline shot, and the elevation at the centerline is recorded in a vertical direction just above the centerline shot.

The heavy lines on the paper are used to represent an elevation ending in 0 or 5 ft. With the elevation of the centerline recorded, these heavy lines can be identified as the elevation they represent.

The notes can be plotted by first reducing the level shots to elevation and then plotting by elevation. Alternatively, the rod shots can be plotted directly from a line on the paper representing the HI. As an example, if the HI is 447.6 ft (rounded off) and the rod shot is 5.4 ft, subtracting 5.4 ft from 7.6 ft gives an elevation of 442.2 ft.

Figure 38.6 Plotting Cross Sections

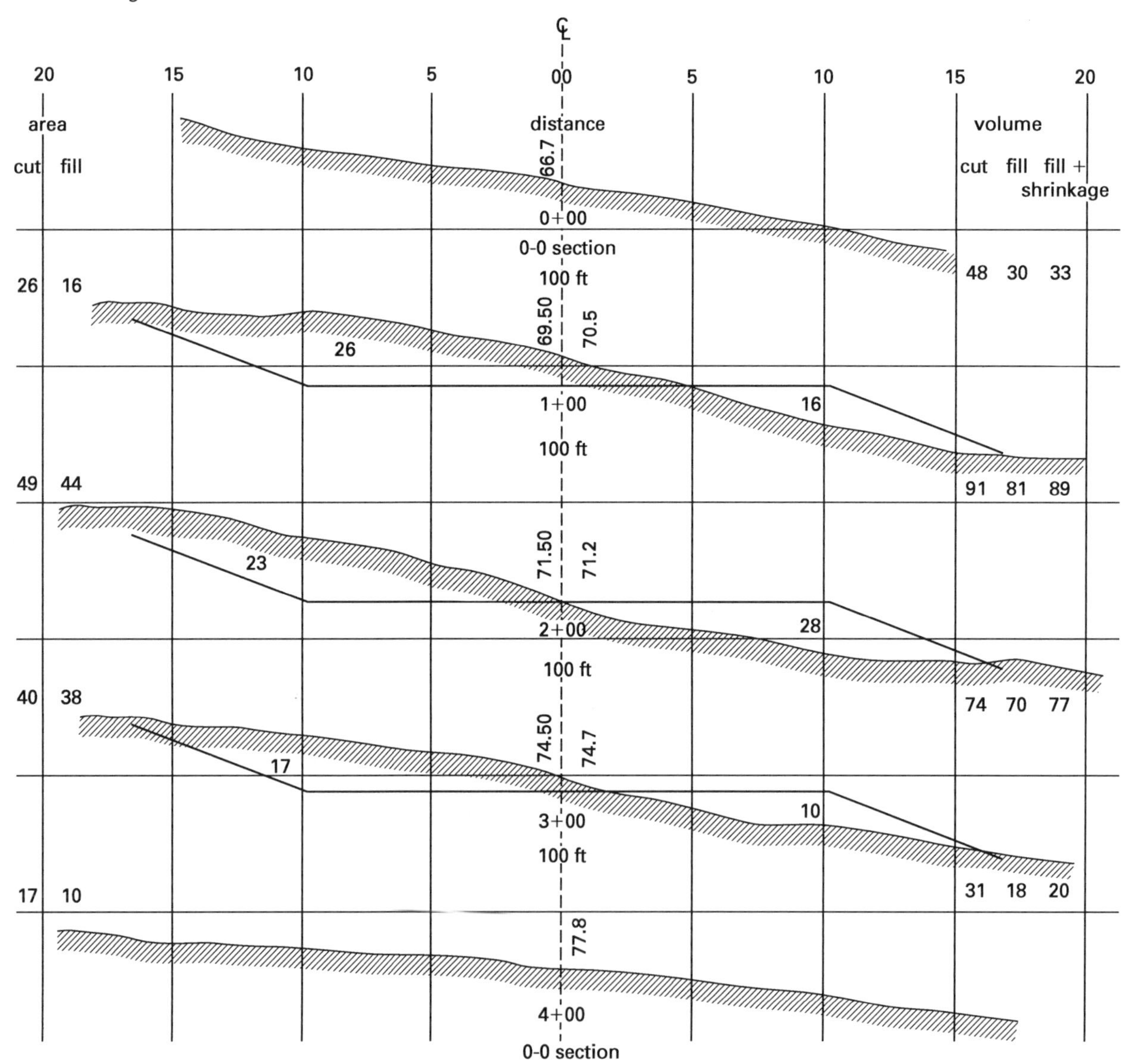

17. DETERMINING END AREAS

End areas are commonly determined by planimetery or by dividing the area into triangles and trapezoids.

Cut areas and fill areas must be kept separate. After the areas have been determined, the sum of each two adjacent areas is placed in a column. The distance between two sections is recorded, and the volume for each sum is computed from Eq. 38.1.

After the volume has been computed, shrinkage must be added to fill quantities to balance with cut quantities. Shrinkage will vary from 30% for light cuts and fills to 10% for heavy cuts and fills.

Excavated rock will occupy larger volume when placed in a fill, and this swell will be subtracted from the fill quantity.

18. VOLUMES FROM PROFILES

For preliminary estimates of earthwork, volumes can be computed from the centerline profiles. After the ground profile and finish grade profile are plotted, the area of cut can be planimetered and the average determined by dividing by the length of cut. Using the average cut, a template can be drawn, and the end area can also be planimetered. This area times the length of the cut will give the volume.

19. BORROW PIT

As mentioned previously, it is often necessary to borrow earth from an adjacent area to construct embankments.

Normally, the *borrow pit* area is laid out in a rectangular grid with 10 ft, 50 ft, or even 100 ft squares. Elevations are determined at the corners of each square by leveling before and after excavation so that the cut at each corner can be computed.

Points outside the cut area are established on the grid lines so that the lines can be reestablished after excavation is completed.

As an example, volumes for two of the prisms shown in Fig. 38.7 are computed by multiplying the average cut by the area of the figure. The volume of the prism A0-B0-B1-A1 is

$$V = \left(\frac{(50\text{ ft})(50\text{ ft})}{27\ \frac{\text{ft}^3}{\text{yd}^3}}\right) \times \left(\frac{3.2\text{ ft} + 3.4\text{ ft} + 3.0\text{ ft} + 2.6\text{ ft}}{4}\right)$$
$$= 282\text{ yd}^3$$

The volume of the triangle E2-F2-E3 is

$$V = \left(\frac{(50\text{ ft})(15\text{ ft})}{(2)\left(27\ \frac{\text{ft}^3}{\text{yd}^3}\right)}\right)\left(\frac{2.3\text{ ft} + 2.4\text{ ft} + 2.4\text{ ft}}{3}\right)$$
$$= 33\text{ yd}^3$$

Figure 38.7 Borrow Pit Areas

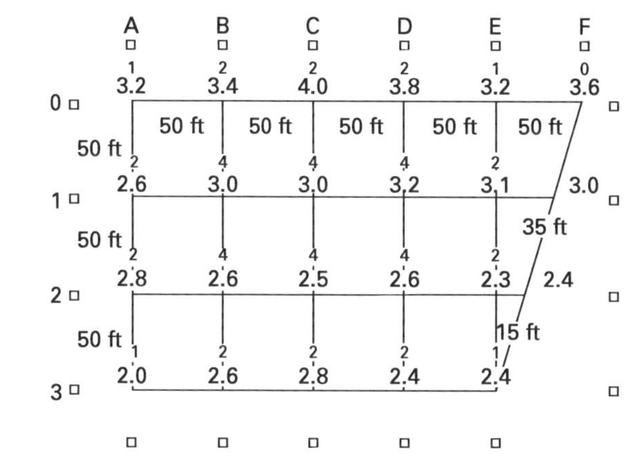

Instead of computing volumes of prisms represented by squares separately, all square-based prisms can be computed collectively by multiplying the area of one square by the sum of the cut at each corner times the number of times that cut appears in any square, divided by 4. For instance, on the second line from the top in Fig. 38.7, which is line 1, 2.6 appears in two squares, 3.0 appears in four squares, 3.0 appears in four squares, 3.2 appears in four squares, and 3.1 appears in two squares. In the figure, the small number above the cut indicates the number of times the cut is used in averaging the cuts for the prisms.

Example 38.1

Calculate the volume of earth excavated from the borrow pit shown in Fig. 38.7.

Solution

The volume of the squares is

$$V = \left(\frac{50\text{ ft}}{\left(27\ \frac{\text{ft}^3}{\text{yd}^3}\right)(4)}\right)(50\text{ ft}) \times \left(\begin{array}{l} 3.2\text{ ft} + (2)(3.4\text{ ft}) + (2)(4.0\text{ ft}) \\ \quad + (2)(3.8\text{ ft}) + 3.2\text{ ft} + (2)(2.6\text{ ft}) \\ \quad + (4)(3.0\text{ ft}) + (4)(3.0\text{ ft}) + (4)(3.2\text{ ft}) \\ \quad + (2)(3.1\text{ ft}) + (2)(2.8\text{ ft}) + (4)(2.6\text{ ft}) \\ \quad + (4)(2.5\text{ ft}) + (4)(2.6\text{ ft}) + (2)(2.3\text{ ft}) \\ \quad + 2.0 + (2)(2.6\text{ ft}) + (2)(2.8\text{ ft}) \\ \quad + (2)(2.4\text{ ft}) + 2.4\text{ ft} \end{array}\right)$$
$$= 3194\text{ yd}^3$$

The volume of the trapezoids is

$$V = \left(\frac{50 \text{ ft} + 35 \text{ ft}}{(2)\left(27 \ \frac{\text{ft}^3}{\text{yd}^3}\right)} \right)(50 \text{ ft}) \times \left(\frac{3.2 \text{ ft} + 3.6 \text{ ft} + 3.1 \text{ ft} + 3.0 \text{ ft}}{4} \right) = 254 \text{ yd}^3$$

$$V = \left(\frac{35 \text{ ft} + 15 \text{ ft}}{(2)\left(27 \ \frac{\text{ft}^3}{\text{yd}^3}\right)} \right)(50 \text{ ft}) \times \left(\frac{3.1 \text{ ft} + 3.0 \text{ ft} + 2.3 \text{ ft} + 2.4 \text{ ft}}{4} \right) = 125 \text{ yd}^3$$

The volume of the triangle is

$$V = \left(\frac{(15 \text{ ft})}{(2)\left(27 \ \frac{\text{ft}^3}{\text{yd}^3}\right)} \right)(50 \text{ ft}) \times \left(\frac{2.3 \text{ ft} + 2.4 \text{ ft} + 2.4 \text{ ft}}{3} \right) = 33 \text{ yd}^3$$

The total volume of earth excavated is

$$3194 \text{ yd}^3 + 254 \text{ yd}^3 + 125 \text{ yd}^3 + 33 \text{ yd}^3 = 3606 \text{ yd}^3$$

20. HAUL

In some contracts for highways and railroads, the contractor is paid per cubic yard for excavation (which includes the cost of excavation, hauling, placing in embankment, and compaction of embankment). However, the cost of hauling one cubic yard of earth over a long distance can easily become greater than the cost of excavation, so that it is often practical to pay a contractor for excavating and hauling earth.

21. FREE HAUL

It is common not to pay for hauling if the material is hauled less than a certain distance, usually 500 ft to 1000 ft. An additional price is paid for hauling the earth beyond the prescribed limit. The haul distance for which no pay is received is known as *free haul.*

22. OVERHAUL

The hauling of material beyond the free haul limit is known as *overhaul.* The unit of overhaul measure is yard-stations or yard-quarters. A *yard-quarter* is the hauling of one cubic yard of earth a distance of 1/4 mi. For example, if six yards of earth were hauled 4 mi, the overhaul would be twenty-four yard-quarters.

Thus, the word "haul" may have two meanings. It may mean linear distance or volume times distance.

It should be mentioned that the distance is measured along the centerline. Distance from the extremity of the right-of-way to the centerline is not considered.

23. BALANCE POINTS

It is important in planning and construction to know the points along the centerline a particular section of cut that will balance a particular section of fill. For example, assume that a cut section extends from sta 12+25 to sta 18+65, and a fill section extends from sta 18+65 to sta 26+80. Also, assume that the excavated material will exactly provide the material needed to make the embankment. Then, cut balances fill, and sta 12+25 and 26+80 are *balance points.*

24. MASS DIAGRAMS

A method of determining economical handling of material, quantities of overhaul, and location of balance points is the mass-diagram method.

The *mass diagram* is a graph that has distance in stations as the abscissa and the cumulative earthwork (i.e., the algebraic sums of cut and fill) as ordinate. The x-axis parallels the centerline, and the cut and fill (plus shrinkage) quantities are taken from the cross-section sheets. Often, the mass diagram is plotted below the centerline profile so that like stations are vertically in line.

To add cut and fill algebraically, cut is given a plus sign, and fill is given a minus sign.

25. PLOTTING THE MASS DIAGRAM

After volumes of cut and fill between stations have been computed, they are tabulated as shown in Table 38.1. The cuts and fills are then added, and the cumulative yardage at each station is recorded in the table. It is this cumulative yardage that is plotted as an ordinate. In Fig. 38.8, the baseline serves as the x-axis and cumulative yardage that has a plus sign is plotted above the baseline. Cumulative yardage that has a minus sign is plotted below the baseline.

The scale is not important. In Fig. 38.8, the horizontal scale is 1 in = 5 stations, and the vertical scale is 1 in = 5000 yd^3. A larger scale would be more practical in actual computations.

Figure 38.8 *Baseline and Centerline Profile Mass Diagram*

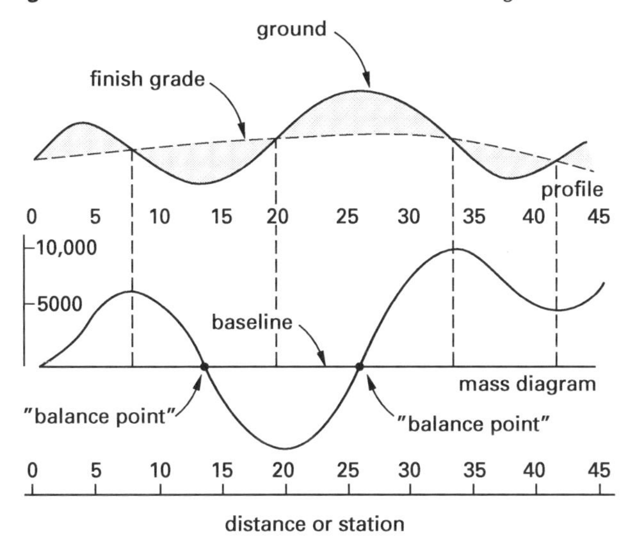

In Fig. 38.8, the mass diagram is plotted on the lower half of the sheet, and the centerline profile of the project is plotted on the upper half.

26. BALANCE LINE

Any horizontal line cutting off a loop of the mass curve intersects the curve at two points, between which the cut is equal to the fill. Figure 38.9 shows a portion of the mass diagram of Fig. 38.8 with an enlarged scale.

27. SUB-BASES

Sub-bases are horizontal balance lines that divide an area of the mass diagram between two balance points into trapezoids for the purpose of more accurately computing overhaul. In Fig. 38.9, the top sub-base is less than 600 ft in length. Therefore, the volume of earth represented by the area above this line will be hauled a distance less than the free-haul distance, and no payment will be made for overhaul. All the volume represented by the area below this top sub-base will receive payment for overhaul.

The area between two sub-bases is very nearly trapezoidal. The average length of the bases of a trapezoid can be measured in feet, and the altitude of the trapezoid can be measured in cubic yards of earth. The product of these two quantities can be expressed in yard-quarters. If free haul is subtracted from the length, the quantity can be expressed as overhaul.

Sub-bases are drawn at distinct breaks in the mass curve. Distinct breaks in Fig. 38.9 can be seen at sta 1+00(184), 2+00(806), and 5+00(4340). After sub-bases are drawn, a horizontal line is drawn midway between the sub-bases. This line represents the average haul for the volume of earth between the two sub-bases. If a horizontal scale of 1 in = 100 ft is used, the length of the average haul can be determined by scaling. This line, shown as a dashed line in Fig. 38.9, scales 875 ft for the area between the top sub-bases. The free haul is subtracted from this in Table 38.1.

Table 38.1 *Typical Cut and Fill Calculations (all volumes are in cubic yards)*

sta	cut +	fill −	cum sum	sta	cut +	fill −	cum sum
0			0	23			−4710
	184				1377		
1			+184	24			−3034
	622				1676		
2			+806	25			−1358
	1035				1860		
3			+1841	26			+502
	1268				1917		
4			+3109	27			+2419
	1231				1839		
5			+4340	28			+4258
	919				1611		
6			+5259	29			+5869
	503				1338		
7			+5762	30			+7207
	164	21			1029		
8			+5905	31			+8236
	12	190			652		
9			+5727	32			+8888
		616			357		
10			+5111	33			+9245
		942			150	39	
11			+4169	34			+9356
		1150			52	236	
12			+3019	35			+9172
		1500				465	
13			+1519	36			+8707
		1773				712	
14			−254	37			+7995
		1755				904	
15			−2009	38			+7091
		1540				904	
16			−3549	39			+6187
		1262				757	
17			−4811	40			+5430
		932				516	
18			−5743	41			+4914
		546				280	
19			−6289	42			+4634
		203				127	
20			−6461	43			+4507
		101				98	
21			−6283	44			+4409
		18				20	
22			−5715	45			+4389

The volume of earth between the two sub-bases is found by subtracting the ordinate of the lower sub-base from the ordinate of the upper sub-base. These ordinates are found in Table 38.1.

Figure 38.9 Mass Diagram Showing Sub-Bases

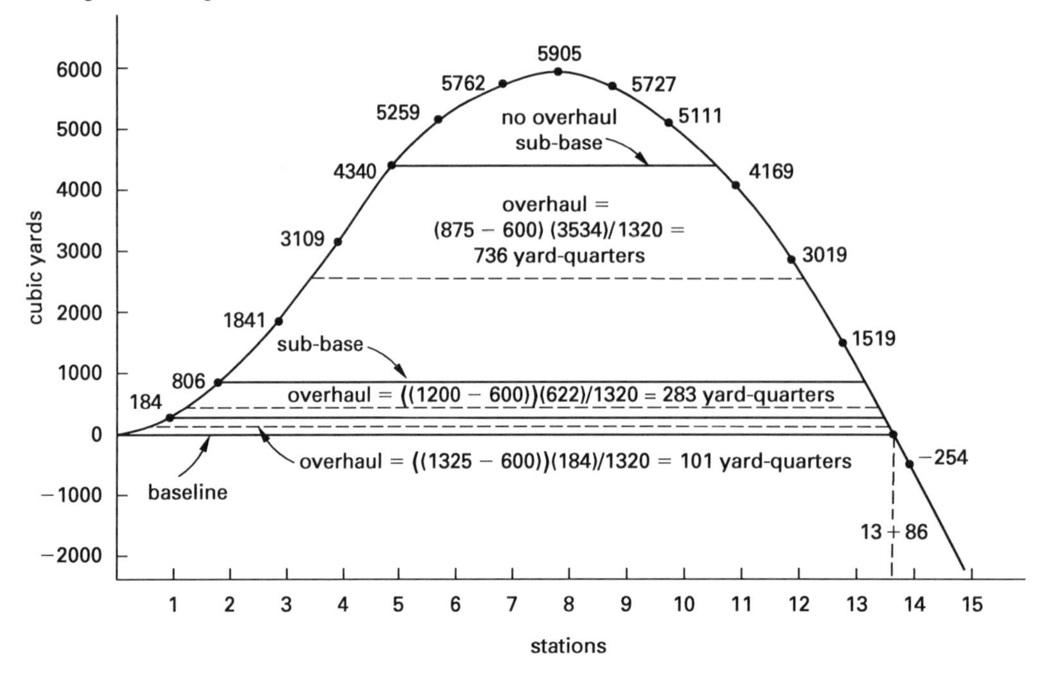

Multiplying average haul minus free haul in feet by volume of earth in cubic yards gives overhaul in yard-feet. Dividing by 1320 ft gives yard-quarters.

28. LOCATING BALANCE POINTS

A balance point occurs where the mass curve crosses the baseline. It can be seen that a balance point falls between sta 13 and 14. The ordinate of 13 is +1519; the ordinate of 14 is −254. Therefore, the curve fell $1519 + 254 = 1773\ \text{yd}^3$ in 100 ft or $17.73\ \text{yd}^3/\text{ft}$. The curve crosses the baseline at a distance of $1519/17.73 = 86$ ft from sta 13 (13+86).

29. CHARACTERISTICS OF THE MASS DIAGRAM

Important characteristics that should be considered in using the mass diagram in planning the economical hauling of earth are as follows.

- A horizontal line connecting two points on the mass curve cuts off a loop in which the cut equals the fill.
- A loop that rises and then falls from left to right indicates that the haul from cut to fill will be from left to right.
- A loop that falls and then rises from left to right indicates the haul will be from right to left.
- A high point on a mass curve indicates a change from cut to fill.
- A low point on a mass curve indicates a change from fill to cut.
- High and low points on the mass curve occur at or near grade points on the profile.

39 Hydrographic Surveying

Nomenclature

d	depth	ft	m
t	time	sec	s
v	velocity	ft/sec	m/s

1. INTRODUCTION

Hydrographic surveying measures and records the shape, location, and contour of underwater terrain. It is similar to topographic surveying, but more complex because the terrain being mapped is not directly viewable. Whereas in topographic surveying, land elevation is recording height in relation to a fixed datum, hydrographic surveying measures depth below a fixed datum.

Bathymetry, the study of underwater depth, uses contour lines to map floor terrain. These maps use the water surface as a reference plane, but because the surface is constantly changing from tidal actions or metrological conditions, specialized surveying equipment and practices must be used.

2. HYDROGRAPHIC EQUIPMENT

Water depth and water level can be measured using a wide range of equipment from simple lead lines and graduated staffs, to complex echo sounding systems and acoustical gauges. Equipment choices depend on the purpose of the survey, what equipment is available, the size of the project, the water depth, and a project's budget.

Water Depth Measurement

A *lead line* is the traditional means of measuring water depths. It consists of a lead weight attached to a graduated line. Although the line may have a soft exterior to allow ease of handling, a stranded wire center should be used to avoid problems due to stretching and shrinking. Waterproof, solid-braid cotton rope with a phosphor-bronze wire center is typically used. Lead lines are traditionally marked with a color-coded thread at even units and a standard color at odd units. They are often used in small projects when it would not be cost effective to use more complex measuring equipment. Graduated staffs and lead lines are also used to confirm least depths over shoals or sunken rocks, to confirm echo sounding in grassy areas, and to determine echo sounding corrections.

In the early 1900s, *echo sounders (fathometers)* revolutionized hydrographic surveying and made surveying in deeper water practical. Echo sounders measure, or "sound," the time required for a sound wave to travel from its point of origin and back, and convert the measured time to depth. Echo sounders are typically designed with an electronic component that generates an electrical pulse at a specified repetition rate. A *transducer* mounted on or beside the hull of the survey vessel converts the electrical pulse to acoustic energy and then reconverts the returning echo to an electrical pulse to measure the elapsed time. Echo sounders are designed with different acoustical pulse frequencies, durations, and widths depending on how and where they will be used.

Choosing a pulse frequency is often a compromise. *Lower frequencies* in the audible range (below 15 kHz) are used for sounding in deeper waters because low-frequency pulses have a lower absorption rate and higher penetrating power. However, low-frequency pulses are less accurate at measuring shoal depths because the pulses carry the same frequency as the noise created by the sounding vessel as it moves through the water. *Ultrasonic frequencies* (20 kHz) are unaffected by noise interference and, therefore, allow more accurate measurements of shoal depths, more detailed profiles of irregular bottoms, and higher directivities of sound waves. However, their higher absorption rates make them ineffective in deep water. They also suffer greater attenuation in turbulent water and in areas with variable water temperature or density.

The *duration of the acoustical pulse* varies. Deep-water echo sounders use longer pulses of 0.001–0.04 sec to assure the pulse reaches the floor. However, the accuracy of the pulse deteriorates at greater depths and provides a relatively poor definition of the floor. General-purpose echo sounders transmit shorter pulses of less than 0.0002 sec. These shorter pulses do not have sufficient power for reaching deeper waters, but they do provide good definition in shallow waters.

The *transducer beam width* of echo sounders generally varies between 2° and 50° depending on the equipment's purpose. Beam widths are inversely proportional to the size of the transducer so that the larger the transducer diameter, the narrower the beam. For the greatest accuracy in echo sounding, the beam should be extremely narrow. However, practical considerations also govern the selection of beam widths since at some frequencies the transducer would need to be very large to produce a narrow beam signal.

Considering the variables of frequency, duration of acoustical pulse, and transducer beam width, echo sounders are generally produced in three classes: (1) deep-water sounders with low frequencies (less than 15 kHz) and wide beam widths (as much as 50°), (2) medium depth sounders designed to operate in depths less than 200 ft (600 m), with medium frequencies (15–20 kHz) and relatively narrow beam widths, and (3) shallow depth sounders designed to operate in waters less than 30 ft (100 m), with high frequencies (above 20 kHz), short wave lengths, and narrow beam widths for detailed bottom definition.

Two modern innovations in echo sounders are used with more advanced systems: the *multiple-beam fathometer* and side-scan sonar. The multiple-beam fathometer is equipped with an array of transducers oriented to varying angles off the vertical axis. This type of system allows for coverage of a wide swath, and is used when 100% coverage of the bottom is required. The *side-scan sonar* uses a beam oriented slightly below the horizontal axis and can locate underwater features protruding above the floor better than vertically oriented echo sounders.

Water Level Measurement

Water level is measured using a wide range of equipment. For short-term observations, a *graduated staff* is often used. It requires little equipment and a minimum amount of installation time and effort since the staff can either be driven into the water floor or secured to a piling. Manually read staffs are also more practical for areas like inter-tidal marshlands where it is desirable to determine a tidal datum, as other gauges (such as float-type gauges) experience problems in areas that are dry for a portion of the tidal cycle.

Most tidal observations lasting longer than a few hours are made using gauges that record the water level either continuously or at fixed intervals. The simplest of these are *float-type gauges*, which are operated by a float that moves in a stilling well with the rise and fall of the water level. A *stilling well* is a vertical tube that extends below the lowest possible tide. Water enters the well through an intake opening near the bottom of the well and rises to the average level of the water surface outside the well. The size of the opening controls the amount of damping that takes place, and must be large enough to allow sufficient flow of water for tidal measurements, while still dampening the rapid water level fluctuation caused by heavy seas. Stilling wells should always be used with float-type gauges so that short period waves (such as wind waves) are dampened, and only the longer-period tidal waves are allowed to move the float. While early float-type gauges used the vertical action of the water to draw a graph of the water level versus time, more recent gauges convert the motion of the float to digital signals that are recorded by some other device. Because float-type gauges cannot be directly leveled to a benchmark or ground elevation, a tidal staff is usually installed near the gauge. Comparisons are made between the gauge and the staff to determine the constant difference between the two readings.

Acoustic gauges also measure water level. They use sensors that measure the water height by timing sound (acoustic) waves or radar beams that bounce off the water surface. There are a variety of output devices. More modern acoustic gauges have a leveling point on the transducer to level against.

A recent trend in tidal measurement has equipped gauges with *telemetry devices* to allow near-real time monitoring of tides in remote locations. These telemetry devices collect information at a remote location and transmit it to another location. The National Ocean Service (NOS) has equipped most of its permanent tidal stations with telemetry devices that relay the data to their headquarters via satellite.

3. SOUNDING PATTERNS

With most bathymetric surveys, it is impractical to develop 100% coverage of the floor within the survey area. It is also typically impossible (unlike with topographic surveys) to see significant breaks in the terrain that should be mapped for an accurate floor representation. Therefore, a systematic sampling pattern must be used to produce a representative floor model. As a result, bathymetric or depth-measurement surveys are generally made along a series of pre-planned sounding lines. A series of evenly spaced parallel lines is generally considered optimal for a systematic delineation of hydrographic features. The spacing of the sounding lines depends on a number of factors including the purpose of the survey, the depth of the water, the topographic configuration of the bottom, and the beam width of the fathometer being used.

For best definition, the sounding lines should run perpendicular to the depth contours, although for steep features such as ridges or trenches, the two lines should cross at 45°. In general, cross lines should supplement the sounding lines at angles of 45° to 90°. When floor topography has significant features, such as a shoal areas or pinnacles, an additional pattern of closely spaced lines, usually parallel to the axis of the feature being mapped, should be run to adequately develop the feature.

4. CORRECTIONS TO DEPTH MEASUREMENTS

It is necessary to make a number of corrections when using echo sounders so that the soundings reflect the actual depth relative to the selected reference datum. Figure 39.1 illustrates such corrections.

Figure 39.1 *Depth Measurement Corrections*

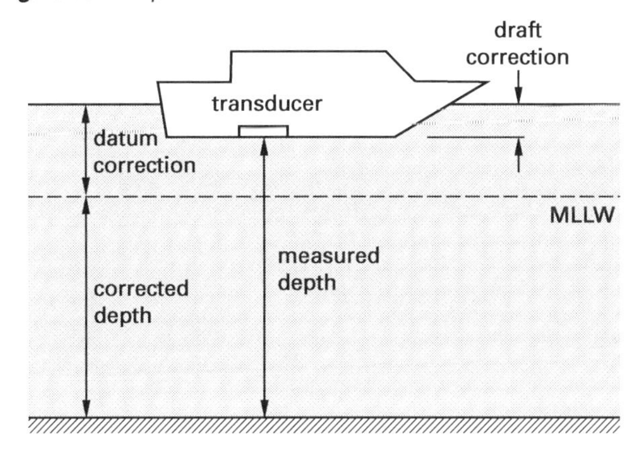

Velocity of Sound Correction

Echo sounders measure depth based on the time a sound pulse takes to travel to the bottom and return. The measured depth, d, is calculated using Eq. 39.1, where v is the velocity of sound in water, and t is the time for the sound pulse to travel to the floor and back.

$$d = \tfrac{1}{2}\mathrm{v}t \qquad 39.1$$

To perform measurements, most echo sounders are calibrated to a velocity of 800 fathoms per second (4800 ft/sec, 1440 m/s), which is the approximate average velocity of sound in the sea. However, the actual velocity varies due to the water's differing salinities and temperatures. Therefore, except for surveys in very shallow waters, a velocity correction should be used.

There are two commonly used methods for determining *velocity corrections*. The first method directly compares readings at various depths using a horizontally suspended bar below the vessel that is raised and lowered to various depths. (A less desirable version of this method is to vertically cast a lead line and compare the readings.) The second method calculates the velocity of sound at various depths using temperature and salinity measurements taken from oceanographic sensors. Some echo sounding systems detect the actual velocity of sound and automatically correct the soundings, making these two correction methods unnecessary.

Draft Corrections

Echo sounders measure the depth below the transducer, not the surface of the water; therefore, corrections must be made for the difference. *Dynamic drafts* correct soundings for the height difference between the surface of the water and the face of the transducer. This correction is the algebraic sum of two corrections: static draft and settlement and squat. *Static draft* is the height difference between the water level and the transducer when the sounding vessel is stationary. Static draft varies with the load aboard the vessel, so this correction needs to be determined for various loads. *Settlement* is the general difference between the elevations of the stationary and moving vessel, while *squat* is the change in trim in the moving vessel—generally it is a lowering of the stern and a rise of the bow. All factors vary with the speed of the vessel, so they should be determined for various speeds.

Instrument Corrections

Although generally not present in soundings obtained by digital methods, soundings scaled from graphic recorders may require corrections to compensate for various errors associated with the depth recorder. Typically errors occur when setting the index (initial pulse) or with other adjustments to the recorder. Generally, errors can be avoided by properly operating the recorder.

Datum of Reference Correction

Since the depth of water at a specific location in a tidally affected water body can vary with time due to the tides and meteorological conditions, depths measured in hydrographic surveys must be adjusted to some common plane of reference. Techniques and datums used for this purpose will be discussed in Sec. 39.6.

5. HORIZONTAL CONTROL OF HYDROGRAPHIC MEASUREMENTS

The horizontal location of the soundings is as important as the accuracy of the depth measurements. Before GPS technology, determining the horizontal location was often problematic, since soundings are generally taken from a moving vessel. GPS technology largely overcame this problem using real-time, differentially corrected code observations to provide submeter precision. *Submeter precision* requires a base station at a known point to receive a correction signal. In the United States, correction signals from U.S. Coast Guard beacons, or from companies offering such services, are commonly used. Where such services are not available, a second GPS receiver may be set on a nearby known point and its observations used for real-time correction.

For small hydrographic projects, it can be more efficient to provide horizontal control using conventional surveying techniques, such as theodolite intersection. *Theodolite intersection* determines the position of the sounding vessel by simultaneously turning angles to the vessel from two known positions on the shore. The resulting position can then be calculated using the bearing-bearing method. (See Ch. 15, Sec. 38.)

Regardless of the method used for horizontal positioning, time must be associated with both the soundings and the horizontal positions for coordinating between the two. It is additionally important that the sounding vessel's point be located directly over the echo sounder transducer. For example, when using GPS positioning, the GPS antenna should be located directly over the transducer or at a know azimuth and distance offset.

6. VERTICAL CONTROL OF HYDROGRAPHIC MEASUREMENTS

Depth measurements are traditionally referenced from the floor to the surface of the water. Since the surface can vary in height depending on the tide, the surface elevation must be monitored during the course of a hydrographic survey to correct soundings to a common, stable plane of reference.

Traditionally, hydrographic soundings in tidally affected waters are reduced to a low-water tidal datum. However, for engineering applications and in non-tidal waters, the soundings can be compared to a geodetic datum. Regardless of the datum used, the basic procedure for correcting soundings involves monitoring the water level during the course of the survey. The monitoring information, combined with the time the depth measurements were taken, can be used to correct the soundings.

Recently, a new process utilizing GPS provides vertical control for hydrographic measurements that bypasses the need to separately measure the water level. GPS provides both horizontal and vertical positions in reference to the ellipsoid for a specific area. If the relationship between the desired vertical reference datum and the ellipsoid is known, GPS can provide the relationship to that desired datum. Such a relationship between the desired vertical datum and the ellipsoid can be determined by static GPS observations at a benchmark, such as a tidal benchmark, where the elevation in reference to the desired vertical datum is known. The resulting difference, combined with the height difference between the fathometer transducer and the phase center of the GPS antenna, can then be used to provide corrections between the observed soundings and the desired vertical datum. One significant advantage of using GPS observations for the vertical datum correction is that it also accounts for the dynamic draft correction since the GPS measures the instantaneous position of the sounding vessel.

7. TIDAL DATUM CALCULATIONS

Hydrographic soundings are traditionally reduced to a low-water tidal datum, since mariners prefer that hydrographic charts show minimum depths. Therefore, a low-water datum is generally used when a tidal datum is used as a reference. Unfortunately, there is a lack of international uniformity in picking a low-water datum. In the United States, the national charting agency uses *mean lower low water* (MLLW) for hydrographic surveys along all coasts. Other nations use a datum closer to the *lowest possible low water.*

Correct application of corrections to a tidal datum requires some understanding of tidal datum theory. A visible tidal cycle is a composite of numerous constituent cycles of various periods that reflect the relationships between the earth, moon, and sun. The longest of these constituent cycles, that associated with the regression of the moon's nodes,[1] has a period of 18.6 years. To develop a statistically significant tidal value, a tidal datum should include all of the periodic variations in tidal height. Therefore, a tidal datum is usually considered the average of all occurrences of a certain tidal extreme for a period of 19 years.[2] Such a period is called a *tidal epoch*.

As shown in Fig. 39.2, *mean high water* (MHW) is an example of tidal datum planes, and is the average height of all the high waters occurring over a period of 19 years. Likewise, *mean low water* (MLW) is defined as the average of all of the low tides over a 19 year tidal epoch. *Mean tide level* (MTL) or *half-tide level* is the plane located halfway between mean high and mean low water, and it used for datum computation purposes. Mean tide level should not be confused with *mean sea level* (MSL), which is the average level of the sea as measured from hourly heights over a tidal epoch. The relationship between mean sea level and mean tide level varies from location to location, depending on the phase and amplitude relations of the location's various tidal constituents.

Figure 39.2 *Tidal Datums*

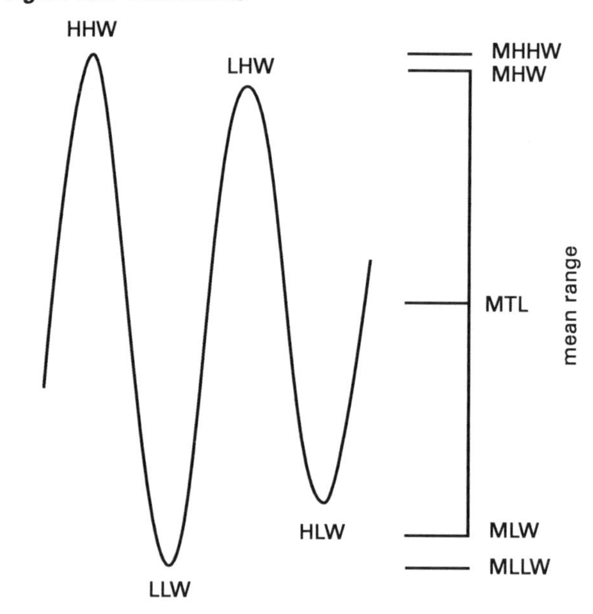

[1]Regression of the moon's nodes refers to the movement of the intersection of the moon's orbital plane and the plane of the earth's equator, which completes a 360° circuit in 18.6 years.
[2]18.6 years rounded to the nearest whole year to include a multiple of the annual cycle associated with the declination of the sun.

Also significant to hydrographic surveying are the mean higher high water and the mean lower low water datum planes. *Mean higher high water* (MHHW) is the average of the higher of the high tides occurring each day. *Mean lower low water* (MLLW) is the average of the lower of the low tides occurring each day. Both of these averages are calculated over a tidal epoch. The height difference between the mean higher high water and mean high water is called *diurnal high water inequality* (DHQ). The difference between mean lower low water and mean low water is called *diurnal low water inequality* (DLQ).

The primary determination of a tidal datum involves the relatively simple determination of the arithmetic mean, or average, of all the occurrences of a certain tidal extreme over a 19-year tidal epoch. In practice, this is usually accomplished by computing the mean values of the various tidal extremes for each calendar month, and then computing the annual mean values by averaging the 12 monthly means for each extreme for each calendar year. Then, the mean values for the tidal epoch are determined by averaging the annual mean values for the 19 years comprising the epoch.

Traditionally, all tidal datum values published in the United States are referred to a specific epoch called the *National Tidal Datum Epoch*. A specific 19-year period is used since apparently non-periodic variation in mean sea level is noted from one 19-year period to the next. A new epoch has historically been adopted every two or three decades when a significant change has occurred. At such times, adjustments are made to all datum elevations. In effect, a quantum jump occurs in the elevations of all tidal datum planes for stations published by the National Ocean Service at those times. The most current epoch of 1983–2001 was adopted in 2003. Previously, the epoch of 1960–1978 was used.

The Range Ratio Method

There are a number of different topographic factors shaping incoming tidal waves that result in significant elevation differences from point-to-point in a tidal datum—even for points in the same general vicinity. Therefore, a tidal datum must be determined in the immediate area of its intended use. Since it is impractical to observe tides for a full 19-year tidal epoch at every location where a tidal datum is needed, most tidal datum elevations are determined from observations of less than 19 years. The average of the tidal extremes observed over a shorter period can be reduced to a value equivalent to a 19-year mean by a correlation process using a ratio of tidal ranges observed at the station and at a control station with a known 19-year value. This correlation process is known as the range ratio method and is based on the following equations, using the nomenclature as given in Table 39.1.

Table 39.1 *Special Nomenclature*

c	control station
HW	high water
LW	low water
MLH	mean diurnal low water inequality
MHHW	19 yr mean higher high water
MHW	19 yr mean high water
MTL	19 yr mean tide level
MLW	19 yr mean low water
MLLW	19 yr mean lower low water
MR	19 yr mean range
HHW	mean higher high water for observation period
LLW	mean lower low water for observation period
R	mean range for observation period
s	subordinate station
TL	mean tide level for observation period

Equation 39.2 is used to find the mean range for an observation period and Eq. 39.3 is used to find the mean tide level for that period.

$$R = \text{HW} - \text{LW} \qquad 39.2$$

$$\text{TL} = \frac{\text{HW} + \text{LW}}{2} \qquad 39.3$$

Equation 39.4 calculates the equivalent 19-year mean range at the subordinate station.

$$\text{MR}_s = \frac{(\text{MR}_c)R_s}{R_c} \qquad 39.4$$

Equation 39.5 calculates the equivalent 19-year mean tide level at the subordinate station.

$$\text{MTL}_s = \text{MTL}_c + \text{TL}_s - \text{TL}_c \qquad 39.5$$

The equivalent 19-year mean high water at the subordinate station can be determined by adding one-half of the 19-year mean range to the 19-year mean tide level.

$$\text{MHW}_s = \text{MTL}_s + \frac{\text{MR}_s}{2} \qquad 39.6$$

Likewise, the equivalent 19-year mean low water at the subordinate station can be determined by subtracting one-half of the 19-year mean range from the 19-year mean tide level.

$$\text{MLW}_s = \text{MTL}_s - \frac{\text{MR}_s}{2} \qquad 39.7$$

The values for 19-year mean higher high water and lower low water can be determined using Eqs. 39.8 and 39.9.

$$\text{MHHW}_s = \text{MHW}_s + \frac{(\text{MHHW}_c - \text{MHW}_c) \times (\text{HHW}_s - \text{HW}_s)}{\text{HHW}_c - \text{HW}_c} \qquad 39.8$$

$$\text{MLLW}_s = \text{MLW}_s - \frac{(\text{MLW}_c - \text{MLLW}_c) \times (\text{LW}_s - \text{LLW}_s)}{\text{LW}_c - \text{LLW}_c} \qquad 39.9$$

$$\text{MLLW}_c = \text{MLW}_c - \text{DLQ}_c \qquad 39.10$$

Example 39.1

Tidal observations were made for one month at a survey site resulting in the following monthly mean values from a gauge datum.

high water (HW) 6.21 ft
low water (LW) 2.62 ft

Simultaneous observations were made at a nearby control tidal station that reported the following 19-year mean values from a gauge datum.

mean tidal level (MTL) 4.55 ft
mean tidal range (MR) 3.40 ft

The monthly mean values on the gauge datum at the control site were

high water (HW) 5.20 ft
low water (LW) 2.00 ft

What is the calculated value of the survey site's mean high water (MHW)?

Solution

Using Eq. 39.2, the observed subordinate range, R_s, is

$$R_s = \mathrm{HW}_s - \mathrm{LW}_s = 6.21 \text{ ft} - 2.62 \text{ ft} = 3.59 \text{ ft}$$

The control range, R_c, is

$$R_c = \mathrm{HW}_c - \mathrm{LW}_c = 5.20 \text{ ft} - 2.00 \text{ ft} = 3.20 \text{ ft}$$

Using Eq. 39.3, the observed tide level, TL_s, is

$$\mathrm{TL}_s = \frac{\mathrm{HW}_s + \mathrm{LW}_s}{2} = \frac{6.21 \text{ ft} + 2.62 \text{ ft}}{2} = 4.415 \text{ ft}$$

The control observed tide level, TL_c, is

$$\mathrm{TL}_c = \frac{\mathrm{HW}_c + \mathrm{LW}_c}{2} = \frac{5.20 \text{ ft} + 2.00 \text{ ft}}{2} = 3.60 \text{ ft}$$

Using Eq. 39.4, the subordinate station mean range is

$$\mathrm{MR}_s = \frac{(\mathrm{MR}_c)R_s}{R_c} = \frac{(3.40 \text{ ft})(3.59 \text{ ft})}{3.20 \text{ ft}} = 3.81 \text{ ft}$$

Using Eq. 39.5, the subordinate mean tide level is

$$\begin{aligned} \mathrm{MTL}_s &= \mathrm{MTL}_c + \mathrm{TL}_s - \mathrm{TL}_c \\ &= 4.55 \text{ ft} + 4.415 \text{ ft} - 3.60 \text{ ft} \\ &= 5.365 \text{ ft} \end{aligned}$$

Therefore, using Eq. 39.6, the survey site's mean high water level is

$$\mathrm{MHW}_s = \mathrm{MTL}_s + \frac{\mathrm{MR}_s}{2} = 5.365 \text{ ft} + \frac{3.81 \text{ ft}}{2} = 7.27 \text{ ft}$$

Datum Calculations

The process for datum calculations uses a strong network of control tidal stations that have been established in most of the coastal areas of the United States, as well as in a number of other countries. Typically, these networks have base primary stations that have been operated for a full 19-year tidal epoch. Such networks are then typically densified with a larger number of secondary and tertiary stations where observations are conducted for a shorter period simultaneously with the observations at the primary stations. The data at such densified stations are corrected to their equivalent 19-year mean values using the previously described range ratio method. At each station, a series of benchmarks are established to perpetuate the resulting data. That is, the elevations are preserved by setting benchmarks and leveling to them. Then, even if the gauge is removed, the elevations are preserved.

When a hydrographic survey uses a tidal datum as its reference datum, any existing local tidal stations must first be located. In the United States, the National Ocean Service is the national repository for such data. If the survey is in the immediate area of an existing control station, a water level staff or gauge may be installed at the station and leveled to the benchmarks for that station. Water level observations made during the survey are used to reduce the soundings to the reference data.

If the survey is not in the immediate area of an existing tidal station, a new tidal datum must be established by installing a gauge both at the most appropriate existing control station and in the survey area. Simultaneous observations taken at the gauges allow the equivalent 19-year mean values to be calculated for the survey area. The observation period necessary for a valid determination will vary depending on the distance between gauges. It is typically recommended that such simultaneous observations be conducted for a minimum of one month to capture the dominant 28-day lunar cycle. However, if the control station is within a few miles and in the same hydrographic regime, as few as three repeated tidal cycles may be sufficient. For greater distances, at least one year of observations may be necessary.

Tidal datums may also serve as a basis for mapping the limits of the water body being surveyed. Most jurisdictions use tidal datum lines to delineate the boundary between the publicly owned submerged lands and the bordering uplands, which are subject to private ownership. Additionally, tidal datum lines are used as a basis for offshore maritime boundaries, making them an essential part of hydrographic surveys.

PRACTICE PROBLEMS

1. The Marianas Trench has a maximum depth of 35,800 ft. Assuming the standard velocity of sound in sea water, most nearly how long would a fathometer sound pulse take to travel to the floor and back?

(A) 3.5 sec
(B) 7.5 sec
(C) 15 sec
(D) 16 sec

2. Bar checks determine that the velocity of sound at a survey site is 4600 ft/sec and the draft correction for the sounding vessel is 2.5 ft. The uncorrected sounding is 59.8 ft. Accounting for sound velocity and draft, what is most nearly the corrected sounding?

(A) 57.3 ft
(B) 58.6 ft
(C) 59.8 ft
(D) 61.2 ft

3. Table A provides depth measurements for a channel cross section. Table B provides tidal height measurements for the same day at a tide station near the channel. What is most nearly the measured depth above the mean lower low water at 100 ft distance?

Table A
depth measurements

distance (ft)	time	depth (ft)
0	4:00	0.5
20	5:00	2.0
40	6:00	10.0
60	7:00	20.0
80	8:00	50.0
100	9:00	50.0
120	10:00	51.0
140	11:00	35.0
160	12:00	15.0
180	13:00	5.0
200	14:00	5.0

Table B
tidal heights

time	height with MLLW (ft)
4:00	2.80
5:00	5.29
6:00	7.38
7:00	8.82
8:00	9.57
9:00	9.17
10:00	7.75
11:00	5.94
12:00	4.28
13:00	2.84
14:00	1.68

(A) 9.2 ft
(B) 40.8 ft
(C) 50.0 ft
(D) 59.2 ft

Use the following information for Probs. 4–6. Tidal observations were made for one month at a survey site resulting in the following mean values from the gauge datum.

higher high water (HHW)	7.45 ft
high water (HW)	7.11 ft
low water (LW)	1.72 ft
lower low water (LLW)	1.55 ft

Simultaneous observations were made at a nearby control tide station with the following known mean values.

mean tide level (MTL)	4.75 ft
mean tidal range (MR)	5.96 ft
mean diurnal high water inequality (DHQ)	0.35 ft
mean diurnal low water inequality (DLQ)	0.19 ft

The monthly mean values on the gauge datum at the control site were

higher high water (HHW)	8.26 ft
high water (HW)	7.91 ft
low water (LW)	1.89 ft
lower low water (LLW)	1.70 ft

4. What is most nearly the mean high water (MHW) at the survey site?

(A) 3.26 ft
(B) 4.26 ft
(C) 5.34 ft
(D) 6.93 ft

5. What is most nearly the mean low water (MLW) at the survey site?

(A) 1.43 ft
(B) 1.55 ft
(C) 1.60 ft
(D) 1.68 ft

6. What is most nearly the mean lower low water (MLLW) at the survey site?

(A) 1.33 ft
(B) 1.43 ft
(C) 1.58 ft
(D) 1.77 ft

SOLUTIONS

1. Rearrange Eq. 39.1 to find how long it would take a fathometer sound pulse to travel to the ocean floor and back.

$$t = \frac{2d}{\text{v}} = \frac{(2)(35{,}800\ \text{ft})}{4800\ \frac{\text{ft}}{\text{sec}}} = 14.9\ \text{sec} \quad (15\ \text{sec})$$

The answer is (C).

2. With the standard velocity of sound (4800 ft/sec) used by the fathometer and a recorded depth of 59.8 ft, the time it takes to measure the depth may be calculated by rearranging Eq. 39.1 as follows.

$$t = \frac{2d}{\text{v}} = \frac{(2)(59.8\ \text{ft})}{4800\ \frac{\text{ft}}{\text{sec}}} = 0.0249\ \text{sec}$$

With that time and the actual velocity of sound (4600 ft/sec), the actual depth below the transducer may then be calculated using Eq. 39.1 as follows.

$$d = \tfrac{1}{2}\text{v}t = \left(\tfrac{1}{2}\right)\left(4600\ \frac{\text{ft}}{\text{sec}}\right)(0.0249\ \text{sec}) = 57.3\ \text{ft}$$

Depth from the water surface may then be calculated by adding the draft correction.

$$\begin{array}{c}\text{corrected}\\ \text{depth}\end{array} = \begin{array}{c}\text{depth below}\\ \text{the transducer}\end{array} + \begin{array}{c}\text{draft}\\ \text{correction}\end{array}$$
$$= 57.3\ \text{ft} + 2.5\ \text{ft} = 59.8\ \text{ft}$$

The answer is (C).

3. At the sounding made at 100 ft, the uncorrected depth is 50.0 ft. At the time of that sounding (9:00), the tidal height above MLLW is 9.17 ft. Therefore, the corrected depth is 50 ft − 9.17 ft = 40.83 ft (40.8 ft).

The answer is (B).

4. Using Eq. 39.2, the observed subordinate range, R_s, is

$$R_s = \text{HW}_s - \text{LW}_s = 7.11\ \text{ft} - 1.72\ \text{ft} = 5.39\ \text{ft}$$

The observed control range, R_c, is

$$R_c = \text{HW}_c - \text{LW}_c = 7.91\ \text{ft} - 1.89\ \text{ft} = 6.02\ \text{ft}$$

From Eq. 39.3, the observed subordinate tide level, TL_s, is

$$\text{TL}_s = \frac{\text{HW}_s + \text{LW}_s}{2} = \frac{7.11\ \text{ft} + 1.72\ \text{ft}}{2} = 4.415\ \text{ft}$$

The observed control tide level, TL_c, is

$$\text{TL}_c = \frac{\text{HW}_c + \text{LW}_c}{2} = \frac{7.91\ \text{ft} + 1.89\ \text{ft}}{2} = 4.90\ \text{ft}$$

Using Eq. 39.4, the subordinate station mean range is

$$\text{MR}_s = \frac{(\text{MR}_c)R_s}{R_c} = \frac{(5.96\ \text{ft})(5.39\ \text{ft})}{6.02\ \text{ft}} = 5.336\ \text{ft}$$

Using Eq. 39.5, the subordinate mean tide level is

$$\begin{aligned}\text{MTL}_s &= \text{MTL}_c + \text{TL}_s - \text{TL}_c\\ &= 4.75\ \text{ft} + 4.415\ \text{ft} - 4.90\ \text{ft}\\ &= 4.265\ \text{ft}\end{aligned}$$

Using Eq. 39.6, the subordinate mean high water level is

$$\begin{aligned}\text{MHW}_s &= \text{MTL}_s + \frac{\text{MR}_s}{2}\\ &= 4.265\ \text{ft} + \frac{5.336\ \text{ft}}{2}\\ &= 6.933\ \text{ft} \quad (6.93\ \text{ft})\end{aligned}$$

The answer is (D).

5. Using Eq. 39.7, the subordinate mean low water is

$$\begin{aligned}\text{MLW}_s &= \text{MTL}_s - \frac{\text{MR}_s}{2}\\ &= 4.265\ \text{ft} - \frac{5.336\ \text{ft}}{2}\\ &= 1.597\ \text{ft} \quad (1.60\ \text{ft})\end{aligned}$$

The answer is (C).

6. Using Eq. 39.7, mean low water at the control station is

$$\text{MLW}_c = \text{MTL}_c - \frac{\text{MR}_c}{2} = 4.75\ \text{ft} - \frac{5.96\ \text{ft}}{2} = 1.77\ \text{ft}$$

From Eq. 39.10, the control mean lower low water is

$$\text{MLLW}_c = \text{MLW}_c - \text{DLQ}_c = 1.77\ \text{ft} - 0.19\ \text{ft} = 1.58\ \text{ft}$$

Using Eq. 39.9, the subordinate mean lower low water is

$$\begin{aligned}\text{MLLW}_s &= \text{MLW}_s - \frac{(\text{MLW}_c - \text{MLLW}_c)(\text{LW}_s - \text{LLW}_s)}{\text{LW}_c - \text{LLW}_c}\\ &= 1.597\ \text{ft} - \frac{(1.77\ \text{ft} - 1.58\ \text{ft})(1.72\ \text{ft} - 1.55\ \text{ft})}{(1.89\ \text{ft} - 1.70\ \text{ft})}\\ &= 1.427\ \text{ft} \quad (1.43\ \text{ft})\end{aligned}$$

The answer is (B).

Topic IX: Computer Operations and Programming

Chapter

40 Computer Hardware

1. EVOLUTION OF COMPUTER HARDWARE

The term *hardware* encompasses the equipment and devices that perform data preparation, input, computation, control, primary and secondary storage, and output functions, but it does not include the programs, routines, and applications (i.e., computer *software*) that control the computer.[1]

Digital computers are generally acknowledged to have gone through five major evolutionary stages.[2,3]

- *first generation:* electromechanical calculators
- *second generation:* vacuum tube computers
- *third generation:* transistor computers
- *fourth generation:* integrated circuit computers
- *fifth generation:* VLSI (very large-scale integration) computers

The term *fifth generation* also is used to refer to the efforts (largely on the part of Japanese researchers) to produce computers that are easier to program and use. These efforts have generally been unsuccessful and have been replaced by research into the *sixth generation* of computers, devices that rely on parallel processing in order to appear more human-like in their programming and processing.

[1]The term "software program" is redundant.

[2]The categorization depends on who is counting and what characteristics are considered evolutionary. Some writers omit electromechanical calculators from the evolution.

[3]Before the days of calculators and computers, some companies were large enough to have employees who did nothing but crank through calculations for engineers and designers. These people were called the company's "computers."

2. COMPUTER SIZE

Computers used for data processing (excluding process control devices) are classified into four categories depending on size and cost.

- *Microcomputers* (*personal computers*, PCs) are small, generally single-user computers without extensive peripherals or storage.
- *Minicomputers* are larger computers, usually dedicated to business data processing at a single site. They have the ability to support multiple terminals and a wide range of peripherals.
- *Mainframe computers* are general-purpose computers used in large, centralized data processing complexes and departments in which many programs are running simultaneously.
- *Supercomputers* are extremely powerful computers that usually have specific functions (e.g., engineering design, number crunching, analysis of strategies).

The distinction between these categories is rapidly becoming indistinct. Some current microcomputers have the same capabilities as minicomputers and mainframes of five years ago. The categorization must be made on the basis of physical size and cost, not on memory size or processing speed as was done in the past.

3. COMPUTER ARCHITECTURE

All digital computers, from giant mainframes to the smallest microcomputers, contain three main components—a central processing unit (CPU), main memory, and external (peripheral) devices. Figure 40.1 illustrates a typical integration of these functions.

Figure 40.1 Simplified Computer Architecture

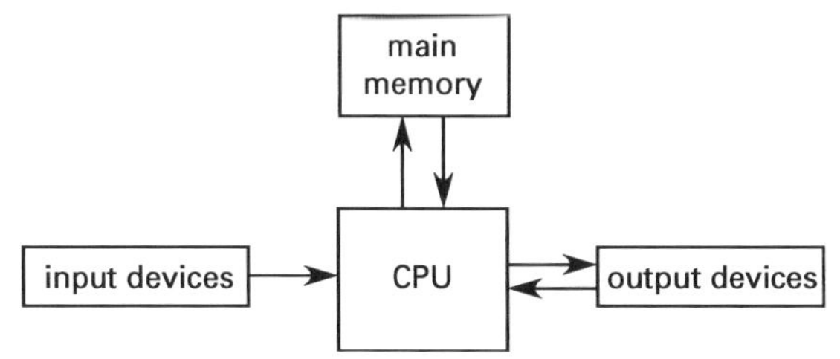

4. MICROPROCESSORS

A *microprocessor* is a central processing unit (CPU) on a single chip. With *large-scale integration* (LSI), most microprocessors are contained on one chip, although other chips in the set can be used for memory and input/output control. The most popular microprocessor families are produced by Intel (Pentium, Celeron, Xeon, Itanium, etc.) and Motorola (680X0 family), although work-alike clones of these chips may be produced by other companies (e.g., AMD's Athalon and K7 processors, and Cyrix's GX and M2 processors).

Microprocessor CPUs consist of an arithmetic and logic unit, several accumulators, one or more registers, stacks, and a control unit. The *control unit* fetches and decodes the incoming instructions and generates the signals necessary for the arithmetic and logic unit to perform the intended function. The *arithmetic and logic unit* (ALU) executes commands and manipulates data.

Accumulators hold data and instructions for further manipulation in the ALU. Registers are used for temporary storage of instructions or data. The *program counter* (PC) is a special register that always points to (contains) the address of the next instruction to be executed. Another special register is the *instruction register* (IR), which holds the current instruction during its execution. *Stacks* provide temporary data storage in sequential order—usually on a last-in, first-out (LIFO) basis. Because their operation is analogous to spring-loaded tray holders in cafeterias, the name *pushdown stack* is also used.

Microprocessors communicate with support chips and peripherals through connections in a *bus* or *channel*, which is logically subdivided into three different functions.[4] The *address bus* directs memory and input/output device transfers. The *data bus* carries the actual data and is the busiest bus. The *control bus* communicates control and status information. The number of lines in the address bus determines the amount of random access memory that can be directly addressed. When there are n address lines in the bus, 2^n words of memory can be addressed.

Microprocessors can be designed to operate on 4-bit, 8-bit, 16-bit, 32-bit, and 64-bit words, although microprocessors with 4- and 8-bit words are now used primarily only in process control applications. Some microprocessors can be combined, and the resulting larger unit is known as a *bit-slice microprocessor*. (For example, four 4-bit microprocessors might be combined into a 16-bit-slice microprocessor.)

[4]The term *bus* refers to the physical path (i.e., wires or circuit board traces) along which the signal travels. The term *channel* refers to the logical path.

Figure 40.2 *Microprocessor Architecture*

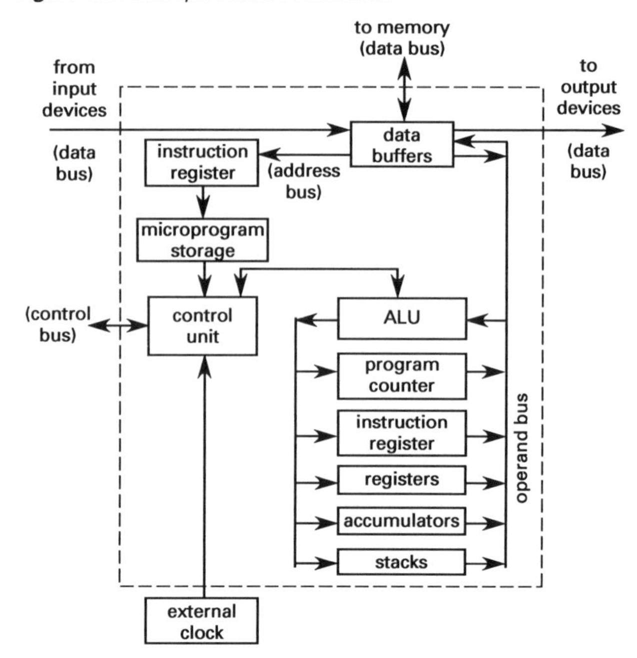

All microprocessors use a crystal-controlled *clock* to control instruction and data movements. The *clock rate* is specified in microprocessor cycles per second (e.g., 2 GHz). Ideally, the clock rate is the number of instructions the microprocessor can execute per second. However, one or more cycles may be required for complex instructions (*macrocommands*). For example, executing a complex instruction may require one or more cycles each to fetch the instruction from memory, decode the instruction to see what to do, execute the instruction, and store (write) the result.[5] Since operations on floating point numbers are macrocommands, the speed of a microprocessor can also be specified in *flops*, the number of floating point operations it can perform per second. Similar units of processing speed are *mips* (millions of instructions per second).

Most microprocessors are rich in complex executable commands (i.e., the *command set*) and are known as *complex instruction-set computing* (CISC) microprocessors. In order to increase the operating speed, however, *reduced instruction-set computing* (RISC) microprocessors are limited to performing simple, standardized format instructions but are otherwise fully featured. Unlike CISC units, RISC microprocessors require four separate instructions for the common fetch-decode-execute-write sequence.

Some microprocessors can emulate the operation of other microprocessors. For example, an 80486 chip could operate in virtual 8086 mode. *Emulation mode* is also referred to as *virtual mode.*

[5]In some cases, other chips (e.g., *memory management units*) can perform some of these tasks.

5. CONTROL OF COMPUTER OPERATION

The user interface and basic operation of a computer are controlled by the *operating system* (OS), also known as the *monitor program.* The operating system is a program that controls the computer at its most basic level and provides the environment for application programs. The operating system manages the memory, schedules processing operations, accesses peripheral devices, communicates with the user/operator, and (in multi-tasking environments) resolves conflicting requirements for resources. Since one of the functions of the operating system is to coordinate use of the peripheral devices (disk drives, keyboards, etc.), the term *basic input/output system* (BIOS) is also used.

All or part of the operating system can be stored in read-only memory (ROM). In early computers, a small part of executable code that was used to initiate data transfers and logical operations when the computer was first started was known as a *bootstrap loader.* Although modern start-up operations are more sophisticated, the phrase "booting the computer" is still used today.

During program operation, peripherals and other parts of the computer signal the operating system through interrupts. An *interrupt* is a signal that stops the execution of the current instruction (or, in some cases, the current program) and transfers control to another memory location, subroutine, or program. Other interrupts signal error conditions such as division by zero, overflow and underflow, and syntax errors. The operating system intercepts, decodes, and acts on these interrupts.

6. COMPUTER MEMORY

Computer memory consists of many equally sized storage locations, each of which has an associated address. The contents of a storage location may change, but the address does not.

The total number of storage locations in a computer can be measured in various ways. A *bit* (binary digit) is the smallest changeable data unit. Bits can only have values of 1 or 0. Bits are combined into *nibbles* (4 bits), *bytes* (8 bits, the smallest number of bits that can represent one alphanumeric character), *half-words* (8 and 16 bits), *words* (8, 16, and 32 bits) and *doublewords* (16, 32, and 64 bits).[6]

The number of memory storage locations is always a multiple of two. The abbreviations K, M, and G are used to designate the quantities 2^{10} (1024), 2^{20} (1,048,576), and 2^{30} (1,073,741,824), respectively. For example, a 6M memory would contain 6×2^{20} bytes. (K and M do not mean one thousand and one million exactly.)

[6]The distinction between doublewords, words, and half-words depends on the computer. Sixteen bits would be a word in a 16-bit computer but would be a half-word in a 32-bit computer. Furthermore, *double-precision (double-length) words* double the number of bytes normally used. The abbreviations KB (*kilobytes*) and KW (*kilowords*) used by some manufacturers do not help much to clarify the ambiguity.

Most of the memory locations are used for user programs and data. However, portions of the memory may be used for video memory, I/O cache memory, the BIOS, and other purposes. *Video memory* (known as VRAM) contains the text displayed on the screen of a terminal. Since the screen is refreshed many times per second, the screen information must be repeatedly read from video memory. *Cache memory* holds the most recently read and frequently read data in memory, making subsequent retrieval of that data much faster than reading from a tape or disk drive, or even from main memory.[7] *OS memory* contains the BIOS that is read in when the computer is first started. *Scratchpad memory* is high-speed memory used to store a small amount of data temporarily so that the data can be retrieved quickly.

Modern memory hardware is semiconductor based.[8] Memory is designated as RAM (random access memory), ROM (read-only memory), PROM (programmable read-only memory), and EPROM (erasable programmable read-only memory). While data in RAM is easily changed, data in ROM cannot be altered. PROMs are initially blank but once filled, they cannot be changed. EPROMs are initially blank but can be filled, erased, and refilled repeatedly.[9] The term *firmware* is used to describe programs stored in ROMs and EPROMs.

The contents of a *volatile memory* are lost when the power is turned off. RAM is usually volatile, while ROM, PROM, and EPROM are *non-volatile.* With *static memory,* data does not need to be refreshed and remains as long as the power stays on. With *dynamic memory,* data must be continually refreshed. Static and dynamic RAM (i.e., DRAM) are both volatile.

Virtual memory (storage) (VS) is a technique by which programs and data larger than main memory can be accessed by the computer. (Virtual memory is not synonymous with *virtual machine,* described in Sec. 12.) In virtual memory systems, some of the disk space is used as an extension of the semiconductor memory. A large application or program is divided into modules of equal size called *pages.* Each page is switched into (and out of) RAM from (and back to) disk storage as needed, a process known as *paging.* This interchange is largely transparent to the user. Of course, access to data stored on a disk drive is much slower than semiconductor memory access. *Thrashing* is a deadlock situation that occurs when a program references a different page for almost every instruction, and there is not even enough real memory to hold most of the virtual memory.

[7]A high-speed mainframe computer may require 200–500 nanoseconds to access main memory but only 20–50 nanoseconds to access cache memory.

[8]The term core, derived from the ferrite cores used in early computers, is seldom used today.

[9]Most EPROMs can be erased by exposing them to ultraviolet light.

Most memory locations are filled and managed by the CPU. However, *direct memory access* (DMA) is a powerful I/O technique that allows peripherals (e.g., tape and disk drives) to transfer data directly into and out of memory without affecting the CPU. Although special DMA hardware is required, DMA does not require explicit program instructions, making data transfer faster.

7. PARITY

Parity is a technique used to ensure that the bits within a memory byte are correct. For every eight data bits, there is a ninth bit—the parity bit—that serves as a *check bit.* The nine bits together constitute a *frame.* In *odd-parity recording,* the parity bit will be set so there is an odd number of one-bits. In *even-parity recording,* the parity bit will be set so that there is an even number of one-bits. When the data are read, the nine bits are checked to ensure valid data.[10]

8. INPUT/OUTPUT DEVICES

Devices that feed data to, or receive data from, the computer are known as *input/output* (I/O) *devices.* Terminals, light pens, digitizers, printers and plotters, and tape and disk drives are common peripherals.[11] Point-of-sale (POS) devices, bar code readers, and magnetic ink character recognition (MICR) and optical character recognition (OCR) readers are less common devices.

Peripherals are connected to their computer through multi-line cables. With a *parallel interface* (used in a *parallel device*), there are as many separate lines in the cable as there are bits (typically seven, eight, or nine) in the code representing a character. An additional line is used as the *strobe signal* to carry a timing signal. With a *serial interface* (used in a *serial device*), all bits pass one at a time along a single line in the cable. The *transmission speed (baud rate)* in bps is the number of bits that pass through the data line per second.[12]

Peripherals such as terminals and printers typically do not have large memories. They only need memories large enough to store the information before the data are displayed or printed. The small memories are known as *buffers.* The peripheral can send the status (i.e., full, empty, off-line, etc.) of its buffer to the computer in several different ways. This is known as *flow control* or *handshaking.*

If the computer and peripheral are configured so that each can send and receive data, the peripheral can send a single character (e.g., the XOFF *character* for transmission off) to the computer when its buffer is full. Similarly, a different character (e.g., the XON *character* for transmission on) can be sent when the peripheral is ready for more data. The computer must monitor the incoming data line for these characters. This is known as *software flow control* or *software handshaking.*

If there are enough separate lines between the computer and the peripheral, one or more of them can be used for *hardware handshaking.*[13] In this method of flow control, the peripheral keeps the voltage on one of the lines high (or low) when it is able to accept more data. The computer monitors the voltage on this line.

Most peripheral devices are connected to the computer by a dedicated channel (cable). However, a pair of *multiplexers* (*statistical multiplexers* or *concentrators*) can be used to carry data for several peripherals along a single cable known as the *composite link.* There are two methods of achieving multiplexed transmission: *frequency division multiplexing* (FDM) and *time division multiplexing* (TDM). With FDM, the available transmission band is divided into narrower bands, each used for a separate channel. In TDM, the connecting channel is operated at a much higher clock rate (proportional to the capacity of the multiplexer), and each peripheral shares equally in the available cycles.

The Electronics Industries Association (EIA) RS-232 standard was developed in an attempt to standardize the connectors and pin uses in serial device cables.

Figure 40.3 *Multiplexed Peripherals*

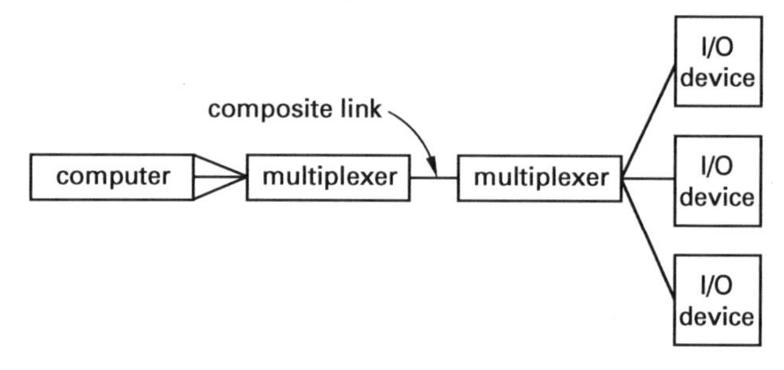

9. RANDOM SECONDARY STORAGE DEVICES

Random access (direct access) storage devices, also known as *mass storage devices,* include magnetic and optical disk drive units. They are random access because individual records can be accessed without having to read through the entire file.

Magnetic disk drives (*hard drives*) are composed of several *platters,* each with one or more read/write heads. The platters typically turn at 4500–7200 rpm, although

[10]This does not detect two of the bits in the frame being incorrect, however.

[11]CRT, the abbreviation for *cathode ray tube,* is often used to mean a terminal.

[12]The name *baud rate* is derived from the use of the *Baudot code* (see Ch. 27). One *baud* is one modulation per second. If there is a one-to-one correspondence between modulations and bits, one baud unit is the same as one bit per second (bps). In general, the unit bps should always be used.

[13]Terminals and printers usually require only two or three lines—data in, data out, and ground. Most computer cables contain more lines than this, and one of the extra lines can be used for DTR (*data terminal ready*) or CTS (*clear to send*) handshaking.

high-performance drives typically turn at 10,000 rpm. Data on a surface are organized into tracks, sectors, and cylinders. *Tracks* are the concentric storage areas. *Sectors* are pie-shaped subdivisions of each track. A *cylinder* consists of the same numbered track on all drive platters. Figure 40.4 illustrates a typical magnetic disk drive. Some platters and *disk packs* are removable, but most hard drives are fixed (i.e., non-removable).

Depending on the media, *optical disk drives* can be *read only* (R/O) or *read/write* (R/W) in nature. WORM drives (write once, read many) can be written by the user, while others such as CD-ROM (compact disk read-only memory) can only be read.[14]

Figure 40.4 *Tracks, Sectors, and Cylinders*

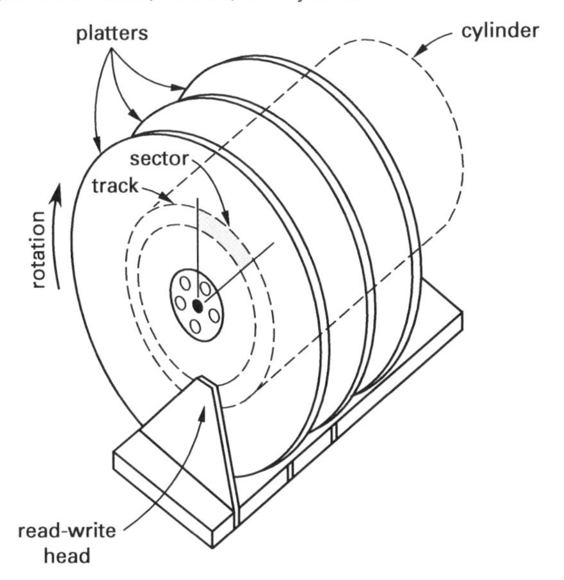

In addition to *storage capacity* (usually specified in megabytes—MB) there are several parameters that describe the performance of a disk drive. *Areal density* is a measure of the number of data bits stored per square inch of disk surface. It is calculated by multiplying the number of bits per track by the number of tracks per (radial) inch. The *average seek time* is the average time it takes to move a head from one location to a new location. The *track-to-track seek time* is the time required to move a head from one track to an adjacent track. *Latency* or *rotational delay* is the time it takes for a disk to spin a particular sector under the head for reading. On the average, latency is one-half of the time to spin a full revolution. The *average access time* is the time to move to a new sector and read the data. Access time is the sum of average latency and average seek time.

Floppy disks (diskettes) are suitable for low-capacity random-access storage. Although their capacities are comparatively low (typically less than one million characters for magnetic media), they are used to transfer programs and data between computers (primarily microcomputers). The capacity of a diskette depends on its size, the recording density, number of tracks, and number of sides.

10. SEQUENTIAL SECONDARY STORAGE DEVICES

Tape units are *sequential access devices* because a computer cannot access information stored at the end of a tape without first reading or passing by the information stored at the front of the tape. Some tapes use *indexed sequential formats* in which a directory of files on the tape is placed at the start of the tape. The tape can be rapidly wound to (near) the start of the target file without having to read everything in between.

There are a number of tape formats in use. One of the few standardized commercial formats is the *nine-track* format.[15] The tape is divided into nine tracks running the length of the tape. The width of the tape is divided into frames (characters). Eight tracks are used to record the data in either ASCII or EBCDIC format. (See Ch. 41.) The remaining track is used to record the parity bit. 1600 bpi (bits or frames per inch) is still a common *recording density* for sharing data, although densities of 9600 bpi and above are in use.

Frames are grouped into fixed-length *blocks* separated by *interblock gaps* (IBG). Reflective spots for photoelectric detection are used to indicate the beginnings and ends of magnetic tapes. These spots are called *load point* and *end-of-file* (EOF) *markers*, respectively.

Magnetic tape is often used to back up hard disks. A *streaming tape* operates in a continuously running—or streaming—mode, with data being written or read while the tape is running.

11. REAL-TIME AND BATCH PROCESSING

Programs run on a computer in one of two main ways: batch mode and real-time mode. In some data processing environments, programs are held (either by the operator or by the operating system) and eventually grouped into efficient categories requiring the same peripherals and resources. This is called *batch mode processing* since all jobs of a particular type are batched together for subsequent processing. There is usually no interaction between the user and the computer once batch processing begins. In *real-time (interactive) processing*, a program runs when it is submitted, often with user interaction during processing.

12. MULTI-TASKING AND TIME-SHARING

If a computer's main memory is large enough and the CPU is fast enough, it may be possible to allocate the main memory among several users running applications

[14]The WORM acronym is also interpreted as write once—read mostly.

[15]Other formats that have received some acceptance include *quarter inch cartridge* (QIC) and *digital audio tape* (DAT).

simultaneously. This is the concept of a *virtual machine* (VM)—each user appears to have his own computer. This is also known as *multi-tasking* and *multi-programming* since multiple tasks can be performed simultaneously. A *multi-user system* is similar to a multi-tasking system in that several users can use the computer simultaneously, although that term also means that all users are using the same program.

Time-sharing (*swapping*) is a technique where each user takes turns (under the control of the operating system) using the entire computer main memory for a certain length of time (usually less than a second). At the end of that time period, all of the active memory is written to a private area, and the memory for the next user is loaded. The swapping occurs so frequently that all users are able to accomplish useful work on a real-time basis.

13. BACKGROUND AND FOREGROUND PROCESSING

A program running in real time is an example of *foreground processing.* There are times, however, when it is convenient to start a long program running while the same computer is used for a second program. The first program continues to run unseen in the background. *Background processing* can be accomplished by segmenting the main computer memory (i.e., establishing virtual machines as defined in Sec. 12) or by time-sharing (i.e., allowing the background application to have all cycles not used by the foreground application).

14. TELEPROCESSING

Teleprocessing is the access of a computer from a remote station, usually over a telephone line (although fiber optic, coaxial, and microwave links can also be used). Since these media transmit analog signals, a *modem* (*modulator-demodulator*) is used to convert to/from the digital signals required by a computer.

With *simplex communication*, transmission is only in one direction. With *half-duplex communication*, data can be transmitted in both directions, but only in one direction at a time. With *duplex* or *full duplex communication*, data can be transmitted in both directions simultaneously. 56,000 bps is the maximum practical transmission speed over voice-grade lines without data compression. Much higher rates, however, are possible over dedicated data lines and wide-band lines.

In *asynchronous* or *start-stop transmission*, each character is preceded and followed by special signals (i.e., *start* and *stop bits*). Thus, every 8-bit character is actually transmitted as 10 bits, and the character transmission rate is one-tenth of the transmission speed in bits per second.[16] With asynchronous transmission, it is possible to distinguish the beginning and end of each character from the bit stream itself.

Synchronous equipment transmits a block of data continuously without pause and requires a built-in clock to maintain synchronization. Synchronous transmission is preceded and interwoven with special clock-synchronizing characters, and the separation of a bit stream into individual characters is done by counting bits from the start of the previous character. Since start and stop bits are not used, synchronous communication is approximately 20 percent faster than asynchronous communication.

There are three classes of communication lines—narrow-band, voice-grade, and wide-band—depending on the bandwidth (i.e., range of frequencies) available for signaling.

Narrow-band may only support a single channel of communication, as the bandwidth is too narrow for modulation. *Wide-band channels* support the highest transfer rates, since the bandwidth can be divided into individual channels. *Voice-grade lines*, supporting frequencies between 300 Hz and 3000–3300 Hz, are midrange in bandwidth.

Errors in transmission can easily occur over voice grade lines at the rate of 1 in 10,000. In general, methods of ensuring the accuracy of transmitted and received data are known as *communications protocols* and *transmission standards.* A simple way of checking the transmission is to have the receiver send each block of data back to the sender. This process is known as *loop checking* or *echo checking.* If the characters in a block do not match, they are re-sent. While accurate, this method requires sending each block of data twice.

Another method of checking the accuracy of transmitted data is for both the receiver and sender to calculate a *check digit* or *block check character* derived from each block of *characters* sent. (A common transmission block size is 128 characters.) With CRC (*cycling redundancy checking*), the block check character is the remainder after dividing all the serialized bits in a transmission block by a predetermined binary number. Then, the block check character is sent and compared after each block of data.

15. DISTRIBUTED SYSTEMS AND LOCAL-AREA NETWORKS

Distributed data processing systems assume many configurations. In the traditional situation, a centrally located main computer interacts with, and is fed by, smaller computers in other locations. In a second configuration, many identical computers are linked together in order to share storage and printing resources. This latter case is known as a *local-area network* (LAN). Local-area networks typically communicate at speeds of 10 kbps or 100 kbps (Ethernet), 54 Mbps (wireless), and even 1 Gbps (Gigabit-Ethernet).

[16]For historical reasons, a second stop bit is used when data are sent at ten characters per second. This is referred to as 110 bps.

PRACTICE PROBLEMS

1. A disk has 1000 tracks with 10,000 bits per track. The disk diameter is 6 inches. The average seek time is 20 ms, and the rotational speed is 2000 rpm. (a) What is the areal density of the disk? (b) Estimate the average access time.

2. What length of nine-track tape with density 1600 bpi is required to store 1 Mb of data?

3. The entire contents of a full 20-Mb disk are sent in an asynchronous transmission at a 2400 baud rate. What is the time required for transmission?

SOLUTIONS

1. (a) areal density

$$\begin{aligned} &= (\text{no. bits per track}) \\ &\quad \times (\text{no. tracks per radial inch}) \\ &= \left(10{,}000\ \frac{\text{bits}}{\text{track}}\right)\left(\frac{1000\ \text{tracks}}{\frac{D}{2}}\right) \\ &= \left(10{,}000\ \frac{\text{bits}}{\text{track}}\right)\left(\frac{2000\ \text{tracks}}{6\ \text{in}}\right) \\ &= \boxed{3.33 \times 10^6\ \text{bits/radial in}} \end{aligned}$$

(b) access time = average latency + average seek time

The average latency can be approximated as one-half time for the full revolution.

$$\left(\frac{1}{2}\ \text{rev}\right)\left(\frac{1\ \text{min}}{2000\ \text{rev}}\right)\left(60\ \frac{\text{sec}}{\text{min}}\right)\left(1000\ \frac{\text{ms}}{\text{sec}}\right) = 15\ \text{ms}$$

$$\begin{aligned} \text{average access time} &= 15\ \text{ms} + 20\ \text{ms} \\ &= \boxed{35\ \text{ms}} \end{aligned}$$

2. Since eight tracks are used for data recording, 1 byte of data is stored across the width of the tape. Therefore, 1600 bytes of data are stored per inch of length. The length required for 1 megabyte is

$$\frac{1{,}048{,}576\ \text{bytes}}{\left(1600\ \frac{\text{bytes}}{\text{in}}\right)\left(12\ \frac{\text{in}}{\text{ft}}\right)} = \boxed{54.6\ \text{ft}}$$

This neglects any interblock gaps.

3. The time required is

$$\begin{aligned} t &= \frac{(20\ \text{Mb})\left(1{,}048{,}576\ \frac{\text{bytes}}{\text{Mb}}\right)\left(10\ \frac{\text{bits}}{\text{byte}}\right)}{\left(2400\ \frac{\text{bits}}{\text{sec}}\right)\left(60\ \frac{\text{sec}}{\text{min}}\right)\left(60\ \frac{\text{min}}{\text{hr}}\right)} \\ &= \boxed{24.3\ \text{hrs}} \end{aligned}$$

41 Data Structure and Programming

1. CHARACTER CODING

Alphanumeric data refers to characters that can be displayed or printed, including numerals and symbols ($, %, &, etc.) but excluding *control characters* (tab, carriage return, form feed, etc.). Since computers can handle binary numbers only, all symbolic data must be represented by binary codes. *Coding* refers to the manner in which alphanumeric data and control characters are represented by sequences of bits.

The standard method for coding data on 80-column, 12-row cards is the *Hollerith code*.[1]

The *Baudot code* is a five-bit code that has long been used in Telex and teletypewriter (TWX and TTY) communications. By shifting to an alternate character set (numerals versus letters), it has a maximum of 64 (2×2^5) characters. Primarily due to its slow transmission speed, but also due to its limited character set, the Baudot code is no longer favored by new equipment manufacturers.

In some early computers, characters were represented by six-bit combinations known as *Binary Coded Decimal* (BCD).[2] The 64 (2^6) different combinations, however, proved insufficient to represent all necessary characters.

The *American Standard Code for Information Interchange* (ASCII) is a seven-bit code permitting 128 (2^7) different combinations. It is commonly used in microcomputers, although use of the high order (eighth) bit is not standardized. ASCII-coded magnetic tape and disk files are used to transfer data and documents between computers of all sizes that would otherwise be unable to share data structures.

The *Extended Binary Coded Decimal Interchange Code* (EBCDIC) is in widespread use in mainframe computers.[3] It uses eight bits (a byte) for each character, allowing a maximum of 256 (2^8) different characters.

Since strings of bits are difficult to read, the *packed decimal* format is used to simplify working with EBCDIC data. Each byte is converted into two strings of four bits each. The two strings are then converted to hexadecimal format. Since $(1111)_2 = (15)_{10} = (F)_{16}$, the largest possible EBCDIC character is coded FF in packed decimal.

Example 41.1

The number $(7)_{10}$ is represented as 11110111 in EBCDIC. What is this in packed decimal?

Solution

The first four bits are 1111, which is $(15)_{10}$ or $(F)_{16}$. The last four bits are 0111, which is $(7)_{10}$ or $(7)_{16}$. The packed decimal representation is F7.

2. PROGRAM DESIGN

A *program* is a sequence of computer instructions that performs some function. The program is designed to implement an algorithm, which is a procedure consisting of a finite set of well-defined steps. Each step in the algorithm usually is implemented by one or more instructions (e.g., READ, GOTO, OPEN, etc.) entered by the programmer. These original "human-readable" instructions are known as source code statements.

Except in rare cases, a computer will not understand source code statements. Therefore, the source code is

[1]Punch cards are now seldom used.

[2]BCD was reintroduced when IBM started using 96-column cards.

[3]EBCDIC is pronounced eb′-sih-dik.

translated into machine-readable object code and absolute memory locations. Eventually, an executable program is produced.

If the executable program is kept on disk or tape, it is normally referred to as *software*. If the program is placed in ROM or EPROM, it is referred to as *firmware*. The computer mechanism itself is known as the *hardware*.

3. FLOWCHARTING SYMBOLS

A *flowchart* is a step-by-step drawing representing a specific procedure or algorithm. Figure 41.1 illustrates the most common flowcharting symbols. The terminal symbol begins and ends a flowchart. The input/output symbol defines an I/O operation, including those to and from keyboard, printer, memory, and permanent data storage. The processing symbol and predefined process symbol refer to calculations or data manipulation. The decision symbol indicates a point where a decision must be made or two items are compared. The connector symbol indicates that the flowchart continues elsewhere. The off-page symbol indicates that the flowchart continues on the following page. Comments can be added in an annotation symbol.

Figure 41.1 Flowcharting Symbols

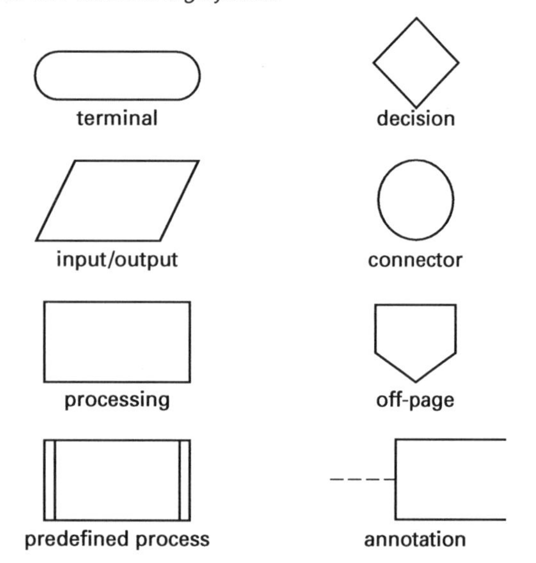

4. LOW-LEVEL LANGUAGES

Programs are written in specific languages, of which there are two general types: low-level and high-level. Low-level languages include machine language and assembly language.

Machine language instructions are intrinsically compatible with and understood by the computer's CPU. They are the CPU's native language. An instruction normally consists of two parts: the operation to be performed (*op-code*) and the operand expressed as a storage location. Each instruction must ultimately be expressed as a series of bits, a form known as *intrinsic machine code*. However, octal and hexadecimal coding are more convenient. In either case, coding a machine language program is tedious and seldom done by hand.

Table 41.1 Comparison of Typical ADD Commands

language	instruction
intrinsic machine code	1111 0001
machine language	1A
assembly language	AR
FORTRAN	+

Assembly language is more sophisticated (i.e., is more symbolic) than machine language. Mnemonic codes are used to specify the operations. The operands are referred to by variable names rather than the addresses. Blocks of code that are to be repeated verbatim at multiple locations in the program are known as *macros* (*macro instructions*). Macros are written only once and are referred to by a symbolic name in the source code.

Assembly language code is translated into machine language by an *assembler* (*macro-assembler* if macros are supported). After assembly, portions of other programs or function libraries may be combined by a *linker*. In order to run, the program must be placed in the computer's memory by a *loader*. Assembly language programs are preferred for highly efficient programs. However, the coding inconvenience outweighs this advantage for most applications.

5. HIGH-LEVEL LANGUAGES

High-level languages are easier to use than low-level languages because the instructions resemble English. High-level statements are translated into machine language by either an interpreter or a compiler. A *compiler* performs the checking and conversion functions on all instructions only when the compiler is invoked. A true stand-alone executable program is created. An *interpreter*, however, checks the instructions and converts them line by line into machine code during execution but produces no stand-alone program capable of being used without the interpreter.[4,5]

There are many high-level languages, although only a few are in widespread use. Some of these languages are listed in Table 41.2.

[4]Some interpreters check syntax as each statement is entered by the programmer.

[5]Some languages and implementations of other languages blur the distinction between interpreters and compilers. Terms such as *pseudo-compiler* and *incremental compiler* are used in these cases.

Table 41.2 High-Level Languages

name	description
BASIC	Beginner's All-purpose Symbolic Instruction Code; English-like instructions; dialects, such as Visual Basic, are popular for user interface programming
C	a structured high-level, function-oriented language capable of low-level machine control
C++	a popular enhancement to the C programming language that allows for procedural and object-oriented programming
COBOL	COmmon Business-Oriented Language; used in business programs; English-like
Forth	a language prevalent in instrument controls; subroutines are called words
FORTRAN	FORmula TRANslation; rich in scientific functions
Java	an object-oriented programming language similar to C++, known for its portability, extensibility, and straightforwardness
PERL	Practical Extraction and Report Language; originally designed for text manipulation; common in website programming
SQL	Structured Query Language; used to create, modify, and retrieve data from relational database management systems
XML	eXtensible Markup Language; a general-purpose language capable of describing other kinds of data

6. SPECIAL PURPOSE LANGUAGES

There are many special purpose languages, some of which are listed in Table 41.3.

Special applications include AI (artificial intelligence—see Sec. 17), CAD (computer-aided design), CAM (computer-aided manufacturing), CAI (computer-aided instruction), DB (database), DBMS (database management system), EIS (executive information system), JCL (job control language), and MIS (management information system).

Table 41.3 Special Purpose Languages

name	use
Discipulus	genetic-programming software
LISP	list-processing; artificial intelligence
Maple	computer algebra system
PostScript	a portable language for printed output
SAS	statistical analysis
SPICE	electric circuit simulation

7. RELATIVE COMPUTATIONAL SPEED

Certain languages are more efficient (i.e., execute faster) than others.[6] While it is impossible to be specific, and exceptions abound, assembly language programs are fastest, followed in order of decreasing speed by compiled, pseudo-compiled, and interpreted programs.

Similarly, certain program structures are more efficient than others. For example, when performing a repetitive operation, the most efficient structure will be a single equation, followed in order of decreasing speed by a stand-alone loop and a loop within a subroutine. Incrementing the loop variables and managing the exit and entry points is known as *overhead* and takes time during execution.

8. STRUCTURE, DATA TYPING, AND PORTABILITY

A language is said to be *structured* if subroutines and other procedures each have one specific entry point and one specific return point.[7] A language has *strong data types* if integer and real numbers cannot be combined in arithmetic statements.

A *portable language* can be implemented on different machines. Most portable languages are either sufficiently rigidly defined (as in the cases of ADA and C) to eliminate variants and extensions, or (as in the case of Pascal) are compiled into an intermediate, machine-independent form. This so-called *pseudo-code* (*p-code*) is neither source nor object code. The language is said to have been "ported to a new machine" when an interpreter is written that converts p-code to the appropriate machine code and supplies specific drivers for input, output, printers, and disk use.[8]

9. STRUCTURED PROGRAMMING

Structured programming (also known as top-down programming, procedure-oriented programming, and GOTO-less programming) divides a procedure or algorithm into parts known as subprograms, subroutines, modules, blocks, or procedures.[9] Internal subprograms are written by the programmer; external subprograms are supplied in a library by another source. Ideally, the mainline program will consist entirely of a series of calls (references) to these subprograms. Liberal use is made of FOR/NEXT, DO/WHILE, and DO/UNTIL

[6]Efficiency can also, but seldom does, refer to the size of the program.
[7]Contrast this with BASIC, which permits (1) a GOSUB to a specific subroutine with a return from anywhere within the subroutine and (2) unlimited GOTO statements to anywhere in the main program.
[8]Some companies have produced Pascal engines that run p-code directly.
[9]The format and readability of the source code—improved by indenting nested structures, for example—do not define structured programming.

commands. Labels and GOTO commands are avoided as much as possible.

Very efficient programs can be constructed in languages that support *recursive calls* (i.e., permit a subprogram to call itself). Some languages (e.g., Pascal and PL/I) permit recursion; others do not.

Variables whose values are accessible strictly within the subprogram are *local variables. Global variables* can be referred to by the main program and all other subprograms.

10. FIELDS, RECORDS, AND FILE TYPES

A collection of *fields* is known as a *record.* For example, name, age, and address might be fields in a personnel record. Groups of records are stored in a *file.*

A *sequential file* structure (e.g., typical of data on magnetic tape) contains consecutive records and must be read starting at the beginning. An *indexed sequential file* is one for which a separate index file (see Sec. 11) is maintained to help locate records.

With a *random* (*direct access*) *file structure*, any record can be accessed without starting at the beginning of the file.

11. FILE INDEXING

It is usually inefficient to place the records of an entire file in order. (A good example is a mailing list with thousands of names. It is more efficient to keep the names in the order of entry than to sort the list each time names are added or deleted.) Indexing is a means of specifying the order of the records without actually changing the order of those records.

An index (key or keyword) file is analogous to the index at the end of this book. It is an ordered list of items with references to the complete record. One field in the data record is selected as the key field (record index).[10] The sorted keys are usually kept in a file separate from the data file. One of the standard search techniques is used to find a specific key.

12. SORTING

Sorting routines place data in ascending or descending numerical or alphabetical order.

With the method of *successive minima*, a list is searched sequentially until the smallest element is found and brought to the top of the list. That element is then ignored, and the remaining elements are searched for the smallest element, which, when found, is placed after previous minimum, and so on. A total of $n(n-1)/2$ comparisons will be required.[11]

[10] More than one field can be indexed. However, each field will require its own index file.

[11] When n is large, $n^2/2$ is sometimes given as the number of comparisons.

Figure 41.2 *Key and Data Files*

key file

key	ref.
ADAMS	3
JONES	2
SMITH	1
THOMAS	4

data file

record	last name	first name	age
1	SMITH	JOHN	27
2	JONES	WANDA	39
3	ADAMS	HENRY	58
4	THOMAS	SUSAN	18

In a *bubble sort*, each element in the list is compared with the element immediately following it. If the first element is larger, the positions of the two elements are reversed (swapped). In effect, the smaller element "bubbles" to the top of the list. The comparisons continue to be made until the bottom of the list is reached. If no swaps are made in a pass, the list is sorted. A total of approximately $n^2/2$ comparisons are needed, on the average, to sort a list in this manner.[12]

In an *insertion sort*, the elements are ordered by rewriting them in the proper sequence. After the proper position of an element is found, all elements below that position are bumped down one place in the sequence. The resulting vacancy is filled by the inserted element. At worst, approximately $n^2/2$ comparisons will be required. On the average, there will be approximately $n^2/4$ comparisons.

Disregarding the number of swaps, the number of comparisons required by the successive minima, bubble, and insertion sorts is on the order of n^2. When n is large, these methods are two slow. The *quicksort* is more complex but reduces the average number of comparisons (with random data) to approximately $n\times\log(n)/\log(2)$, generally considered as being on the order of $n\times\log(n)$.[13] The maximum number of comparisons for a heap sort is $n \times \log(n)/\log(2)$, but it is likely that even fewer comparisons will be needed.

13. SEARCHING[14]

If a group of records (i.e., a list) is randomly organized, a particular element in the list can be found only by a linear search (sequential search). At best, only one comparison and, at worst, n comparisons will be required

[12] This is the same as for the successive minima approach. However, swapping occurs more frequently in the bubble sort, slowing it down.

[13] However, the quicksort falters (in speed) when the elements are in near-perfect order.

[14] The term *probing* is synonymous with searching.

to find something (an event known as a *hit*) in a list of n elements. The average is $n/2$ comparisons, described as being on the order of n.

If the records are in ascending or descending order, a binary search will be superior.[15] The search begins by looking at the middle element in the list. If the middle element is the sought-for element, the search is over. If not, half of the list can be disregarded in further searching since elements in that portion will be either too large or too small. The middle element in the remaining part of the list is investigated, and the procedure continues until a hit occurs or the list is exhausted. The number of required comparisons in a list of n elements will be $\log(n)/\log(2)$ (i.e., on the order of $\log(n)$).

14. HASHING

An index file is not needed if the record number (i.e., the storage location for a read or write operation) can be calculated directly from the key, a technique known as *hashing*.[16] The procedure by which a numeric or nonnumeric key (e.g., a last name) is converted into a record number is called the hashing function or hashing algorithm. Most hashing algorithms use a remaindering modulus—the remainder after dividing the key by the number of records, n, in the list. Excellent results are obtained if n is a prime number; poor results occur if n is a power of 2.

Not all hashed record numbers will be correct. A *collision* occurs when an attempt is made to use a record number that is already in use. Chaining, linear probing, and double hashing are techniques used to resolve such collisions.

15. DATABASE STRUCTURES

Databases can be implemented as indexed files, linked lists, and tree structures; in all three cases, the records are written and remain in the order of entry.

An indexed file such as that shown in Fig. 41.2 keeps the data in one file and maintains separate index files (usually in sorted order) for each key field. The index file must be recreated each time records are added to the file. A *flat file* has only one key field by which records can be located. Searching techniques (see Sec. 13) are used to locate a particular record. In a *linked list* (*threaded list*), each record has an associated *pointer* (usually a record number or memory address) to the next record in key sequence. Only two pointers are changed when a record is added or deleted. Generally, a linear search following the links through the records is used.

Pointers are also used in *tree structures*. Each record has one or more pointers to other records that meet certain criteria. In a binary tree structure, each record has two pointers—usually to records that are lower and higher, respectively, in key sequence. In general, records in a tree structure are referred to as *nodes*. The first record in the file is called the *root node*. A particular node will have one node above it (the *parent* or *ancestor*) and one or more nodes below it (the *daughters* or *offspring*). Records are found in a tree by starting at the root node and moving sequentially according to the tree structure. The number of comparisons required to find a particular element is $1 + (\log(n)/\log(2))$, which is on the order of $\log(n)$.

Figure 41.3 *Database Structures*

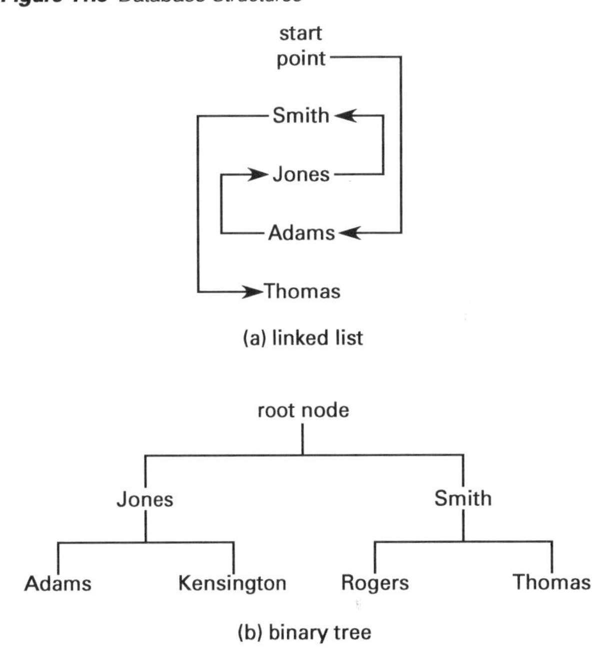

16. HIERARCHICAL AND RELATIONAL DATA STRUCTURES

A *hierarchical database* contains records in an organized, structured format. Records are organized according to one or more of the indexing schemes. However, each field within a record is not normally accessible.

A *relational database* stores all information in the equivalent of a matrix. Nothing else (no index files, pointers, etc.) is needed to find, read, edit, or select information. Any piece of information can be accessed directly by referring to the field name and field value.

Figure 41.4 *A Hierarchical Personnel File*

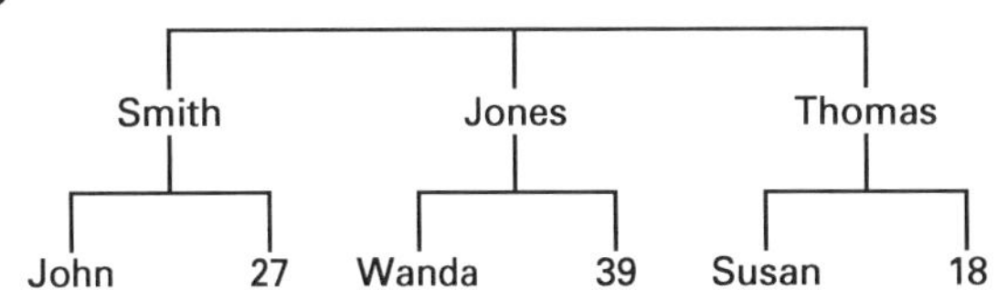

[15]A binary search is unrelated to a binary tree. A binary tree structure (see Sec. 15) greatly reduces search time but does not use a sorted list.

[16]Of course, finding a record in this manner requires it to have been written in a location determined by the same hashing routine.

Figure 41.5 *A Relational Personnel File*

rec. no.	last	first	age
1	Smith	John	27
2	Jones	Wanda	39
3	Thomas	Susan	18

17. ARTIFICIAL INTELLIGENCE

Artificial intelligence (AI) in a machine implies that the machine is capable of absorbing and organizing new data, learning new concepts, reasoning logically, and responding to inquiries. AI is implemented in a category of programs known as *expert systems* that "learn" rules from sets of events that are entered whenever they occur. (The manner in which the entry is made depends on the particular system.) Once the rules are learned, an expert system can participate in a dialogue to give advice, make predictions and diagnoses, or draw conclusions.

18. HIERARCHY OF OPERATIONS

Operations in an arithmetic statement are performed in the order of exponentiation first, multiplication and division second, and addition and subtraction third. In the event there are two consecutive operations with the same hierarchy (e.g., a multiplication followed by a division), the operations are performed in the order encountered, normally left to right (except for exponentiation, which is right to left).[17] Parentheses can modify this order; operations within parentheses are always evaluated before operations outside. If nested parentheses are present in an expression, the expression is evaluated outward starting from the innermost pair.

Example 41.2

Evaluate J in the following expression. Mixed-mode arithmetic is permitted, and expressions are scanned left to right.

$$J = (6.0 + 3.0)*3.0/6.0 + 5.0 - 6.0**2.0$$

Solution

The expression within the parentheses is evaluated first.

$$9.0*3.0/6.0 + 5.0 - 6.0**2.0$$

The exponentiation is performed next.

$$9.0*3.0/6.0 + 5.0 - 36.0$$

[17]In most implementations, a statement will be scanned from left to right. Once a left-to-right scan is complete, some implementations then scan from right to left; others return to the equals sign and start a second left-to-right scan. Parentheses should be used to define the intended order of operations.

The multiplication and division are performed next.

$$4.5 + 5.0 - 36.0$$

The addition and subtraction are performed last.

$$-26.5$$

However, J is an integer variable, and the assignment of J = −26.5 results in truncating the fractional part. Ultimately, J has the value of −26.

19. LOGIC GATES

A *gate* performs a logical operation on one or more inputs. The inputs, labeled A, B, C, etc., and output are limited to the values of 0 or 1. A listing of the output value for all possible input values is known as a *truth table.* Table 41.4 combines the symbols, names, and truth tables for the most common gates.

Table 41.4 *Logic Gates*

inputs		not	and	or	nand	nor	exclusive or
A	A	$-A$ or $\overline{A}$	AB	$A+B$	$\overline{AB}$	$\overline{A+B}$	$A \oplus B$
0	0	1	0	0	1	1	0
0	1	1	0	1	1	0	1
1	0	0	0	1	1	0	1
1	1	0	1	1	0	0	0

20. BOOLEAN ALGEBRA

The rules of *Boolean Algebra* are used to write and simplify expressions of binary variables (i.e., variables constrained to two values). The basic laws governing Boolean variables are listed here.

commutative: $A + B = B + A$

$A \cdot B = B \cdot A$

associative: $A + (B + C) = (A + B) + C$

$A \cdot (B \cdot C) = (A \cdot B) \cdot C$

distributive: $A \cdot (B + C) = (A \cdot B) + (A \cdot C)$

$A + (B \cdot C) = (A + B) \cdot (A + C)$

absorptive: $A + (A \cdot B) = A$

$A \cdot (A + B) = A$

De Morgan's theorems are

$$\overline{(A + B)} = \overline{A} \cdot \overline{B}$$

$$\overline{A \cdot B} = \overline{A} + \overline{B}$$

The following basic identities are used to simplify Boolean expressions.

$$0+0=0$$
$$0+1=1$$
$$1+0=1$$
$$1+1=1$$
$$0\cdot 0=0$$
$$0\cdot 1=1$$
$$1\cdot 0=0$$
$$1\cdot 1=1$$
$$A+0=A$$
$$A+1=1$$
$$A+A=A$$
$$A+\overline{A}=1$$
$$A\cdot 0=0$$
$$A\cdot 1=A$$
$$A\cdot A=A$$
$$A\cdot\overline{A}=0$$
$$-0=1$$
$$-1=0$$
$$-\overline{A}=A$$

Example 41.3

Simplify and write the truth tables for the following network of logic gates.

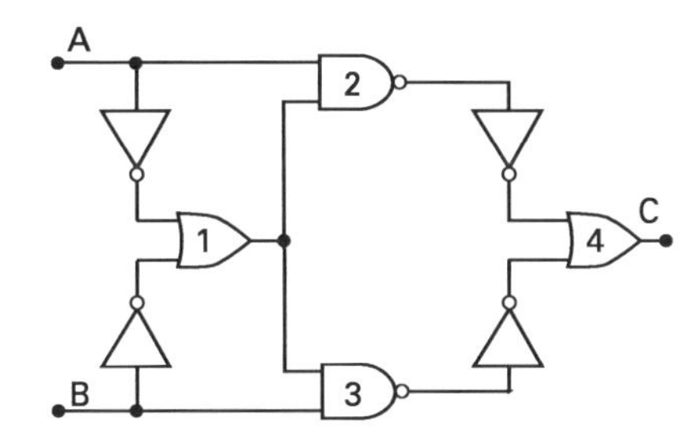

Solution

Determine the inputs and output of each gate in turn.

gate	inputs	output
1	$\overline{A}$, $\overline{B}$	$(\overline{A}+\overline{B})$
2	A, $(\overline{A}+\overline{B})$	$\overline{A\cdot(\overline{A}+\overline{B})}$
3	B, $(\overline{A}+\overline{B})$	$\overline{B\cdot(\overline{A}+\overline{B})}$
4	$A\cdot(\overline{A}+\overline{B})$, $B\cdot(\overline{A}+\overline{B})$	$A\cdot(\overline{A}+\overline{B})+B\cdot(\overline{A}+\overline{B})$

Simplify the output of gate 4.

$A\cdot(\overline{A}+\overline{B})+B\cdot(\overline{A}+\overline{B})$	[original]
$A\cdot\overline{A}+A\cdot\overline{B}+B\cdot\overline{A}+B\cdot\overline{B}$	[distributive]
$A\cdot\overline{B}+B\cdot\overline{A}$	[since $A\cdot\overline{A}=0$]
$A\oplus B$	[definition]

The truth table is

A	B	C
0	0	0
0	1	1
1	0	1
1	1	0

PRACTICE PROBLEMS

1. Flowchart the following procedure: If A is greater than B and A is greater than C, then add B to C and go to PLACE1. If B is greater than A and B is greater than C, then add A and C and go to PLACE2. If B is greater than A but less than C, then subtract A and B and go to PLACE1.

2. Flowchart the following procedure: If A is greater than 10 and if A is less than 14, subtract X from A. If A is less than 10 or if A is greater than 14, exit the program.

3. Draw a flowchart representing an algorithm that finds the real roots of a second-degree equation with real coefficients. The coefficients are to be read from an external peripheral, and the results are to be printed.

4. Draw a flowchart representing the bubble sort algorithm. (Sort into ascending order.)

5. For the binary search procedure, derive the number of required comparisons in a list of n elements.

SOLUTIONS

1.

start
input A, B, C
A > B and A > C
y
B = B + C
place1
n
B > A and B > C
y
A = A + C
place2
n
B > A and B < C
y
B = B − A
n
end

2.

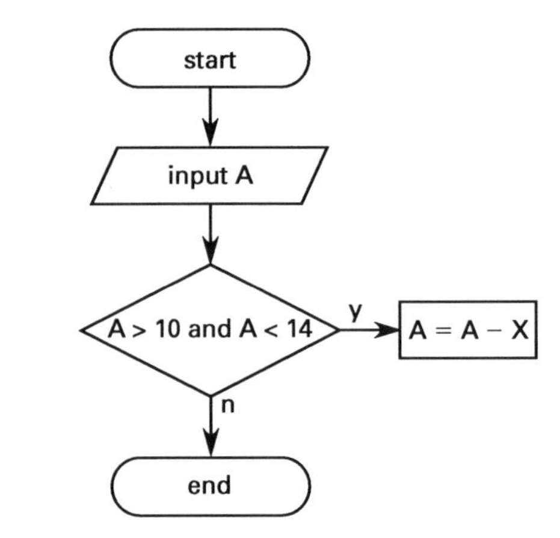

3.

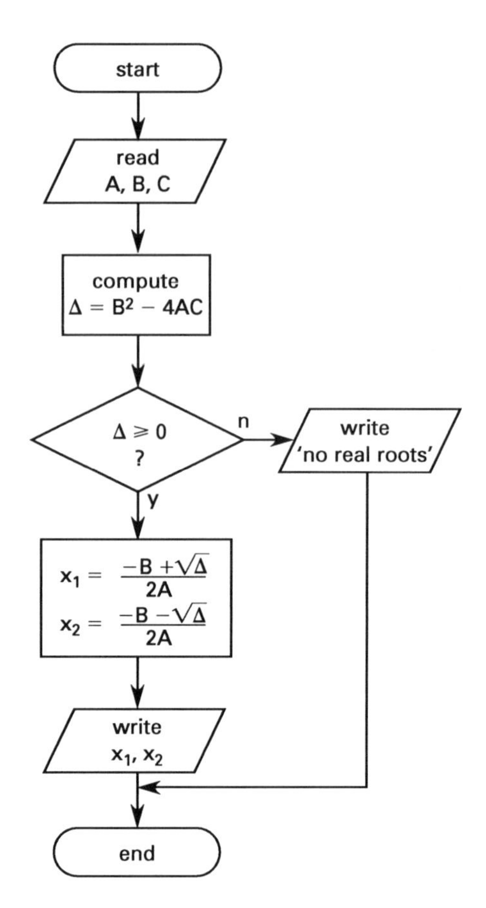

4.

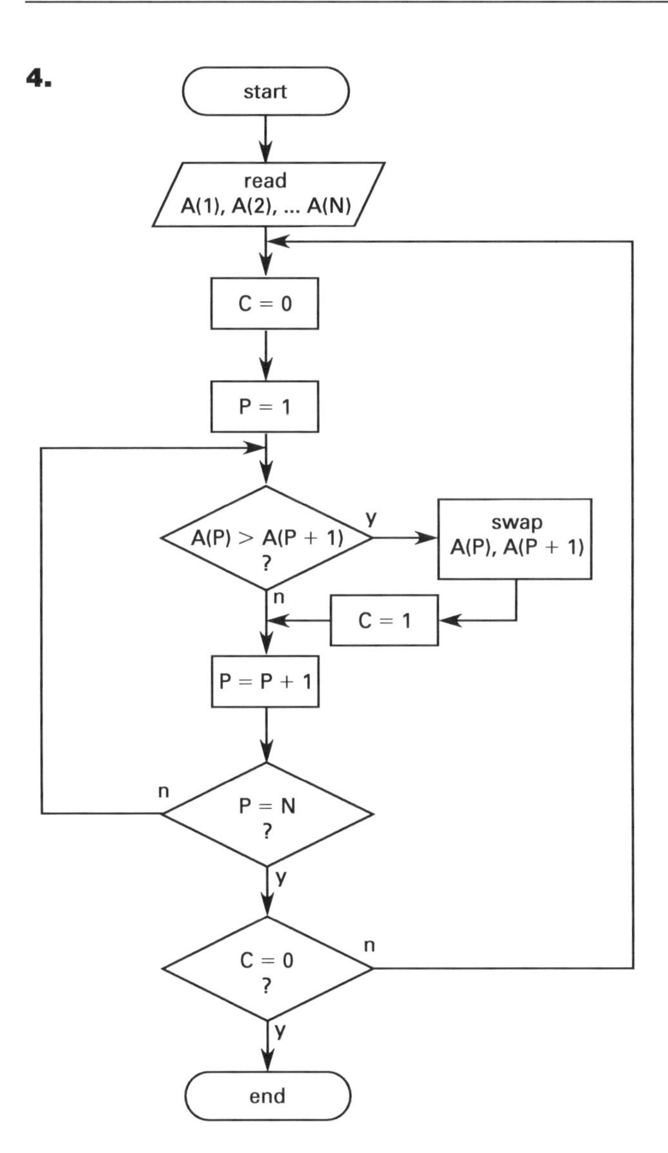

5. To simplify, assume that the number of elements in the list is some integer power of 2.

$$n = 2^p$$

Assume that the sought-for element is the first element in the list. This will produce a worst-case search. Let $a(n)$ be the number of comparisons needed to find the element in the list. Since the binary search procedure discards (disregards) half of the list after each comparison,

$$a(n) = 1 + a\left(\frac{n}{2}\right) \qquad \text{[I]}$$

But, $n = 2^p$.

$$a(n) = a\left(2^p\right) = 1 + a\left(\frac{2^p}{2}\right) = 1 + a\left(2^{p-1}\right) \qquad \text{[II]}$$

Equation I can be applied to the second term of Eq. II.

$$a\left(2^{p-1}\right) = 1 + a\left(\frac{2^{p-1}}{2}\right) = 1 + a\left(2^{p-2}\right) \qquad \text{[III]}$$

Combining Eqs. II and III,

$$a(n) = a\left(2^p\right) = 1 + 1 + a\left(2^{p-2}\right)$$

Equation I can continue to be applied to the last term until the list contains only two elements, after which the search becomes trivial. This results in a string of p 1's.

$$a(n) = 1 + 1 + 1 + \cdots + 1 = p$$

(An argument for $a(n) = p + 1$ can be made depending on the how the search logic handles a miss with 2 elements in the list. If it merely selects the remaining element without comparing it, then $a(n) = p$. If a singular element is investigated, then $a(n) = p + 1$.)

Now,

$$n = 2^p$$

$$\log_2(n) = \log_2\left(2^p\right) = p$$

But $p = a(n) = \log_2(n)$.

Using logarithm identities,

$$p = \log_2(n) = \frac{\log_{10}(n)}{\log_{10}(2)}$$

Topic X: Business Management Practices

Chapter

42 Job Costing

1. INTRODUCTION

Job costing is one of the most critical functions of successfully managing a surveying company. It is the ability to develop a reliable estimate of the cost of a survey project before it starts and to determine the actual cost of that project after it ends. Estimating the costs of a prospective project allows the surveyor to negotiate fair fees and to manage the project. Determining the actual cost of the project after completion allows the surveyor to analyze the profitability of the project for business planning purposes, and to generate feedback that will improve the estimate process. A good understanding of job costing is essential to surveying professionals involved in marketing and business management as well to those involved in hands-on surveying services.

Knowledge of the principles of job costing is especially important when a surveyor deals with government agencies, which require their contractors to strictly adhere to appropriate job costing procedures. For example, the Brooks Act (40 USC 759(f), 41USC 541) requires that all federal agencies negotiate contracts for professional services, including surveying and mapping, by taking into account the estimated value of the services. Under the Federal Acquisition Regulations, that estimate process must involve job costing as described in this chapter. Under this process, the cost of surveying services for a specific project is usually considered to be the sum of the direct costs together with a proportional share of the overhead costs, or indirect costs, of the surveying company. This chapter will define these costs and provide procedures for determining them.

2. DETERMINING THE QUANTITY OF DIRECT LABOR

In estimating project cost, the first and probably most important step is determining which professional classes of personnel will be needed to accomplish the project and how much time will be required for each class. This requires a thoughtful analysis of the project to develop a detailed *scope of services*. The scope of services document should detail each task and product necessary to complete the work. Following the development of the scope of services, each task should be analyzed to identify the class of personnel required and estimate the amount of time it will take. This is typically called a *staff hour estimate* or a *task analysis*. For completed projects, the quantity of direct labor may be obtained from a ledger generated by a job cost accounting system that contains a total of the time expended on the project.

3. DIRECT COSTS

Direct labor costs are the salary or wages paid to employees for work on specific projects. With a staff hour estimate as described in the previous section, an estimate of direct labor costs may be obtained by multiplying the quantity of labor time for each job class by the wage rate to get the labor cost for that class, and then adding the costs for all classes involved in the project. *Other direct costs* is a category of expenses other than labor costs that are associated with a specific project. Examples of these direct costs include travel expenses, and costs of material or supplies that are purchased for the project.

4. INDIRECT COSTS

Indirect labor costs are any salary and wages that are spread over several projects, rather than dedicated to only one. These costs include salaries paid to management and marketing personnel as well as wages paid to other personnel for duties that might be divided among several projects. *Other indirect costs* is a category of indirect costs other than indirect labor costs. Examples of typical costs in this category include office rent and utilities.

5. OVERHEAD RATES

The typical method for determining what portion of the surveying organization's indirect costs, or *overhead*, applies to a specific project is to first calculate an *overhead rate* from the previous year's performance. The overhead rate on a direct labor basis is easily calculated using Eq. 42.1 given the direct and indirect costs for a typical year. The overhead rate can then be multiplied by the direct labor costs for an individual project to yield the overhead for the project as in Eq. 42.2.

$$\text{overhead rate} = \frac{\text{total indirect costs}}{\text{total direct labor costs}} \qquad 42.1$$

$$\begin{aligned}\text{overhead} &= (\text{overhead rate}) \\ &\quad \times (\text{project direct labor costs}) \qquad 42.2\end{aligned}$$

An example of a financial statement used for the determination of overhead rate is provided in Table 42.1. The actual calculation is provided at the bottom of the table, using data from the financial statement. Note that on these types of statements, certain indirect costs (bank loan interest costs, entertainment costs, and country club dues in this example) must be excluded from the total of indirect costs in accordance with Federal Acquisition Regulations. The regulations do not allow such costs to be considered in the calculation of overhead rate for government contracting purposes, even though the costs may be legitimate business expenses. In the example, the overhead rate is 170.63%. This indicates that for every dollar the surveying firm spent on direct labor costs, the firm spent just over \$1.70 on indirect costs.

6. DETERMINING COST OF SERVICES

The total cost of a particular project, or *job cost*, is the sum of the direct costs and a proportional share of the indirect costs. With a known overhead rate, the cost of a project may be determined using the following equation.

$$\begin{aligned}\text{job cost} &= \text{direct labor costs} \\ &\quad + (\text{direct labor costs})(\text{overhead rate}) \\ &\quad + \text{other direct costs} \qquad 42.3\end{aligned}$$

Example 42.1

The vice president of a surveying firm would like to estimate the job cost of a potential project. The firm has a total overhead rate of 170.63%, as shown in Table 42.1. A task analysis has indicated the project will require 80 hr of work from a three-person survey crew, 40 hr of office technician time, and 40 hr of professional supervision time. Other direct costs are estimated to come to \$1000. The hourly wages paid the staff are as follows.

professional surveyor	\$35
office technician	\$15
crew	
crew chief	\$17
instrument technician	\$13
rod technician	\$10

Solution

The direct labor costs would be calculated as follows. For the crew,

$$(80 \text{ hr})\left(\frac{\$17}{1 \text{ hr}} + \frac{\$13}{1 \text{ hr}} + \frac{\$10}{1 \text{ hr}}\right) = \$3200$$

For the office technician,

$$(40 \text{ hr})\left(\frac{\$15}{1 \text{ hr}}\right) = \$600$$

For the professional surveyor,

$$(40 \text{ hr})\left(\frac{\$35}{1 \text{ hr}}\right) = \$1400$$

The total direct labor cost is

$$\$3200 + \$600 + \$1400 = \$5200$$

Use Eq. 42.3.

$$\begin{aligned}\text{job cost} &= \text{direct labor costs} \\ &\quad + (\text{direct labor costs})(\text{overhead rate}) \\ &\quad + \text{other direct costs} \\ &= \$5200 + (\$5200)(1.7063) + \$1000 \\ &= \$15{,}073\end{aligned}$$

7. DETERMINING OPERATING MARGIN

Operating margin is an amount above a surveying firm's costs for a project that the firm adds as part of the total fee charged the client. As used in this context, total fee is the total amount to be charged the client, including all direct costs, a prorated amount of indirect costs, and operating margin. (The Federal Acquisition Regulations assume a fair operating margin to be in the range of 6–15% of the total fee.) The operating margin represents the return on an investment of the surveying firm. It is also a means of recovering any indirect costs excluded in the determination of the overhead rate.

Operating margin may be determined by the following equation.

$$\text{operating margin} = \frac{\text{total fee} - \text{job cost}}{\text{job cost}} \qquad 42.4$$

Some government agencies consider other direct expenses and subcontractor costs as direct pass-through costs, not subject to operating margin. Under that approach, such costs would be excluded from both the fee and job cost. However, since such expenses do represent an investment by the surveying company, the operating margin is more typically applied to these expenses as well as to direct labor costs and overhead.

Table 42.1 Sample Financial Statement for Overhead Rate Determination

	total costs ($)	unallowable costs ($)	allowable costs ($)
direct labor costs			
employee wages	$126,000		$126,000[1]
indirect costs			
employee fringe benefits			
vacation and holidays	16,000		6,000
payroll taxes	17,000		17,000
group insurance	20,000		20,000
workers' comp insurance	16,000		16,000
general and administrative expenses			
indirect salaries	76,000		76,000
office rent	12,000		12,000
insurance	15,000		15,000
dues	5000	5000[a]	
depreciation	15,000		15,000
marketing travel	5000		5000
entertainment	5000	5000[b]	
office supplies	8000		8000
communications	12,000		12,000
miscellaneous	3000		3000
interest expenses	7500	7500[c]	
total indirect costs	$232,500	$17,500	$215,000[2]

unallowable costs (per Federal Acquisition Regulations)
[a]31.205-12 country club dues
[b]31.205-14 entertainment costs
[c]31.205-20 bank loan interest

overhead rate calculation

$$\text{overhead rate} = \frac{^2\text{total indirect costs}}{^1\text{total direct labor costs}} = \frac{\$215{,}000}{\$126{,}000} = 1.7063 \quad (170.63\%)$$

8. JOB COST ACCOUNTING

A surveying company needs a job cost accounting system to accumulate data that will allow it to determine a valid overhead rate and track project costs. The system may be as informal as a series of manually maintained ledgers, it may be based on computer spread sheets, or it may take advantage of readily available formal job cost accounting systems. The key requirements of such an accounting system is that it must segregate direct and indirect costs, and it must accumulate direct costs for each project separately. The basis for such accounting is an accurate time sheet system that identifies the quantity of hours worked and the project.

Most job cost accounting systems have the capability to produce periodic reports of costs associated with all active projects. Such progress reports are invaluable for project management and for assessment of the accuracy of estimates. Annual reports from such a system also provide the cost information, such as that shown in Table 42.1, to allow a surveying company to determine its overhead rate.

PRACTICE PROBLEMS

1. What is the overhead rate, on a direct labor basis, for a firm with direct labor costs of $150,000, fringe benefit costs of $75,000, and other indirect expenses of $150,000?

(A) 135%
(B) 140%
(C) 145%
(D) 150%

2. What operating margin was achieved for a project for which a fee of $50,000 was received if the project costs were $46,300?

(A) 7%
(B) 8%
(C) 9%
(D) 10%

3. What is the total estimated cost of a project for a firm with a 150% overhead rate if the direct labor cost is estimated to be $25,000 and other direct expenses to be $5000?

(A) $42,500
(B) $62,500
(C) $67,500
(D) $72,225

4. For the project in Prob. 3, what is most nearly the fee that should be charged by the firm if it wants an operating margin of 10%?

(A) $46,300
(B) $68,800
(C) $74,300
(D) $79,400

5. A task analysis has indicated that a prospective project will require two 40 hr weeks of crew time and four 8 hr days of professional surveyor time. If the combined wage rate for the survey crew is $45/hr and the wage rate for the professional surveyor is $40/hr, what is most nearly the estimated direct labor cost for the project?

(A) $4900
(B) $5000
(C) $5100
(D) $5700

6. For the project in Prob. 5, if the firm estimates other direct expenses at $1000, has an overhead rate of 150% and desires an operating margin of 10%, what is most nearly the fee the firm should propose?

(A) $8320
(B) $13,200
(C) $13,800
(D) $14,600

SOLUTIONS

1.

$$\text{overhead rate} = \frac{\text{total indirect costs}}{\text{total direct labor costs}} = \frac{\$75{,}000 + \$150{,}000}{\$150{,}000} = 150\%$$

The answer is (D).

2.

$$\text{operating margin} = \frac{\text{fee received} - \text{job cost}}{\text{job cost}} = \frac{\$50{,}000 - \$46{,}300}{\$46{,}300} = 0.08 \quad (8\%)$$

The answer is (B).

3.

$$\begin{aligned}\text{job cost} &= \text{direct labor costs} \\ &\quad + (\text{direct labor costs})(\text{overhead rate}) \\ &\quad + \text{other direct costs} \\ &= \$25{,}000 + (\$25{,}000)(1.50) + \$5000 \\ &= \$67{,}500\end{aligned}$$

The answer is (C).

4. If a 10% operating margin is desired,

$$\begin{aligned}\text{fee} &= (1 + \text{operating margin})(\text{job cost}) \\ &= (1.10)(\$67{,}500) \\ &= \$74{,}250 \quad (\$74{,}300)\end{aligned}$$

The answer is (C).

5. The direct labor costs would be calculated as follows. For the crew,

$$(80 \text{ hr})\left(\frac{\$45}{1 \text{ hr}}\right) = \$3600$$

For the professional surveyor,

$$(32 \text{ hr})\left(\frac{\$40}{1 \text{ hr}}\right) = \$1280$$

The total direct labor costs are

$$\$3600 + \$1280 = \$4880 \quad (\$4900)$$

The answer is (A).

6. The total estimated cost for the project would be calculated as follows.

$$\begin{aligned}\text{job cost} &= \text{direct labor costs}\\ &\quad + (\text{direct labor costs})(\text{overhead rate})\\ &\quad + \text{other direct costs}\\ &= \$4900 + (\$4900)(1.50) + \$1000\\ &= \$13{,}250\end{aligned}$$

If a 10% operating margin is desired, the proposed fee should be

$$\begin{aligned}\text{fee} &= (1 + \text{operating margin})(\text{job cost})\\ &= (1.10)(\$13{,}250)\\ &= \$14{,}575 \quad (\$14{,}600)\end{aligned}$$

The answer is (D).

43 Economic Analysis

Nomenclature

A	annual amount	\$
B	present worth of all benefits	\$
BV_j	book value at end of the jth year	\$
C	cost or present worth of all costs	\$
d	declining balance depreciation rate	decimal
D	demand	various
D	depreciation	\$
DR	present worth of after-tax depreciation recovery	\$
e	constant inflation rate	decimal
E_0	initial amount of an exponentially growing cash flow	\$
EAA	equivalent annual amount	\$
EUAC	equivalent uniform annual cost	\$
F	future worth	\$
i	effective rate per period (usually per year)	decimal per unit time
i'	effective interest rate corrected for inflation	decimal
k	number of compounding periods per year	–
m	an integer	–
n	number of compounding periods or life of asset	–
P	present worth	\$
r	nominal rate per year (rate per annum)	decimal per unit time
ROI	return on investment	\$
ROR	rate of return	decimal per unit time
t	time	yr (typical)
T	a quantity equal to $^1/_2 n(n+1)$	–

Subscripts

0	initial
j	at time j
n	at time n
t	at time t

1. INTRODUCTION

In its simplest form, an *economic analysis* is a study of the desirability of making an investment. There is very little, if any, true economics in this subject. The decision-making principles in this chapter can be applied by individuals as well as by companies. The nature of the spending opportunity or industry is not important. Land, personal investments, and multimillion dollar infrastructure improvements can all be evaluated using the same principles.

Similarly, the applicable principles are largely insensitive to the monetary units. Although *dollars* are used in this chapter, it is equally convenient to use pounds, yen, or euros.

Finally, this chapter may give the impression that investment alternatives must be evaluated on a year-by-year basis. Actually, the *effective period* can be defined as a day, month, century, or any other convenient period of time.

There is a wide variety of problem types that, collectively, are considered to be economic analysis problems.

By far, the majority of economic analysis problems are *alternative comparisons*. In these problems, two or more mutually exclusive investments compete for limited funds. A variation of this is a *replacement/retirement analysis*, which is repeated each year to determine if an existing asset should be replaced. Finding the percentage return on an investment is a *rate of return problem*, one of the alternative comparison solution methods.

Investigating interest and principal amounts in loan payments is a *loan repayment problem.* An *economic life analysis* will determine when an asset should be retired. In addition, there are miscellaneous problems involving economic order quantity, learning curves, break-even points, product costs, and so on.

2. TYPES OF CASH FLOWS

To evaluate a real-world project, it is necessary to present the project's cash flows in terms of standard cash flows that can be handled by economic analysis techniques. Although there are other, more complex types, the two types of cash flows covered in this chapter are single payment and uniform series.

A *single payment cash flow* can occur at the beginning of the time line (designated as $t = 0$), at the end of the time line (designated as $t = n$), or at any time in between.

The *uniform series cash flow* consists of a series of equal transactions starting at $t = 1$ and ending at $t = n$. The symbol A is typically given to the magnitude of each individual cash flow.[1]

3. SINGLE-PAYMENT EQUIVALENCE

The equivalence of any present amount, P, at $t = 0$, to any future amount, F, at $t = n$, is called the *future worth* and can be calculated from Eq. 43.1.

$$F = P(1+i)^n \qquad 43.1$$

The factor $(1+i)^n$ is known as the single payment *compound amount factor* and has been tabulated in the factor table in App. L for various combinations of i and n.

Similarly, the equivalence of any future amount to any present amount is called the *present worth* and can be calculated from Eq. 43.2.

$$P = F(1+i)^{-n} = \frac{F}{(1+i)^n} \qquad 43.2$$

The factor $(1+i)^{-n}$ is known as the *single payment present worth factor.*[2]

The interest rate used in Eqs. 43.1 and 43.2 must be the effective rate per period. Also, the basis of the rate (annually, monthly, etc.) must agree with the type of period used to count n. Thus, it would be incorrect to use an effective annual interest rate if n was the number of compounding periods in months.

[1]Notice that the cash flows do not begin at $t = 0$. This is an important concept with all of the series cash flows. This convention has been established to accommodate the timing of annual maintenance (and similar) cash flows for which the year-end convention is applicable.

[2]The *present worth* is also called the *present value* and *net present value.* These terms are used interchangeably and no significance should be attached to the terms *value, worth,* and *net.*

Example 43.1

How much should you put into a 10% (effective annual rate) savings account in order to have \$10,000 in 5 yr?

Solution

This problem could also be stated: What is the equivalent present worth of \$10,000 5 yr from now if money is worth 10% per year?

$$P = F(1+i)^{-n} = (\$10{,}000)(1+0.10)^{-5} = \$6209$$

The factor $(1 + 0.10)^{-5}$, or 0.6209, would usually be obtained from the factor table in App. L.

4. STANDARD CASH FLOW FACTORS AND SYMBOLS

Equations 43.1 and 43.2 may give the impression that solving economic analysis problems involves a lot of calculator use, and, in particular, a lot of exponentiation. Such calculations may be necessary from time to time, but most problems are simplified by the use of tabulated values of the factors.

Rather than actually writing the formula for the compound amount factor (which converts a present amount to a future amount), it is common convention to substitute the standard functional notation of $(F/P, i\%, n)$. Thus, the future value in n periods of a present amount would be symbolically written as

$$F = P(F/P, i\%, n) \qquad 43.3$$

Similarly, the present worth factor has a functional notation of $(P/F, i\%, n)$. The present worth of a future amount n periods hence would be symbolically written as

$$P = F(P/F, i\%, n) \qquad 43.4$$

Values of these *cash flow (discounting) factors* are tabulated in App. L. There is often initial confusion about whether the (F/P) or (P/F) column should be used in a particular problem. There are several ways of remembering what the functional notations mean.

One method of remembering which factor should be used is to think of the factors as conditional probabilities. The conditional probability of event **A** given that event **B** has occurred is written as $p\{\mathbf{A}|\mathbf{B}\}$, where the given event comes after the vertical bar. In the standard notational form of discounting factors, the given amount is similarly placed after the slash. What you want comes before the slash. (F/P) would be a factor to find F given P.

Another method of remembering the notation is to interpret the factors algebraically. Thus, the (F/P) factor could be thought of as the fraction F/P. Algebraically, Eq. 43.3 would be

$$F = P(F/P) \tag{43.5}$$

This algebraic approach is actually more than an interpretation. The numerical values of the discounting factors are consistent with this algebraic manipulation. Thus, the (F/A) factor could be calculated as $(F/P) \times (P/A)$.

Example 43.2

If you put $10,000 into a money market account at a 5% interest rate, what would it be worth in 10 yr?

Solution

This problem could be stated as follows.

$$F = P(F/P, 5\%, 10 \text{ yr})$$

Using the factor table in App. L,

$$F = (\$10{,}000)(1.6289) = \$16{,}289$$

Example 43.3

How much would you have to invest at 5% to have $10,000 in 10 yr?

Solution

This problem could be stated as follows.

$$P = F(P/F, 5\%, 10 \text{ yr})$$

Using the factor table in App. L,

$$P = (\$10{,}000)(0.6139) = \$6139$$

5. CALCULATING UNIFORM SERIES EQUIVALENCE

A cash flow that repeats each year for n years without change in amount is known as an *annual amount* and is given the symbol A. As an example, a piece of equipment may require annual maintenance, and the maintenance cost will be an annual amount. Although the equivalent value for each of the n annual amounts could be calculated and then summed, it is more expedient to use one of the uniform series factors. For example, it is possible to convert from an annual amount to a future amount by use of the (F/A) factor.

$$F = A(F/A, i\%, n) \tag{43.6}$$

Business Mgmt. Practices

Table 43.1 *Discount Factors for Discrete Compounding*

factor name	converts	symbol	formula
single payment compound amount	P to F	$(F/P, i\%, n)$	$(1+i)^n$
single payment present worth	F to P	$(P/F, i\%, n)$	$(1+i)^{-n}$
uniform series sinking fund	F to A	$(A/F, i\%, n)$	$\frac{i}{(1+i)^n - 1}$
capital recovery	P to A	$(A/P, i\%, n)$	$\frac{i(1+i)^n}{(1+i)^n - 1}$
uniform series compound amount	A to F	$(F/A, i\%, n)$	$\frac{(1+i)^n - 1}{i}$
uniform series present worth	A to P	$(P/A, i\%, n)$	$\frac{(1+i)^n - 1}{i(1+i)^n}$

A *sinking fund* is a fund or account into which annual deposits of A are made in order to accumulate F at $t = n$ in the future. Since the annual deposit is calculated as $A = F(A/F, i\%, n)$, the (A/F) factor is known as the *sinking fund factor*. An *annuity* is a series of equal payments (A) made over a period of time.[3] Usually, it is necessary to "buy into" an investment (a bond, and insurance policy, etc.) in order to ensure the annuity. In the simplest case of an annuity that starts at the end of the first year and continues for n years, the purchase price (P) is

$$P = A(P/A, i\%, n) \qquad 43.7$$

The present worth of an *infinite (perpetual) series* of annual amounts is known as a *capitalized cost*. There is no $(P/A, i\%, \infty)$ factor in the tables, but the capitalized cost can be calculated simply as

$$P = \frac{A}{i} \qquad [i \text{ in decimal form}] \qquad 43.8$$

Alternatives with different lives will generally be compared by way of *equivalent uniform annual cost* (EUAC). An EUAC is the annual amount that is equivalent to all of the cash flows in the alternative. The EUAC differs in sign from all of the other cash flows. Costs and expenses expressed as EUACs, which would normally be considered negative, are actually positive. The term *cost* in the designation EUAC serves to make clear the meaning of a positive number.

Example 43.4

What would be the monthly payment on a \$100,000 mortgage for the purchase of land if the interest rate is 6% with a 5 yr term?

Solution

This problem could be stated as follows.

$$\begin{aligned} A &= P\left(A/P, \frac{6\%}{12\ \frac{\text{mo}}{\text{yr}}}, (5\text{ yr})\left(12\ \frac{\text{mo}}{\text{yr}}\right)\right) \\ &= P(A/P, 0.5\%, 60\text{ mo}) \end{aligned}$$

Using the factor table in App. L,

$$A = (\$100{,}000)(0.0193) = \$1930$$

Example 43.5

What would be the annual return from a \$300,000 annuity with a 30 yr payout term at 6% interest?

[3]An annuity may also consist of a lump sum payment made at some future time. However, this interpretation is not considered in this chapter.

Solution

This problem could be stated as follows.

$$A = P(A/P, 6\%, 30\text{ yr})$$

Using the factor table in App. L,

$$A = (\$300{,}000)(0.0726) = \$21{,}780$$

6. FINDING PAST VALUES

From time to time, it will be necessary to determine an amount in the past equivalent to some current (or future) amount. For example, you might have to calculate the original investment made 15 yr ago given a current annuity payment.

Such problems are solved by placing the $t = 0$ point at the time of the original investment, and then calculating the past amount as a P value. For example, the original investment, P, can be extracted from the annuity, A, by using the standard cash flow factors.

$$P = A(P/A, i\%, n) \qquad 43.9$$

The choice of $t = 0$ is flexible. As a general rule, the $t = 0$ point should be selected for convenience in solving a problem.

Example 43.6

You are currently paying \$250 per month to lease your office phone equipment. You have 3 yr (36 mo) left on the 5 yr (60 mo) lease. What would have been an equivalent purchase price 2 yr ago? The effective interest rate per month is 1%.

Solution

The solution of this example is not affected by the fact that investigation is being performed in the middle of the horizon. This is a simple calculation of present worth.

$$\begin{aligned} P &= A(P/A, 1\%, 60\text{ mo}) \\ &= (-\$250)(44.9550) = -\$11{,}239 \end{aligned}$$

7. THE MEANING OF PRESENT WORTH AND *i*

If \$100 is invested in a 5% bank account (using annual compounding), you can remove \$105 a year from now; if this investment is made, you will receive a *return on investment* (ROI) of \$5. The cash flow diagram and the present worth of the two transactions are

$$P = -\$100 + (\$105)(P/F, 5\%, 1)$$
$$= -\$100 + (\$105)(0.9524) = 0$$

Figure 43.1 *Cash Flow Diagram*

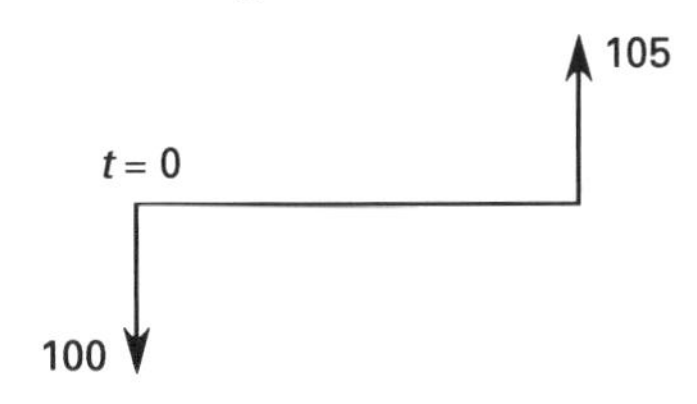

Notice that the present worth is zero even though you will receive a \$5 return on your investment.

However, if you are offered \$120 for the use of \$100 over a 1 yr period, the cash flow diagram and present worth (at 5%) would be

$$P = -\$100 + (\$120)(P/F, 5\%, 1)$$
$$= -\$100 + (\$120)(0.9524) = \$14.29$$

Figure 43.2 *Cash Flow Diagram*

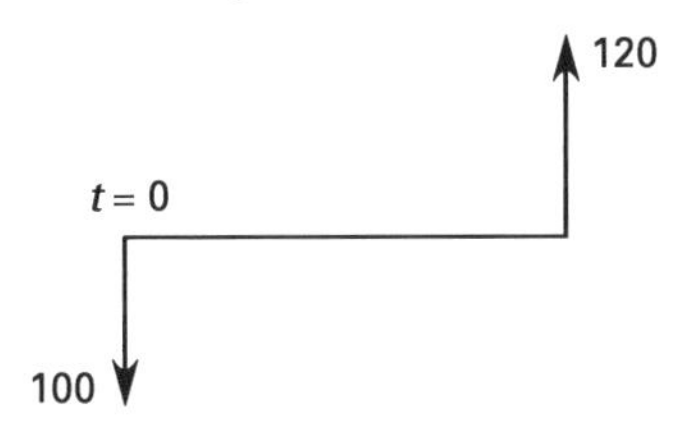

Therefore, the present worth of an alternative is seen to be equal to the equivalent value at $t = 0$ of the increase in return above that which you would be able to earn in an investment offering $i\%$ per period. In the previous case, \$14.29 is the present worth of (\$20 − \$5), the difference in the two ROIs.

The present worth is also the amount that you would have to be given to dissuade you from making an investment, since placing the initial investment amount along with the present worth into a bank account earning $i\%$ will yield the same eventual return on investment. Relating this to the previous paragraphs, you could be dissuaded from investing \$100 in an alternative that would return \$120 in 1 yr by a $t = 0$ payment of \$14.29. Clearly, (\$100 + \$14.29) invested at $t = 0$ will also yield \$120 in 1 yr at 5%.

Income-producing alternatives with negative present worths are undesirable, and alternatives with positive present worths are desirable because they increase the average earning power of invested capital. (In some cases, such as municipal and public works projects, the present worths of all alternatives are negative, in which case, the least negative alternative is best.)

The selection of the interest rate is difficult in engineering economics problems. Usually, it is taken as the average rate of return that an individual or business organization has realized in past investments. Alternatively, the interest rate may be associated with a particular level of risk. Usually, i for individuals is the interest rate that can be earned in relatively *risk-free investments.*

8. SIMPLE AND COMPOUND INTEREST

If \$100 is invested at 5%, it will grow to \$105 in 1 yr. During the second year, 5% interest continues to be accrued, but on \$105, not on \$100. This is the principle of *compound interest*: The interest accrues interest.[4]

If only the original principal accrues interest, the interest is said to be *simple interest.* Simple interest is rarely encountered in economic analyses, but the concept may be incorporated into short-term transactions.

9. RATE OF RETURN VERSUS RETURN ON INVESTMENT

Rate of return (ROR) is an effective annual interest rate, typically stated in percent per year. *Return on investment* (ROI) is a dollar amount. Thus, *rate of return* and *return on investment* are not synonymous.

Return on investment can be calculated in two different ways. The accounting method is to subtract the total of all investment costs from the total of all net profits (i.e., revenues less expenses). The time value of money is not considered.

In economic analysis, the return on investment is calculated from equivalent values. Specifically, the present worth (at $t = 0$) of all investment costs is subtracted from the future worth (at $t = n$) of all net profits.

When there are only two cash flows, a single investment amount and a single payback, the two definitions of return on investment yield the same numerical value. When there are more than two cash flows, the returns on investment will be different depending on which definition is used.

10. CHOICE OF ALTERNATIVES: COMPARING ONE ALTERNATIVE WITH ANOTHER ALTERNATIVE

Several methods exist for selecting a superior alternative from among a group of proposals. Each method has its own merits and applications.

[4]This assumes, of course, that the interest remains in the account. If the interest is removed and spent, only the remaining funds accumulate interest.

Present Worth Method

When two or more alternatives are capable of performing the same functions, the superior alternative will have the largest present worth. The *present worth method* is restricted to evaluating alternatives that are mutually exclusive and that have the same lives. This method is suitable for ranking the desirability of alternatives.

Example 43.7

Investment A costs \$10,000 today and pays back \$11,500 2 yr from now. Investment B costs \$8000 today and pays back \$4500 each year for 2 yr. If an interest rate of 5% is used, which alternative is superior?

Solution

$$\begin{aligned} P(\text{A}) &= -\$10{,}000 + (\$11{,}500)(P/F, 5\%, 2\ \text{yr}) \\ &= -\$10{,}000 + (\$11{,}500)(0.9070) \\ &= \$431 \\ P(\text{B}) &= -\$8000 + (\$4500)(P/A, 5\%, 2\ \text{yr}) \\ &= -\$8000 + (\$4500)(1.8594) \\ &= \$367 \end{aligned}$$

Alternative A is superior and should be chosen.

Capitalized Cost Method

The present worth of a project with an infinite life is known as the *capitalized cost* or *life cycle cost.* Capitalized cost is the amount of money at $t = 0$ needed to perpetually support the project on the earned interest only. Capitalized cost is a positive number when expenses exceed income.

In comparing two alternatives, each of which is infinitely lived, the superior alternative will have the lowest capitalized cost.

Normally, it would be difficult to work with an infinite stream of cash flows since most economics tables do not list factors for periods in excess of 100 yr. However, the (A/P) discounting factor approaches the interest rate as n becomes large. Since the (P/A) and (A/P) factors are reciprocals of each other, it is possible to divide an infinite series of equal cash flows by the interest rate in order to calculate the present worth of the infinite series. This is the basis of Eq. 43.10.

$$\text{capitalized cost} = \text{initial cost} + \frac{\text{annual costs}}{i} \qquad 43.10$$

Equation 43.10 can be used when the annual costs are equal in every year. If the operating and maintenance costs occur irregularly instead of annually, or if the costs vary from year to year, it will be necessary to somehow determine a cash flow of equal annual amounts (EAA) that is equivalent to the stream of original costs.

The equal annual amount may be calculated in the usual manner by first finding the present worth of all the actual costs and then multiplying the present worth by the interest rate (the (A/P) factor for an infinite series). However, it is not even necessary to convert the present worth to an equal annual amount since Eq. 43.11 will convert the equal amount back to the present worth.

$$\begin{aligned} \text{capitalized cost} &= \text{initial cost} + \frac{\text{EAA}}{i} \\ &= \text{initial cost} + \begin{array}{c}\text{present worth} \\ \text{of all expenses}\end{array} \end{aligned} \qquad 43.11$$

Example 43.8

What is the capitalized cost of a public works project that will cost \$25,000,000 now and will require \$2,000,000 in maintenance annually? The effective annual interest rate is 12%.

Solution

Worked in millions of dollars, from Eq. 43.10, the capitalized cost is

$$\begin{aligned} \text{capitalized cost} &= 25 + (2)(P/A, 12\%, \infty) \\ &= 25 + \frac{2}{0.12} = 41.67 \end{aligned}$$

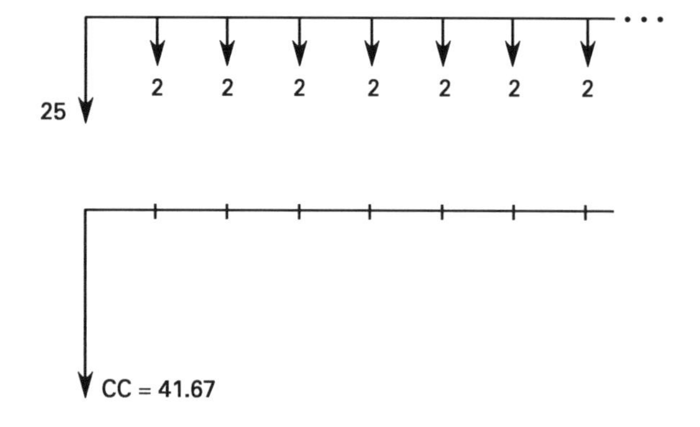

Annual Cost Method

Alternatives that accomplish the same purpose but that have unequal lives must be compared by the *annual cost method.*[5] The annual cost method assumes that each alternative will be replaced by an identical twin at the end of its useful life (infinite renewal). This method, which may also be used to rank alternatives according to their desirability, is also called the *annual return method* or *capital recovery method.*

Restrictions are that the alternatives must be mutually exclusive and repeatedly renewed up to the duration of the longest-lived alternative. The calculated

[5]Of course, the annual cost method can be used to determine the superiority of assets with identical lives as well.

annual cost is known as the *equivalent uniform annual cost* (EUAC) or just *equivalent annual cost.* Cost is a positive number when expenses exceed income.

Example 43.9

Which of the following alternatives is superior over a 30 yr period if the interest rate is 7%?

	alternative A	alternative B
type	brick	wood
life	30 yr	10 yr
initial cost	$1800	$450
maintenance	$5/yr	$20/yr

Solution

$$\begin{aligned}\text{EUAC(A)} &= (\$1800)(A/P, 7\%, 30\text{ yr}) + \$5\\ &= (\$1800)(0.0806) + \$5\\ &= \$150\\ \text{EUAC(B)} &= (\$450)(A/P, 7\%, 10\text{ yr}) + \$20\\ &= (\$450)(0.1424) + \$20\\ &= \$84\end{aligned}$$

Alternative B is superior since its annual cost of operation is the lowest. It is assumed that three wood facilities, each with a life of 10 yr and a cost of $450, will be built to span the 30 yr period.

11. CHOICE OF ALTERNATIVES: COMPARING AN ALTERNATIVE WITH A STANDARD

With specific economic performance criteria, it is possible to qualify an investment as acceptable or unacceptable without having to compare it with another investment. Two such performance criteria are the benefit-cost ratio and the minimum attractive rate of return.

Benefit-Cost Ratio Method

The *benefit-cost ratio* method is often used in municipal project evaluations where benefits and costs accrue to different segments of the community. With this method, the present worth of all benefits (irrespective of the beneficiaries) is divided by the present worth of all costs. The project is considered acceptable if the ratio equals or exceeds 1.0, that is, if $B/C \geq 1.0$.

When the benefit-cost ratio method is used, disbursements by the initiators or sponsors are *costs.* Disbursements by the users of the project are known as *disbenefits.* It is often difficult to determine whether a cash flow is a cost or a disbenefit (whether to place it in the numerator or denominator of the benefit-cost ratio calculation).

Regardless of where the cash flow is placed, an acceptable project will always have a benefit-cost ratio greater than or equal to 1.0, although the actual numerical result will depend on the placement. For this reason, the benefit-cost ratio method should not be used to rank competing projects.

The benefit-cost ratio method of comparing alternatives is used extensively in transportation engineering where the ratio is often (but not necessarily) written in terms of annual benefits and annual costs instead of present worths. Another characteristic of highway benefit-cost ratios is that the route (road, highway, etc.) is usually already in place and that various alternative upgrades are being considered. There will be existing benefits and costs associated with the current route. Therefore, the *change* (usually an increase) in benefits and costs is used to calculate the benefit-cost ratio.[6]

$$B/C = \frac{\Delta_{\text{benefits}}^{\text{user}}}{\Delta_{\text{cost}}^{\text{investment}} + \Delta\text{maintenance} - \Delta_{\text{value}}^{\text{residual}}} \qquad 43.12$$

Notice that the change in *residual value (terminal value)* appears in the denominator as a negative item. An increase in the residual value would decrease the denominator.

Example 43.10

By building a bridge over a ravine, a state department of transportation can shorten the time it takes to drive through a mountainous area. Estimates of costs and benefits (due to decreased travel time, fewer accidents, reduced gas usage, etc.) have been prepared. Should the bridge be built? Use the benefit-cost ratio method of comparison.

	millions
initial cost	40
capitalized cost of perpetual annual maintenance	12
capitalized value of annual user benefits	49
residual value	0

Solution

If Eq. 43.12 is used, the benefit-cost ratio is

$$B/C = \frac{49}{40 + 12 + 0} = 0.942$$

[6]This discussion of highway benefit-cost ratios is not meant to imply that everyone agrees with Eq. 43.12. In *Economic Analysis for Highways* (International Textbook Company, Scranton, PA, 1969), author Robley Winfrey takes a strong stand on one aspect of the benefits versus disbenefits issue: highway maintenance. According to Winfrey, regular highway maintenance costs should be placed in the numerator as a subtraction from the user benefits. Some have called this mandate the *Winfrey method.*

Since the benefit-cost ratio is less than 1.00, the bridge should not be built.

If the maintenance costs are placed in the numerator (per Ftn. 6), the benefit-cost ratio value will be different but the conclusion will not change.

$$B/C_{\text{alternate method}} = \frac{49 - 12}{40} = 0.925$$

Rate of Return Method

The minimum attractive rate of return (MARR) has already been introduced as a standard of performance against which an investment's actual *rate of return* (ROR) is compared. If the rate of return is equal to or exceeds the minimum attractive rate of return, the investment is qualified. This is the basis for the *rate of return method* of alternative selection.

Finding the rate of return can be a long, iterative process. Usually, the actual numerical value of rate of return is not needed; it is sufficient to know whether or not the rate of return exceeds the minimum attractive rate of return. This *comparative analysis* can be accomplished without calculating the rate of return simply by finding the present worth of the investment using the minimum attractive rate of return as the effective interest rate (i.e., i = MARR). If the present worth is zero or positive, the investment is qualified. If the present worth is negative, the rate of return is less than the minimum attractive rate of return.

12. RANKING MUTUALLY EXCLUSIVE MULTIPLE PROJECTS

Ranking of multiple investment alternatives is required when there is sufficient funding for more than one investment. Since the best investments should be selected first, it is necessary to be able to place all investments into an ordered list.

Ranking is relatively easy if the present worths, future worths, capitalized costs, or equivalent uniform annual costs have been calculated for all the investments. The highest ranked investment will be the one with the largest present or future worth, or the smallest capitalized or annual cost. Present worth, future worth, capitalized cost, and equivalent uniform annual cost can all be used to rank multiple investment alternatives.

However, neither rates of return nor benefit-cost ratios should be used to rank multiple investment alternatives. Specifically, if two alternatives both have rates of return exceeding the minimum acceptable rate of return, it is not sufficient to select the alternative with the highest rate of return.

An *incremental analysis*, also known as a *rate of return on added investment study*, should be performed if rate of return is used to select between investments. An incremental analysis starts by ranking the alternatives in order of increasing initial investment. Then, the cash flows for the investment with the lower initial cost are subtracted from the cash flows for the higher-priced alternative on a year-by-year basis. This produces, in effect, a third alternative representing the costs and benefits of the added investment. The added expense of the higher priced investment is not warranted unless the rate of return of this third alternative exceeds the minimum attractive rate of return as well. The choice criterion is to select the alternative with the higher initial investment if the incremental rate of return exceeds the minimum attractive rate of return.

An incremental analysis is also required if ranking is to be done by the benefit-cost ratio method. The incremental analysis is accomplished by calculating the ratio of differences in benefits to differences in costs for each possible pair of alternatives. If the ratio exceeds 1.0, alternative 2 is superior to alternative 1. Otherwise, alternative 1 is superior.[7]

$$\frac{B_2 - B_1}{C_2 - C_1} \geq 1 \quad \text{[alternative 2 superior]} \qquad 43.13$$

13. ALTERNATIVES WITH DIFFERENT LIVES

Comparison of two alternatives is relatively simple when both alternatives have the same life. For example, a problem might be stated: "Which would you rather have: car A with a life of 3 yr, or car B with a life of 5 yr?"

However, care must be taken to understand what is going on when the two alternatives have different lives. If car A has a life of 3 yr and car B has a life of 5 yr, what happens at $t = 3$ if the 5 yr car is chosen? If a car is needed for 5 yr, what happens at $t = 3$ if the 3 yr car is chosen?

In this type of situation, it is necessary to distinguish between the length of the need (the *analysis horizon*) and the lives of the alternatives or assets intended to meet that need. The lives do not have to be the same as the horizon.

Finite Horizon with Incomplete Asset Lives

If an asset with a 5 yr life is chosen for a 3 yr need, the disposition of the asset at $t = 3$ must be known in order to evaluate the alternative. If the asset is sold at $t = 3$, the salvage value is entered into the analysis (at $t = 3$) and the alternative is evaluated as a 3 yr investment. The fact that the asset is sold when it has some useful life remaining does not affect the analysis horizon.

Similarly, if a 3 yr asset is chosen for a 5 yr need, something about how the need is satisfied during the last 2 yr must be known. Perhaps a rental asset will be

[7]It goes without saying that the benefit-cost ratios for all investment alternatives by themselves must also be equal to or greater than 1.0.

used. Or, perhaps the function will be "farmed out" to an outside firm. In any case, the costs of satisfying the need during the last 2 yr enter the analysis, and the alternative is evaluated as a 5 yr investment.

If both alternatives are "converted" to the same life, any of the alternative selection criteria (present worth method, annual cost method, etc.) can be used to determine which alternative is superior.

Finite Horizon with Integer Multiple Asset Lives

It is common to have a long-term horizon (need) that must be met with short-lived assets. In special instances, the horizon will be an integer number of asset lives. For example, a company may be making a 12 yr transportation plan and may be evaluating two cars: one with a 3 yr life, and another with a 4 yr life.

In this example, four of the first car or three of the second car are needed to reach the end of the 12 yr horizon.

If the horizon is an integer number of asset lives, any of the alternative selection criteria can be used to determine which is superior. If the present worth method is used, all alternatives must be evaluated over the entire horizon. (In this example, the present worth of 12 yr of car purchases and use must be determined for both alternatives.)

If the equivalent uniform annual cost method is used, it may be possible to base the calculation of annual cost on one lifespan of each alternative only. It may not be necessary to incorporate all of the cash flows into the analysis. (In the running example, the annual cost over 3 yr would be determined for the first car; the annual cost over 4 yr would be determined for the second car.) This simplification is justified if the subsequent asset replacements (renewals) have the same cost and cash flow structure as the original asset. This assumption is typically made implicitly when the annual cost method of comparison is used.

Infinite Horizon

If the need horizon is infinite, it is not necessary to impose the restriction that asset lives of alternatives be integer multiples of the horizon. The superior alternative will be replaced (renewed) whenever it is necessary to do so, forever.

Infinite horizon problems are almost always solved with either the annual cost or capitalized cost method. It is common to (implicitly) assume that the cost and cash flow structure of the asset replacements (renewals) are the same as the original asset.

14. CAPITALIZED ASSETS VERSUS EXPENSES

High expenses reduce profit, which in turn reduces income tax. It seems logical to label each and every expenditure, even an asset purchase, as an expense. As an alternative to this *expensing the asset*, it may be decided to capitalize the asset. *Capitalizing the asset* means that the cost of the asset is divided into equal or unequal parts, and only one of these parts is taken as an expense each year. Expensing is clearly the more desirable alternative, since the after-tax profit is increased early in the asset's life.

There are long-standing accounting conventions as to what can be expensed and what must be capitalized.[8] Some companies capitalize everything—regardless of cost—with expected lifetimes greater than 1 yr. Most companies, however, expense items whose purchase costs are below a cut-off value. A cut-off value in the range of $250 to $500, depending on the size of the company, is chosen as the maximum purchase cost of an expensed asset. Assets costing more than this are capitalized.

It is not necessary for a large corporation to keep track of every lamp, desk, and chair for which the purchase price is greater than the cut-off value. Such assets, all of which have the same lives and have been purchased in the same year, can be placed into groups or *asset classes*. A group cost, equal to the sum total of the purchase costs of all items in the group, is capitalized as though the group was an identifiable and distinct asset itself.

15. PURPOSE OF DEPRECIATION

Depreciation is an *artificial expense* that spreads the purchase price of an asset or other property over a number of years.[9] Depreciating an asset is an example of capitalization, as previously defined. The inclusion of depreciation in economic analysis problems will increase the after-tax present worth (profitability) of an asset. The larger the depreciation, the greater will be the profitability. Therefore, individuals and companies eligible to utilize depreciation want to maximize and accelerate the depreciation available to them.

Although the entire property purchase price is eventually recognized as an expense, the net recovery from the expense stream never equals the original cost of the asset. That is, depreciation cannot realistically be thought of as a fund (an annuity or sinking fund) that accumulates capital to purchase a replacement at the end of the asset's life. The primary reason for this is that the depreciation expense is reduced significantly by the impact of income taxes, as will be seen in later sections.

[8]For example, purchased vehicles must be capitalized; payments for leased vehicles can be expensed. Repainting a building with paint that will last five years is an expense, but the replacement cost of a leaking roof must be capitalized.

[9]The IRS tax regulations allow depreciation on almost all forms of *business property* except land. The following types of property are distinguished: *real* (e.g., buildings used for business), *residential* (e.g., buildings used as rental property), and *personal* (e.g., equipment used for business). Personal property does *not* include items for personal use (such as a personal residence), despite its name. *Tangible personal property* is distinguished from *intangible property* (goodwill, copyrights, patents, trademarks, franchises, agreements not to compete, etc.).

16. DEPRECIATION BASIS OF AN ASSET

The *depreciation basis* of an asset is the part of the asset's purchase price that is spread over the *depreciation period*, also known as the *service life*.[10] Usually, the depreciation basis and the purchase price are not the same.

A common depreciation basis is the difference between the purchase price and the expected salvage value at the end of the depreciation period. That is,

$$\text{depreciation basis} = C - S_n \qquad 43.14$$

There are several methods of calculating the year-by-year depreciation of an asset. Equation 43.14 is not universally compatible with all depreciation methods. Some methods do not consider the salvage value. This is known as an *unadjusted basis.* When the depreciation method is known, the depreciation basis can be rigorously defined.[11]

17. DEPRECIATION METHODS

Generally, tax regulations do not allow the cost of an asset to be treated as a deductible expense in the year of purchase. Rather, portions of the depreciation basis must be allocated to each of the n years of the asset's depreciation period. The amount that is allocated each year is called the *depreciation.*

Various methods exist for calculating an asset's depreciation each year.[12] Although the depreciation calculations may be considered independently (for the purpose of determining book value or as an academic exercise), it is important to recognize that depreciation has no effect on economic analyses unless income taxes are also considered.

Straight Line Method

With the *straight line method*, depreciation is the same each year. The depreciation basis $(C - S_n)$ is allocated uniformly to all of the n years in the depreciation period. Each year, the depreciation will be

$$D = \frac{C - S_n}{n} \qquad 43.15$$

[10]The *depreciation period* is selected to be as short as possible within recognized limits. This depreciation will not normally coincide with the *economic life* or *useful life* of an asset. For example, a car may be capitalized over a depreciation period of 3 yr. It may become uneconomical to maintain and use at the end of an economic life of 9 yr. However, the car may be capable of operation over a useful life of 25 yr.

[11]For example, with the Accelerated Cost Recovery System (ACRS) the *depreciation basis* is the total purchase cost, regardless of the expected salvage value. With declining balance methods, the depreciation basis is the purchase cost less any previously taken depreciation.

[12]This discussion gives the impression that any form of depreciation may be chosen regardless of the nature and circumstances of the purchase. In reality, the IRS tax regulations place restrictions on the higher-rate (accelerated) methods, such as declining balance and sum-of-the-years' digits methods. Furthermore, the *Economic Recovery Act of 1981* and the *Tax Reform Act of 1986* substantially changed the laws relating to personal and corporate income taxes.

Constant Percentage Method

The *constant percentage method*[13] is similar to the straight line method in that the depreciation is the same each year. If the fraction of the basis used as depreciation is $1/n$, there is no difference between the constant percentage and straight line methods. The two methods differ only in what information is available. (With the straight line method, the life is known. With the constant percentage method, the depreciation fraction is known.)

Each year, the depreciation will be

$$\begin{aligned} D &= (\text{depreciation fraction})(\text{depreciation basis}) \\ &= (\text{depreciation fraction})(C - S_n) \end{aligned} \qquad 43.16$$

Sum-of-the-Years' Digits Method

In *sum-of-the-years' digits* (SOYD) depreciation, the digits from 1 to n inclusive are summed. The total, T, can also be calculated from

$$T = \tfrac{1}{2}n(n+1) \qquad 43.17$$

The depreciation in year j can be found from Eq. 43.18. Notice that the depreciation in year j, D_j, decreases by a constant amount each year.

$$D_j = \frac{(C - S_n)(n - j + 1)}{T} \qquad 43.18$$

Double Declining Balance Method[14]

Double declining balance[15] (DDB) depreciation is independent of salvage value. Furthermore, the book value never stops decreasing, although the depreciation decreases in magnitude. Usually, any book value in excess of the salvage value is written off in the last year of the asset's depreciation period. Unlike any of the other depreciation methods, double declining balance depends on accumulated depreciation.

$$D_{\text{first year}} = \frac{2C}{n} \qquad 43.19$$

$$D_j = \frac{2\left(C - \sum\limits_{m=1}^{j-1} D_m\right)}{n} \qquad 43.20$$

[13]The *constant percentage method* should not be confused with the declining balance method, which used to be known as the *fixed percentage on diminishing balance method.*

[14]In the past, the *declining balance method* has also been known as the *fixed percentage of book value* and *fixed percentage on diminishing balance method.*

[15]Double declining balance depreciation is a particular form of *declining balance depreciation,* as defined by the IRS tax regulations. Declining balance depreciation includes 125% declining balance and 150% declining balance depreciations that can be calculated by substituting 1.25 and 1.50, respectively, for the 2 in Eq. 43.19.

Calculating the depreciation in the middle of an asset's life appears particularly difficult with double declining balance, since all previous years' depreciation amounts seem to be required. It appears that the depreciation in the sixth year, for example, cannot be calculated unless the values of depreciation for the first 5 yr are calculated. However, this is not true.

Depreciation in the middle of an asset's life can be found from the following equations. (d is known as the *depreciation rate.*)

$$d = \frac{2}{n} \quad \text{43.21}$$

$$D_j = dC(1-d)^{j-1} \quad \text{43.22}$$

Statutory Depreciation Systems

In the United States, property placed into service in 1981 and thereafter must use the *Accelerated Cost Recovery System* (ACRS), and after 1986, the *Modified Accelerated Cost Recovery System* (MACRS) or other statutory method. Other methods (straight line, declining balance, etc.) cannot be used except in special cases.

Property placed into service in 1980 or before must continue to be depreciated according to the method originally chosen (e.g., straight line, declining balance, or sum-of-the-years' digits). ACRS and MACRS cannot be used.

Under ACRS and MACRS, the cost recovery amount in the jth year of an asset's cost recovery period is calculated by multiplying the initial cost by a factor.

$$D_j = C \times \text{factor} \quad \text{43.23}$$

The initial cost used is not reduced by the asset's salvage value for ACRS and MACRS calculations. The factor used depends on the asset's cost recovery period. Such factors are subject to continuing legislation changes. Current tax publications should be consulted before using this method.

Table 43.2 *Representative MACRS Depreciation Factors**

	depreciation rate for recovery period, n			
year, j	3 yr	5 yr	7 yr	10 yr
1	33.33%	20.00%	14.29%	10.00%
2	44.45%	32.00%	24.49%	18.00%
3	14.81%	19.20%	17.49%	14.40%
4	7.41%	11.52%	12.49%	11.52%
5		11.52%	8.93%	9.22%
6		5.76%	8.92%	7.37%
7			8.93%	6.55%
8			4.46%	6.55%
9				6.56%
10				6.55%
11				3.28%

*Values are for the "half-year" convention. This table gives typical values only. Since these factors are subject to continuing revision, they should not be used without consulting an accounting professional.

Production or Service Output Method

If an asset has been purchased for a specific task and that task is associated with a specific lifetime amount of output or production, the depreciation may be calculated by the fraction of total production produced during the year. The depreciation is not expected to be the same each year.

$$D_j = (C - S_n)\left(\frac{\text{actual output in year } j}{\text{estimated lifetime output}}\right) \quad \text{43.24}$$

Sinking Fund Method

The *sinking fund method* is seldom used in industry because the initial depreciation is low. The formula for sinking fund depreciation (which increases each year) is

$$D_j = (C - S_n)(A/F, i\%, n)(F/P, i\%, j-1) \quad \text{43.25}$$

Disfavored Methods

Three other depreciation methods are mentioned here, not because they are currently accepted or in widespread use, but because they are still occasionally encountered in the literature.[16]

The *sinking fund plus interest on first cost* depreciation method, like the following two methods, is an attempt to include the *opportunity interest cost* on the purchase price with the depreciation. That is, the purchasing company not only incurs an annual expense due to the drop in book value, but it also loses the interest on the purchase price. The formula for this method is

$$D = (C - S_n)(A/F, i\%, n) + Ci \quad \text{43.26}$$

The *straight line plus interest on first cost* method is similar. Its formula is

$$D = \left(\frac{1}{n}\right)(C - S_n) + Ci \quad \text{43.27}$$

The *straight line plus average interest method* assumes that the opportunity interest cost should be based on the book value only, not on the full purchase price. Since the book value changes each year, an average value is used. The depreciation formula is

$$D = \left(\frac{C - S_n}{n}\right)\left(1 + \frac{i(n+1)}{2}\right) + iS_n \quad \text{43.28}$$

Example 43.11

An asset is purchased for \$9000. Its estimated economic life is 10 yr, after which it will be sold for \$200. Find

[16]These three depreciation methods should not be used in the usual manner (e.g., in conjunction with the income tax rate). These methods are attempts to calculate a more accurate annual cost of an alternative. Sometimes they give misleading answers. Their use cannot be recommended. They are included in this chapter only for the sake of completeness.

the depreciation in the first 3 yr using straight line, double declining balance, and sum-of-the-years' digits depreciation methods.

Solution

$$\text{SL:}\quad D = \frac{\$9000 - \$200}{10} = \$880 \text{ each year}$$

$$\text{DDB:}\quad D_1 = \frac{(2)(\$9000)}{10} = \$1800 \text{ in year 1}$$

$$D_2 = \frac{(2)(\$9000 - \$1800)}{10} = \$1440 \text{ in year 2}$$

$$D_3 = \frac{(2)(\$9000 - \$3240)}{10} = \$1152 \text{ in year 3}$$

$$\text{SOYD:}\quad T = \left(\tfrac{1}{2}\right)(10)(11) = 55$$

$$D_1 = \left(\tfrac{10}{55}\right)(\$9000 - \$200) = \$1600 \text{ in year 1}$$

$$D_2 = \left(\tfrac{9}{55}\right)(\$8800) = \$1440 \text{ in year 2}$$

$$D_3 = \left(\tfrac{8}{55}\right)(\$8800) = \$1280 \text{ in year 3}$$

18. ACCELERATED DEPRECIATION METHODS

An *accelerated depreciation method* is one that calculates a depreciation amount greater than a straight line amount. Double declining balance and sum-of-the-years' digits methods are accelerated methods. The ACRS and MACRS methods are explicitly accelerated methods. Straight line and sinking fund methods are not accelerated methods.

Use of an accelerated depreciation method may result in unexpected tax consequences when the depreciated asset or property is disposed of. Professional tax advice should be obtained in this area.

19. BOOK VALUE

The difference between original purchase price and accumulated depreciation is known as *book value*.[17] At the end of each year, the book value (which is initially equal to the purchase price) is reduced by the depreciation in that year.

It is important to properly synchronize depreciation calculations. It is difficult to answer the question, "What is the book value in the fifth year?" unless the timing of the book value change is mutually agreed upon. It is better to be specific about an inquiry by identifying when the book value change occurs. For example, the following question is unambiguous: "What is the book value at the end of year 5, after subtracting depreciation in the fifth year?" or "What is the book value after five years?"

Unfortunately, this type of care is seldom taken in book value inquiries, and it is up to the respondent to exercise reasonable care in distinguishing between beginning-of-year book value and end-of-year book value. To be consistent, the book value equations in this chapter have been written in such a way that the year subscript (j) has the same meaning in book value and depreciation calculations. That is, BV_5 means the book value at the end of the fifth year, after 5 yr of depreciation, including D_5, has been subtracted from the original purchase price.

There can be a great difference between the book value of an asset and the *market value* of that asset. There is no legal requirement for the two values to coincide, and no intent for book value to be a reasonable measure of market value.[18] Therefore, it is apparent that book value is merely an accounting convention with little practical use. Even when a depreciated asset is disposed of, the book value is used to determine the consequences of disposal, not the price the asset should bring at sale.

The calculation of book value is relatively easy, even for the case of the declining balance depreciation method.

For the straight line depreciation method, the book value at the end of the jth year, after the jth depreciation deduction has been made, is

$$\text{BV}_j = C - \frac{j(C - S_n)}{n} = C - jD \qquad 43.29$$

For the sum-of-the-years' digits method, the book value is

$$\text{BV}_j = (C - S_n)\left(1 - \frac{j(2n + 1 - j)}{n(n + 1)}\right) + S_n \qquad 43.30$$

For the declining balance method, including double declining balance (see Ftn. 15), the book value is

$$\text{BV}_j = C(1 - d)^j \qquad 43.31$$

For the sinking fund method, the book value is calculated directly as

$$\text{BV}_j = C - (C - S_n)(A/F, i\%, n)(F/A, i\%, j) \qquad 43.32$$

[17]The balance sheet of a corporation usually has two asset accounts: the *equipment account* and the *accumulated depreciation account*. There is no book value account on this financial statement, other than the implicit value obtained from subtracting the accumulated depreciation account from the equipment account. The book values of various assets, as well as their original purchase cost, date of purchase, salvage value, and so on, and accumulated depreciation appear on detail sheets or other peripheral records for each asset.

[18]Common examples of assets with great divergences of book and market values are buildings (rental houses, apartment complexes, factories, etc.) and company luxury automobiles (Porsches, Mercedes, etc.) during periods of inflation. Book values decrease, but actual values increase.

Of course, the book value at the end of year j can always be calculated for any method by successive subtractions (i.e., subtraction of the accumulated depreciation), as Eq. 43.33 illustrates.

$$\text{BV}_j = C - \sum_{m=1}^{j} D_m \qquad 43.33$$

Figure 43.3 illustrates the book value of a hypothetical asset depreciated using several depreciation methods. Notice that the double declining balance method initially produces the fastest write-off, while the sinking fund method produces the slowest write-off. Note also that the book value does not automatically equal the salvage value at the end of an asset's depreciation period with the double declining balance method.[19]

Figure 43.3 *Book Value with Different Depreciation Methods*

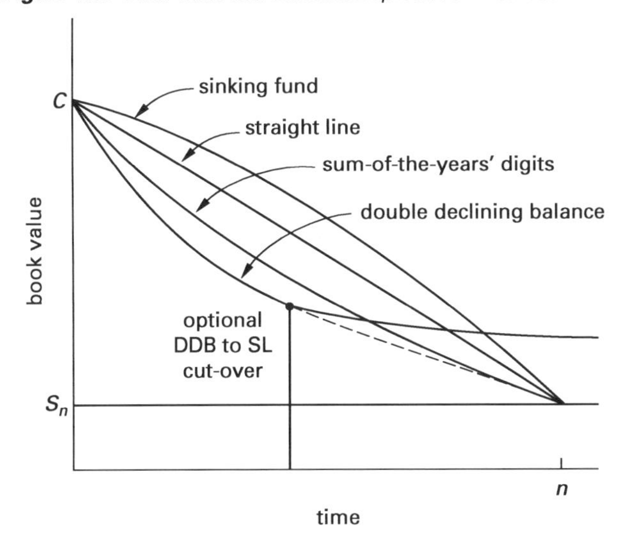

Example 43.12

For the asset described in Ex. 43.11, calculate the book value at the end of the first 3 yr if sum-of-the-years' digits depreciation is used. The book value at the beginning of year 1 is $9000.

Solution

From Eq. 43.33,

$$\begin{aligned}\text{BV}_1 &= \$9000 - \$1600 = \$7400\\ \text{BV}_2 &= \$7400 - \$1440 = \$5960\\ \text{BV}_3 &= \$5960 - \$1280 = \$4680\end{aligned}$$

[19]This means that the straight line method of depreciation may result in a lower book value at some point in the depreciation period than if double declining balance is used. A *cut-over* from double declining balance to straight line may be permitted in certain cases. Finding the *cut-over point*, however, is usually done by comparing book values determined by both methods. The analytical method is complicated.

20. AMORTIZATION

Amortization and depreciation are similar in that they both divide up the cost basis or value of an asset. In fact, in certain cases, the term "amortization" may be used in place of the term "depreciation." However, depreciation is a specific form of amortization.

Amortization spreads the cost basis or value of an asset over some base. The base can be time, units of production, number of customers, and so on. The asset can be tangible (e.g., a delivery truck or building) or intangible (e.g., goodwill or a patent).

If the asset is tangible, if the base is time, and if the length of time is consistent with accounting standards and taxation guidelines, then the term "depreciation" is appropriate. However, if the asset is intangible, if the base is some other variable, or if some length of time other than the customary period is used, then the term "amortization" is more appropriate.[20]

Example 43.13

A company purchases complete and exclusive patent rights to an invention for $1,200,000. It is estimated that once commercially produced, the invention will have a specific but limited market of 1200 units. For the purpose of allocating the patent right cost to production cost, what is the amortization rate in dollars per unit?

Solution

The patent should be amortized at the rate of

$$\frac{\$1,200,000}{1200 \text{ units}} = \$1000 \text{ per unit}$$

21. ACCOUNTING PRINCIPLES

Basic Bookkeeping

An accounting or *bookkeeping system* is used to record historical financial transactions. The resultant records are used for product costing, satisfaction of statutory requirements, reporting of profit for income tax purposes, and general company management.

Bookkeeping consists of two main steps: recording the transactions, followed by categorization of the transactions.[21] The transactions (receipts and disbursements) are recorded in a *journal (book of original entry)* to complete the first step. Such a journal is organized in

[20]From time to time, the U.S. Congress has allowed certain types of facilities (e.g., emergency, grain storage, and pollution control) to be written off more rapidly than would otherwise be permitted in order to encourage investment in such facilities. The term "amortization" has been used with such write-off periods.
[21]These two steps are not to be confused with the *double-entry bookkeeping method.*

a simple chronological and sequential manner.[22] The transactions are then categorized (into interest income, advertising expense, etc.) and posted (i.e., entered or written) into the appropriate *ledger account.*

The ledger accounts together constitute the *general ledger* or *ledger.* All ledger accounts can be classified into one of three types: *asset accounts, liability accounts,* and *owners' equity accounts.* Strictly speaking, income and expense accounts, kept in a separate journal, are included within the classification of owners' equity accounts.

Together, the journal and ledger are known simply as "the books" of the company.

Balancing the Books

In a business environment, *balancing the books* means more than reconciling the checkbook and bank statements. All accounting entries must be posted in such a way as to maintain the equality of the *basic accounting equation,*

$$\text{assets} = \text{liability} + \text{owners' equity} \qquad 43.34$$

In a *double-entry bookkeeping system,* the equality is maintained within the ledger system by entering each transaction into two balancing ledger accounts. For example, paying a utility bill would decrease the cash account (an asset account) and decrease the utility expense account (a liability account) by the same amount.

Transactions are either *debits* or *credits,* depending on their sign. Increases in asset accounts are debits; decreases are credits. For liability and equity accounts, the opposite is true: Increases are credits, and decreases are debits.[23]

Cash and Accrual Systems[24]

The simplest form of bookkeeping is based on the *cash system.* The only transactions that are entered into the journal are those that represent cash receipts and disbursements. In effect, a checkbook register or bank deposit book could serve as the journal.

During a given period (e.g., month or quarter), expense liabilities may be incurred even though the payments for those expenses have not been made. For example, an invoice (bill) may have been received but not paid. Under the *accrual system,* the obligation is posted into the appropriate expense account before it is paid.[25] Analogous to expenses, under the accrual system, income will be claimed before payment is received. Specifically, a sales transaction can be recorded as income when the customer's order is received, when the outgoing invoice is generated, or when the merchandise is shipped.

Financial Statements

Each period, two types of corporate financial statements are typically generated: the *balance sheet* and *profit and loss (P&L) statements.*[26] The profit and loss statement, also known as a *statement of income and retained earnings,* is a summary of sources of *income* or *revenue* (interest, sales, fees charged, etc.) and *expenses* (utilities, advertising, repairs, etc.) for the period. The expenses are subtracted from the revenues to give a *net income* (generally, before taxes).[27] Figure 43.4 illustrates a simple profit and loss statement.

The *balance sheet* presents the *basic accounting equation* in tabular form. The balance sheet lists the major categories of assets and outstanding liabilities. The difference between asset values and liabilities is the *equity,* as defined in Eq. 43.34. This equity represents what would be left over after satisfying all debts by liquidating the company.

Figure 43.4 *Simplified Profit and Loss Statement*

revenue			
interest	2000		
sales	237,000		
returns	(23,000)		
net revenue		216,000	
expenses			
salaries	149,000		
utilities	6000		
advertising	28,000		
insurance	4000		
supplies	1000		
net expenses		188,000	
period net income		28,000	
beginning retained earnings		63,000	
net year-to-date earnings			91,000

[22]The two-step process is more typical of a *manual bookkeeping system* than a computerized *general ledger system.* However, even most computerized systems produce reports in journal entry order, as well as account summaries.

[23]There is a difference in sign between asset and liability accounts. Thus, an increase in an expense account is actually a decrease. The accounting profession, apparently, is comfortable with the common confusion that exists between debits and credits.

[24]There is also a distinction made between cash flows that are known and those that are expected. It is a *standard accounting principle* to record losses in full, at the time they are recognized, even before their occurrence. In the construction industry, for example, losses are recognized in full and projected to the end of a project as soon as they are foreseeable. Profits, on the other hand, are recognized only as they are realized (typically, as a percentage of project completion). The difference between cash and accrual systems is a matter of *bookkeeping.* The difference between loss and profit recognition is a matter of *accounting convention.* Surveyors seldom need to be concerned with the accounting tradition.

[25]The expense for an item or service might be accrued even *before* the invoice is received. It might be recorded when the purchase order for the item or service is generated, or when the item or service is received.

[26]Other types of financial statements (*statements of changes in financial position, cost of sales statements,* inventory and asset reports, etc.) also will be generated, depending on the needs of the company.

[27]Financial statements also can be prepared with percentages (of total assets and net revenue) instead of dollars, in which case they are known as *common size financial statements.*

There are several terms that appear regularly on balance sheets.

- *current assets:* cash and other assets that can be converted quickly into cash, such as accounts receivable, notes receivable, and merchandise (inventory). Also known as *liquid assets.*
- *fixed assets:* relatively permanent assets used in the operation of the business and relatively difficult to convert into cash. Examples are land, buildings, and equipment. Also known as *nonliquid assets.*
- *current liabilities:* liabilities due within a short period of time (e.g., within 1 yr) and typically paid out of current assets. Examples are accounts payable, notes payable, and other accrued liabilities.
- *long-term liabilities:* obligations that are not totally payable within a short period of time (e.g., within 1 yr).

Figure 43.5 is a simplified balance sheet.

Figure 43.5 *Simplified Balance Sheet*

ASSETS			
current assets			
cash	14,000		
accounts receivable	36,000		
notes receivable	20,000		
inventory	89,000		
prepaid expenses	3000		
total current assets		162,000	
plant, property, and equipment			
land and buildings	217,000		
motor vehicles	31,000		
equipment	94,000		
accumulated depreciation	(52,000)		
total fixed assets		290,000	
total assets			452,000
LIABILITIES AND OWNERS' EQUITY			
current liabilities			
accounts payable	66,000		
accrued income taxes	17,000		
accrued expenses	8000		
total current liabilities		91,000	
long-term debt			
notes payable	117,000		
mortgage	23,000		
total long-term debt		140,000	
owners' and stockholders' equity			
stock	130,000		
retained earnings	91,000		
total owners' equity		221,000	
total liabilities and owners' equity			452,000

Analysis of Financial Statements

Financial statements are evaluated by management, lenders, stockholders, potential investors, and many other groups for the purpose of determining the *health of the company.* The health can be measured in terms of *liquidity* (ability to convert assets to cash quickly), *solvency* (ability to meet debts as they become due), and *relative risk* (of which one measure is *leverage*—the portion of total capital contributed by owners).

The analysis of financial statements involves several common ratios, usually expressed as percentages. The following are some frequently encountered ratios.

- *current ratio:* an index of short-term paying ability.

$$\text{current ratio} = \frac{\text{current assets}}{\text{current liabilities}}$$

- *quick* (or *acid-test*) *ratio:* a more stringent measure of short-term debt-paying ability. The *quick assets* are defined to be current assets minus inventories and prepaid expenses.

$$\text{quick ratio} = \frac{\text{quick assets}}{\text{current liabilities}}$$

- *receivable turnover:* a measure of the average speed with which accounts receivable are collected.

$$\text{receivable turnover} = \frac{\text{net credit sales}}{\text{average net receivables}}$$

- *average age of receivables:* number of days, on the average, in which receivables are collected.

$$\text{average age of receivables} = \frac{365 \text{ days}}{\text{receivable turnover}}$$

- *inventory turnover:* a measure of the speed with which inventory is sold, on the average.

$$\text{inventory turnover} = \frac{\text{cost of goods sold}}{\text{average cost of inventory on hand}}$$

- *days supply of inventory on hand:* number of days, on the average, that the current inventory would last.

$$\text{days supply of inventory on hand} = \frac{365 \text{ days}}{\text{inventory turnover}}$$

- *book value per share of common stock:* number of dollars represented by the balance sheet owners' equity for each share of common stock outstanding.

$$\text{book value per share of common stock} = \frac{\text{common shareholders' equity}}{\text{number of outstanding shares}}$$

- *gross margin:* gross profit as a percentage of sales. (Gross profit is sales less cost of goods sold.)

$$\text{gross margin} = \frac{\text{gross profit}}{\text{net sales}}$$

- *profit margin ratio:* percentage of each dollar of sales that is net income.

$$\text{profit margin} = \frac{\text{net income before taxes}}{\text{net sales}}$$

- *return on investment ratio:* shows the percent return on owners' investment.

$$\text{return on investment} = \frac{\text{net income}}{\text{owners' equity}}$$

- *price-earnings ratio:* indication of relationship between earnings and market price per share of common stock, useful in comparisons between alternative investments.

$$\text{price-earnings} = \frac{\text{market price per share}}{\text{earnings per share}}$$

22. INFLATION

It is important to perform economic studies in terms of *constant value dollars.* One method of converting all cash flows to constant value dollars is to divide the flows by some annual *economic indicator* or price index.

If indicators are not available, cash flows can be adjusted by assuming that inflation is constant at a decimal rate (e) per year. Then, all cash flows can be converted to $t = 0$ dollars by dividing by $(1 + e)^n$, where n is the year of the cash flow.

An alternative is to replace the effective annual interest rate (i) with a value corrected for inflation. This corrected value (i') is

$$i' = i + e + ie \qquad 43.35$$

This method has the advantage of simplifying the calculations. However, precalculated factors are not available for the non-integer values of i'. Therefore, Table 43.1 must be used to calculate the factors.

Example 43.14

What is the uninflated present worth of a $2000 future value in 2 yr if the average inflation rate is 6% and i is 10%?

Solution

$$P = \frac{\$2000}{(1 + 0.10)^2(1 + 0.06)^2} = \$1471$$

Example 43.15

Repeat Ex. 43.14 using i'.

Solution

$$i' = i + e + ie = 0.10 + 0.06 + (0.10)(0.06) = 0.166$$

$$P = \frac{\$2000}{(1 + 0.166)^2} = \$1471$$

23. CONSUMER LOANS

Special Nomenclature

BAL_j	balance after the jth payment
LV	principal total value loaned (cost minus down payment)
j	payment or period number
N	total number of payments to pay off the loan
PI_j	jth interest payment
PP_j	jth principal payment
PT_j	jth total payment
ϕ	effective rate per period (r/k)

Many different arrangements can be made between a borrower and a lender. With the advent of creative financing concepts, it often seems that there are as many variations of loans as there are loans made. Nevertheless, there are several traditional types of transactions. Real estate or investment texts, or a financial consultant, should be consulted for more complex problems.

Simple Interest

Interest due does not compound with a *simple interest loan.* The interest due is merely proportional to the length of time that the principal is outstanding. Because of this, simple interest loans are seldom made for long periods (e.g., more than 1 yr). (For loans less than 1 yr, it is commonly assumed that a year consists of 12 mo of 30 days each.)

Example 43.16

A $12,000 simple interest loan is taken out at 16% per annum interest rate. The loan matures in 2 yr with no intermediate payments. How much will be due at the end of the second year?

Solution

The interest each year is

$$\text{PI} = (0.16)(\$12{,}000) = \$1920$$

The total amount due in 2 yr is

$$\text{PT} = \$12{,}000 + (2)(\$1920) = \$15{,}840$$

Example 43.17

\$4000 is borrowed for 75 days at 16% per annum simple interest. How much will be due at the end of 75 days?

Solution

$$\begin{aligned}\text{amount due} &= \$4000 + (0.16)\left(\frac{75 \text{ days}}{360 \ \frac{\text{days}}{\text{bank yr}}}\right)(\$4000) \\ &= \$4133\end{aligned}$$

Loans with Constant Amount Paid Toward Principal

With this loan type, the payment is not the same each period. The amount paid toward the principal is constant, but the interest varies from period to period. The equations that govern this type of loan are

$$\text{BAL}_j = \text{LV} - (j)(\text{PP}) \quad \textit{43.36}$$

$$\text{PI}_j = \phi(\text{BAL})_{j-1} \quad \textit{43.37}$$

$$\text{PT}_j = \text{PP} + \text{PI}_j \quad \textit{43.38}$$

$$\text{PP} = \frac{\text{LV}}{N} \quad \textit{43.39}$$

$$N = \frac{\text{LV}}{\text{PP}} \quad \textit{43.40}$$

$$\begin{aligned}\text{LV} &= (\text{PP} + \text{PI}_1)(P/A,\phi,N) \\ &\quad - \text{PI}_N(P/G,\phi,N)\end{aligned} \quad \textit{43.41}$$

$$\begin{aligned}1 &= \left(\frac{1}{N} + \phi\right)(P/A,\phi,N) \\ &\quad - \left(\frac{\phi}{N}\right)(P/G,\phi,N)\end{aligned} \quad \textit{43.42}$$

Figure 43.6 *Loan with Constant Amount Paid Toward Principal*

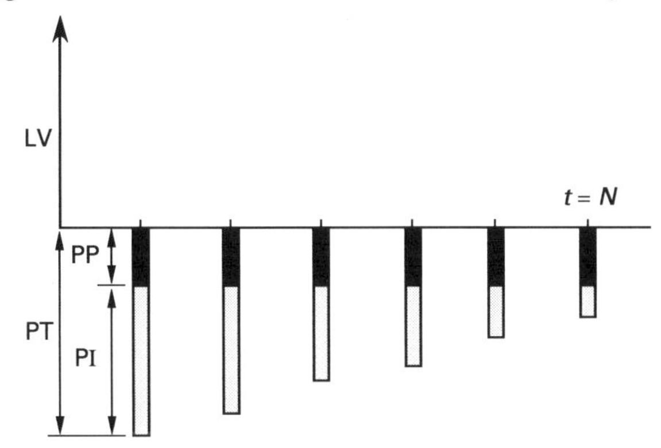

Example 43.18

A \$12,000 6 yr loan is taken from a bank that charges 15% effective annual interest. Payments toward the principal are uniform, and repayments are made at the end of each year. Tabulate the interest, total payments, and the balance remaining after each payment is made.

Solution

The amount of each principal payment is

$$\text{PP} = \frac{\$12{,}000}{6} = \$2000$$

At the end of the first year (before the first payment is made), the principal balance is \$12,000 (i.e., $BAL_0 =$ \$12,000). From Eq. 43.37, the interest payment is

$$\text{PI}_1 = (0.15)(\$12{,}000) = \$1800$$

The total first payment is

$$\begin{aligned}\text{PT}_1 &= \text{PP} + \text{PI} = \$2000 + \$1800 \\ &= \$3800\end{aligned}$$

The following table is similarly constructed.

j	BAL_j (\$)	PP_j (\$)	PI_j (\$)	PT_j (\$)
0	12,000	–	–	–
1	10,000	2000	1800	3800
2	8000	2000	1500	3500
3	6000	2000	1200	3200
4	4000	2000	900	2900
5	2000	2000	600	2600
6	0	2000	300	2300

Direct Reduction Loans

This is the typical "interest paid on unpaid balance" loan. The amount of the periodic payment is constant, but the amounts paid toward the principal and interest both vary.

$$\text{BAL}_{j-1} = \text{PT}\left(\frac{1-(1+\phi)^{j-1-N}}{\phi}\right) \quad \textit{43.43}$$

$$\text{PI}_j = \phi(\text{BAL})_{j-1} \quad \textit{43.44}$$

$$\text{PP}_j = \text{PT} - \text{PI}_j \quad \textit{43.45}$$

$$\text{BAL}_j = \text{BAL}_{j-1} - \text{PP}_j \quad \textit{43.46}$$

$$N = \frac{-\ln\left(1 - \frac{\phi(\text{LV})}{\text{PT}}\right)}{\ln(1+\phi)} \quad \textit{43.47}$$

Equation 43.47 calculates the number of payments necessary to pay off a loan. This equation can be solved with effort for the total periodic payment (PT) or the initial value of the loan (LV). It is easier, however, to use the $(A/P, i\%, n)$ factor to find the payment and loan value.

$$\mathrm{PT} = \mathrm{LV}(A/P, \phi\%, N) \qquad 43.48$$

For monthly payments, N is the number of months in the loan period. If the loan is repaid in yearly installments, then i is the effective annual rate. If the loan is paid off monthly, then i should be replaced by the effective rate per month. The effective rate per month, or for any compounding period, may be calculated with the following equation, in which r is the nominal interest rate and k is the number of compounding periods per year.

$$\phi = \frac{r}{k} \qquad 43.49$$

Figure 43.7 *Direct Reduction Loan*

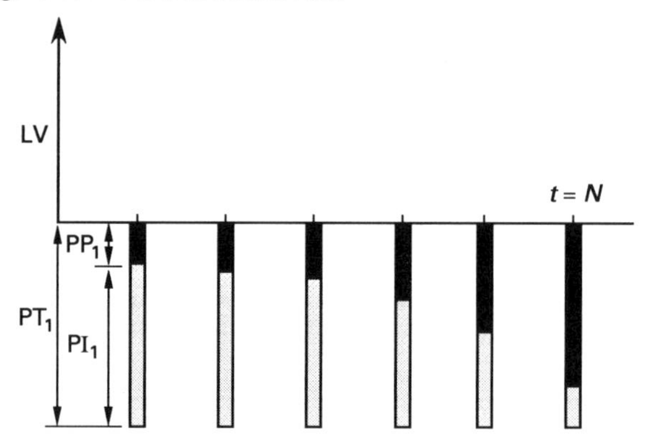

Example 43.19

A $45,000 loan is financed at 9.25% per annum. The monthly payment is $385. What are the amounts paid toward interest and principal in the 14th period? What is the remaining principal balance after the 14th payment has been made?

Solution

The effective rate per month is

$$\phi = \frac{r}{k} = \frac{0.0925}{12} = 0.0077083 \quad (0.007708)$$

$$N = \frac{-\ln\left(1 - \frac{(0.007708)(45{,}000)}{385}\right)}{\ln(1 + 0.007708)} = 301$$

$$\mathrm{BAL}_{13} = (\$385)\left(\frac{1 - (1 + 0.007708)^{14-1-301}}{0.007708}\right) = \$44{,}476.39$$

$$\mathrm{PI}_{14} = (0.007708)(\$44{,}476.39) = \$342.82$$

$$\mathrm{PP}_{14} = \$385 - \$342.82 = \$42.18$$

$$\mathrm{BAL}_{14} = \$44{,}476.39 - \$42.18 = \$44{,}434.21$$

Direct Reduction Loans with Balloon Payments

This type of loan has a constant periodic payment, but the duration of the loan is insufficient to completely pay back the principal (i.e, the loan is not fully amortized). Therefore, all remaining unpaid principal must be paid back in a lump sum when the loan matures. This large payment is known as a *balloon payment.*[28]

Equations 43.44 through 43.48 also can be used with this type of loan. The remaining balance after the last payment is the balloon payment. This balloon payment must be repaid along with the last regular payment calculated.

Figure 43.8 *Direct Reduction Loan with Balloon Payment*

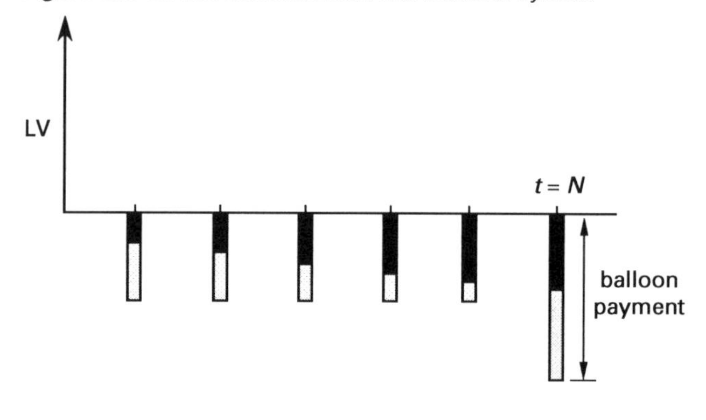

[28]The term *balloon payment* may include the final interest payment as well. Generally, the problem statement will indicate whether the balloon payment is inclusive or exclusive of the regular payment made at the end of the loan period.

44 Ethics for Surveyors

1. CREEDS, CODES, CANONS, STATUTES, AND RULES

It is generally conceded that an individual acting on his or her own cannot be counted on to always act in a proper and moral manner. Creeds, statutes, rules, and codes all attempt to complete the guidance needed for a surveyor to do "the correct thing."

A *creed* is a statement or oath, often religious in nature, taken or assented to by an individual in ceremonies. A *code* is a system of nonstatutory, nonmandatory canons of personal conduct. A *canon* is a fundamental belief that usually encompasses several rules.

The National Society of Professional Surveyors (NSPS) of the American Congress on Surveying and Mapping (ACSM) has the following creed and canons.

> As a Professional Surveyor, I dedicate my professional knowledge and skills to the advancement and betterment of human welfare.
>
> I pledge:
>
> To give the utmost of performance;
>
> To participate in none but honest enterprise;
>
> To live and work according to the laws of humankind and the highest standards of professional conduct;
>
> To place service before profit, honor and standing of the profession before personal advantage, and the public welfare above all other considerations;
>
> In humility and with need for Divine Guidance, I make this pledge.
>
> Canon 1. A Professional Surveyor should refrain from conduct that is detrimental to the public.
>
> Canon 2. A Professional Surveyor should abide by the rules and regulations pertaining to the practice of surveying within the licensing jurisdiction.
>
> Canon 3. A Professional Surveyor should accept assignments only in one's area of professional competence and expertise.
>
> Canon 4. A Professional Surveyor should develop and communicate a professional analysis and opinion without bias or personal interest.
>
> Canon 5. A Professional Surveyor should maintain the confidential nature of the surveyor-client relationship.
>
> Canon 6. A Professional Surveyor should use care to avoid advertising or solicitation that is misleading or otherwise contrary to the public interest.
>
> Canon 7. A Professional Surveyor should maintain professional integrity when dealing with members of other professions.

A *rule* is a guide (principle, standard, or norm) for conduct and action in a certain situation. A *statutory rule* is enacted by the legislative branch of state or federal government and carries the weight of law. Some U.S. engineering and surveying registration boards have statutory *rules of professional conduct*.

2. PURPOSE OF A CODE OF ETHICS

Many different sets of *codes of ethics* (*canons of ethics*, *rules of professional conduct*, etc.) have been produced by various professional societies, registration boards, and other organizations.[1] The purpose of these ethical guidelines is to guide the conduct and decision making of engineers. Most codes are primarily educational. Nevertheless, from time to time they have been used by the societies and regulatory agencies as the basis for disciplinary actions.

Fundamental to ethical codes is the requirement that surveyors render faithful, honest, professional service. In providing such service, surveyors must represent the interests of their employers or clients and, at the same time, protect public health, safety, and welfare.

[1]All of the major engineering technical and professional societies in the United States (ASCE, IEEE, ASME, AIChE, NSPE, etc.) and throughout the world have adopted codes of ethics. Most U.S. societies have endorsed the *Code of Ethics of Engineers* developed by the Accreditation Board for Engineering and Technology (ABET), formerly the Engineers' Council for Professional Development (ECPD). The National Council of Examiners for Engineering and Surveying (NCEES) has developed its *Model Rules of Professional Conduct* as a guide for state registration boards in developing guidelines for the professional engineers in those states.

There is an important distinction between what is legal and what is ethical. Many legal actions can be violations of codes of ethical or professional behavior.[2] For example, a surveyor's contract with a client may give the surveyor the right to assign the surveyor's responsibilities, but doing so without informing the client would be unethical.

Ethical guidelines can be categorized on the basis of who is affected by the surveyor's actions—the client, other surveyors, or the public at large.[3]

3. ETHICAL PRIORITIES

There are frequently conflicting demands on surveyors. While it is impossible to use a single decision-making process to solve every ethical dilemma, it is clear that ethical considerations will force surveyors to subjugate their own self-interests. Specifically, the ethics of surveyors dealing with others need to be considered in the following order from highest to lowest priority.

- society and the public
- the law
- the surveying profession
- the surveyor's client
- the surveyor's firm
- other involved surveyors
- the surveyor personally

4. DEALING WITH CLIENTS AND EMPLOYERS

The most common ethical guidelines affecting surveyors' interactions with their employer (the *client*) can be summarized as follows.

- Surveyors should not accept assignments for which they do not have the skill, knowledge, or time.
- Surveyors must recognize their own limitations. They should use associates and other experts when the survey requirements exceed their abilities.
- The client's interests must be protected. The extent of this protection exceeds normal business relationships and transcends the legal requirements of the surveyor-client contract.
- Surveyors must not be bound by what the client wants in instances where such desires would be unsuccessful, dishonest, unethical, unhealthy, or unsafe.
- Confidential client information remains the property of the client and must be kept confidential.
- Surveyors must avoid conflicts of interest and should inform the client of any business connections or interests that might influence their judgment. Surveyors should also avoid the *appearance* of a conflict of interest when such an appearance would be detrimental to the profession, their client, or themselves.
- The surveyors' sole source of income for a particular project should be the fee paid by their client. Surveyors should not accept compensation in any form from more than one party for the same services.
- If the client rejects the surveyor's recommendations, the surveyor should fully explain the consequences to the client.
- Surveyors must freely and openly admit to the client any errors made.

All courts of law have required a surveyor to perform in a manner consistent with normal professional standards. This is not the same as saying a surveyor's work must be error free. If a surveyor completes a survey, has the calculations checked by another competent surveyor, and an error is subsequently shown to have been made, the surveyor may be held responsible, but will probably not be considered negligent.

5. DEALING WITH OTHER SURVEYORS

Surveyors should try to protect the surveying profession as a whole, to strengthen it, and to enhance its public stature. The following ethical guidelines apply.

- A surveyor should not attempt to maliciously injure the professional reputation, business practice, or employment position of another surveyor. However, if there is proof that another surveyor has acted unethically or illegally, the surveyor should advise the proper authority.
- A surveyor should not review another surveyor's work while the other surveyor is still employed unless the other surveyor is made aware of the review.
- A surveyor should not try to replace another surveyor once the other surveyor has received employment.
- A surveyor should not use the advantages of a salaried position to compete unfairly (i.e., moonlight) with other surveyors who have to charge more for the same consulting services.
- Subject to legal and proprietary restraints, a surveyor should freely report, publish, and distribute information that would be useful to other surveyors.

[2]Whether the guidelines emphasize ethical behavior or professional conduct is a matter of wording. The intention is the same: to provide guidelines that transcend the requirements of the law.
[3]Some authorities also include ethical guidelines for dealing with the employees of a surveyor. However, these guidelines are no different for a surveying employer than they are for a supermarket, automobile assembly line, or airline employer. Ethics is not a unique issue when it comes to employees.

6. DEALING WITH (AND AFFECTING) THE PUBLIC

In regard to the social consequences of surveying, the relationship between a surveyor and the public is essentially straightforward. Responsibilities to the public demand that the surveyor place service to humankind above personal gain. Furthermore, proper ethical behavior requires that a surveyor avoid association with projects that are contrary to public health and welfare or that are of questionable legal character.

- Surveyors must consider the safety, health, and welfare of the public in all work performed.
- Surveyors must uphold the honor and dignity of their profession by refraining from self-laudatory advertising, by explaining (when required) their work to the public, and by expressing opinions only in areas of knowledge.
- When surveyors issue a public statement, they must clearly indicate if the statement is being made on anyone's behalf (i.e., if anyone is benefitting from their position).
- Surveyors must keep their skills at a state-of-the-art level.
- Surveyors should develop public knowledge and appreciation of the surveying profession and its achievements.
- Surveyors must notify the proper authorities when decisions adversely affecting public safety and welfare are made.[4]

7. CONFLICTS OF INTEREST

Conflict of interest is one of the more commonly encountered ethical problems and as such deserves consideration. A *conflict of interest* is generally defined as a situation in which one's professional judgment is likely to be compromised due to a private or personal interest. It is important to avoid apparent and potential conflicts as well as actual conflicts. Some examples of conflicts of interest that might be encountered by a surveyor include the following.

1. a surveyor who uses a position, such as working for a public agency, to secure work for a private business affiliated with the surveyor
2. a surveyor who receives some benefit in exchange for influencing a process, such as the award of a contract
3. a surveyor who uses an employer's equipment or software in personal consulting work
4. a surveyor who uses confidential information, such as knowledge about a client's plans, for personal gain
5. a surveyor who moonlights in direct competition with his or her employer
6. a surveyor who resigns from a company and then deliberately secures the clients and staff of the former employer

8. COMPETITIVE BIDDING

The ethical guidelines for dealing with other surveyors presented here and in more detailed codes of ethics no longer include a prohibition on *competitive bidding.* Until 1971, most codes of ethics considered competitive bidding detrimental to public welfare, since cost cutting normally results in a lower quality design.

However, in a 1971 case against the National Society of Professional Engineers that went all the way to the U.S. Supreme Court, the prohibition against competitive bidding was determined to be a violation of the Sherman Antitrust Act (i.e., it was an unreasonable restraint of trade).

The opinion of the Supreme Court does not *require* competitive bidding—it merely forbids a prohibition against competitive bidding in NSPE's code of ethics. The following points must be considered.

- Surveyors may individually continue to refuse to bid competitively on surveying services.
- Clients are not required to seek competitive bids for survey services.
- Federal, state, and local statutes governing the procedures for procuring surveying services, even those statutes that prohibit competitive bidding, are not affected.
- Any prohibitions against competitive bidding in individual state surveying registration laws remain unaffected.
- Surveyors and their societies may actively and aggressively lobby for legislation that would prohibit competitive bidding for surveying services by public agencies.

[4]This practice has come to be known as *whistle-blowing.*

Topic XI: Support Material

Appendices Table of Contents

APPENDIX A

Tangents and Externals for Horizontal Curves

Δ = interior angle $\quad T^* = 5729.578 \tan \dfrac{\Delta}{2} \quad E^* = 5729.578 \left(\sec \dfrac{\Delta}{2} - 1\right)$

For arc definition, calculate T and E for any degree of curve D by dividing both table values by D.

$$T = \frac{T^*}{D} \qquad E = \frac{E^*}{D}$$

	$\Delta = 0°$		$\Delta = 1°$		$\Delta = 2°$		$\Delta = 3°$		$\Delta = 4°$		$\Delta = 5°$		$\Delta = 6°$		
minutes	T^*	E^*	T^*	E^*	T^*	E^*	T^*	E^*	T^*	E^*	T^*	E^*	T^*	E^*	minutes
0	0.00	0.00	50.00	0.22	100.01	0.87	150.03	1.96	200.08	3.49	250.16	5.46	300.27	7.86	0
1	0.83	0.00	50.83	0.23	100.84	0.89	150.87	1.99	200.92	3.52	250.99	5.50	301.11	7.91	1
2	1.67	0.00	51.67	0.23	101.68	0.90	151.70	2.01	201.75	3.55	251.83	5.53	301.95	7.95	2
3	2.50	0.00	52.50	0.24	102.51	0.92	152.54	2.03	202.58	3.58	252.66	5.57	302.78	7.99	3
4	3.33	0.00	53.33	0.25	103.34	0.93	153.37	2.05	203.42	3.61	253.50	5.60	303.62	8.04	4
5	4.17	0.00	54.17	0.26	104.18	0.95	154.20	2.08	204.25	3.64	254.33	5.64	304.45	8.08	5
6	5.00	0.00	55.00	0.26	105.01	0.96	155.04	2.10	205.09	3.67	255.17	5.68	305.29	8.13	6
7	5.83	0.00	55.84	0.27	105.85	0.98	155.87	2.12	205.92	3.70	256.00	5.72	306.12	8.17	7
8	6.67	0.00	56.67	0.28	106.68	0.99	156.71	2.14	206.76	3.73	256.84	5.75	306.96	8.22	8
9	7.50	0.00	57.50	0.29	107.51	1.01	157.54	2.17	207.59	3.76	257.67	5.79	307.80	8.26	9
10	8.33	0.01	58.34	0.30	108.35	1.02	158.37	2.19	208.43	3.79	258.51	5.83	308.63	8.31	10
11	9.17	0.01	59.17	0.31	109.18	1.04	159.21	2.21	209.26	3.82	259.34	5.87	309.47	8.35	11
12	10.00	0.01	60.00	0.31	110.01	1.06	160.04	2.23	210.09	3.85	260.18	5.90	310.30	8.40	12
13	10.83	0.01	60.84	0.32	110.85	1.07	160.88	2.26	210.93	3.88	261.01	5.94	311.14	8.44	13
14	11.67	0.01	61.67	0.33	111.68	1.09	161.71	2.28	211.76	3.91	261.85	5.98	311.97	8.49	14
15	12.50	0.01	62.50	0.34	112.51	1.10	162.54	2.30	212.60	3.94	262.68	6.02	312.81	8.53	15
16	13.33	0.02	63.34	0.35	113.35	1.12	163.38	2.33	213.43	3.97	263.52	6.06	313.65	8.58	16
17	14.17	0.02	64.17	0.36	114.18	1.14	164.21	2.35	214.27	4.00	264.35	6.09	314.48	8.62	17
18	15.00	0.02	65.00	0.37	115.02	1.15	165.05	2.38	215.10	4.04	265.19	6.13	315.32	8.67	18
19	15.83	0.02	65.84	0.38	115.85	1.17	165.88	2.40	215.94	4.07	266.02	6.17	316.15	8.72	19
20	16.67	0.02	66.67	0.39	116.68	1.19	166.71	2.42	216.77	4.10	266.86	6.21	316.99	8.76	20
21	17.50	0.03	67.50	0.40	117.52	1.20	167.55	2.45	217.60	4.13	267.69	6.25	317.83	8.81	21
22	18.33	0.03	68.34	0.41	118.35	1.22	168.38	2.47	218.44	4.16	268.53	6.29	318.66	8.85	22
23	19.17	0.03	69.17	0.42	119.18	1.24	169.22	2.50	219.27	4.19	269.36	6.33	319.50	8.90	23
24	20.00	0.03	70.00	0.43	120.02	1.26	170.05	2.52	220.11	4.23	270.20	6.37	320.33	8.95	24
25	20.83	0.04	70.84	0.44	120.85	1.27	170.88	2.55	220.94	4.26	271.04	6.41	321.17	8.99	25
26	21.67	0.04	71.67	0.45	121.68	1.29	171.72	2.57	221.78	4.29	271.87	6.45	322.01	9.04	26
27	22.50	0.04	72.50	0.46	122.52	1.31	172.55	2.60	222.61	4.32	272.71	6.49	322.84	9.09	27
28	23.33	0.05	73.34	0.47	123.35	1.33	173.39	2.62	223.45	4.36	273.54	6.53	323.68	9.14	28
29	24.17	0.05	74.17	0.48	124.19	1.34	174.22	2.65	224.28	4.39	274.38	6.57	324.51	9.18	29
30	25.00	0.05	75.00	0.49	125.02	1.36	175.05	2.67	225.12	4.42	275.21	6.61	325.35	9.23	30
31	25.83	0.06	75.84	0.50	125.85	1.38	175.89	2.70	225.95	4.45	276.05	6.65	326.19	9.28	31
32	26.67	0.06	76.67	0.51	126.69	1.40	176.72	2.72	226.78	4.49	276.88	6.69	327.02	9.32	32
33	27.50	0.07	77.50	0.52	127.52	1.42	177.56	2.75	227.62	4.52	277.72	6.73	327.86	9.37	33
34	28.33	0.07	78.34	0.54	128.35	1.44	178.39	2.78	228.45	4.55	278.55	6.77	328.69	9.42	34
35	29.17	0.07	79.17	0.55	129.19	1.46	179.23	2.80	229.29	4.59	279.39	6.81	329.53	9.47	35
36	30.00	0.08	80.01	0.56	130.02	1.48	180.06	2.83	230.12	4.62	280.22	6.85	330.37	9.52	36
37	30.83	0.08	80.84	0.57	130.86	1.49	180.89	2.86	230.96	4.65	281.06	6.89	331.20	9.56	37
38	31.67	0.09	81.67	0.58	131.69	1.51	181.73	2.88	231.79	4.69	281.89	6.93	332.04	9.61	38
39	32.50	0.09	82.51	0.59	132.52	1.53	182.56	2.91	232.63	4.72	282.73	6.97	332.87	9.66	39
40	33.33	0.10	83.34	0.61	133.36	1.55	183.40	2.93	233.46	4.75	283.56	7.01	333.71	9.71	40
41	34.17	0.10	84.17	0.62	134.19	1.57	184.23	2.96	234.30	4.79	284.40	7.05	334.55	9.76	41
42	35.00	0.11	85.01	0.63	135.02	1.59	185.06	2.99	235.13	4.82	285.24	7.09	335.38	9.81	42
43	35.83	0.11	85.84	0.64	135.86	1.61	185.90	3.01	235.97	4.86	286.07	7.14	336.22	9.86	43
44	36.67	0.12	86.67	0.66	136.69	1.63	186.73	3.04	236.80	4.89	286.91	7.18	337.05	9.91	44
45	37.50	0.12	87.51	0.67	137.53	1.65	187.57	3.07	237.64	4.93	287.74	7.22	337.89	9.95	45
46	38.33	0.13	88.34	0.68	138.36	1.67	188.40	3.10	238.47	4.96	288.58	7.26	338.73	10.00	46
47	39.17	0.13	89.17	0.69	139.19	1.69	189.24	3.12	239.31	4.99	289.41	7.30	339.56	10.05	47
48	40.00	0.14	90.01	0.71	140.03	1.71	190.07	3.15	240.14	5.03	290.25	7.35	340.40	10.10	48
49	40.83	0.15	90.84	0.72	140.86	1.73	190.90	3.18	240.98	5.07	291.08	7.39	341.24	10.15	49
50	41.67	0.15	91.67	0.73	141.70	1.75	191.74	3.21	241.81	5.10	291.92	7.43	342.07	10.20	50
51	42.50	0.16	92.51	0.75	142.53	1.77	192.57	3.23	242.64	5.13	292.75	7.47	342.91	10.25	51
52	43.33	0.16	93.34	0.76	143.36	1.79	193.41	3.26	243.48	5.17	293.59	7.52	343.74	10.30	52
53	44.17	0.17	94.18	0.77	144.20	1.81	194.24	3.29	244.31	5.21	294.43	7.56	344.58	10.35	53
54	45.00	0.18	95.01	0.79	145.03	1.84	195.08	3.32	245.15	5.24	295.26	7.60	345.42	10.40	54
55	45.83	0.18	95.84	0.80	145.86	1.86	195.91	3.35	245.98	5.28	296.10	7.65	346.25	10.45	55
56	46.67	0.19	96.68	0.82	146.70	1.88	196.74	3.38	246.82	5.31	296.93	7.69	347.09	10.50	56
57	47.50	0.20	97.51	0.83	147.53	1.90	197.58	3.41	247.65	5.35	297.77	7.73	347.93	10.55	57
58	48.33	0.20	98.34	0.84	148.37	1.92	198.41	3.43	248.49	5.39	298.60	7.78	348.76	10.61	58
59	49.17	0.21	99.18	0.86	149.20	1.94	199.25	3.46	249.32	5.42	299.44	7.82	349.60	10.66	59

(continued)

APPENDIX A *(continued)*
Tangents and Externals for Horizontal Curves

Δ = interior angle $\quad T^* = 5729.578 \tan \frac{\Delta}{2} \quad E^* = 5729.578 \left(\sec \frac{\Delta}{2} - 1\right)$

For arc definition, calculate T and E for any degree of curve D by dividing both table values by D.

$$T = \frac{T^*}{D} \qquad E = \frac{E^*}{D}$$

	$\Delta = 7°$		$\Delta = 8°$		$\Delta = 9°$		$\Delta = 10°$		$\Delta = 11°$		$\Delta = 12°$		$\Delta = 13°$		
minutes	T^*	E^*	T^*	E^*	T^*	E^*	T^*	E^*	T^*	E^*	T^*	E^*	T^*	E^*	minutes
0	350.44	10.71	400.65	13.99	450.93	17.72	501.27	21.89	551.70	26.50	602.20	31.56	652.80	37.07	0
1	351.27	10.76	401.49	14.05	451.77	17.78	502.11	21.96	552.54	26.58	603.05	31.65	653.65	37.16	1
2	352.11	10.81	402.33	14.11	452.60	17.85	502.95	22.03	553.38	26.66	603.89	31.74	654.49	37.26	2
3	352.95	10.86	403.16	14.17	453.44	17.91	503.79	22.11	554.22	26.74	604.73	31.82	655.34	37.36	3
4	353.78	10.91	404.00	14.23	454.28	17.98	504.63	22.18	555.06	26.82	605.57	31.91	656.18	37.45	4
5	354.62	10.96	404.84	14.28	455.12	18.05	505.47	22.25	555.90	26.90	606.42	32.00	657.02	37.55	5
6	355.45	11.02	405.68	14.34	455.96	18.11	506.31	22.33	556.74	26.99	607.26	32.09	657.87	37.64	6
7	356.29	11.07	406.51	14.40	456.80	18.18	507.15	22.40	557.58	27.07	608.10	32.18	658.71	37.74	7
8	357.13	11.12	407.35	14.46	457.64	18.25	507.99	22.48	558.42	27.15	608.94	32.27	659.56	37.84	8
9	357.96	11.17	408.19	14.52	458.47	18.31	508.83	22.55	559.27	27.23	609.79	32.36	660.40	37.93	9
10	358.80	11.22	409.03	14.58	459.31	18.38	509.67	22.62	560.11	27.31	610.63	32.45	661.25	38.03	10
11	359.64	11.28	409.86	14.64	460.15	18.45	510.51	22.70	560.95	27.39	611.47	32.54	662.09	38.13	11
12	360.47	11.33	410.70	14.70	460.99	18.51	511.35	22.77	561.79	27.48	612.32	32.63	662.93	38.22	12
13	361.31	11.38	411.54	14.76	461.83	18.58	512.19	22.85	562.63	27.56	613.16	32.72	663.78	38.32	13
14	362.15	11.43	412.38	14.82	462.67	18.65	513.03	22.92	563.47	27.64	614.00	32.80	664.62	38.42	14
15	362.98	11.49	413.21	14.88	463.51	18.72	513.87	23.00	564.31	27.72	614.84	32.89	665.47	38.52	15
16	363.82	11.54	414.05	14.94	464.35	18.79	514.71	23.07	565.16	27.81	615.69	32.98	666.31	38.61	16
17	364.66	11.59	414.89	15.00	465.18	18.85	515.55	23.15	566.00	27.89	616.53	33.08	667.16	38.71	17
18	365.49	11.65	415.73	15.06	466.02	18.92	516.39	23.22	566.84	27.97	617.37	33.17	668.00	38.81	18
19	366.33	11.70	416.56	15.12	466.86	18.99	517.23	23.30	567.68	28.05	618.22	33.26	668.85	38.91	19
20	367.17	11.75	417.40	15.18	467.70	19.06	518.07	23.37	568.52	28.14	619.06	33.35	669.69	39.00	20
21	368.00	11.81	418.24	15.24	468.54	19.13	518.91	23.45	569.36	28.22	619.90	33.44	670.54	39.10	21
22	368.84	11.86	419.08	15.31	469.38	19.19	519.75	23.53	570.20	28.30	620.75	33.53	671.38	39.20	22
23	369.68	11.91	419.92	15.37	470.22	19.26	520.59	23.60	571.05	28.39	621.59	33.62	672.23	39.30	23
24	370.52	11.97	420.75	15.43	471.06	19.33	521.43	23.68	571.89	28.47	622.43	33.71	673.07	39.40	24
25	371.35	12.02	421.59	15.49	471.90	19.40	522.27	23.75	572.73	28.55	623.27	33.80	673.92	39.50	25
26	372.19	12.08	422.43	15.55	472.73	19.47	523.11	23.83	573.57	28.64	624.12	33.89	674.76	39.60	26
27	373.03	12.13	423.27	15.61	473.57	19.54	523.95	23.91	574.41	28.72	624.96	33.98	675.61	39.69	27
28	373.86	12.18	424.11	15.67	474.41	19.61	524.79	23.98	575.25	28.81	625.80	34.08	676.45	39.79	28
29	374.70	12.24	424.94	15.74	475.25	19.68	525.63	24.06	576.10	28.89	626.65	34.17	677.30	39.89	29
30	375.54	12.29	425.78	15.80	476.09	19.75	526.47	24.14	576.94	28.97	627.49	34.26	678.14	39.99	30
31	376.37	12.35	426.62	15.86	476.93	19.82	527.31	24.21	577.78	29.06	628.33	34.35	678.99	40.09	31
32	377.21	12.40	427.46	15.92	477.77	19.89	528.16	24.29	578.62	29.14	629.18	34.44	679.83	40.19	32
33	378.05	12.46	428.30	15.99	478.61	19.96	529.00	24.37	579.46	29.23	630.02	34.52	680.68	40.29	33
34	378.88	12.51	429.13	16.05	479.45	20.02	529.84	24.45	580.31	29.31	630.86	34.63	681.52	40.39	34
35	379.72	12.57	429.97	16.11	480.29	20.10	530.68	24.52	581.15	29.40	631.71	34.72	682.37	40.49	35
36	380.56	12.62	430.81	16.17	481.13	20.17	531.52	24.60	581.99	29.48	632.55	34.81	683.21	40.59	36
37	381.40	12.68	431.65	16.24	481.97	20.24	532.36	24.68	582.83	29.57	633.39	34.90	684.06	40.69	37
38	382.23	12.74	432.49	16.30	482.80	20.31	533.20	24.76	583.67	29.65	634.24	35.00	684.90	40.79	38
39	383.07	12.79	433.32	16.36	483.64	20.38	534.04	24.83	584.52	29.74	635.08	35.09	685.75	40.89	39
40	383.91	12.85	434.16	16.43	484.48	20.45	534.88	24.91	585.36	29.82	635.93	35.18	686.59	40.99	40
41	384.74	12.90	435.00	16.49	485.32	20.52	535.72	24.99	586.20	29.91	636.77	35.28	687.44	41.09	41
42	385.58	12.96	435.84	16.55	486.16	20.59	536.56	25.07	587.04	29.99	637.61	35.37	688.28	41.19	42
43	386.42	13.02	436.68	16.62	487.00	20.66	537.40	25.15	587.88	30.08	638.46	35.46	689.13	41.29	43
44	387.25	13.07	437.51	16.68	487.84	20.73	538.24	25.23	588.73	30.17	639.30	35.56	689.97	41.39	44
45	388.09	13.13	438.35	16.74	488.68	20.80	539.08	25.30	589.57	30.25	640.14	35.65	690.82	41.50	45
46	388.93	13.18	439.19	16.81	489.52	20.87	539.92	25.38	590.41	30.34	640.99	35.74	691.66	41.60	46
47	389.77	13.24	440.03	16.87	490.36	20.94	540.76	25.46	591.25	30.43	641.83	35.84	692.51	41.70	47
48	390.60	13.30	440.87	16.94	491.20	21.02	541.60	25.54	592.09	30.51	642.68	35.93	693.36	41.80	48
49	391.44	13.36	441.71	17.00	492.04	21.09	542.45	25.62	592.94	30.60	643.52	36.03	694.20	41.90	49
50	392.28	13.41	442.54	17.06	492.88	21.16	543.29	25.70	593.78	30.69	644.36	36.12	695.05	42.00	50
51	393.12	13.47	443.38	17.13	493.72	21.23	544.13	25.78	594.62	30.77	645.21	36.21	695.89	42.11	51
52	393.95	13.53	444.22	17.19	494.56	21.30	544.97	25.86	595.46	30.86	646.05	36.31	696.74	42.21	52
53	394.79	13.59	445.06	17.26	495.40	21.38	545.81	25.94	596.31	30.95	646.89	36.40	697.58	42.31	53
54	395.63	13.64	445.90	17.32	496.24	21.45	546.65	26.02	597.15	31.03	647.74	36.50	698.43	42.41	54
55	396.46	13.70	446.74	17.39	497.07	21.52	547.49	26.10	597.99	31.12	648.58	36.59	699.27	42.51	55
56	397.30	13.76	447.57	17.46	497.91	21.59	548.33	26.18	598.83	31.21	649.43	36.69	700.12	42.62	56
57	398.14	13.82	448.41	17.52	498.75	21.67	549.17	26.26	599.68	31.30	650.27	36.78	700.97	42.72	57
58	398.98	13.87	449.25	17.59	499.59	21.74	550.01	26.34	600.52	31.38	651.11	36.88	701.81	42.82	58
59	399.81	13.93	450.09	17.65	500.43	21.81	550.85	26.42	601.36	31.47	651.96	36.97	702.66	42.92	59

(continued)

APPENDIX A *(continued)*

Tangents and Externals for Horizontal Curves

Δ = interior angle $\quad T^* = 5729.578 \tan \frac{\Delta}{2} \quad E^* = 5729.578 \left(\sec \frac{\Delta}{2} - 1 \right)$

For arc definition, calculate T and E for any degree of curve D by dividing both table values by D.

$$T = \frac{T^*}{D} \qquad E = \frac{E^*}{D}$$

	$\Delta = 14°$		$\Delta = 15°$		$\Delta = 16°$		$\Delta = 17°$		$\Delta = 18°$		$\Delta = 19°$		$\Delta = 20°$		
minutes	T^*	E^*	T^*	E^*	T^*	E^*	T^*	E^*	T^*	E^*	T^*	E^*	T^*	E^*	minutes
0	703.50	43.03	754.31	49.44	805.24	56.31	856.29	63.63	907.48	71.42	958.80	79.67	1010.28	88.39	0
1	704.35	43.13	755.16	49.55	806.09	56.43	857.14	63.76	908.33	71.55	959.66	79.81	1011.14	88.54	1
2	705.20	43.23	756.01	49.66	806.94	56.55	858.00	63.88	909.18	71.69	960.52	79.95	1012.00	88.69	2
3	706.04	43.34	756.86	49.77	807.79	56.66	858.85	64.01	910.04	71.82	961.37	80.09	1012.86	88.84	3
4	706.89	43.44	757.70	49.88	808.64	56.78	859.70	64.14	910.89	71.96	962.23	80.24	1013.72	88.99	4
5	707.73	43.55	758.55	49.99	809.49	56.90	860.55	64.26	911.75	72.09	963.09	80.38	1014.58	89.14	5
6	708.58	43.65	759.40	50.11	810.34	57.02	861.40	64.39	912.60	72.22	963.94	80.52	1015.44	89.29	6
7	709.43	43.75	760.25	50.22	811.19	57.14	862.26	64.52	913.46	72.36	964.80	80.66	1016.29	89.44	7
8	710.27	43.86	761.10	50.33	812.04	57.26	863.11	64.64	914.31	72.49	965.66	80.81	1017.15	89.59	8
9	711.12	43.96	761.94	50.44	812.89	57.38	863.96	64.77	915.17	72.63	966.51	80.95	1018.01	89.74	9
10	711.96	44.07	762.79	50.55	813.74	57.50	864.81	64.90	916.02	72.76	967.37	81.09	1018.87	89.89	10
11	712.81	44.17	763.64	50.67	814.59	57.62	865.66	65.03	916.88	72.90	968.23	81.23	1019.73	90.04	11
12	713.66	44.27	764.49	50.78	815.44	57.74	866.52	65.15	917.73	73.03	969.09	81.38	1020.59	90.19	12
13	714.50	44.38	765.34	50.89	816.29	57.86	867.37	65.28	918.58	73.17	969.94	81.52	1021.45	90.34	13
14	715.35	44.48	766.19	51.00	817.14	57.98	868.22	65.41	919.44	73.30	970.80	81.66	1022.31	90.49	14
15	716.20	44.59	767.03	51.11	817.99	58.10	869.07	65.54	920.29	73.44	971.66	81.81	1023.17	90.64	15
16	717.04	44.69	767.88	51.23	818.84	58.22	869.93	65.66	921.15	73.57	972.51	81.95	1024.03	90.79	16
17	717.89	44.80	768.73	51.34	819.69	58.34	870.78	65.79	922.00	73.71	973.37	82.09	1024.89	90.94	17
18	718.73	44.90	769.58	51.45	820.54	58.46	871.63	65.92	922.86	73.85	974.23	82.24	1025.75	91.09	18
19	719.58	45.01	770.43	51.57	821.39	58.58	872.48	66.05	923.71	73.98	975.09	82.38	1026.61	91.25	19
20	720.43	45.11	771.28	51.68	822.24	58.70	873.34	66.18	924.57	74.12	975.94	82.52	1027.47	91.40	20
21	721.27	45.22	772.12	51.79	823.09	58.82	874.19	66.31	925.42	74.25	976.80	82.67	1028.33	91.55	21
22	722.12	45.33	772.97	51.91	823.94	58.94	875.04	66.43	926.28	74.39	977.66	82.81	1029.19	91.70	22
23	722.97	45.43	773.82	52.02	824.79	59.06	875.90	66.56	927.13	74.53	978.52	82.96	1030.05	91.85	23
24	723.81	45.54	774.67	52.13	825.64	59.18	876.75	66.69	927.99	74.66	979.37	83.10	1030.91	92.01	24
25	724.66	45.64	775.52	52.25	826.50	59.30	877.60	66.82	928.84	74.80	980.23	83.25	1031.77	92.16	25
26	725.51	45.75	776.37	52.36	827.35	59.43	878.45	66.95	929.70	74.94	981.09	83.39	1032.63	92.31	26
27	726.35	45.86	777.22	52.47	828.20	59.55	879.31	67.08	930.56	75.08	981.95	83.53	1033.49	92.46	27
28	727.20	45.96	778.06	52.59	829.05	59.67	880.16	67.21	931.41	75.21	982.81	83.68	1034.36	92.62	28
29	728.05	46.07	778.91	52.70	829.90	59.79	881.01	67.34	932.27	75.35	983.66	83.83	1035.22	92.77	29
30	728.89	46.18	779.76	52.82	830.75	59.91	881.87	67.47	933.12	75.49	984.52	83.97	1036.08	92.92	30
31	729.74	46.28	780.61	52.93	831.60	60.03	882.72	67.60	933.98	75.62	985.38	84.12	1036.94	93.08	31
32	730.59	46.39	781.46	53.05	832.45	60.16	883.57	67.73	934.83	75.76	986.24	84.26	1037.80	93.23	32
33	731.44	46.50	782.31	53.16	833.30	60.28	884.43	67.86	935.69	75.90	987.10	84.41	1038.66	93.38	33
34	732.28	46.61	783.16	53.28	834.15	60.40	885.28	67.99	936.54	76.04	987.95	84.55	1039.52	93.54	34
35	733.13	46.71	784.01	53.39	835.00	60.53	886.13	68.12	937.40	76.18	988.81	84.70	1040.38	93.69	35
36	733.98	46.82	784.85	53.51	835.86	60.65	886.99	68.25	938.25	76.31	989.67	84.84	1041.24	93.84	36
37	734.82	46.93	785.70	53.62	836.71	60.77	887.84	68.38	939.11	76.45	990.53	84.99	1042.10	94.00	37
38	735.67	47.04	786.55	53.74	837.56	60.89	888.69	68.51	939.97	76.59	991.39	85.14	1042.96	94.15	38
39	736.52	47.14	787.40	53.85	838.41	61.02	889.55	68.64	940.82	76.73	992.24	85.28	1043.82	94.31	39
40	737.36	47.25	788.25	53.97	839.26	61.14	890.40	68.77	941.68	76.87	993.10	85.43	1044.68	94.46	40
41	738.21	47.36	789.10	54.08	840.11	61.26	891.25	68.90	942.53	77.01	993.96	85.58	1045.55	94.62	41
42	739.06	47.47	789.95	54.20	840.96	61.39	892.11	69.03	943.39	77.15	994.82	85.72	1046.41	94.77	42
43	739.91	47.58	790.80	54.32	841.81	61.51	892.96	69.17	944.25	77.28	995.68	85.87	1047.27	94.93	43
44	740.75	47.69	791.65	54.43	842.66	61.63	893.81	69.30	945.10	77.42	996.54	86.02	1048.13	95.08	44
45	741.60	47.79	792.50	54.55	843.52	61.76	894.67	69.43	945.96	77.56	997.40	86.16	1048.99	95.23	45
46	742.45	47.90	793.35	54.66	844.37	61.88	895.52	69.56	946.81	77.70	998.25	86.31	1049.85	95.39	46
47	743.29	48.01	794.20	54.78	845.22	62.01	896.37	69.69	947.67	77.84	999.11	86.46	1050.71	95.54	47
48	744.14	48.12	795.04	54.90	846.07	62.13	897.23	69.83	948.53	77.98	999.97	86.61	1051.57	95.70	48
49	744.99	48.23	795.89	55.01	846.92	62.26	898.08	69.96	949.38	78.12	1000.83	86.75	1052.44	95.86	49
50	745.84	48.34	796.74	55.13	847.77	62.38	898.94	70.09	950.24	78.26	1001.69	86.90	1053.30	96.01	50
51	746.68	48.45	797.59	55.25	848.63	62.51	899.79	70.22	951.09	78.40	1002.55	87.05	1054.16	96.17	51
52	747.53	48.56	798.44	55.37	849.48	62.63	900.64	70.35	951.95	78.54	1003.41	87.20	1055.02	96.32	52
53	748.38	48.67	799.29	55.48	850.33	62.76	901.50	70.49	952.81	78.68	1004.27	87.35	1055.88	96.48	53
54	749.23	48.78	800.14	55.60	851.18	62.88	902.35	70.62	953.66	78.82	1005.12	87.49	1056.74	96.64	54
55	750.07	48.89	800.99	55.72	852.03	63.01	903.21	70.75	954.52	78.97	1005.98	87.64	1057.61	96.79	55
56	750.92	49.00	801.84	55.84	852.88	63.13	904.06	70.89	955.38	79.11	1006.84	87.79	1058.47	96.95	56
57	751.77	49.11	802.69	55.95	853.74	63.26	904.91	71.02	956.23	79.25	1007.70	87.94	1059.33	97.11	57
58	752.62	49.22	803.54	56.07	854.59	63.38	905.77	71.15	957.09	79.39	1008.56	88.09	1060.19	97.26	58
59	753.47	49.33	804.39	56.19	855.44	63.51	906.62	71.29	957.95	79.53	1009.42	88.24	1061.05	97.42	59

(continued)

APPENDIX A *(continued)*

Tangents and Externals for Horizontal Curves

Δ = interior angle $\quad T^* = 5729.578 \tan \frac{\Delta}{2} \quad E^* = 5729.578 \left(\sec \frac{\Delta}{2} - 1 \right)$

For arc definition, calculate T and E for any degree of curve D by dividing both table values by D.

$$T = \frac{T^*}{D} \qquad E = \frac{E^*}{D}$$

	$\Delta = 21°$		$\Delta = 22°$		$\Delta = 23°$		$\Delta = 24°$		$\Delta = 25°$		$\Delta = 26°$		$\Delta = 27°$		
minutes	T^*	E^*	T^*	E^*	T^*	E^*	T^*	E^*	T^*	E^*	T^*	E^*	T^*	E^*	minutes
0	1061.91	97.58	1113.72	107.24	1165.70	117.38	1217.86	128.00	1270.22	139.11	1322.78	150.71	1375.55	162.81	0
1	1062.78	97.73	1114.58	107.40	1166.56	117.55	1218.73	128.18	1271.09	139.30	1323.66	150.91	1376.43	163.01	1
2	1063.64	97.89	1115.45	107.57	1167.43	117.73	1219.60	128.36	1271.97	139.49	1324.53	151.11	1377.31	163.22	2
3	1064.50	98.05	1116.31	107.73	1168.30	117.90	1220.47	128.55	1272.84	139.68	1325.41	151.30	1378.19	163.42	3
4	1065.36	98.21	1117.18	107.90	1169.17	118.07	1221.34	128.73	1273.71	139.87	1326.29	151.50	1379.08	163.63	4
5	1066.22	98.36	1118.04	108.07	1170.04	118.25	1222.22	128.91	1274.59	140.06	1327.17	151.70	1379.96	163.84	5
6	1067.09	98.52	1118.91	108.23	1170.90	118.42	1223.09	129.09	1275.46	140.25	1328.04	151.90	1380.84	164.04	6
7	1067.95	98.68	1119.77	108.40	1171.77	118.59	1223.96	129.27	1276.34	140.44	1328.92	152.10	1381.72	164.25	7
8	1068.81	98.84	1120.64	108.56	1172.64	118.77	1224.83	129.45	1277.21	140.63	1329.80	152.30	1382.60	164.46	8
9	1069.67	99.00	1121.50	108.73	1173.51	118.94	1225.70	129.64	1278.09	140.82	1330.68	152.49	1383.48	164.66	9
10	1070.54	99.15	1122.37	108.90	1174.38	119.12	1226.57	129.82	1278.96	141.01	1331.56	152.69	1384.37	164.87	10
11	1071.40	99.31	1123.23	109.06	1175.25	119.29	1227.44	130.00	1279.84	141.20	1332.44	152.89	1385.25	165.08	11
12	1072.26	99.47	1124.10	109.23	1176.11	119.46	1228.32	130.18	1280.71	141.39	1333.31	153.09	1386.13	165.29	12
13	1073.12	99.63	1124.96	109.40	1176.98	119.64	1229.19	130.37	1281.59	141.58	1334.19	153.29	1387.01	165.49	13
14	1073.99	99.79	1125.83	109.56	1177.85	119.81	1230.06	130.55	1282.46	141.77	1335.07	153.49	1387.90	165.70	14
15	1074.85	99.95	1126.69	109.73	1178.72	119.99	1230.93	130.73	1283.34	141.96	1335.95	153.69	1388.78	165.91	15
16	1075.71	100.11	1127.56	109.90	1179.59	120.16	1231.80	130.92	1284.21	142.16	1336.83	153.89	1389.66	166.12	16
17	1076.57	100.27	1128.43	110.06	1180.46	120.34	1232.67	131.10	1285.09	142.35	1337.71	154.09	1390.54	166.33	17
18	1077.44	100.42	1129.29	110.23	1181.33	120.52	1233.55	131.28	1285.96	142.54	1338.59	154.29	1391.42	166.53	18
19	1078.30	100.58	1130.16	110.40	1182.19	120.69	1234.42	131.47	1286.84	142.73	1339.47	154.49	1392.31	166.74	19
20	1079.16	100.74	1131.02	110.57	1183.06	120.87	1235.29	131.65	1287.71	142.92	1340.34	154.69	1393.19	166.95	20
21	1080.03	100.90	1131.89	110.73	1183.93	121.04	1236.16	131.83	1288.59	143.12	1341.22	154.89	1394.07	167.16	21
22	1080.89	101.06	1132.76	110.90	1184.80	121.22	1237.03	132.02	1289.47	143.31	1342.10	155.09	1394.96	167.37	22
23	1081.75	101.22	1133.62	111.07	1185.67	121.39	1237.91	132.20	1290.34	143.50	1342.98	155.29	1395.84	167.58	23
24	1082.61	101.38	1134.49	111.24	1186.54	121.57	1238.78	132.39	1291.22	143.69	1343.86	155.49	1396.72	167.79	24
25	1083.48	101.54	1135.35	111.41	1187.41	121.75	1239.65	132.57	1292.09	143.88	1344.74	155.69	1397.60	167.99	25
26	1084.34	101.70	1136.22	111.57	1188.28	121.92	1240.52	132.76	1292.97	144.08	1345.62	155.89	1398.49	168.20	26
27	1085.20	101.87	1137.09	111.74	1189.15	122.10	1241.40	132.94	1293.84	144.27	1346.50	156.09	1399.37	168.41	27
28	1086.07	102.03	1137.95	111.91	1190.02	122.28	1242.27	133.13	1294.72	144.46	1347.38	156.29	1400.25	168.62	28
29	1086.93	102.19	1138.82	112.08	1190.88	122.45	1243.14	133.31	1295.60	144.66	1348.26	156.50	1401.14	168.83	29
30	1087.79	102.35	1139.68	112.25	1191.75	122.63	1244.01	133.50	1296.47	144.85	1349.14	156.70	1402.02	169.04	30
31	1088.66	102.51	1140.55	112.42	1192.62	122.81	1244.89	133.68	1297.35	145.04	1350.02	156.90	1402.90	169.25	31
32	1089.52	102.67	1141.42	112.59	1193.49	122.98	1245.76	133.87	1298.22	145.24	1350.90	157.10	1403.79	169.46	32
33	1090.38	102.83	1142.28	112.76	1194.36	123.16	1246.63	134.05	1299.10	145.43	1351.78	157.30	1404.67	169.67	33
34	1091.25	102.99	1143.15	112.93	1195.23	123.34	1247.50	134.24	1299.98	145.62	1352.66	157.50	1405.55	169.88	34
35	1092.11	103.15	1144.02	113.10	1196.10	123.52	1248.38	134.42	1300.85	145.82	1353.53	157.71	1406.44	170.09	35
36	1092.98	103.32	1144.88	113.27	1196.97	123.69	1249.25	134.61	1301.73	146.01	1354.41	157.91	1407.32	170.30	36
37	1093.84	103.48	1145.75	113.44	1197.84	123.87	1250.12	134.80	1302.60	146.21	1355.29	158.11	1408.20	170.51	37
38	1094.70	103.64	1146.62	113.61	1198.71	124.05	1251.00	134.98	1303.48	146.40	1356.17	158.31	1409.09	170.73	38
39	1095.57	103.80	1147.48	113.78	1199.58	124.23	1251.87	135.17	1304.36	146.59	1357.05	158.52	1409.97	170.94	39
40	1096.43	103.97	1148.35	113.95	1200.45	124.41	1252.74	135.35	1305.23	146.79	1357.93	158.72	1410.85	171.15	40
41	1097.29	104.13	1149.22	114.12	1201.32	124.59	1253.62	135.54	1306.11	146.98	1358.81	158.92	1411.74	171.36	41
42	1098.16	104.29	1150.08	114.29	1202.19	124.76	1254.49	135.73	1306.99	147.18	1359.70	159.13	1412.62	171.57	42
43	1099.02	104.45	1150.95	114.46	1203.06	124.94	1255.36	135.91	1307.86	147.37	1360.58	159.33	1413.51	171.78	43
44	1099.89	104.62	1151.82	114.63	1203.93	125.12	1256.24	136.10	1308.74	147.57	1361.46	159.53	1414.39	171.99	44
45	1100.75	104.78	1152.68	114.80	1204.80	125.30	1257.11	136.29	1309.62	147.76	1362.34	159.74	1415.27	172.21	45
46	1101.61	104.94	1153.55	114.97	1205.67	125.48	1257.98	136.48	1310.49	147.96	1363.22	159.94	1416.16	172.42	46
47	1102.48	105.10	1154.42	115.14	1206.54	125.66	1258.86	136.66	1311.37	148.16	1364.10	160.14	1417.04	172.63	47
48	1103.34	105.27	1155.29	115.31	1207.41	125.84	1259.73	136.85	1312.25	148.35	1364.98	160.35	1417.93	172.84	48
49	1104.21	105.43	1156.15	115.48	1208.28	126.02	1260.60	137.04	1313.13	148.55	1365.86	160.55	1418.81	173.06	49
50	1105.07	105.60	1157.02	115.66	1209.15	126.20	1261.48	137.23	1314.00	148.74	1366.74	160.76	1419.70	173.27	50
51	1105.94	105.76	1157.89	115.83	1210.02	126.38	1262.35	137.41	1314.88	148.94	1367.62	160.96	1420.58	173.48	51
52	1106.80	105.92	1158.76	116.00	1210.89	126.56	1263.22	137.60	1315.76	149.14	1368.50	161.17	1421.47	173.69	52
53	1107.66	106.09	1159.62	116.17	1211.76	126.74	1264.10	137.79	1316.63	149.33	1369.38	161.37	1422.35	173.91	53
54	1108.53	106.25	1160.49	116.34	1212.63	126.92	1264.97	137.98	1317.51	149.53	1370.26	161.57	1423.23	174.12	54
55	1109.39	106.41	1161.36	116.52	1213.51	127.10	1265.85	138.17	1318.39	149.73	1371.14	161.78	1424.12	174.33	55
56	1110.26	106.58	1162.22	116.69	1214.38	127.28	1266.72	138.36	1319.27	149.92	1372.02	161.99	1425.00	174.55	56
57	1111.12	106.74	1163.09	116.86	1215.25	127.46	1267.59	138.54	1320.14	150.12	1372.91	162.19	1425.89	174.76	57
58	1111.99	106.91	1163.96	117.03	1216.12	127.64	1268.47	138.73	1321.02	150.32	1373.79	162.40	1426.77	174.98	58
59	1112.85	107.07	1164.83	117.21	1216.99	127.82	1269.34	138.92	1321.90	150.51	1374.67	162.60	1427.66	175.19	59

(continued)

APPENDIX A *(continued)*

Tangents and Externals for Horizontal Curves

Δ = interior angle $\quad T^* = 5729.578 \tan \frac{\Delta}{2} \quad E^* = 5729.578 \left(\sec \frac{\Delta}{2} - 1\right)$

For arc definition, calculate T and E for any degree of curve D by dividing both table values by D.

$$T = \frac{T^*}{D} \qquad E = \frac{E^*}{D}$$

	$\Delta = 28°$		$\Delta = 29°$		$\Delta = 30°$		$\Delta = 31°$		$\Delta = 32°$		$\Delta = 33°$		$\Delta = 34°$		
minutes	T^*	E^*	T^*	E^*	T^*	E^*	T^*	E^*	T^*	E^*	T^*	E^*	T^*	E^*	minutes
0	1428.54	175.40	1481.77	188.51	1535.24	202.12	1588.95	216.25	1642.93	230.90	1697.18	246.08	1751.71	261.79	0
1	1429.43	175.62	1482.66	188.73	1536.13	202.35	1589.85	216.49	1643.83	231.15	1698.08	246.34	1752.62	262.06	1
2	1430.31	175.83	1483.55	188.95	1537.02	202.58	1590.75	216.73	1644.73	231.40	1698.99	246.59	1753.53	262.33	2
3	1431.20	176.05	1484.44	189.17	1537.92	202.81	1591.65	216.97	1645.64	231.64	1699.90	246.85	1754.44	262.59	3
4	1432.09	176.26	1485.33	189.40	1538.81	203.04	1592.54	217.21	1646.54	231.89	1700.80	247.11	1755.35	262.86	4
5	1432.97	176.48	1486.22	189.62	1539.77	203.27	1593.44	217.45	1647.44	232.14	1701.71	247.37	1756.26	263.13	5
6	1433.86	176.69	1487.11	189.84	1540.60	203.51	1594.34	217.69	1648.34	232.39	1702.62	247.63	1757.18	263.40	6
7	1434.74	176.91	1487.99	190.07	1541.49	203.74	1595.24	217.93	1649.24	232.64	1703.53	247.88	1758.09	263.66	7
8	1435.63	177.12	1488.88	190.29	1542.38	203.97	1596.13	218.17	1650.15	232.89	1704.43	248.14	1759.00	263.93	8
9	1436.51	177.34	1489.77	190.51	1543.28	204.20	1597.03	218.41	1651.05	233.14	1705.34	248.40	1759.91	264.20	9
10	1437.40	177.55	1490.66	190.74	1544.17	204.44	1597.93	218.65	1651.95	233.39	1706.25	248.66	1760.82	264.47	10
11	1438.28	177.77	1491.55	190.96	1545.06	204.67	1598.83	218.89	1652.85	233.64	1707.15	248.92	1761.74	264.73	11
12	1439.17	177.98	1492.44	191.19	1545.96	204.90	1599.73	219.14	1653.76	233.89	1708.06	249.18	1762.65	265.00	12
13	1440.06	178.20	1493.33	191.41	1546.85	205.13	1600.63	219.38	1654.66	234.14	1708.97	249.44	1763.56	265.27	13
14	1440.94	178.41	1494.22	191.64	1547.75	205.37	1601.52	219.62	1655.56	234.39	1709.88	249.70	1764.47	265.54	14
15	1441.83	178.63	1495.11	191.86	1548.64	205.60	1602.42	219.86	1656.47	234.64	1710.78	249.96	1765.39	265.81	15
16	1442.71	178.85	1496.00	192.08	1549.54	205.83	1603.32	220.10	1657.37	234.89	1711.69	250.22	1766.30	266.08	16
17	1443.60	179.06	1496.89	192.31	1550.43	206.07	1604.22	220.35	1658.27	235.15	1712.60	250.48	1767.21	266.35	17
18	1444.49	179.28	1497.78	192.53	1551.32	206.30	1605.12	220.59	1659.18	235.40	1713.51	250.74	1768.12	266.61	18
19	1445.37	179.50	1498.67	192.76	1552.22	206.54	1606.02	220.83	1660.08	235.65	1714.42	251.00	1769.04	266.88	19
20	1446.26	179.71	1499.56	192.98	1553.11	206.77	1606.92	221.07	1660.98	235.90	1715.32	251.26	1769.95	267.15	20
21	1447.15	179.93	1500.45	193.21	1554.01	207.00	1607.81	221.32	1661.89	236.15	1716.23	251.52	1770.86	267.42	21
22	1448.03	180.15	1501.35	193.44	1554.90	207.24	1608.71	221.56	1662.79	236.40	1717.14	251.78	1771.77	267.69	22
23	1448.92	180.36	1502.24	193.66	1555.80	207.47	1609.61	221.80	1663.69	236.66	1718.05	252.04	1772.69	267.96	23
24	1449.81	180.58	1503.13	193.89	1556.69	207.71	1610.51	222.04	1664.60	236.91	1718.96	252.30	1773.60	268.23	24
25	1450.69	180.80	1504.02	194.11	1557.59	207.94	1611.41	222.29	1665.50	237.16	1719.86	252.56	1774.51	268.50	25
26	1451.58	181.02	1504.91	194.34	1558.48	208.18	1612.31	222.53	1666.40	237.41	1720.77	252.82	1775.43	268.77	26
27	1452.47	181.24	1505.80	194.57	1559.38	208.41	1613.21	222.77	1667.31	237.66	1721.68	253.08	1776.34	269.04	27
28	1453.35	181.45	1506.69	194.79	1560.27	208.65	1614.11	223.02	1668.21	237.92	1722.59	253.35	1777.25	269.31	28
29	1454.24	181.67	1507.58	195.02	1561.17	208.88	1615.01	223.26	1669.12	238.17	1723.50	253.61	1778.17	269.58	29
30	1455.13	181.89	1508.47	195.25	1562.06	209.12	1615.91	223.51	1670.02	238.42	1724.41	253.87	1779.08	269.85	30
31	1456.01	182.11	1509.36	195.47	1562.96	209.35	1616.81	223.75	1670.92	238.68	1725.32	254.13	1779.99	270.13	31
32	1456.90	182.33	1510.25	195.70	1563.85	209.59	1617.71	224.00	1671.83	238.93	1726.22	254.39	1780.91	270.40	32
33	1457.79	182.55	1511.15	195.93	1564.75	209.82	1618.61	224.24	1672.73	239.18	1727.13	254.66	1781.82	270.67	33
34	1458.68	182.76	1512.04	196.16	1565.64	210.06	1619.51	224.49	1673.64	239.44	1728.04	254.92	1782.74	270.94	34
35	1459.56	182.98	1512.93	196.38	1566.54	210.30	1620.41	224.73	1674.54	239.69	1728.95	255.18	1783.65	271.21	35
36	1460.45	183.20	1513.82	196.61	1567.44	210.53	1621.31	224.97	1675.45	239.94	1729.86	255.44	1784.56	271.48	36
37	1461.34	183.42	1514.71	196.84	1568.33	210.77	1622.21	225.22	1676.35	240.20	1730.77	255.71	1785.48	271.76	37
38	1462.23	183.64	1515.60	197.07	1569.23	211.01	1623.11	255.47	1677.26	240.45	1731.68	255.97	1786.39	272.03	38
39	1463.11	183.86	1516.49	197.29	1570.12	211.24	1624.01	225.71	1678.16	240.71	1732.59	256.23	1787.31	272.30	39
40	1464.00	184.08	1517.39	197.52	1571.02	211.48	1624.91	225.96	1679.06	240.96	1733.50	256.50	1788.22	272.57	40
41	1464.89	184.30	1518.28	197.75	1571.91	211.72	1625.81	226.20	1679.97	241.21	1734.41	256.76	1789.14	272.84	41
42	1465.78	184.52	1519.17	197.98	1572.81	211.95	1626.71	226.45	1680.88	241.47	1735.32	257.02	1790.05	273.12	42
43	1466.67	184.74	1520.06	198.21	1573.71	212.19	1627.61	226.69	1681.78	241.72	1736.23	257.29	1790.97	273.39	43
44	1467.55	184.96	1520.95	198.44	1574.60	212.43	1628.51	226.94	1682.69	241.98	1737.14	257.55	1791.88	273.66	44
45	1468.44	185.18	1521.85	198.67	1575.50	212.67	1629.41	227.19	1683.59	242.23	1738.05	257.82	1792.80	273.94	45
46	1469.33	185.40	1522.74	198.90	1576.40	212.90	1630.31	227.43	1684.50	242.49	1738.96	258.08	1793.71	274.21	46
47	1470.22	185.62	1523.63	199.12	1577.29	213.14	1631.21	227.68	1685.40	242.74	1739.87	258.34	1794.63	274.48	47
48	1471.11	185.84	1524.52	199.35	1578.19	213.38	1632.11	227.93	1686.31	243.00	1740.78	258.61	1795.54	274.76	48
49	1471.99	186.06	1525.42	199.58	1579.09	213.62	1633.01	228.17	1687.21	243.26	1741.69	258.87	1796.46	275.03	49
50	1472.88	186.29	1526.31	199.81	1579.98	213.86	1633.92	228.42	1688.12	243.51	1742.60	259.14	1797.37	275.30	50
51	1473.77	186.51	1527.20	200.04	1580.88	214.09	1634.82	228.67	1689.02	243.77	1743.51	259.40	1798.29	175.58	51
52	1474.66	186.73	1528.09	200.27	1581.78	214.33	1635.72	228.92	1689.93	244.02	1744.42	259.67	1799.20	275.85	52
53	1475.55	186.95	1528.99	200.50	1582.67	214.57	1636.62	229.16	1690.84	244.28	1745.33	259.93	1800.12	276.13	53
54	1476.44	187.17	1529.88	200.73	1583.57	214.81	1637.52	229.41	1691.74	244.54	1746.24	260.20	1801.03	276.40	54
55	1477.32	187.39	1530.77	200.96	1584.47	215.05	1638.42	229.66	1692.65	244.79	1747.15	260.46	1801.95	276.68	55
56	1478.21	187.62	1531.66	201.19	1585.36	215.29	1639.32	229.91	1693.55	245.05	1748.06	260.73	1802.87	276.95	56
57	1479.10	187.84	1532.56	201.42	1586.26	215.53	1640.22	230.15	1694.46	245.31	1748.97	261.00	1803.78	277.23	57
58	1479.99	188.06	1533.45	201.66	1587.16	215.77	1641.13	230.40	1695.37	245.56	1749.89	261.26	1804.70	277.50	58
59	1480.88	188.28	1534.34	201.89	1588.06	216.01	1642.03	230.65	1696.27	245.82	1750.80	261.53	1805.61	277.78	59

(continued)

Support Material

APPENDIX A *(continued)*
Tangents and Externals for Horizontal Curves

Δ = interior angle $\quad T^* = 5729.578 \tan \frac{\Delta}{2} \quad E^* = 5729.578 \left(\sec \frac{\Delta}{2} - 1 \right)$

For arc definition, calculate T and E for any degree of curve D by dividing both table values by D.

$$T = \frac{T^*}{D} \qquad E = \frac{E^*}{D}$$

	$\Delta = 35°$		$\Delta = 36°$		$\Delta = 37°$		$\Delta = 38°$		$\Delta = 39°$		$\Delta = 40°$		$\Delta = 41°$		
minutes	T^*	E^*	T^*	E^*	T^*	E^*	T^*	E^*	T^*	E^*	T^*	E^*	T^*	E^*	minutes
0	1806.53	278.05	1861.65	294.86	1917.09	312.22	1972.85	330.14	2028.96	348.64	2085.40	367.71	2142.20	387.37	0
1	1807.45	278.33	1862.57	295.14	1918.02	312.51	1973.78	330.45	2029.89	348.95	2086.34	368.03	2143.15	387.71	1
2	1808.36	278.60	1863.50	295.43	1918.94	312.81	1974.72	330.75	2030.83	349.26	2087.28	368.36	2144.10	388.04	2
3	1809.28	278.88	1864.42	295.71	1919.87	313.10	1975.65	331.05	2031.76	349.58	2088.23	368.68	2145.05	388.37	3
4	1810.19	279.15	1865.34	296.00	1920.80	313.40	1976.58	331.36	2032.70	349.89	2089.17	369.00	2146.00	388.70	4
5	1811.11	279.43	1866.26	296.28	1921.72	313.69	1977.51	331.66	2033.64	350.20	2090.12	369.33	2146.95	389.04	5
6	1812.03	279.71	1867.18	296.57	1922.65	313.98	1978.45	331.97	2034.58	350.52	2091.06	369.65	2147.90	389.37	6
7	1812.94	279.98	1868.10	296.85	1923.58	314.28	1979.38	332.27	2035.52	350.83	2092.00	369.97	2148.85	389.71	7
8	1813.86	280.26	1869.03	297.14	1924.51	314.58	1980.31	332.57	2036.46	351.15	2092.95	370.30	2149.80	390.04	8
9	1814.78	280.54	1869.95	297.43	1925.43	314.87	1981.24	332.88	2037.39	351.46	2093.89	370.62	2150.75	390.37	9
10	1815.70	280.81	1870.87	297.71	1926.36	315.17	1982.18	333.18	2038.33	351.78	2094.84	370.95	2151.70	390.71	10
11	1816.61	281.09	1871.79	298.00	1927.29	315.46	1983.11	333.49	2039.27	352.09	2095.78	371.27	2152.66	391.04	11
12	1817.53	281.37	1872.71	298.28	1928.22	315.76	1984.04	333.80	2040.21	352.41	2096.73	371.60	2153.61	391.38	12
13	1818.45	281.65	1873.64	298.57	1929.14	316.05	1984.98	334.10	2041.15	352.72	2097.67	371.92	2154.56	391.71	13
14	1819.36	281.92	1874.56	298.86	1930.07	316.35	1985.91	334.41	2042.09	353.04	2098.62	372.25	2155.51	392.05	14
15	1820.28	282.20	1875.48	299.14	1931.00	316.65	1986.84	334.71	2043.03	353.35	2099.56	372.57	2156.46	392.38	15
16	1821.20	282.48	1876.41	299.43	1931.93	316.94	1987.78	335.02	2043.97	353.67	2100.51	372.90	2157.41	392.72	16
17	1822.12	282.76	1877.33	299.72	1931.86	317.24	1988.71	335.32	2044.91	353.98	2101.45	373.22	2158.36	393.05	17
18	1823.03	283.03	1878.25	300.01	1933.78	317.53	1989.65	335.63	2045.85	354.30	2102.40	373.55	2159.31	393.39	18
19	1823.95	283.31	1879.17	300.29	1934.71	317.83	1990.58	335.94	2046.79	354.61	2103.34	373.87	2160.27	393.72	19
20	1824.87	283.59	1880.10	300.58	1935.64	318.13	1991.51	336.24	2047.73	354.93	2104.29	374.20	2161.22	394.06	20
21	1825.79	283.87	1881.02	300.87	1936.57	318.43	1992.45	336.55	2048.67	355.25	2105.24	374.53	2162.17	394.39	21
22	1826.71	284.15	1881.94	301.16	1937.50	318.72	1993.38	336.86	2049.61	355.56	2106.18	374.85	2163.12	394.73	22
23	1827.62	284.43	1882.87	301.45	1938.43	319.02	1994.32	337.16	2050.55	355.88	2107.13	375.18	2164.07	395.07	23
24	1828.54	284.71	1883.79	301.73	1939.36	319.32	1995.25	337.47	2051.49	356.20	2108.07	375.50	2165.03	395.40	24
25	1829.46	284.99	1884.71	302.02	1940.28	319.62	1996.18	337.78	2052.43	356.51	2109.02	375.83	2165.98	395.74	25
26	1830.38	285.27	1885.64	302.31	1941.21	319.92	1997.12	338.09	2053.37	356.83	2109.97	376.16	2166.93	396.08	26
27	1831.30	285.54	1886.56	302.60	1942.14	320.21	1998.05	338.39	2054.31	357.15	2110.91	376.49	2167.88	396.41	27
28	1832.22	285.83	1887.48	302.89	1943.07	320.51	1998.99	338.70	2055.25	357.47	2111.86	376.81	2168.84	396.75	28
29	1833.13	286.10	1888.41	303.18	1944.00	320.81	1999.92	339.01	2056.19	357.78	2112.81	377.14	2169.79	397.09	29
30	1834.05	286.39	1889.33	303.47	1944.93	321.11	2000.86	339.32	2057.13	358.10	2113.75	377.47	2170.74	397.43	30
31	1834.97	286.67	1890.26	303.76	1945.86	321.41	2001.79	339.63	2058.07	358.42	2114.70	377.80	2171.70	397.76	31
32	1835.89	286.95	1891.18	304.05	1946.79	321.71	2002.73	339.93	2059.01	358.74	2115.65	378.12	2172.65	398.10	32
33	1836.81	287.23	1892.10	304.34	1947.72	322.01	2003.66	340.24	2059.95	359.06	2116.59	378.45	2173.60	398.44	33
34	1837.73	287.51	1893.03	304.63	1948.65	322.31	2004.60	340.55	2060.89	359.37	2117.54	378.78	2174.56	398.78	34
35	1838.65	287.79	1893.95	304.92	1949.58	322.60	2005.53	340.86	2061.83	359.69	2118.49	379.11	2175.51	399.12	35
36	1839.57	288.07	1894.88	305.21	1950.51	322.90	2006.47	341.17	2062.78	360.01	2119.44	379.44	2176.46	399.46	36
37	1840.49	288.35	1895.80	305.50	1951.44	323.20	2007.41	341.48	2063.72	360.33	2120.38	379.77	2177.42	399.79	37
38	1841.41	288.63	1896.73	305.79	1952.37	323.50	2008.34	341.79	2064.66	360.65	2121.33	380.10	2178.37	400.13	38
39	1842.32	288.91	1897.65	306.08	1953.30	323.80	2009.28	342.10	2065.60	360.97	2122.28	380.42	2179.32	400.47	39
40	1843.24	289.19	1898.58	306.37	1954.23	324.10	2010.21	342.41	2066.54	361.29	2123.23	380.75	2180.28	400.81	40
41	1844.16	289.48	1899.50	306.66	1955.16	324.40	2011.15	342.72	2067.48	361.61	2124.17	381.08	2181.23	401.15	41
42	1845.08	289.76	1900.43	306.95	1956.09	324.71	2012.08	343.03	2068.42	361.93	2125.12	381.41	2182.19	401.49	42
43	1846.00	290.04	1901.35	307.24	1957.02	325.01	2013.02	343.34	2069.37	362.25	2126.07	381.74	2183.14	401.83	43
44	1846.92	290.32	1902.28	307.53	1957.95	325.31	2013.96	343.65	2070.31	362.57	2127.02	382.07	2184.09	402.17	44
45	1847.84	290.60	1903.20	307.83	1958.88	325.61	2014.89	343.96	2071.25	362.89	2127.97	382.40	2185.05	402.51	45
46	1848.76	290.89	1904.13	308.12	1959.81	325.91	2015.83	344.27	2072.19	363.21	2128.91	382.73	2186.00	402.85	46
47	1849.68	291.17	1905.05	308.41	1960.74	326.21	2016.77	344.58	3072.14	363.53	2129.86	383.06	2186.96	403.19	47
48	1850.60	291.45	1905.98	308.70	1961.67	326.51	2017.70	344.89	2074.08	363.85	2130.81	383.39	2187.91	403.53	48
49	1851.52	291.73	1906.90	308.99	1962.60	326.81	2018.64	345.20	2075.02	364.17	2131.76	383.72	2188.87	403.87	49
50	1852.44	292.02	1907.83	309.29	1963.54	327.11	2019.58	345.51	2075.96	364.49	2132.71	384.05	2189.82	404.21	50
51	1853.36	292.30	1908.75	309.58	1964.47	327.42	2020.51	345.83	2076.91	364.81	2133.66	384.39	2190.78	404.55	51
52	1854.29	292.58	1909.68	309.87	1965.40	327.72	2021.45	346.14	2077.85	365.13	2134.61	384.72	2191.73	404.89	52
53	1855.21	292.87	1910.61	310.16	1966.33	328.02	2022.39	346.45	2078.79	365.46	2135.56	385.05	2192.69	405.24	53
54	1856.13	293.15	1911.53	310.46	1967.26	328.32	2023.32	346.76	2079.74	365.78	2136.50	385.38	2193.64	405.58	54
55	1857.05	293.44	1912.46	310.75	1968.19	328.63	2024.26	347.07	2080.68	366.10	2137.45	385.71	2194.60	405.92	55
56	1857.97	293.72	1913.38	311.04	1969.12	328.93	2025.20	347.39	2081.62	366.42	2138.40	386.04	2195.56	406.26	56
57	1858.89	294.00	1914.31	311.34	1970.06	329.23	2026.14	347.70	2082.57	366.74	2139.35	386.38	2196.51	406.60	57
58	1859.81	294.29	1915.24	311.63	1970.99	329.53	2027.07	348.01	2083.51	367.07	2140.30	386.71	2197.47	406.95	58
59	1860.73	294.57	1916.16	311.92	1971.92	329.84	2028.01	348.32	2084.45	367.39	2141.25	387.04	2198.42	407.29	59

(continued)

APPENDIX A *(continued)*
Tangents and Externals for Horizontal Curves

Δ = interior angle $\quad T^* = 5729.578 \tan \dfrac{\Delta}{2} \quad E^* = 5729.578 \left(\sec \dfrac{\Delta}{2} - 1\right)$

For arc definition, calculate T and E for any degree of curve D by dividing both table values by D.

$$T = \frac{T^*}{D} \qquad E = \frac{E^*}{D}$$

	$\Delta = 42°$		$\Delta = 43°$		$\Delta = 44°$		$\Delta = 45°$		
minutes	T^*	E^*	T^*	E^*	T^*	E^*	T^*	E^*	minutes
0	2199.38	407.63	2256.94	428.49	2314.90	449.97	2373.27	472.07	0
1	2200.34	407.97	2257.90	428.85	2315.87	450.33	2374.25	472.45	1
2	2201.29	408.32	2258.87	429.20	2316.84	450.70	2375.22	472.82	2
3	2202.25	408.66	2259.83	429.55	2317.81	451.06	2376.20	473.19	3
4	2203.20	409.00	2260.79	429.91	2318.78	451.42	2377.18	473.57	4
5	2204.16	409.35	2261.76	430.26	2319.75	451.79	2378.15	473.94	5
6	2205.12	409.69	2262.72	430.61	2320.72	452.15	2379.13	474.32	6
7	2206.07	410.03	2263.68	430.97	2321.69	452.52	2380.11	474.69	7
8	2207.03	410.38	2264.65	431.32	2322.66	452.88	2381.08	475.07	8
9	2207.99	410.72	2265.61	431.68	2323.63	453.25	2382.06	475.44	9
10	2208.95	411.06	2266.57	432.03	2324.60	453.61	2383.04	475.82	10
11	2209.90	411.41	2267.54	432.38	2325.57	453.98	2384.02	476.19	11
12	2210.86	411.75	2268.50	432.74	2326.54	454.34	2384.99	476.57	12
13	2211.82	412.10	2269.46	433.09	2327.51	454.71	2385.97	476.94	13
14	2212.78	412.44	2270.43	433.45	2328.48	455.07	2386.95	477.32	14
15	2213.73	412.79	2271.39	433.81	2329.45	455.44	2387.93	477.70	15
16	2214.69	413.13	2272.36	434.16	2330.42	455.80	2388.90	478.07	16
17	2215.65	413.48	2273.32	434.52	2331.40	456.17	2389.88	478.45	17
18	2216.61	413.83	2274.29	434.87	2332.37	456.54	2390.86	478.83	18
19	2217.56	414.17	2275.25	435.23	2333.34	456.90	2391.84	479.20	19
20	2218.52	414.52	2276.22	435.58	2334.31	457.27	2392.82	479.58	20
21	2219.48	414.86	2277.18	435.94	2335.28	457.63	2393.80	479.96	21
22	2220.44	415.21	2278.15	436.30	2336.25	458.00	2394.78	480.34	22
23	2221.40	415.56	2279.11	436.65	2337.23	458.37	2395.76	480.71	23
24	2222.36	415.90	2280.08	437.01	2338.20	458.74	2396.73	481.09	24
25	2223.32	416.25	2281.04	437.37	2339.17	459.10	2397.71	481.47	25
26	2224.27	416.60	2282.01	437.72	2340.14	459.47	2398.69	481.85	26
27	2225.23	416.94	2282.97	438.08	2341.11	459.84	2399.67	482.22	27
28	2226.19	417.29	2283.94	438.44	2342.09	460.21	2400.65	482.60	28
29	2227.15	417.64	2284.90	438.80	2343.06	460.57	2401.63	482.98	29
30	2228.11	417.99	2285.87	439.15	2344.03	460.94	2402.61	483.36	30
31	2229.07	418.33	2286.84	439.51	2345.01	461.31	2403.59	483.74	31
32	2230.03	418.68	2287.80	439.87	2345.98	461.68	2404.57	484.12	32
33	2230.99	419.03	2288.77	440.23	2346.95	462.05	2405.55	484.50	33
34	2231.95	419.38	2289.73	440.59	2347.92	462.42	2406.53	484.88	34
35	2232.91	419.73	2290.70	440.95	2348.90	462.79	2407.51	485.26	35
36	2233.87	420.08	2291.67	441.31	2349.87	463.16	2408.49	485.64	36
37	2234.83	420.42	2292.63	441.66	2350.84	463.53	2409.47	486.02	37
38	2235.79	420.77	2293.60	442.02	2351.82	463.89	2410.45	486.40	38
39	2236.75	421.12	2294.57	442.38	2352.79	464.27	2411.44	486.78	39
40	2237.71	421.47	2295.54	442.74	2353.77	464.64	2412.42	487.16	40
41	2238.67	421.82	2296.50	443.10	2354.74	465.00	2413.40	487.54	41
42	2239.63	422.17	2297.47	443.46	2355.71	465.38	2414.38	487.92	42
43	2240.59	422.52	2298.44	443.82	2356.69	465.75	2415.36	488.30	43
44	2241.55	422.87	2299.40	444.18	2357.66	466.12	2416.34	488.68	44
45	2242.51	423.22	2300.37	444.54	2358.64	466.49	2417.32	489.06	45
46	2243.47	423.57	2301.34	444.90	2359.61	466.86	2418.30	489.45	46
47	2244.44	423.92	2302.31	445.26	2360.59	467.23	2419.29	489.83	47
48	2245.40	424.27	2303.28	445.63	2361.56	467.60	2420.27	490.21	48
49	2246.36	424.62	2304.24	445.99	2362.54	467.97	2421.25	490.59	49
50	2247.32	424.97	2305.21	446.35	2363.51	468.34	2422.23	490.97	50
51	2248.28	425.32	2306.18	446.71	2364.49	468.72	2423.22	491.36	51
52	2249.24	425.68	2307.15	447.07	2365.46	469.09	2424.20	491.74	52
53	2250.21	426.03	2308.12	447.43	2366.44	469.46	2425.18	492.12	53
54	2251.17	426.38	2309.09	447.79	2367.41	469.83	2426.16	492.51	54
55	2252.13	426.73	2310.05	448.16	2368.39	470.21	2427.15	492.89	55
56	2253.09	427.08	2311.02	448.52	2369.36	470.58	2428.13	493.27	56
57	2254.05	427.44	2311.99	448.88	2370.34	470.95	2429.11	493.66	57
58	2255.02	427.79	2312.96	449.24	2371.32	471.32	2430.09	494.04	58
59	2255.98	428.14	2313.93	449.61	2372.29	471.70	2431.08	494.42	59

APPENDIX B
Radius When Degree of Curve Is Known

$$R = \frac{5729.578}{D}$$

D	R (ft)	D	R (ft)	D	R (ft)	D	R (ft)	D	R (ft)
0°00′		9°00′	636.62	18°00′	318.31	27°00′	212.21	36°00′	159.15
15′	22918.31	15′	619.41	15′	313.95	15′	210.26	15′	158.06
30′	11459.16	30′	603.11	30′	309.71	30′	208.35	30′	156.97
45′	7639.44	45′	587.65	45′	305.58	45′	206.47	45′	155.91
1°00′	5729.58	10°00′	572.96	19°00′	301.56	28°00′	204.63	37°00′	154.85
15′	4583.66	15′	558.98	15′	297.64	15′	202.82	15′	153.81
30′	3819.72	30′	545.67	30′	293.82	30′	201.04	30′	152.79
45′	3274.04	45′	532.98	45′	290.11	45′	199.29	45′	151.78
2°00′	2864.79	11°00′	520.87	20°00′	286.48	29°00′	197.57	38°00′	150.78
15′	2546.48	15′	509.30	15′	282.94	15′	195.88	15′	149.79
30′	2291.83	30′	498.22	30′	279.49	30′	194.22	30′	148.82
45′	2083.48	45′	487.62	45′	276.12	45′	192.59	45′	147.86
3°00′	1909.86	12°00′	477.46	21°00′	272.84	30°00′	190.99	39°00′	146.91
15′	1762.95	15′	467.72	15′	269.63	15′	189.41	15′	145.98
30′	1637.02	30′	458.37	30′	266.49	30′	187.86	30′	145.05
45′	1527.89	45′	449.38	45′	263.43	45′	186.33	45′	144.14
4°00′	1432.39	13°00′	440.74	22°00′	260.44	31°00′	184.83	40°00′	143.24
15′	1348.14	15′	432.42	15′	257.51	15′	183.35	15′	142.35
30′	1273.24	30′	424.41	30′	254.65	30′	181.89	30′	141.47
45′	1206.23	45′	416.70	45′	251.85	45′	180.46	45′	140.60
5°00′	1145.92	14°00′	409.26	23°00′	249.11	32°00′	179.05	41°00′	139.75
15′	1091.35	15′	402.08	15′	246.43	15′	177.66	15′	138.90
30′	1041.74	30′	395.14	30′	243.81	30′	176.29	30′	138.06
45′	996.45	45′	388.45	45′	241.25	45′	174.95	45′	137.24
6°00′	954.93	15°00′	381.97	24°00′	238.73	33°00′	173.62	42°00′	136.42
15′	916.73	15′	375.71	15′	236.27	15′	172.32	15′	135.61
30′	881.47	30′	369.65	30′	233.86	30′	171.03	30′	134.81
45′	848.83	45′	363.78	45′	231.50	45′	169.77	45′	134.03
7°00′	818.51	16°00′	358.10	25°00′	229.18	34°00′	168.52	43°00′	133.25
15′	790.29	15′	352.59	15′	226.91	15′	167.29	15′	132.48
30′	763.94	30′	347.25	30′	224.69	30′	166.07	30′	131.71
45′	739.30	45′	342.06	45′	222.51	45′	164.88	45′	130.96
8°00′	716.20	17°00′	337.03	26°00′	220.37	35°00′	163.70	44°00′	130.22
15′	694.49	15′	332.15	15′	218.27	15′	162.54	15′	129.48
30′	674.07	30′	327.40	30′	216.21	30′	161.40	30′	128.75
45′	654.81	45′	322.79	45′	214.19	45′	160.27	45′	128.04
								45°00′	127.32
								15′	126.62
								30′	125.92
								45′	125.24

(continued)

APPENDIX B *(continued)*
Radius When Degree of Curve Is Known

$$R = \frac{5729.578}{D}$$

D	R (ft)	D	R (ft)	D	R (ft)	D	R (ft)	D	R (ft)
46°00′	124.56	55°00′	104.17	64°00′	89.52	73°00′	78.49	82°00′	69.87
15′	123.88	15′	103.70	15′	89.18	15′	78.22	15′	69.66
30′	123.22	30′	103.24	30′	88.83	30′	77.95	30′	69.45
45′	122.56	45′	102.77	45′	88.49	45′	77.69	45′	69.24
47°00′	121.91	56°00′	102.31	65°00′	88.15	74°00′	77.43	83°00′	69.03
15′	121.26	15′	101.86	15′	87.81	15′	77.17	15′	68.82
30′	120.62	30′	101.41	30′	87.47	30′	76.91	30′	68.62
45′	119.99	45′	100.96	45′	87.14	45′	76.65	45′	68.41
48°00′	119.37	57°00′	100.52	66°00′	86.81	75°00′	76.39	84°00′	68.21
15′	118.75	15′	100.08	15′	86.48	15′	76.14	15′	68.01
30′	118.14	30′	99.64	30′	86.16	30′	75.89	30′	67.81
45′	117.53	45′	99.21	45′	85.84	45′	75.64	45′	67.61
49°00′	116.93	58°00′	98.79	67°00′	85.52	76°00′	75.39	85°00′	67.41
15′	116.34	15′	98.36	15′	85.20	15′	75.14	15′	67.21
30′	115.75	30′	97.94	30′	84.88	30′	74.90	30′	67.01
45′	115.17	45′	97.52	45′	84.57	45′	74.65	45′	66.82
50°00′	114.59	59°00	97.11	68°00′	84.26	77°00′	74.41	86°00′	66.62
15′	114.02	15′	96.70	15′	83.95	15′	74.17	15′	66.43
30′	113.46	30′	96.30	30′	83.64	30′	73.93	30′	66.24
45′	112.90	45′	95.89	45′	83.34	45′	73.69	45′	66.05
51°00′	112.34	60°00′	95.49	69°00′	83.04	78°00′	73.46	87°00′	65.86
15′	111.80	15′	95.10	15′	82.74	15′	73.22	15′	65.67
30′	111.25	30′	94.70	30′	82.44	30′	72.99	30′	65.48
45′	110.72	45′	94.31	45′	82.14	45′	72.76	45′	65.29
52°00′	110.18	61°00′	93.93	70°00′	81.85	79°00′	72.53	88°00′	65.11
15′	109.66	15′	93.54	15′	81.56	15′	72.30	15′	64.92
30′	109.13	30′	93.16	30′	81.27	30′	72.07	30′	64.74
45′	108.62	45′	92.79	45′	80.98	45′	71.84	45′	64.56
53°00′	108.11	62°00′	92.41	71°00′	80.70	80°00′	71.62	89°00	64.38
15′	107.60	15′	92.04	15′	80.42	15′	71.40	15′	64.20
30′	107.09	30′	91.67	30′	80.13	30′	71.17	30′	64.02
45′	106.60	45′	91.31	45′	79.85	45′	70.95	45′	63.84
54°00′	106.10	63°00′	90.95	72°00′	79.58	81°00′	70.74	90°00′	63.66
15′	105.61	15′	90.59	15′	79.30	15′	70.52	15′	63.49
30′	105.13	30′	90.23	30′	79.03	30′	70.30	30′	63.31
45′	104.65	45′	89.88	45′	78.76	45′	70.09	45′	63.14

APPENDIX C
Chord Lengths of Circular Arcs (Arc Definition)

$$C = 2R\sin\left(\frac{D}{2}\right) = \frac{(2)(360°)(100\text{ ft})}{2\pi D}\sin\left(\frac{D}{2}\right) = \frac{5729.578}{D}\sin\left(\frac{D}{2}\right)$$

(all distances in feet)

degree of curve	for arc of 100	95	90	85	80	75	70	60	degree of curve
1°	100	95	90	85	80	75	70	60	1°
2°	100	95	90	85	80	75	70	60	2°
3°	99.99	94.99	89.99	85	80	75	70	60	3°
4°	99.98	94.98	89.98	84.99	79.99	74.99	70	60	4°
5°	99.97	94.97	89.98	84.98	79.98	74.99	69.99	59.99	5°
6°	99.95	94.96	89.97	84.97	79.98	74.98	69.98	59.99	6°
7°	99.94	94.95	89.96	84.96	79.96	74.97	69.98	59.99	7°
8°	99.92	94.93	89.94	84.95	79.96	74.97	69.97	59.98	8°
9°	99.90	94.91	89.93	84.94	79.95	74.96	69.96	59.98	9°
10°	99.87	94.89	89.91	84.92	79.94	74.95	69.96	59.97	10°

degree of curve	for arc of 100	60	55	50	45	40	35	30	degree of curve
11°	99.85	59.97	54.97	49.98	44.99	39.99	34.99	29.99	11°
12°	99.82	59.96	54.97	49.98	44.98	39.99	34.99	29.99	12°
13°	99.79	59.95	54.96	49.97	44.98	39.99	34.99	29.99	13°
14°	99.75	59.95	54.96	49.97	44.98	39.98	34.99	29.99	14°
15°	99.71	59.94	54.95	49.96	44.97	39.98	34.99	29.99	15°
16°	99.68	59.93	54.95	49.96	44.97	39.98	34.99	29.99	16°
17°	99.63	59.92	54.94	49.95	44.97	39.98	34.98	29.99	17°
18°	99.59	59.91	54.93	49.95	44.96	39.97	34.98	29.99	18°
19°	99.54	59.90	54.92	49.94	44.96	39.97	34.98	29.99	19°
20°	99.49	59.89	54.92	49.94	44.95	39.97	34.98	29.99	20°
21°	99.44	59.88	54.91	49.93	44.95	39.96	34.98	29.98	21°
22°	99.39	59.87	54.90	49.92	44.94	39.96	34.97	29.98	22°
23°	99.33	59.85	54.89	49.92	44.94	39.96	34.97	29.98	23°
24°	99.27	59.84	54.88	49.91	44.93	39.95	34.97	29.98	24°
25°	99.21	59.83	54.87	49.90	44.93	39.95	34.97	29.98	25°

Support Material

APPENDIX D

Conversion Factors

multiply	by	to obtain
acres	0.4047	hectares
	43,560.0	square feet
	1.5625×10^{-3}	square miles
ampere-hours	3600.0	coulombs
angstrom units	3.937×10^{-9}	inches
	1×10^{-4}	microns
astronomical units	1.496×10^{8}	kilometers
atmospheres	76.0	centimeters of mercury
atomic mass unit	9.316×10^{8}	electron-volts
	1.492×10^{-10}	joules
	1.66×10^{-27}	kilograms
BeV (also GeV)	1×10^{9}	electron-volts
Btu	3.93×10^{-4}	horsepower-hours
	778.3	foot-pounds
	2.93×10^{-4}	kilowatt hours
	1.0×10^{-5}	therms
Btu/hr	0.2161	foot-pounds/sec
	3.929×10^{-4}	horsepower
	0.293	watts
bushels	2150.4	cubic inches
calories, gram (mean)	3.9683×10^{-3}	Btu (mean)
centares	1.0	square meters
centimeters	1×10^{-5}	kilometers
	1×10^{-2}	meters
	10.0	millimeters
	3.281×10^{-2}	feet
	0.3937	inches
chains	792.0	inches
coulombs	1.036×10^{-5}	faradays
cubic centimeters	0.06102	cubic inches
	2.113×10^{-3}	pints (U.S. liquid)
cubic feet	0.02832	cubic meters
	7.4805	gallons
cubic feet/min	62.43	pounds H_2O/min
cubic feet/sec	448.831	gallons/min
	0.64632	millions of gallons per day
cubits	18.0	inches
days	86,400.0	seconds
degrees (angle)	1.745×10^{-2}	radians
degrees/sec	0.1667	revolutions/min
dynes	1×10^{-5}	newtons
electron-volts	1.074×10^{-9}	atomic mass units
	1×10^{-9}	BeV (also GeV)
	1.602×10^{-19}	joules
	1.783×10^{-36}	kilograms
	1×10^{-6}	MeV
faradays/sec	96,500	amperes (absolute)
fathoms	6.0	feet
feet (international)	30.48	centimeters
	0.3048	meters
feet (U.S. Survey)	1200/39.37	centimeters
	12/39.37	meters
	1.645×10^{-4}	miles (nautical)
	1.894×10^{-4}	miles (statute)
feet/min	0.5080	centimeters/sec
feet/sec	0.592	knots
	0.6818	miles/hr
foot-pounds	1.285×10^{-3}	Btu
	5.051×10^{-7}	horsepower-hours
	3.766×10^{-7}	kilowatt-hours
foot-pound/sec	4.6272	Btu/hr
	1.818×10^{-3}	horsepower
	1.356×10^{-3}	kilowatts
furlongs	660.0	feet
	0.125	miles (statute)
gallons	0.1337	cubic feet
	3.785	liters
gallons H_2O	8.3453	pounds H_2O
gallons/min	8.0208	cubic feet/hr
	0.002228	cubic feet/sec
GeV (also BeV)	1×10^{9}	electron-volts
grams	1×10^{-3}	kilograms
	3.527×10^{-2}	ounces (avoirdupois)
	3.215×10^{-2}	ounces (troy)
	2.205×10^{-3}	pounds
hectares	2.471	acres
	1.076×10^{5}	square feet
horsepower	2545.0	Btu/hr
	42.44	Btu/min
	550	foot-pounds/sec
	0.7457	kilowatts
	745.7	watts
horsepower-hours	2545.0	Btu
	1.976×10^{-6}	foot-pounds
	0.7457	kilowatt-hours
hours	4.167×10^{-2}	days
	5.952×10^{-3}	weeks
inches	2.540	centimeters
	1.578×10^{-5}	miles
inches, H_2O	5.199	pounds force/ft^2
	0.0361	psi
	0.0735	inches, mercury
inches, mercury	70.7	pounds force/ft^2
	0.491	pounds force/in^2
	13.60	inches, H_2O
joules	6.705×10^{9}	atomic mass units
	9.480×10^{-4}	Btu
	1×10^{7}	ergs
	6.242×10^{18}	electron-volts
	1.113×10^{-17}	kilograms
kilograms	6.025×10^{26}	atomic mass units
	5.610×10^{35}	electron-volts
	8.987×10^{16}	joules
	2.205	pounds
kilometers	3281.0	feet
	1000.0	meters
	0.6214	miles
kilometers/hr	0.5396	knots

(continued)

APPENDIX D *(continued)*
Conversion Factors

multiply	by	to obtain
kilowatts	3412.9	Btu/hr
	737.6	foot-pounds/sec
	1.341	horsepower
kilowatt-hours	3413.0	Btu
knots	6076.0	feet/hr
	1.0	nautical miles/hr
	1.151	statute miles/hr
light years	5.9×10^{12}	miles
links (surveyor)	7.92	inches
liters	1000.0	cubic centimeters
	61.02	cubic inches
	0.2642	gallons (U.S. liquid)
	1000.0	milliliters
	2.113	pints
MeV	1×10^{6}	electron-volts
meters	100.0	centimeters
	1/0.3048	feet (international)
	39.37/12	feet (U.S. Survey)
	1×10^{-3}	kilometers
	5.396×10^{-4}	miles (nautical)
	6.214×10^{-4}	miles (statute)
	1000.0	millimeters
microns	1×10^{-6}	meters
miles (nautical)	6076	feet
	1.853	kilometers
	1.1516	miles (statute)
miles (statute)	5280.0	feet
	1.609	kilometers
	0.8684	miles (nautical)
miles/hr	88.0	feet/min
milligrams/liter	1.0	parts/million
milliliters	1×10^{-3}	liters
millimeters	3.937×10^{-2}	inches
newtons	1×10^{5}	dynes
ohms (international)	1.0005	ohms (absolute)
ounces	28.349527	grams
	6.25×10^{-2}	pounds
ounces (troy)	1.09714	ounces (avoirdupois)
parsecs	3.086×10^{13}	kilometers
	1.9×10^{13}	miles
pascal-sec	1000	centipoise
	10	poise
	0.02089	pound force-sec/ft^2
	0.6720	pound mass/ft-sec
	0.02089	slug/ft-sec
pints (liquid)	473.2	cubic centimeters
	28.87	cubic inches
	0.125	gallons
	0.5	quarts (liquid)
poise	0.002089	pound-sec/ft^2
pounds	0.4536	kilograms
	16.0	ounces
	14.5833	ounces (troy)
	1.21528	pounds (troy)
pounds/ft^2	0.006944	pounds/in^2
pounds/in^2	2.308	feet, H_2O
	27.7	inches, H_2O
	2.037	inches, mercury
	144	pounds/ft^2
quarts (dry)	67.20	cubic inches
quarts (liquid)	57.75	cubic inches
	0.25	gallons
	0.9463	liters
radians	57.30	degrees
	3438.0	minutes
revolutions	360.0	degrees
revolutions/min	6.0	degrees/sec
rods	16.5	feet
	5.029	meters
rods (surveyor)	5.5	yards
seconds	1.667×10^{-2}	minutes
square meters/sec	1×10^{6}	centistokes
	10.76	square feet/sec
	1×10^{4}	stokes
slugs	32.174	pounds mass
stokes	0.0010764	square feet/sec
tons (long)	1016.0	kilograms
	2240.0	pounds
	1.120	tons (short)
tons (short)	907.1848	kilograms
	2000.0	pounds
	0.89287	tons (long)
volts (absolute)	3.336×10^{-3}	statvolts
watts	3.4129	Btu/hr
	1.341×10^{-3}	horsepower
yards	0.9144	meters
	4.934×10^{-4}	miles (nautical)
	5.682×10^{-4}	miles (statute)

APPENDIX E

Surveying Conversion Factors

multiply	by	to obtain
ac	43,560	ft^2
	10	$chains^2$
	4046.87	m^2
ac-ft	43,560	ft^3
	1233.49	m^3
chain	66	ft
	22	yd
	4	rods
day (mean solar)	86,400	sec
day (sidereal)	86,164.09	sec
deg (angle)	0.0174533	rad
	17.77778	mils
engineer's link	1	ft
ft (U.S. Survey)	0.3048006	m
grads	0.9	degrees (angle)
	0.01570797	rad
hectare	2.47104	ac
	10,000	m^2
in	25.4	mm
labors	177.14	ac
leagues	4428.40	ac
link (see engineer's link and surveyor's link)		
mils	0.05625	degrees (angle)
	3,037,500	min
mi (statute)	5280	ft
	80	chains (surveyor's)
	320	rods
	0.86839	mi (nautical)
mi^2	640	ac
	27,878,400	ft^2
min (angle)	0.29630	mils
	0.000290888	rad
min (mean solar)	60	sec
min (sidereal)	59.83617	sec
outs	330	ft
	10	33 ft chains
rad	57.2957795	degrees (angle)
	$57°17'44.806''$	degrees (angle)
rods	16.5	ft
	1	perches
	1	poles
sec (angle)	4.848137×10^{-6}	rad
sec (sidereal)	0.9972696	sec (mean solar)
surveyor's link	0.66	ft
	7.92	in
VARA (California)	33	in
VARA (Texas)	$33\frac{1}{3}$	in
yd (U.S.)	0.914402	m

Support Material

APPENDIX F
Miscellaneous Constants and Conversions

0.0000001 per °F	= coefficient of expansion for invar tape
0.00000645 per °F	= coefficient of expansion for steel tape
0.6745	= coefficient for 50% standard deviation
1.15 mi	= 1 minute of latitude
1.6449	= coefficient for 90% standard deviation
6 mi	= length and width of township
10 square chains	= 1 ac
15° longitude	= width of one time zone
23°26.5′	= maximum declination of the sun at solstice
24 hr	= 360° of longitude
36	= number of sections in a township
69.1 mi	= 1° latitude
100	= usual stadia ratio
101 ft	= 1 second of latitude
400 grads	= 360°
480 chains	= width and length of township
640 ac	= 1 normal section
4046.9 mi^2	= 1 ac
6400 mils	= 360°
43,560 ft^2	= 1 ac
20,906,000 ft	= mean radius of earth

APPENDIX G
Glossary

Not all words are covered in this glossary.
Refer to the index for additional words that are discussed in the text.

Abstract – A summary of facts.

Abstract of title – A condensed history of the title to land.

Accessory to corner – A physical object that is adjacent to a corner. An accessory is usually considered part of the monument.

Acclivity – An upward slope of ground.

Accretion – The gradual accumulation of land by natural causes.

Acknowledgment – A declaration by a person before an official (usually a notary public) that he or she executed a legal document.

Acquiescence – Implied consent to a transaction, to the accrual of a right, or to any act, by one's silence (or without express assent).

Adjudication – The giving or pronouncing of a judgment or decree.

Adverse possession – A method of acquiring property by holding it for a period of time under certain conditions.

Affidavit – A written declaration under oath before an authorized official (usually a notary public).

Alienation – The transfer of property and/or possessions from one person to another.

Aliquot – A portion contained in something else a whole number of times.

Alluvium – Sand or soil deposited by streams.

Appellant – The party that takes an appeal from one court or jurisdiction to another.

Appurtenance – A right, privilege, or improvement belonging to and passing with a piece of property when it is conveyed.

Assigns – Those to whom property is transferred.

Avulsion – A sudden and perceptible change of shoreline by the violent action of water.

Bayou – An outlet from a swamp or lagoon to the sea.

Bed of stream – The depression between the banks of a water course worn by the regular and usual flow of the water.

Bequest – A gift by will of personal property.

Bounty lands – Portions of the public domain given or donated as a bounty for services rendered.

Chain of title – A chronological list of documents that comprise the record history of title of real property.

Civil law – That part of the law pertaining to civil rights, as distinguished from criminal law. Civil law and Roman law have the same meaning. In contradistinction to English common law, civil law is enacted by legislative bodies.

Clear title – Good title. Title free from encumbrances.

Cloud on title – A claim or encumbrance on a title to land that may or may not be valid.

Color of title – Any written instrument that appears to convey title, even though it does not.

Common law – Principles and rules of action determined by court decisions that have been accepted by generation after generation, and that are distinguished from laws enacted by legislative bodies.

Consideration – Something of value given to make an agreement binding.

Conveyance – Any instrument in writing by which an interest in real property is transferred.

Covenant – When used in deeds, restrictions imposed on the grantee as to the use of land conveyed.

Crown – The sovereign power in a monarchy.

Cut bank – The watershed and relatively permanent elevation or acclivity that separates the bed of a river from its adjacent upland.

Decree – A judgment by the court in a legal proceeding.

Dedication – An appropriation of land to some public use made by the owner, and accepted for such use by or on behalf of the public.

Deed – Evidence in writing of the transfer of real property.

Deed of trust – An instrument taking the place of a mortgage, by which the legal title to real property is placed in one or more trustees to secure repayment of a sum of money.

Demurrer – In legal pleading, the formal mode of disputing the sufficiency of the pleading of the other side.

Devise – A gift of real property by the last will and testament of the donor.

Easement – The right that the public, an individual, or individuals have in the lands of another.

Egress – The right or permission to go out from a place; right of exit.

Eminent domain – The right or power of government or certain other agencies to take private property for public use on payment of just compensation to the owner.

Support Material

APPENDIX G *(continued)*
Glossary

Encroachment – An obstruction that intrudes upon the land of another. The gradual, stealthy, illegal acquisition of property.

Encumbrance – Any burden or claim on property, such as a mortgage or delinquent taxes.

Equity – The excess of the market value over any indebtedness.

Erosion – The process by which the surface of the earth is worn away by the action of waters, glaciers, wind, or waves.

Escheat – Reversion of property to the state where there is no competent or available person to inherit it.

Escrow – Something placed in the keeping of a third person for delivery to a given party upon fulfillment of some condition.

Estate – An interest in property, real or personal.

Estoppel – A bar or impediment that precludes allegation or denial of a certain fact or state of facts in consequence of a final adjudication.

Et al. – An abbreviation for "and others."

Et Mode Ad Hune Diem – An abbreviation for "and now at this day."

Et ux – An abbreviation for "and wife."

Evidence aliunde – Evidence from outside or from another source.

Extrinsic evidence – Evidence NOT contained in the deed, but offered to clear up an ambiguity found to exist when applying the description to the ground.

Grant – A transfer of property.

Grantee – The person to whom a grant is made.

Grantor – The person by whom a grant is made.

Good faith – An honest intention to abstain from taking advantage of another.

Gradient – An inclined surface. The change in elevation per unit of horizontal distance.

Hereditament – Something capable of being inherited, be it real or personal property.

Hiatus – An area between two surveys of record described as having one or more common boundary lines with no omission.

Holograph – A will written entirely by the testator in his or her own handwriting.

Incumbrance – A right, interest in, or legal liability upon real property that does not prohibit passing title to the land but that diminishes its value.

Ingress – The right or permission to go upon a place; right of entrance.

Intent – The true meaning (from the written words of an instrument).

Intestate – Without making a will.

Judgment – The official and authentic decision of a court of justice.

Leasehold – An estate in realty held under a lease; an estate for a fixed term of years.

Lessee – The person to whom a lease is made.

Lessor – The person who grants a lease.

Lien – A claim or charge on property for payment of some debt, obligation, or duty.

Lis pendens – A pending suit. A notice of lis pendens is filed for the purpose of warning all persons that a suit is pending.

Litigation – Contest in a court of justice for the purpose of enforcing a right.

Littoral – Belonging to the shore, as of seas and lakes.

Logical relevancy – A relationship in logic between the fact for which evidence is offered and a fact in issue such that the existence of the former renders probable or improbable the existence of the latter.

Mean – Intermediate; the middle between two extremes.

Memorial – That which contains the particulars of a deed, and so on. In practice, a memorial is a short note, abstract, memorandum, or rough draft of the orders of the court, from which the records thereof may at any time be fully made up.

Mortgage – A conditional conveyance of an estate as a pledge for the security of a debt.

Muniment – Documentary evidence of title.

Option – The right as granted in a contract or by an initial payment of acquiring something in the future.

Parcel – A part of a piece of land that cannot be identified by a lot or tract number.

Parol evidence – Evidence that is given verbally.

Patent – A government grant of land. The instrument by which a government conveys title to land.

Plat – A scaled diagram showing boundaries of a tract of land or subdivisions. May constitute a legal description of the land and be used in lieu of a written description.

Power of attorney – A written document given by one person to another authorizing the latter to act for the former.

Prescription – Creation of an easement under claim of right by use of land that has been open, continuous, and exclusive for a period of time prescribed by law.

APPENDIX G *(continued)*
Glossary

Prima facie evidence – Facts presumed to be true unless disproved by evidence to the contrary.

Privity – The relationship that exists between parties to a contract. Mutual or successive relationship to the same rights of property, such as the relationship of heir with ancestor or donee with donor.

Privy – A person who is in privity with another.

Probate – The act or process of validating a will.

Quiet title – Action of law to remove an adverse claim or cloud on title.

Quitclaim deed – A conveyance that passes any title, interest, or claim that the grantor may have.

Reliction – A gradual and imperceptible recession of water, resulting in increased shoreline, beach, or property.

Relinquishment – The forsaking, abandonment, renouncement, or gift of a right.

Remand – To send a cause back to the same court out of which it came for the purpose of having some action taken upon it there.

Riparian – Belonging or relating to the bank of a river.

Royalty – A share of the profit from sale of minerals paid to the owner of the property by the lessee.

Said – Refers to one previously mentioned.

Scrivener – A person whose occupation is to draw up contracts, write deeds and mortgages, and prepare other written instruments.

Shore – The place lying between the line of ordinary high tide and the line of lowest tide.

Sovereign – A person, body, or state in which independent and supreme authority is vested.

Squatter – One who settles on another's land without legal authority.

Statute – A particular law established by the legislative branch of government.

Statutory – Relating to a statute.

Submerged land – In tidal areas, land that extends seaward from the shore and is continuously covered during the ebb and flow of the tide.

Substantive evidence – Evidence used to prove a fact (as opposed to evidence given for the purpose of discrediting a claim).

Tenancy by the entirety – Husband and wife each possesses the entire estate in order that, upon the death of either spouse, the survivor is entitled to the estate in its entirety.

Tenancy, Joint – The holding of property by two or more persons, each of whom has an undivided interest. After the death of one of the joint tenants, the surviving tenant(s) receive the descendent's share.

Tenant – One who has the temporary use and occupation of real property owned by another person (the landlord).

Tenements – Property held by a tenant. Everything of a permanent nature. In a more restrictive sense, a house or dwelling.

Testament – A will of personal property.

Testator – One who makes a testament or will. One who dies leaving a will.

Thalweg – The deepest part of a channel.

Thence – From that place; the following course is continuous from the one before it.

Title policy – Insurance against loss or damage resulting from defects or failure of title to a particular parcel of land.

To wit – That is to say; namely.

Upland – Land above mean high water and subject to private ownership (as distinguished from tidelands, which are in the state). Also used as meaning nonriparian.

Watercourse – A running stream of water fed from permanent or natural sources running in a particular direction and having a channel formed by a well-defined bed and banks (though it need not flow continuously).

Warranty deed – A deed in which the grantor proclaims that he or she is the lawful owner of real property and will forever defend the grantee against any claim on the property.

Will – The legal declaration of a person's wishes as to the disposition of his or her property after his or her death.

Witness mark – A mark placed at a known location to aid in recovery and identification of a monument or corner.

Writ – A mandatory order issued from a court of justice.

Writ of coram nobis – A common law writ, the purpose of which is to correct an error in a judgment in the same court in which it was rendered.

APPENDIX H
Areas under the Standard Normal Curve
(0 to z)

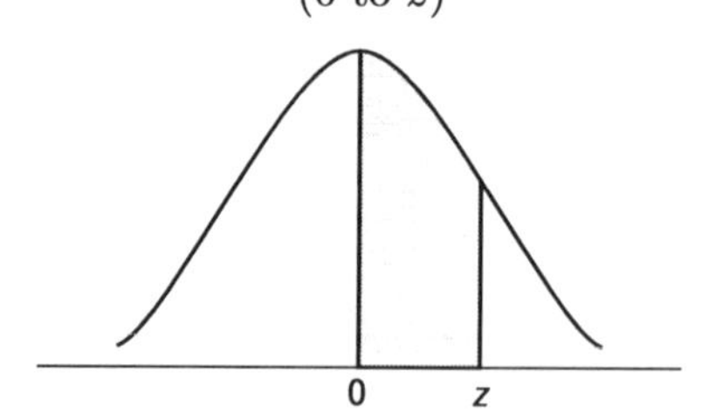

z	0	1	2	3	4	5	6	7	8	9
0.0	0.0000	0.0040	0.0080	0.0120	0.0160	0.0199	0.0239	0.0279	0.0319	0.0359
0.1	0.0398	0.0438	0.0478	0.0517	0.0557	0.0596	0.0636	0.0675	0.0714	0.0754
0.2	0.0793	0.0832	0.0871	0.0910	0.0948	0.0987	0.1026	0.1064	0.1103	0.1141
0.3	0.1179	0.1217	0.1255	0.1293	0.1331	0.1368	0.1406	0.1443	0.1480	0.1517
0.4	0.1554	0.1591	0.1628	0.1664	0.1700	0.1736	0.1772	0.1808	0.1844	0.1879
0.5	0.1915	0.1950	0.1985	0.2019	0.2054	0.2088	0.2123	0.2157	0.2190	0.2224
0.6	0.2258	0.2291	0.2324	0.2357	0.2389	0.2422	0.2454	0.2486	0.2518	0.2549
0.7	0.2580	0.2612	0.2642	0.2673	0.2704	0.2734	0.2764	0.2794	0.2823	0.2852
0.8	0.2881	0.2910	0.2939	0.2967	0.2996	0.3023	0.3051	0.3078	0.3106	0.3133
0.9	0.3159	0.3186	0.3212	0.3238	0.3264	0.3289	0.3315	0.3340	0.3365	0.3389
1.0	0.3413	0.3438	0.3461	0.3485	0.3508	0.3531	0.3554	0.3577	0.3599	0.3621
1.1	0.3643	0.3665	0.3686	0.3708	0.3729	0.3749	0.3770	0.3790	0.3810	0.3830
1.2	0.3849	0.3869	0.3888	0.3907	0.3925	0.3944	0.3962	0.3980	0.3997	0.4015
1.3	0.4032	0.4049	0.4066	0.4082	0.4099	0.4115	0.4131	0.4147	0.4162	0.4177
1.4	0.4192	0.4207	0.4222	0.4236	0.4251	0.4265	0.4279	0.4292	0.4306	0.4319
1.5	0.4332	0.4345	0.4357	0.4370	0.4382	0.4394	0.4406	0.4418	0.4429	0.4441
1.6	0.4452	0.4463	0.4474	0.4484	0.4495	0.4505	0.4515	0.4525	0.4535	0.4545
1.7	0.4554	0.4564	0.4573	0.4582	0.4591	0.4599	0.4608	0.4616	0.4625	0.4633
1.8	0.4641	0.4649	0.4656	0.4664	0.4671	0.4678	0.4686	0.4693	0.4699	0.4706
1.9	0.4713	0.4719	0.4726	0.4732	0.4738	0.4744	0.4750	0.4756	0.4761	0.4767
2.0	0.4772	0.4778	0.4783	0.4788	0.4793	0.4798	0.4803	0.4808	0.4812	0.4817
2.1	0.4821	0.4826	0.4830	0.4834	0.4838	0.4842	0.4846	0.4850	0.4854	0.4857
2.2	0.4861	0.4864	0.4868	0.4871	0.4875	0.4878	0.4881	0.4884	0.4887	0.4890
2.3	0.4893	0.4896	0.4898	0.4901	0.4904	0.4906	0.4909	0.4911	0.4913	0.4916
2.4	0.4918	0.4920	0.4922	0.4925	0.4927	0.4929	0.4931	0.4932	0.4934	0.4936
2.5	0.4938	0.4940	0.4941	0.4943	0.4945	0.4946	0.4948	0.4949	0.4951	0.4952
2.6	0.4953	0.4955	0.4956	0.4957	0.4959	0.4960	0.4961	0.4962	0.4963	0.4964
2.7	0.4965	0.4966	0.4967	0.4968	0.4969	0.4970	0.4971	0.4972	0.4973	0.4974
2.8	0.4974	0.4975	0.4976	0.4977	0.4977	0.4978	0.4979	0.4979	0.4980	0.4981
2.9	0.4981	0.4982	0.4982	0.4983	0.4984	0.4984	0.4985	0.4985	0.4986	0.4986
3.0	0.4987	0.4987	0.4987	0.4988	0.4988	0.4989	0.4989	0.4989	0.4990	0.4990
3.1	0.4990	0.4991	0.4991	0.4991	0.4992	0.4992	0.4992	0.4992	0.4993	0.4993
3.2	0.4993	0.4993	0.4994	0.4994	0.4994	0.4994	0.4994	0.4995	0.4995	0.4995
3.3	0.4995	0.4995	0.4996	0.4996	0.4996	0.4996	0.4996	0.4996	0.4996	0.4997
3.4	0.4997	0.4997	0.4997	0.4997	0.4997	0.4997	0.4997	0.4997	0.4997	0.4998
3.5	0.4998	0.4998	0.4998	0.4998	0.4998	0.4998	0.4998	0.4998	0.4998	0.4998
3.6	0.4998	0.4998	0.4999	0.4999	0.4999	0.4999	0.4999	0.4999	0.4999	0.4999
3.7	0.4999	0.4999	0.4999	0.4999	0.4999	0.4999	0.4999	0.4999	0.4999	0.4999
3.8	0.4999	0.4999	0.4999	0.4999	0.4999	0.4999	0.4999	0.4999	0.4999	0.4999
3.9	0.5000	0.5000	0.5000	0.5000	0.5000	0.5000	0.5000	0.5000	0.5000	0.5000

APPENDIX I

Representative Plane Coordinate Projection Tables for Texas*

Constants for Texas zones

Constant	Zone	
	North	North Central
C	2,000,000.00	2,000,000.00
Central Meridian	101° 30' 00".000	97° 30' 00".000
R_D	29,972,959.94	32,691,654.54
y_o	516,052.65	503,844.96
ℓ	0.57953 58654	0.54539 44146
$\dfrac{1}{2\rho_o^2 \sin 1''}$	2.360 × 10^{-10}	2.362 × 10^{-10}
$\log \dfrac{1}{2\rho_o^2 \sin 1''}$	0.373 0025 – 10	0.373 2288 – 10
$\log \ell$	9.76308 03181 – 10	9.73671 06860 – 10
log K	7.63475 78652	7.65172 89823

Constant	Zone
	Central
C	2,000,000.00
Central Meridian	100° 20' 00".000
R_D	35,337,121.23
y_o	485,417.77
ℓ	0.51505 88857
$\dfrac{1}{2\rho_o^2 \sin 1''}$	2.363 × 10^{-10}
$\log \dfrac{1}{2\rho_o^2 \sin 1''}$	0.373 4182 – 10
$\log \ell$	9.71185 68921 – 10
log K	7.66885 39642

Lambert Projection for Texas (North Central)
Table I (Cont'd)

Lat.	R (feet)	Y' y value on central meridian (feet)	Tabular difference for 1 sec. of lat. (feet)	Scale in units of 7th place of logs	Scale expressed as a ratio
33° 26'	32048935.45	642719.09	101.06550	–457.1	0.9998947
27	32042871.52	648783.02	101.06617	–448.6	0.9998967
28	32036807.55	654846.99	101.06650	–439.6	0.9998988
29	32030743.56	660910.98	101.06700	–430.3	0.9999009
30	32024679.54	666975.00	101.06750	–420.6	0.9999032
33° 31'	32018615.49	673039.05	101.06817	–410.6	0.9999055
32	32012551.40	679103.14	101.06850	–400.2	0.9999079
33	32006487.29	685167.25	101.06917	–389.4	0.9999103
34	32000423.14	691231.40	101.06967	–378.2	0.9999129
35	31994358.96	697295.58	101.07017	–366.7	0.9999156
33° 36'	31988294.75	703359.79	101.07067	–354.8	0.9999183
37	31982230.51	709424.03	101.07133	–342.6	0.9999211
38	31976166.23	715488.31	101.07183	–329.9	0.9999240
39	31970101.92	721552.62	101.07250	–316.9	0.9999270
40	31965037.57	727616.97	101.07317	–303.6	0.9999301
33° 41'	31957973.18	733681.36	101.07350	–289.8	0.9999333
42	31951908.77	739745.77	101.07433	–275.7	0.9999365
43	31945844.31	745810.23	101.07483	–261.3	0.9999398
44	31939779.82	751874.72	101.07550	–246.4	0.9999433
45	31933715.29	757939.25	101.07617	–231.2	0.9999468
33° 46'	31927650.72	764003.82	101.07683	–215.7	0.9999503
47	31921586.11	770068.43	101.07750	–199.7	0.9999540
48	31915521.46	776133.08	101.07800	–183.4	0.9999578
49	31909456.78	782197.76	101.07883	–166.7	0.9999616
50	31903392.05	788262.49	101.07950	–149.7	0.9999655
33° 51'	31897327.28	794327.26	101.08017	–132.3	0.9999695
52	31891262.47	800392.07	101.08083	–114.5	0.9999736
53	31885197.62	806456.92	101.08150	–96.3	0.9999778
54	31879132.73	812521.81	101.08217	–77.8	0.9999821
55	31873067.80	818586.74	101.08300	–58.9	0.9999864
33° 56'	31867002.82	824651.72	101.08367	–39.6	0.9999909
57	31860937.80	830716.74	101.08450	–20.0	0.9999954
58	31854872.73	836781.81	101.08517	0	1.0000000
59	31848807.62	842846.92	101.08600	+20.4	1.0000047
34° 00'	31842742.46	848912.08	101.08667	+41.1	1.0000095

Lambert Projection for Texas (North Central)
Table II (Cont'd)
1" of Long. = 0".54539441 of Θ

Long.	Θ	Long.	Θ	Long.	Θ
98° 31'	–0° 33' 16".1436	99° 06'	–0° 52' 21".4718	99° 41'	–1° 11' 26".8001
32	–0 33 48.8672	07	–0 52 54.1955	42	–1 11 59.5238
33	–0 34 21.5909	08	–0 53 26.9192	43	–1 12 32.2474
34	–0 34 54.3146	09	–0 53 59.6428	44	–1 13 04.9711
35	–0 35 27.0382	10	–0 54 32.3665	45	–1 13 37.6948
98° 36'	–0 35 59.7619	99° 11'	–0 55 05.0902	99° 46'	–1 14 10.4184
37	–0 36 32.4855	12	–0 55 37.8138	47	–1 14 43.1421
38	–0 37 05.2092	13	–0 56 10.5375	48	–1 15 15.8658
39	–0 37 37.9329	14	–0 56 43.2611	49	–1 15 48.5894
40	–0 38 10.6565	15	–0 57 15.9848	50	–1 16 21.3131
98° 41'	–0 38 43.3802	99° 16'	–0 57 48.7085	99° 51'	–1 16 54.0367
42	–0 39 16.1039	17	–0 58 21.4321	52	–1 17 26.7604
43	–0 39 48.8275	18	–0 58 54.1558	53	–1 17 59.4841
44	–0 40 21.5512	19	–0 59 26.8795	54	–1 18 32.2077
45	–0 40 54.2749	20	–0 59 59.6031	55	–1 19 04.9314
98° 46'	–0 41 26.9985	99° 21'	–1 00 32.3268	99° 56'	–1 19 37.6551
47	–0 41 59.7222	22	–1 01 05.0505	57	–1 20 10.3787
48	–0 42 32.4459	23	–1 01 37.7741	58	–1 20 43.1024
49	–0 43 05.1695	24	–1 02 10.4978	59	–1 21 15.8261
50	–0 43 37.8932	25	–1 02 43.2215	100° 00	–1 21 48.5497
98° 51'	–0 44 10.6169	99° 26'	–1 03 15.9451	100° 01'	–1 22 21.2734
52	–0 44 43.3405	27	–1 03 48.6688	02	–1 22 53.9971
53	–0 45 16.0642	28	–1 04 21.3925	03	–1 23 26.7207
54	–0 45 48.7878	29	–1 04 54.1161	04	–1 23 59.4444
55	–0 46 21.5115	30	–1 05 26.8398	05	–1 24 32.1681
98° 56'	–0 46 54.2352	99° 31'	–1 05 59.5635	100° 06'	–1 25 04.8917
57	–0 47 26.9588	32	–1 06 32.2871	07	–1 25 37.6154
58	–0 47 59.6825	33	–1 07 05.0108	08	–1 26 10.3391
59	–0 48 32.4062	34	–1 07 37.7344	09	–1 26 43.0627
99° 00'	–0 49 05.1298	35	–1 08 10.4581	10	–1 27 15.7864
99° 01'	–0 49 37.8535	99° 36'	–1 08 43.1818	100° 11'	–1 27 48.5100
02	–0 50 10.5772	37	–1 09 15.9054	12	–1 28 21.2337
03	–0 50 43.3008	38	–1 09 48.6291	13	–1 28 53.9574
04	–0 51 16.0245	39	–1 10 21.3528	14	–1 29 26.6810
05	–0 51 48.7482	40	–1 10 54.0764	15	–1 29 59.4047

Lambert Projection for Texas (Central)
Table I (Cont'd)

Lat.	R (feet)	Y' y value on central meridian (feet)	Tabular difference for 1 sec. of lat. (feet)	Scale in units of 7th place of logs	Scale expressed as a ratio
31° 26'	34694604.31	642516.92	101.03467	–390.6	0.9999101
27	34688542.23	648579.00	101.03500	–380.9	0.9999123
28	34682480.13	654641.10	101.03567	–370.9	0.9999146
29	34676417.99	660703.24	101.03617	–360.4	0.9999170
30	34670355.82	666765.41	101.03667	–349.7	0.9999195
31° 31'	34664293.62	672827.61	101.03733	–338.5	0.9999221
32	34658231.38	678889.85	101.03767	–327.0	0.9999247
33	34652169.12	684952.11	101.03833	–315.1	0.9999274
34	34646106.52	691014.41	101.03883	–302.5	0.9999303
35	34640044.49	697076.74	101.03950	–290.2	0.9999332
31° 36'	34633982.12	703139.11	101.04000	–277.2	0.9999362
37	34627919.72	709201.51	101.04050	–263.9	0.9999392
38	34621857.29	715263.94	101.04117	–250.1	0.9999424
39	34615794.82	721326.41	101.04183	–236.0	0.9999457
40	34609732.31	727388.92	101.04233	–221.6	0.9999490
31° 41'	34603669.77	733451.46	101.04300	–206.7	0.9999524
42	34597607.19	739514.04	101.04367	–191.5	0.9999559
43	34591544.57	745576.66	101.04417	–176.0	0.9999595
44	34585481.92	751639.31	101.04483	–160.0	0.9999632
45	34579419.23	757702.00	101.04567	–143.7	0.9999669
31° 46'	34573356.49	763764.74	101.04617	–127.1	0.9999707
47	34567293.72	769827.51	101.04683	–110.0	0.9999747
48	37561230.91	775890.32	101.04750	–92.6	0.9999787
49	34555168.06	781953.17	101.04833	–74.8	0.9999825
50	34549105.16	788016.07	101.04883	–56.7	0.9999869
31° 51'	34543042.23	794079.00	101.04967	–38.1	0.9999912
52	34536979.25	800141.98	101.05033	–19.3	0.9999956
53	34530916.23	806205.00	101.05100	0.0	1.0000000
54	34524853.17	812268.06	101.05183	+19.6	1.0000045
55	34518790.06	818331.17	101.05250	+39.6	1.0000091
31° 56'	34512726.91	824394.32	101.05333	+60.0	1.0000138
57	34506663.71	830457.52	101.05400	+80.7	1.0000186
58	34500600.47	836520.76	101.05433	+101.8	1.0000234
59	34494537.18	842584.05	101.05550	+123.3	1.0000284
32° 00'	34488473.85	848647.38	101.05650	+145.2	1.0000334

*Not for actual use.

(continued)

Support Material

APPENDIX I *(continued)*

Representative Plane Coordinate Projection Tables for Texas*

Lambert Projection for Texas (Central)
Table II (Cont'd)
1″ of Long. = 0″.51505889

Long.	θ	Long.	θ	Long.	θ
96° 51'	+1° 47' 38″.8384	97° 26'	+1° 29' 37″.2148	98° 01'	+1° 11' 35″.5911
52	+1 47 07.9349	27	+1 29 06.3112	02	+1 11 04.6876
53	+1 46 37.0314	28	+1 28 35.4077	03	+1 10 33.7840
54	+1 46 06.1278	29	+1 28 04.5042	04	+1 10 02.8805
55	+1 45 35.2243	30	+1 27 33.6006	05	+1 09 31.9770
96° 56'	+1 45 04.3208	97° 31'	+1 27 02.6971	98° 06'	+1 09 01.0734
57	+1 44 33.4172	32	+1 26 31.7936	07	+1 08 30.1699
58	+1 44 02.5137	33	+1 26 00.8900	08	+1 07 59.2664
59	+1 43 31.6102	34	+1 25 29.9865	09	+1 07 28.3628
97° 00	+1 43 00.7066	35	+1 24 59.0830	10	+1 06 57.4593
97° 01'	+1 42 29.8031	97° 36'	+1 24 28.1794	98° 11'	+1 06 26.5558
02	+1 41 58.8996	37	+1 23 57.2759	12	+1 05 55.6522
03	+1 41 27.9960	38	+1 23 26.3724	13	+1 05 24.7487
04	+1 40 57.0925	39	+1 22 55.4688	14	+1 04 53.8452
05	+1 40 26.1890	40	+1 22 24.5653	15	+1 04 22.9416
97° 06'	+1 39 55.2854	97° 41'	+1 21 53.6618	98° 16'	+1 03 52.0381
07	+1 39 24.3819	42	+1 21 22.7582	17	+1 03 21.1346
08	+1 38 53.4784	43	+1 20 51.8547	18	+1 02 50.2310
09	+1 38 22.5748	44	+1 20 20.9512	19	+1 02 19.3275
10	+1 37 51.6713	45	+1 19 50.0476	20	+1 01 48.4240
97° 11'	+1 37 20.7678	97° 46'	+1 19 19.1441	98° 21'	+1 01 17.5204
12	+1 36 49.8642	47	+1 18 48.2406	22	+1 00 46.6169
13	+1 36 18.9607	48	+1 18 17.3370	23	+1 00 15.7134
14	+1 35 48.0572	49	+1 17 46.4335	24	+0 59 44.8098
15	+1 35 17.1536	50	+1 17 15.5300	25	+0 59 13.9063
97° 16'	+1 34 46.2501	97° 51'	+1 16 44.6264	98° 26'	+0 58 43.0028
17	+1 34 15.3466	52	+1 16 13.7229	27	+0 58 12.0992
18	+1 33 44.4430	53	+1 15 42.8194	28	+0 57 41.1957
19	+1 33 13.5395	54	+1 15 11.9158	29	+0 57 10.2922
20	+1 32 42.6360	55	+1 14 41.0123	30	+0 56 39.3886
97° 21'	+1 32 11.7324	97° 56'	+1 14 10.1088	98° 31'	+0 56 08.4851
22	+1 31 40.8289	57	+1 13 39.2052	32	+0 55 37.5816
23	+1 31 09.9254	58	+1 13 08.3017	33	+0 55 06.6780
24	+1 30 39.0218	59	+1 12 37.3982	34	+0 54 35.7745
25	+1 30 08.1183	98° 00	+1 12 06.4946	35	+0 54 04.8710

Lambert Projection for Texas (Central)
Table II (Cont'd)
1″ of Long. = 0″.51505889

Long.	θ	Long.	θ	Long.	θ
102° 06'	–0° 54' 35″.7745	102° 41'	–1° 12' 37″.3982	103° 16'	–1° 30' 39″.0218
07	–0 55 06.6780	42	–1 13 08.3017	17	–1 31 09.9254
08	–0 55 37.5816	43	–1 13 39.2052	18	–1 31 40.8289
09	–0 56 08.4851	44	–1 14 10.1088	19	–1 32 11.7324
10	–0 56 39.3886	45	–1 14 41.0123	20	–1 32 42.6360
102° 11'	–0 57 10.2922	102° 46'	–1 15 11.9158	103° 21'	–1 33 13.5395
12	–0 57 41.1957	47	–1 15 42.8194	22	–1 33 44.4430
13	–0 58 12.0992	48	–1 16 13.7229	23	–1 34 15.3466
14	–0 58 43.0028	49	–1 16 44.6264	24	–1 34 46.2501
15	–0 59 13.9063	50	–1 17 15.5300	25	–1 35 17.1536
102° 16'	–0 59 44.8098	102° 51'	–1 17 46.4335	103° 26'	–1 35 48.0572
17	–1 00 15.7134	52	–1 18 17.3370	27	–1 36 18.9607
18	–1 00 46.6169	53	–1 18 48.2406	28	–1 36 49.8642
19	–1 01 17.5204	54	–1 19 19.1441	29	–1 37 20.7678
20	–1 01 48.4240	55	–1 19 50.0476	30	–1 37 51.6713
102° 21'	–1 02 19.3275	102° 56'	–1 20 20.9512	103° 31'	–1 38 22.5748
22	–1 02 50.2310	57	–1 20 51.8547	32	–1 38 53.4784
23	–1 03 21.1346	58	–1 21 22.7582	33	–1 39 24.3819
24	–1 03 52.0381	59	–1 21 53.6618	34	–1 39 55.2854
25	–1 04 22.9416	103° 00'	–1 22 24.5653	35	–1 40 26.1890
102° 26'	–1 04 53.8452	103° 01'	–1 22 55.4688	103° 36'	–1 40 57.0925
27	–1 05 24.7487	02	–1 23 26.3724	37	–1 41 27.9960
28	–1 05 55.6522	03	–1 23 57.2759	38	–1 41 58.8996
29	–1 06 26.5558	04	–1 24 28.1794	39	–1 42 29.8031
30	–1 06 57.4593	05	–1 24 59.0830	40	–1 43 00.7066
102° 31'	–1 07 28.3628	103° 06'	–1 25 29.9865	103° 41'	–1 43 31.6102
32	–1 07 59.2664	07	–1 26 00.8900	42	–1 44 02.5137
33	–1 08 30.1669	08	–1 26 31.7936	43	–1 44 33.4172
34	–1 09 01.0734	09	–1 27 02.6971	44	–1 45 04.3208
35	–1 09 31.9770	10	–1 27 33.6006	45	–1 45 35.2243
102° 36'	–1 10 02.8805	103° 11'	–1 28 04.5042	103° 46'	–1 46 06.1278
37	–1 10 33.7840	12	–1 28 35.4077	47	–1 46 37.0314
38	–1 11 04.6876	13	–1 29 06.3112	48	–1 47 07.9349
39	–1 11 35.5911	14	–1 29 37.2148	49	–1 47 38.8384
40	–1 12 06.4946	15	–1 30 08.1183	50	–1 48 09.7420

*Not for actual use.

Used with permission of the U.S. Department of Commerce National Geodetic Survey, Special Publication No. 252, Plane Coordinate Projection Tables—Texas, 1950 (reprinted 1979).

APPENDIX J
Mensuration of Two-Dimensional Areas

Nomenclature

A	total surface area
b	base
c	chord length
d	distance
h	height
L	length
p	perimeter
r	radius
s	side (edge) length, arc length
θ	vertex angle, in radians
ϕ	central angle, in radians

Triangle

equilateral

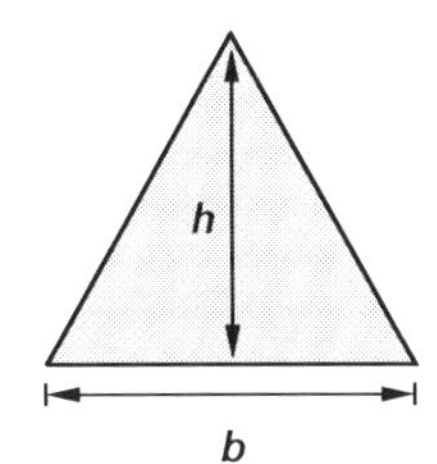

$$A = \tfrac{1}{2}bh = \frac{\sqrt{3}}{4}b^2$$

$$h = \frac{\sqrt{3}}{2}b$$

right

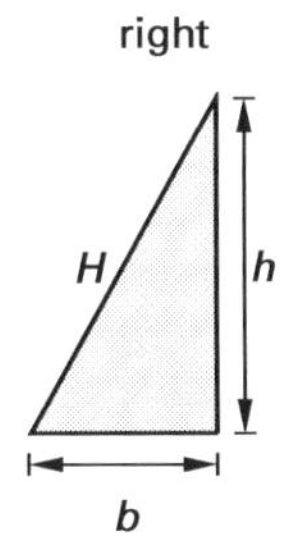

$$A = \tfrac{1}{2}bh$$

$$H^2 = b^2 + h^2$$

oblique

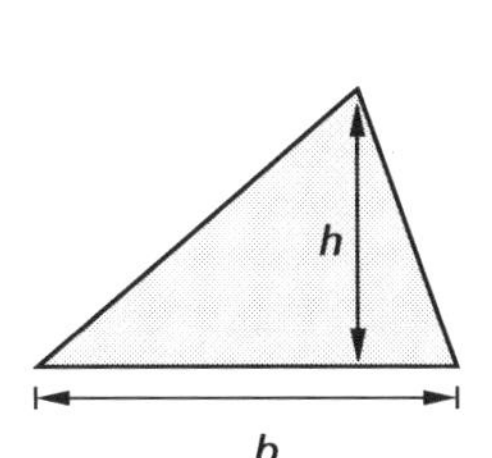

$$A = \tfrac{1}{2}bh$$

Circle

$$p = 2\pi r$$

$$A = \pi r^2 = \frac{p^2}{4\pi}$$

Circular Segment

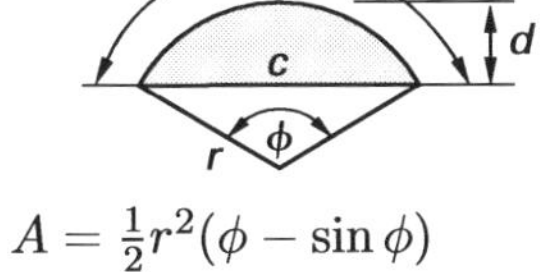

$$A = \tfrac{1}{2}r^2(\phi - \sin\phi)$$

$$\phi = \frac{s}{r} = 2\left(\arccos\frac{r-d}{r}\right)$$

$$c = 2r\sin\left(\frac{\phi}{2}\right)$$

Circular Sector

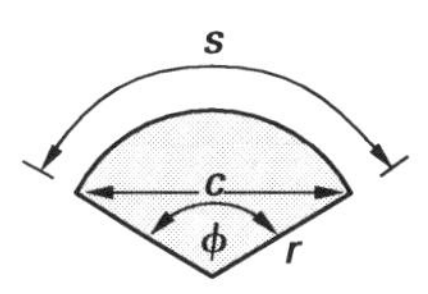

$$A = \tfrac{1}{2}\phi r^2 = \tfrac{1}{2}sr$$

$$\phi = \frac{s}{r}$$

$$s = r\phi$$

$$c = 2r\sin\left(\frac{\phi}{2}\right)$$

Parabola

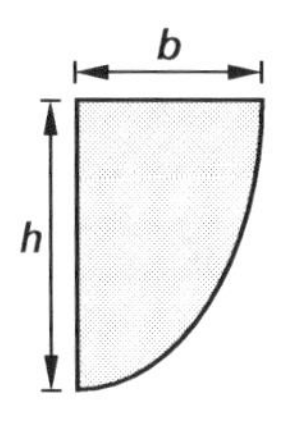

$$A = \tfrac{2}{3}bh$$

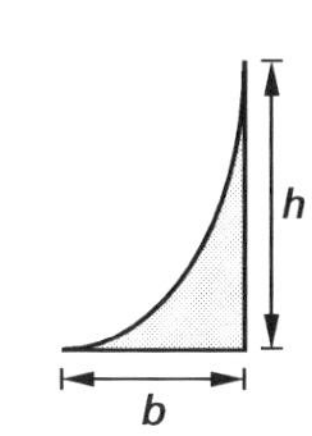

$$A = \tfrac{1}{3}bh$$

Ellipse

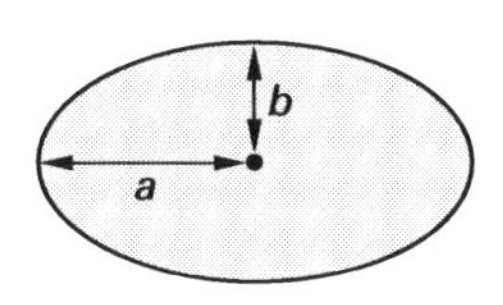

$$A = \pi ab$$

$$p \approx 2\pi\sqrt{\tfrac{1}{2}(a^2 + b^2)} \quad \left[\begin{array}{c}\text{Euler's}\\ \text{upper bound}\end{array}\right]$$

(*continued*)

APPENDIX J *(continued)*
Mensuration of Two-Dimensional Areas

Trapezoid

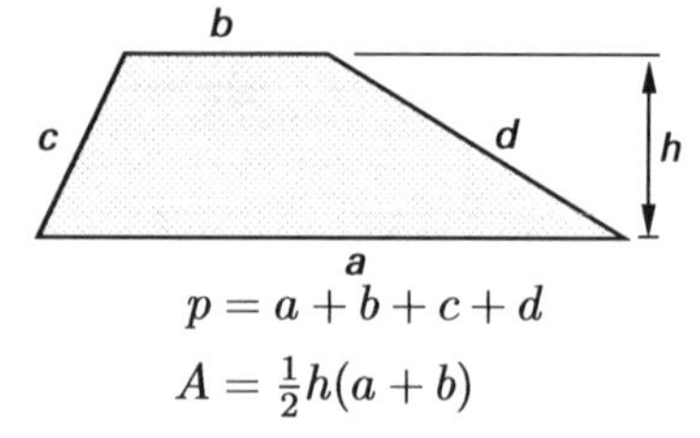

$$p = a + b + c + d$$

$$A = \tfrac{1}{2}h(a+b)$$

The trapezoid is isosceles if $c = d$.

Parallelogram

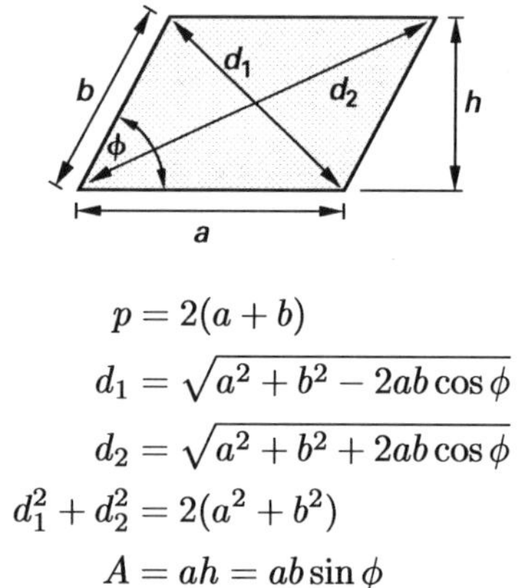

$$p = 2(a+b)$$

$$d_1 = \sqrt{a^2 + b^2 - 2ab\cos\phi}$$

$$d_2 = \sqrt{a^2 + b^2 + 2ab\cos\phi}$$

$$d_1^2 + d_2^2 = 2(a^2 + b^2)$$

$$A = ah = ab\sin\phi$$

If $a = b$, the parallelogram is a rhombus.

Regular Polygon (*n* equal sides)

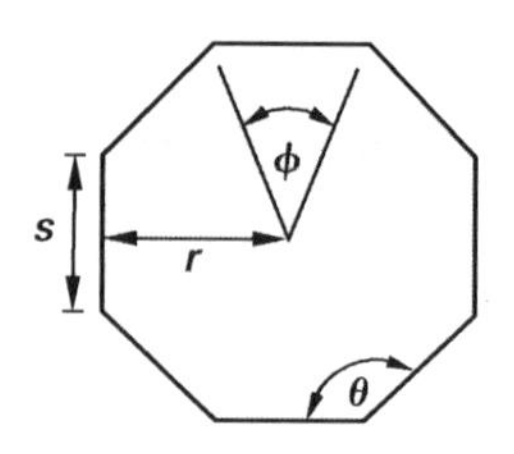

$$\phi = \frac{2\pi}{n}$$

$$\theta = \frac{\pi(n-2)}{n} = \pi - \phi$$

$$p = ns$$

$$s = 2r\tan\left(\frac{\theta}{2}\right)$$

$$A = \tfrac{1}{2}nsr$$

regular polygons							
sides	name	area (A) when diameter of inscribed circle = 1	area (A) when side = 1	radius (r) of circumscribed circle when side = 1	length (L) of side when radius (r) of circumscribed circle = 1	length (L) of side when perpendicular to center = 1	perpendicular (p) to center when side = 1
3	triangle	1.299	0.433	0.577	1.732	3.464	0.289
4	square	1.000	1.000	0.707	1.414	2.000	0.500
5	pentagon	0.908	1.720	0.851	1.176	1.453	0.688
6	hexagon	0.866	2.598	1.000	1.000	1.155	0.866
7	heptagon	0.843	3.634	1.152	0.868	0.963	1.038
8	octagon	0.828	4.828	1.307	0.765	0.828	1.207
9	nonagon	0.819	6.182	1.462	0.684	0.728	1.374
10	decagon	0.812	7.694	1.618	0.618	0.650	1.539
11	undecagon	0.807	9.366	1.775	0.563	0.587	1.703
12	dodecagon	0.804	11.196	1.932	0.518	0.536	1.866

APPENDIX K
Mensuration of Three-Dimensional Volumes

Nomenclature

A	area
b	base
h	height
r	radius
R	radius
s	side (edge) length
V	volume

Sphere

$$V = \frac{4\pi r^3}{3}$$
$$A = 4\pi r^2$$

Right Circular Cone

$$V = \frac{\pi r^2 h}{3}$$
$$A = \pi r\sqrt{r^2 + h^2}$$

(does not include base area)

Right Circular Cylinder

$$V = \pi r^2 h$$
$$A = 2\pi r h$$

(does not include end area)

Spherical Segment (Spherical Cap)

Surface area of a spherical segment of radius r cut out by an angle θ_0 rotated from the center about a radius, r, is

$$A = 2\pi r^2 (1 - \cos\theta_0)$$
$$\omega = \frac{A}{r^2} = 2\pi (1 - \cos\theta_0)$$

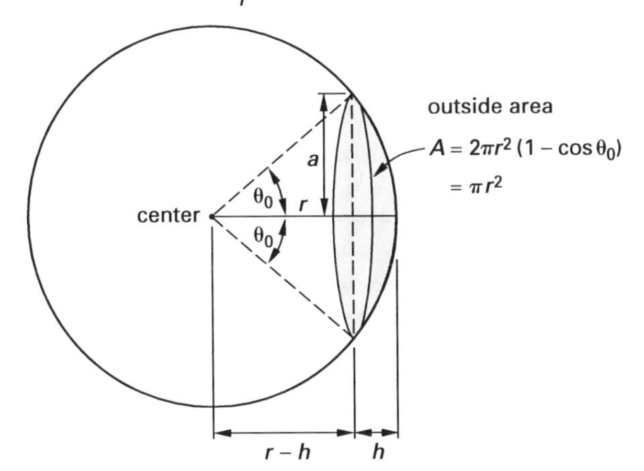

$$V_{\text{cap}} = \frac{\pi}{6}h(3a^2 + h^2)$$
$$= \frac{\pi}{3}h^2(3r - h)$$
$$a = \sqrt{h(2r - h)}$$

Paraboloid of Revolution

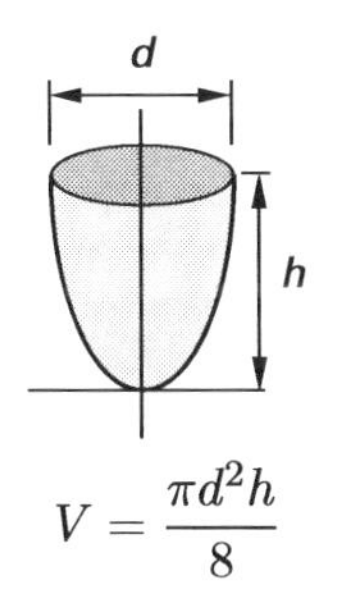

$$V = \frac{\pi d^2 h}{8}$$

Torus

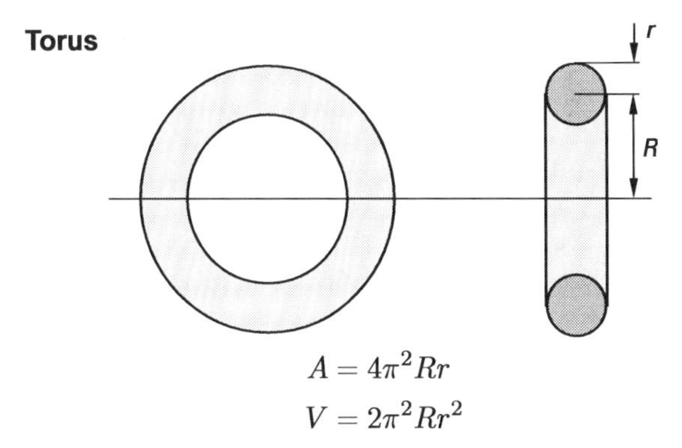

$$A = 4\pi^2 Rr$$
$$V = 2\pi^2 Rr^2$$

Regular Polyhedra (identical faces)

name	number of faces	form of faces	total surface area	volume
tetrahedron	4	equilateral triangle	$1.7321s^2$	$0.1179s^3$
cube	6	square	$6.0000s^2$	$1.0000s^3$
octahedron	8	equilateral triangle	$3.4641s^2$	$0.4714s^3$
dodecahedron	12	regular pentagon	$20.6457s^2$	$7.6631s^3$
isosahedron	20	equilateral triangle	$8.6603s^2$	$2.1817s^3$

The radius of a sphere inscribed within a regular polyhedron is

$$r = \frac{3V_{\text{polyhedron}}}{A_{\text{polyhedron}}}$$

APPENDIX L
Factor Tables

$I = 0.50\%$

n	P/F	P/A	P/G	F/P	F/A	A/P	A/F	A/G	n
1	0.9950	0.9950	0.0000	1.0050	1.0000	1.0050	1.0000	0.0000	1
2	0.9901	0.9851	0.9901	1.0100	2.0050	0.5038	0.4988	0.4988	2
3	0.9851	2.9702	2.9604	1.0151	3.0150	0.3367	0.3317	0.9967	3
4	0.9802	3.9505	5.9011	1.0202	4.0301	0.2531	0.2481	1.4938	4
5	0.9754	4.9259	9.8026	1.0253	5.0503	0.2030	0.1980	1.9900	5
6	0.9705	5.8964	14.6552	1.0304	6.0755	0.1696	0.1646	2.4855	6
7	0.9657	6.8621	20.4493	1.0355	7.1059	0.1457	0.1407	2.9801	7
8	0.9609	7.8230	27.1755	1.0407	8.1414	0.1278	0.1228	3.4738	8
9	0.9561	8.7791	34.8244	1.0459	9.1821	0.1139	0.1089	3.9668	9
10	0.9513	9.7304	43.3865	1.0511	10.2280	0.1028	0.0978	4.4589	10
11	0.9466	10.6770	52.8526	1.0564	11.2792	0.0937	0.0887	4.9501	11
12	0.9419	11.6189	63.2136	1.0617	12.3356	0.0861	0.0811	5.4406	12
13	0.9372	12.5562	74.4602	1.0670	13.3972	0.0796	0.0746	5.9302	13
14	0.9326	13.4887	86.5835	1.0723	14.4642	0.0741	0.0691	6.4190	14
15	0.9279	14.4166	99.5743	1.0777	15.5365	0.0694	0.0644	6.9069	15
16	0.9233	15.3399	113.4238	1.0831	16.6142	0.0652	0.0602	7.3940	16
17	0.9187	16.2586	128.1231	1.0885	17.6973	0.0615	0.0565	7.8803	17
18	0.9141	17.1728	143.6634	1.0939	18.7858	0.0582	0.0532	8.3658	18
19	0.9096	18.0824	160.0360	1.0994	19.8797	0.0553	0.0503	8.8504	19
20	0.9051	18.9874	177.2322	1.1049	20.9791	0.0527	0.0477	9.3342	20
21	0.9006	19.8880	195.2434	1.1104	22.0840	0.0503	0.0453	9.8172	21
22	0.8961	20.7841	214.0611	1.1160	23.1944	0.0481	0.0431	10.2993	22
23	0.8916	21.6757	233.6768	1.1216	24.3104	0.0461	0.0411	10.7806	23
24	0.8872	22.5629	254.0820	1.1272	25.4320	0.0443	0.0393	11.2611	24
25	0.8828	23.4456	275.2686	1.1328	26.5591	0.0427	0.0377	11.7407	25
26	0.8784	24.3240	297.2281	1.1385	27.6919	0.0411	0.0361	12.2195	26
27	0.8740	25.1980	319.9523	1.1442	28.8304	0.0397	0.0347	12.6975	27
28	0.8697	26.0677	343.4332	1.1499	29.9745	0.0384	0.0334	13.1747	28
29	0.8653	26.9330	367.6625	1.1556	31.1244	0.0371	0.0321	13.6510	29
30	0.8610	27.7941	392.6324	1.1614	32.2800	0.0360	0.0310	14.1265	30
31	0.8567	28.6508	418.3348	1.1672	33.4414	0.0349	0.0299	14.6012	31
32	0.8525	29.5033	444.7618	1.1730	34.6086	0.0339	0.0289	15.0750	32
33	0.8482	30.3515	471.9055	1.1789	35.7817	0.0329	0.0279	15.5480	33
34	0.8440	31.1955	499.7583	1.1848	36.9606	0.0321	0.0271	16.0202	34
35	0.8398	32.0354	528.3123	1.1907	38.1454	0.0312	0.0262	16.4915	35
36	0.8356	32.8710	557.5598	1.1967	39.3361	0.0304	0.0254	16.9621	36
37	0.8315	33.7025	587.4934	1.2027	40.5328	0.0297	0.0247	17.4317	37
38	0.8274	34.5299	618.1054	1.2087	41.7354	0.0290	0.0240	17.9006	38
39	0.8232	35.3531	649.3883	1.2147	42.9441	0.0283	0.0233	18.3686	39
40	0.8191	36.1722	681.3347	1.2208	44.1588	0.0276	0.0226	18.8359	40
41	0.8151	36.9873	713.9372	1.2269	45.3796	0.0270	0.0220	19.3022	41
42	0.8110	37.7983	747.1886	1.2330	46.6065	0.0265	0.0215	19.7678	42
43	0.8070	38.6053	781.0815	1.2392	47.8396	0.0259	0.0209	20.2325	43
44	0.8030	39.4082	815.6087	1.2454	49.0788	0.0254	0.0204	20.6964	44
45	0.7990	40.2072	850.7631	1.2516	50.3242	0.0249	0.0199	21.1595	45
46	0.7950	41.0022	886.5376	1.2579	51.5758	0.0244	0.0194	21.6217	46
47	0.7910	41.7932	922.9252	1.2642	52.8337	0.0239	0.0189	22.0831	47
48	0.7871	42.5803	959.9188	1.2705	54.0978	0.0235	0.0185	22.5437	48
49	0.7832	43.3635	997.5116	1.2768	55.3683	0.0231	0.0181	23.0035	49
50	0.7793	44.1428	1035.6966	1.2832	56.6452	0.0227	0.0177	23.4624	50
51	0.7754	44.9182	1074.4670	1.2896	57.9284	0.0223	0.0173	23.9205	51
52	0.7716	45.6897	1113.8162	1.2961	59.2180	0.0219	0.0169	24.3778	52
53	0.7677	46.4575	1153.7372	1.3026	60.5141	0.0215	0.0165	24.8343	53
54	0.7639	47.2214	1194.2236	1.3091	61.8167	0.0212	0.0162	25.2899	54
55	0.7601	47.9814	1235.2686	1.3156	63.1258	0.0208	0.0158	25.7447	55
60	0.7414	51.7256	1448.6458	1.3489	69.7700	0.0193	0.0143	28.0064	60
65	0.7231	55.3775	1675.0272	1.3829	76.5821	0.0181	0.0131	30.2475	65
70	0.7053	58.9394	1913.6427	1.4178	83.5661	0.0170	0.0120	32.4680	70
75	0.6879	62.4136	2163.7525	1.4536	90.7265	0.0160	0.0110	34.6679	75
80	0.6710	65.8023	2424.6455	1.4903	98.0677	0.0152	0.0102	36.8474	80
85	0.6545	69.1075	2695.6389	1.5280	105.5943	0.0145	0.0095	39.0065	85
90	0.6383	72.3313	2976.0769	1.5666	113.3109	0.0138	0.0088	41.1451	90
95	0.6226	75.4757	3265.3298	1.6061	121.2224	0.0132	0.0082	43.2633	95
100	0.6073	78.5426	3562.7934	1.6467	129.3337	0.0127	0.0077	45.3613	100

(continued)

APPENDIX L *(continued)*
Factor Tables
$I = 0.75\%$

n	P/F	P/A	P/G	F/P	F/A	A/P	A/F	A/G	n
1	0.9926	0.9926	0.0000	1.0075	1.0000	1.0075	1.0000	0.0000	1
2	0.9852	1.9777	0.9852	1.0151	2.0075	0.5056	0.4981	0.4981	2
3	0.9778	2.9556	2.9408	1.0227	3.0226	0.3383	0.3308	0.9950	3
4	0.9706	3.9261	5.8525	1.0303	4.0452	0.2547	0.2472	1.4907	4
5	0.9633	4.8894	9.7058	1.0381	5.0756	0.2045	0.1970	1.9851	5
6	0.9562	5.8456	14.4866	1.0459	6.1136	0.1711	0.1636	2.4782	6
7	0.9490	6.7946	20.1808	1.0537	7.1595	0.1472	0.1397	2.9701	7
8	0.9420	7.7366	26.7747	1.0616	8.2132	0.1293	0.1218	3.4608	8
9	0.9350	8.6716	34.2544	1.0696	9.2748	0.1153	0.1078	3.9502	9
10	0.9280	9.5996	42.6064	1.0776	10.3443	0.1042	0.0967	4.4384	10
11	0.9211	10.5207	51.8174	1.0857	11.4219	0.0951	0.0876	4.9253	11
12	0.9142	11.4349	61.8740	1.0938	12.5076	0.0875	0.0800	5.4110	12
13	0.9074	12.3423	72.7632	1.1020	13.6014	0.0810	0.0735	5.8954	13
14	0.9007	13.2430	84.4720	1.1103	14.7034	0.0755	0.0680	6.3786	14
15	0.8940	14.1370	96.9876	1.1186	15.8137	0.0707	0.0632	6.8606	15
16	0.8873	15.0243	110.2973	1.1270	16.9323	0.0666	0.0591	7.3413	16
17	0.8807	15.9050	124.3887	1.1354	18.0593	0.0629	0.0554	7.8207	17
18	0.8742	16.7792	139.2494	1.1440	19.1947	0.0596	0.0521	8.2989	18
19	0.8676	17.6468	154.8671	1.1525	20.3387	0.0567	0.0492	8.7759	19
20	0.8612	18.5080	171.2297	1.1612	21.4912	0.0540	0.0465	9.2516	20
21	0.8548	19.3628	188.3253	1.1699	22.6524	0.0516	0.0441	9.7261	21
22	0.8484	20.2112	206.1420	1.1787	23.8223	0.0495	0.0420	10.1994	22
23	0.8421	21.0533	224.6682	1.1875	25.0010	0.0475	0.0400	10.6714	23
24	0.8358	21.8891	243.8923	1.1964	26.1885	0.0457	0.0382	11.1422	24
25	0.8296	22.7188	263.8029	1.2054	27.3849	0.0440	0.0365	11.6117	25
26	0.8234	23.5422	284.3888	1.2144	28.5903	0.0425	0.0350	12.0800	26
27	0.8173	24.3595	305.6387	1.2235	29.8047	0.0411	0.0336	12.5470	27
28	0.8112	25.1707	327.5416	1.2327	31.0282	0.0397	0.0322	13.0128	28
29	0.8052	25.9759	350.0867	1.2420	32.2609	0.0385	0.0310	13.4774	29
30	0.7992	26.7751	373.2631	1.2513	33.5029	0.0373	0.0298	13.9407	30
31	0.7932	27.5683	397.0602	1.2607	34.7542	0.0363	0.0288	14.4028	31
32	0.7873	28.3557	421.4675	1.2701	36.0148	0.0353	0.0278	14.8636	32
33	0.7815	29.1371	446.4746	1.2796	37.2849	0.0343	0.0268	15.3232	33
34	0.7757	29.9128	472.0712	1.2892	38.5646	0.0334	0.0259	15.7816	34
35	0.7699	30.6827	498.2471	1.2989	39.8538	0.0326	0.0251	16.2387	35
36	0.7641	31.4468	524.9924	1.3086	41.1527	0.0318	0.0243	16.6946	36
37	0.7585	32.2053	552.2969	1.3185	42.4614	0.0311	0.0236	17.1493	37
38	0.7528	32.9581	580.1511	1.3283	43.7798	0.0303	0.0228	17.6027	38
39	0.7472	33.7053	608.5451	1.3383	45.1082	0.0297	0.0222	18.0549	39
40	0.7416	34.4469	637.4693	1.3483	46.4465	0.0290	0.0215	18.5058	40
41	0.7361	35.1831	666.9144	1.3585	47.7948	0.0284	0.0209	18.9556	41
42	0.7306	35.9137	696.8709	1.3686	49.1533	0.0278	0.0203	19.4040	42
43	0.7252	36.6389	727.3297	1.3789	50.5219	0.0273	0.0198	19.8513	43
44	0.7198	37.3587	758.2815	1.3893	51.9009	0.0268	0.0193	20.2973	44
45	0.7145	38.0732	789.7173	1.3997	53.2901	0.0263	0.0188	20.7421	45
46	0.7091	38.7823	821.6283	1.4102	54.6898	0.0258	0.0183	21.1856	46
47	0.7039	39.4862	854.0056	1.4207	56.1000	0.0253	0.0178	21.6280	47
48	0.6986	40.1848	886.8404	1.4314	57.5207	0.0249	0.0174	22.0691	48
49	0.6934	40.8782	920.1243	1.4421	58.9521	0.0245	0.0170	22.5089	49
50	0.6883	41.5664	953.8486	1.4530	60.3943	0.0241	0.0166	22.9476	50
51	0.6831	42.2496	988.0050	1.4639	61.8472	0.0237	0.0162	23.3850	51
52	0.6780	42.9276	1022.5852	1.4748	63.3111	0.0233	0.0158	23.8211	52
53	0.6730	43.6006	1057.5810	1.4859	64.7859	0.0229	0.0154	24.2561	53
54	0.6680	44.2686	1092.9842	1.4970	66.2718	0.0226	0.0151	24.6898	54
55	0.6630	44.9316	1128.7869	1.5083	67.7688	0.0223	0.0148	25.1223	55
60	0.6387	48.1734	1313.5189	1.5657	75.4241	0.0208	0.0133	27.2665	60
65	0.6153	51.2963	1507.0910	1.6253	83.3709	0.0195	0.0120	29.3801	65
70	0.5927	54.3046	1708.6065	1.6872	91.6201	0.0184	0.0109	31.4634	70
75	0.5710	57.2027	1917.2225	1.7514	100.1833	0.0175	0.0100	33.5163	75
80	0.5500	59.9944	2132.1472	1.8180	109.0725	0.0167	0.0092	35.5391	80
85	0.5299	62.6838	2352.6375	1.8873	118.3001	0.0160	0.0085	37.5318	85
90	0.5104	65.2746	2577.9961	1.9591	127.8790	0.0153	0.0078	39.4946	90
95	0.4917	67.7704	2807.5694	2.0337	137.8225	0.0148	0.0073	41.4277	95
100	0.4737	70.1746	3040.7453	2.1111	148.1445	0.0143	0.0068	43.3311	100

(continued)

Support Material

APPENDIX L *(continued)*
Factor Tables

$I = 1.00\%$

n	P/F	P/A	P/G	F/P	F/A	A/P	A/F	A/G	n
1	0.9901	0.9901	0.0000	1.0100	1.0000	1.0100	1.0000	0.0000	1
2	0.9803	1.9704	0.9803	1.0201	2.0100	0.5075	0.4975	0.4975	2
3	0.9706	2.9410	2.9215	1.0303	3.0301	0.3400	0.3300	0.9934	3
4	0.9610	3.9020	5.8044	1.0406	4.0604	0.2563	0.2463	1.4876	4
5	0.9515	4.8534	9.6103	1.0510	5.1010	0.2060	0.1960	1.9801	5
6	0.9420	5.7955	14.3205	1.0615	6.1520	0.1725	0.1625	2.4710	6
7	0.9327	6.7282	19.9168	1.0721	7.2135	0.1486	0.1386	2.9602	7
8	0.9235	7.6517	26.3812	1.0829	8.2857	0.1307	0.1207	3.4478	8
9	0.9143	8.5660	33.6959	1.0937	9.3685	0.1167	0.1067	3.9337	9
10	0.9053	9.4713	41.8435	1.1046	10.4622	0.1056	0.0956	4.4179	10
11	0.8963	10.3676	50.8067	1.1157	11.5668	0.0965	0.0865	4.9005	11
12	0.8874	11.2551	60.5687	1.1268	12.6825	0.0888	0.0788	5.3815	12
13	0.8787	12.1337	71.1126	1.1381	13.8093	0.0824	0.0724	5.8607	13
14	0.8700	13.0037	82.4221	1.1495	14.9474	0.0769	0.0669	6.3384	14
15	0.8613	13.8651	94.4810	1.1610	16.0969	0.0721	0.0621	6.8143	15
16	0.8528	14.7179	107.2734	1.1726	17.2579	0.0679	0.0579	7.2886	16
17	0.8444	15.5623	120.7834	1.1843	18.4304	0.0643	0.0543	7.7613	17
18	0.8360	16.3983	134.9957	1.1961	19.6147	0.0610	0.0510	8.2323	18
19	0.8277	17.2260	149.8950	1.2081	20.8109	0.0581	0.0481	8.7017	19
20	0.8195	18.0456	165.4664	1.2202	22.0190	0.0554	0.0454	9.1694	20
21	0.8114	18.8570	181.6950	1.2324	23.2392	0.0530	0.0430	9.6354	21
22	0.8034	19.6604	198.5663	1.2447	24.4716	0.0509	0.0409	10.0998	22
23	0.7954	20.4558	216.0660	1.2572	25.7163	0.0489	0.0389	10.5626	23
24	0.7876	21.2434	234.1800	1.2697	26.9735	0.0471	0.0371	11.0237	24
25	0.7798	22.0232	252.8945	1.2824	28.2432	0.0454	0.0354	11.4831	25
26	0.7720	22.7952	272.1957	1.2953	29.5256	0.0439	0.0339	11.9409	26
27	0.7644	23.5596	292.0702	1.3082	30.8209	0.0424	0.0324	12.3971	27
28	0.7568	24.3164	312.5047	1.3213	32.1291	0.0411	0.0311	12.8516	28
29	0.7493	25.0658	333.4863	1.3345	33.4504	0.0399	0.0299	13.3044	29
30	0.7419	25.8077	355.0021	1.3478	34.7849	0.0387	0.0287	13.7557	30
31	0.7346	26.5423	377.0394	1.3613	36.1327	0.0377	0.0277	14.2052	31
32	0.7273	27.2696	399.5858	1.3749	37.4941	0.0367	0.0267	14.6532	32
33	0.7201	27.9897	422.6291	1.3887	38.8690	0.0357	0.0257	15.0995	33
34	0.7130	28.7027	446.1572	1.4026	40.2577	0.0348	0.0248	15.5441	34
35	0.7059	29.4086	470.1583	1.4166	41.6603	0.0340	0.0240	15.9871	35
36	0.6989	30.1075	494.6207	1.4308	43.0769	0.0332	0.0232	16.4285	36
37	0.6920	30.7995	519.5329	1.4451	44.5076	0.0325	0.0225	16.8682	37
38	0.6852	31.4847	544.8835	1.4595	45.9527	0.0318	0.0218	17.3063	38
39	0.6784	32.1630	570.6616	1.4741	47.4123	0.0311	0.0211	17.7428	39
40	0.6717	32.8347	596.8561	1.4889	48.8864	0.0305	0.0205	18.1776	40
41	0.6650	33.4997	623.4562	1.5038	50.3752	0.0299	0.0199	18.6108	41
42	0.6584	34.1581	650.4514	1.5188	51.8790	0.0293	0.0193	19.0424	42
43	0.6519	34.8100	677.8312	1.5340	53.3978	0.0287	0.0187	19.4723	43
44	0.6454	35.4555	705.5853	1.5493	54.9318	0.0282	0.0182	19.9006	44
45	0.6391	36.0945	733.7037	1.5648	56.4811	0.0277	0.0177	20.3273	45
46	0.6327	36.7272	762.1765	1.5805	58.0459	0.0272	0.0172	20.7524	46
47	0.6265	37.3537	790.9938	1.5963	59.6263	0.0268	0.0168	21.1758	47
48	0.6203	37.9740	820.1460	1.6122	61.2226	0.0263	0.0163	21.5976	48
49	0.6141	38.5881	849.6237	1.6283	62.8348	0.0259	0.0159	22.0178	49
50	0.6080	39.1961	879.4176	1.6446	64.4632	0.0255	0.0155	22.4363	50
51	0.6020	39.7981	909.5186	1.6611	66.1078	0.0251	0.0151	22.8533	51
52	0.5961	40.3942	939.9175	1.6777	67.7689	0.0248	0.0148	23.2686	52
53	0.5902	40.9844	970.6057	1.6945	69.4466	0.0244	0.0144	23.6823	53
54	0.5843	41.5687	1001.5743	1.7114	71.1410	0.0241	0.0141	24.0945	54
55	0.5785	42.1472	1032.8148	1.7285	72.8525	0.0237	0.0137	24.5049	55
60	0.5504	44.9550	1192.8061	1.8167	81.6697	0.0222	0.0122	26.5333	60
65	0.5237	47.6266	1358.3903	1.9094	90.9366	0.0210	0.0110	28.5217	65
70	0.4983	50.1685	1528.6474	2.0068	100.6763	0.0199	0.0099	30.4703	70
75	0.4741	52.5871	1702.7340	2.1091	110.9128	0.0190	0.0090	32.3793	75
80	0.4511	54.8882	1879.8771	2.2167	121.6715	0.0182	0.0082	34.2492	80
85	0.4292	57.0777	2059.3701	2.3298	132.9790	0.0175	0.0075	36.0801	85
90	0.4084	59.1609	2240.5675	2.4486	144.8633	0.0169	0.0069	37.8724	90
95	0.3886	61.1430	2422.8811	2.5735	157.3538	0.0164	0.0064	39.6265	95
100	0.3697	63.0289	2605.7758	2.7048	170.4814	0.0159	0.0059	41.3426	100

(continued)

APPENDIX L *(continued)*
Factor Tables

$I = 1.50\%$

n	P/F	P/A	P/G	F/P	F/A	A/P	A/F	A/G	n
1	0.9852	0.9852	0.0000	1.0150	1.0000	1.0150	1.0000	0.0000	1
2	0.9707	1.9559	0.9707	1.0302	2.0150	0.5113	0.4963	0.4963	2
3	0.9563	2.9122	2.8833	1.0457	3.0452	0.3434	0.3284	0.9901	3
4	0.9422	3.8544	5.7098	1.0614	4.0909	0.2594	0.2444	1.4814	4
5	0.9283	4.7826	9.4229	1.0773	5.1523	0.2091	0.1941	1.9702	5
6	0.9145	5.6972	13.9956	1.0934	6.2296	0.1755	0.1605	2.4566	6
7	0.9010	6.5982	19.4018	1.1098	7.3230	0.1516	0.1366	2.9405	7
8	0.8877	7.4859	25.6157	1.1265	8.4328	0.1336	0.1186	3.4219	8
9	0.8746	8.3605	32.6125	1.1434	9.5593	0.1196	0.1046	3.9008	9
10	0.8617	9.2222	40.3675	1.1605	10.7027	0.1084	0.0934	4.3772	10
11	0.8489	10.0711	48.8568	1.1779	11.8633	0.0993	0.0843	4.8512	11
12	0.8364	10.9075	58.0571	1.1956	13.0412	0.0917	0.0767	5.3227	12
13	0.8240	11.7315	67.9454	1.2136	14.2368	0.0852	0.0702	5.7917	13
14	0.8118	12.5434	78.4994	1.2318	15.4504	0.0797	0.0647	6.2582	14
15	0.7999	13.3432	89.6974	1.2502	16.6821	0.0749	0.0599	6.7223	15
16	0.7880	14.1313	101.5178	1.2690	17.9324	0.0708	0.0558	7.1839	16
17	0.7764	14.9076	113.9400	1.2880	19.2014	0.0671	0.0521	7.6431	17
18	0.7649	15.6726	126.9435	1.3073	20.4894	0.0638	0.0488	8.0997	18
19	0.7536	16.4262	140.5084	1.3270	21.7967	0.0609	0.0459	8.5539	19
20	0.7425	17.1686	154.6154	1.3469	23.1237	0.0582	0.0432	9.0057	20
21	0.7315	17.9001	169.2453	1.3671	24.4705	0.0559	0.0409	9.4550	21
22	0.7207	18.6208	184.3798	1.3876	25.8376	0.0537	0.0387	9.9018	22
23	0.7100	19.3309	200.0006	1.4084	27.2251	0.0517	0.0367	10.3462	23
24	0.6995	20.0304	216.0901	1.4295	28.6335	0.0499	0.0349	10.7881	24
25	0.6892	20.7196	232.6310	1.4509	30.0630	0.0483	0.0333	11.2276	25
26	0.6790	21.3986	249.6065	1.4727	31.5140	0.0467	0.0317	11.6646	26
27	0.6690	22.0676	267.0002	1.4948	32.9867	0.0453	0.0303	12.0992	27
28	0.6591	22.7267	284.7958	1.5172	34.4815	0.0440	0.0290	12.5313	28
29	0.6494	23.3761	302.9779	1.5400	35.9987	0.0428	0.0278	12.9610	29
30	0.6398	24.0158	321.5310	1.5631	37.5387	0.0416	0.0266	13.3883	30
31	0.6303	24.6461	340.4402	1.5865	39.1018	0.0406	0.0256	13.8131	31
32	0.6210	25.2671	359.6910	1.6103	40.6883	0.0396	0.0246	14.2355	32
33	0.6118	25.8790	379.2691	1.6345	42.2986	0.0386	0.0236	14.6555	33
34	0.6028	26.4817	399.1607	1.6590	43.9331	0.0378	0.0228	15.0731	34
35	0.5939	27.0756	419.3521	1.6839	45.5921	0.0369	0.0219	15.4882	35
36	0.5851	27.6607	439.8303	1.7091	47.2760	0.0362	0.0212	15.9009	36
37	0.5764	28.2371	460.5822	1.7348	48.9851	0.0354	0.0204	16.3112	37
38	0.5679	28.8051	481.5954	1.7608	50.7199	0.0347	0.0197	16.7191	38
39	0.5595	29.3646	502.8576	1.7872	52.4807	0.0341	0.0191	17.1246	39
40	0.5513	29.9158	524.3568	1.8140	54.2679	0.0334	0.0184	17.5277	40
41	0.5431	30.4590	546.0814	1.8412	56.0819	0.0328	0.0178	17.9284	41
42	0.5351	30.9941	568.0201	1.8688	57.9231	0.0323	0.0173	18.3267	42
43	0.5272	31.5212	590.1617	1.8969	59.7920	0.0317	0.0167	18.7227	43
44	0.5194	32.0406	612.4955	1.9253	61.6889	0.0312	0.0162	19.1162	44
45	0.5117	32.5523	635.0110	1.9542	63.6142	0.0307	0.0157	19.5074	45
46	0.5042	33.0565	657.6979	1.9835	65.5684	0.0303	0.0153	19.8962	46
47	0.4967	33.5532	680.5462	2.0133	67.5519	0.0298	0.0148	20.2826	47
48	0.4894	34.0426	703.5462	2.0435	69.5652	0.0294	0.0144	20.6667	48
49	0.4821	34.5247	726.6884	2.0741	71.6087	0.0290	0.0140	21.0484	49
50	0.4750	34.9997	749.9636	2.1052	73.6828	0.0286	0.0136	21.4277	50
51	0.4680	35.4677	773.3629	2.1368	75.7881	0.0282	0.0132	21.8047	51
52	0.4611	35.9287	796.8774	2.1689	77.9249	0.0278	0.0128	22.1794	52
53	0.4543	36.3830	820.4986	2.2014	80.0938	0.0275	0.0125	22.5517	53
54	0.4475	36.8305	844.2184	2.2344	82.2952	0.0272	0.0122	22.9217	54
55	0.4409	37.2715	868.0285	2.2679	84.5296	0.0268	0.0118	23.2894	55
60	0.4093	39.3803	988.1674	2.4432	96.2147	0.0254	0.0104	25.0930	60
65	0.3799	41.3378	1109.4752	2.6320	108.8028	0.0242	0.0092	26.8393	65
70	0.3527	43.1549	1231.1658	2.8355	122.3638	0.0232	0.0082	28.5290	70
75	0.3274	44.8416	1352.5600	3.0546	136.9728	0.0223	0.0073	30.1631	75
80	0.3039	46.4073	1473.0741	3.2907	152.7109	0.0215	0.0065	31.7423	80
85	0.2821	47.8607	1592.2095	3.5450	169.6652	0.0209	0.0059	33.2676	85
90	0.2619	49.2099	1709.5439	3.8189	187.9299	0.0203	0.0053	34.7399	90
95	0.2431	50.4622	1824.7224	4.1141	207.6061	0.0198	0.0048	36.1602	95
100	0.2256	51.6247	1937.4506	4.4320	228.8030	0.0194	0.0044	37.5295	100

(continued)

Support Material

APPENDIX L *(continued)*
Factor Tables

$I = 2.00\%$

n	P/F	P/A	P/G	F/P	F/A	A/P	A/F	A/G	n
1	0.9804	0.9804	0.0000	1.0200	1.0000	1.0200	1.0000	0.0000	1
2	0.9612	1.9416	0.9612	1.0404	2.0200	0.5150	0.4950	0.4950	2
3	0.9423	2.8839	2.8458	1.0612	3.0604	0.3468	0.3268	0.9868	3
4	0.9238	3.8077	5.6173	1.0824	4.1216	0.2626	0.2426	1.4752	4
5	0.9057	4.7135	9.2403	1.1041	5.2040	0.2122	0.1922	1.9604	5
6	0.8880	5.6014	13.6801	1.1262	6.3081	0.1785	0.1585	2.4423	6
7	0.8706	6.4720	18.9035	1.1487	7.4343	0.1545	0.1345	2.9208	7
8	0.8535	7.3255	24.8779	1.1717	8.5830	0.1365	0.1165	3.3961	8
9	0.8368	8.1622	31.5720	1.1951	9.7546	0.1225	0.1025	3.8681	9
10	0.8203	8.9826	38.9551	1.2190	10.9497	0.1113	0.0913	4.3367	10
11	0.8043	9.7868	46.9977	1.2434	12.1687	0.1022	0.0822	4.8021	11
12	0.7885	10.5753	55.6712	1.2682	13.4121	0.0946	0.0746	5.2642	12
13	0.7730	11.3484	64.9475	1.2936	14.6803	0.0881	0.0681	5.7231	13
14	0.7579	12.1062	74.7999	1.3195	15.9739	0.0826	0.0626	6.1786	14
15	0.7430	12.8493	85.2021	1.3459	17.2934	0.0778	0.0578	6.6309	15
16	0.7284	13.5777	96.1288	1.3728	18.6393	0.0737	0.0537	7.0799	16
17	0.7142	14.2919	107.5554	1.4002	20.0121	0.0700	0.0500	7.5256	17
18	0.7002	14.9920	119.4581	1.4282	21.4123	0.0667	0.0467	7.9681	18
19	0.6864	15.6785	131.8139	1.4568	22.8406	0.0638	0.0438	8.4073	19
20	0.6730	16.3514	144.6003	1.4859	24.2974	0.0612	0.0412	8.8433	20
21	0.6598	17.0112	157.7959	1.5157	25.7833	0.0588	0.0388	9.2760	21
22	0.6468	17.6580	171.3795	1.5460	27.2990	0.0566	0.0366	9.7055	22
23	0.6342	18.2922	185.3309	1.5769	28.8450	0.0547	0.0347	10.1317	23
24	0.6217	18.9139	199.6305	1.6084	30.4219	0.0529	0.0329	10.5547	24
25	0.6095	19.5235	214.2592	1.6406	32.0303	0.0512	0.0312	10.9745	25
26	0.5976	20.1210	229.1987	1.6734	33.6709	0.0497	0.0297	11.3910	26
27	0.5859	20.7069	244.4311	1.7069	35.3443	0.0483	0.0283	11.8043	27
28	0.5744	21.2813	259.9392	1.7410	37.0512	0.0470	0.0270	12.2145	28
29	0.5631	21.8444	275.7064	1.7758	38.7922	0.0458	0.0258	12.6214	29
30	0.5521	22.3965	291.7164	1.8114	40.5681	0.0446	0.0246	13.0251	30
31	0.5412	22.9377	307.9538	1.8476	42.3794	0.0436	0.0236	13.4257	31
32	0.5306	23.4683	324.4035	1.8845	44.2270	0.0426	0.0226	13.8230	32
33	0.5202	23.9886	341.0508	1.9222	46.1116	0.0417	0.0217	14.2172	33
34	0.5100	24.4986	357.8817	1.9607	48.0338	0.0408	0.0208	14.6083	34
35	0.5000	24.9986	374.8826	1.9999	49.9945	0.0400	0.0200	14.9961	35
36	0.4902	25.4888	392.0405	2.0399	51.9944	0.0392	0.0192	15.3809	36
37	0.4806	25.9695	409.3424	2.0807	54.0343	0.0385	0.0185	15.7625	37
38	0.4712	26.4406	426.7764	2.1223	56.1149	0.0378	0.0178	16.1409	38
39	0.4619	26.9026	444.3304	2.1647	58.2372	0.0372	0.0172	16.5163	39
40	0.4529	27.3555	461.9931	2.2080	60.4020	0.0366	0.0166	16.8885	40
41	0.4440	27.7995	479.7535	2.2522	62.6100	0.0360	0.0160	17.2576	41
42	0.4353	28.2348	497.6010	2.2972	64.8622	0.0354	0.0154	17.6237	42
43	0.4268	28.6616	515.5253	2.3432	67.1595	0.0349	0.0149	17.9866	43
44	0.4184	29.0800	533.5165	2.3901	69.5027	0.0344	0.0144	18.3465	44
45	0.4102	29.4902	551.5652	2.4379	71.8927	0.0339	0.0139	18.7034	45
46	0.4022	29.8923	569.6621	2.4866	74.3306	0.0335	0.0135	19.0571	46
47	0.3943	30.2866	587.7985	2.5363	76.8172	0.0330	0.0130	19.4079	47
48	0.3865	30.6731	605.9657	2.5871	79.3535	0.0326	0.0126	19.7556	48
49	0.3790	31.0521	624.1557	2.6388	81.9406	0.0322	0.0122	20.1003	49
50	0.3715	31.4236	642.3606	2.6916	84.5794	0.0318	0.0118	20.4420	50
51	0.3642	31.7878	660.5727	2.7454	87.2710	0.0315	0.0115	20.7807	51
52	0.3571	32.1449	678.7849	2.8003	90.0164	0.0311	0.0111	21.1164	52
53	0.3501	32.4950	696.9900	2.8563	92.8167	0.0308	0.0108	21.4491	53
54	0.3432	32.8383	715.1815	2.9135	95.6731	0.0305	0.0105	21.7789	54
55	0.3365	33.1748	733.3527	2.9717	98.5865	0.0301	0.0101	22.1057	55
60	0.3048	34.7609	823.6975	3.2810	114.0515	0.0288	0.0088	23.6961	60
65	0.2761	36.1975	912.7085	3.6225	131.1262	0.0276	0.0076	25.2147	65
70	0.2500	37.4986	999.8343	3.9996	149.9779	0.0267	0.0067	26.6632	70
75	0.2265	38.6771	1084.6393	4.4158	170.7918	0.0259	0.0059	28.0434	75
80	0.2051	39.7445	1166.7868	4.8754	193.7720	0.0252	0.0052	29.3572	80
85	0.1858	40.7113	1246.0241	5.3829	219.1439	0.0246	0.0046	30.6064	85
90	0.1683	41.5869	1322.1701	5.9431	247.1567	0.0240	0.0040	31.7929	90
95	0.1524	42.3800	1395.1033	6.5617	278.0850	0.0236	0.0036	32.9189	95
100	0.1380	43.0984	1464.7527	7.2446	312.2323	0.0232	0.0032	33.9863	100

(continued)

APPENDIX L *(continued)*
Factor Tables

$I = 3.00\%$

n	P/F	P/A	P/G	F/P	F/A	A/P	A/F	A/G	n
1	0.9709	0.9709	0.0000	1.0300	1.0000	1.0300	1.0000	0.0000	1
2	0.9426	1.9135	0.9426	1.0609	2.0300	0.5226	0.4926	0.4926	2
3	0.9151	2.8286	2.7729	1.0927	3.0909	0.3535	0.3235	0.9803	3
4	0.8885	3.7171	5.4383	1.1255	4.1836	0.2690	0.2390	1.4631	4
5	0.8626	4.5797	8.8888	1.1593	5.3091	0.2184	0.1884	1.9409	5
6	0.8375	5.4172	13.0762	1.1941	6.4684	0.1846	0.1546	2.4138	6
7	0.8131	6.2303	17.9547	1.2299	7.6625	0.1605	0.1305	2.8819	7
8	0.7894	7.0197	23.4806	1.2668	8.8923	0.1425	0.1125	3.3450	8
9	0.7664	7.7861	29.6119	1.3048	10.1591	0.1284	0.0984	3.8032	9
10	0.7441	8.5302	36.3088	1.3439	11.4639	0.1172	0.0872	4.2565	10
11	0.7224	9.2526	43.5330	1.3842	12.8078	0.1081	0.0781	4.7049	11
12	0.7014	9.9540	51.2482	1.4258	14.1920	0.1005	0.0705	5.1485	12
13	0.6810	10.6350	59.4196	1.4685	15.6178	0.0940	0.0640	5.5872	13
14	0.6611	11.2961	68.0141	1.5126	17.0863	0.0885	0.0585	6.0210	14
15	0.6419	11.9379	77.0002	1.5580	18.5989	0.0838	0.0538	6.4500	15
16	0.6232	12.5611	86.3477	1.6047	20.1569	0.0796	0.0496	6.8742	16
17	0.6050	13.1661	96.0280	1.6528	21.7616	0.0760	0.0460	7.2936	17
18	0.5874	13.7535	106.0137	1.7024	23.4144	0.0727	0.0427	7.7081	18
19	0.5703	14.3238	116.2788	1.7535	25.1169	0.0698	0.0398	8.1179	19
20	0.5537	14.8775	126.7987	1.8061	26.8704	0.0672	0.0372	8.5229	20
21	0.5375	15.4150	137.5496	1.8603	28.6765	0.0649	0.0349	8.9231	21
22	0.5219	15.9369	148.5094	1.9161	30.5368	0.0627	0.0327	9.3186	22
23	0.5067	16.4436	159.6566	1.9736	32.4529	0.0608	0.0308	9.7093	23
24	0.4919	16.9355	170.9711	2.0328	34.4265	0.0590	0.0290	10.0954	24
25	0.4776	17.4131	182.4336	2.0938	36.4593	0.0574	0.0274	10.4768	25
26	0.4637	17.8768	194.0260	2.1566	38.5530	0.0559	0.0259	10.8535	26
27	0.4502	18.3270	205.7309	2.2213	40.7096	0.0546	0.0246	11.2255	27
28	0.4371	18.7641	217.5320	2.2879	42.9309	0.0533	0.0233	11.5930	28
29	0.4243	19.1885	229.4137	2.3566	45.2189	0.0521	0.0221	11.9558	29
30	0.4120	19.6004	241.3613	2.4273	47.5754	0.0510	0.0210	12.3141	30
31	0.4000	20.0004	253.3609	2.5001	50.0027	0.0500	0.0200	12.6678	31
32	0.3883	20.3888	265.3993	2.5751	52.5028	0.0490	0.0190	13.0169	32
33	0.3770	20.7658	277.4642	2.6523	55.0778	0.0482	0.0182	13.3616	33
34	0.3660	21.1318	289.5437	2.7319	57.7302	0.0473	0.0173	13.7018	34
35	0.3554	21.4872	301.6267	2.8139	60.4621	0.0465	0.0165	14.0375	35
36	0.3450	21.8323	313.7028	2.8983	63.2759	0.0458	0.0158	14.3688	36
37	0.3350	22.1672	325.7622	2.9852	66.1742	0.0451	0.0151	14.6957	37
38	0.3252	22.4925	337.7956	3.0748	69.1594	0.0445	0.0145	15.0182	38
39	0.3158	22.8082	349.7942	3.1670	72.2342	0.0438	0.0138	15.3363	39
40	0.3066	23.1148	361.7499	3.2620	75.4013	0.0433	0.0133	15.6502	40
41	0.2976	23.4124	373.6551	3.3599	78.6633	0.0427	0.0127	15.9597	41
42	0.2890	23.7014	385.5024	3.4607	82.0232	0.0422	0.0122	16.2650	42
43	0.2805	23.9819	397.2852	3.5645	85.4839	0.0417	0.0117	16.5660	43
44	0.2724	24.2543	408.9972	3.6715	89.0484	0.0412	0.0112	16.8629	44
45	0.2644	24.5187	420.6325	3.7816	92.7199	0.0408	0.0108	17.1556	45
46	0.2567	24.7754	432.1856	3.8950	96.5015	0.0404	0.0104	17.4441	46
47	0.2493	25.0247	443.6515	4.0119	100.3965	0.0400	0.0100	17.7285	47
48	0.2420	25.2667	455.0255	4.1323	104.4084	0.0396	0.0096	18.0089	48
49	0.2350	25.5017	466.3031	4.2562	108.5406	0.0392	0.0092	18.2852	49
50	0.2281	25.7298	477.4803	4.3839	112.7969	0.0389	0.0089	18.5575	50
51	0.2215	25.9512	488.5535	4.5154	117.1808	0.0385	0.0085	18.8258	51
52	0.2150	26.1662	499.5191	4.6509	121.6962	0.0382	0.0082	19.0902	52
53	0.2088	26.3750	510.3742	4.7904	126.3471	0.0379	0.0079	19.3507	53
54	0.2027	26.5777	521.1157	4.9341	131.1375	0.0376	0.0076	19.6073	54
55	0.1968	26.7744	531.7411	5.0821	136.0716	0.0373	0.0073	19.8600	55
60	0.1697	27.6756	583.0526	5.8916	163.0534	0.0361	0.0061	21.0674	60
65	0.1464	28.4529	631.2010	6.8300	194.3328	0.0351	0.0051	22.1841	65
70	0.1263	29.1234	676.0869	7.9178	230.5941	0.0343	0.0043	23.2145	70
75	0.1089	29.7018	717.6978	9.1789	272.6309	0.0337	0.0037	24.1634	75
80	0.0940	30.2008	756.0865	10.6409	321.3630	0.0331	0.0031	25.0353	80
85	0.0811	30.6312	791.3529	12.3357	377.8570	0.0326	0.0026	25.8349	85
90	0.0699	31.0024	823.6302	14.3005	443.3489	0.0323	0.0023	26.5667	90
95	0.0603	31.3227	853.0742	16.5782	519.2720	0.0319	0.0019	27.2351	95
100	0.0520	31.5989	879.8540	19.2186	607.2877	0.0316	0.0016	27.8444	100

(continued)

APPENDIX L *(continued)*
Factor Tables

$I = 4.00\%$

n	P/F	P/A	P/G	F/P	F/A	A/P	A/F	A/G	n
1	0.9615	0.9615	0.0000	1.0400	1.0000	1.0400	1.0000	0.0000	1
2	0.9246	1.8861	0.9246	1.0816	2.0400	0.5302	0.4902	0.4902	2
3	0.8890	2.7751	2.7025	1.1249	3.1216	0.3603	0.3203	0.9739	3
4	0.8548	3.6299	5.2670	1.1699	4.2465	0.2755	0.2355	1.4510	4
5	0.8219	4.4518	8.5547	1.2167	5.4163	0.2246	0.1846	1.9216	5
6	0.7903	5.2421	12.5062	1.2653	6.6330	0.1908	0.1508	2.3857	6
7	0.7599	6.0021	17.0657	1.3159	7.8983	0.1666	0.1266	2.8433	7
8	0.7307	6.7327	22.1806	1.3686	9.2142	0.1485	0.1085	3.2944	8
9	0.7026	7.4353	27.8013	1.4233	10.5828	0.1345	0.0945	3.7391	9
10	0.6756	8.1109	33.8814	1.4802	12.0061	0.1233	0.0833	4.1773	10
11	0.6496	8.7605	40.3772	1.5395	13.4864	0.1141	0.0741	4.6090	11
12	0.6246	9.3851	47.2477	1.6010	15.0258	0.1066	0.0666	5.0343	12
13	0.6006	9.9856	54.4546	1.6651	16.6268	0.1001	0.0601	5.4533	13
14	0.5775	10.5631	61.9618	1.7317	18.2919	0.0947	0.0547	5.8659	14
15	0.5553	11.1184	69.7355	1.8009	20.0236	0.0899	0.0499	6.2721	15
16	0.5339	11.6523	77.7441	1.8730	21.8245	0.0858	0.0458	6.6720	16
17	0.5134	12.1657	85.9581	1.9479	23.6975	0.0822	0.0422	7.0656	17
18	0.4936	12.6593	94.3498	2.0258	25.6454	0.0790	0.0390	7.4530	18
19	0.4746	13.1339	102.8933	2.1068	27.6712	0.0761	0.0361	7.8342	19
20	0.4564	13.5903	111.5647	2.1911	29.7781	0.0736	0.0336	8.2091	20
21	0.4388	14.0292	120.3414	2.2788	31.9692	0.0713	0.0313	8.5779	21
22	0.4220	14.4511	129.2024	2.3699	34.2480	0.0692	0.0292	8.9407	22
23	0.4057	14.8568	138.1284	2.4647	36.6179	0.0673	0.0273	9.2973	23
24	0.3901	15.2470	147.1012	2.5633	39.0826	0.0656	0.0256	9.6479	24
25	0.3751	15.6221	156.1040	2.6658	41.6459	0.0640	0.0240	9.9925	25
26	0.3607	15.9828	165.1212	2.7725	44.3117	0.0626	0.0226	10.3312	26
27	0.3468	16.3296	174.1385	2.8834	47.0842	0.0612	0.0212	10.6640	27
28	0.3335	16.6631	183.1424	2.9987	49.9676	0.0600	0.0200	10.9909	28
29	0.3207	16.9837	192.1206	3.1187	52.9663	0.0589	0.0189	11.3120	29
30	0.3083	17.2920	201.0618	3.2434	56.0849	0.0578	0.0178	11.6274	30
31	0.2965	17.5885	209.9556	3.3731	59.3283	0.0569	0.0169	11.9371	31
32	0.2851	17.8736	218.7924	3.5081	62.7015	0.0559	0.0159	12.2411	32
33	0.2741	18.1476	227.5634	3.6484	66.2095	0.0551	0.0151	12.5396	33
34	0.2636	18.4112	236.2607	3.7943	69.8579	0.0543	0.0143	12.8324	34
35	0.2534	18.6646	244.8768	3.9461	73.6522	0.0536	0.0136	13.1198	35
36	0.2437	18.9083	253.4052	4.1039	77.5983	0.0529	0.0129	13.4018	36
37	0.2343	19.1426	261.8399	4.2681	81.7022	0.0522	0.0122	13.6784	37
38	0.2253	19.3679	270.1754	4.4388	85.9703	0.0516	0.0116	13.9497	38
39	0.2166	19.5845	278.4070	4.6164	90.4091	0.0511	0.0111	14.2157	39
40	0.2083	19.7928	286.5303	4.8010	95.0255	0.0505	0.0105	14.4765	40
41	0.2003	19.9931	294.5414	4.9931	99.8265	0.0500	0.0100	14.7322	41
42	0.1926	20.1856	302.4370	5.1928	104.8196	0.0495	0.0095	14.9828	42
43	0.1852	20.3708	310.2141	5.4005	110.0124	0.0491	0.0091	15.2284	43
44	0.1780	20.5488	317.8700	5.6165	115.4129	0.0487	0.0087	15.4690	44
45	0.1712	20.7200	325.4028	5.8412	121.0294	0.0483	0.0083	15.7047	45
46	0.1646	20.8847	332.8104	6.0748	126.8706	0.0479	0.0079	15.9356	46
47	0.1583	21.0429	340.0914	6.3178	132.9454	0.0475	0.0075	16.1618	47
48	0.1522	21.1951	347.2446	6.5705	139.2632	0.0472	0.0072	16.3832	48
49	0.1463	21.3415	354.2689	6.8333	145.8337	0.0469	0.0069	16.6000	49
50	0.1407	21.4822	361.1638	7.1067	152.6671	0.0466	0.0066	16.8122	50
51	0.1353	21.6175	367.9289	7.3910	159.7738	0.0463	0.0063	17.0200	51
52	0.1301	21.7476	374.5638	7.6866	167.1647	0.0460	0.0060	17.2232	52
53	0.1251	21.8727	381.0686	7.9941	174.8513	0.0457	0.0057	17.4221	53
54	0.1203	21.9930	387.4436	8.3138	182.8454	0.0455	0.0055	17.6167	54
55	0.1157	22.1086	393.6890	8.6464	191.1592	0.0452	0.0052	17.8070	55
60	0.0951	22.6235	422.9966	10.5196	237.9907	0.0442	0.0042	18.6972	60
65	0.0781	23.0467	449.2014	12.7987	294.9684	0.0434	0.0034	19.4909	65
70	0.0642	23.3945	472.4789	15.5716	364.2905	0.0427	0.0027	20.1961	70
75	0.0528	23.6804	493.0408	18.9453	448.6314	0.0422	0.0022	20.8206	75
80	0.0434	23.9154	511.1161	23.0498	551.2450	0.0418	0.0018	21.3718	80
85	0.0357	24.1085	526.9384	28.0436	676.0901	0.0415	0.0015	21.8569	85
90	0.0293	24.2673	540.7369	34.1193	827.9833	0.0412	0.0012	22.2826	90
95	0.0241	24.3978	552.7307	41.5114	1012.7846	0.0410	0.0010	22.6550	95
100	0.0198	24.5050	563.1249	50.5049	1237.6237	0.0408	0.0008	22.9800	100

(continued)

APPENDIX L *(continued)*
Factor Tables
$I = 5.00\%$

n	P/F	P/A	P/G	F/P	F/A	A/P	A/F	A/G	n
1	0.9524	0.9524	0.0000	1.0500	1.0000	1.0500	1.0000	0.0000	1
2	0.9070	1.8594	0.9070	1.1025	2.0500	0.5378	0.4878	0.4878	2
3	0.8638	2.7232	2.6347	1.1576	3.1525	0.3672	0.3172	0.9675	3
4	0.8227	3.5460	5.1028	1.2155	4.3101	0.2820	0.2320	1.4391	4
5	0.7835	4.3295	8.2369	1.2763	5.5256	0.2310	0.1810	1.9025	5
6	0.7462	5.0757	11.9680	1.3401	6.8019	0.1970	0.1470	2.3579	6
7	0.7107	5.7864	16.2321	1.4071	8.1420	0.1728	0.1228	2.8052	7
8	0.6768	6.4632	20.9700	1.4775	9.5491	0.1547	0.1047	3.2445	8
9	0.6446	7.1078	26.1268	1.5513	11.0266	0.1407	0.0907	3.6758	9
10	0.6139	7.7217	31.6520	1.6289	12.5779	0.1295	0.0795	4.0991	10
11	0.5847	8.3064	37.4988	1.7103	14.2068	0.1204	0.0704	4.5144	11
12	0.5568	8.8633	43.6241	1.7959	15.9171	0.1128	0.0628	4.9219	12
13	0.5303	9.3936	49.9879	1.8856	17.7130	0.1065	0.0565	5.3215	13
14	0.5051	9.8986	56.5538	1.9799	19.5986	0.1010	0.0510	5.7133	14
15	0.4810	10.3797	63.2880	2.0789	21.5786	0.0963	0.0463	6.0973	15
16	0.4581	10.8378	70.1597	2.1829	23.6575	0.0923	0.0423	6.4736	16
17	0.4363	11.2741	77.1405	2.2920	25.8404	0.0887	0.0387	6.8423	17
18	0.4155	11.6896	84.2043	2.4066	28.1324	0.0855	0.0355	7.2034	18
19	0.3957	12.0853	91.3275	2.5270	30.5390	0.0827	0.0327	7.5569	19
20	0.3769	12.4622	98.4884	2.6533	33.0660	0.0802	0.0302	7.9030	20
21	0.3589	12.8212	105.6673	2.7860	35.7193	0.0780	0.0280	8.2416	21
22	0.3418	13.1630	112.8461	2.9253	38.5052	0.0760	0.0260	8.5730	22
23	0.3256	13.4886	120.0087	3.0715	41.4305	0.0741	0.0241	8.8971	23
24	0.3101	13.7986	127.1402	3.2251	44.5020	0.0725	0.0225	9.2140	24
25	0.2953	14.0939	134.2275	3.3864	47.7271	0.0710	0.0210	9.5238	25
26	0.2812	14.3752	141.2585	3.5557	51.1135	0.0696	0.0196	9.8266	26
27	0.2678	14.6430	148.2226	3.7335	54.6691	0.0683	0.0183	10.1224	27
28	0.2551	14.8981	155.1101	3.9201	58.4026	0.0671	0.0171	10.4114	28
29	0.2429	15.1411	161.9126	4.1161	62.3227	0.0660	0.0160	10.6936	29
30	0.2314	15.3725	168.6226	4.3219	66.4388	0.0651	0.0151	10.9691	30
31	0.2204	15.5928	175.2333	4.5380	70.7608	0.0641	0.0141	11.2381	31
32	0.2099	15.8027	181.7392	4.7649	75.2988	0.0633	0.0133	11.5005	32
33	0.1999	16.0025	188.1351	5.0032	80.0638	0.0625	0.0125	11.7566	33
34	0.1904	16.1929	194.4168	5.2533	85.0670	0.0618	0.0118	12.0063	34
35	0.1813	16.3742	200.5807	5.5160	90.3203	0.0611	0.0111	12.2498	35
36	0.1727	16.5469	206.6237	5.7918	95.8363	0.0604	0.0104	12.4872	36
37	0.1644	16.7113	212.5434	6.0814	101.6281	0.0598	0.0098	12.7186	37
38	0.1566	16.8679	218.3378	6.3855	107.7095	0.0593	0.0093	12.9440	38
39	0.1491	17.0170	224.0054	6.7048	114.0950	0.0588	0.0088	13.1636	39
40	0.1420	17.1591	229.5452	7.0400	120.7998	0.0583	0.0083	13.3775	40
41	0.1353	17.2944	234.9564	7.3920	127.8398	0.0578	0.0078	13.5857	41
42	0.1288	17.4232	240.2389	7.7616	135.2318	0.0574	0.0074	13.7884	42
43	0.1227	17.5459	245.3925	8.1497	142.9933	0.0570	0.0070	13.9857	43
44	0.1169	17.6628	250.4175	8.5572	151.1430	0.0566	0.0066	14.1777	44
45	0.1113	17.7741	255.3145	8.9850	159.7002	0.0563	0.0063	14.3644	45
46	0.1060	17.8801	260.0844	9.4343	168.6852	0.0559	0.0059	14.5461	46
47	0.1009	17.9810	264.7281	9.9060	178.1194	0.0556	0.0056	14.7226	47
48	0.0961	18.0772	269.2467	10.4013	188.0254	0.0553	0.0053	14.8943	48
49	0.0916	18.1687	273.6418	10.9213	198.4267	0.0550	0.0050	15.0611	49
50	0.0872	18.2559	277.9148	11.4674	209.3480	0.0548	0.0048	15.2233	50
51	0.0831	18.3390	282.0673	12.0408	220.8154	0.0545	0.0045	15.3808	51
52	0.0791	18.4181	286.1013	12.6428	232.8562	0.0543	0.0043	15.5337	52
53	0.0753	18.4934	290.0184	13.2749	245.4990	0.0541	0.0041	15.6823	53
54	0.0717	18.5651	293.8208	13.9387	258.7739	0.0539	0.0039	15.8265	54
55	0.0683	18.6335	297.5104	14.6356	272.7126	0.0537	0.0037	15.9664	55
60	0.0535	18.9293	314.3432	18.6792	353.5837	0.0528	0.0028	16.6062	60
65	0.0419	19.1611	328.6910	23.8399	456.7980	0.0522	0.0022	17.1541	65
70	0.0329	19.3427	340.8409	30.4264	588.5285	0.0517	0.0017	17.6212	70
75	0.0258	19.4850	351.0721	38.8327	756.6537	0.0513	0.0013	18.0176	75
80	0.0202	19.5965	359.6460	49.5614	971.2288	0.0510	0.0010	18.3526	80
85	0.0158	19.6838	366.8007	63.2544	1245.0871	0.0508	0.0008	18.6346	85
90	0.0124	19.7523	372.7488	80.7304	1597.6073	0.0506	0.0006	18.8712	90
95	0.0097	19.8059	377.6774	103.0347	2040.6935	0.0505	0.0005	19.0689	95
100	0.0076	19.8479	381.7492	131.5013	2610.0252	0.0504	0.0004	19.2337	100

(continued)

APPENDIX L *(continued)*
Factor Tables

$I = 6.00\%$

n	P/F	P/A	P/G	F/P	F/A	A/P	A/F	A/G	n
1	0.9434	0.9434	0.0000	1.0600	1.0000	1.0600	1.0000	0.0000	1
2	0.8900	1.8334	0.8900	1.1236	2.0600	0.5454	0.4854	0.4854	2
3	0.8396	2.6730	2.5692	1.1910	3.1836	0.3741	0.3141	0.9612	3
4	0.7921	3.4651	4.9455	1.2625	4.3746	0.2886	0.2286	1.4272	4
5	0.7473	4.2124	7.9345	1.3382	5.6371	0.2374	0.1774	1.8836	5
6	0.7050	4.9173	11.4594	1.4185	6.9753	0.2034	0.1434	2.3304	6
7	0.6651	5.5824	15.4497	1.5036	8.3938	0.1791	0.1191	2.7676	7
8	0.6274	6.2098	19.8416	1.5938	9.8975	0.1610	0.1010	3.1952	8
9	0.5919	6.8017	24.5768	1.6895	11.4913	0.1470	0.0870	3.6133	9
10	0.5584	7.3601	29.6023	1.7908	13.1808	0.1359	0.0759	4.0220	10
11	0.5268	7.8869	34.8702	1.8983	14.9716	0.1268	0.0668	4.4213	11
12	0.4970	8.3838	40.3369	2.0122	16.8699	0.1193	0.0593	4.8113	12
13	0.4688	8.8527	45.9629	2.1329	18.8821	0.1130	0.0530	5.1920	13
14	0.4423	9.2950	51.7128	2.2609	21.0151	0.1076	0.0476	5.5635	14
15	0.4173	9.7122	57.5546	2.3966	23.2760	0.1030	0.0430	5.9260	15
16	0.3936	10.1059	63.4592	2.5404	25.6725	0.0990	0.0390	6.2794	16
17	0.3714	10.4773	69.4011	2.6928	28.2129	0.0954	0.0354	6.6240	17
18	0.3503	10.8276	75.3569	2.8543	30.9057	0.0924	0.0324	6.9597	18
19	0.3305	11.1581	81.3062	3.0256	33.7600	0.0896	0.0296	7.2867	19
20	0.3118	11.4699	87.2304	3.2071	36.7856	0.0872	0.0272	7.6051	20
21	0.2942	11.7641	93.1136	3.3996	39.9927	0.0850	0.0250	7.9151	21
22	0.2775	12.0416	98.9412	3.6035	43.3923	0.0830	0.0230	8.2166	22
23	0.2618	12.3034	104.7007	3.8197	46.9958	0.0813	0.0213	8.5099	23
24	0.2470	12.5504	110.3812	4.0489	50.8156	0.0797	0.0197	8.7951	24
25	0.2330	12.7834	115.9732	4.2919	54.8645	0.0782	0.0182	9.0722	25
26	0.2198	13.0032	121.4684	4.5494	59.1564	0.0769	0.0169	9.3414	26
27	0.2074	13.2105	126.8600	4.8223	63.7058	0.0757	0.0157	9.6029	27
28	0.1956	13.4062	132.1420	5.1117	68.5281	0.0746	0.0146	9.8568	28
29	0.1846	13.5907	137.3096	5.4184	73.6398	0.0736	0.0136	10.1032	29
30	0.1741	13.7648	142.3588	5.7435	79.0582	0.0726	0.0126	10.3422	30
31	0.1643	13.9291	147.2864	6.0881	84.8017	0.0718	0.0118	10.5740	31
32	0.1550	14.0840	152.0901	6.4534	90.8898	0.0710	0.0110	10.7988	32
33	0.1462	14.2302	156.7681	6.8406	97.3432	0.0703	0.0103	11.0166	33
34	0.1379	14.3681	161.3192	7.2510	104.1838	0.0696	0.0096	11.2276	34
35	0.1301	14.4982	165.7427	7.6861	111.4348	0.0690	0.0090	11.4319	35
36	0.1227	14.6210	170.0387	8.1473	119.1209	0.0684	0.0084	11.6298	36
37	0.1158	14.7368	174.2072	8.6361	127.2681	0.0679	0.0079	11.8213	37
38	0.1092	14.8460	178.2490	9.1543	135.9042	0.0674	0.0074	12.0065	38
39	0.1031	14.9491	182.1652	9.7035	145.0585	0.0669	0.0069	12.1857	39
40	0.0972	15.0463	185.9568	10.2857	154.7620	0.0665	0.0065	12.3590	40
41	0.0917	15.1380	189.6256	10.9029	165.0477	0.0661	0.0061	12.5264	41
42	0.0865	15.2245	193.1732	11.5570	175.9505	0.0657	0.0057	12.6883	42
43	0.0816	15.3062	196.6017	12.2505	187.5076	0.0653	0.0053	12.8446	43
44	0.0770	15.3832	199.9130	12.9855	199.7580	0.0650	0.0050	12.9956	44
45	0.0727	15.4558	203.1096	13.7646	212.7435	0.0647	0.0047	13.1413	45
46	0.0685	15.5244	206.1938	14.5905	226.5081	0.0644	0.0044	13.2819	46
47	0.0647	15.5890	209.1681	15.4659	241.0986	0.0641	0.0041	13.4177	47
48	0.0610	15.6500	212.0351	16.3939	256.5645	0.0639	0.0039	13.5485	48
49	0.0575	15.7076	214.7972	17.3775	272.9584	0.0637	0.0037	13.6748	49
50	0.0543	15.7619	217.4574	18.4202	290.3359	0.0634	0.0034	13.7964	50
51	0.0512	15.8131	220.0181	19.5254	308.7561	0.0632	0.0032	13.9137	51
52	0.0483	15.8614	222.4823	20.6969	328.2814	0.0630	0.0030	14.0267	52
53	0.0456	15.9070	224.8525	21.9387	348.9783	0.0629	0.0029	14.1355	53
54	0.0430	15.9500	227.1316	23.2550	370.9170	0.0627	0.0027	14.2402	54
55	0.0406	15.9905	229.3222	24.6503	394.1720	0.0625	0.0025	14.3411	55
60	0.0303	16.1614	239.0428	32.9877	533.1282	0.0619	0.0019	14.7909	60
65	0.0227	16.2891	246.9450	44.1450	719.0829	0.0614	0.0014	15.1601	65
70	0.0169	16.3845	253.3271	59.0759	967.9322	0.0610	0.0010	15.4613	70
75	0.0126	16.4558	258.4527	79.0569	1300.9487	0.0608	0.0008	15.7058	75
80	0.0095	16.5091	262.5493	105.7960	1746.5999	0.0606	0.0006	15.9033	80
85	0.0071	16.5489	265.8096	141.5789	2342.9817	0.0604	0.0004	16.0620	85
90	0.0053	16.5787	268.3946	189.4645	3141.0752	0.0603	0.0003	16.1891	90
95	0.0039	16.6009	270.4375	253.5463	4209.1042	0.0602	0.0002	16.2905	95
100	0.0029	16.6175	272.0471	339.3021	5638.3681	0.0602	0.0002	16.3711	100

(continued)

APPENDIX L *(continued)*
Factor Tables

$I = 7.00\%$

n	P/F	P/A	P/G	F/P	F/A	A/P	A/F	A/G	n
1	0.9346	0.9346	0.0000	1.0700	1.0000	1.0700	1.0000	0.0000	1
2	0.8734	1.8080	0.8734	1.1449	2.0700	0.5531	0.4831	0.4831	2
3	0.8163	2.6243	2.5060	1.2250	3.2149	0.3811	0.3111	0.9549	3
4	0.7629	3.3872	4.7947	1.3108	4.4399	0.2952	0.2252	1.4155	4
5	0.7130	4.1002	7.6467	1.4026	5.7507	0.2439	0.1739	1.8650	5
6	0.6663	4.7665	10.9784	1.5007	7.1533	0.2098	0.1398	2.3032	6
7	0.6227	5.3893	14.7149	1.6058	8.6540	0.1856	0.1156	2.7304	7
8	0.5820	5.9713	18.7889	1.7182	10.2598	0.1675	0.0975	3.1465	8
9	0.5439	6.5152	23.1404	1.8385	11.9780	0.1535	0.0835	3.5517	9
10	0.5083	7.0236	27.7156	1.9672	13.8164	0.1424	0.0724	3.9461	10
11	0.4751	7.4987	32.4665	2.1049	15.7836	0.1334	0.0634	4.3296	11
12	0.4440	7.9427	37.3506	2.2522	17.8885	0.1259	0.0559	4.7025	12
13	0.4150	8.3577	42.3302	2.4098	20.1406	0.1197	0.0497	5.0648	13
14	0.3878	8.7455	47.3718	2.5785	22.5505	0.1143	0.0443	5.4167	14
15	0.3624	9.1079	52.4461	2.7590	25.1290	0.1098	0.0398	5.7583	15
16	0.3387	9.4466	57.5271	2.9522	27.8881	0.1059	0.0359	6.0897	16
17	0.3166	9.7632	62.5923	3.1588	30.8402	0.1024	0.0324	6.4110	17
18	0.2959	10.0591	67.6219	3.3799	33.9990	0.0994	0.0294	6.7225	18
19	0.2765	10.3356	72.5991	3.6165	37.3790	0.0968	0.0268	7.0242	19
20	0.2584	10.5940	77.5091	3.8697	40.9955	0.0944	0.0244	7.3163	20
21	0.2415	10.8355	82.3393	4.1406	44.8652	0.0923	0.0223	7.5990	21
22	0.2257	11.0612	87.0793	4.4304	49.0057	0.0904	0.0204	7.8725	22
23	0.2109	11.2722	91.7201	4.7405	53.4361	0.0887	0.0187	8.1369	23
24	0.1971	11.4693	96.2545	5.0724	58.1767	0.0872	0.0172	8.3923	24
25	0.1842	11.6536	100.6765	5.4274	63.2490	0.0858	0.0158	8.6391	25
26	0.1722	11.8258	104.9814	5.8074	68.6765	0.0846	0.0146	8.8773	26
27	0.1609	11.9867	109.1656	6.2139	74.4838	0.0834	0.0134	9.1072	27
28	0.1504	12.1371	113.2264	6.6488	80.6977	0.0824	0.0124	9.3289	28
29	0.1406	12.2777	117.1622	7.1143	87.3465	0.0814	0.0114	9.5427	29
30	0.1314	12.4090	120.9718	7.6123	94.4608	0.0806	0.0106	9.7487	30
31	0.1228	12.5318	124.6550	8.1451	102.0730	0.0798	0.0098	9.9471	31
32	0.1147	12.6466	128.2120	8.7153	110.2182	0.0791	0.0091	10.1381	32
33	0.1072	12.7538	131.6435	9.3253	118.9334	0.0784	0.0084	10.3219	33
34	0.1002	12.8540	134.9507	9.9781	128.2588	0.0778	0.0078	10.4987	34
35	0.0937	12.9477	138.1353	10.6766	138.2369	0.0772	0.0072	10.6687	35
36	0.0875	13.0352	141.1990	11.4239	148.9135	0.0767	0.0067	10.8321	36
37	0.0818	13.1170	144.1441	12.2236	160.3374	0.0762	0.0062	10.9891	37
38	0.0765	13.1935	146.9730	13.0793	172.5610	0.0758	0.0058	11.1398	38
39	0.0715	13.2649	149.6883	13.9948	185.6403	0.0754	0.0054	11.2845	39
40	0.0668	13.3317	152.2928	14.9745	199.6351	0.0750	0.0050	11.4233	40
41	0.0624	13.3941	154.7892	16.0227	214.6096	0.0747	0.0047	11.5565	41
42	0.0583	13.4524	157.1807	17.1443	230.6322	0.0743	0.0043	11.6842	42
43	0.0545	13.5070	159.4702	18.3444	247.7765	0.0740	0.0040	11.8065	43
44	0.0509	13.5579	161.6609	19.6285	266.1209	0.0738	0.0038	11.9237	44
45	0.0476	13.6055	163.7559	21.0025	285.7493	0.0735	0.0035	12.0360	45
46	0.0445	13.6500	165.7584	22.4726	306.7518	0.0733	0.0033	12.1435	46
47	0.0416	13.6916	167.6714	24.0457	329.2244	0.0730	0.0030	12.2463	47
48	0.0389	13.7305	169.4981	25.7289	353.2701	0.0728	0.0028	12.3447	48
49	0.0363	13.7668	171.2417	27.5299	378.9990	0.0726	0.0026	12.4387	49
50	0.0339	13.8007	172.9051	29.4570	406.5289	0.0725	0.0025	12.5287	50
51	0.0317	13.8325	174.4915	31.5190	435.9860	0.0723	0.0023	12.6146	51
52	0.0297	13.8621	176.0037	33.7253	467.5050	0.0721	0.0021	12.6967	52
53	0.0277	13.8898	177.4447	36.0861	501.2303	0.0720	0.0020	12.7751	53
54	0.0259	13.9157	178.8173	38.6122	537.3164	0.0719	0.0019	12.8500	54
55	0.0242	13.9399	180.1243	41.3150	575.9286	0.0717	0.0017	12.9215	55
60	0.0173	14.0392	185.7677	57.9464	813.5204	0.0712	0.0012	13.2321	60
65	0.0123	14.1099	190.1452	81.2729	1146.7552	0.0709	0.0009	13.4760	65
70	0.0088	14.1604	193.5185	113.9894	1614.1342	0.0706	0.0006	13.6662	70
75	0.0063	14.1964	196.1035	159.8760	2269.6574	0.0704	0.0004	13.8136	75
80	0.0045	14.2220	198.0748	224.2344	3189.0627	0.0703	0.0003	13.9273	80
85	0.0032	14.2403	199.5717	314.5003	4478.5761	0.0702	0.0002	14.0146	85
90	0.0023	14.2533	200.7042	441.1030	6287.1854	0.0702	0.0002	14.0812	90
95	0.0016	14.2626	201.5581	618.6697	8823.8535	0.0701	0.0001	14.1319	95
100	0.0012	14.2693	202.2001	867.7163	12381.6618	0.0701	0.0001	14.1703	100

(continued)

APPENDIX L *(continued)*
Factor Tables

$I = 8.00\%$

n	P/F	P/A	P/G	F/P	F/A	A/P	A/F	A/G	n
1	0.9259	0.9259	0.0000	1.0800	1.0000	1.0800	1.0000	0.0000	1
2	0.8573	1.7833	0.8573	1.1664	2.0800	0.5608	0.4808	0.4808	2
3	0.7938	2.5771	2.4450	1.2597	3.2464	0.3880	0.3080	0.9487	3
4	0.7350	3.3121	4.6501	1.3605	4.5061	0.3019	0.2219	1.4040	4
5	0.6806	3.9927	7.3724	1.4693	5.8666	0.2505	0.1705	1.8465	5
6	0.6302	4.6229	10.5233	1.5869	7.3359	0.2163	0.1363	2.2763	6
7	0.5835	5.2064	14.0242	1.7138	8.9228	0.1921	0.1121	2.6937	7
8	0.5403	5.7466	17.8061	1.8509	10.6366	0.1740	0.0940	3.0985	8
9	0.5002	6.2469	21.8081	1.9990	12.4876	0.1601	0.0801	3.4910	9
10	0.4632	6.7101	25.9768	2.1589	14.4866	0.1490	0.0690	3.8713	10
11	0.4289	7.1390	30.2657	2.3316	16.6455	0.1401	0.0601	4.2395	11
12	0.3971	7.5361	34.6339	2.5182	18.9771	0.1327	0.0527	4.5957	12
13	0.3677	7.9038	39.0463	2.7196	21.4953	0.1265	0.0465	4.9402	13
14	0.3405	8.2442	43.4723	2.9372	24.2149	0.1213	0.0413	5.2731	14
15	0.3152	8.5595	47.8857	3.1722	27.1521	0.1168	0.0368	5.5945	15
16	0.2919	8.8514	52.2640	3.4259	30.3243	0.1130	0.0330	5.9046	16
17	0.2703	9.1216	56.5883	3.7000	33.7502	0.1096	0.0296	6.2037	17
18	0.2502	9.3719	60.8426	3.9960	37.4502	0.1067	0.0267	6.4920	18
19	0.2317	9.6036	65.0134	4.3157	41.4463	0.1041	0.0241	6.7697	19
20	0.2145	9.8181	69.0898	4.6610	45.7620	0.1019	0.0219	7.0369	20
21	0.1987	10.0168	73.0629	5.0338	50.4229	0.0998	0.0198	7.2940	21
22	0.1839	10.2007	76.9257	5.4365	55.4568	0.0980	0.0180	7.5412	22
23	0.1703	10.3711	80.6726	5.8715	60.8933	0.0964	0.0164	7.7786	23
24	0.1577	10.5288	84.2997	6.3412	66.7648	0.0950	0.0150	8.0066	24
25	0.1460	10.6748	87.8041	6.8485	73.1059	0.0937	0.0137	8.2254	25
26	0.1352	10.8100	91.1842	7.3964	79.9544	0.0925	0.0125	8.4352	26
27	0.1252	10.9352	94.4390	7.9881	87.3508	0.0914	0.0114	8.6363	27
28	0.1159	11.0511	97.5687	8.6271	95.3388	0.0905	0.0105	8.8289	28
29	0.1073	11.1584	100.5738	9.3173	103.9659	0.0896	0.0096	9.0133	29
30	0.0994	11.2578	103.4558	10.0627	113.2832	0.0888	0.0088	9.1897	30
31	0.0920	11.3498	106.2163	10.8677	123.3459	0.0881	0.0081	9.3584	31
32	0.0852	11.4350	108.8575	11.7371	134.2135	0.0875	0.0075	9.5197	32
33	0.0789	11.5139	111.3819	12.6760	145.9506	0.0869	0.0069	9.6737	33
34	0.0730	11.5869	113.7924	13.6901	158.6267	0.0863	0.0063	9.8208	34
35	0.0676	11.6546	116.0920	14.7853	172.3168	0.0858	0.0058	9.9611	35
36	0.0626	11.7172	118.2839	15.9682	187.1021	0.0853	0.0053	10.0949	36
37	0.0580	11.7752	120.3713	17.2456	203.0703	0.0849	0.0049	10.2225	37
38	0.0537	11.8289	122.3579	18.6253	220.3159	0.0845	0.0045	10.3440	38
39	0.0497	11.8786	124.2470	20.1153	238.9412	0.0842	0.0042	10.4597	39
40	0.0460	11.9246	126.0422	21.7245	259.0565	0.0839	0.0039	10.5699	40
41	0.0426	11.9672	127.7470	23.4625	280.7810	0.0836	0.0036	10.6747	41
42	0.0395	12.0067	129.3651	25.3395	304.2435	0.0833	0.0033	10.7744	42
43	0.0365	12.0432	130.8998	27.3666	329.5830	0.0830	0.0030	10.8692	43
44	0.0338	12.0771	132.3547	29.5560	356.9496	0.0828	0.0028	10.9592	44
45	0.0313	12.1084	133.7331	31.9204	386.5056	0.0826	0.0026	11.0447	45
46	0.0290	12.1374	135.0384	34.4741	418.4261	0.0824	0.0024	11.1258	46
47	0.0269	12.1643	136.2739	37.2320	452.9002	0.0822	0.0022	11.2028	47
48	0.0249	12.1891	137.4428	40.2106	490.1322	0.0820	0.0020	11.2758	48
49	0.0230	12.2122	138.5480	43.4274	530.3427	0.0819	0.0019	11.3451	49
50	0.0213	12.2335	139.5928	46.9016	573.7702	0.0817	0.0017	11.4107	50
51	0.0197	12.2532	140.5799	50.6537	620.6718	0.0816	0.0016	11.4729	51
52	0.0183	12.2715	141.5121	54.7060	671.3255	0.0815	0.0015	11.5318	52
53	0.0169	12.2884	142.3923	59.0825	726.0316	0.0814	0.0014	11.5875	53
54	0.0157	12.3041	143.2229	63.8091	785.1141	0.0813	0.0013	11.6403	54
55	0.0145	12.3186	144.0065	68.9139	848.9232	0.0812	0.0012	11.6902	55
60	0.0099	12.3766	147.3000	101.2571	1253.2133	0.0808	0.0008	11.9015	60
65	0.0067	12.4160	149.7387	148.7798	1847.2481	0.0805	0.0005	12.0602	65
70	0.0046	12.4428	151.5326	218.6064	2720.0801	0.0804	0.0004	12.1783	70
75	0.0031	12.4611	152.8448	321.2045	4002.5566	0.0802	0.0002	12.2658	75
80	0.0021	12.4735	153.8001	471.9548	5886.9354	0.0802	0.0002	12.3301	80
85	0.0014	12.4820	154.4925	693.4565	8655.7061	0.0801	0.0001	12.3772	85
90	0.0010	12.4877	154.9925	1018.9151	12723.9386	0.0801	0.0001	12.4116	90
95	0.0007	12.4917	155.3524	1497.1205	18701.5069	0.0801	0.0001	12.4365	95
100	0.0005	12.4943	155.6107	2199.7613	27484.5157	0.0800	0.0000	12.4545	100

(continued)

APPENDIX L *(continued)*
Factor Tables

$I = 9.00\%$

n	P/F	P/A	P/G	F/P	F/A	A/P	A/F	A/G	n
1	0.9174	0.9174	0.0000	1.0900	1.0000	1.0900	1.0000	0.0000	1
2	0.8417	1.7591	0.8417	1.1881	2.0900	0.5685	0.4785	0.4785	2
3	0.7722	2.5313	2.3860	1.2950	3.2781	0.3951	0.3051	0.9426	3
4	0.7084	3.2397	4.5113	1.4116	4.5731	0.3087	0.2187	1.3925	4
5	0.6499	3.8897	7.1110	1.5386	5.9847	0.2571	0.1671	1.8282	5
6	0.5963	4.4859	10.0924	1.6771	7.5233	0.2229	0.1329	2.2498	6
7	0.5470	5.0330	13.3746	1.8280	9.2004	0.1987	0.1087	2.6574	7
8	0.5019	5.5348	16.8877	1.9926	11.0285	0.1807	0.0907	3.0512	8
9	0.4604	5.9952	20.5711	2.1719	13.0210	0.1668	0.0768	3.4312	9
10	0.4224	6.4177	24.3728	2.3674	15.1929	0.1558	0.0658	3.7978	10
11	0.3875	6.8052	28.2481	2.5804	17.5603	0.1469	0.0569	4.1510	11
12	0.3555	7.1607	32.1590	2.8127	20.1407	0.1397	0.0497	4.4910	12
13	0.3262	7.4869	36.0731	3.0658	22.9534	0.1336	0.0436	4.8182	13
14	0.2992	7.7862	39.9633	3.3417	26.0192	0.1284	0.0384	5.1326	14
15	0.2745	8.0607	43.8069	3.6425	29.3609	0.1241	0.0341	5.4346	15
16	0.2519	8.3126	47.5849	3.9703	33.0034	0.1203	0.0303	5.7245	16
17	0.2311	8.5436	51.2821	4.3276	36.9737	0.1170	0.0270	6.0024	17
18	0.2120	8.7556	54.8860	4.7171	41.3013	0.1142	0.0242	6.2687	18
19	0.1945	8.9501	58.3868	5.1417	46.0185	0.1117	0.0217	6.5236	19
20	0.1784	9.1285	61.7770	5.6044	51.1601	0.1095	0.0195	6.7674	20
21	0.1637	9.2922	65.0509	6.1088	56.7645	0.1076	0.0176	7.0006	21
22	0.1502	9.4424	68.2048	6.6586	62.8733	0.1059	0.0159	7.2232	22
23	0.1378	9.5802	71.2359	7.2579	69.5319	0.1044	0.0144	7.4357	23
24	0.1264	9.7066	74.1433	7.9111	76.7898	0.1030	0.0130	7.6384	24
25	0.1160	9.8226	76.9265	8.6231	84.7009	0.1018	0.0118	7.8316	25
26	0.1064	9.9290	79.5863	9.3992	93.3240	0.1007	0.0107	8.0156	26
27	0.0976	10.0266	82.1241	10.2451	102.7231	0.0997	0.0097	8.1906	27
28	0.0895	10.1161	84.5419	11.1671	112.9682	0.0989	0.0089	8.3571	28
29	0.0822	10.1983	86.8422	12.1722	124.1354	0.0981	0.0081	8.5154	29
30	0.0754	10.2737	89.0280	13.2677	136.3076	0.0973	0.0073	8.6657	30
31	0.0691	10.3428	91.1024	14.4618	149.5752	0.0967	0.0067	8.8083	31
32	0.0634	10.4062	93.0690	15.7633	164.0370	0.0961	0.0061	8.9436	32
33	0.0582	10.4644	94.9314	17.1820	179.8003	0.0956	0.0056	9.0718	33
34	0.0534	10.5178	96.6935	18.7284	196.9823	0.0951	0.0051	9.1933	34
35	0.0490	10.5668	98.3590	20.4140	215.7108	0.0946	0.0046	9.3083	35
36	0.0449	10.6118	99.9319	22.2512	236.1247	0.0942	0.0042	9.4171	36
37	0.0412	10.6530	101.4162	24.2538	258.3759	0.0939	0.0039	9.5200	37
38	0.0378	10.6908	102.8158	26.4367	282.6298	0.0935	0.0035	9.6172	38
39	0.0347	10.7255	104.1345	28.8160	309.0665	0.0932	0.0032	9.7090	39
40	0.0318	10.7574	105.3762	31.4094	337.8824	0.0930	0.0030	9.7957	40
41	0.0292	10.7866	106.5445	34.2363	369.2919	0.0927	0.0027	9.8775	41
42	0.0268	10.8134	107.6432	37.3175	403.5281	0.0925	0.0025	9.9546	42
43	0.0246	10.8380	108.6758	40.6761	440.8457	0.0923	0.0023	10.0273	43
44	0.0226	10.8605	109.6456	44.3370	481.5218	0.0921	0.0021	10.0958	44
45	0.0207	10.8812	110.5561	48.3273	525.8587	0.0919	0.0019	10.1603	45
46	0.0190	10.9002	111.4103	52.6767	574.1860	0.0917	0.0017	10.2210	46
47	0.0174	10.9176	112.2115	57.4176	626.8628	0.0916	0.0016	10.2780	47
48	0.0160	10.9336	112.9625	62.5852	684.2804	0.0915	0.0015	10.3317	48
49	0.0147	10.9482	113.6661	68.2179	746.8656	0.0913	0.0013	10.3821	49
50	0.0134	10.9617	114.3251	74.3575	815.0836	0.0912	0.0012	10.4295	50
51	0.0123	10.9740	114.9420	81.0497	889.4411	0.0911	0.0011	10.4740	51
52	0.0113	10.9853	115.5193	88.3442	970.4908	0.0910	0.0010	10.5158	52
53	0.0104	10.9957	116.0593	96.2951	1058.8349	0.0909	0.0009	10.5549	53
54	0.0095	11.0053	116.5642	104.9617	1155.1301	0.0909	0.0009	10.5917	54
55	0.0087	11.0140	117.0362	114.4083	1260.0918	0.0908	0.0008	10.6261	55
60	0.0057	11.0480	118.9683	176.0313	1944.7921	0.0905	0.0005	10.7683	60
65	0.0037	11.0701	120.3344	270.8460	2998.2885	0.0903	0.0003	10.8702	65
70	0.0024	11.0844	121.2942	416.7301	4619.2232	0.0902	0.0002	10.9427	70
75	0.0016	11.0938	121.9646	641.1909	7113.2321	0.0901	0.0001	10.9940	75
80	0.0010	11.0998	122.4306	986.5517	10950.5741	0.0901	0.0001	11.0299	80
85	0.0007	11.1038	122.7533	1517.9320	16854.8003	0.0901	0.0001	11.0551	85
90	0.0004	11.1064	122.9758	2335.5266	25939.1842	0.0900	0.0000	11.0726	90
95	0.0003	11.1080	123.1287	3593.4971	39916.6350	0.0900	0.0000	11.0847	95
100	0.0002	11.1091	123.2335	5529.0408	61422.6755	0.0900	0.0000	11.0930	100

(continued)

Support Material

APPENDIX L *(continued)*
Factor Tables

$I = 10.00\%$

n	P/F	P/A	P/G	F/P	F/A	A/P	A/F	A/G	n
1	0.9091	0.9091	0.0000	1.1000	1.0000	1.1000	1.0000	0.0000	1
2	0.8264	1.7355	0.8264	1.2100	2.1000	0.5762	0.4762	0.4762	2
3	0.7513	2.4869	2.3291	1.3310	3.3100	0.4021	0.3021	0.9366	3
4	0.6830	3.1699	4.3781	1.4641	4.6410	0.3155	0.2155	1.3812	4
5	0.6209	3.7908	6.8618	1.6105	6.1051	0.2638	0.1638	1.8101	5
6	0.5645	4.3553	9.6842	1.7716	7.7156	0.2296	0.1296	2.2236	6
7	0.5132	4.8684	12.7631	1.9487	9.4872	0.2054	0.1054	2.6216	7
8	0.4665	5.3349	16.0287	2.1436	11.4359	0.1874	0.0874	3.0045	8
9	0.4241	5.7590	19.4215	2.3579	13.5795	0.1736	0.0736	3.3724	9
10	0.3855	6.1446	22.8913	2.5937	15.9374	0.1627	0.0627	3.7255	10
11	0.3505	6.4951	26.3963	2.8531	18.5312	0.1540	0.0540	4.0641	11
12	0.3186	6.8137	29.9012	3.1384	21.3843	0.1468	0.0468	4.3884	12
13	0.2897	7.1034	33.3772	3.4523	24.5227	0.1408	0.0408	4.6988	13
14	0.2633	7.3667	36.8005	3.7975	27.9750	0.1357	0.0357	4.9955	14
15	0.2394	7.6061	40.1520	4.1772	31.7725	0.1315	0.0315	5.2789	15
16	0.2176	7.8237	43.4164	4.5950	35.9497	0.1278	0.0278	5.5493	16
17	0.1978	8.0216	46.5819	5.0545	40.5447	0.1247	0.0247	5.8071	17
18	0.1799	8.2014	49.6395	5.5599	45.5992	0.1219	0.0219	6.0526	18
19	0.1635	8.3649	52.5827	6.1159	51.1591	0.1195	0.0195	6.2861	19
20	0.1486	8.5136	55.4069	6.7275	57.2750	0.1175	0.0175	6.5081	20
21	0.1351	8.6487	58.1095	7.4002	64.0025	0.1156	0.0156	6.7189	21
22	0.1228	8.7715	60.6893	8.1403	71.4027	0.1140	0.0140	6.9189	22
23	0.1117	8.8832	63.1462	8.9543	79.5430	0.1126	0.0126	7.1085	23
24	0.1015	8.9847	65.4813	9.8497	88.4973	0.1113	0.0113	7.2881	24
25	0.0923	9.0770	67.6964	10.8347	98.3471	0.1102	0.0102	7.4580	25
26	0.0839	9.1609	69.7940	11.9182	109.1818	0.1092	0.0092	7.6186	26
27	0.0763	9.2372	71.7773	13.1100	121.0999	0.1083	0.0083	7.7704	27
28	0.0693	9.3066	73.6495	14.4210	134.2099	0.1075	0.0075	7.9137	28
29	0.0630	9.3696	75.4146	15.8631	148.6309	0.1067	0.0067	8.0489	29
30	0.0573	9.4269	77.0766	17.4494	164.4940	0.1061	0.0061	8.1762	30
31	0.0521	9.4790	78.6395	19.1943	181.9434	0.1055	0.0055	8.2962	31
32	0.0474	9.5264	80.1078	21.1138	201.1378	0.1050	0.0050	8.4091	32
33	0.0431	9.5694	81.4856	23.2252	222.2515	0.1045	0.0045	8.5152	33
34	0.0391	9.6086	82.7773	25.5477	245.4767	0.1041	0.0041	8.6149	34
35	0.0356	9.6442	83.9872	28.1024	271.0244	0.1037	0.0037	8.7086	35
36	0.0323	9.6765	85.1194	30.9127	299.1268	0.1033	0.0033	8.7965	36
37	0.0294	9.7059	86.1781	34.0039	330.0395	0.1030	0.0030	8.8789	37
38	0.0267	9.7327	87.1673	37.4043	364.0434	0.1027	0.0027	8.9562	38
39	0.0243	9.7570	88.0908	41.1448	401.4478	0.0125	0.0025	9.0285	39
40	0.0221	9.7791	88.9525	45.2593	442.5926	0.1023	0.0023	9.0962	40
41	0.0201	9.7991	89.7560	49.7852	487.8518	0.1020	0.0020	9.1596	41
42	0.0183	9.8174	90.5047	54.7637	537.6370	0.1019	0.0019	9.2188	42
43	0.0166	9.8340	91.2019	60.2401	592.4007	0.1017	0.0017	9.2741	43
44	0.0151	9.8491	91.8508	66.2641	652.6408	0.1015	0.0015	9.3258	44
45	0.0137	9.8628	92.4544	72.8905	718.9048	0.1014	0.0014	9.3740	45
46	0.0125	9.8753	93.0157	80.1795	791.7953	0.1013	0.0013	9.4190	46
47	0.0113	9.8866	93.5372	88.1975	871.9749	0.1011	0.0011	9.4610	47
48	0.0103	9.8969	94.0217	97.0172	960.1723	0.1010	0.0010	9.5001	48
49	0.0094	9.9063	94.4715	106.7190	1057.1896	0.1009	0.0009	9.5365	49
50	0.0085	9.9148	94.8889	117.3909	1163.9085	0.1009	0.0009	9.5704	50
51	0.0077	9.9226	95.2761	129.1299	1281.2994	0.1008	0.0008	9.6020	51
52	0.0070	9.9296	95.6351	142.0429	1410.4293	0.1007	0.0007	9.6313	52
53	0.0064	9.9360	95.9679	156.2472	1552.4723	0.1006	0.0006	9.6586	53
54	0.0058	9.9418	96.2763	171.8719	1708.7195	0.1006	0.0006	9.6840	54
55	0.0053	9.9471	96.5619	189.0591	1880.5914	0.1005	0.0005	9.7075	55
60	0.0033	9.9672	97.7010	304.4816	3034.8164	0.1003	0.0003	9.8023	60
65	0.0020	9.9796	98.4705	490.3707	4893.7073	0.1002	0.0002	9.8672	65
70	0.0013	9.9873	98.9870	789.7470	7887.4696	0.1001	0.0001	9.9113	70
75	0.0008	9.9921	99.3317	1271.8954	12708.9537	0.1001	0.0001	9.9410	75
80	0.0005	9.9951	99.5606	2048.4002	20474.0021	0.1000	0.0000	9.9609	80
85	0.0003	9.9970	99.7120	3298.9690	32979.6903	0.1000	0.0000	9.9742	85
90	0.0002	9.9981	99.8118	5313.0226	53120.2261	0.1000	0.0000	9.9831	90
95	0.0001	9.9988	99.8773	8556.6760	85556.7605	0.1000	0.0000	9.9889	95
100	0.0001	9.9993	99.9202	13780.6123	137796.1234	0.1000	0.0000	9.9927	100

(continued)

APPENDIX L *(continued)*
Factor Tables

$I = 12.00\%$

n	P/F	P/A	P/G	F/P	F/A	A/P	A/F	A/G	n
1	0.8929	0.8929	0.0000	1.1200	1.0000	1.1200	1.0000	0.0000	1
2	0.7972	1.6901	0.7972	1.2544	2.1200	0.5917	0.4717	0.4717	2
3	0.7118	2.4018	2.2208	1.4049	3.3744	0.4163	0.2963	0.9246	3
4	0.6355	3.0373	4.1273	1.5735	4.7793	0.3292	0.2092	1.3589	4
5	0.5674	3.6048	6.3970	1.7623	6.3528	0.2774	0.1574	1.7746	5
6	0.5066	4.1114	8.9302	1.9738	8.1152	0.2432	0.1232	2.1720	6
7	0.4523	4.5638	11.6443	2.2107	10.0890	0.2191	0.0991	2.5515	7
8	0.4039	4.9676	14.4714	2.4760	12.2997	0.2013	0.0813	2.9131	8
9	0.3606	5.3282	17.3563	2.7731	14.7757	0.1877	0.0677	3.2574	9
10	0.3220	5.6502	20.2541	3.1058	17.5487	0.1770	0.0570	3.5847	10
11	0.2875	5.9377	23.1288	3.4785	20.6546	0.1684	0.0484	3.8953	11
12	0.2567	6.1944	25.9523	3.8960	24.1331	0.1614	0.0414	4.1897	12
13	0.2292	6.4235	28.7024	4.3635	28.0291	0.1557	0.0357	4.4683	13
14	0.2046	6.6282	31.3624	4.8871	32.3926	0.1509	0.0309	4.7317	14
15	0.1827	6.8109	33.9202	5.4736	37.2797	0.1468	0.0268	4.9803	15
16	0.1631	6.9740	36.3670	6.1304	42.7533	0.1434	0.0234	5.2147	16
17	0.1456	7.1196	38.6973	6.8660	48.8837	0.1405	0.0205	5.4353	17
18	0.1300	7.2497	40.9080	7.6900	55.7497	0.1379	0.0179	5.6427	18
19	0.1161	7.3658	42.9979	8.6128	63.4397	0.1358	0.0158	6.8375	19
20	0.1037	7.4694	44.9676	9.6463	72.0524	0.1339	0.0139	6.0202	20
21	0.0926	7.5620	46.8188	10.8038	81.6987	0.1322	0.0122	6.1913	21
22	0.0826	7.6446	48.5543	12.1003	92.5026	0.1308	0.0108	6.3514	22
23	0.0738	7.7184	50.1776	13.5523	104.6029	0.1296	0.0096	6.5010	23
24	0.0659	7.7843	51.6929	15.1786	118.1552	0.1285	0.0085	6.6406	24
25	0.0588	7.8431	53.1046	17.0001	133.3339	0.1275	0.0075	6.7708	25
26	0.0525	7.8957	54.4177	19.0401	150.3339	0.1267	0.0067	6.8921	26
27	0.0469	7.9426	55.6369	21.3249	169.3740	0.1259	0.0059	7.0049	27
28	0.0419	7.9844	56.7674	23.8839	190.6989	0.1252	0.0052	7.1098	28
29	0.0374	8.0218	57.8141	26.7499	214.5828	0.1247	0.0047	7.2071	29
30	0.0334	8.0552	58.7821	29.9599	241.3327	0.1241	0.0041	7.2974	30
31	0.0298	8.0850	59.6761	33.5551	271.2926	0.1237	0.0037	7.3811	31
32	0.0266	8.1116	60.5010	37.5817	304.8477	0.1233	0.0033	7.4586	32
33	0.0238	8.1354	61.2612	42.0915	342.4294	0.1229	0.0029	7.5302	33
34	0.0212	8.1566	61.9612	47.1425	384.5210	0.1226	0.0026	7.5965	34
35	0.0189	8.1755	62.6052	52.7996	431.6635	0.1223	0.0023	7.6577	35
36	0.0169	8.1924	63.1970	59.1356	484.4631	0.1221	0.0021	7.7141	36
37	0.0151	8.2075	63.7406	66.2318	543.5987	0.1218	0.0018	7.7661	37
38	0.0135	8.2210	64.2394	74.1797	609.8305	0.1216	0.0016	7.8141	38
39	0.0120	8.2330	64.6967	83.0812	684.0102	0.1215	0.0015	7.8582	39
40	0.0107	8.2438	65.1159	93.0510	767.0914	0.1213	0.0013	7.8988	40
41	0.0096	8.2534	65.4997	104.2171	860.1424	0.1212	0.0012	7.9361	41
42	0.0086	8.2619	65.8509	116.7231	964.3595	0.1210	0.0010	7.9704	42
43	0.0076	8.2696	66.1722	130.7299	1081.0826	0.1209	0.0009	8.0019	43
44	0.0068	8.2764	66.4659	146.4175	1211.8125	0.1208	0.0008	8.0308	44
45	0.0061	8.2825	66.7342	163.9876	1358.2300	0.1207	0.0007	8.0572	45
46	0.0054	8.2880	66.9792	183.6661	1522.2176	0.1207	0.0007	8.0815	46
47	0.0049	8.2928	67.2028	205.7061	1705.8838	0.1206	0.0006	8.1037	47
48	0.0043	8.2972	67.4068	230.3908	1911.5898	0.1205	0.0005	8.1241	48
49	0.0039	8.3010	67.5929	258.0377	2141.9806	0.1205	0.0005	8.1427	49
50	0.0035	8.3045	67.7624	289.0022	2400.0182	0.1204	0.0004	8.1597	50
51	0.0031	8.3076	67.9169	323.6825	2689.0204	0.1204	0.0004	8.1753	51
52	0.0028	8.3103	68.0576	362.5243	3012.7029	0.1203	0.0003	8.1895	52
53	0.0025	8.3128	68.1856	406.0273	3375.2272	0.1203	0.0003	8.2025	53
54	0.0022	8.3150	68.3022	454.7505	3781.2545	0.1203	0.0003	8.2143	54
55	0.0020	8.3170	68.4082	509.3206	4236.0050	0.1202	0.0002	8.2251	55
60	0.0011	8.3240	68.8100	897.5969	7471.6411	0.1201	0.0001	8.2664	60
65	0.0006	8.3281	69.0581	1581.8725	13173.9374	0.1201	0.0001	8.2922	65
70	0.0004	8.3303	69.2103	2787.7998	23223.3319	0.1200	0.0000	8.3082	70
75	0.0002	8.3316	69.3031	4913.0558	40933.7987	0.1200	0.0000	8.3181	75
80	0.0001	8.3324	69.3594	8658.4831	72145.6925	0.1200	0.0000	8.3241	80
85	0.0001	8.3328	69.3935	15259.2057	127151.7140	0.1200	0.0000	8.3278	85
90	0.0000	8.3330	69.4140	26891.9342	224091.1185	0.1200	0.0000	8.3300	90
95	0.0000	8.3332	69.4263	47392.7766	394931.4719	0.1200	0.0000	8.3313	95
100	0.0000	8.3332	69.4336	83522.2657	696010.5477	0.1200	0.0000	8.3321	100

(continued)

Support Material

APPENDIX L *(continued)*
Factor Tables

$I = 15.00\%$

n	P/F	P/A	P/G	F/P	F/A	A/P	A/F	A/G	n
1	0.8696	0.8696	0.0000	1.1500	1.0000	1.1500	1.0000	0.0000	1
2	0.7561	1.6257	0.7561	1.3225	2.1500	0.6151	0.4651	0.4651	2
3	0.6575	2.2832	2.0712	1.5209	3.4725	0.4380	0.2880	0.9071	3
4	0.5718	2.8550	3.7864	1.7490	4.9934	0.3503	0.2003	1.3263	4
5	0.4972	3.3522	5.7751	2.0114	6.7424	0.2983	0.1483	1.7228	5
6	0.4323	3.7845	7.9368	2.3131	8.7537	0.2642	0.1142	2.0972	6
7	0.3759	4.1604	10.1924	2.6600	11.0668	0.2404	0.0904	2.4498	7
8	0.3269	4.4873	12.4807	3.0590	13.7268	0.2229	0.0729	2.7813	8
9	0.2843	4.7716	14.7548	3.5179	16.7858	0.2096	0.0596	3.0922	9
10	0.2472	5.0188	16.9795	4.0456	20.3037	0.1993	0.0493	3.3832	10
11	0.2149	5.2337	19.1289	4.6524	24.3493	0.1911	0.0411	3.6549	11
12	0.1869	5.4206	21.1849	5.3503	29.0017	0.1845	0.0345	3.9082	12
13	0.1625	5.5831	23.1352	6.1528	34.3519	0.1791	0.0291	4.1438	13
14	0.1413	5.7245	24.9725	7.0757	40.5047	0.1747	0.0247	4.3624	14
15	0.1229	5.8474	26.9630	8.1371	47.5804	0.1710	0.0210	4.5650	15
16	0.1069	5.9542	28.2960	9.3576	55.7175	0.1679	0.0179	4.7522	16
17	0.0929	6.0472	29.7828	10.7613	65.0751	0.1654	0.0154	4.9251	17
18	0.0808	6.1280	31.1565	12.3755	75.8364	0.1632	0.0132	5.0843	18
19	0.0703	6.1982	32.4213	14.2318	88.2118	0.1613	0.0113	5.2307	19
20	0.0611	6.2593	33.5822	16.3665	102.4436	0.1598	0.0098	5.3651	20
21	0.0531	6.3125	34.6448	18.8215	118.8101	0.1584	0.0084	5.4883	21
22	0.0462	6.3587	35.6150	21.6447	137.6316	0.1573	0.0073	5.6010	22
23	0.0402	6.3988	36.4988	24.8915	159.2764	0.1563	0.0063	5.7040	23
24	0.0349	6.4338	37.3023	28.6252	184.1678	0.1554	0.0054	5.7979	24
25	0.0304	6.4641	38.0314	32.9190	212.7930	0.1547	0.0047	5.8834	25
26	0.0264	6.4906	38.6918	37.8568	245.7120	0.1541	0.0041	5.9612	26
27	0.0230	6.5135	39.2890	43.5353	283.5688	0.1535	0.0035	6.0319	27
28	0.0200	6.5335	39.8283	50.0656	327.1041	0.1531	0.0031	6.0960	28
29	0.0174	6.5509	40.3146	57.5755	377.1697	0.1527	0.0027	6.1541	29
30	0.0151	6.5660	40.7526	66.2118	434.7451	0.1523	0.0023	6.2066	30
31	0.0131	6.5791	41.1466	76.1435	500.9569	0.1520	0.0020	6.2541	31
32	0.0114	6.5905	41.5006	87.5651	577.1005	0.1517	0.0017	6.2970	32
33	0.0099	6.6005	41.8184	100.6998	664.6655	0.1515	0.0015	6.3357	33
34	0.0086	6.6091	42.1033	115.8048	765.3654	0.1513	0.0013	6.3705	34
35	0.0075	6.6166	42.3586	133.1755	881.1702	0.1511	0.0011	6.4019	35
36	0.0065	6.6231	42.5872	153.1519	1014.3457	0.1510	0.0010	6.4301	36
37	0.0057	6.6288	42.7916	176.1246	1167.4975	0.1509	0.0009	6.4554	37
38	0.0049	6.6338	42.9743	202.5433	1343.6222	0.1507	0.0007	6.4781	38
39	0.0043	6.6380	43.1374	232.9248	1546.1655	0.1506	0.0006	6.4985	39
40	0.0037	6.6418	43.2830	267.8635	1779.0903	0.1506	0.0006	6.5168	40
41	0.0032	6.6450	43.4128	308.0431	2046.9539	0.1505	0.0005	6.5331	41
42	0.0028	6.6478	43.5286	354.2495	2354.9969	0.1504	0.0004	6.5478	42
43	0.0025	6.6503	43.6317	407.3870	2709.2465	0.1504	0.0004	6.5609	43
44	0.0021	6.6524	43.7235	468.4950	3116.6334	0.1503	0.0003	6.5725	44
45	0.0019	6.6543	43.8051	538.7693	3585.1285	0.1503	0.0003	6.5830	45
46	0.0016	6.6559	43.8778	619.5847	4123.8977	0.1502	0.0002	6.5923	46
47	0.0014	6.6573	43.9423	712.5224	4743.4824	0.1502	0.0002	6.6006	47
48	0.0012	6.6585	43.9997	819.4007	5456.0047	0.1502	0.0002	6.6080	48
49	0.0011	6.6596	44.0506	942.3108	6275.4055	0.1502	0.0002	6.6146	49
50	0.0009	6.6605	44.0958	1083.6574	7217.7163	0.1501	0.0001	6.6205	50
51	0.0008	6.6613	44.1360	1246.2061	8301.3737	0.1501	0.0001	6.6257	51
52	0.0007	6.6620	44.1715	1433.1370	9547.5798	0.1501	0.0001	6.6304	52
53	0.0006	6.6626	44.2031	1648.1075	10980.7167	0.1501	0.0001	6.6345	53
54	0.0005	6.6631	44.2311	1895.3236	12628.8243	0.1501	0.0001	6.6382	54
55	0.0005	6.6636	44.2558	2179.6222	14524.1479	0.1501	0.0001	6.6414	55
60	0.0002	6.6651	44.3431	4383.9987	29219.9916	0.1500	0.0000	6.6530	60
65	0.0001	6.6659	44.3903	8817.7874	58778.5826	0.1500	0.0000	6.6593	65
70	0.0001	6.6663	44.4156	17735.7200	118231.4669	0.1500	0.0000	6.6627	70
75	0.0000	6.6665	44.4292	35672.8680	237812.4532	0.1500	0.0000	6.6646	75
80	0.0000	6.6666	44.4364	71750.8794	478332.5293	0.1500	0.0000	6.6656	80
85	0.0000	6.6666	44.4402	144316.6470	962104.3133	0.1500	0.0000	6.6661	85
90	0.0000	6.6666	44.4422	290272.3252	1935142.1680	0.1500	0.0000	6.6664	90
95	0.0000	6.6667	44.4433	583841.3276	3892268.8509	0.1500	0.0000	6.6665	95
100	0.0000	6.6667	44.4438	1174313.4507	7828749.6713	0.1500	0.0000	6.6666	100

(continued)

APPENDIX L *(continued)*
Factor Tables
$I = 20.00\%$

n	P/F	P/A	P/G	F/P	F/A	A/P	A/F	A/G	n
1	0.8333	0.8333	0.0000	1.2000	1.0000	1.2000	1.0000	0.0000	1
2	0.6944	1.5278	0.6944	1.4400	2.2000	0.6545	0.4545	0.4545	2
3	0.5787	2.1065	1.8519	1.7280	3.6400	0.4747	0.2747	0.8791	3
4	0.4823	2.5887	3.2986	2.0736	5.3680	0.3863	0.1863	1.2742	4
5	0.4019	2.9906	4.9061	2.4883	7.4416	0.3344	0.1344	1.6405	5
6	0.3349	3.3255	6.5806	2.9860	9.9299	0.3007	0.1007	1.9788	6
7	0.2791	3.6046	8.2551	3.5832	12.9159	0.2774	0.0774	2.2902	7
8	0.2326	3.8372	9.8831	4.2998	16.4991	0.2606	0.0606	2.5756	8
9	0.1938	4.0310	11.4335	5.1598	20.7989	0.2481	0.0481	2.8364	9
10	0.1615	4.1925	12.8871	6.1917	25.9587	0.2385	0.0385	3.0739	10
11	0.1346	4.3271	14.2330	7.4301	32.1504	0.2311	0.0311	3.2893	11
12	0.1122	4.4392	15.4667	8.9161	39.5805	0.2253	0.0253	3.4841	12
13	0.0935	4.5327	16.5883	10.6993	48.4966	0.2206	0.0206	3.6597	13
14	0.0779	4.6106	17.6008	12.8392	59.1959	0.2169	0.0169	3.8175	14
15	0.0649	4.6755	18.5095	15.4070	72.0351	0.2139	0.0139	3.9588	15
16	0.0541	4.7296	19.3208	18.4884	87.4421	0.2114	0.0114	4.0851	16
17	0.0451	4.7746	20.0419	22.1861	105.9306	0.2094	0.0094	4.1976	17
18	0.0376	4.8122	20.6805	26.6233	128.1167	0.2078	0.0078	4.2975	18
19	0.0313	4.8435	21.2439	31.9480	154.7400	0.2065	0.0065	4.3861	19
20	0.0261	4.8696	21.7395	38.3376	186.6880	0.2054	0.0054	4.4643	20
21	0.0217	4.8913	22.1742	46.0051	225.0256	0.2044	0.0044	4.5334	21
22	0.0181	4.9094	22.5546	55.2061	271.0307	0.2037	0.0037	4.5941	22
23	0.0151	4.9245	22.8867	66.2474	326.2369	0.2031	0.0031	4.6475	23
24	0.0126	4.9371	23.1760	79.4968	392.4842	0.2025	0.0025	4.6943	24
25	0.0105	4.9476	23.4276	95.3962	471.9811	0.2021	0.0021	4.7352	25
26	0.0087	4.9563	23.6460	114.4755	567.3773	0.2018	0.0018	4.7709	26
27	0.0073	4.9636	23.8353	137.3706	681.8528	0.2015	0.0015	4.8020	27
28	0.0061	4.9697	23.9991	164.8447	819.2233	0.2012	0.0012	4.8291	28
29	0.0051	4.9747	24.1406	197.8136	984.0680	0.2010	0.0010	4.8527	29
30	0.0042	4.9789	24.2628	237.3763	1181.8816	0.2008	0.0008	4.8731	30
31	0.0035	4.9824	24.3681	284.8516	1419.2579	0.2007	0.0007	4.8908	31
32	0.0029	4.9854	24.4588	341.8219	1704.1095	0.2006	0.0006	4.9061	32
33	0.0024	4.9878	24.5368	410.1863	2045.9314	0.2005	0.0005	4.9194	33
34	0.0020	4.9898	24.6038	492.2235	2456.1176	0.2004	0.0004	4.9308	34
35	0.0017	4.9915	24.6614	590.6682	2948.3411	0.2003	0.0003	4.9406	35
36	0.0014	4.9929	24.7108	708.8019	3539.0094	0.2003	0.0003	4.9491	36
37	0.0012	4.9941	24.7531	850.5622	4247.8112	0.2002	0.0002	4.9564	37
38	0.0010	4.9951	24.7894	1020.6747	5098.3735	0.2002	0.0002	4.9627	38
39	0.0008	4.9959	24.8204	1224.8096	6119.0482	0.2002	0.0002	4.9681	39
40	0.0007	4.9966	24.8469	1469.7716	7343.8578	0.2001	0.0001	4.9728	40
41	0.0006	4.9972	24.8696	1763.7259	8813.6294	0.2001	0.0001	4.9767	41
42	0.0005	4.9976	24.8890	2116.4711	10577.3553	0.2001	0.0001	4.9801	42
43	0.0004	4.9980	24.9055	2539.7653	12693.8263	0.2001	0.0001	4.9831	43
44	0.0003	4.9984	24.9196	3047.7183	15233.5916	0.2001	0.0001	4.9856	44
45	0.0003	4.9986	24.9316	3657.2620	18281.3099	0.2001	0.0001	4.9877	45
46	0.0002	4.9989	24.9419	4388.7144	21938.5719	0.2000	0.0000	4.9895	46
47	0.0002	4.9991	24.9506	5266.4573	26327.2863	0.2000	0.0000	4.9911	47
48	0.0002	4.9992	24.9581	6319.7487	31593.7436	0.2000	0.0000	4.9924	48
49	0.0001	4.9993	24.9644	7583.6985	37913.4923	0.2000	0.0000	4.9935	49
50	0.0001	4.9995	24.9698	9100.4382	45497.1908	0.2000	0.0000	4.9945	50
51	0.0001	4.9995	24.9744	10920.5258	54597.6289	0.2000	0.0000	4.9953	51
52	0.0001	4.9996	24.9783	13104.6309	65518.1547	0.2000	0.0000	4.9960	52
53	0.0001	4.9997	24.9816	15725.5571	78622.7856	0.2000	0.0000	4.9966	53
54	0.0001	4.9997	24.9844	18870.6685	94348.3427	0.2000	0.0000	4.9971	54
55	0.0000	4.9998	24.9868	22644.8023	113219.0113	0.2000	0.0000	4.9976	55
60	0.0000	4.9999	24.9942	56347.5144	281732.5718	0.2000	0.0000	4.9989	60
65	0.0000	5.0000	24.9975	140210.6469	701048.2346	0.2000	0.0000	4.9995	65
70	0.0000	5.0000	24.9989	348888.9569	1744439.7847	0.2000	0.0000	4.9998	70
75	0.0000	5.0000	24.9995	868147.3693	4340731.8466	0.2000	0.0000	4.9999	75

(continued)

Support Material

APPENDIX L *(continued)*
Factor Tables

$I = 25.00\%$

n	P/F	P/A	P/G	F/P	F/A	A/P	A/F	A/G	n
1	0.8000	0.8000	0.0000	1.2500	1.0000	1.2500	1.0000	0.0000	1
2	0.6400	1.4400	0.6400	1.5625	2.2500	0.6944	0.0444	0.4444	2
3	0.5120	1.9520	1.6640	1.9531	3.8125	0.5123	0.2623	0.8525	3
4	0.4096	2.3616	2.8928	2.4414	5.7656	0.4234	0.1734	1.2249	4
5	0.3277	2.6893	4.2035	3.0518	8.2070	0.3718	0.1218	1.5631	5
6	0.2621	2.9514	5.5142	3.8147	11.2588	0.3383	0.0888	1.8683	6
7	0.2097	3.1611	6.7725	4.7684	15.0735	0.3163	0.0663	2.1424	7
8	0.1678	3.3289	7.9469	5.9605	19.8419	0.3004	0.0504	2.3872	8
9	0.1342	3.4631	9.0207	7.4506	25.8023	0.2888	0.0388	2.6048	9
10	0.1074	3.5705	9.9870	9.3132	33.2529	0.2801	0.0301	2.7971	10
11	0.0859	3.6564	10.8460	11.6415	42.5661	0.2735	0.0235	2.9663	11
12	0.0687	3.7251	11.6020	14.5519	54.2077	0.2684	0.0184	3.1145	12
13	0.0550	3.7801	12.2617	18.1899	68.7596	0.2645	0.0145	3.2437	13
14	0.0440	3.8241	12.8334	22.7374	86.9495	0.2615	0.0115	3.3559	14
15	0.0352	3.8593	13.3260	28.4217	109.6868	0.2591	0.0091	3.4530	15
16	0.0281	3.8874	13.7482	35.5271	138.1085	0.2572	0.0072	3.5366	16
17	0.0225	3.9099	14.1085	44.4089	173.6357	0.2558	0.0058	3.6084	17
18	0.0180	3.9279	14.4147	55.5112	218.0446	0.2546	0.0046	3.6698	18
19	0.0144	3.9424	14.6741	69.3889	273.5558	0.2537	0.0037	3.7222	19
20	0.0115	3.9539	14.8932	86.7362	342.9447	0.2529	0.0029	3.7667	20
21	0.0092	3.9631	15.0777	108.4202	429.6809	0.2523	0.0023	3.8045	21
22	0.0074	3.9705	15.2326	135.5253	538.1011	0.2519	0.0019	3.8365	22
23	0.0059	3.9764	15.3625	169.4066	673.6264	0.2515	0.0015	3.8634	23
24	0.0047	3.9811	15.4711	211.7582	843.0329	0.2512	0.0012	3.8861	24
25	0.0038	3.9849	15.5618	264.6978	1054.7912	0.2509	0.0009	3.9052	25
26	0.0030	3.9879	15.6373	330.8722	1319.4890	0.2508	0.0008	3.9212	26
27	0.0024	3.9903	15.7002	413.5903	1650.3612	0.2506	0.0006	3.9346	27
28	0.0019	3.9923	15.7524	516.9879	2063.9515	0.2505	0.0005	3.9457	28
29	0.0015	3.9938	15.7957	646.2349	2580.9394	0.2504	0.0004	3.9551	29
30	0.0012	3.9950	15.8316	807.7936	3227.1743	0.2503	0.0003	3.9628	30
31	0.0010	3.9960	15.8614	1009.7420	4034.9678	0.2502	0.0002	3.9693	31
32	0.0008	3.9968	15.8859	1262.1774	5044.7098	0.2502	0.0002	3.9746	32
33	0.0006	3.9975	15.9062	1577.7218	6306.8872	0.2502	0.0002	3.9791	33
34	0.0005	3.9980	15.9229	1972.1523	7884.6091	0.2501	0.0001	3.9828	34
35	0.0004	3.9984	15.9367	2465.1903	9856.7613	0.2501	0.0001	3.9858	35
36	0.0003	3.9987	15.9481	3081.4879	12321.9516	0.2501	0.0001	3.9883	36
37	0.0003	3.9990	15.9574	3851.8599	15403.4396	0.2501	0.0001	3.9904	37
38	0.0002	3.9992	15.9651	4814.8249	19255.2994	0.2501	0.0001	3.9921	38
39	0.0002	3.9993	15.9714	6018.5311	24070.1243	0.2500	0.0000	3.9935	39
40	0.0001	3.9995	15.9766	7523.1638	30088.6554	0.2500	0.0000	3.9947	40
41	0.0001	3.9996	15.9809	9403.9548	37611.8192	0.2500	0.0000	3.9956	41
42	0.0001	3.9997	15.9843	11754.9435	47015.7740	0.2500	0.0000	3.9964	42
43	0.0001	3.9997	15.9872	14693.6794	58770.7175	0.2500	0.0000	3.9971	43
44	0.0001	3.9998	15.9895	18367.0992	73464.3969	0.2500	0.0000	3.9976	44
45	0.0000	3.9998	15.9915	22958.8740	91831.4962	0.2500	0.0000	3.9980	45
46	0.0000	3.9999	15.9930	28698.5925	114790.3702	0.2500	0.0000	3.9984	46
47	0.0000	3.9999	15.9943	35873.2407	143488.9627	0.2500	0.0000	3.9987	47
48	0.0000	3.9999	15.9954	44841.5509	179362.2034	0.2500	0.0000	3.9989	48
49	0.0000	3.9999	15.9962	56051.9386	224203.7543	0.2500	0.0000	3.9991	49
50	0.0000	3.9999	15.9969	70064.9232	280255.6929	0.2500	0.0000	3.9993	50
51	0.0000	4.0000	15.9975	87581.1540	350320.6161	0.2500	0.0000	3.9994	51
52	0.0000	4.0000	15.9980	109476.4425	437901.7701	0.2500	0.0000	3.9995	52
53	0.0000	4.0000	15.9983	136845.5532	547378.2126	0.2500	0.0000	3.9996	53
54	0.0000	4.0000	15.9986	171056.9414	684223.7658	0.2500	0.0000	3.9997	54
55	0.0000	4.0000	15.9989	213821.1768	855280.7072	0.2500	0.0000	3.9997	55
60	0.0000	4.0000	15.9996	652530.4468	2610117.7872	0.2500	0.0000	3.9999	60

(continued)

APPENDIX L *(continued)*
Factor Tables

$I = 30.00\%$

n	P/F	P/A	P/G	F/P	F/A	A/P	A/F	A/G	n
1	0.7692	0.7692	0.0000	1.3000	1.0000	1.3000	1.0000	0.000	1
2	0.5917	1.3609	0.5917	1.6900	2.3000	0.7348	0.4348	0.434	2
3	0.4552	1.8161	1.5020	2.1970	3.9900	0.5506	0.2506	0.827	3
4	0.3501	2.1662	2.5524	2.8561	6.1870	0.4616	0.1616	1.178	4
5	0.2693	2.4356	3.6297	3.7129	9.0431	0.4106	0.1106	1.490	5
6	0.2072	2.6427	4.6656	4.8268	12.7560	0.3784	0.0784	1.765	6
7	0.1594	2.8021	5.6218	6.2749	17.5828	0.3569	0.0569	2.006	7
8	0.1226	2.9247	6.4800	8.1573	23.8577	0.3419	0.0419	2.215	8
9	0.0943	3.0190	7.2343	10.6045	32.0150	0.3312	0.0312	2.396	9
10	0.0725	3.0915	7.8872	13.7858	42.6195	0.3235	0.0235	2.551	10
11	0.0558	3.1473	8.4452	17.9216	56.4053	0.3177	0.0177	2.683	11
12	0.0429	3.1903	8.9173	23.2981	74.3270	0.3135	0.0135	2.795	12
13	0.0330	3.2233	9.3135	30.2875	97.6250	0.3102	0.0102	2.889	13
14	0.0254	3.2487	9.6437	39.3738	127.9125	0.3078	0.0078	2.968	14
15	0.0195	3.2682	9.9172	51.1859	167.2863	0.3060	0.0060	3.034	15
16	0.0150	3.2832	10.1426	66.5417	218.4722	0.3046	0.0046	3.089	16
17	0.0116	3.2948	10.3276	86.5042	285.0139	0.3035	0.0035	3.134	17
18	0.0089	3.3037	10.4788	112.4554	371.5180	0.3027	0.0027	3.171	18
19	0.0068	3.3105	10.6019	146.1920	483.9734	0.3021	0.0021	3.202	19
20	0.0053	3.3158	10.7019	190.0496	630.1655	0.3016	0.0016	3.227	20
21	0.0040	3.3198	10.7828	247.0645	820.2151	0.3012	0.0012	3.248	21
22	0.0031	3.3230	10.8482	321.1839	1067.2796	0.3009	0.0009	3.264	22
23	0.0024	3.3254	10.9009	417.5391	1388.4635	0.3007	0.0007	3.278	23
24	0.0018	3.3272	10.9433	542.8008	1806.0026	0.3006	0.0006	3.289	24
25	0.0014	3.3286	10.9773	705.6410	2348.8033	0.3004	0.0004	3.297	25
26	0.0011	3.3297	11.0045	917.3333	3054.4443	0.3003	0.0003	3.305	26
27	0.0008	3.3305	11.0263	1192.5333	3971.7776	0.3003	0.0003	3.310	27
28	0.0006	3.3312	11.0437	1550.2933	5164.3109	0.3002	0.0002	3.315	28
29	0.0005	3.3317	11.0576	2015.3813	6714.6042	0.3001	0.0001	3.318	29
30	0.0004	3.3321	11.0687	2619.9956	8729.9855	0.3001	0.0001	3.321	30
31	0.0003	3.3324	11.0775	3405.9943	11349.9811	0.3001	0.0001	3.324	31
32	0.0002	3.3326	11.0845	4427.7926	14755.9755	0.3001	0.0001	3.326	32
33	0.0002	3.3328	11.0901	5756.1304	19183.7681	0.3001	0.0001	3.327	33
34	0.0001	3.3329	11.0945	7482.9696	24939.8985	0.3000	0.0000	3.328	34
35	0.0001	3.3330	11.0980	9727.8604	32422.8681	0.3000	0.0000	3.329	35
36	0.0001	3.3331	11.1007	12646.2186	42150.7285	0.3000	0.0000	3.330	36
37	0.0001	3.3331	11.1029	16440.0841	54796.9471	0.3000	0.0000	3.331	37
38	0.0000	3.3332	11.1047	21372.1094	71237.0312	0.3000	0.0000	3.331	38
39	0.0000	3.3332	11.1060	27783.7422	92609.1405	0.3000	0.0000	3.331	39
40	0.0000	3.3332	11.1071	36118.8648	120392.8827	0.3000	0.0000	3.332	40
41	0.0000	3.3333	11.1080	46954.5243	156511.7475	0.3000	0.0000	3.332	41
42	0.0000	3.3333	11.1086	61040.8815	203466.2718	0.3000	0.0000	3.332	42
43	0.0000	3.3333	11.1092	79353.1460	264507.1533	0.3000	0.0000	3.332	43
44	0.0000	3.3333	11.1096	103159.0898	343860.2993	0.3000	0.0000	3.332	44
45	0.0000	3.3333	11.1099	134106.8167	447019.3890	0.3000	0.0000	3.333	45
46	0.0000	3.3333	11.1102	174338.8617	581126.2058	0.3000	0.0000	3.333	46
47	0.0000	3.3333	11.1104	226640.5202	755465.0675	0.3000	0.0000	3.333	47
48	0.0000	3.3333	11.1105	294632.6763	982105.5877	0.3000	0.0000	3.333	48
49	0.0000	3.3333	11.1107	383022.4792	1276738.2640	0.3000	0.0000	3.333	49
50	0.0000	3.3333	11.1108	497929.2230	1659760.7433	0.3000	0.0000	3.333	50

(continued)

APPENDIX L *(continued)*
Factor Tables

$I = 40.00\%$

n	P/F	P/A	P/G	F/P	F/A	A/P	A/F	A/G	n
1	0.7143	0.7143	0.0000	1.4000	1.0000	1.4000	1.0000	0.000	1
2	0.5102	1.2245	0.5102	1.9600	2.4000	0.8167	0.4167	0.416	2
3	0.3644	1.5889	1.2391	2.7440	4.3600	0.6294	0.2294	0.779	3
4	0.2603	1.8492	2.0200	3.8416	7.1040	0.5408	0.1408	1.092	4
5	0.1859	2.0352	2.7637	5.3782	10.9456	0.4914	0.0914	1.358	5
6	0.1328	2.1680	3.4278	7.5295	16.3238	0.4613	0.0613	1.581	6
7	0.0949	2.2628	3.9970	10.5414	23.8534	0.4419	0.0419	1.766	7
8	0.0678	2.3306	4.4713	14.7579	34.3947	0.4291	0.0291	1.918	8
9	0.0484	2.3790	4.8585	20.6610	49.1526	0.4203	0.0203	2.042	9
10	0.0346	2.4136	5.1696	28.9255	69.8137	0.4143	0.0143	2.141	10
11	0.0247	2.4383	5.4166	40.4957	98.7391	0.4101	0.0101	2.221	11
12	0.0176	2.4559	5.6106	56.6939	139.2348	0.4072	0.0072	2.284	12
13	0.0126	2.4685	5.7618	79.3715	195.9287	0.4051	0.0051	2.334	13
14	0.0090	2.4775	5.8788	111.1201	275.3002	0.4036	0.0036	2.372	14
15	0.0064	2.4839	5.9688	155.5681	386.4202	0.4026	0.0026	2.403	15
16	0.0046	2.4885	6.0376	217.7953	541.9883	0.4018	0.0018	2.426	16
17	0.0033	2.4918	6.0901	304.9135	759.7837	0.4013	0.0013	2.444	17
18	0.0023	2.4941	6.1299	426.8789	1064.6971	0.4009	0.0009	2.457	18
19	0.0017	2.4958	6.1601	597.6304	1491.5760	0.4007	0.0007	2.468	19
20	0.0012	2.4970	6.1828	836.6826	2089.2064	0.4005	0.0005	2.476	20
21	0.0009	2.4979	6.1998	1171.3556	2925.8889	0.4003	0.0003	2.482	21
22	0.0006	2.4985	6.2127	1639.8978	4097.2445	0.4002	0.0002	2.486	22
23	0.0004	2.4989	6.2222	2295.8569	5737.1423	0.4002	0.0002	2.490	23
24	0.0003	2.4992	6.2294	3214.1997	8032.9993	0.4001	0.0001	2.492	24
25	0.0002	2.4994	6.2347	4499.8796	11247.1990	0.4001	0.0001	2.494	25
26	0.0002	2.4996	6.2387	6299.8314	15747.0785	0.4001	0.0001	2.495	26
27	0.0001	2.4997	6.2416	8819.7640	22046.9099	0.4000	0.0000	2.496	27
28	0.0001	2.4998	6.2438	12347.6696	30866.6739	0.4000	0.0000	2.497	28
29	0.0001	2.4999	6.2454	17286.7374	43214.3435	0.4000	0.0000	2.498	29
30	0.0000	2.4999	6.2466	24201.4324	60501.0809	0.4000	0.0000	2.498	30
31	0.0000	2.4999	6.2475	33882.0053	84702.5132	0.4000	0.0000	2.499	31
32	0.0000	2.4999	6.2482	47434.8074	118584.5185	0.4000	0.0000	2.499	32
33	0.0000	2.5000	6.2487	66408.7304	166019.3260	0.4000	0.0000	2.499	33
34	0.0000	2.5000	6.2490	92972.2225	232428.0563	0.4000	0.0000	2.499	34
35	0.0000	2.5000	6.2493	130161.1116	325400.2789	0.4000	0.0000	2.499	35
36	0.0000	2.5000	6.2495	182225.5562	455561.3904	0.4000	0.0000	2.499	36
37	0.0000	2.5000	6.2496	255115.7786	637786.9466	0.4000	0.0000	2.499	37
38	0.0000	2.5000	6.2497	357162.0901	892902.7252	0.4000	0.0000	2.499	38
39	0.0000	2.5000	6.2498	500026.9261	1250064.8153	0.4000	0.0000	2.499	39
40	0.0000	2.5000	6.2498	700037.6966	1750091.7415	0.4000	0.0000	2.499	40
41	0.0000	2.5000	6.2499	980052.7752	2450129.4381	0.4000	0.0000	2.500	41
42	0.0000	2.5000	6.2499	1372073.8853	3430182.2133	0.4000	0.0000	2.500	42
43	0.0000	2.5000	6.2499	1920903.4394	4802256.0986	0.4000	0.0000	2.500	43
44	0.0000	2.5000	6.2500	2689264.8152	6723159.5381	0.4000	0.0000	2.500	44
45	0.0000	2.5000	6.2500	3764970.7413	9412424.3533	0.4000	0.0000	2.500	45

Index

Italicized page numbers represent sources of data in tables, figures, and appendices

Index

Italicized page numbers represent sources of data in tables, figures, and appendices

Italicized page numbers represent sources of data in tables, figures, and appendices

Italicized page numbers represent sources of data in tables, figures, and appendices

Italicized page numbers represent sources of data in tables, figures, and appendices

Index

Italicized page numbers represent sources of data in tables, figures, and appendices

Italicized page numbers represent sources of data in tables, figures, and appendices